AF814225

Test Drive

India's Most Trusted Practice Resource

15 Practice Sets

JEE MAIN 2025

Test Drive

India's Most Trusted Practice Resource

15 Practice Sets JEE MAIN 2025

PHYSICS

CHEMISTRY

MATHEMATICS

Authors

Sumit Shukla *(Physics)*

Swati Chauhan *(Chemistry)*

Karan Mishra *(Mathematics)*

arihant

SUCCESS AND NOTHING LESS

arihant
SUCCESS AND NOTHING LESS

ARIHANT PRAKASHAN (SERIES), MEERUT

ADMINISTRATIVE & PRODUCTION OFFICES

Regd. Office
'Ramchhaya' 4577/15, Agarwal Road, Darya Ganj, New Delhi -110002
Tele: 011- 47630600, 43518550; Fax: 011- 23280316

Head Office
Kalindi, TP Nagar, Meerut (UP) - 250002, Tel: 0121-7156203, 7156204

SALES & SUPPORT OFFICES
Agra, Ahmedabad, Bengaluru, Bareilly, Chennai, Delhi, Guwahati, Hyderabad, Jaipur, Jhansi, Kolkata, Lucknow, Nagpur & Pune.

PRICE

PO No : TXT-XX-XXXXXXX-X-XX

Published By Arihant Publications (India) Ltd.

For further information about the books published by Arihant, log on to www.arihantbooks.com or e-mail at info@arihantbooks.com

Follow us on

7 Reasons

Why Practice Sets are Crucial for JEE Main Preparation?

Gain Insight into the Exam

Practice Sets mirror the actual exam, providing you with a glimpse of what to anticipate on the crucial exam day. By solving these sets, you familiarize yourself with the exam syllabus, pattern, and level of difficulty.

Evaluate Your Preparation and Performance

Practice Sets serve as a reality check to assess the effectiveness of your preparation. They reinforce your understanding of concepts and help prevent forgetting what you have learned and mastered.

Experiment and Test Strategies

Through various Practice Sets, you can experiment and implement different strategies. This allows you to determine the most effective approach to utilize during the actual exam, enabling you to plan and strategize effectively.

Efficient Time Management

Practice Sets provide insights into your speed and efficiency. By solving these sets, you can identify sections that require more time and those that can be completed faster. It helps you allocate your time wisely during the exam,

Identify Strengths and Weaknesses

By consistently solving Practice Sets, you gain a clear understanding of your strengths, weaknesses, and even specific topics within each section that pose challenges. Armed with this knowledge, you can prioritize your focus and allocate more time to areas that need improvement.

Reduce Pre-Exam Nervousness

Practice Sets simulate the actual exam environment. As you solve an ample number of them, you become familiar with the exam format and feel more at ease on the day of the exam. Achieving good scores in these sets also boosts your confidence in performing well on the actual exam.

Track Your Progress

Regularly solving Practice Sets allows you to track your progress over time. By monitoring your scores as you solve more sets, you can observe how your performance has improved or identify instances where you fell short. This feedback helps you identify areas that require further attention and measure your overall progress.

Scan this QR Code for PYQs Practice

CONTENTS

Section I (Foundation Builders) 1-65

- **Practice Set 1** 3-10
  Explanations 11-22
- **Practice Set 2** 23-30
  Detailed Solutions 31-45
- **Practice Set 3** 46-53
  Detailed Solutions 54-65

Section II (Mastery Builders) 67-285

- **Practice Set 4** 69-76
  Detailed Solutions 77-90
- **Practice Set 5** 91-98
  Detailed Solutions 99-112
- **Practice Set 6** 113-120
  Detailed Solutions 121-133
- **Practice Set 7** 134-141
  Detailed Solutions 142-155
- **Practice Set 8** 156-164
  Detailed Solutions 165-177
- **Practice Set 9** 178-185
  Detailed Solutions 186-198
- **Practice Set 10** 199-206
  Detailed Solutions 207-219
- **Practice Set 11** 220-228
  Detailed Solutions 229-241
- **Practice Set 12** 242-249
  Detailed Solutions 250-263
- **Practice Set 13** 264-271
  Detailed Solutions 272-285

Section III (Achiever's Section) 287-333

- **Practice Set 14** 289-296
  Answers & Explanations 297-310
- **Practice Set 15** 311-318
  Answers & Explanations 319-333

JEE Main 2024 Analysis

Physics	Average		Chemistry	Average		Mathematics	Average	
	Jan.	April		Jan.	April		Jan.	April
Physics & Measurements	1-2	0-1	Some Basic Concepts in Chemistry	1-2	1-2	Sets	0-1	0-1
Motion in a Straight Line	0-1	0-1	Atomic Structure	0-1	1-2	Relations & Functions	1-2	1-2
Motion in a Plane	0-1	1-2	Classification of Elements and Periodicity in Properties	0-1	1-2	Trigonometry	0-1	1-2
Laws of Motion	2-3	1-2	Chemical Bonding and Molecular Structure	1-2	2-3	Complex Numbers & Quadratic Equations	1-2	1-2
Work, Energy & Power	0-1	1-2	Thermodynamics	0-1	1-2	Permutations and Combinations	1-2	1-2
System of Particles & Rotational Motion	0-1	0-1	Equilibrium	0-1	1-2	Binomial Theorem and its Applications	1-2	1-2
Gravitation	0-1	1-2	Redox Reactions	0-1	0-1	Sequence and Series	1-2	1-2
Mechanical Properties of Solids	0-1	0-1	*p*-Block Elements-1	0-1	0-1	Straight Lines and Pair of Straight Lines	1-2	1-2
Mechanical Properties of Fluids	1-2	1-2	Purification & Characterisation of Organic Compounds	0-1	0-1	Circle	0-1	0-1
Thermal Properties of Matter	0-1	0-1	Some Basic Principles of Organic Chemistry	2-3	2-3	Conic Section	1-2	1-2
Thermodynamics	0-1	0-1	Hydrocarbons	1-2	1-2	Matrices and Determinants	1-2	1-2
Kinetic Theory of Gases	1-2	0-1	Solutions	0-1	0-1	Limit, Continuity and Differentiability	2-3	2-3
Oscillations	0-1	0-1	Electrochemistry	1-2	1-2	Differentiation	1-2	1-2
Waves	0-1	0-1	Chemical Kinetics	0-1	0-1	Application of Derivatives	0-1	0-1
Electric Charges and Fields	1-2	1-2	*p*-Block Elements - 2	1-2	0-1	Integrals	2-3	2-3
Electrostatics Potential and Capacitance	0-1	0-1	*d* & *f*-Block Elements	1-2	1-2	Application of Intergrals	1-2	1-2
Current Electricity	2-3	2-3	Coordination Compounds	1-2	2-3	Differential Equations	1-2	1-2
Moving Charges and Magnetism	1-2	1-2	Organic Compounds Containing Halogens	1-2	1-2	Vector Algebra	1-2	1-2
Magnetism and Matter	0-1	0-1	Alcohols, Phenols & Ethers	1-2	2-3	Three Dimentional Geometry	1-2	1-2
Electromagnetic Induction	0-1	0-1	Aldehydes, Ketones and Carboxylic Acids	1-2	1-2	Statistics	0-1	0-1
Alternating Current	1-3	1-2	Organic Compounds Containing Nitrogen	1-2	0-1	Probability	1-2	1-2
Electromagnetic Waves	1-2	0-1	Biomolecules	0-1	0-1			
Ray Optics & Optical Instruments	1-2	0-1	Principles Related to Practical Chemistry	1-2	1-2			
Wave Optics	1-2	1-2						
Dual Nature of Matter & Radiation	0-1	1-2						
Atoms	1-2	1-2						
Nuclei	0-1	0-1						
Semiconductor Electronics	1	1-2						
Experimental Physics	0-1	1-2						

SYLLABUS

Physics

UNIT 1 Physics and Measurement

Units of measurements, System of Units, S I Units, fundamental and derived units, least count, significant figures, Errors in measurements, Dimensions of Physics quantities, dimensional analysis, and its applications.

UNIT 2 Kinematics

The frame of reference, motion in a straight line, Position- time graph, speed and velocity; Uniform and non-uniform motion, average speed and instantaneous velocity, uniformly accelerated motion, velocity-time, position-time graph, relations for uniformly accelerated motion, Scalars and Vectors, Vector. Addition and subtraction, scalar and vector products, Unit Vector, Resolution of a Vector. Relative Velocity, Motion in a plane, Projectile Motion, Uniform Circular Motion.

UNIT 3 Laws of Motion

Force and inertia, Newton's First law of motion; Momentum, Newton's Second law of motion, Impulses; Newton's Third law of motion. Law of conservation of linear momentum and its applications. Equilibrium of concurrent forces.

Static and Kinetic friction, laws of friction, rolling friction.

Dynamics of uniform circular motion: centripetal force and its applications: vehicle on a level circular road, vehicle on a banked road.

UNIT 4 Work, Energy and Power

Work done by a constant force and a variable force; kinetic and potential energies, work-energy theorem, power.

The potential energy of spring conservation of mechanical energy, conservative and non-conservative forces; motion in a vertical circle: Elastic and inelastic collisions in one and two dimensions.

UNIT 5 Rotational Motion

Centre of the mass of a two-particles system, Centre of the mass of a rigid body; Basic concepts of rotational motion; moment of a force; torque, angular momentum, conservation of angular momentum and its applications;

The moment of inertia, the radius of gyration, values of moments of inertia for simple geometrical objects, parallel and perpendicular axes theorems and their applications.

Equilibrium of rigid bodies, rigid body rotation and equations of rotational motion, comparison of linear and rotational motions.

UNIT 6 Gravitation

The universal law of gravitation. Acceleration due to gravity and its variation with altitude and depth. Kepler's law of planetary motion. Gravitational potential energy; gravitational potential. Escape velocity, Motion of a satellite, orbital velocity, time period and energy of satellite.

UNIT 7 Properties of Solids and Liquids

Elastic behaviour, Stress-strain relationship, Hooke's Law. Young's modulus, bulk modulus, and modulus of rigidity. Pressure due to a fluid column; Pascal's law and its applications. Effect of gravity on fluid pressure.
Viscosity, Stokes' law, terminal velocity, streamline, and turbulent flow, critical velocity. Bernoulli's principle and its applications. Surface energy and surface tension, angle of contact, excess of pressure across a curved surface, application of surface tension - drops, bubbles, and capillary rise. Heat, temperature, thermal expansion; specific heat capacity, calorimetry; change of state, latent heat. Heat transfer conduction, convection and radiation.

UNIT 8 Thermodynamics

Thermal equilibrium, zeroth law of thermodynamics, the concept of temperature. Heat, work and internal energy. The first law of thermodynamics, isothermal and adiabatic processes. The second law of thermodynamics: reversible and irreversible processes.

UNIT 9 Kinetic Theory of Gases

Equation of state of a perfect gas, work done on compressing a gas, Kinetic theory of gases -assumptions, the concept of pressure. Kinetic interpretation of temperature: RMS speed of gas molecules: Degrees of freedom. Law of equipartition of energy and applications to specific heat capacities of gases; Mean free path. Avogadro's number.

UNIT 10 Oscillations and Waves

Oscillations and periodic motion – time period, frequency, displacement as a function of time. Periodic functions. Simple harmonic motion (SHM) and its equation; phase: oscillations of a spring -restoring force and force constant: energy in SHM - Kinetic and potential energies; Simple pendulum - derivation of expression for its time period:
Wave motion. Longitudinal and transverse waves, speed of the travelling wave. Displacement relation for a progressive wave. Principle of superposition of waves, reflection of waves. Standing waves in strings and organ pipes, fundamental mode and harmonics. Beats.

UNIT 11 Electrostatics

Electric charges: Conservation of charge. Coulomb's law forces between two point charges, forces between multiple charges: superposition principle and continuous charge distribution.
Electric field: Electric field due to a point charge, Electric field lines. Electric dipole, Electric field due to a dipole. Torque on a dipole in a uniform electric field
Electric flux. Gauss's law and its applications to find field due to infinitely long uniformly charged straight wire uniformly charged infinite plane sheet and uniformly charged thin spherical shell. Electric potential and its calculation for a point charge, electric dipole and system of charges; potential difference, Equipotential surfaces, Electrical potential energy of a system of two point charges and of electric dipole in an electrostatic field.
Conductors and insulators. Dielectrics and electric polarization, capacitors and capacitances, the combination of capacitors in series and parallel, and capacitance of a parallel plate capacitor with and without dielectric medium between the plates. Energy stored in a capacitor.

UNIT 12 Current Electricity

Electric current. Drift velocity, mobility, and their relation with electric current. Ohm's law, Electrical resistance. *V-I* characteristics of Ohmic and non-ohmic conductors. Electrical energy and power. Electrical resistivity and conductivity. Series and parallel combinations of resistors; Temperature dependence of resistance.

Internal resistance, potential difference, and emf of a cell, a combination of cells in series and parallel. Kirchhoff's laws and their applications. Wheatstone bridge. Metre bridge.

UNIT 13 Magnetic Effects of Current and Magnetism

Biot - Savart' law and its application to the current-carrying circular loop. Ampere's law and its applications to infinitely long current carrying straight wire and solenoid. Force on a moving charge in uniform magnetic and electric fields.

Force on a current-carrying conductor in a uniform magnetic field. The force between two parallel currents carrying conductors-definition of ampere. Torque experienced by a current loop in a uniform magnetic field: Moving coil galvanometer, its sensitivity, and conversion to ammeter and voltmeter.

Current loop as a magnetic dipole and its magnetic dipole moment. Bar magnet as an equivalent solenoid, magnetic field lines; Magnetic field due to a magnetic dipole (bar magnet) along its axis and perpendicular to its axis. Torque on a magnetic dipole in a uniform magnetic field. Para, dia and ferromagnetic substances with examples, the effect of temperature on magnetic properties.

UNIT 14 Electromagnetic Induction and Alternating Currents

Electromagnetic induction: Faraday's law. Induced emf and current: Lenz's Law, Eddy currents. Self and mutual inductance. Alternating currents, peak and RMS value of alternating current/ voltage: reactance and impedance: *LCR* series circuit, resonance: power in AC circuits, wattless current. AC generator and transformer.

UNIT 15 Electromagnetic Waves

Displacement current. Electromagnetic waves and their characteristics, Transverse nature of electromagnetic waves, Electromagnetic spectrum (radio waves, microwaves, infrared, visible, ultraviolet, X-rays, Gamma rays), Applications of EM waves.

UNIT 16 Optics

Reflection of light, spherical mirrors, mirror formula. Refraction of light at plane and spherical surfaces, thin lens formula, and lens maker formula. Total internal reflection and its applications. Magnification. Power of a Lens. Combination of thin lenses in contact. Refraction of light through a prism. Microscope and Astronomical Telescope (reflecting and refracting) and their magnifying powers.

Wave optics: wavefront and Huygens' principle. Laws of reflection and refraction using Huygens principle. Interference, Young's double-slit experiment, and expression for fringe width, coherent sources, and sustained interference of light. Diffraction due to a single slit, width of central maximum. Polarization, plane-polarized light: Brewster's law, uses of plane-polarized light and Polaroid.

UNIT 17 Dual Nature of Matter and Radiation

Dual nature of radiation. Photoelectric effect. Hertz and Lenard's observations; Einstein's photoelectric equation: particle nature of light. Matter waves-wave nature of particle, de Broglie relation.

UNIT 18 Atoms and Nuclei

Alpha-particle scattering experiment; Rutherford's model of atom; Bohr model, energy levels, hydrogen spectrum. Composition and size of nucleus, atomic masses, Mass-energy relation, mass defect; binding energy per nucleon and its variation with mass number, nuclear fission, and fusion.

UNIT 19 Electronic Devices

Semiconductors; semiconductor diode: *I-V* characteristics in forward and reverse bias; diode as a rectifier; *I-V* characteristics of LED, the photodiode, solar cell, and Zener diode; Zener diode as a voltage regulator. Logic gates (OR, AND, NOT, NAND and NOR).

UNIT 20 Experimental Skills

Familiarity with the basic approach and observations of the experiments and activities:

1. Vernier calipers -its use to measure the internal and external diameter and depth of a vessel.
2. Screw gauge - its use to determine the thickness/ diameter of thin sheet/wire.
3. Simple Pendulum-dissipation of energy by plotting a graph between the square of amplitude and time.
4. Metre Scale - the mass of a given object by the principle of moments.
5. Young's modulus of elasticity of the material of a metallic wire.
6. Surface tension of water by capillary rise and effect of detergents.
7. Co-efficient of Viscosity of a given viscous liquid by measuring the terminal velocity of a given spherical body.
8. Speed of sound in air at room temperature using a resonance tube.
9. Specific heat capacity of a given (i) solid and (ii) liquid by method of mixtures.
10. The resistivity of the material of a given wire using a metre bridge.
11. The resistance of a given wire using Ohm's law.
12. Resistance and figure of merit of a galvanometer by half deflection method.
13. The focal length of;

(i) Convex mirror

(ii) Concave mirror

(iii) Convex lens, using the parallax method.

14. The plot of the angle of deviation vs angle of incidence for a triangular prism.
15. The refractive index of a glass slab using a travelling microscope.
16. Characteristic curves of a p-n junction diode in forward and reverse bias.
17. Characteristic curves of a Zener diode and finding reverse breakdown voltage.
18. Identification of Diode. LED, Resistor. A capacitor from a mixed collection of such items.

Chemistry
Physical Chemistry

UNIT 1 Some Basic Concepts in Chemistry

Matter and its nature, Dalton's atomic theory: Concept of atom, molecule, element, and compound:: Laws of chemical combination; Atomic and molecular masses, mole concept, molar mass, percentage composition, empirical and molecular formulae: Chemical equations and stoichiometry.

UNIT 2 Atomic Structure

Nature of electromagnetic radiation, photoelectric effect; Spectrum of the hydrogen atom. Bohr model of a hydrogen atom - its postulates, derivation of the relations for the energy of the electron and radii of the different orbits, limitations of Bohr's model; Dual nature of matter, de Broglie's relationship. Heisenberg uncertainty principle. Elementary ideas of quantum mechanics, quantum mechanics, the quantum mechanical model of the atom, and its important features. Concept of atomic orbitals as one-electron wave functions: Variation of ψ and ψ^2 with r for $1s$ and $2s$ orbitals; various quantum numbers (principal, angular momentum, and magnetic quantum numbers) and their significance; shapes of *s*, *p*, and *d* - orbitals, electron spin and spin quantum number: Rules for filling electrons in orbitals – Aufbau principle. Pauli's exclusion principle and Hund's rule, electronic configuration of elements, and extra stability of half-filled and completely filled orbitals.

UNIT 3 Chemical Bonding and Molecular Structure

Kossel-Lewis approach to chemical bond formation, the concept of ionic and covalent bonds.

Ionic Bonding: Formation of ionic bonds, factors affecting the formation of ionic bonds; calculation of lattice enthalpy.

Covalent Bonding: Concept of electronegativity. Fajan's rule, dipole moment: Valence Shell Electron Pair Repulsion (VSEPR) theory and shapes of simple molecules.

Quantum mechanical approach to covalent bonding: Valence bond theory - its important features, the concept of hybridization involving *s*, *p*, and *d* orbitals; Resonance.

Molecular Orbital Theory - Its important features. LCAOs, types of molecular orbitals (bonding, antibonding), sigma and pi-bonds, molecular orbital electronic configurations of homonuclear diatomic molecules, the concept of bond order, bond length, and bond energy. Elementary idea of metallic bonding. Hydrogen bonding and its applications.

UNIT 4 Chemical Thermodynamics

Fundamentals of thermodynamics: System and surroundings, extensive and intensive properties, state functions, entropy, types of processes.

The first law of thermodynamics - Concept of work, heat internal energy and enthalpy, heat capacity, molar heat capacity; Hess's law of constant heat summation; Enthalpies of bond dissociation, combustion, formation, atomization, sublimation, phase transition, hydration, ionization, and solution.

The second law of thermodynamics - Spontaneity of processes; ΔS of the universe and ΔG of the system as criteria for spontaneity. $\Delta G°$ (Standard Gibbs energy change) and equilibrium constant.

UNIT 5 Solutions

Different methods for expressing the concentration of solution - molality, molarity, mole fraction, percentage (by volume and mass both), the vapour pressure of solutions and Raoult's Law - Ideal and non-ideal solutions, vapour pressure - composition, plots for ideal and non-ideal solutions; Colligative properties of dilute solutions - a relative lowering of vapour pressure, depression of freezing point, the elevation of boiling point and osmotic pressure; Determination of molecular mass using colligative properties; Abnormal value of molar mass, Van't Hoff factor and its significance.

UNIT 6 Equilibrium

Meaning of equilibrium is the concept of dynamic equilibrium.

Equilibria involving physical processes: Solid-liquid, liquid-gas - gas and solid-gas equilibria, Henry's law. General characteristics of equilibrium involving physical processes.

Equilibrium involving chemical processes: Law of chemical equilibrium, equilibrium constants (K_p and K_c) and their significance, the significance of ΔG and $\Delta G°$ in chemical equilibrium, factors affecting equilibrium concentration, pressure, temperature, the effect of catalyst; Le Chatelier's principle.

Ionic equilibrium: Weak and strong electrolytes, ionization of electrolytes, various concepts of acids and bases (Arrhenius. Bronsted - Lowry and Lewis) and their ionization, acid-base equilibria (including multistage ionization) and ionization constants, ionization of water. pH scale, common ion effect, hydrolysis of salts and pH of their solutions, the solubility of sparingly soluble salts and solubility products, and buffer solutions.

UNIT 7 Redox Reactions and Electrochemistry

Electronic concepts of oxidation and reduction, redox reactions, oxidation number, rules for assigning oxidation number and balancing of redox reactions.

Electrolytic and metallic conduction, conductance in electrolytic solutions, molar conductivities and their variation with concentration: Kohlrausch's law and its applications.

Electrochemical cells - Electrolytic and Galvanic cells, different types of electrodes, electrode potentials including standard electrode potential, half-cell and cell reactions, emf of a Galvanic cell and its measurement: Nernst equation and its applications; Relationship between cell potential and Gibbs' energy change: Dry cell and lead accumulator; Fuel cells.

UNIT 8 Chemical Kinetics

Rate of a chemical reaction, factors affecting the rate of reactions: concentration, temperature, pressure, and catalyst; elementary and complex reactions, order and molecularity of reactions, rate law, rate constant and its units, differential and integral forms of zero and first-order reactions, their characteristics and half-lives, the effect of temperature on the rate of reactions, Arrhenius theory, activation energy and its calculation, collision theory of bimolecular gaseous reactions (no derivation).

Inorganic Chemistry

UNIT 9 Classification of Elements and Periodicity in Properties

Modem periodic law and present form of the periodic table, *s, p. d* and *f* block elements, periodic trends in properties of elements, atomic and ionic radii, ionisation enthalpy, electron gain enthalpy, valence, oxidation states, and chemical reactivity.

UNIT 10 *p*-block Elements

Group-13 to Group 18 Elements

General Introduction: Electronic configuration and general trends in physical and chemical properties of elements across the periods and down the groups; unique behaviour of the first element in each group.

UNIT 11 *d*- and *f*- Block Elements

Transition Elements

General introduction, electronic configuration, occurrence and characteristics, general trends in properties of the first-row transition elements - physical properties, ionization enthalpy, oxidation states, atomic radii, colour, catalytic behaviour, magnetic properties, complex formation, interstitial compounds, alloy formation; Preparation, properties, and uses of $K_2Cr_2O_7$, and $KMnO_4$.

Inner Transition Elements

Lanthanoids - Electronic configuration, oxidation states, and lanthanoid contraction.

Actinoids - Electronic configuration and oxidation states.

UNIT 12 Co-ordination Compounds

Introduction to coordination compounds. Werner's theory; ligands, coordination number, denticity. chelation; IUPAC nomenclature of mononuclear co-ordination compounds, isomerism; Bonding-Valence bond approach and basic ideas of Crystal field theory, colour and magnetic properties; Importance of co-ordination compounds (in qualitative analysis, extraction of metals, and in biological systems).

Organic Chemistry

UNIT 13 Purification and Characterisation of Organic Compounds

Purification - Crystallization, sublimation, distillation, differential extraction, and chromatography - principles and their applications.

Qualititative analysis Detection of nitrogen, sulphur, phosphorus, and halogens.

Quantitative analysis (basic principles only) - Estimation of carbon, hydrogen, nitrogen, halogens, sulphur, and phosphorus. Calculations of empirical formulae and molecular formulae: Numerical problems in organic quantitative analysis.

UNIT 14 Some Basic Principles of Organic Chemistry

Tetravalency of carbon: Shapes of simple molecules - hybridization (*s* and *p*): Classification of organic compounds based on functional groups: and those containing halogens, oxygen, nitrogen, and sulphur; Homologous series: Isomerism - structural and stereoisomerism.

Nomenclature (Trivial and IUPAC)

Covalent bond fission - Homolytic and heterolytic: free radicals, carbocations and carbanions; stability of carbocations and free radicals, electrophiles and nucleophiles.

Electronic displacement in a covalent bond - Inductive effect, electromeric effect, resonance and hyperconjugation.

Common types of organic reactions - Substitution, addition, elimination, and rearrangement

UNIT 15 Hydrocarbons

Classification, isomerism, IUPAC nomenclature, general methods of preparation, properties, and reactions.

Alkanes - Conformations: Sawhorse and Newman projections (of ethane): Mechanism of halogenation of alkanes.

Alkenes - Geometrical isomerism: Mechanism of electrophilic addition: addition of hydrogen, halogens, water, hydrogen halides (Markownikoffs and peroxide effect): Ozonolysis and polymerization.

Alkynes - Acidic character: Addition of hydrogen, halogens, water, and hydrogen halides: Polymerisation.

Aromatic hydrocarbons - Nomenclature, benzene - structure and aromaticity: Mechanism of electrophilic substitution: halogenation, nitration.

Friedel-Craft's alkylation and acylation, directive influence of the functional group in monosubstituted benzene.

UNIT 16 Organic Compounds Containing Halogens

General methods of preparation, properties, and reactions: Nature of C—*X* bond; Mechanisms of substitution reactions. Uses: Environmental effects of chloroform, iodoform, freons and DDT.

UNIT 17 Organic Compounds Containing Oxygen

General methods of preparation, properties, reactions, and uses.

ALCOHOLS, PHENOLS, AND ETHERS

Alcohols: Identification of primary, secondary, and tertiary alcohols: mechanism of dehydration.

Phenols: Acidic nature, electrophilic substitution reactions: halogenation. nitration and sulphonation. Reimer - Tiemann reaction.

Ethers: Structure.

Aldehyde and Ketones: Nature of carbonyl group; Nucleophilic addition to $>C{=}O$ group, relative reactivities of aldehydes and ketones; Important reactions such as - Nucleophilic addition reactions (addition of HCN. NH_3, and its derivatives), Grignard reagent; oxidation: reduction (Wolf Kishner and Clemmensen); the acidity of α-hydrogen. aldol condensation, Cannizzaro reaction. Haloform reaction, Chemical tests to distinguish between aldehydes and ketones.

Carboxylic Acids

Acidic strength and factors affecting it.

UNIT 18 Organic Compounds Containing Nitrogen

General methods of preparation. Properties, reactions, and uses.

Amines: Nomenclature, classification structure, basic character, and identification of primary, secondary, and tertiary amines and their basic character.

Diazonium Salts: Importance in synthetic organic chemistry.

UNIT 19 Biomolecules

General introduction and importance of biomolecules.

CARBOHYDRATES - Classification; aldoses and ketoses: monosaccharides (glucose and fructose) and constituent monosaccharides of oligosaccharides (sucrose, lactose, and maltose).

PROTEINS - Elementary Idea of α-amino acids, peptide bond, polypeptides. Proteins: primary, secondary, tertiary, and quaternary structure (qualitative idea only), denaturation of proteins, enzymes.

VITAMINS – Classification and functions.

NUCLEIC ACIDS – Chemical constitution of DNA and RNA.

Biological functions of nucleic acids.

Hormones (General introduction)

UNIT 20 Principles Related To Practical Chemistry

Detection of extra elements (Nitrogen, Sulphur, halogens) in organic compounds; Detection of the following functional groups; hydroxyl (alcoholic and phenolic), carbonyl (aldehyde and ketones) carboxyl, and amino groups in organic compounds.

- The chemistry involved in the preparation of the following:

  Inorganic compounds; Mohr's salt, potash alum.

  Organic compounds: Acetanilide, p-nitro acetanilide, aniline yellow, iodoform.

- The chemistry involved in the titrimetric exercises – Acids, bases, and the use of indicators, oxalic-acid vs $KMnO_4$, Mohr's salt vs $KMnO_4$.
- Chemical principles involved in the qualitative salt analysis:

Cations — Pb^{2+}, Cu^{2+}, Al^{3+}, Fe^{3+}, Zn^{2+}, Ni^{2+}, Ca^{2+}, Ba^{2+}, Mg^{2+} NH^{4+}.

Anions – CO_3^{2-}, S^{2-}, SO_4^{2-}, NO_2, NO_3, Cl^-, Br^-, I^- (Insoluble salts excluded).

Chemical principles involved in the following experiments:

1. Enthalpy of solution of $CuSO_4$.
2. Enthalpy of neutralization of strong acid and strong base.
3. Preparation of lyophilic and lyophobic sols.
4. Kinetic study of the reaction of iodide ions with hydrogen peroxide at room temperature.

MATHEMATICS

UNIT 1 Sets, Relations And Functions

Sets and their representation: Union, intersection, and complement of sets and their algebraic properties; Power set; Relation, Type of relations, equivalence relations, functions; one-one, into and onto functions, the composition of functions

UNIT 2 Complex Numbers And Quadratic Equations

Complex numbers as ordered pairs of reals, Representation of complex numbers in the form $a + ib$ and their representation in a plane, Argand diagram, algebra of complex number, modulus, and argument (or amplitude) of a complex number, Quadratic equations in real and complex number system and their solutions, Relations between roots and co-efficient, nature of roots, the formation of quadratic equations with given roots

UNIT 3 Matrices And Determinants

Matrices, algebra of matrices, type of matrices, determinants, and matrices of order two and three, evaluation of determinants, area of triangles using determinants, Adjoint and evaluation of inverse of a square matrix using determinates and Test of consistency and solution of simultaneous linear equations in two or three variables using matrices

UNIT 4 Permutations And Combinations

The fundamental principle of counting, permutation as an arrangement and combination as section, Meaning of $P(n,r)$ and $C(n,r)$, simple applications

UNIT 5 Binomial Theorem And Its Simple Applications

Binomial theorem for a positive integral index, general term and middle term, and simple applications

UNIT 6 Sequence And Series

Arithmetic and Geometric progressions, insertion of arithmetic, geometric means between two given numbers, Relation between A.M and G.M

UNIT 7 Limit, Continuity And Differentiability

Real–valued functions, algebra of functions, polynomials, rational, trigonometric, logarithmic, and exponential functions, inverse function. Graphs of simple functions. Limits, continuity and differentiability. Differentiation of the sum, difference, product, and quotient of two functions. Differentiation of trigonometric, inverse trigonometric, logarithmic, exponential, composite, and implicit functions; derivatives of order up to two, Applications of derivatives:
Rate of change of quantities, monotonic-Increasing and decreasing functions, Maxima and minima of functions of one variable,

UNIT 8 Integral Calculus

Integral as an anti-derivative, Fundamental integral involving algebraic, trigonometric, exponential, and logarithmic functions. Integrations by substitution, by parts and by partial functions. Integration using trigonometricidentities. Evaluation of simple integrals of the type

$$\int \frac{dx}{x^2 \pm a^2},\ \int \frac{dx}{\sqrt{x^2 \pm a^2}},\ \int \frac{dx}{a^2 - x^2},\ \int \frac{dx}{\sqrt{a^2 - x^2}},\ \int \frac{dx}{ax^2+bx+c},\ \int \frac{dx}{\sqrt{ax^2+bx+c}},\ \int \frac{(px+q)dx}{ax^2+bx+c},\ \int \frac{(px+q)dx}{\sqrt{ax^2+bx+c}},\ \int \sqrt{a^2 \pm x^2}\,dx,\ \int \sqrt{x^2 - a^2}\,dx$$

The fundamental theorem of calculus, properties of definite integrals. Evaluation of definite integrals, determining areas of the regions bounded by simple curves in standard form

UNIT 9 Diffrential Equations

Ordinary differential equations, their order, and degree, the solution of differential equation by the method of separation of variables, solution of a homogeneous and linear differential equation of the type $\frac{dy}{dx}+p(x)y=q(x)$

UNIT 10 Co-ordinate Geometry

Cartesian system of rectangular coordinates in a plane, distance formula, sections formula, locus and its equation, the slope of a line, parallel and perpendicular lines, intercepts of a line on the co-ordinate axis.

Straight line - Various forms of equations of a line, intersection of lines, angles between two lines, conditions for concurrence of three lines, the distance of a point form a line, co-ordinate of the centroid, orthocentre, and circumcentre of a triangle,

Circle, conic sections - A standard form of equations of a circle, the general form of the equation of a circle, its radius and central, equation of a circle when the endpoints of a diameter are given, points of intersection of a line and a circle with the centre at the origin and sections of conics, equations of conic sections (parabola, ellipse, and hyperbola) in standard forms,

UNIT 11 Three Dimensional Geometry

Coordinates of a point in space, the distance between two points, section formula, directions ratios, and direction cosines, and the angle between two intersecting lines. Skew lines, the shortest distance between them, and its equation. Equations of a line

UNIT 12 Vector Algebra

Vectors and scalars, the addition of vectors, components of a vector in two dimensions and three-dimensional space, scalar and vector products

UNIT 13 Statistics And Probability

Measures of discretion; calculation of mean, median, mode of grouped and ungrouped data calculation of standard deviation, variance, and mean deviation for grouped and ungrouped data.

Probability: Probability of an event, addition and multiplication theorems of probability, Baye's theorem, probability distribution of a random variate,

UNIT 14 Trigonometry

Trigonometrical identities and trigonometrical functions, inverse trigonometrical functions, and their properties,

JEE Main
TEST DRIVE

SECTION I
Foundation Builders

This section contains **3 Practice Sets** of beginning level, helping you to strengthen your foundational knowledge and build foundation for higher level practice.

SECTION I

Foundation Builders

PRACTICE SET 1

Total Marks : 300	Total Time : 3 hrs	No. of Questions : 75

Instructions

1. There are **three** parts in the questions paper consisting of **Physics**, **Chemistry** and **Mathematics** having 25 questions in each part of equal weightage. Each part (subject) has two sections.
2. **Section A** This section contains 20 multiple choice questions which have only one correct answer. Each question carries **4 marks** for correct answer and **– 1 mark** for wrong answer.
3. **Section B** This section contains 5 questions. The answer to each of the questions is a numerical value. Each question carries **4 marks** for correct answer and **– 1 mark** for wrong answer. For Section B, the answer should be rounded off to the nearest integer.

PHYSICS

Section A : Objective Type Questions

1. A stone is thrown vertically up from the top of a tower which reaches the ground in time t_1 and another stone is thrown vertically down with the same speed which reaches ground in time t_2. A third stone is released from rest from the same location reaches the ground in time t_3, then

(a) $t_3 = \dfrac{t_1 - t_2}{2}$ (b) $t_3 = \sqrt{t_1 t_2}$

(c) $\dfrac{1}{t_3} = \dfrac{1}{t_2 - t_1}$ (d) $t_3^2 = t_1^2 - t_2^2$

2. Two massive bodies of masses m_1 and m_2 are at rest and separated by infinite distance. Now, they move towards each other due to their mutual attraction. The velocity of approach, when distance between m_1 and m_2 was r, is

(a) $\sqrt{\dfrac{2G}{r}(m_1 + m_2)}$ (b) $\sqrt{\dfrac{2G}{r}(m_1 - m_2)}$

(c) $\sqrt{\dfrac{2Gm_1 m_2}{r(m_1 + m_2)}}$ (d) $\sqrt{\dfrac{2G}{r} m_1 m_2}$

3. **Assertion** The net dipole moment of a given volume containing large number of polar molecules is zero.

Reason The dielectric constant of polar molecules is independent of temperature.

In the light of above statements, choose the most appropriate answer from the options given below.

(a) Both Assertion and Reason are correct and Reason is the correct explanation of Assertion.
(b) Both Assertion and Reason are correct but Reason is not the correct explanation of Assertion.
(c) Assertion is correct but Reason is incorrect.
(d) Reason is correct but Assertion is incorrect.

4. A boy is in a lift which is moving downwards with an acceleration $g/5$. At an instant when the velocity of the lift is 10 m/s, then boy throws a ball upwards with a velocity 10 m/s. The time after which, ball will return into the hand of the boy is

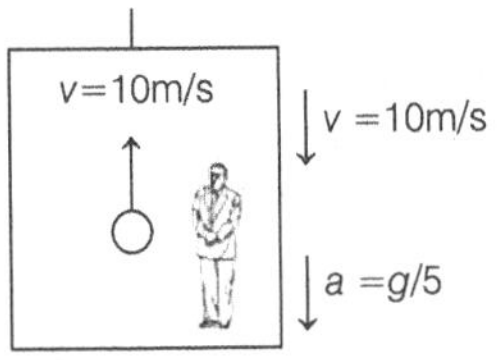

(a) 3 s (b) 4 s
(c) 5 s (d) Not possible

5. **Statement I** If the centre of mass of a system is at rest, the KE of the system must be zero.

Statement II The change in velocity depends on the external force.

In the light of above statements, choose the most appropriate answer from the options given below.

(a) Statement I is true but Statement II is false.
(b) Both Statement I and Statement II are true.
(c) Statement I is false but Statement II is true.
(d) Both Statement I and Statement II are false.

6. Two resistances are expressed as $R_1 = (3 \pm 0.2)\Omega$ and $R_2 = (6 \pm 0.3)\Omega$. The equivalent resistance of their series and parallel combination are, respectively.

(a) $(9 \pm 0.1)\Omega, (2 \pm 0.12)\Omega$
(b) $(9 \pm 0.5)\Omega, (2 \pm 0.12)\Omega$
(c) $(9 \pm 0.8)\Omega, (2 \pm 0.1)\Omega$
(d) $(9 \pm 0.6)\Omega, (2 \pm 0.02)\Omega$

7. The curve represents the distribution of potential along the straight line joining two charges Q_1 and Q_2 separated by distance r, then which of the following statements are correct ?

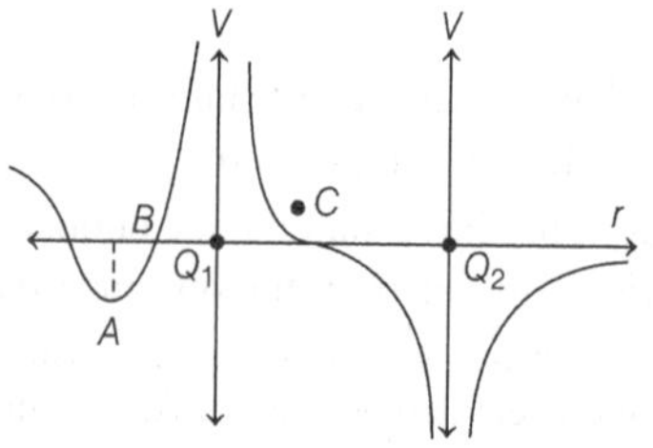

I. Magnitude of charge Q_1 is less than that of charge Q_2
II. Q_1 is positive charge and Q_2 is negative charge
III. A is a point of stable equilibrium
IV. B and C are points of unstable equilibrium.

(a) I and II
(b) I, II, III and IV
(c) I, II and III
(d) I and III

8. An electron with kinetic energy 3.4 eV is incident on a hydrogen atom in its ground state. The collision

(a) must be elastic
(b) may be partially elastic
(c) must be completely inelastic
(d) may be completely inelastic

9. The length of an open organ pipe is 60 cm and that of a closed organ pipe is 40 cm, then the ratio of the third overtone of closed organ pipe and the second overtone of open organ pipe is

(a) 4 : 3
(b) 3 : 4
(c) 7 : 4
(d) 4 : 7

10. A lamp emits monochromatic green light uniformly in all directions. The lamp is 3% efficient in converting electric power to electromagnetic waves and consume 100W of power. The amplitude of the electric field associated with the electromagnetic radiation at a distance of 5 m from the lamp will be

(a) $2.68\,Vm^{-1}$
(b) $3.15\,Vm^{-1}$
(c) $2.01\,Vm^{-1}$
(d) 0

11. The specific heat at constant pressure is C_p and specific heat at constant volume is C_V, then the ratio of specific heats $\left(\frac{C_p}{C_V}\right)$ in terms of degree of freedom (f) is given by

(a) $\left(1+\frac{3}{f}\right)$ (b) $\left(1+\frac{2}{f}\right)$ (c) $\left(1+\frac{f}{3}\right)$ (d) $\left(1+\frac{1}{f}\right)$

12. A pure germanium plate of area $3.5\times 10^{-4}\ m^2$ and of thickness 1.5×10^{-3} m is connected across a battery of potential 5 V. Find the amount of heat generated in the plate in 120 s. Given that, the concentration of carriers in germanium at room temperature is 1.6×10^6 per cubic metre. The mobilities of electrons and holes are $0.4\ m^2 V^{-1} s^{-1}$ and $0.2\ m^2 V^{-1} s^{-1}$, respectively.

(a) 10.74×10^{-11}
(b) 9×10^{-6}
(c) 10.48×10^{-9}
(d) 6.26×10^{-10}

13. In an experiment of coefficient of viscosity, two drops of water of same size are falling through air with terminal velocity of 10 cm/s. If the two drops combine to form a single drop, then the terminal velocity is

(a) 15.9 cm/s (b) 22.8 cm/s (c) 12.8 cm/s (d) 18.4 cm/s

14. For a projectile thrown into space with a speed v, the horizontal range is $\frac{\sqrt{3}v^2}{2g}$. The vertical range is $\frac{v^2}{8g}$. The angle which the projectile makes with the horizontal initially is

(a) 15° (b) 30° (c) 45° (d) 60°

15. A body of m kg slides with velocity 30 m/s along the curve of vertical circle from point A to B in frictionless path. The velocity of the body at B is (Given, $R = 30$ m, $g = 10\,ms^{-2}$ and $\sqrt{2} = 1.4$)

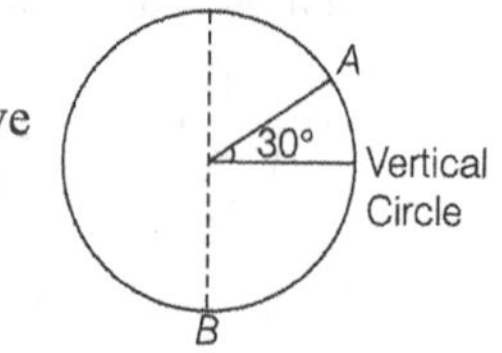

(a) 30 m/s
(b) 20 m/s
(c) 42 m/s
(d) 19.6 m/s

16. In changing the state of a gas adiabatically from an equilibrium state A to another equilibrium state B, an amount of work equal to 22.3 J is done on the system. If the gas is taken from state A to B *via* a process in which the net heat absorbed by the system is 9.35 cal, how much is the net work done by the system in the latter case ?

(a) 15.6 J (b) 11.2 J (c) 14.9 J (d) 16.8 J

17. Which of the following logic state is represented by the combination of logic gates ?

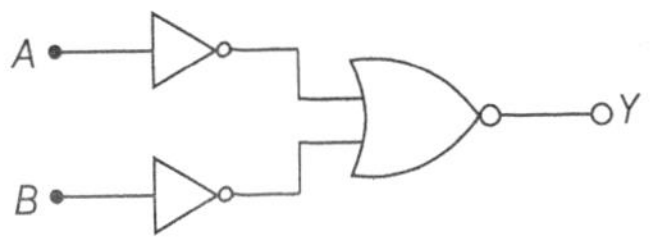

(a) NAND gate (b) NOR gate
(c) AND gate (d) OR gate

18. Match the Column I with Column II and select the correct option from the codes given below.

	Column I		Column II
A.	$\oint \mathbf{B}\cdot d\mathbf{l} = \mu_0 I$	1.	Gauss's law
B.	$\oint \mathbf{B}\cdot d\mathbf{l} = \mu_0\left(I + \varepsilon_0 \frac{d\phi_E}{dt}\right)$	2.	Faraday's laws of electromagnetic induction
C.	$\oint \mathbf{E}\cdot d\mathbf{l} = -\frac{d}{dt}\oint_s \mathbf{B}\cdot d\mathbf{S}$	3.	Ampere-Maxwell's law
D.	$\oint \mathbf{E}\cdot d\mathbf{S} = \frac{1}{\varepsilon_0}\oint_s \rho dS$	4.	Ampere's circuital law

Codes

	A	B	C	D		A	B	C	D
(a)	1	2	3	4	(b)	4	3	2	1
(c)	4	2	1	3	(d)	3	1	2	4

19. Two waves are travelling in the opposite direction, on superposition, produces standing wave. The transverse displacement is given by

$$y(x,t) = 2\sin\left(\frac{8\pi}{9}x\right)\cos(400\pi t)$$

Speed of the travelling wave moving in the positive x-direction is (x and t are in metre and second, respectively)

(a) 160 m/s (b) 450 m/s
(c) 180 m/s (d) 220 m/s

20. A load of 4.0 kg is suspended from a ceiling through a steel wire of length 2.0 m and radius 2.0 mm. It is found that, the length of the wire increases by 0.031 mm as equilibrium is achieved. Taking, $g = 3.1\pi\ \text{ms}^{-2}$, the Young's modulus of steel is

(a) $2.0\times10^{8}\ \text{Nm}^{-2}$ (b) $2.0\times10^{9}\ \text{Nm}^{-2}$
(c) $2.0\times10^{11}\ \text{Nm}^{-2}$ (d) $2.0\times10^{13}\ \text{Nm}^{-2}$

Section B : Numerical Value Type Questions

21. A mixture of gas consists of 3 moles of O_2 and 5 moles of Ar at temperature T. Considering only rotational and translation modes. The total internal energy of the system is found to be xRT. The value of x is ....

22. The surface of some metal is radiated by waves of $\lambda_1 = 3.5\times10^{-7}$ m and $\lambda_2 = 6.2\times10^{-7}$ m, respectively. The ratio of the stopping potential in two cases is 2 : 1. The work function of the material is $\frac{1}{x}$ eV, then the value of x is ....

23. A transparent cube contains a small air bubble. Its apparent distance is 2 cm when seen through one face and 5 cm when seen through other face. If the refractive index of the material of the cube is 1.5, then the real length of the edge of the cube must be $\frac{x}{2}$ cm. The value of x is ....

24. The resistance of a 10 m long wire is 10 Ω. Its length is increased by 25% by stretching the wire uniformly. The resistance of wire will change to (approximately) ....Ω.

25. A rectangular loop with a sliding conductor of length $l = 1.0$ m is situated in a uniform magnetic field $B = 2$T, perpendicular to the plane of loop. Resistance of conductor is $R = 2\Omega$. Two resistances of 6Ω and 3Ω are connected as shown in figure. The external force required to keep the conductor moving with a constant velocity $v = 2\ \text{ms}^{-1}$ is .... N.

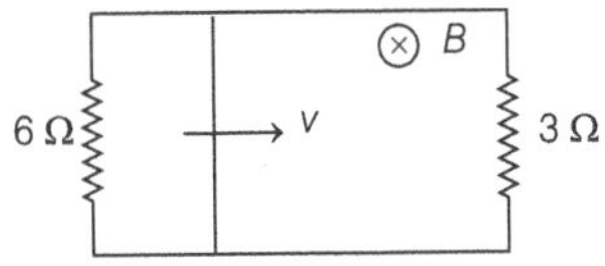

CHEMISTRY

Section A : Objective Type Questions

26. Bromine and fluorine react to form more than one compound. When an atom of bromine in its second excited state, reacts with fluorine to form a binary compound. The formula and shape of the compound is

(a) BrF and linear
(b) BrF_3 and pyramidal
(c) BrF_3 and T-shaped
(d) BrF_5 and square pyramidal

27. The major product of the given reaction is

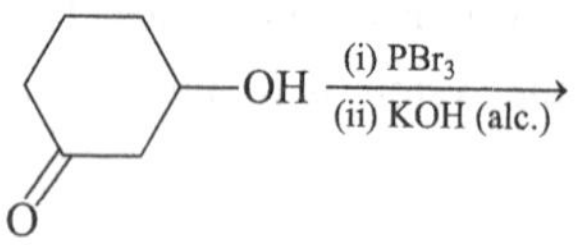

(a) 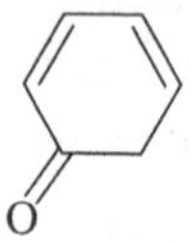
(b) 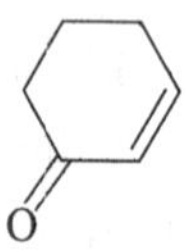
(c) 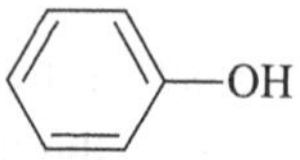
(d) 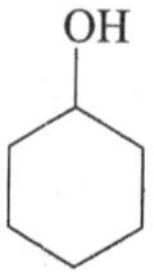

28. Arrange the following in correct order of increasing dipole moment.

I. 1, 2, 3-trichlorobenzene
II. 1, 4-dichlorobenzene
III. 1, 2, 4-trichlorobenzene
IV. 1, 3, 5-trichlorobenzene

(a) II < IV < III < I
(b) II < IV < I < III
(c) II < III < IV < I
(d) I < III < II < IV

29. An organic compound having molecular formula $C_6H_4NO_2F$ is a cyclic compound which when treated with DMF and $(CH_3)_2NH$ produces A and on further treatment with H_2/Pt produces B. The possible molecular structure of B is

(a) 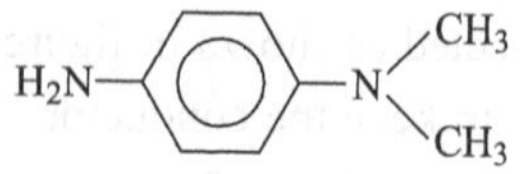
(b) 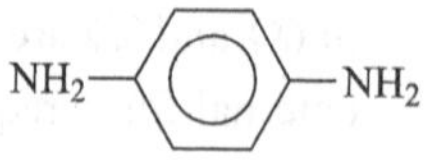
(c) 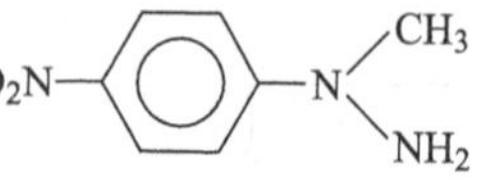
(d) 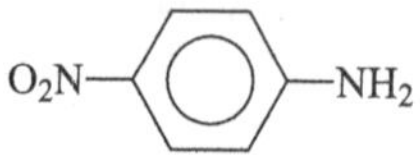

30. Match List-I with List-II.

List-I	List-II
A. 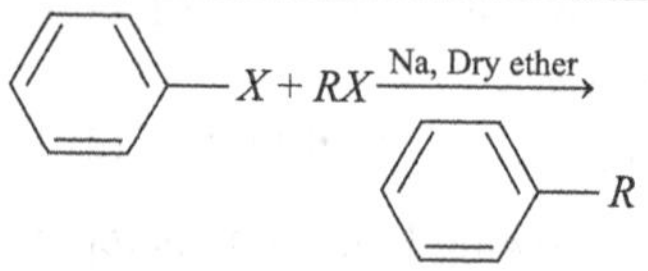	I. Fittig reaction
B. 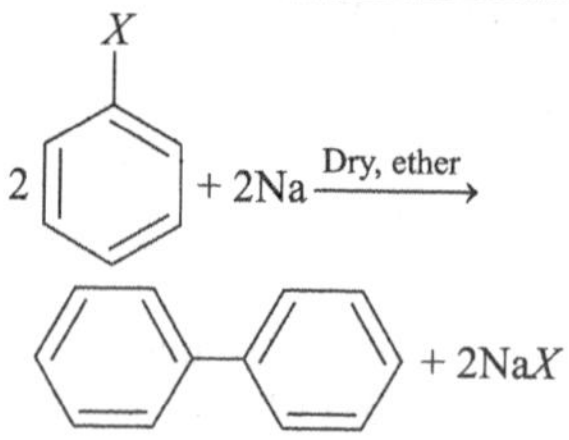	II. Wurtz-Fittig reaction
C. 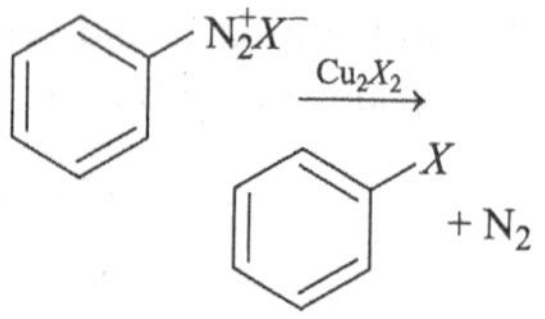	III. Hunsdiecker reaction
D. $C_6H_5COOAg \xrightarrow{Br_2} C_6H_5Br + CO_2 + Ag$	IV. Sandmeyer reaction

Choose the correct answer from the options given below.

(a) A-III, B-II, C-I, D-IV
(b) A-IV, B-III, C-II, D-I
(c) A-II, B-I, C-IV, D-III
(d) A-I, B-III, C-II, D-IV

31. Match List -I with List-II.

List -I	List-II
A. Myosin	I. Ascorbic acid
B. Vitamin C	II. Fibrous protein
C. Haemoglobin	III. Steroids
D. Androgens	IV. Globular protein

Choose the correct answer from the options given below.

(a) A-II, B-III, C-IV, D-I
(b) A-II, B-I, C-IV, D-III
(c) A-I, B-IV, C-II, D-III
(d) A-I, B-IV, C-III, D-II

32. On the basis of following reaction, choose the correct option for A and B, respectively.

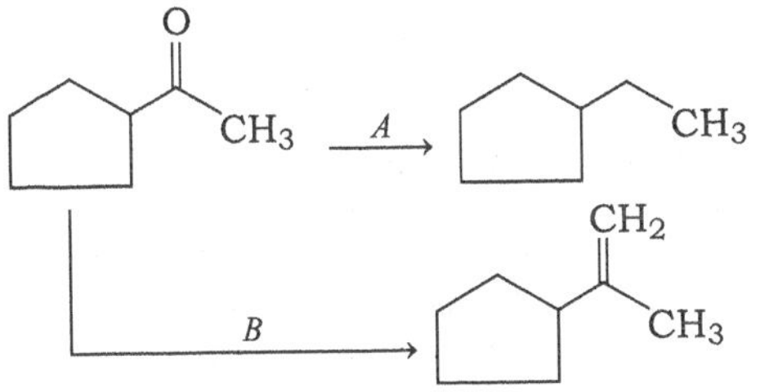

(a) $PPh_3 = CH_2$, Zn - Hg/HCl
(b) $PPh_3 = CH_2$, $LiAlH_4$
(c) $NaBH_4$, Zn - Hg / HCl
(d) Zn - Hg/HCl, $PPh_3 = CH_2$

33. A solution containing sulphate and sulphite ion can be easily distinguished by
(a) $FeSO_4$
(b) $Na_2[Fe(CN)_5NO]$
(c) $BaCl_2$
(d) $Na_3[Co(NO_2)_6]$

34. Formaldehyde + $NH_3 \longrightarrow X$
Here, 'X' is
(a) nitro compound used as sedative.
(b) acid used as an antipyretic.
(c) tertiary amine used as an antiseptic.
(d) heterocyclic compound used as an antiseptic.

35. Phosphorus on reaction with NaOH produces a colourless gas with rotten fish smell. The gas gives a vortex ring. The colourless gas is
(a) PH_3 (b) P_2O_3
(c) P_2O_5 (d) P_2S_5

36. Elements A and D react to form more than one type of compounds. The solubility product values of three such compounds are as follows

Compound	K_{sp}
AD	5×10^{-8}
AD_2	3×10^{-14}
AD_3	2.8×10^{-15}

I. Solubility of compound AD is more than that of compound AD_3.
II. Solubility of compound AD_2 is more than that of compound AD_3.
III. Solubility of compound AD_3 is more than that of compound AD and less than compound AD_2.
IV. Solubility of compound AD_2 is less than AD and AD_3.

Choose the most appropriate answer from the options given below.
(a) I, II and IV are correct.
(b) II, III and IV are correct.
(c) I, III and IV are correct.
(d) I and IV are correct.

37. Which of the following complex/ion has 3 geometrical isomers?
(a) $[Co(NH_3)_4Cl_2]^+$
(b) $[Co(H_2O)_2(NH_3)_2(en)]^{3+}$
(c) $[Pt(NH_3)_2Cl_2]$
(d) $[Cr(H_2O)_2(NH_3)_4]^{3+}$

38. Consider the following statements.
Statement I Na^+ and Al^{3+} are isoelectronic ions but the magnitude of ionic radius of Na^+ is less than that of Al^{3+}.
Statement II In isoelectronic ions, greater the negative charge lower is the ionisation energy.
In the light of the above statements, choose the correct answer from the options given below:
(a) Both statement I and statement II are false.
(b) Statement I is true but statement II is false.
(c) Statement I is false but statement II is true.
(d) Both statement I and statement II are true.

39. Two solutions of glucose have osmotic pressure 1.5 atm and 2.5 atm respectively. 1 L of first solution is mixed with 2 L of second solution at the same temperature. Then, the osmotic pressure of resultant solution will be
(a) 2.16 atm (b) 3.5 atm
(c) 5.12 atm (d) 7.2 atm

40. The bond order of N_2^+ and N_2^- and N_2 is
(a) 2.5, 3 and 3 (b) 3, 2.5 and 3
(c) 2.5, 2.5 and 3 (d) 3, 3 and 2.5

41. Resistance of 0.2 M solution of an electrolyte is 60 Ω. The specific conductance of the solution is $1.5\ Sm^{-1}$. The resistance of 0.5 M solution of the same electrolyte is 280 Ω. The molar conductivity of 0.5 M solution of the electrolyte in Scm^2mol^{-1} is
(a) 6.420 (b) 6.235 (c) 5.381 (d) 5.833

42. Identify the hormone that increase the glucose level in the blood.
(a) Oxytocin (b) Glucagon
(c) Insulin (d) Thyroxine

43. The number of species that are colourless in aqueous solutions is

$$Ti^{4+}, V^{4+}, Mn^{2+}, Fe^{3+}, Ni^{2+}$$

(a) 2 (b) 1 (c) 5 (d) 0

44. The incorrect statement about the 'Hinsberg reagent' is

(a) It is known as *p*-toluenesulphonyl chloride.
(b) On treatment with secondary, it leads to a product, that is soluble in alkali.
(c) It doesn't react with tertiary amines.
(d) It is used to distinguish primary and secondary amines.

45. Given below are two statements, one is labelled as Assertion (A) and the other is labelled as Reason (R).

Assertion (A) 1-chloro-3, 3-di-methyl-1-butene do not undergo nucleophilic substitution reaction easily.

Reason (R) Even though the intermediate carbocation is stabilised by loosely held π-electrons, the cleavage is difficult because of strong bonding.

In the light of the above statements, choose the correct answer from the options given below.

(a) Both (A) and (R) are true and (R) is the true explanation of (A).
(b) Both (A) and (R) are true but (R) is not the true explanation of (A).
(c) (A) is true but (R) is false.
(d) (A) is false but (R) is true.

Section B : Numerical Value Type Questions

46. Number of electrons will have magnetic quantum number ($m=0$) in Fe^{3+} ion is ....

47. A radioactive substance decays in such a way that its amount decreases from initial value of 80 dpm to 40 dpm in 40 minutes. The number of atoms of radioactive substance present initially in the sample are approximately ....

48. 8.4 g of pure aniline is treated with bromine water at room temperature to give a white precipitate of the product '*P*'. The mass of product '*P*' obtained is 26.4 g. The percentage yield is ....%

49. If the standard molar enthalpy change for combustion of graphite powder is -3.45×10^{2} kJ mol^{-1}, the amount of heat generated on combustion of 1g of graphite powder is .... kJ. (Nearest integer)

50. The total number of equations that involve reduction of metal is ....

A. $Fe_2O_3 + 3CO \longrightarrow 2Fe + 3CO_2$
B. $TiO_2 + 2C + 2Cl_2 \longrightarrow TiCl_4 + 2CO$
C. $Cr_2O_3 + 2Al \longrightarrow 2Cr + Al_2O_3$
D. $TiCl_4 + 4Na \longrightarrow Ti + 4NaCl$

MATHEMATICS

Section A : Objective Type Questions

51. Let $A=\int_0^1 \frac{e^x}{x+1}dx$, then the value of $\int_0^1 \frac{xe^{x^2}}{x^2+1}dx$ is

(a) A^2 (b) $\frac{1}{2}A$
(c) $2A$ (d) $\frac{1}{2}A^2$

52. Let $z=1+ai$ be a complex number, $a>0$, such that z^3 is a real number. Then, the sum of $1+z+z^2+\ldots\ldots+z^{11}$ is equal to

(a) $1215\sqrt{3}\,i$ (b) $-1215\sqrt{3}\,i$
(c) $-1365\sqrt{3}\,i$ (d) $1365\sqrt{3}\,i$

53. The function $g(x)=\lim\limits_{n\to\infty}\frac{f(x)+x^n h(x)}{1+x^n}$ shall be continuous everywhere, if

Statement I $f(1)=h(1)$

Statement II $f(-1)=h(-1)$

Then, which of the following option is correct?

(a) Statement I is true and Statement II is false
(b) Statement I is false and Statement II is true
(c) Statement I and Statement II both are true
(d) Statement I and Statement II both are false

54. The sum of $1\cdot 2C_2+2\cdot 3C_3+\ldots+(n-1)nC_n$ is equal to

(a) $n(n-1)2^{n-1}$ (b) $n(n-1)2^{n-2}$
(c) $(n-1)2^{n-2}$ (d) $n\,2^{n-2}$

55. The area bounded by $y=x^3$, $3y+x=4$ and X-axis is equal to

(a) 2 (b) 1.9 (c) 1.75 (d) 1.5

56. A rod of length l moves such that its end A and B always lie on the lines $3x-y+5=0$ and $y+5=0$, then locus of the point P, which divide AB internally in the ratio $2:1$, is

(a) $l^2 = \frac{1}{4}(3x+3y-5)^2 + (3y+5)^2$

(b) $l^2 = \frac{1}{4}(3x-3y+5)^2 + (3y-5)^2$

(c) $l^2 = \frac{1}{4}(3x-3y-5)^2 + (3y-5)^2$

(d) None of the above

57. The distance of point of intersection of lines $\frac{x-4}{1} = \frac{y+3}{-4} = \frac{z+1}{7}$ and $\frac{x-1}{2} = \frac{y+1}{-3} = \frac{z+10}{8}$ from (1, – 4, 7), is

(a) $\sqrt{15}$ (b) $\sqrt{27}$
(c) $\sqrt{26}$ (d) $\sqrt{14}$

58. Let $h(x)$ be a function such that $h'(x) = h(x)$, $h(0) = 1$ and $g(x)$ be a function such that $h(x) + g(x) = 2x$. Then, $\int_0^2 h(x)\,g(x)dx$ is equal to

(a) $(e^2-1)(e^2-3)$ (b) $\frac{(e^2+1)(e^2-3)}{2}$
(c) $\frac{(e^2+1)(e^2-1)}{2}$ (d) $\frac{(e^2+1)(5-e^2)}{2}$

59. If $\sin^{-1} x + \sin^{-1} y + \sin^{-1} z = \frac{\pi}{2}$, then the value of $x^2 + y^2 + z^2 + 2xyz$ is equal to

(a) 8 (b) 1
(c) 0 (d) 3

60. Let $|\mathbf{a}| = |\mathbf{c}| = 1$ and $|\mathbf{b}| = 4$ with $\mathbf{a} \times \mathbf{b} = 2\mathbf{a} \times \mathbf{c}$. The angle between $\mathbf{a}$ and $\mathbf{c}$ is $\cos^{-1}\left(\frac{1}{4}\right)$. If $\mathbf{b} - 2\mathbf{c} = \lambda\mathbf{a}$ $(\lambda > 0)$, then λ is equal to

(a) 1 (b) 3
(c) $\frac{3}{2}$ (d) $\frac{2}{3}$

61. The total number of local maxima and local minima of the function $f(x) = \begin{cases} (3+x)^3, & -3 < x \le -1 \\ 3x^{\frac{2}{3}}, & -1 < x < 2 \end{cases}$ is

(a) 3 (b) 0
(c) 4 (d) 2

62. The number of positive integral solutions of the equation $abcde = 1050$ is 375 k, when $k \in N$. Then, k is equal to

(a) 2 (b) 5
(c) 6 (d) 11

63. The equation of curve which satisfy the equation $\frac{dy}{dx} = \frac{y(x+y^3)}{x(y^3-x)}$ and passing through the point (1, 1), is

(a) $y^3 + 2x + 3x^2 y = 0$
(b) $y^3 + 2x - 3x^2 y = 0$
(c) $y^3 + x + x^2 y = 0$
(d) $y^3 + 3x + 2x^2 y = 0$

64. Let p, q and r be three non-zero real numbers such that the equation $\sqrt{3}p\cos x + 2q\sin x = r$, x is in first and fourth quadrant, has two distinct real roots α and β with $\alpha + \beta = \frac{\pi}{3}$. Then, $q : p$ is equal to

(a) 2 : 1 (b) 3 : 2
(c) 1 : 2 (d) 2 : 3

65. The number of points with integral coordinates lying on or within the common region bounded by the curves $y^2 = 4|x|$ and the figure formed by joining the extremeties of latus rectum with straight lines is:

(a) 9 (b) 11
(c) 13 (d) 17

66. The range of the function $f(x) = [16^x + 2^{2x+1} + 1]$ is where, $[\cdot]$ denotes greatest integer function.

(a) Natural numbers (b) Whole numbers
(c) Even integers (d) Integers

67. If $x^3 + 2x^2 - x + \lambda = 0$ and $x^2 + 5x - 3\lambda = 0$ has a common root, then λ can be

(a) 1 (b) 4 (c) –2 (d) 2

68. If $D_k = \begin{vmatrix} 1 & n & n \\ 2k & n^2+n+1 & n^2+n \\ 2k-1 & n^2 & n^2+n+1 \end{vmatrix}$ and $\sum_{k=1}^{n} D_k = 56$, then n equals to

(a) 7 (b) 6 (c) 5 (d) 4

69. If C_i denotes ${}^4C_i$, then the value of $\sum_{i=1}^{4}\left(\frac{i \cdot C_i}{C_i + C_{4-i}}\right)^2$ is equal to

(a) $\frac{20}{3}$ (b) $\frac{15}{7}$
(c) $\frac{30}{7}$ (d) $\frac{15}{2}$

70. Let $p, q, r \in Z$ such that $p^2 + q^2 + r^2 = 2$,

let $A = \begin{bmatrix} p & q & r \\ q & r & p \\ r & p & q \end{bmatrix}$. Then, probability of $|A| \neq 0$ is

(a) $\frac{1}{4}$ (b) $\frac{1}{2}$ (c) $\frac{2}{3}$ (d) $\frac{1}{3}$

Section B : Numerical Value Type Questions

71. Let $g(x) = \sin hx$ and if $g[f(x)] = x$, then $\frac{1}{501} f\left(\frac{e^{1002} - 1}{2e^{501}}\right)$ is ....

72. If ratio of second and third term in the expansion of $(x + y)^n$ is equal to the ratio of third and fourth term in the expansion of $(x + y)^{n+3}$, then n is equal to ....

73. Let $a_1 = \tan^{-1} 2$, and $\{a_n\}_{n=1}^{\infty}$ be decreasing sequence of positive real numbers satisfying the condition $\sin(a_{n+1} - a_n) + \frac{\sin(a_n)\sin(a_{n+1})}{2^{n+1}} = 0$ for $n \geq 1$. Then, the value of $\lim_{n \to \infty} \cot a_n$ is equal to .....

74. An equilateral triangle is inscribed in an ellipse whose equation is $\frac{x^2}{9} + y^2 = 1$. If one vertex of triangle is (0, 1) and perimeter of triangle is k, then the value of $7\sqrt{3}\, k$ is ....

75. The mean and standard deviation of 20 observations are found to be 10 and 2, respectively. On review, it was found that an observation 10 was incorrect. Then, correct standard deviation if the wrong item is omitted, is ....

DETAILED SOLUTIONS

Physics

1. *(b)*

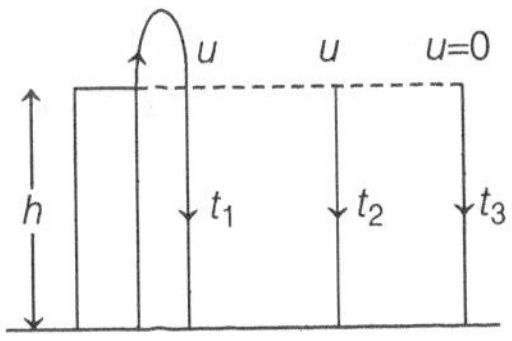

For stone 1,

$$h = -ut_1 + \frac{1}{2}gt_1^2 \quad \ldots(i)$$

For stone 2,

$$h = ut_2 + \frac{1}{2}gt_2^2 \quad \ldots(ii)$$

For stone 3,

$$h = \frac{1}{2}gt_3^2 \quad \ldots(iii)$$

Multiplying Eq. (i) by t_2 and Eq (ii) by t_1 and adding, we get

$$h(t_1 + t_2) = -ut_1t_2 + \frac{1}{2}gt_1^2t_2 + ut_2t_1 + \frac{1}{2}gt_2^2t_1$$

$$h(t_1 + t_2) = \frac{1}{2}gt_1t_2(t_1 + t_2)$$

$$h = \frac{1}{2}gt_1t_2 \quad \ldots(iv)$$

From Eqs. (iii) and (iv), we get

$$t_3^2 = t_1t_2$$

$$t_3 = \sqrt{t_1t_2}$$

2. *(a)* Consider v_1 and v_2 are velocities of approach at a distance r.

Apply conservation of momentum,

$$m_1v_1 - m_2v_2 = 0$$

$$m_1v_1 = m_2v_2 \quad \ldots(i)$$

Now, apply conservation of energy,

Change in PE = Change in KE

$$\frac{Gm_1m_2}{r} = \frac{1}{2}m_1v_1^2 + \frac{1}{2}m_2v_2^2$$

$$\frac{m_1^2v_1^2}{m_1} + \frac{m_2^2v_2^2}{m_2} = \frac{2Gm_1m_2}{r} \quad \ldots(ii)$$

Substitute value of m_1v_1 from Eq. (i)

$$\frac{m_2^2v_2^2}{m_1} + \frac{m_2^2v_2^2}{m_2} = \frac{2Gm_1m_2}{r}$$

$$(m_2v_2)^2 = \frac{2G}{r}\frac{(m_1m_2)^2}{(m_1 + m_2)}$$

$$v_2 = \sqrt{\frac{2Gm_1^2}{r(m_1 + m_2)}}$$

Again from Eq. (i),

$$v_1 = \sqrt{\frac{2Gm_2^2}{r(m_1 + m_2)}}$$

Approach velocity $= v_1 - (-v_2) = v_1 + v_2$

$$= \sqrt{\frac{2G}{r}(m_1 + m_2)}$$

3. *(c)* For polar molecules, the net dipole moment of a given volume containing large number of molecules is zero, because of the dipole moments of molecules are oriented in different direction due to thermal agitation. The dielectric constant of polar molecules depends on its temperature. Hence, option (c) is correct.

4. *(c)* This question is based on one-dimensional motion. Here, one thing should be noticed that the final displacement of the boy and the ball is same.

Boy	Ball
$u = +10$ m/s,	$u = -10$ m/s,
$a = +2\,\text{m/s}^2$	$a = +10\,\text{m/s}^2$

$\therefore \quad s_{\text{Boy}} = s_{\text{Ball}}$

$$10t + \frac{1}{2} \times 2 \times t^2 = -10t + \frac{1}{2} \times 10 \times t^2$$

$$\Rightarrow \quad 20t = 5t^2 - t^2$$

$$\Rightarrow \quad 20t = 4t^2$$

$$\Rightarrow \quad t = 5\,\text{s}$$

5. *(c)* Statement I is false but the Statement II is true.

If a body at rest explodes into two parts due to internal force, then its centre of mass will be at rest but the KE of the system is non-zero.

The change in velocity, depends on the external force.

6. *(b)* In series combination,

$$R = R_1 + R_2$$

$\therefore \quad R_{\text{eq}} = 3 + 6 = 9\Omega$

and $\quad \Delta R_{\text{eq}} = \Delta R_1 + \Delta R_2$

$$= 0.2 + 0.3 = 0.5$$

Thus $\quad R = (9 \pm 0.5)\Omega$

In parallel combination,

$$\frac{1}{R_{\text{eq}}} = \frac{1}{R_1} + \frac{1}{R_2} = \frac{1}{3} + \frac{1}{6} = \frac{1}{2}$$

$$R_{\text{eq}} = 2\Omega$$

Again $\frac{1}{R_{eq}} = \frac{1}{R_1} + \frac{1}{R_2}$

$$\frac{\Delta R}{R_{eq}^2} = \frac{\Delta R_1}{R_1^2} + \frac{\Delta R_2}{R_2^2}$$

$$\frac{\Delta R}{4} = \frac{0.2}{9} + \frac{0.3}{36}$$

$\Rightarrow$ $\Delta R = 0.12\ \Omega$

Thus, $R = (2 \pm 0.12)\Omega$

Hence, option (b) is correct.

7. *(a)* Clearly, Q_1 is +ve and Q_2 is –ve because potential due to + ve charge, is + ve and due to –ve charge, is –ve.

Secondly, $|Q_2|$ is more than $|Q_1|$ as seen from graph.

Also potential energy at point A would be $U = qV$.

For stable equilibrium, U should be minimum and U would be minimum only when charge placed at A is +ve.

Thus, stable equilibrium would depend on the sign of charge placed at A and thus statement III is not always true.

Points B and C are not equilibrium points.

8. *(a)* The energy levels of a hydrogen atom are given by the formula

$$E_n = \frac{-13.6}{n^2}\ \text{eV}$$

for $(n = 1)$

$$E_1 = \frac{-13.6}{(1)^2} = -13.6\ \text{eV}$$

For $(n = 2)$

$$E_2 = \frac{-13.6}{(2)^2} = -3.4\ \text{eV}$$

$$\Delta E = E_2 - E_1 = (-3.4) - (-13.6) = 10.2\ \text{eV}$$

Since, the energy required to excite the electron to the first excited state (10.2 eV) is greater than the kinetic energy of the incident electron (3.4 eV). Therefore, there is no energy loss in process, thus collision is elastic.

9. *(c)* Given, length of open organ pipe,

$$l_1 = 60\ \text{cm} = 0.6\ \text{m}$$

length of closed organ pipe,

$$l_2 = 40\ \text{cm} = 0.4\ \text{m}$$

As we know, for 3rd overtone of closed organ pipe,

$$(\nu_3)_c = \frac{7v}{4l_2} = \frac{7v}{4(0.4)} \quad \ldots \text{(i)}$$

Here, v is velocity of sound.

and for 2nd overtone of open organ pipe,

$$(\nu_2)_o = \frac{3v}{2l_1} = \frac{3v}{2 \times (0.6)} \quad \ldots \text{(ii)}$$

From Eqs. (i) and (ii), we get

$$\frac{(\nu_3)_c}{(\nu_2)_o} = \frac{\frac{7v}{4 \times 0.4}}{\frac{3v}{2 \times 0.6}} = \frac{7v \times 2 \times 0.6}{3v \times 4 \times 0.4} = \frac{7}{4}$$

Therefore, the ratio of third overtone of closed organ pipe and second overtone of open organ pipe is 7 : 4.

10. *(a)* Average intensity

$$= \frac{P}{4\pi R^2} = \frac{1}{2}\varepsilon_0 E_0^2 c$$

or

$$E_0 = \sqrt{\frac{P}{2\pi R^2 \varepsilon_0 c}}$$

$$= \sqrt{\frac{3}{2 \times 3.14 \times 25 \times 8.85 \times 10^{-12} \times 3 \times 10^8}}$$

$$= 2.68\ \text{Vm}^{-1}$$

11. *(b)* As we know that,

$$C_V = \frac{f}{2}R \text{ and } C_p = C_V + R$$

$$\therefore \quad C_p = \frac{f}{2}R + R = \left(1 + \frac{f}{2}\right)R$$

Now

$$\frac{C_p}{C_V} = \frac{\left(1 + \frac{f}{2}\right)R}{\frac{f}{2}(R)} = \frac{2+f}{f} = \left(1 + \frac{2}{f}\right)$$

Hence, option (b) is correct.

12. *(a)* Conductivity of semiconductor is given by

$$\sigma = n_e e \mu_e + n_h e \mu_h$$

Since semiconductor is intrinsic

$\Rightarrow$ $n_e = n_h = n_i$

or

$$\sigma = n_i e(\mu_e + \mu_h)$$

$$= 1.6 \times 10^6 \times 1.6 \times 10^{-19}\ (0.4 + 0.2)$$

$$= 1.536 \times 10^{-13}\ \Omega^{-1}\text{m}^{-1}$$

Current flowing, $I = jA$ $\quad \ldots$(i)

where, j = current density $= \sigma E = \sigma\left(\frac{V}{d}\right)$

From Eq. (i), we get

$$I = 1.536 \times 10^{-13}\left(\frac{5}{1.5 \times 10^{-3}}\right)(3.5 \times 10^{-4})$$

$$= 1.79 \times 10^{-13}\ \text{A}$$

Heat produced, $H = VIt$

$$= 5 \times 1.79 \times 10^{-13} \times 120$$

$$= 10.74 \times 10^{-11}\ \text{J}$$

13. *(a)* Here, volume of new drop = volume of two small drops.

$$\frac{4}{3}\pi R^3 = 2\times\frac{4}{3}\pi r^3$$

$$\Rightarrow \quad R^3 = 2r^3 \quad \ldots \text{(i)}$$

where, R = radius of new drop

and r = radius of each small drop.

Each drop acquires a velocity equal to its terminal velocity, when viscous force acting on it equals to its effective weight. Let v_1 and v_2 are the velocities of new drop and each small drop, respectively.

Then, $6\pi\eta R v_1 = \frac{4}{3}\pi R^3(\rho-\sigma)g \quad \ldots \text{(ii)}$

and $6\pi\eta r v_2 = \frac{4}{3}\pi r^3(\rho-\sigma)g \quad \ldots \text{(iii)}$

From Eqs. (ii) and (iii), we get

$$\frac{Rv_1}{rv_2} = \frac{R^3}{r^3} \Rightarrow \frac{Rv_1}{rv_2} = \left(\frac{R}{r}\right)^3$$

Using Eqs. (i),

$$(2)^{\frac{1}{3}}\frac{v_1}{v_2} = 2 \Rightarrow \frac{v_1}{v_2} = 2^{2/3}$$

$$v_1 = 2^{\frac{2}{3}}\cdot v_2 = 2^{\frac{2}{3}}\times 10 = 15.9\text{ cm/s}$$

14. *(b)* As, $R = \frac{\sqrt{3}v^2}{2g}$

$$\Rightarrow \frac{v^2\sin 2\theta}{g} = \frac{\sqrt{3}v^2}{2g}$$

or $\quad \sin 2\theta = \frac{\sqrt{3}}{2}$

or $\quad 2\theta = 60°$

$\therefore \quad \theta = 30°$

15. *(c)* Apply work-energy theorem from A to B,

$$W_{mg} = K_B - K_A$$

$$mg\times(R + R\sin 30°) = \frac{1}{2}mv_B^2 - \frac{1}{2}mv_A^2$$

$$10\times\frac{3R}{2} = \frac{v_B^2}{2} - \frac{1}{2}\times(30)^2$$

$$30R = v_B^2 - 900$$

$$v_B^2 = 30\times 30 + 900$$

$$v_B = \sqrt{1800}$$

$$v_B = 30\sqrt{2} = 30\times 1.4 = 42\text{ m/s}$$

16. *(d)* Given, work done, $W = -22.3$ J

Work done is taken negative as work is done on the system.

In an adiabatic change, $\Delta Q = 0$

Using first law of thermodynamics,

$$\Delta U = \Delta Q - W = 0 - (-22.3) = 22.3\text{ J}$$

For another process between state A and state B,

Heat absorbed $(\Delta Q) = +9.35$ cal

$$= +(9.35\times 4.18)\text{ J} = +39.08\text{ J}$$

Change in internal energy between two states *via* different paths are equal.

$\therefore \quad \Delta U = 22.3$ J

$\therefore$ From first law of thermodynamics,

$$\Delta U = \Delta Q - W$$

or $\quad W = \Delta Q - \Delta U$

$$= 39.08 - 22.3$$

$$= 16.78\text{ J} \approx 16.8\text{ J}$$

17. *(c)* From the figure, the truth table is given below :

A	*B*	*Y*
0	0	0
0	1	0
1	0	0
1	1	1

i.e., AND gate

Study Tactics

$Y = \overline{\overline{A}+\overline{B}} = \overline{\overline{A}}\cdot\overline{\overline{B}} = AB$ (AND gate)

18. *(b)* According to Gauss's law

$$\oint \mathbf{E}\cdot d\mathbf{S} = \frac{1}{\varepsilon_0}\oint_s \rho d\mathbf{S}$$

$$[\because Q = \oint_s \rho d\mathbf{S}, \rho = \text{surface charge density}]$$

According to Faraday's law of electromagnetic induction,

$$\oint \mathbf{E}\cdot d\mathbf{l} = \frac{-d\phi}{dt} = \frac{-d}{dt}\oint \mathbf{B}\cdot d\mathbf{S}$$

$$\because \quad \phi = \oint \mathbf{B}\cdot d\mathbf{S}$$

According to Ampere-Maxwell's law,

$$\oint \mathbf{B}\cdot d\mathbf{l} = \mu_0\left(I + \varepsilon_0\frac{d\phi_E}{dt}\right)$$

According to Ampere's circuital law,

$$\oint \mathbf{B} \cdot d\mathbf{l} = \mu_0 I$$

Hence, option (b) is correct.

19. *(b)* Compare, $y(x,t) = 2\sin\left(\frac{8\pi x}{9}\right)\cos(400\pi t)$

with $y(x,t) = 2A \sin kx \cos \omega t$

$$k = \frac{8\pi}{9}, \omega = 400\pi$$

Now, speed of travelling wave, $v = \frac{\omega}{k}$

$$= \frac{400\pi}{\frac{8\pi}{9}} = 450 \text{ m/s}$$

20. *(c)* As we know that,

$$Y = \frac{MgL}{\pi r^2 \times l} = \frac{4 \times 3.1 \times \pi \times 2.0}{\pi \times (2 \times 10^{-3})^2 \times (0.031 \times 10^{-3})}$$

$$= \frac{4 \times 3.1 \times 2}{4 \times 10^{-6} \times 3.1 \times 10^{-5}} = 2 \times 10^{11} \text{ N/m}^2$$

21. (15) As we know that, internal energy of gas with f degree of freedom,

$$U = \frac{nfRT}{2}$$

$\therefore$ $U_{O_2} = 3 \times \frac{5}{2} RT$ $\quad [\because n = 3, f = 5]$

$U_{Ar} = 5 \times \frac{3}{2} RT$ $\quad [\because n = 5, f = 3]$

Total internal energy,

$$U = U_{O_2} + U_{Ar} = 15RT = xRT$$

So, $x = 15$

22. (2) According to Einstein photoelectric equation,

$$(\text{KE})_{\max} = \frac{hc}{\lambda} - W$$

$\therefore$ Stopping potential,

$$V = \frac{1}{e}\left(\frac{hc}{\lambda} - W\right) \quad [\because (\text{KE})_{\max} = eV]$$

$$\therefore \quad \frac{V_1}{V_2} = \frac{\frac{hc}{\lambda_1} - W}{\frac{hc}{\lambda_2} - W}$$

$$\Rightarrow \quad \frac{2}{1} = \frac{\frac{12400}{3500} - W}{\frac{12400}{6200} - W} \quad [\because hc = 12400 \text{ eV-Å}]$$

$$\Rightarrow \quad \frac{2}{1} = \frac{3.5 - W}{2 - W} \Rightarrow 4 - 2W = 3.5 - W$$

$$\Rightarrow \quad W = 0.5 \text{ eV}$$

$$= \frac{1}{2} \text{ eV} = \frac{1}{x} \text{ eV} \quad \text{(according to question)}$$

So, $x = 2$

23. (21) Now, $\mu = \frac{\text{real depth}}{\text{apparent depth}}$

or, real depth $= \mu \times$ apparent depth

Therefore, real length of the edge of the cube

$= \mu \times$ (apparent distance of the bubble from one face + apparent distance of the bubble from second face).

$$= 1.5 \times (2 + 5) = 1.5 \times 7 = \frac{21}{2} \text{ cm} = \frac{x}{2} \text{ cm.}$$

So, $x = 21$

24. (16) Given, $l_2 = l + \frac{25}{100} l = \frac{5l}{4}$

$\therefore$ Volume remain unchanged.

$$A_1 l_1 = A_2 l_2$$

$$Al = A_2 \frac{5}{4} l$$

$$A_2 = \frac{4A}{5}$$

Now, $R_2 = \frac{\rho l_2}{A_2} = \frac{\rho\, 5l/4}{4A/5} = \frac{25\rho l}{16A}$

$$= \frac{25}{16} \times R = \frac{25 \times 10}{16} \quad [\because R = 10\Omega]$$

$$= 15.6\Omega \simeq 16\Omega$$

25. (2) Motional emf induced in the conductor,

$$e = Blv = 2(1)(2) = 4 \text{ V}$$

This acts as a cell of emf 4V and internal resistance 2Ω. 6Ω and 3Ω resistors are in parallel.

$$\therefore \quad \frac{1}{R_p} = \frac{1}{6} + \frac{1}{3} = \frac{1+2}{6} = \frac{3}{6} = \frac{1}{2}$$

$$R_p = 2\Omega$$

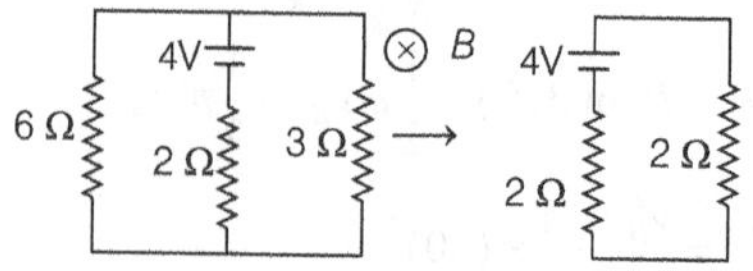

$\therefore$ Current through the conductor (i)

$$i = \frac{E}{R_p + r} = \frac{4}{2+2} = 1\text{A}$$

Magnetic force on the conductor

$$F = Bil = 2(1)(1) = 2\text{N}$$

Therefore, to keep the conductor moving with a constant velocity, a force of 2 N to be applied to the right side.

Chemistry

26. *(d)* Bromine in its second excited state, reacts with fluorine to form BrF_5. Its shape is square pyramidal.

	4s	4p			4d				
Br (Ground state) =	⇅	⇅	⇅	1					
Br (First excited state) =	⇅	⇅	1	1	1				
Br (Second excited state) =	⇅	1	1	1	1	1			

∴ Br in second excited state will react with F to form BrF_5 having square pyramidal shape.

27. *(b)* The reaction takes place as follows

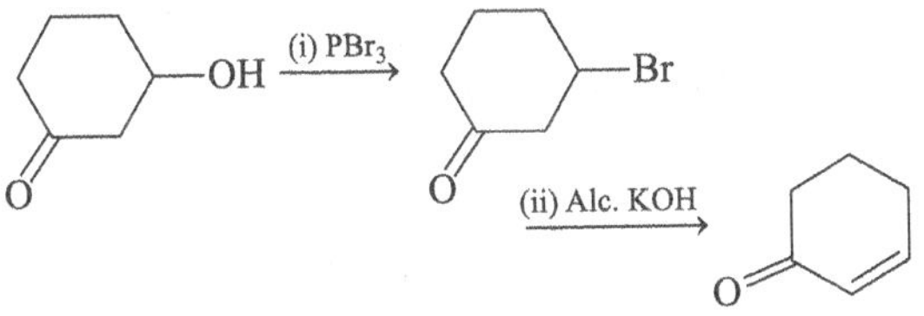

28. *(a)* Structure of given compounds are as follows

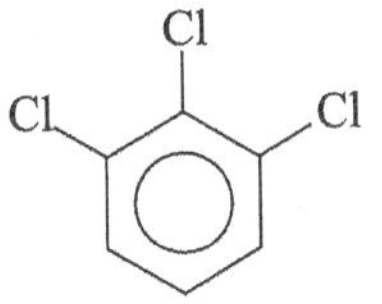

$\mu = 2.39D$
1,2,3-trichlorobenzene
(I)

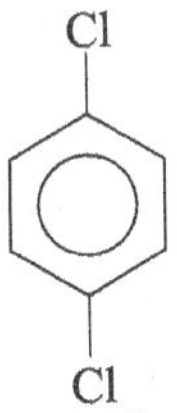

$\mu = 0$
1,4-dichlorobenzene
(II)

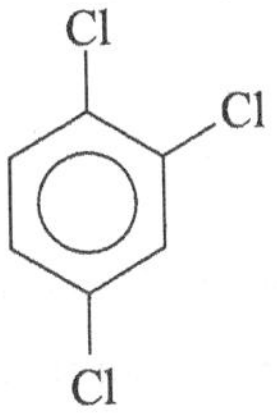

$\mu = 1.26D$
1,2,4-trichlorobenzene
(III)

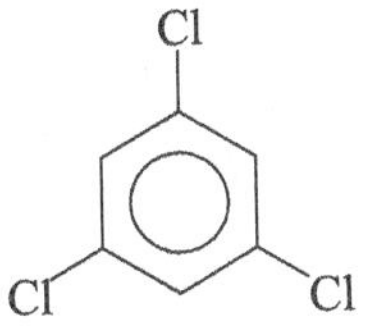

$\mu = 0.28D$
1,3,5-trichlorobenzene
(IV)

Thus, the increasing order of dipole moment is II < IV < III < I.

29. *(a)* **Determination of molecular structure** Structure of given compound may be determined as molecular formula

[Degree of unsaturation (u)]

$$= (C + 1) - \frac{H}{2} - \frac{X}{2} + \frac{N}{2}$$

$$= (6 + 1) - \frac{4}{2} - \frac{1}{2} + \frac{1}{2} = 7 - 2 = 5$$

Since, the organic compound is cyclic, it must contain benzene ring as final product which is also aromatic in nature.

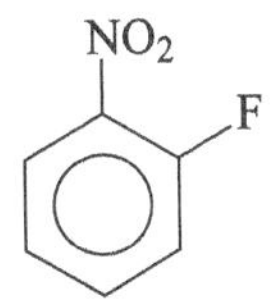

ortho-form

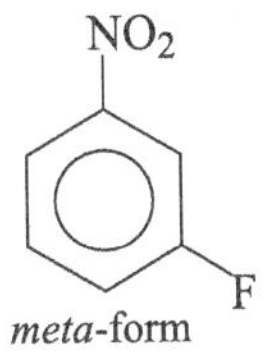

meta-form

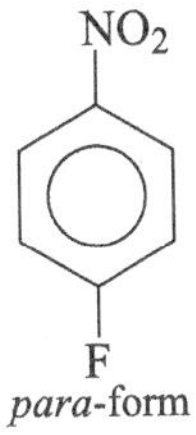

para-form

From the given options, we will choose *para* form as a starting material and further sequence of chemical reaction will be as follows

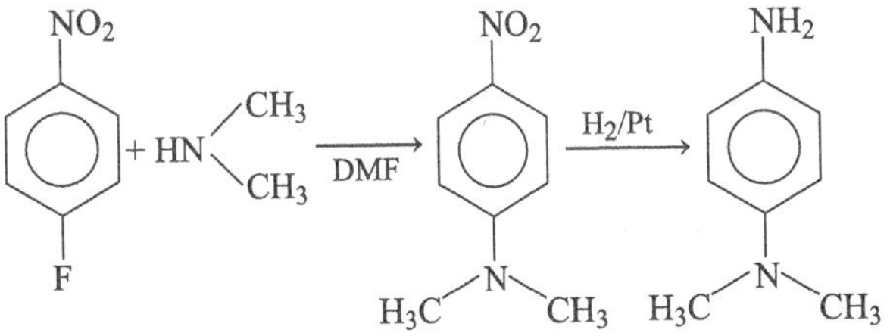

(A) *(B)*

Hence, the possible structure of compound *B* is

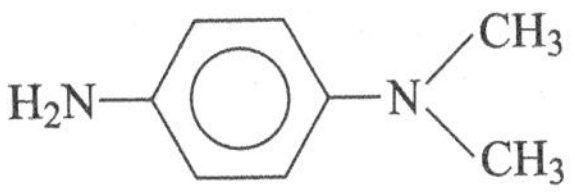

Study Tactics

The problem includes conceptual mixing of determination of molecular structure of compound and nucleophilic substitution reaction on substituted benzene ring.

30. *(c)* The correct match is A-II, B-I, C-IV, D-III.

31. *(b)* The correct match is A-II, B-I, C-IV, D-III.

- **Myosin** is an example of fibrous protein.
- **Vitamin C** Ascorbic acid is also known as vitamin-C.
- **Haemoglobin** It is an example of globular protein.
- **Androgens** is an example of steroids.

32. *(d)* **Wittig reaction** The conversion of cyclopentyl methylketone to 2 cyclopentyl propene is done by the use of Wittig reagent, i.e. phosphorus ylide. Therefore, the reagent *B* is

$$PPh_3 = CH_2 \text{ or } \overset{\oplus}{P}Ph_3 - \overset{\ominus}{C}H_2$$

Reaction involved by using is as follows

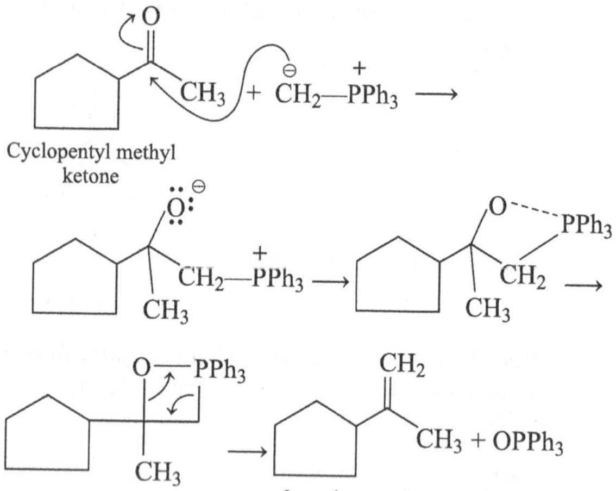

Cyclopentyl methyl ketone

2-cyclopentyl propene

Clemmensen reduction When carbonyl compound is treated with amalgamated zinc in presence of concentrated HCl, it converts $\rangle C{=}O$ to $\rangle CH_2$.

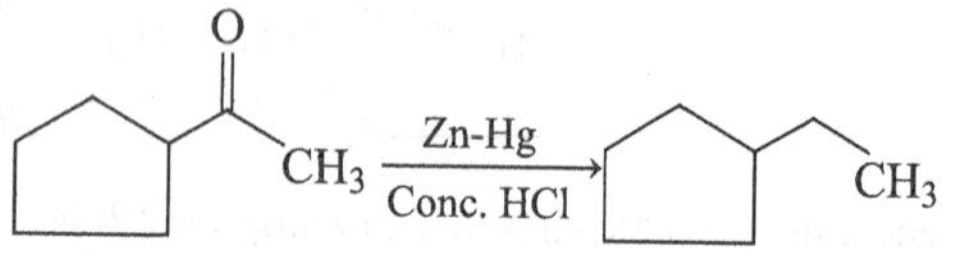

This is a direct method for conversion of ketone to hydrocarbon. Thus, the reagent *A* used is Zn-Hg in conc. HCl.

Study Tactics

This problem includes conceptual mixing of Wittig reaction and Clemmensen reduction.

33. *(c)* $SO_3^{2-}(aq) + SO_4^{2-}(aq) \xrightarrow{BaCl_2} \underset{\text{White ppt.}}{BaSO_4} + 2Cl^- + SO_3^{2-}$

Therefore, using $BaCl_2$ sulphite and sulphate get separated easily.

34. *(d)* $\underset{\text{Formaldehyde}}{6HCHO} + 4NH_3 \longrightarrow \underset{\text{Hexamethylene - tetramine}}{(CH_2)_6N_4} + 6H_2O$

When formaldehyde reacts with ammonia, hexamethylene tetramine is formed which is used as urinary antiseptic (urotropine).

35. *(a)* The reaction is given as

$$\underset{\text{(Tetraphosphorus)}}{P_4} + 3NaOH + 3H_2O \longrightarrow \underset{\text{(Sodium hypophosphite)}}{3NaH_2PO_2} + \underset{\text{(Phosphine)}}{PH_3\uparrow}$$

Thus, phosphorus (P_4) on reaction with NaOH produces phosphine (PH_3) gas.

36. *(d)* Solubility of salt *AD*

$$AD \rightleftharpoons \underset{(S)}{A^+} + \underset{(S)}{D^-}$$

$$K_{sp} = \underset{S}{[A^+]}\underset{S}{[D^-]}$$

$$\therefore \quad K_{sp} = S^2$$

or $\quad S = \sqrt{K_{sp}} = \sqrt{5 \times 10^{-8}} = 2.23 \times 10^{-4}$

Solubility of salt AD_2

$$AD_2 \rightleftharpoons \underset{S}{A^{+2}} + \underset{2S}{2D^-}$$

$$K_{sp} = [A^{2+}][D^-]^2 = S(2S)^2 = 4S^3$$

$$S = \sqrt[3]{K_{sp}/4} = \sqrt[3]{\frac{3 \times 10^{-14}}{4}} = 1.95 \times 10^{-5}$$

Solubility of salt AD_3

$$AD_3 \rightleftharpoons \underset{S}{A^{3+}} + \underset{3S}{3D^-}$$

$$K_{sp} = \underset{S}{[A^{3+}]}\underset{(3S)^3}{[D^-]^3} = S\,(3S)^3 = 27S^4$$

$$S = \sqrt[4]{K_{sp}/27} = \sqrt[4]{\frac{2.8 \times 10^{-15}}{27}} = 1.0 \times 10^{-4}$$

It means solubility order is $AD > AD_3 > AD_2$.

$\therefore$ Only statements (I) and (IV) are correct.

37. *(b)* Among the given complexes $[Co(H_2O)_2(NH_3)_2(en)]^{3+}$ has 3 geometrical isomers.

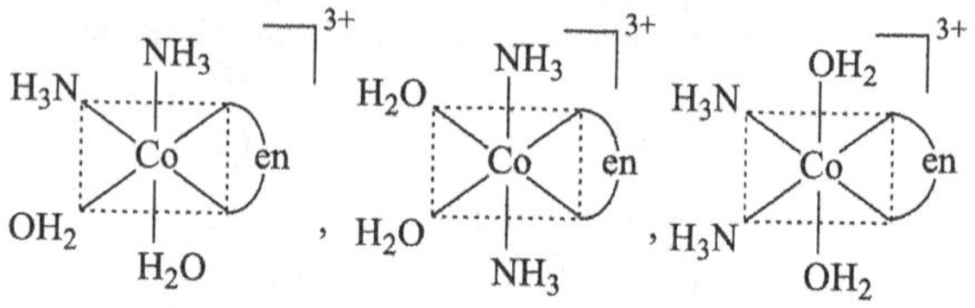

while, all other have 2 geometrical isomers.

38. *(c)* Statement I is false but statement II is true.

Na^+ and Al^{3+} are isoelectronic ions but the magnitude of ionic radius of Na^+ is more than that of Al^{3+}.

39. *(a)* Given, osmotic pressure

(π_1) for one solution = 1.5 atm

(π_2) for other solution = 2.5 atm

$$V_1 = 1\,L$$

$$V_2 = 2\,L$$

The osmotic pressure of resultant solution is

$$\pi_R(V_1 + V_2) = \pi_1 V_1 + \pi_2 V_2$$

$$3\pi_R = 1.5 \times 1 + 2.5 \times 2$$

$$\pi_R = \frac{6.5}{3} = 2.16 \text{ atm}$$

40. *(c)*

$$N_2^+ = \frac{9-4}{2} = \frac{5}{2} = 2.5$$

$$N_2^- = \frac{10-5}{2} = \frac{5}{2} = 2.5$$

$$N_2 = \frac{10-4}{2} = 3$$

Study Tactics

N_2 has 14 electrons and its bond order is 3. Every electron added or subtracted to 14, reduces the bond order by 0.5.

41. *(a)* Given for 0.2 M solution,

$R = 60\Omega$

$\kappa = 1.5\ \text{Sm}^{-1} = 1.5 \times 10^{-2}\ \text{S cm}^{-1}$

Now,

$$R = \rho \frac{l}{A} = \frac{1}{\kappa} \times \frac{l}{A}$$

$$\frac{l}{A} = R \times \kappa = 60 \times 1.5 \times 10^{-2}\ \text{cm}^{-1}$$

For, 0.5 M solution

$R = 280\Omega$

$\kappa = ?$

As we know,

$$R = \rho \frac{l}{A} = \frac{1}{\kappa} \times \frac{l}{A}$$

$$\kappa = \frac{1}{R} \times \frac{l}{A} = \frac{1}{280} \times 60 \times 1.5 \times 10^{-2}\ \text{S cm}^{-1}$$

$$= 0.321 \times 10^{-2}\ \text{S cm}^{-1} = 3.21 \times 10^{-3}\ \text{S cm}^{-1}$$

Now, $\Lambda_m = \frac{\kappa \times 1000}{M}$

$$= \frac{3.21 \times 10^{-3}\ \text{Scm}^{-1} \times 1000\ \text{cm}^3\text{L}^{-1}}{0.5\ \text{mol L}^{-1}}$$

$$= 6.420\ \text{S cm}^2\ \text{mol}^{-1}$$

42. *(b)* Among the given hormones, glucagon increase the glucose level in the blood. While insulin keeping the blood glucose level within the narrow limit.

43. *(b)* The species which have no unpaired electron in their valence shell possess no colour.

$Ti^{4+} = [Ar]\,3d^0$ (0 unpaired electrons)

$Mn^{2+} = [Ar]\,3d^5$ (5 unpaired electrons)

$Fe^{3+} = [Ar]\,3d^5$ (5 unpaired electrons)

$Ni^{2+} = [Ar]\,3d^8$ (2 unpaired electrons)

Hence, only Ti^{4+} is colourless in their aqueous solution.

44. *(b)* *p*-toluenesulphonyl chloride is the derivative of Benzene sulphonyl chloride also known as Hinsberg's reagent.

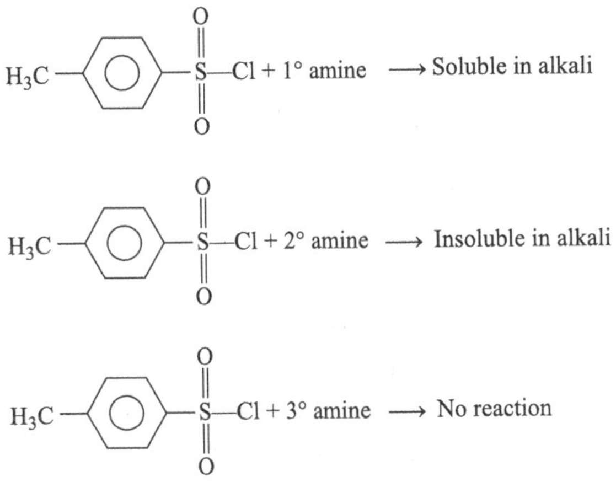

45. *(c)* (A) is true but (R) is false.

Vinyl halide $[(CH_3)_3C—CH{=}CH—Cl]$ do not undergo nucleophilic substitution reaction.

This is due to the formation of highly unstable carbocation $(CH_3)_3C—CH{=}\overset{\oplus}{C}H$; which cannot be delocalised by the π-electrons, also C—Cl has partial double bond character which is difficult to break.

46. *(11)* For *s*-orbital $l = 0$, $m_l = 0$

For *p*-orbital $l = 1, m_l = -1, 0, +1$

For *d*-orbital $l = 2, m_l = -2, -1, 0, +1, +2$

$\therefore$ For every orbital, 2 electrons will contribute for $m_l = 0$.

$Fe = 1s^2, 2s^2, 2p^6, 3s^2, 3p^6, 4s^2, 3d^6$ [Atomic no. 26]

$Fe^{3+} = 1s^2\, 2s^2\, 2p^6\, 3s^2\, 3p^6\, 3d^5$

For $m = 0$, No. of electrons = ⥮ (1s) ⥮ (2s) ⥮ (2p) ⥮ (3s) ⥮ (3p) 1 (3d)

Hence, electrons with $m = 0$ are 11.

47. *(4624)* $t_{1/2} = 40$ minutes

$$\lambda = \frac{0.693}{t_{1/2}} = \frac{0.693}{40} = 0.0173$$

Number of atoms present initially

$$= \frac{\text{Initial rate of decay}}{\lambda}$$

$$= \frac{80}{0.0173} = 4624.2 \simeq 4624$$

48. *(89)* The reaction take place is given by

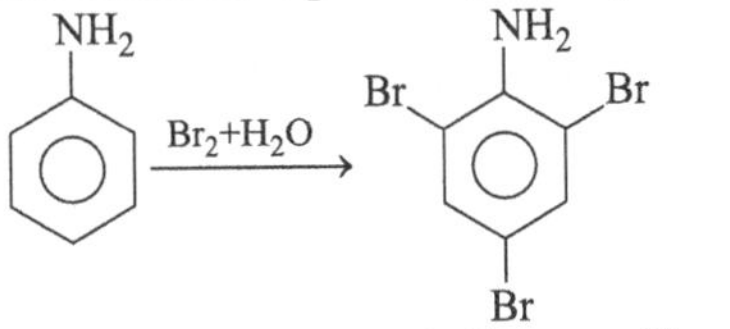

2, 4, 6 tribromoaniline

93 g of aniline produces 330 g of 2, 4, 6 tribromoaniline

Hence, 8.4 g of aniline will give

$$\frac{330}{93} \times 8.4 = 29.8$$

$$\text{Percentage yield} = \frac{26.4 \times 100}{29.8} = 88.59\% = 88.6\% \simeq 89$$

49. *(29)* Molecular weight of graphite = 12 g/mol

Heat of combustion per mol for graphite $= -3.45 \times 10^2$ kJ

Heat generated for 1g of graphite

$$= -\frac{3.45 \times 10^2}{12} = -0.2875 \times 10^2 \text{ kJ}$$

$$= 28.75 \text{ kJ} \approx 29 \text{ kJ}$$

50. *(3)* Among the given equations only three i.e. A, C and D involve reduction while B does not involve reduction.

$$\underset{(+4)}{TiO_2} \longrightarrow \underset{(+4)}{TiCl_4}$$

Mathematics

51. *(b)* Let $I = \int_0^1 \frac{xe^{x^2}}{x^2+1} dx$

Put $x^2 = t \Rightarrow 2x\,dx = dt$

At $x = 0, t = 0$

$x = 1, t = 1$

$$\therefore \quad I = \frac{1}{2}\int_0^1 \frac{e^t\,dt}{t+1} \Rightarrow I = \frac{1}{2}A$$

52. *(c)* Here, $z = 1 + ai$

$\Rightarrow \quad z^2 = 1 - a^2 + 2ai$

So, $z^2 \cdot z = \{(1-a^2) + 2ai\}\{1 + ai\}$

$= (1-a^2) + 2ai + (1-a^2)ai - 2a^2$

$\because z^3$ is real.

$\therefore \quad 2a + (1-a^2)a = 0$

$\Rightarrow \quad a(3-a^2) = 0 \Rightarrow a = \sqrt{3} \quad [\because a > 0]$

Now, $1 + z + z^2 + \ldots\ldots + z^{11}$

$$= \frac{(1+\sqrt{3}i)^{12} - 1}{1 + \sqrt{3}i - 1} = \frac{(1+\sqrt{3}i)^{12} - 1}{\sqrt{3}i}$$

$$\because \quad (1+\sqrt{3}\,i)^{12} = 2^{12}\left(\frac{1}{2} + \frac{\sqrt{3}}{2}i\right)^{12}$$

$$= 2^{12}\left(\cos\frac{\pi}{3} + i\sin\frac{\pi}{3}\right)^{12}$$

$$= 2^{12}(\cos 4\pi + i\sin 4\pi) = 2^{12}$$

$$\because 1 + z + z^2 + \ldots\ldots + z^{11} = \frac{2^{12} - 1}{\sqrt{3}\,i} = \frac{4095}{\sqrt{3}\,i}$$

$$= -\frac{4095}{3}\sqrt{3}\,i = -1365\sqrt{3}\,i$$

53. *(c)* $g(x) = \lim\limits_{x \to \infty} \frac{f(x) + x^n h(x)}{1 + x^n}$

If $|x| = 1$, then $g(x) = \frac{f(x) + h(x)}{2}$ ...(i)

If $|x| < 1$, then $\lim\limits_{n \to \infty} x^n = 0$

$\therefore \quad g(x) = f(x)$ ...(ii)

If $|x| > 1$, then $\lim\limits_{n \to \infty} x^{-n} = 0$

$$g(x) = \lim_{n \to \infty} \frac{x^{-n} f(x) + h(x)}{x^{-n} + 1} = h(x) \quad \text{...(iii)}$$

The function $g(x)$ shall be continuous everywhere, if $f(x)$ and $h(x)$ are continuous everywhere and if $g(x)$ is continuous at $x = \pm 1$.

Now, from Eqs. (i), (ii) and (iii), we get

$$f(1) = \frac{f(1) + h(1)}{2} = h(1)$$

$\Rightarrow \quad f(1) = h(1)$

and $f(-1) = \frac{f(-1) + h(-1)}{2} = h(-1)$

$\Rightarrow \quad f(-1) = h(-1)$

Statement I and Statement II both are true.

54. *(b)* $\because \ (1+x)^n = C_0 + C_1x + C_2x^2 + \ldots + C_nx^n$

On differentiating above expansion two times, we get

$$n(n-1)(1+x)^{n-2} = 1\cdot 2C_2 + 2\cdot 3C_3x + \ldots + (n-1)nC_n \cdot x^{n-2}$$

On putting $x = 1$ in above equation, we get

$$1\cdot 2C_2 + 2\cdot 3C_3 + \ldots + (n-1)nC_n = n(n-1)2^{n-2}$$

55. *(c)*

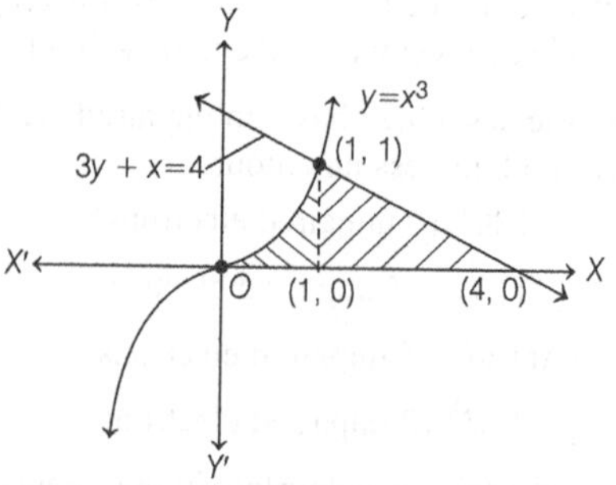

$$\text{Required area} = \int_0^4 y\,dx = \int_0^1 x^3\,dx + \int_1^4 \frac{4-x}{3}\,dx$$

$$=\left(\frac{x^4}{4}\right)_0^1+\frac{1}{3}\left(4x-\frac{x^2}{2}\right)_1^4$$

$$=\frac{1}{4}+\frac{1}{3}\left[(16-8)-\left(4-\frac{1}{2}\right)\right]=\frac{1}{4}+\frac{3}{2}=\frac{7}{4}$$

$$=1.75$$

56. *(d)* Given, equation of lines

$$3x-y+5=0 \quad \text{...(i)}$$

$$y+5=0 \quad \text{...(ii)}$$

Let any point on line (i) be $A(\alpha, 3\alpha+5)$ and on line (ii) be $B(\beta, -5)$.

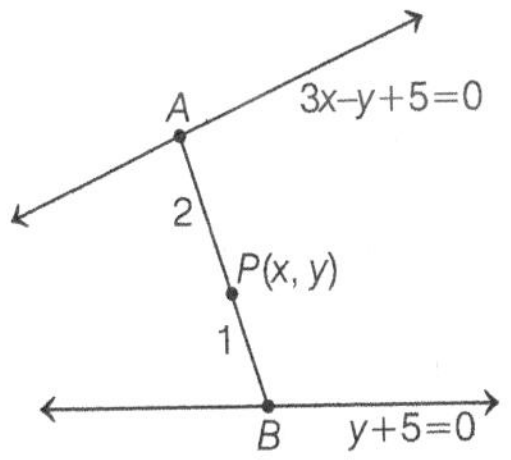

Using section formula,

$$x=\frac{\alpha+2\beta}{3}\Rightarrow 3x=\alpha+2\beta \quad \text{...(iii)}$$

and $\quad y=\dfrac{-10+3\alpha+5}{3}$

$\Rightarrow \quad 3y=3\alpha-5 \quad$ ...(iv)

From Eqs. (iii) and (iv), on eliminating α and β, we get

$$\alpha=\frac{3y+5}{3},\ \beta=\frac{9x-3y-5}{6}$$

$\therefore \quad l^2=AB^2$

$$=(\alpha-\beta)^2+(3\alpha+10)^2$$

$$=\frac{1}{4}(3x-3y-5)^2+(3y+15)^2$$

57. *(c)* Given lines, $\dfrac{x-4}{1}=\dfrac{y+3}{-4}=\dfrac{z+1}{7}=\lambda\text{ (say)}$ ...(i)

and $\quad \dfrac{x-1}{2}=\dfrac{y+1}{-3}=\dfrac{z+10}{8}=\mu\text{ (say)}$ ...(ii)

From Eqs. (i) and (ii), we get

$$\lambda+4=2\mu+1$$

$\Rightarrow \quad \lambda-2\mu=-3$ ...(iii)

and $\quad -3-4\lambda=-1-3\mu$

$\Rightarrow \quad 4\lambda-3\mu=-2$ ....(iv)

From Eqs. (iii) and (iv), we get

$$\lambda=1 \text{ and } \mu=2$$

$\therefore$ Point of intersection $(4+1, -3-4, -1+7)$

i.e. $(5, -7, 6)$.

Distance of $(5, -7, 6)$ from $(1, -4, 7)=\sqrt{16+9+1}$

$=\sqrt{26}$ units

58. *(d)* $h'(x)=h(x)\Rightarrow \dfrac{h'(x)}{h(x)}=1$

On integrating both sides w.r.t x, we get

$$\log h(x)=\log A+x$$

$\Rightarrow \quad h(x)=Ae^x$

$\because \quad h(0)=1$

$\Rightarrow \quad A=1$

$\therefore \quad h(x)=e^x$

and $\quad g(x)=2x-e^x$

$\therefore \quad I=\int_0^2 e^x(2x-e^x)dx$

$$=\int_0^2(2xe^x-e^{2x})dx$$

$$=2[xe^x-e^x]_0^2-\left[\frac{e^{2x}}{2}\right]_0^2$$

$$=2[(2e^2-e^2)-(0-1)]-\left[\frac{e^4-1}{2}\right]$$

$$=2[e^2+1]-\frac{(e^2-1)(e^2+1)}{2}$$

$$=\frac{(e^2+1)(4-e^2+1)}{2}$$

$$=\frac{(e^2+1)(5-e^2)}{2}$$

59. *(b)* Given, $\sin^{-1}x+\sin^{-1}y+\sin^{-1}z=\dfrac{\pi}{2}$

On taking $\sin^{-1}x=\alpha$, $\sin^{-1}y=\beta$ and $\sin^{-1}z=\gamma$, we get

$\therefore \quad \alpha+\beta+\gamma=\dfrac{\pi}{2}$

$$\alpha+\beta=\frac{\pi}{2}-\gamma$$

$$\cos(\alpha+\beta)=\cos\left(\frac{\pi}{2}-\gamma\right)$$

$$\cos\alpha\cos\beta-\sin\alpha\sin\beta=\sin\gamma \quad \text{...(i)}$$

We have, $\sin\alpha=x$, $\sin\beta=y$ and $\sin\gamma=z$

$\therefore \quad \cos\alpha=\sqrt{1-x^2}$, $\cos\beta=\sqrt{1-y^2}$

$\therefore$From Eq. (i), we get

$$\sqrt{1-x^2}\cdot\sqrt{1-y^2}-xy=z$$

$$\sqrt{1-x^2}\cdot\sqrt{1-y^2}=(z+xy)$$

On squaring both sides, we get

$$x^2+y^2+z^2+2xyz=1$$

60. *(b)* $\because \mathbf{a}\cdot\mathbf{c}=|\mathbf{a}||\mathbf{c}|\cos\theta$

and given that $\theta=\cos^{-1}\left(\dfrac{1}{4}\right)$

$\Rightarrow \quad \cos\theta = \dfrac{1}{4}$

$\therefore \quad \mathbf{a}\cdot\mathbf{c} = |\mathbf{a}||\mathbf{c}| \times \dfrac{1}{4} = \dfrac{1}{4} \quad [\because |\mathbf{a}| = |\mathbf{c}| = 1]$

On taking dot product by **a**, **b** and **c** with $\mathbf{b} - 2\mathbf{c} = \lambda\mathbf{a}$, we have

$$\mathbf{a}\cdot\mathbf{b} - 2(\mathbf{a}\cdot\mathbf{c}) = \lambda(\mathbf{a}\cdot\mathbf{a})$$

$$\Rightarrow \quad \mathbf{a}\cdot\mathbf{b} = \lambda + \frac{1}{2} \quad \text{...(i)}$$

Similarly, $\quad \mathbf{b}\cdot\mathbf{c} = 8 - \dfrac{\lambda^2}{2} - \dfrac{\lambda}{4} \quad$ ...(ii)

and $\quad \mathbf{b}\cdot\mathbf{c} - 2 = \lambda(\mathbf{a}\cdot\mathbf{c}) \quad$ ...(iii)

From Eqs. (i), (ii) and (iii), we get

$$8 - \frac{\lambda^2}{2} - \frac{\lambda}{4} - 2 = \lambda\left(\frac{1}{4}\right) \Rightarrow (\lambda - 3)(\lambda + 4) = 0$$

$\because \quad \lambda > 0 \Rightarrow \lambda = 3$

61. *(d)* $f'(x) = \begin{cases} 3(3+x)^2, & -3 < x \le -1 \\ 2x^{\frac{-1}{3}}, & -1 < x < 2 \end{cases}$

Clearly, $f'(x)$ changes its sign at $x = -1$ from positive to negative and so $f(x)$ has local maxima at $x = -1$.

Also, $f'(0)$ does not exist but $f'(0^-) < 0$ and $f'(0^+) > 0$. It can only inferred that $f(x)$ has a possibility of minimum at $x = 0$.

Hence, it has one local maxima at $x = -1$ and one local minima at $x = 0$.

Thus, there are total number of 2 local maxima and local minima.

62. *(b)* $abcde = 1050 = 2 \times 3 \times 5^2 \times 7$

We can assign 2, 3 or 7 to any of 5 variables *a, b, c d* and *e*.

We can assign 5^2 to just one variable in 5 ways or can assign $5^2 = 5 \times 5$ to two variables in ${}^5C_2$ ways, then 5 can be assigned in ${}^5C_1 + {}^5C_2 = 5 + 10 = 15$ ways

Hence, required solutions $= 5 \times 5 \times 5 \times 15 = 1875$.

Now, according to the question, we get

$$375\,k = 1875 \Rightarrow k = \frac{1875}{375} = 5$$

63. *(b)* Given, $\quad \dfrac{dy}{dx} = \dfrac{y(x + y^3)}{x(y^3 - x)}$

$$\Rightarrow \quad x(y^3 - x)dy = y(x + y^3)dx$$

$$\Rightarrow \quad (xy^3 - x^2)dy = (xy + y^4)dx$$

$$\Rightarrow \quad xydx + y^4dx - xy^3dy + x^2dy = 0$$

$$\Rightarrow x(ydx + xdy) + y^3(ydx - xdy) = 0$$

$$\Rightarrow xd(xy) - y^3\left(\frac{xdy - ydx}{x^2}\right)\cdot x^2 = 0$$

$$\Rightarrow xd(xy) - x^2y^3 d\left(\frac{y}{x}\right) = 0 \Rightarrow \frac{d(xy)}{x^2y^2} = \frac{y}{x}d\left(\frac{y}{x}\right)$$

On integrating both sides, we get

$$\int \frac{d(xy)}{(xy)^2} = \int \frac{y}{x}d\left(\frac{y}{x}\right)$$

$$\Rightarrow \quad -\frac{1}{xy} = \left(\frac{y}{x}\right)^2 \cdot \frac{1}{2} + C$$

$$\Rightarrow \quad y^3 + 2x + 2Cx^2y = 0$$

At (1, 1), $\quad 1 + 2 + 2C = 0 \Rightarrow C = -\dfrac{3}{2}$

Then, $\quad y^3 + 2x - 3x^2y = 0$

64. *(c)* Given, $\sqrt{3}p\cos x + 2q\sin x = r \quad$ ...(i)

It has two roots such that $\alpha + \beta = \dfrac{\pi}{3}$

$\therefore$ From Eq. (i), we get

$$\sqrt{3}p\cos\left(\frac{\pi}{3} - x\right) + 2q\sin\left(\frac{\pi}{3} - x\right) = r$$

$$\Rightarrow \quad \sqrt{3}p\left(\cos x\cos\frac{\pi}{3} + \sin x\sin\frac{\pi}{3}\right) + 2q\left(\sin\frac{\pi}{3}\cos x - \cos\frac{\pi}{3}\sin x\right) = r$$

$$\Rightarrow \quad \sqrt{3}p\left(\cos x\cdot\frac{1}{2} + \sin x\cdot\frac{\sqrt{3}}{2}\right) + 2q\left(\frac{\sqrt{3}}{2}\cos x - \frac{1}{2}\sin x\right) = r$$

$$\Rightarrow \left(\frac{\sqrt{3}}{2}p + \sqrt{3}q\right)\cos x + \left(\frac{3}{2}p - q\right)\sin x = r \quad \text{...(ii)}$$

On subtracting Eq. (i) from Eq. (ii), we get

$$\left(\sqrt{3}q - \frac{\sqrt{3}}{2}p\right)\cos x + \left(\frac{3}{2}p - 3q\right)\sin x = 0$$

$$\Rightarrow \quad \sqrt{3}q - \frac{\sqrt{3}}{2}p = 0 \Rightarrow \frac{q}{p} = \frac{1}{2}$$

$$\Rightarrow \quad q : p = 1 : 2$$

Study Tactics

Given, $\sqrt{3}p\cos x + 2q\sin x = r \quad$ ...(i)

$\because$ Eq. (i) has two distinct real roots α and β, with $\alpha + \beta = \dfrac{\pi}{3}$.

So, replace x by $\left(\dfrac{\pi}{3} - x\right)$ in given equation, solve both equations and find $p : q$.

65. *(b)* The common region is shown shaded. AB and CD are the latusrectum of the parabolas.

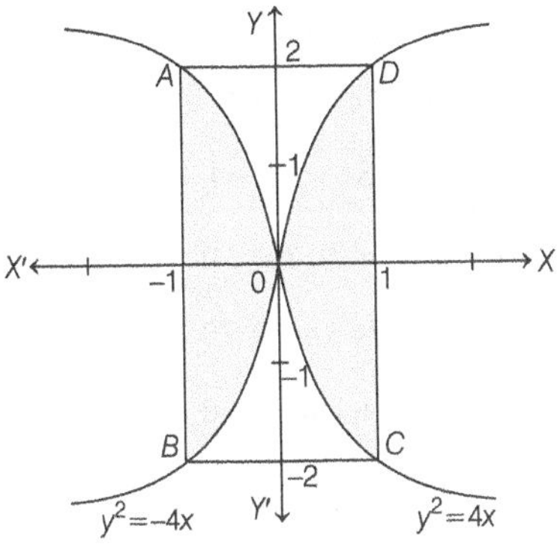

Thus, the integral points on or within the shaded region are
(1, 0), (1, 1), (1, 2), (1, –1), (1, –2), (–1, 0), (–1, 1), (–1, 2), (–1, –1), (–1, –2) and (0, 0).
Hence, there will be 11 such points.

66. *(a)* $f(x)=[16^x+2^{2x+1}+1]=[4^{2x}+2\cdot 4^x+1]$

$$f(x)=[(4^x+1)^2]$$

$\Rightarrow\ (4^x+1)^2>1\ \forall\ x$

$\therefore\ [(4^x+1)^2]=$ Integral values which is greater than equal to one.

Hence, natural numbers are range.

67. *(c)* Let α be the common root.

$$\alpha^3+2\alpha^2-\alpha+\lambda=0 \Rightarrow \lambda=-\alpha^3-2\alpha^2+\alpha$$

and $\quad \alpha^2+5\alpha-3\lambda=0$

$$\alpha^2+5\alpha-3(-\alpha^3-2\alpha^2+\alpha)=0$$

$\Rightarrow\quad 3\alpha^3+7\alpha^2+2\alpha=0$

$\Rightarrow\quad \alpha(3\alpha^2+7\alpha+2)=0$

$\Rightarrow\quad \alpha[3\alpha(\alpha+2)+1(\alpha+2)]=0$

$\Rightarrow\quad \alpha(3\alpha+1)(\alpha+2)=0$

$$\alpha=0,-2,\frac{-1}{3}$$

$$\lambda=0,-2,\frac{-14}{27}$$

68. *(a)* $\sum_{k=1}^{n} D_k=56$

$$\Rightarrow \begin{vmatrix} \sum_{k=1}^{n}1 & n & n \\ \sum_{k=1}^{n}2k & n^2+n+1 & n^2+n \\ \sum_{k=1}^{n}(2k-1) & n^2 & n^2+n+1 \end{vmatrix}=56$$

$$\Rightarrow \begin{vmatrix} n & n & n \\ n(n+1) & n^2+n+1 & n^2+n \\ n^2 & n^2 & n^2+n+1 \end{vmatrix}=56$$

$\Rightarrow\quad n(n+1)=56$

$\Rightarrow\quad n=7$

69. *(d)* $\dfrac{i\cdot C_i}{C_i+C_{4-i}}=\dfrac{i\cdot {}^4C_i}{{}^4C_i+{}^4C_{4-i}}=\dfrac{i\ {}^4C_i}{2\ {}^4C_i}=\dfrac{i}{2}$

$$\therefore \sum_{i=1}^{4}\left(\frac{i\cdot C_i}{C_i+C_{4-i}}\right)^2=\sum_{i=1}^{4}\left(\frac{i}{2}\right)^2$$

$$=\frac{1}{4}(1^2+2^2+3^2+4^2)$$

$$=\frac{1}{4}(30)=\frac{15}{2}$$

70. *(b)* $\because\ p^2+q^2+r^2=2$

$\Rightarrow\quad p=0, q=\pm 1, r=\pm 1$

$p=\pm 1, q=0, r=\pm 1$

$p=\pm 1, q=\pm 1, r=0$

Total solutions $=12$

Now, $\quad |A|\neq 0 \Rightarrow p+q+r\neq 0$

$\therefore$ Required solution $=6$

$\therefore$ Probability of system has unique solution

$$=\frac{6}{12}=\frac{1}{2}$$

71. *(1)* Let $y=\sin hx=\dfrac{e^x-e^{-x}}{2}$

$\Rightarrow\quad e^{2x}-1=2ye^x$

$\Rightarrow\quad e^{2x}-2ye^x-1=0$

$\Rightarrow\quad e^x=y\pm\sqrt{y^2+1}$

$\Rightarrow\quad g^{-1}(x)=\log(x+\sqrt{x^2+1}) \qquad [\because e^x>0]$

$\because\quad f(x)=g^{-1}(x)$

$$=\log(y+\sqrt{y^2+1})$$

$\Rightarrow\quad f\left(\dfrac{e^{1002}-1}{2e^{501}}\right)=\log\left(\dfrac{e^{1002}-1}{2e^{501}}+\dfrac{e^{1002}+1}{2e^{501}}\right)$

$$=\log e^{501}=501$$

$\Rightarrow\quad \dfrac{1}{501}f\left(\dfrac{e^{1002}-1}{2e^{501}}\right)=1$

72. *(5)* General term T_{r+1} in the expansion of $(x+y)^n$ is given by

$$T_{r+1}={}^nC_r\,x^{n-r}\,y^r$$

and in the expansion of $(x+y)^{n+3}$

$$T'_{r+1}={}^{n+3}C_r x^{n+3-r}\ y^r$$

Given, $\frac{T_2}{T_3}=\frac{T_3'}{T_4'}$

$\Rightarrow \frac{{}^nC_1x^{n-1}y^1}{{}^nC_2x^{n-2}y^2}=\frac{{}^{n+3}C_2\,x^{n+1}y^2}{{}^{n+3}C_3x^ny^3}$

$\Rightarrow \frac{2n}{n(n-1)}=\frac{\frac{(n+3)(n+2)}{2}}{\frac{(n+3)(n+2)(n+1)}{6}}$

$\Rightarrow \frac{2}{n-1}=\frac{3}{n+1}\Rightarrow n=5$

73. *(1)* $\because \sin(a_{n+1}-a_n)+\frac{\sin(a_n)\sin(a_{n+1})}{2^{n+1}}=0$

$\Rightarrow \sin(a_{n+1})\cos(a_n)-\cos(a_{n+1})\sin(a_n)$

$+\frac{\sin(a_n)\sin(a_{n+1})}{2^{n+1}}=0$

On dividing by $\sin(a_n)\sin(a_{n+1})$ to both sides, we get

$\cot a_{n+1}-\cot a_n=\frac{1}{2^{n+1}}$

$\Rightarrow \cot a_2-\cot a_1=\frac{1}{2^2}$

$\cot a_3-\cot a_2=\frac{1}{2^3}$

$\cot a_4-\cot a_3=\frac{1}{2^4}$ and so on

But $a_1=\tan^{-1}2$

$\Rightarrow \tan a_1=2\Rightarrow \cot a_1=\frac{1}{2}$

$\therefore \cot a_n=\frac{1}{2}+\frac{1}{2^2}+\frac{1}{2^3}+\ldots+\frac{1}{2^n}$

$\lim_{n\to\infty}\cot a_n=\lim_{n\to\infty}\left(\frac{1}{2}+\frac{1}{2^2}+\frac{1}{2^3}+\ldots+\frac{1}{2^n}\right)$

$=\frac{\frac{1}{2}}{1-\frac{1}{2}}=1$

74. *(81)* Given, equation of ellipse is

$\frac{x^2}{9}+\frac{y^2}{1}=1 \quad \ldots\text{(i)}$

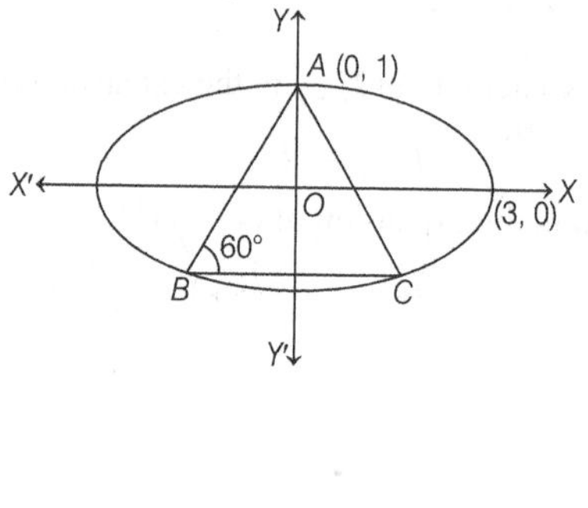

$\therefore$ Equation of side AB is

$y-1=(\tan 60^\circ)(x-0)$

$\Rightarrow y=\sqrt{3}x+1 \quad \ldots\text{(ii)}$

On solving Eqs. (i) and (ii), we get

$\frac{x^2}{9}+(\sqrt{3}x+1)^2=1$

$\Rightarrow x^2+9(3x^2+1+2\sqrt{3}x)=9$

$\Rightarrow 28x^2+18\sqrt{3}x+9=9$

$\Rightarrow 28x^2+18\sqrt{3}x=0$

$\Rightarrow 14x^2+9\sqrt{3}x=0$

$\Rightarrow x(14x+9\sqrt{3})=0$

$\Rightarrow x=0$ or $x=-\frac{9\sqrt{3}}{14}$

On putting $x=-\frac{9\sqrt{3}}{14}$ in Eq. (ii), we get

$y=\sqrt{3}\left(-\frac{9\sqrt{3}}{14}\right)+1=\frac{-27}{14}+1=\frac{-13}{14}$

Hence, $B\equiv\left(\frac{-9\sqrt{3}}{14},\frac{-13}{14}\right)$

$\therefore AB=\sqrt{\left(-\frac{9\sqrt{3}}{14}\right)^2+\left(\frac{-13}{14}-1\right)^2}=\frac{9\sqrt{3}}{7}$ units

$\Rightarrow$ Perimeter of $\Delta ABC=\frac{3\times 9\sqrt{3}}{7}=\frac{27}{7}\sqrt{3}$ units $=k$

$\therefore 7\sqrt{3}k=81$

75. *(2)* Given, $n=20$, $\sigma=2$ and $\bar{x}=10$

$\therefore \Sigma x=n\bar{x}=20\times 10=200$

Incorrect $\Sigma x=200$

Also, $\frac{1}{n}\Sigma x^2-(\bar{x})^2=\sigma^2$

$\Rightarrow \frac{\Sigma x^2}{20}=4+100\Rightarrow \Sigma x^2=2080$

If wrong observation 10 is omitted

Correct $\Sigma x=200-10=190$

Correct $\Sigma x^2=2080-10^2=1980$

and $n=20-1=19$

Correct mean $\bar{x}=\frac{190}{19}=10$

Correct standard deviation $=\sqrt{\frac{\Sigma x^2}{n}-(\bar{x})^2}$

$=\sqrt{\frac{1980}{19}-(10)^2}=2.05\approx 2$

SECTION I

Foundation Builders

PRACTICE SET 2

Total Marks : 300 | Total Time : 3 hrs | No. of Questions : 75

Instructions

1. There are **three** parts in the questions paper consisting of **Physics**, **Chemistry** and **Mathematics** having 25 questions in each part of equal weightage. Each part (subject) has two sections.
2. **Section A** This section contains 20 multiple choice questions which have only one correct answer. Each question carries **4 marks** for correct answer and **−1 mark** for wrong answer.
3. **Section B** This section contains 5 questions. The answer to each of the questions is a numerical value. Each question carries **4 marks** for correct answer and **−1 mark** for wrong answer. For Section B, the answer should be rounded off to the nearest integer.

PHYSICS

Section A : Objective Type Questions

1. A series *L-C-R* circuit with $R = 120\Omega$ has resonance angular frequency $4 \times 10^5 \text{ rad s}^{-1}$. At resonance, voltage across resistor and inductor are 60 V and 40 V, respectively. If angular frequency at which the circuit current lags behind the voltage by a phase of $\frac{\pi}{4}$ rad is $k \times 10^5$ rad / s, then the value of k is

(a) 4 (b) 2
(c) 8 (d) 12

2. The resistance of hexagon circuit between *A* and *B* represented in figure is (Take, resistance of each resistor 1Ω)

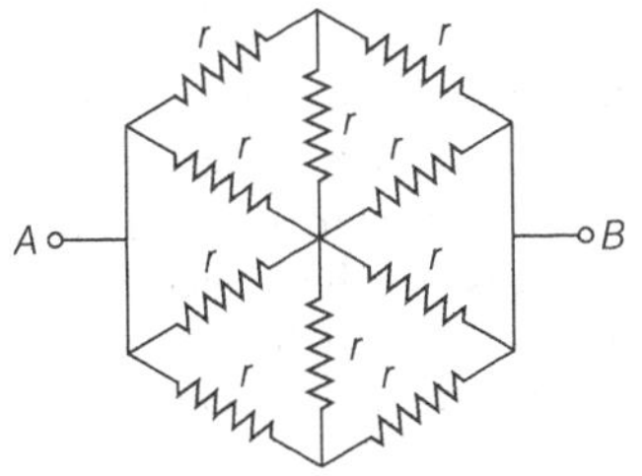

(a) 1Ω (b) 0.5Ω
(c) 2Ω (d) 3Ω

3. The ratio of maximum acceleration to maximum velocity in a simple harmonic motion is 20 s^{-1}. At $t = 0$, the displacement is 10 m. What is the maximum acceleration? The initial phase is $\frac{\pi}{4}$.

(a) $4000\sqrt{2} \text{ ms}^{-2}$ (b) $400\sqrt{2} \text{ ms}^{-2}$
(c) $4\sqrt{2} \text{ ms}^{-2}$ (d) $10\sqrt{2} \text{ ms}^{-2}$

4. A proton of mass 1.67×10^{-27} kg and charge 1.6×10^{-19} C is projected with a speed of 2×10^6 m/s at an angle of 60° to the *X*-axis. If a uniform magnetic field of 0.104 T is applied along *Y*-axis, the path of proton is

(a) a circle of radius = 0.2 m and time period $\pi \times 10^{-7}$ s
(b) a circle of radius = 0.1 m and time period $2\pi \times 10^{-7}$ s
(c) a helix of radius = 0.1 m and time period $2\pi \times 10^{-7}$ s
(d) a helix of radius = 0.2 m and time period $4\pi \times 10^{-7}$ s

5. If the velocity v of a particle moving along a straight line decreases linearly with its displacement s from 20 ms^{-1} to a value approaching zero at $s = 20$ m, then acceleration of the particle at $s = 10$ m is

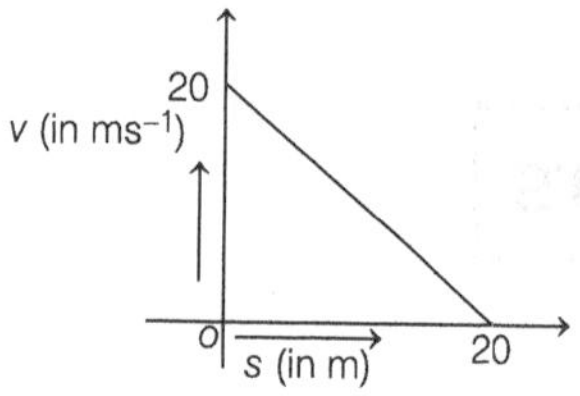

(a) $-20\,m/s^2$ (b) $-10\,m/s^2$
(c) $10\,m/s^2$ (d) $20\,m/s^2$

6. A billiard ball C of mass m moving with velocity u collides two identical balls A and B in contact and at rest. After the collision, ball C is stopped dead and balls A and B move along directions shown in figure with the same speed v. Then,

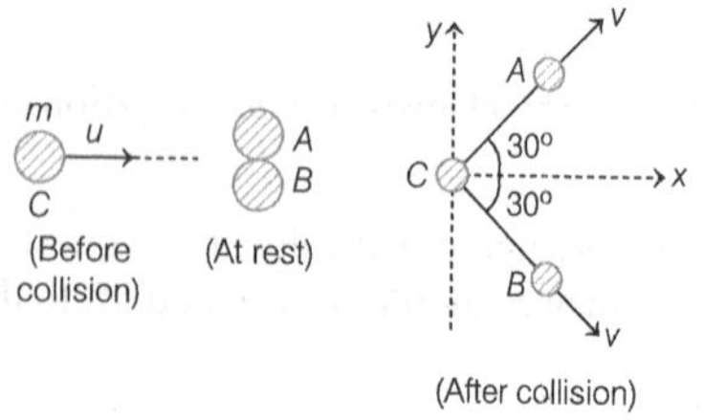

(a) $v=\frac{u}{\sqrt{3}}$ (b) $v=\frac{u}{\sqrt{2}}$
(c) loss in KE $=\frac{1}{3}mu^2$ (d) loss in KE $=\frac{1}{8}mu^2$

7. Assertion When a charge Q is taken from centre of the charged spherical shell to its surface, then the external work done is zero.
Reason The electric potential within the charged spherical shell remains constant.
In the light of above statements, choose the most appropriate answer from the options given below.
(a) Both Assertion and Reason are correct and Reason is the correct explanation of Assertion.
(b) Both Assertion and Reason are correct but Reason is not the correct explanation of Assertion.
(c) Assertion is correct but Reason is incorrect.
(d) Assertion is incorrect but Reason is correct.

8. Two blocks A and B of masses 1 kg and 2 kg are being pressed against a wall by a force F as shown in the figure below. If the coefficient of friction between the blocks is 0.15 and the wall is 0.2, then the frictional force applied by the wall on the block B is

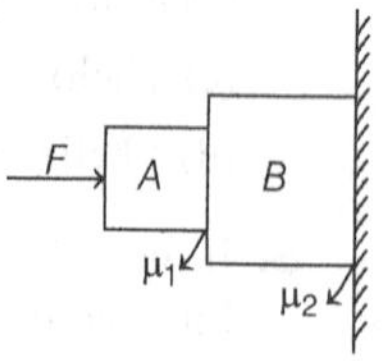

(a) 20 N (b) 30 N
(c) 50 N (d) 40 N

9. Two point masses A and B having masses in the ratio of 4 : 3 are separated by a distance of 1 m. When another point mass C of mass M is placed in between A and B, the force between A and C is $\frac{1}{3}$rd of the force between B and C. Then, distance of C from A is
(a) $\frac{2}{3}$ m (b) $\frac{1}{3}$ m (c) $\frac{1}{4}$ m (d) $\frac{2}{7}$ m

10. I. Stress and Young's modulus have same dimensions.
II. The dimensions of thermal conductivity of a substance is $[MLT^{-3}K]$.
III. Stress and strains are dimensionless physical quantities.
Which of the following statements is/are correct?
(a) I and II (b) II and III
(c) Only I (d) All of these

11. A gasoline engine takes in 5 mol of air at 20°C and 1 atm and compresses it adiabatically to (1/10) th of the original volume. Assume, air to be diatomic. The work done and change in internal energy is
(a) 46 kJ, -46 kJ (b) 36 kJ, -36 kJ
(c) -46 kJ, 46 kJ (d) 36 kJ, -46 kJ

12. The maximum and minimum distance of a comet from the sun are 2×10^{12} m and 8×10^{10} m, respectively. If the speed of the comet at the nearest point is $8\times10^{4}\,ms^{-1}$, then the speed at the farthest point is.
(a) $3.2\times10^{5}\,ms^{-1}$ (b) $3.2\times10^{3}\,ms^{-1}$
(c) $1.6\times10^{5}\,ms^{-1}$ (d) $4.2\times10^{3}\,ms^{-1}$

13. Statement I The root mean square speed of a given gas molecule is more than the average speed of that gas molecule.
Statement II In each degree of freedom, energy of one molecule of an ideal gas is $\frac{3}{2}K_BT$.
In the light of above statements, choose the most appropriate answer from the options given below.
(a) Statement I is true but Statement II is false.
(b) Statement I is false but Statement II is true.
(c) Both Statement I and Statement II are true.
(d) Both Statement I and Statement II are false.

14. A plane electromagnetic wave of frequency 60 MHz travels in vacuum along the positive x-direction. The electric field **E** at a particular space point x at an instant of time t is $9.6\,\hat{j}\,Vm^{-1}$.

The magnetic field **B** at time t is

(a) $3.5\times 10^{-8}\,\hat{\mathbf{k}}\,\text{T}$ (b) $3.0\times 10^{-8}\,\hat{\mathbf{k}}\,\text{T}$
(c) $2.8\times 10^{-8}\,\hat{\mathbf{k}}\,\text{T}$ (d) $3.2\times 10^{-8}\,\hat{\mathbf{k}}\,\text{T}$

15. A particle is moving n times as fast as an electron. The ratio of de-Broglie wavelength of the particle to that of electron is x, if mass of the electron is y, then the mass of the particle is

(a) $\frac{y}{nx}$ (b) $\frac{x}{ny}$ (c) $\frac{n}{xy}$ (d) $\frac{xy}{n}$

16. In the circuit shown below, the barrier voltage of diode is 0.7 V. Match the physical quantities given in Column I to the results given in Column II and select the correct answer from the codes.

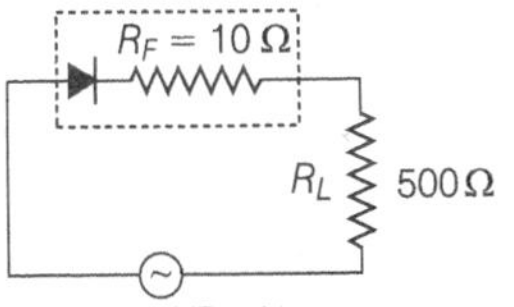

	Column I		Column II
A.	Peak current (in mA) in diode	1.	37.8
B.	Peak voltage (in V) across the ends of load	2.	40.0
C.	Peak current (in mA), if diode is ideal	3.	20.0
D.	Peak voltage (in V) across the ends of load, if diode is ideal	4.	18.9

Codes

	A	B	C	D		A	B	C	D
(a)	1	4	3	2	(b)	4	3	1	2
(c)	1	4	2	3	(d)	3	1	2	4

17. The screw on a screw gauge is given 10 rotations and it moves by 1 mm on the main scale. If there are 100 divisions on the circular scale, then the least count of the screw gauge is

(a) 0.001 mm (b) 0.01 cm
(c) 0.001 cm (d) 0.1 cm

18. In Bohr model of hydrogen atom, the force on the electron depends on the principle quantum number as

(a) $F \propto \frac{1}{n^3}$ (b) $F \propto \frac{1}{n^4}$
(c) $F \propto \frac{1}{n^5}$ (d) Does not depend on n

19. Find out the value of Y.

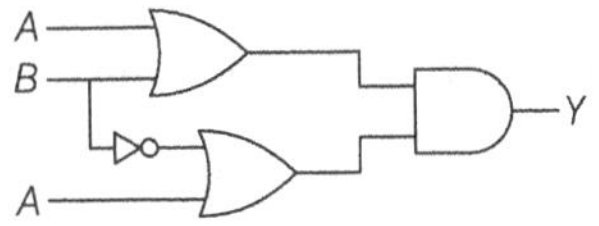

(a) A (b) B (c) $A+B$ (d) $A\cdot B$

20. A parallel plate capacitor of capacitance 120 pF is connected to a battery of emf 30 V. If a dielectric material of dielectric constant $K=\frac{3}{2}$ is inserted between the plates, then magnitude of induced charge will be

(a) 1.5 nC (b) 1.6 nC (c) 1.8 nC (d) 2.4 nC

Section B : Numerical Value Type Questions

21. When a solid sphere is rolling without slipping along a level surface, then the percentage of its translational kinetic energy is x %. The value of x is …… .

22. A non-isotropic solid metal cube has coefficient of linear expansion as $4\times 10^{-4}/^\circ\text{C}$ along X-axis and $5\times 10^{-5}/^\circ\text{C}$ along Y-axis. If coefficient of volumetric expansion of the solid is $57\times 10^{-5}/^\circ\text{C}$, then the coefficient of linear expansion along Z-axis is $n\times 10^{-5}/^\circ\text{C}$. The value of n is …… .

23. The focal length of a convex lens is 20 cm ($\mu_{\text{glass}} = 1.5$). Now, the lens in submerged in water ($\mu_{\text{water}} = \frac{4}{3}$). Find the ratio of focal length in medium to focal length in air.

24. In YDSE, a monochromatic light of wavelength λ is used. The intensity of light at a point on the screen, where the path difference is 2λ is c units. The intensity of light at a point where the path difference is $\frac{\lambda}{4}$ is given by $\frac{nc}{10}$, then the value of n is …… .

25. An ideal cell of emf 20 V is connected in a circuit shown in figure below. Each resistance is 5 Ω. The potential difference (in V) across capacitor when it is fully charged will be ……… .

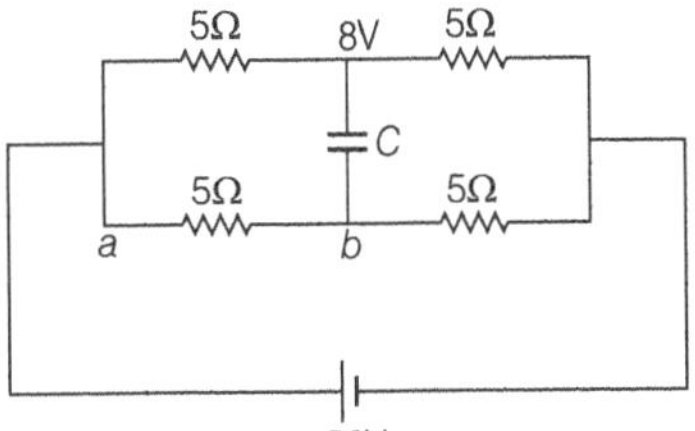

CHEMISTRY

Section A : Objective Type Questions

26. Consider the following compounds of nitrogen.

I. NH_4NO_3 II. $Pb(NO_3)_2$

III. N_2O_4 IV. $(NH_4)_2Cr_2O_7$

V. HNO_3

The total number of compounds that will give nitrogen gas upon heating.

(a) 4 (b) 3

(c) 1 (d) 2

27. The formulas of *A* and *B* for the following reaction sequence are

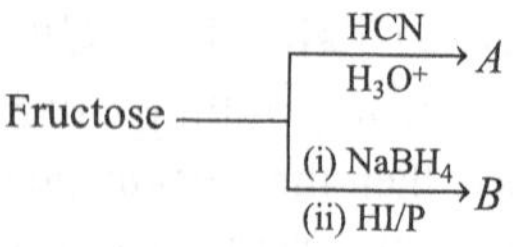

(a) $A = C_7H_{14}O_8, B = C_6H_{14}$

(b) $A = C_7H_{13}O_7, B = C_7H_{14}O$

(c) $A = C_7H_{12}O_8, B = C_6H_{14}$

(d) $A = C_7H_{14}O_8, B = C_6H_{14}O_6$

28. Λ°_{m} for NaCl, HCl and Na*A* are 126.4, 425.9 and 100.5 S cm^2 mol^{-1}, respectively. If the conductivity of 0.001 M H*A* is 6×10^{-5} S cm^{-1}, degree of dissociation of H*A* is

(a) 0.75 (b) 0.25

(c) 0.150 (d) 0.50

29. Phenol on reaction with a mixture of conc. HNO_3 and conc. H_2SO_4 produces a compound. The degree of unsaturation and the nature of compound is

(a) 5, acidic (b) 7, basic

(c) 6, neutral (d) 7, acidic

30. When C_2H_2, CH_4 and C_2H_4 passes through a test tube which have ammoniacal Cu_2Cl_2. Find out which gas comes out unaffected from test tube?

(a) C_2H_2 and CH_4 (b) C_2H_2 and C_2H_4

(c) C_2H_4 and CH_4 (d) Only C_2H_2

31. Which of the following will not produce the benzyl amine?

(a) Benzamide $\xrightarrow{LiAlH_4}$

(b) Phenyl nitrile $\xrightarrow{LiAlH_4}$

(c) 2-phenylethyl isonitrile $\xrightarrow{LiAlH_4}$

(d) 2-phenyl ethanamide $\xrightarrow{Br_2 + KOH}$

32. "Ruhemann's purple" is a corroborative test for the presence of

(a) cupric particle

(b) protein

(c) starch

(d) reducing sugar

33. Given below are two statements.

Statement I The condition for any solution to be ideal is that it should obey Raoult's law over entire range of concentration.

Statement II A solution with $\Delta_{mix} S < 0, \Delta_{mix} V = 0, \Delta_{mix} H = 0, \Delta_{mix} G = 0$ is said to be an ideal solution.

In the light of the above statements, choose the correct answer from the options given below.

(a) Both Statement I and Statement II are correct.

(b) Both Statement I and Statement II all incorrect.

(c) Statement I is correct but Statement II is incorrect.

(d) Statement I is incorrect but Statement II is correct.

34. $CH_3CH{=}CH_2 \xrightarrow{A} CH_3{-}CH_2{-}CH_2OH$

The reagent *A* is

(a) H_2O/ H_2SO_4

(b) $Hg(OAc)_2 / H_2O$ followed by $NaBH_4$

(c) B_2H_6 followed by H_2O_2

(d) CH_3CO_2H/H_2SO_4

35. IUPAC name and degree of unsaturation of the following compound respectively is

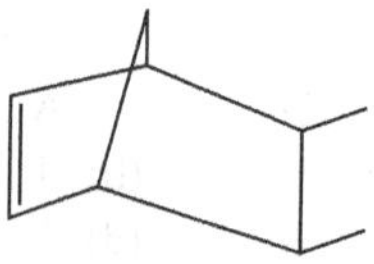

(a) 2, 3- dimethyl bicyclo [2,2,1] hept-5 ene,2

(b) 1, 2- dimethyl bicyclo [2,2,1] hept-4 ene,3

(c) 5, 6- dimethyl bicyclo [2,2,1] hept-2 ene,3

(d) 4, 5- dimethyl bicyclo [2,2,1] hept-1 ene,2

36. Match List-I with List-II.

List-I	List-II
A. BF_4^-	I. Square planar
B. BrF_4^-	II. Bent shape
C. SF_4	III. Tetrahedral
D. BrF_2^+	IV. *See-saw*

Choose the correct answer from the options given below.
(a) A-II, B-III, C-I, D-IV
(b) A-III, B-IV, C-II, D-I
(c) A-III, B-I, C-IV, D-II
(d) A-IV, B-III, C-II, D-I

37. Which of the following statement(s) is/are true about product obtained by treatment of CCl_3CHO with chlorobenzene in presence of H_2SO_4?

A. It has one chiral centre.
B. It is used as an insecticide.
C. It is not easily metabolised by animals.
D. Its name is 1, 1, 1-trichloro-2, 2-*bis*-(*p*-chlorophenyl) ethane

Choose the correct answer from the options given below.
(a) A and B (b) B and C
(c) B, C and D (d) Only D

38. Match List-I with List-II.

	List-I		List-II
A.	$[Cr(NH_3)_4Cl_2]Cl$	I.	Paramagnetic and exhibits ionisation isomerism
B.	$[Ti(H_2O)_5Cl](NO_3)_2$	II.	Diamagnetic and exhibits *cis-trans* isomerism
C.	$[Pt(en)(NH_3)Cl]NO_3$	III.	Paramagnetic and exhibits *cis-trans* isomerism
D.	$[Co(NH_3)_4(NO_3)_2]NO_3$	IV.	Diamagnetic and exhibits ionisation isomerism

Choose the correct answer from the options given below.
(a) A-IV, B-II, C-III, D-I (b) A-III, B-I, C-IV, D-II
(c) A-II, B-I, C-III, D-IV (d) A-I, B-III, C-IV, D-II

39. Consider the following statements.

Statement I 2 L of an ideal gas expands isothermally into vacuum to a total volume of 6 L, then change in internal energy is 4 J.

Statement II Internal energy is a function of temperature.

In the following questions, two statements are given choose the most appropriate answer from the options given below.
(a) Statement I is correct but Statement II is incorrect.
(b) Both Statement I and Statement II are correct.
(c) Statement I is incorrect but Statement II is correct.
(d) Both Statement I and Statement II are incorrect.

40. Which of the following contains a coordinate covalent bond?
(a) $N_2H_5^+$ (b) $BaCl_2$
(c) HCl (d) H_2O

41. Identify *A*, *B* and *C* in the following sequence.

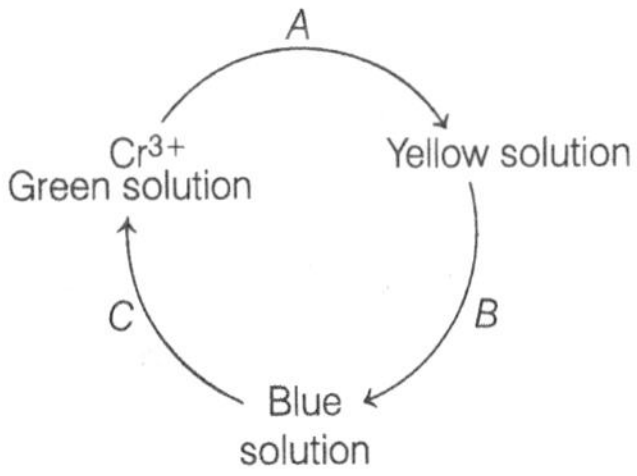

(a) Alkaline H_2O_2, acidified H_2O_2, on standing
(b) Alkaline O_3, acidified O_3, Zn/HCl
(c) Acidified H_2O_2, alkaline H_2O_2, heat
(d) Alkaline O_3, heat, NH_4OH

42. Consider the following compound.
A : Tetracyanomethane B : Carbon dioxide
C : Benzene D : 1, 3-butadiene
Ratio of σ-and π-bonds is in order.
(a) $A = B < C < D$ (b) $A = B < D < C$
(c) $A = B = C = D$ (d) $C < D < A < B$

43. For a solution formed by mixing liquids *L* and *M*, the vapour pressure of *L* plotted against the mole fraction of *M* in solution is shown in the following figure. Here, X_L and X_M represent mole fraction of *L* and *M*, respectively in the solution. The correct statement applicable to this system is

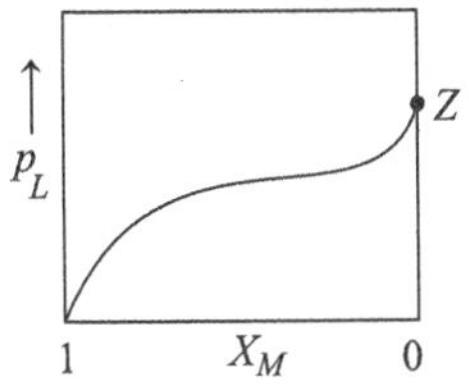

(a) The point *Z* represents vapour pressure of pure liquid *M* and Raoult's law is obeyed from $X_L = 0$ to $X_L = 1$.

(b) The point Z represents vapour pressure of pure liquid L and Raoult's law is not obeyed from $X_L \rightarrow 1$.
(c) The point Z represents vapour pressure of pure liquid M and Raoult's law is obeyed when $X_L \rightarrow 0$.
(d) Attractive intermolecular interaction between L—L in pure liquid L and M—M in pure liquid M are stronger than those between L—M when mixed in solution.

44. Which of the following is incorrect statement?
(a) Cu^+ and Ag^+ ion can not be precipitated by both HCl and H_2S.
(b) In flame test of a mixture of salt, a green colour flame with blue centre indicates the presence of Cu^{2+} cation.
(c) $Pb(NO_3)_2$, $Zn(NO_3)_2$ and $Bi(NO_3)_2$ give white precipitate with dil HCl and dil. NaOH.
(d) Mn^{2+} produces yellow green colour in flame test.

45. Given below are two statements, one is labelled as Assertion (A) and the other is labelled as Reason (R).
Assertion (A) $[Ti(H_2O)_6]^{3+}$ is coloured while $[Sc(H_2O)_6]^{3+}$ is colourless.
Reason (R) d - d transition is not possible in $[Sc(H_2O)_6]^{3+}$.
In the light of the above statements, choose the correct answer from the options given below.
(a) Both (A) and (R) are true and (R) is the correct explanation of (A).
(b) Both (A) and (R) are true but (R) is not the correct explanation of (A).
(c) (A) is true but (R) is false.
(d) (A) is false but (R) is true.

Section B : Numerical Value Type Questions

46. The chlorate ion can disproportionate in basic solution according to the reaction,

$$2ClO_3^- \rightleftharpoons ClO_2^- + ClO_4^-$$

The equilibrium concentration of perchlorate ions for a solution initially at 0.1M in chlorate ions at 298 K is ...... $\times 10^{-2}$. (Nearest integer)

[Given: $E^\circ_{ClO_4^-/ClO_3^-} = 0.39$ V and $E^\circ_{ClO_3^-/ClO_2^-} = 0.36$ V at 298 K.]

47. $KMnO_4$ reacts with ferrous sulphate as follows,
$$MnO_4^- + 5Fe^{2+} + 8H^+ \longrightarrow Mn^{2+} + 5Fe^{3+} + 4H_2O$$
Here, 10 mL of 0.1 M $KMnO_4$ is equivalent to ...... of 0.1 M $FeSO_4$.

48. For a complex reaction,
$$P \xrightarrow{k} \text{Products}$$
$E_{a_1} = 200$ kJ/mol, $E_{a_2} = 90$ kJ/mol, $E_{a_3} = 80$ kJ/mol and $A = 1$.
Overall rate constant, k is related to individual rate constant by the equation, $k = \left(\frac{k_1 k_2}{k_3}\right)^{2/3}$.

Activation energy (kJ/mol) for the overall reaction is ....

49. The number of correct statement(s) about the given reaction is ....
$$I^- + ClO_3^- + H_2SO_4 \longrightarrow Cl^- + HSO_4^- + I_2$$
A. Stoichiometric coefficient of HSO_4^- is 6.
B. Iodide is oxidised.
C. Sulphur is reduced.
D. H_2O is one of the products.

50. An element undergoes a reaction as follows,
$X + 2e^- \longrightarrow X^{2-}$ and energy released $= -30.86$ eV/atom. If the energy released is used to dissociate 4 g of H_2 molecule and equally into H^+ and H^*, where H^* is excited state of H-atoms where the electron travels in orbit whose circumference is equal to four times of its de-Broglie's wavelength. The least moles of X that would be required is ....

[Given: Ionisation energy of H = 13.6 eV/atom, binding energy of H_2 = 4.52 eV/molecule.]

MATHEMATICS

Section A : Objective Type Questions

51. In given figure, person can move towards right or in upward direction only, how many ways are there to reach the point B from point A?

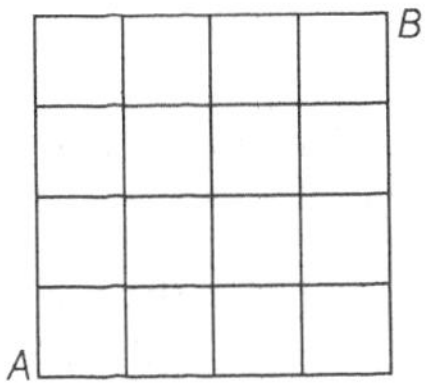

(a) 70 (b) 90
(c) 8 (d) 16

52. Solution of system of equation $2a^4 = b^4 + c^4$, $abc = 27$ knowing that $\log_b a$, $\log_c b$ and $\log_a c$ form a geometric progression, is

(a) $a = b = c = 1$ (b) $a = b = c = 3$
(c) $a = 3$ and $b = c = 2$ (d) $a = b = 2$ and $c = 3$

53. If $f(x, y) = c$ represents the circle having radius r which satisfies the differential equation $\frac{dy}{dx} = \frac{ax+5}{4y+b}$, then

(a) $a = -4,\ r = \frac{1}{4}\sqrt{b^2 + 4c + 25}$

(b) $a = 4,\ r = \frac{1}{4}\sqrt{b^2 + 8c + 100}$

(c) $r = 3,\ a = 3$

(d) $a = -4,\ r = \frac{1}{4}\sqrt{b^2 + 8c + 25}$

54. For complex number z and ω, $|z|^2\omega - |\omega|^2 z = z - \omega$, if and only if

(a) $z = \bar{\omega}$ or $\bar{z}\omega = 1$ (b) $z = \omega$ or $z\omega = 1$
(c) $z = \bar{\omega}$ or $z\bar{\omega} = \frac{1}{2}$ (d) $z = \omega$ or $z\bar{\omega} = 1$

55. If $\int \frac{e^{x-1}}{(x^2 - 5x + 4)} 2x\, dx = AF(x-1) + BF(x-4) + C$ and $F(x) = \int \frac{e^x}{x} dx$, then (A, B) is equal to

(a) $\left(\frac{-2}{3}, \frac{8}{3}\right)$ (b) $\left(\frac{-2}{3}, \frac{8e^3}{3}\right)$
(c) $\left(\frac{8}{3}, \frac{2}{3}\right)$ (d) $\left(\frac{-2}{3}, \frac{8e^{-3}}{3}\right)$

56. If $\left(\frac{1}{a}\right)^b$

$$= \lim_{x \to 0} \frac{8}{x^8}\left(1 - \cos\frac{x^2}{2} - \cos\frac{x^2}{4} + \cos\frac{x^2}{2}\cos\frac{x^2}{4}\right),$$

where a is a prime number, then ab

(a) is greater than $2b$
(b) is less than $2b$
(c) is equal to $2b$
(d) is equal to $5b$

57. Let $p, q, r \in R$, such that $pqr = 5$, if

$A = \begin{bmatrix} 3p & q & r \\ q & 3r & p \\ r & p & 3q \end{bmatrix}$ and $AA^T = 4I$ and $|A|$ is a positive real number, then $3\,(p^3 + q^3 + r^3)$ is equal to

(a) 125 (b) 137 (c) 64 (d) 16

58. If $f(x) = \begin{cases} \dfrac{(3^x - 1)^2}{\sin\left(\frac{x}{c}\right)\log\left(1 + \frac{x}{3}\right)}, & x \neq 0 \\ 10(\log_e 3)^2, & x = 0 \end{cases}$ is continuous at $x = 0$, then c is equal to

(a) $\frac{10}{3}$ (b) $\frac{3}{10}$ (c) $\frac{10}{4}$ (d) $\frac{3}{5}$

59. The coordinates of the focus of the parabola $4x^2 - 4xy + y^2 - 8x - 6y + 5 = 0$ are

(a) $\left(\frac{4}{5}, \frac{3}{5}\right)$ (b) $\left(\frac{3}{5}, \frac{4}{5}\right)$
(c) $\left(\frac{4}{5}, -\frac{3}{5}\right)$ (d) $\left(-\frac{3}{5}, -\frac{4}{5}\right)$

60. Let f be a real valued function defined on the interval $[-10, 10]$ by

$$f(x) = \begin{cases} x - [x], & \text{if } [x] \text{ is odd} \\ 1 + [x] - x, & \text{if } [x] \text{ is even} \end{cases}$$

(where $[\cdot]$ is greatest integer function) and if $\int_{-10}^{10} f(x)\cos \pi x\, dx = \frac{20}{\pi^2}k$, then k is equal to

(a) 1 (b) 2 (c) 3 (d) 4

61. If $f: A \rightarrow B$ is an onto function such that $f(x)=\sqrt{|x|-x}+\dfrac{1}{\sqrt{|x|-x}}$, then A and B are respectively

(a) $(-\infty, 0),(0, \infty)$ (b) $(-\infty, 0),[2, \infty)$
(c) $(0, \infty),(2, \infty)$ (d) $(-\infty, 0],(0, \infty)$

62. A hyperbola having the transverse axis of length $2 \sin \theta$ is confocal with the ellipse $4x^2+6y^2=12$. Then, its equation is

(a) $\dfrac{x^2}{\sin^2 \theta}-\dfrac{y^2}{3\cos^2 \theta}=1$ (b) $\dfrac{x^2}{\operatorname{cosec}^2 \theta}-\dfrac{y^2}{\sec^2 \theta}=1$
(c) $\dfrac{x^2}{\sin^2 \theta}-\dfrac{3y^2}{1+3\cos^2 \theta}=1$ (d) $\dfrac{x^2}{\sin^2 \theta}-\dfrac{y^2}{\cos^2 \theta}=1$

63. A bag contains x blue balls and y red balls $(x \leq 13)$. If two balls chosen simultaneously probability of (one blue, one red) is 0.5. Consider the following statements

I. The possible minimum number of blue balls is 3.

II. The possible maximum number of blue balls is 12.

(a) Statement I is true and Statement II is false
(b) Statement I is false and Statement II is true
(c) Statement I and Statement II both are false
(d) Statement I and Statement II both are true

64. A mirror and a source of light are situated at the origin O and at a point on OX, respectively. A ray of light from the source strikes the mirror and is reflected. If the direction ratios of the normal to the plane are proportional to $1,-1,1$, then direction cosines of the reflected rays are

(a) $\dfrac{1}{3}, \dfrac{2}{3}, \dfrac{2}{3}$ (b) $-\dfrac{1}{3}, \dfrac{2}{3}, \dfrac{2}{3}$
(c) $-\dfrac{1}{3},-\dfrac{2}{3},-\dfrac{2}{3}$ (d) $-\dfrac{1}{3},-\dfrac{2}{3}, \dfrac{2}{3}$

65. $\mathbf{a}$ and $\mathbf{c}$ are unit collinear vectors and $|\mathbf{b}|=6$, then $\mathbf{b}-3\mathbf{c}=\lambda \mathbf{a}$, if λ is

(a) $-9, 3$ (b) $9, 3$
(c) $3,-3$ (d) None of these

66. The value of λ for which the lines

$$\frac{X-2}{1}=\frac{Y-9}{2}=\frac{Z-13}{3} \text{ and } \frac{X-\lambda}{-1}=\frac{Y-7}{2}=\frac{Z+2}{-3}$$

intersect, is

(a) -2 (b) -5 (c) -3 (d) 5

67. The line passing through the points $(5, 1, a)$ and $(3, b, 1)$ crosses the line $X=0$ at the point $\left(0, \dfrac{17}{2}, \dfrac{-13}{2}\right)$, then

(a) $a=2, b=8$
(b) $a=4, b=6$
(c) $a=6, b=4$
(d) $a=8, b=2$

68. The coefficient of x^7 in the expansion of $(1-x-x^2+x^3)^6$ is

(a) 144 (b) 132
(c) -144 (d) -132

69. If $\Delta=\begin{vmatrix}\sin \theta \cos \phi & \sin \theta \sin \phi & \cos \theta \\ \cos \theta \cos \phi & \cos \theta \sin \phi & -\sin \theta \\ -\sin \theta \sin \phi & \sin \theta \cos \phi & 0\end{vmatrix}$, then Δ is independent of

(a) θ (b) ϕ
(c) θ and ϕ (d) None of these

70. The area (in sq units) enclosed by the curve $y=\ln(x+e)$, $x=\ln(1/y)$ and X-axis is

(a) 4 (b) 2
(c) -3 (d) -6

Section B : Numerical Value Type Questions

71. The number of three-digit numbers, whose middle digit is greater than the digit at unit place and hundredth place is ....

72. If $\tan^{-1}(x+1)+\tan^{-1} x+\tan^{-1}(x-1)=\tan^{-1} 3$, then the value of x is equal to ....

73. The number of points in $(-\infty, \infty)$, for which $x^2-x \sin x-\cos x=0$, is ....

74. If the mean deviation about the median of the numbers $k, 2k, 3k, \ldots 50k$ is 100, then $|k|$ is equal to ....

75. A circle has the same centre as an ellipse and passing through the foci F_1 and F_2 of the ellipse, such that the curves intersect in 4 points. Let P be any one of their points of intersection, if the major axis of the ellipse is 17 and the area of the ΔPF_1F_2 is 30, then the distance between the foci is ....

DETAILED SOLUTIONS

Physics

1. *(c)* At resonance condition of series L-C-R circuit,

$$I = \frac{V}{R} = \frac{60}{120} = 0.5 \text{ A}$$

Thus, $V_L = IX_L = I\omega_r L$

$$\Rightarrow \quad L = \frac{V_L}{I\omega_r} = \frac{40}{0.5 \times 4 \times 10^5} = 2 \times 10^{-4} \text{ H}$$

Resonance angular frequency, $\omega_r = \dfrac{1}{\sqrt{LC}}$

$$\Rightarrow \quad C = \frac{1}{L\omega_r^2} = \frac{1}{2 \times 10^{-4} \times (4 \times 10^5)^2} = \frac{1}{32} \times 10^{-6} \text{ F}$$

$$\Rightarrow \quad C = \frac{10^{-6}}{32} \text{ F}$$

Since, current lags behind the voltage by $\dfrac{\pi}{4}$ rad, hence

$$\tan\phi = \frac{X_L - X_C}{R} \Rightarrow \tan 45° = \frac{\omega L - \dfrac{1}{\omega C}}{R}$$

$$\Rightarrow \quad \omega L - \frac{1}{\omega C} - R = 0$$

$$\Rightarrow \quad \omega \times 2 \times 10^{-4} - \frac{1}{\omega \times \dfrac{1}{32} \times 10^{-6}} - 120 = 0$$

$$\Rightarrow \quad \omega^2 - 6 \times 10^5 \omega - 16 \times 10^{10} = 0$$

$$\Rightarrow \quad \omega = \frac{6 \times 10^5 \pm \sqrt{(6 \times 10^5)^2 + (64 \times 10^{10})}}{2}$$

$$= 8 \times 10^5 \text{ rad/s} = k \times 10^5 \text{ rad/s} \quad \text{(given)}$$

$$\therefore \quad k = 8$$

2. *(b)* The given hexagon circuit is symmetric about the line AB, so the modified circuit can be drawn as below.

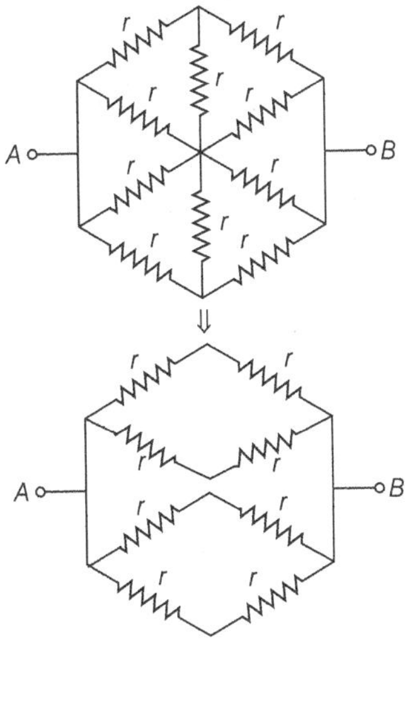

The equivalent resistance between points A and B is given by

$$R_{eq} = (2r) || (2r) || (2r) || (2r)$$

$$\therefore \quad \frac{1}{R_{eq}} = \frac{1}{2r} + \frac{1}{2r} + \frac{1}{2r} + \frac{1}{2r} = \frac{4}{2r}$$

$$\therefore \quad R_{eq} = \frac{r}{2}$$

Given, $r = 1\Omega$

$$\therefore \quad R_{eq} = \frac{1}{2} = 0.5\Omega$$

3. *(a)* For simple harmonic motion,

$$\frac{\text{maximum acceleration}}{\text{maximum velocity}} = 20$$

$$\Rightarrow \quad \frac{\omega^2 a}{\omega a} = 20 \text{ or } \omega = 20 \text{ rad/s}$$

At $t = 0$, displacement, $x = 10$ m

$$x = a\sin(\omega t + \phi)$$

$$10 = a\sin\left(0 + \frac{\pi}{4}\right)$$

or, $$10 = a\sin\frac{\pi}{4} \Rightarrow 10 = a \times \frac{1}{\sqrt{2}}$$

$$a = 10\sqrt{2} \text{ m}$$

Maximum acceleration $= \omega^2 a$

$$= (20)^2 \times 10\sqrt{2} = 400 \times 10\sqrt{2} = 4000\sqrt{2} \text{ ms}^{-2}$$

4. *(c)* Given, $m = 1.67 \times 10^{-27}$ kg,

$q = 1.6 \times 10^{-19}$ C

$v = 2 \times 10^6$ m/s,

$\theta = 60°$, $B = 0.104$ T (along Y-axis)

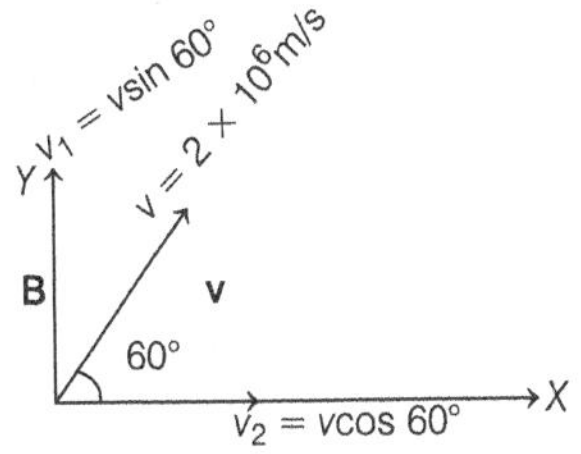

Due to the horizontal component of velocity, the proton will perform circular motion in the XZ-plane.

There is no force due to the velocity in the vertical direction.

So, due to the combination of circular motion in XZ-plane and rectilinear motion in y-direction, the path of proton is helical path.

∴ Radius of the helical path,

$$r = \frac{mv_2}{Bq} = \frac{mv\cos 60°}{Bq}$$

$$\therefore \quad r = \frac{1.67\times 10^{-27}\times 2\times 10^6}{0.104\times 1.6\times 10^{-19}\times 2} = 0.1\text{ m}$$

Time period of revolution,

$$T = \frac{2\pi r}{v_2} = \frac{2\pi m}{Bq}$$

$$\therefore \quad T = \frac{2\pi\times 1.67\times 10^{-27}}{0.104\times 1.6\times 10^{-19}} = 2\pi\times 10^{-7}\text{ s}$$

5. *(b)* Equation of line, $\frac{s}{20} + \frac{v}{20} = 1$

$$\frac{v}{20} = 1 - \frac{s}{20} \quad \text{...(i)}$$

Velocity at $s = 10$ m, i.e.

$$v = \frac{ds}{dt} = 20 - 10 = 10\text{ m/s}$$

Differentiating Eq. (i) with respect to time acceleration,

$$\frac{dv}{dt} = -\frac{ds}{dt}$$

$$a = \left.\frac{dv}{dt}\right|_{S=10\text{ m}} = -\left.\frac{ds}{dt}\right|_{s=10\text{ m}} = -10\text{ m/s}^2$$

6. *(a)* As net external force is zero.

∴Applying conservation of momentum,

$$p_1 = p_2$$

$$\Rightarrow mu + 0 = mv\cos 30° + mv\cos 30°$$

$$\therefore \quad v = \frac{u}{\sqrt{3}} \quad \text{...(i)}$$

Now, initial KE, $K_1 = \frac{1}{2}mu^2$

Final KE, $K_2 = \frac{1}{2}mv^2 + \frac{1}{2}mv^2$

$$= mv^2 = m\left(\frac{u}{\sqrt{3}}\right)^2 = \frac{1}{3}mu^2 \quad \text{[using Eq. (i)]}$$

∴ Loss in KE $= K_1 - K_2$

$$= \frac{1}{2}mu^2 - \frac{1}{3}mu^2 = \frac{1}{6}mu^2$$

7. *(a)*

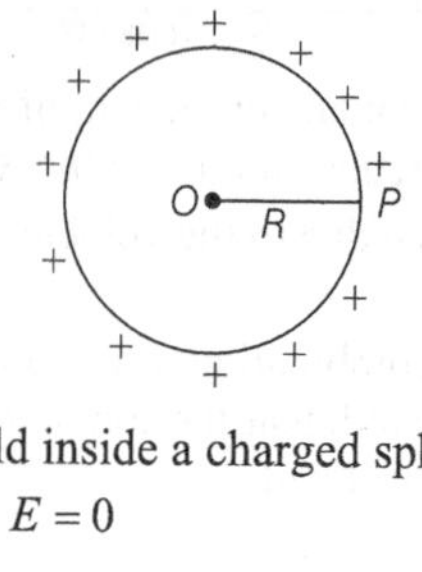

As electric field inside a charged spherical shell is zero.

$$\therefore \quad E = 0$$

$$\therefore \quad dV = -E\,dr = 0$$

$$\therefore \quad V_0 = V_p = \text{constant} = \frac{Q}{4\pi\varepsilon_0 R}$$

∴ The external work done,

$$W = \Delta U = q(V_p - V_0) = 0$$

8. *(b)* Given, $M_A = 1\text{ kg}, M_B = 2\text{ kg}, \mu_1 = 0.15, \mu_2 = 0.2$

As the blocks are at rest, so static friction acts in the system of blocks.

The FBDs of the blocks A and B are shown below.

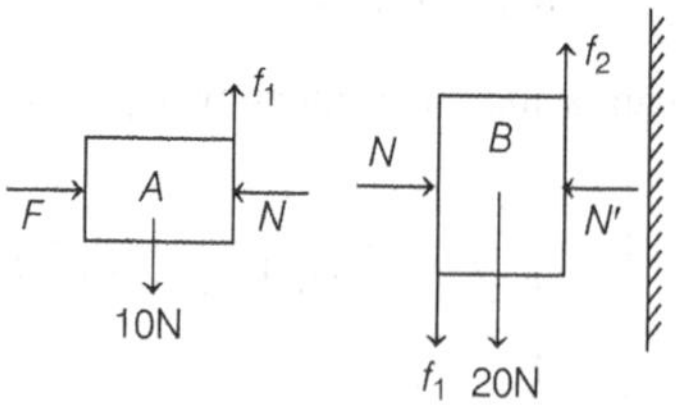

For equilibrium of body A,

$$f_1 = 10\text{ N}$$

For equilibrium of body B,

$$f_2 = f_1 + 20 = 10 + 20 = 30\text{ N}$$

∴Frictional force applied by the wall on the block,

$$f_2 = 30\text{ N}$$

9. *(a)* Let a point mass at C is placed at a distance of x from the point mass at A as shown in figure.

Here, $\frac{M_A}{M_B} = \frac{4}{3}$, Force between A and C is

$$F_{AC} = \frac{GMM_A}{x^2}$$

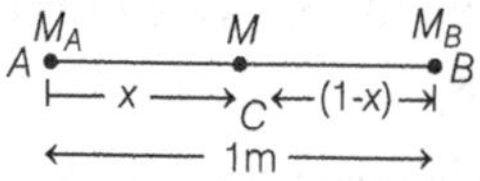

Force between B and C is $F_{BC} = \frac{GMM_B}{(1-x)^2}$

According to question, $F_{AC} = \frac{1}{3}F_{BC}$

$$\therefore \quad \frac{GM_AM}{x^2} = \frac{1}{3}\left(\frac{GM_BM}{(1-x)^2}\right)$$

$$\Rightarrow \quad \frac{M_A}{M_B} = \frac{x^2}{3(1-x)^2}$$

$$\Rightarrow \quad \frac{4}{3} = \frac{x^2}{3(1-x)^2}$$

$$\Rightarrow \quad 4 = \frac{x^2}{(1-x)^2} \Rightarrow 2 = \frac{x}{(1-x)}$$

By solving, we get $x = \frac{2}{3}$ m

10. *(c)* [Stress] = [Young modulus] [Strain]

As, $[\text{Strain}] = \dfrac{[\text{Change in length}]}{[\text{Original length}]} = \dfrac{[M^0LT^0]}{[M^0LT^0]}$

$= [M^0L^0T^0] = \text{Dimensionless}$

$\therefore$ [Stress] = [Young modulus]

Thermal conductivity, $[K]$

$$= \frac{[\text{Energy/Time}][\text{length}]}{[\text{Area}][\text{Temperature difference}]}$$

$$= \frac{[ML^2T^{-3}][L]}{[L^2][K]} = [MLT^{-3}K^{-1}]$$

Strain is a dimensionless physical quantity, whereas stress is not a dimensionless physical quantity.

11. *(c)* Let $p_1 = 1\,\text{atm}, n = 5\,\text{mol}, T_1 = 293\,\text{K}$

and $V_2 = \dfrac{V_1}{10}$

Using, $T_1V_1^{\gamma-1} = T_2V_2^{\gamma-1}$

$$\Rightarrow \quad T_2 = T_1\left(\frac{V_1}{V_2}\right)^{\gamma-1} = 293(10)^{0.4} = 736\,\text{K}$$

[$\because$ for diatomic gas, $\gamma = 1.4$]

Work done, $W = \dfrac{nR(T_1 - T_2)}{\gamma - 1}$

$$= \frac{5 \times 8.3 \times (293 - 736)}{0.4} = -46\,\text{kJ}$$

$$\Delta U = \Delta Q - W = 0 - W = 46\,\text{kJ}$$

12. *(b)* $r_{max} = 2 \times 10^{12}\,\text{m}$

$r_{min} = 8 \times 10^{10}\,\text{m}$

$v_{max} = 8 \times 10^4\,\text{ms}^{-1}$

By conservation of angular momentum,

$mvr = \text{constant}$

$v_{min} \times r_{max} = v_{max} \times r_{min}$

$$v_{min} = \frac{8 \times 10^4 \times 8 \times 10^{10}}{2 \times 10^{12}} = 4 \times 8 \times 10^{14} \times 10^{-12}$$

$$= 32 \times 10^2 = 3.2 \times 10^3\,\text{ms}^{-1}$$

13. *(a)* The rms speed of gas molecules is $v_{rms} = \sqrt{\dfrac{3RT}{M}}$

The average speed of gas molecules is

$$v_{av} = \sqrt{\frac{8RT}{\pi M}}$$

$$\therefore \quad v_{rms} : v_{av} = \sqrt{3} : \sqrt{\frac{8}{\pi}} = 1.732 : 1.595$$

$$\therefore \quad v_{rms} > v_{av}$$

According to law of equipartition of energy, the energy of an ideal gas is distributed equally among its all degree of freedom. The energy associated with each degree of freedom per molecule is $\dfrac{1}{2}K_BT$.

14. *(d)* The magnitude of B is

$$B = \frac{E}{c} = \frac{9.6}{3 \times 10^8} = 3.2 \times 10^{-8}\,\text{T}$$

In EM wave, **E** and **B** oscillates perpendicular to each other. As wave is propagating along $+x$-direction and **E** is along $+y$-direction. So, magnetic field **B** should be along $+z$-direction.

$$\therefore \quad \mathbf{B} = 3.2 \times 10^{-8}\,\hat{\mathbf{k}}\,\text{T}$$

15. *(a)* Given, $v_p = nv_e, m_e = y,$

$$\frac{\lambda_p}{\lambda_e} = x$$

The de-Broglie wavelength of a particle is

$$\lambda = \frac{h}{mv}$$

$$\frac{\lambda_p}{\lambda_e} = \frac{m_e v_e}{m_p v_p} = x$$

$$\Rightarrow \quad \frac{y}{m_p} \times \frac{v_e}{nv_e} = x$$

$$\therefore \quad m_p = \frac{y}{nx}$$

16. *(c)* Diode gets forward biased during positive half-cycle of alternating source, hence current flows. Let the barrier voltage of diode is V_B, its forward resistance be R_F and load resistance be R_L, then peak voltage of applied signal is

$$V_0 = V_B + i_0\,(R_F + R_L)$$

$$\therefore \quad i_0 = \frac{V_0 - V_B}{R_F + R_L}$$

$$= \frac{20 - 0.7}{10 + 500} = 37.8 \times 10^{-3}\,\text{A}$$

$$= 37.8\,\text{mA}$$

Peak voltage across the ends of the load R_L,

$V_L = i_0 \times R_L = 37.8 \times 10^{-3} \times 500 = 18.9\,\text{V}$

For ideal diode, $V_B = 0$ and $R_F = 0$

$$\therefore \quad i_0 = \frac{V_0}{R_L} = \frac{20}{500} = 0.04\,\text{A} = 40\,\text{mA}$$

and $V_L = i_0R_L = 0.04 \times 500 = 20\,\text{V}$

17. *(a)* Pitch of screw of screw gauge

$$= \frac{\text{Distance moved by screw}}{\text{Number of rotations}}$$

$$= \frac{1\,\text{mm}}{10} = 0.1\,\text{mm}$$

and least count of screw gauge

$$= \frac{\text{Pitch}}{\text{Number of circular scale divisons}} = \frac{0.1}{100}$$

$= 0.001\,\text{mm}$

18. *(b)* According to Bohr's model,

$$F \propto \frac{v^2}{r}$$

Also, $v \propto \frac{1}{n}$ and $r \propto n^2$ (n = principal quantum number)

$$\Rightarrow \quad F \propto \frac{1}{n^4}$$

19. *(a)* $Y = (A+B)\cdot(\overline{B}+A)$

$= A\cdot\overline{B} + A\cdot A + B\cdot\overline{B} + B\cdot A$

$= A\cdot\overline{B} + A + 0 + B\cdot A \qquad (\because B\cdot\overline{B} = 0 \text{ and } A\cdot A = A)$

$= A(\overline{B}+1+B) = A \qquad (\because \overline{B}+1+B = 1)$

20. *(c)* $C = 120\text{pF} = 120\times 0^{-12}\,\text{F}$,

$$V = 30\text{V}, K = \frac{3}{2}$$

Let induced charge on the dielectric be Q'

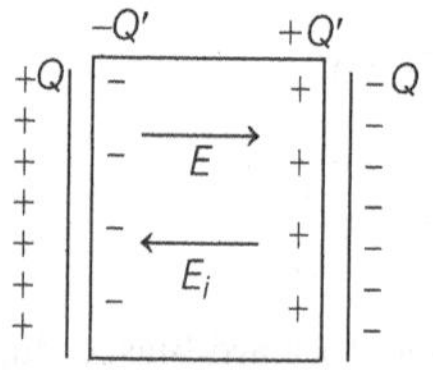

The charge on the capacitor with dielectric medium,

$$Q = C'V = KCV$$

$\therefore$ Electric field inside the plates of capacitor due to charge Q,

$$E = \frac{Q}{A\varepsilon_0}$$

$$E_i = \text{Induced electric field} = \frac{Q'}{\varepsilon_0 A}$$

$\therefore$ Resultant electric field between the plates of capacitor after insertion of dielectric material,

$$E' = E - E_i = \frac{Q-Q'}{\varepsilon_0 A} \quad \ldots\text{(i)}$$

Also, $$E' = \frac{E}{K} = \frac{Q}{\varepsilon_0 AK} \quad \ldots\text{(ii)}$$

From Eqs. (i) and (ii), we get

$$\frac{Q-Q'}{\varepsilon_0 A} = \frac{Q}{\varepsilon_0 AK}$$

$$\therefore \quad Q' = Q\left(1-\frac{1}{K}\right) \qquad [\because Q = KCV]$$

$$\therefore \quad Q' = KCV\left(1-\frac{1}{K}\right) = (K-1)CV$$

$$= \left(\frac{3}{2}-1\right)\times 120\times 10^{-12}\times 30 = 1.8\,\text{nC}$$

Study Tactics

Start from the calculation of electric field inside the plates of capacitor and then induced electric field. Equate both the electric field and finally, calculate the induced charge on dielectric.

21. (71) For a body rolling along a level surface,

$$(\text{KE})_{\text{Total}} = (\text{KE})_T + (\text{KE})_R = \frac{1}{2}mv^2 + \frac{1}{2}I\omega^2$$

$$\Rightarrow (\text{KE})_{\text{Total}} = \frac{1}{2}(mv^2 + I\omega^2) \quad \ldots\text{(i)}$$

Now, moment of inertia for the solid sphere,

$$I = \frac{2}{5}mr^2$$

Substituting this value of moment of inertia in Eq. (i), we get

$$(\text{KE})_{\text{Total}} = \frac{1}{2}\left(mv^2 + \frac{2}{5}mr^2\omega^2\right)$$

Angular velocity of a body is given as

$$\omega = \frac{v}{r} \Rightarrow v = \omega r$$

Substituting value of ωr as v in the above equation,

$$(\text{KE})_{\text{Total}} = \frac{1}{2}\left(mv^2 + \frac{2}{5}mv^2\right)$$

$$\Rightarrow (\text{KE})_{\text{Total}} = \frac{7}{10}mv^2$$

Now, the contribution of translational KE in total KE can be found on dividing the translational KE by total KE.

$$\frac{(\text{KE})_{\text{Trans}}}{(\text{KE})_{\text{Total}}} = \frac{\frac{1}{2}mv^2}{\frac{7}{10}mv^2} = \frac{1}{2}\times\frac{10}{7} = \frac{10}{14} = 0.7142$$

$$\therefore \quad \%(\text{KE})_{\text{Trans}} = 0.7142\times 100 = 71.42\% \simeq 71\% = x\%$$

So, $x = 71$

22. (12) The coefficient of linear expansion, $\alpha = \frac{1}{T}\frac{\Delta l}{l}$

The coefficient of volumetric expansion,

$$\gamma = \frac{1}{T}\cdot\frac{\Delta V}{V}$$

∴For a cube, $V = xyz$

$$\therefore \quad \frac{\Delta V}{V} = \frac{\Delta x}{x} + \frac{\Delta y}{y} + \frac{\Delta z}{z}$$

$$\Rightarrow \quad \frac{1}{T}\cdot\frac{\Delta V}{V} = \frac{1}{T}\cdot\frac{\Delta x}{x} + \frac{1}{T}\cdot\frac{\Delta y}{y} + \frac{1}{T}\cdot\frac{\Delta z}{z}$$

$$\Rightarrow \quad \gamma = \alpha_x + \alpha_y + \alpha_z$$

$$\Rightarrow \quad \alpha_z = \gamma - (\alpha_x + \alpha_y)$$

$$= 57\times10^{-5} - (40+5)\times10^{-5}$$

$$\Rightarrow \quad \alpha_z = 12\times10^{-5}/^\circ\text{C} = n\times10^{-5}/^\circ\text{C}$$

$$\Rightarrow \quad n = 12$$

23. (4) $\frac{1}{f_a} = \left(\frac{\mu_g}{\mu_a} - 1\right)\left(\frac{1}{R_1} - \frac{1}{R_2}\right)$ ...(i)

Similarly, $\frac{1}{f_m} = \left(\frac{\mu_g}{\mu_m} - 1\right)\left(\frac{1}{R_1} - \frac{1}{R_2}\right)$ ......(ii)

Dividing Eq. (i) by Eq. (ii), we get

$$\frac{f_m}{f_a} = \frac{\left(\frac{\mu_g}{\mu_a} - 1\right)}{\left(\frac{\mu_g}{\mu_m} - 1\right)} = \frac{\left(\frac{1.5}{1} - 1\right)}{\left(\frac{1.5}{\frac{4}{3}} - 1\right)} = 4$$

$$\therefore \quad \frac{f_m}{f_a} = 4$$

24. (5) For path difference, $\Delta x = 2\lambda$

Phase difference, $\Delta\phi = \frac{2\pi}{\lambda}\times\Delta x = \frac{2\pi}{\lambda}\times 2\lambda = 4\pi$

Intensity, $I = I_1 + I_2 + 2\sqrt{I_1 I_2}\cos(\Delta\phi)$

As, light is monochromatic,

∴ $I_1 = I_2 = I_0$

At, $\Delta\phi = 4\pi$,

$I = 4I_0$ (given)

Also given, $I = c$

∴ $I_0 = \frac{c}{4}$

For path difference,

$$\Delta x = \frac{\lambda}{4}$$

Phase difference,

$$\Delta\phi = \frac{2\pi}{\lambda}\times\frac{\lambda}{4} = \frac{\pi}{2}$$

$$\therefore \quad I' = I_1 + I_2 + 2\sqrt{I_1 I_2}\cos(\Delta\phi)$$

$$= \frac{c}{4} + \frac{c}{4} + 2\times\frac{c}{4}\times 0$$

$$= \frac{c}{2}$$

According to question, $\frac{nc}{10} = \frac{c}{2}$

$\Rightarrow \quad n = 5$

25. (4) When capacitor is fully charged, then it behaves as open circuit.

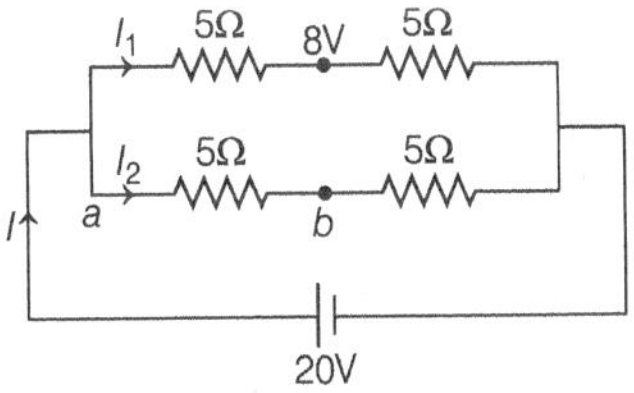

$$\therefore \quad R_{\text{net}} = \frac{10\times10}{10+10} = 5\Omega$$

$$I_1 = \frac{20-8}{5} = \frac{12}{5} = 2.4\text{ A}$$

$$I = \frac{E}{R_{\text{net}}} = \frac{20}{5} = 4\text{ A}$$

$$\therefore \quad I_2 = (4-2.4) = 1.6\text{ A}$$

$$\therefore \quad V_{ab} = I_2\times5 = 1.6\times5 = 8\text{ V}$$

$$V_a = 20\text{ V}$$

$$\therefore \quad V_{ab} = V_a - V_b$$

$$\Rightarrow \quad 8 = 20 - V_b$$

$$\therefore \quad V_b = 12\text{ V}$$

∴Potential difference across the capacitor,

$$V = V_b - 8 = 12 - 8 = 4\text{ V}$$

Chemistry

26. (*c*) Among the given compounds, only one i.e. ammonium dichromate upon heating will give nitrogen gas.

I. $NH_4NO_3 \xrightarrow{\Delta} N_2O + 2H_2O$

II. $2Pb(NO_3)_2 \xrightarrow{673\text{ K}} 4NO_2 + 2PbO + O_2$

III. $N_2O_4 \xrightarrow{\Delta} 2NO_2$

IV. $(NH_4)_2Cr_2O_7 \xrightarrow{\Delta} N_2 + 4H_2O + Cr_2O_3$

V. $4HNO_3 \longrightarrow 2H_2O + 4NO_2 + O_2\uparrow$

27. (*a*) The reaction is represented as

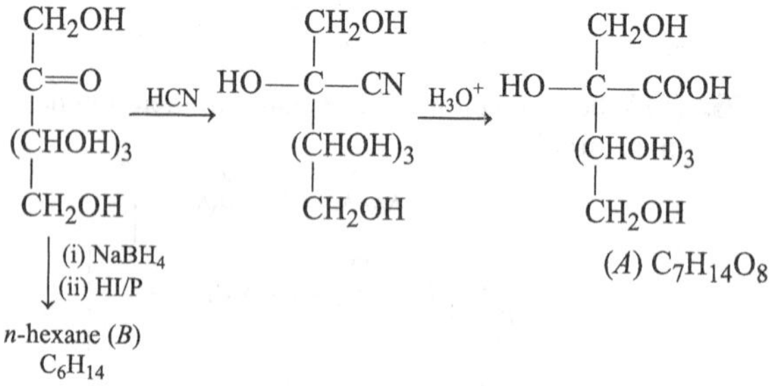

28. (*c*) Given, $\Lambda^\circ_m NaCl = 126.4\text{ S cm}^2\text{mol}^{-1}$

$$\Lambda^\circ_m HCl = 425.9\text{ S cm}^2\text{mol}^{-1}$$

$$\Lambda^\circ_m NaA = 100.5\text{ S cm}^2\text{mol}^{-1}$$

$$\Lambda^\circ_m HA = \Lambda^\circ_m HCl + \Lambda^\circ_m NaA - \Lambda^\circ_m NaCl$$

$$= 425.9 + 100.5 - 126.4$$

$$= 400\text{ S cm}^2\text{mol}^{-1}$$

$$\Lambda_m = \frac{1000\kappa}{C} = 6 \times 10^{-5} \times \frac{1000}{0.001}$$

$$= 60\text{ S cm}^2\text{mol}^{-1}$$

$$\alpha_m = \frac{\Lambda_m}{\Lambda^\circ_m} = \frac{60}{400} = 0.150$$

29. (*d*) Phenol undergoes electrophilic substitution reaction on reaction with a mixture of conc. HNO_3 and conc. H_2SO_4 (known as nitrating mixture). This reaction produces nitro compound and on successive nitration it produces trinitrophenol.

$$\text{Conc. } HNO_3 + \text{Conc. } H_2SO_4 \longrightarrow \overset{+}{N}O_2 \text{ (electrophile)}$$

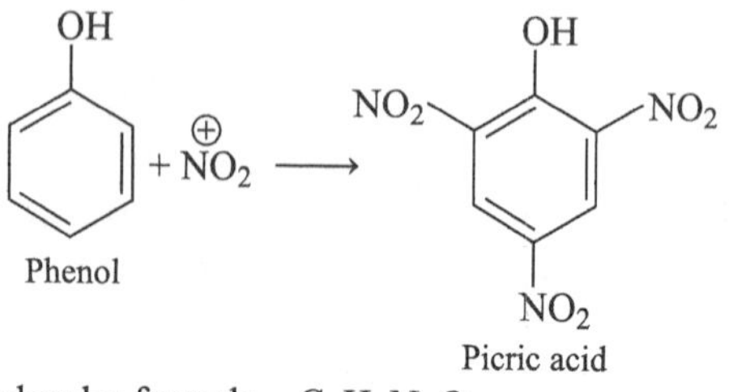

Molecular formula $= C_6H_3N_3O_7$

Degree of unsaturation $= (C+1) - \frac{H}{2} - \frac{X}{2} + \frac{N}{2}$

$$= (6+1) - \frac{3}{2} + \frac{3}{2} = 7 - 0 = 7$$

Nature of compound The compound is acidic in nature due to presence of three strong electron withdrawing groups (NO_2).

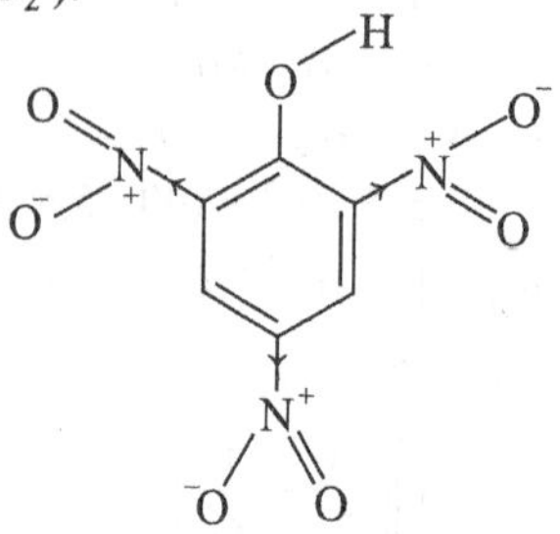

30. (*c*) Copper and silver alkylides are obtained by passing alkynes in the ammoniacal solution of cuprous chloride or silver nitrate respectively. The reaction are used for detecting the presence of acetylenic hydrogen atom is

$$HC \equiv CH + Cu_2Cl_2 + 2NH_4OH \longrightarrow \underset{\substack{\text{Copper acetylide} \\ \text{(Red ppt.)}}}{Cu \cdot C \equiv C \cdot Cu\downarrow} + 2NH_4Cl + 2H_2O$$

While, C_2H_4 and CH_4 comes out unaffected from test tube.

31. (*c*) The given reaction are represented as

(a) 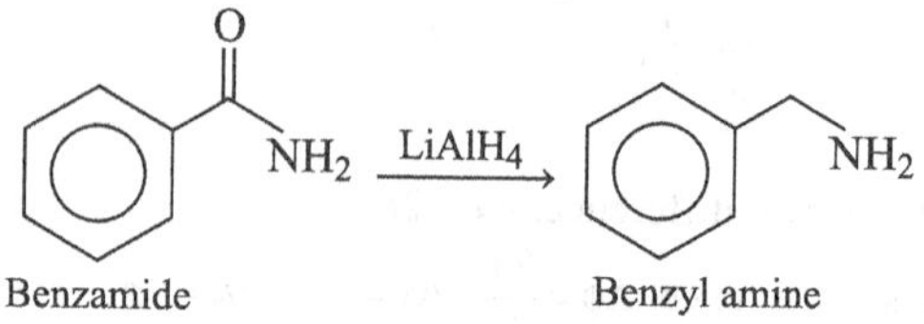

(b) 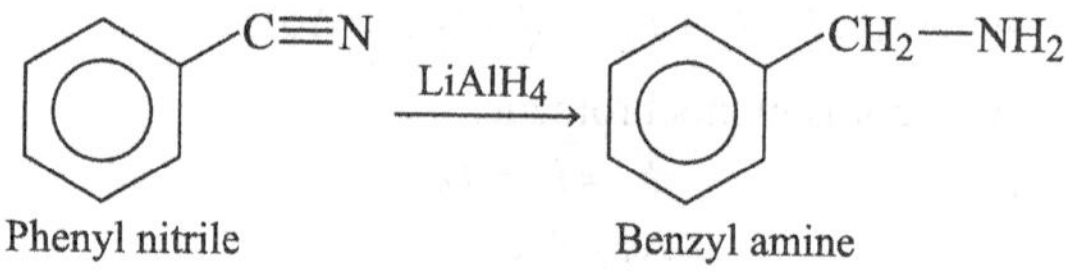

(c) 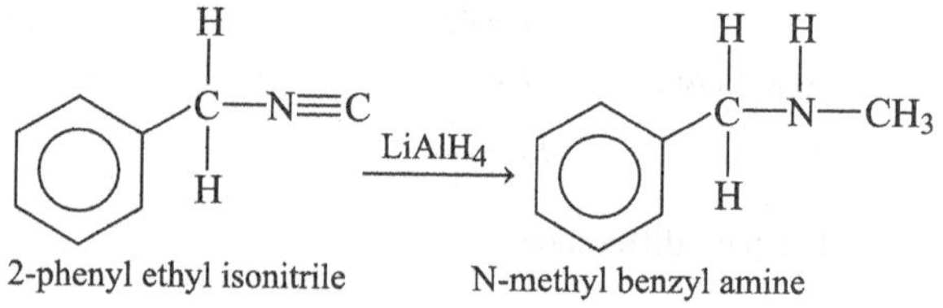

(d) 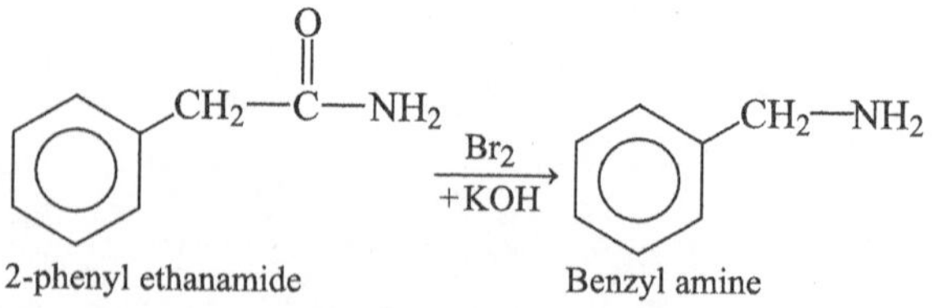

Last process is a good example of Hofmann bromamide reaction in which carboxylamide are converted into amine having one carbon atom less than carboxyl amide.

32. *(b)* All α-amino acids on treatment with ninhydrin (2, 2-dihydroxyindane -1, 3-dione) give purple colouration. The purple coloured complex is Ruhemann's purple. This test is called ninhydrin test. Hence, it is a corroborative test for the presence of protein. The reaction is as follows

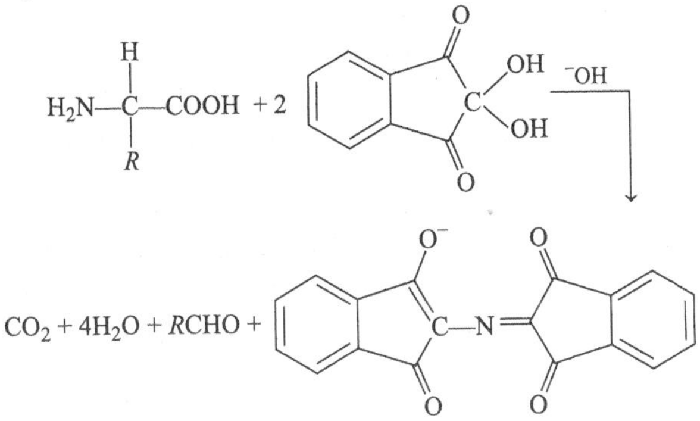

Purple coloured complex

33. *(c)* Statement I is correct but statement II is incorrect.
For a solution to be ideal it should obey Raoult's law over entire range of concentration.
In case of ideal solutions, other conditions are

$$\Delta_{mix} S > 0, \Delta_{mix} V = 0, \Delta_{mix} H = 0, \Delta_{mix} G < 0$$

34. *(c)* $\underset{\text{Propene}}{CH_3—CH{=}CH_2} + B_2H_6 \xrightarrow{H_2O_2} \underset{\text{Propan-1-ol}}{CH_3—CH_2—CH_2OH}$

When propene reacts with diborane followed by H_2O_2, it gives primary alcohols which is in accordance with anti-Markownikoff's rule.

35. *(c)*

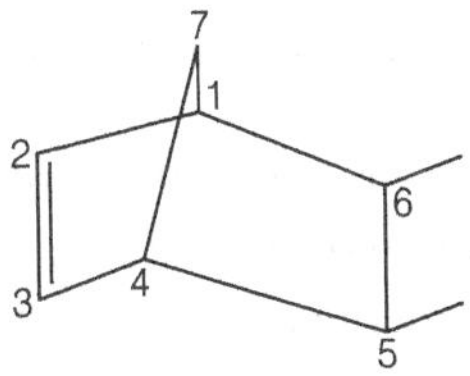

Total carbon atom forming the bicyclic ring = 7 (hept)
Functional group present ⇒ double bond (ene)
Position of double bond ⇒ 2-ene
Substituents ⇒ 2-methyl groups ⇒ dimethyl
Position of substituents = 5, 6 ⟶ 5, 6-dimethyl
Number of cyclic chain = 2 ⟶ Bicyclo
3-bridges are 2 carbons, 2 carbons and one carbon.
IUPAC name = 5, 6-dimethylbicyclo [2, 2, 1] hept- 2-ene
Molecular formula of compound is C_9H_{14}.
Degree of unsaturation can be calculated as

$$u = (C + 1) - \frac{H}{2} - \frac{X}{2} + \frac{N}{2}$$

where, u = degree of unsaturation
C = number of carbons
H = number of hydrogens
X = number of halogens
N = number of nitrogens.

Hence, for a compound having molecular formula C_9H_{14} the degree of unsaturation may be calculated as

$$u = (9 + 1) - \frac{14}{2}$$
$$= 10 - 7$$
$$= 3$$

Study Tactics

The problem can be solved by identifying the parent chain, functional group, position of functional group, substituent and their position one by one and then write the name of compound according to IUPAC names and finally calculate degree of unsaturation by using formula

$$u = (C + 1) - \frac{H}{2} - \frac{X}{2} + \frac{N}{2}$$

36. *(c)* The correct match is A-III, B-I, C-IV, D-II.

The molecules with their shapes are given below.

(A) BF_4^- – Tetrahedral

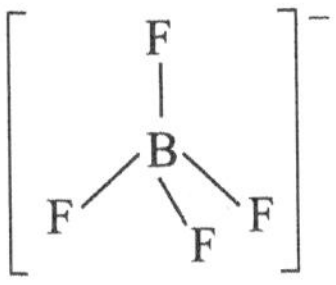

(B) BrF_4^- – Square planar

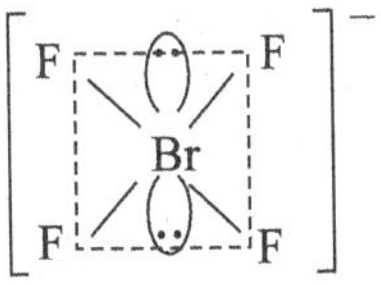

(C) SF_4 – See-saw

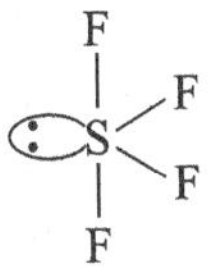

(D) BrF_2^+ – Bent-shape

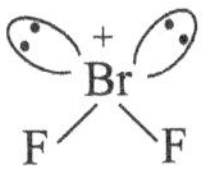

37. *(c)* Among the given statements B, C and D are correct while statement A is incorrect.

Preparation of DDT

DDT is *p*, *p*-Dichlorodiphenyltrichloroethane prepared by reaction of chlorobenzene with $CCl_3—CHO$ in presence of H_2SO_4 through electrophilic substitution reaction.

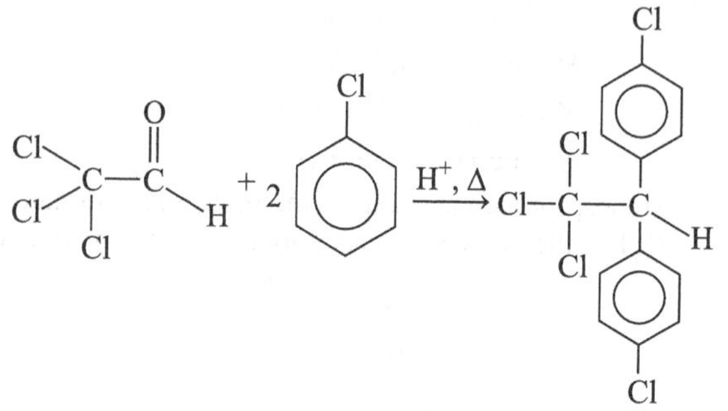

DDT

Characteristics of DDT

- It is used as an insecticide to kill mosquito so that people are prevented by malaria.
- It is highly toxic towards fish.
- It is toxic for animals because it is not easily metabolised by animal.
- It has no chiral centre.

38. (*b*) The correct match is A-III, B-I, C-IV, D-II.

A. Cr^{3+} has $3d^3$ configuration, with 3 unpaired electrons. Hence, it shows paramagnetic behaviour. Complex of the type Ma_4b_2 shows *cis-trans* isomerism.

B. Ti^{3+} has $3d^1$ configuration, hence shows paramagnetic behaviour. Complex gives Cl^- and NO_3^- ions in solution hence, shows ionisation isomerism.

C. Pt^{2+} has $5d^8$ configuration but ligands are strong field ligands hence, it forms square planar complex. thus, all electrons are paired and it also exhibits ionisation isomerism.

D. Co^{3+} has $3d^6$ configuration. But, ligands present are strong enough to cause electron pairing, hence, it shows diamagnetic behaviour and exhibits *cis-trans* isomerism as it is Ma_4b_2 type complex.

39. (*c*) Statement I is incorrect but statement II is correct. The correct form of statement I is,

When 2 L of an ideal gas expands isothermally into vacuum to a total volume of 6 L. So, the change in internal energy is zero.

$$pV = \text{constant.}$$

40. (*a*) $N_2H_4 + H^+ \longrightarrow [N_2H_4 \longrightarrow H]^+$

A coordinate bond is present between N and H and a covalent bond with 4 hydrogens. Thus, it form coordinate covalent bond.

41. (*a*) (*A*) $\underbrace{\underset{\text{Green}}{2Cr^{3+}} + 10OH^- + 3H_2O_2}_{A} \longrightarrow \underset{\text{Yellow}}{2CrO_4^{2-}} + 8H_2O$

(*B*) $\underbrace{CrO_4^{2-} + 2H_2O_2 + 2H^+}_{B} \longrightarrow$ 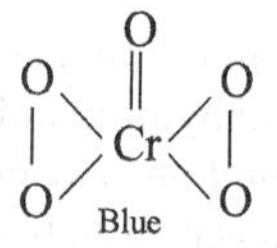 $+ 3H_2O$

In aqueous solution, CrO_5 is unstable and it further decomposes.

(*C*) $2CrO_5 \longrightarrow \underset{\text{(Amphoteric)}}{Cr_2O_3} + \frac{7}{2}O_2$

$Cr_2O_3 + 3H_2SO_4 \longrightarrow Cr_2(SO_4)_3 + 3H_2O$

42. (*a*) A. Tetracyanomethane : $C(CN)_4$ has 8 σ and 8 π-bond. So, ratio is 1.

B. CO_2 has 2 σ and 2 π-bonds. So, ratio is 1.

C. Benzene has 12 σ and 3 π-bonds. So, the ratio is 4.

D. 1, 3-butadiene has 9 σ and 2 π-bonds. So, the ratio is $\frac{9}{2}$.

Hence, the correct order is $A = B < C < D$.

43. (*d*) .

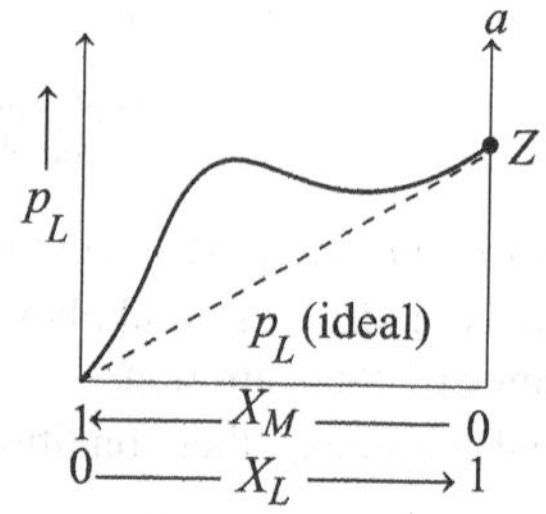

The graph representing the deviation from Raoult's law therefore intermolecular forces of attraction between

$$M—L < M—M \text{ or } L—L$$

$$p_L \geq p_L^\circ X_L$$

but when $X_L = 1$, the mixture has almost pure liquid L so, $p_L = p_L^\circ$

44. (*a*) Ag^+ is precipitated by HCl only while Cu^+ is precipitated by passing H_2S in the presence of HCl.

45. (*a*) Both (A) and (R) are true and (R) is the correct explanation of (A).

$[Sc(H_2O)_6]^{3+}$ has no unpaired electron in its *d*-subshell and thus d - d transition is not possible. $[Ti(H_2O)_6]^{3+}$ has one unpaired electron in its *d*-subshell which gives rise to *d*-*d* transition.

46. (*2*) **At anode**

$H_2O(l) + ClO_3^-(aq) \longrightarrow ClO_4^-(aq) + 2H^+(aq) + 2e^-$

At cathode

$2H^+(aq) + ClO_3^-(aq) + 2e^- \longrightarrow ClO_2^-(aq) + H_2O(l)$

$2ClO_3^-(aq) \longrightarrow ClO_2^-(aq) + ClO_4^-(aq)$

$$E^\circ_{cell} = -0.39 + 0.36 = -0.03$$

$$E^\circ_{cell} = \frac{RT}{2F} \ln K$$

$$-0.03 = \frac{0.06}{2} \log K \text{ or } K = 0.1$$

$$2ClO_3^- \rightleftharpoons ClO_4^- + ClO_2^-$$
$$\quad 0.1-2x \qquad x \qquad x$$

$$\frac{x^2}{(0.1-2x)^2} = 0.1 = \frac{1}{10}$$

$$3.16x = 0.1 - 2x \Rightarrow 5.16x = 0.1$$

$$x = \frac{0.1}{5.16} = 0.0193 \approx 1.93 \times 10^{-2} \approx 2 \times 10^{-2}$$

47. *(50)* $KMnO_4$ reacts with ferrous sulphate according to the given reaction. So, reaction can be represented as

$$2KMnO_4 + 10FeSO_4 + 8H_2SO_4 \longrightarrow K_2SO_4 + 2MnSO_4 + 5Fe_2(SO_4)_3 + 8H_2O$$

2 moles of $KMnO_4$ reacts with 10 moles of $FeSO_4$.

The number of moles of $KMnO_4$ in 10 ml of 0.1 M

$$= 0.1 \times 0.01 = 10^{-3} \text{ moles}$$

Therefore, 10^{-3} moles $KMnO_4$ reacts with 5×10^{-3} moles of $FeSO_4$.

Hence, 10 mL of 0.1 M $KMnO_4$ is equivalent to its 5 times, i.e. 50 mL of 0.1 M $FeSO_4$.

48. *(140)* According to Arrhenius equation,

$$k = Ae^{-E_a/RT}$$

So, $k_1 = Ae^{-\frac{E_{a1}}{RT}}$,

$$k_2 = Ae^{-\frac{E_{a2}}{RT}}$$

and $k_3 = Ae^{-\frac{E_{a3}}{RT}}$

$$k = \left(\frac{k_1 k_2}{k_3}\right)^{2/3} \quad \text{[Given]}$$

$$\therefore \quad k = \left(\frac{Ae^{-\frac{E_{a1}}{RT}} \cdot Ae^{-\frac{E_{a2}}{RT}}}{Ae^{-E_{a3}/RT}}\right)^{2/3} = Ae^{-E_a/RT}$$

$$e^{\frac{1}{RT}\left[\frac{2}{3}(E_{a3} - (E_{a1} + E_{a2}))\right]} = e^{-E_a/RT}$$

$$E_a = \frac{2}{3}[E_{a_1} + E_{a_2} - E_{a_3}]$$

$$E_a = \frac{2}{3}[200 + 90 - 80] = \frac{2}{3}[210] = 140 \text{ kJ / mol}$$

49. *(3)* Statements (A), (B) and (D) are correct.

Balanced chemical equation is

$$6\overset{-1}{I^-} + ClO_3^- + 6H_2SO_4 \longrightarrow Cl^- + 6HSO_4^- + 3\overset{0}{I_2} + 3H_2O$$

The stoichiometric coefficient of HSO_4^- is 6.

Iodide's oxidation state has changed from −1 to 0 oxidation state, i.e. iodide is oxidised.

H_2O is one of the products in the reaction.

Sulphur oxidation state remains the same after reactions, i.e. + 6, hence sulphur is not reduced.

50. *(2)* $X + 2e^- \longrightarrow X^{2-}$;

energy released = − 30.86 eV/atom

Total energy released = number of moles of molecule

× energy released by one mole of molecule

$$= y \times (-\ 30.86)\ N_A \text{eV}$$

Number of moles of $H_2 = \frac{4}{2} = 2$

According to de-Broglie, $2\pi r = n\lambda$

Now, it is given that, $2\pi r = 4\lambda$

$\therefore \quad n = 4$ [Energy level]

$$2H_2 \longrightarrow 2H^+ + 2H^*$$

Total energy required = total energy required to dissociate two moles of H_2 + total energy required in ionisation of two H_2 to two H^+ + total energy required in ionisation of two H to 4th excited energy level.

$$= 2 \times 4.52 \times N_A + 2 \times 13.6 N_A + 2 \times 13.6 \times \left(1 - \frac{1}{16}\right) \times N_A$$

$$= N_A \ (9.04 + 27.2 + 27.2 \times 0.94)$$

$$= N_A \ (61.80) \text{ eV}$$

We know that, during formation of H^+ and H^* in above reaction,

Total energy required = Total energy released

$$\therefore \quad (61.80 \times N_A) \text{ eV} = (-30.86y) N_A \text{ eV}$$

$$y = \frac{61.80}{30.86} \approx 2$$

Hence, number of moles required = 2

Study Tactics

This problem is based upon the conceptual mixing of de-Broglie equation and energy consideration during formation of molecule. So, students are advised to calculate the energy required and energy released during process. Before calculating energy required write the reaction of formation of H^+ and H^* by using concept of de-Broglie's equation.

Mathematics

51. *(a)* We have, shown two possible path in the following figure

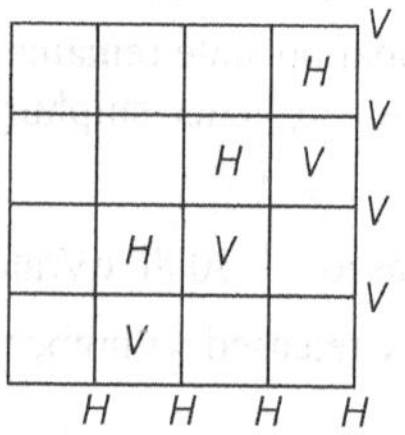

If horizontal movement is written as H and vertical movement is written as V.

Then, these paths can be written as

$HHHHVVVV$ and $HV\ HV\ HV\ HV$

Hence, the possible path is basically arrangement of 4 H and 4 V.

$\therefore$ Total number of ways $= \dfrac{8!}{4!4!} = \dfrac{8\times 7\times 6\times 5\times 4!}{4\times 3\times 2\times 1\times 4!} = 70$

52. *(b)* $\log_b a, \log_c b$ and $\log_a c$ are in GP.

$\Rightarrow \quad (\log_c b)^2 = \log_b a \times \log_a c$

$\Rightarrow \quad (\log_c b)^2 = \log_b c$

$\Rightarrow \quad (\log_c b)^2 = \dfrac{1}{\log_c b}$

$\Rightarrow \quad (\log_c b)^3 = 1$

$\Rightarrow \quad \log_c b = 1$

$\Rightarrow \quad b = c \, (a, b, c > 0)$

Then, $\quad 2a^4 = b^4 + c^4$

$\Rightarrow \quad 2a^4 = 2b^4$

$\Rightarrow \quad a = b = c$

By $\quad abc = 27 \Rightarrow a^3 = 27 \Rightarrow a = 3$

$\therefore a = b = c = 3$

53. *(d)* Given, $\dfrac{dy}{dx} = \dfrac{ax+5}{4y+b}$

$\Rightarrow (4y+b)\,dy = (ax+5)\,dx$

$\Rightarrow \quad 2y^2 + by = \dfrac{ax^2}{2} + 5x + C$

$\Rightarrow \quad 2y^2 - \dfrac{ax^2}{2} + by - 5x - C = 0$

represents a circle, then $2 = -\dfrac{a}{2} \Rightarrow a = -4$

$\therefore 2(x^2+y^2) + by - 5x - C = 0$

$\Rightarrow x^2 + y^2 - \dfrac{5}{2}x + \dfrac{b}{2}y - \dfrac{C}{2} = 0$ is a circle of radius,

$$r = \sqrt{\left(\frac{-5}{4}\right)^2 + \left(\frac{b}{4}\right)^2 + \frac{C}{2}}$$

Hence, $r = \dfrac{1}{4}\sqrt{b^2 + 8C + 25}$

54. *(d)* $|z|^2\omega - z|\omega|^2 - z + \omega = 0 \quad \ldots(i)$

$\Rightarrow \omega(|z|^2+1) - z(|\omega|^2+1) = 0$

$\Rightarrow \quad \dfrac{\omega}{z} = \dfrac{|\omega|^2+1}{|z|^2+1}$

$\Rightarrow \quad \dfrac{\omega}{z}$ is a real number.

$\therefore \quad \dfrac{\omega}{z} = \dfrac{\overline{\omega}}{\overline{z}} \Rightarrow \overline{\omega}z = \overline{z}\omega \quad \ldots(ii)$

From Eq. (i), we get

$z\overline{z}\omega - z\omega\overline{\omega} - z + \omega = 0 \quad [\because z\overline{z} = |z|^2]$

$\Rightarrow \quad z(z\overline{\omega} - 1) - \omega(\overline{\omega}z - 1) = 0 \quad$ [using Eq. (ii)]

$\Rightarrow \quad (z-\omega)(z\overline{\omega} - 1) = 0$

$\Rightarrow \quad z = \omega$ or $z\overline{\omega} = 1$

55. *(b)* Given, $\displaystyle\int \frac{e^{x-1}\,2x\,dx}{x^2 - 5x + 4} = AF(x-1) + BF(x-4) + C$

$$\frac{2x}{x^2-5x+4} = \frac{2x}{(x-4)(x-1)} = \frac{A_1}{x-4} + \frac{A_2}{x-1}$$

$\Rightarrow \quad 2x = A_1(x-1) + A_2(x-4)$

On solving, we get

$$A_1 = \frac{8}{3}, A_2 = \frac{-2}{3}$$

$$\therefore \int \frac{e^{x-1}\,2x\,dx}{x^2-5x+4} = \int e^{x-1}\left(\frac{-\frac{2}{3}}{x-1} + \frac{\frac{8}{3}}{x-4}\right)dx$$

$$= \frac{-2}{3}\int \frac{e^{x-1}}{x-1}\,dx + \frac{8}{3}\int \frac{e^{x-1}}{x-4}\,dx$$

$$= \frac{-2}{3}F(x-1) + \frac{8}{3}e^3F(x-4) + C$$

$$A = \frac{-2}{3} \text{ and } B = \frac{8}{3}e^3$$

$$(A, B) \equiv \left(\frac{-2}{3}, \frac{8}{3}e^3\right)$$

Study Tactics

First, use partial fraction method of integration, then use $F(x) = \int \dfrac{e^x}{x}\,dx$ to get expression in the form of $C \cdot F(x)$ where, C is constant.

56. *(c)* $\left(\frac{1}{a}\right)^b = \lim_{x\to 0} \frac{8}{x^8}\left(1-\cos\frac{x^2}{2}-\cos\frac{x^2}{4}+\cos\frac{x^2}{2}\cos\frac{x^2}{4}\right)$

$$= \lim_{x\to 0} \frac{8}{x^8}\left[\left(1-\cos\frac{x^2}{2}\right)\left(1-\cos\frac{x^2}{4}\right)\right]$$

$$= \lim_{x\to 0} \frac{8}{x^8}\left[\left(2\sin^2\frac{x^2}{4}\right)\left(2\sin^2\frac{x^2}{8}\right)\right]$$

$$= \lim_{x\to 0} 32\left(\frac{\sin\frac{x^2}{4}}{x^2}\right)^2\left(\frac{\sin\frac{x^2}{8}}{x^2}\right)^2$$

$$= \lim_{x\to 0} 32\left(\frac{\sin\frac{x^2}{4}}{\frac{4x^2}{4}}\right)^2\left(\frac{\sin\frac{x^2}{8}}{8\times\frac{x^2}{8}}\right)^2$$

$$= \frac{32}{4^2\times 8^2} = \frac{1}{4\times 8} = \frac{1}{2^5}$$

$\therefore \left(\frac{1}{a}\right)^b = \left(\frac{1}{2}\right)^5$

$a = 2$ and $b = 5$

$\therefore \quad a\cdot b = 10 = 2b$

57. *(b)* $\because$ Given, $pqr = 5$, $AA^T = 4I$

and $A = \begin{bmatrix} 3p & q & r \\ q & 3r & p \\ r & p & 3q \end{bmatrix} = A^T$

$\therefore \quad AA^T = 4I \Rightarrow AA = 4I \Rightarrow A^2 = 4I$

$\Rightarrow \quad |A|^2 = |4I| = 4^3\,|I| \Rightarrow |A| = \pm 8$

But $|A|$ is positive real number.

$\Rightarrow \quad |A| = 8$

Now, $|A| = 8 \Rightarrow \begin{vmatrix} 3p & q & r \\ q & 3r & p \\ r & p & 3q \end{vmatrix} = 8$

$\Rightarrow \quad 29pqr - 3(p^3+q^3+r^3) = 8$

$\Rightarrow \quad 3(p^3+q^3+r^3) = 29\times 5 - 8 = 137$

58. *(a)* $\lim_{x\to 0} f(x) = \lim_{x\to 0} \frac{(3^x-1)^2}{\sin\left(\frac{x}{c}\right)\log\left(1+\frac{x}{3}\right)}$

$$= \lim_{x\to 0}\left(\frac{3^x-1}{x}\right)^2 \times \frac{x^2}{\frac{x\sin\frac{x}{c}}{c\times\frac{x}{c}}} \times \frac{1}{\frac{\log\left(1+\frac{x}{3}\right)}{\frac{x}{3}}\times\frac{x}{3}}$$

$$= \lim_{x\to 0}\left(\frac{3^x-1}{x}\right)^2 \times \lim_{x\to 0}\frac{c}{\left(\frac{\sin\frac{x}{c}}{\frac{x}{c}}\right)} \times \lim_{x\to 0}\frac{3}{\frac{\log\left(1+\frac{x}{3}\right)}{\left(\frac{x}{3}\right)}}$$

$$= (\log_e 3)^2 \times c \times 3 = 3c(\log_e 3)^2$$

Also, $f(0) = 10(\log_e 3)^2$

$\because f(x)$ is continuous at $x = 0$.

$\therefore\ 3c(\log_e 3)^2 = 10(\log_e 3)^2 \Rightarrow c = \frac{10}{3}$

59. *(a)* Given equation is $(2x-y)^2 = 8x+6y-5$

Since, the lines $2x - y = 0$ and $8x + 6y - 5 = 0$ are not at right angles, so adjusting the equations further, we rewrite it as

$4x^2 + y^2 + 1 - 4xy - 4x + 2y = 4x + 8y - 4$

$\Rightarrow \quad (2x-y-1)^2 = 4\,(x+2y-1)$

$\Rightarrow \quad \left(\frac{2x-y-1}{\sqrt{2^2+1^2}}\right)^2 = \frac{4}{\sqrt{5}}\left(\frac{x+2y-1}{\sqrt{1^2+2^2}}\right),$

which is of the form $Y^2 = 4aX$, since the straight lines $2x - y - 1 = 0$ and $x + 2y - 1 = 0$ are perpendicular to each other.

Focus is obtained from $X = a$ and $Y = 0$

$\Rightarrow \quad \frac{x+2y-1}{\sqrt{5}} = \frac{1}{\sqrt{5}}$ and $\frac{2x-y-1}{\sqrt{5}} = 0$

$\Rightarrow x + 2y - 2 = 0$ and $2x - y - 1 = 0$

Solving them, the focus is $\left(\frac{4}{5}, \frac{3}{5}\right)$.

60. *(b)* $f(x) = \begin{cases} x-[x], & \text{if } [x] \text{ is odd} \\ 1+[x]-x, & \text{if } [x] \text{ is even} \end{cases}$

$$= \begin{cases} \{x\}, & 2n-1 \le x < 2n \\ 1-\{x\}, & 2n \le x < 2n+1 \end{cases}$$

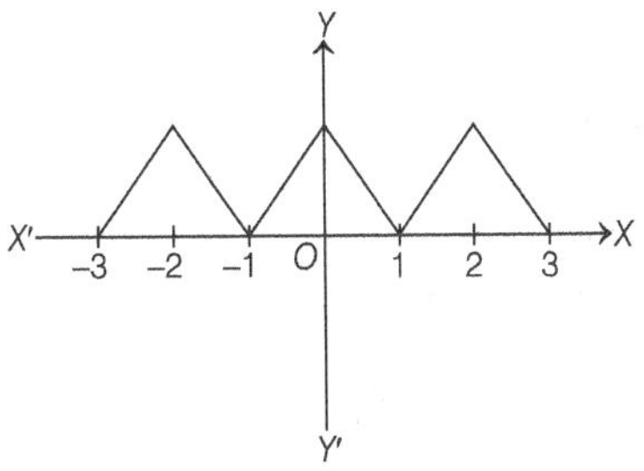

$f(x)$ is periodic.

$$I = \int_{-10}^{10} f(x)\cos \pi x\, dx$$

$$= 2\int_0^{10} f(x)\cos \pi x\, dx$$

$$= 2\times 5\int_0^{2} f(x)\cos \pi x\, dx$$

$$= 10\left[\int_0^{1} (1-x)\cos \pi x\, dx + \int_1^{2} (x-1)\cos \pi x\, dx\right]$$

$$= 10[I_1 + I_2]$$

$$I_1 = \int_0^1 (1-x)\cos \pi x\, dx$$

$$= -\int_0^1 x\cos \pi x\, dx$$

$$I_2 = \int_1^2 (x-1)\cos \pi x\, dx$$

On putting $x-1=t$

$$= -\int_0^1 t\cos \pi t\, dt$$

$$= -\int_0^1 x\cos \pi x\, dx$$

$$\therefore \quad I = 10\left(-2\int_0^1 x\cos \pi x\, dx\right)$$

$$= -20\left[\frac{x\sin \pi x}{\pi} + \frac{\cos \pi x}{\pi^2}\right]_0^1$$

$$= -20\left(-\frac{1}{\pi^2} - \frac{1}{\pi^2}\right)$$

$$= \frac{40}{\pi^2}$$

$$\because \int_{-10}^{10} f(x)\cos \pi x\, dx = \frac{20}{\pi^2} k = \frac{40}{\pi^2}$$

$$= \frac{20}{\pi^2}\times 2$$

$$\Rightarrow \quad k = 2$$

Study Tactics

First, using property of greatest integer function, draw the graph. Then, by properties of definite integration get the required result.

61. *(b)* $f(x) = \sqrt{|x|-x} + \dfrac{1}{\sqrt{|x|-x}}$

If $x \ge 0$, then $|x| - x = 0$

If $x < 0$, then $|x| - x = -2x$

$\therefore f(x) = \sqrt{-2x} + \dfrac{1}{\sqrt{-2x}}$,

where $-2x > 0$

$\Rightarrow \quad A = (-\infty, 0)$

We know that AM $\ge$ GM

$$\Rightarrow \frac{\sqrt{-2x} + \dfrac{1}{\sqrt{-2x}}}{2} \ge \left(\sqrt{-2x}\times\frac{1}{\sqrt{-2x}}\right)^{1/2}$$

$$\Rightarrow \quad f(x) \ge 2$$

$$\Rightarrow \quad B = [2, \infty)$$

Study Tactics

Use property of modulus function, then apply AM $\ge$ GM.

62. *(d)* Given that the length of transverse axis is $2\sin\theta = 2a$

$\Rightarrow a = \sin\theta$

Also, for the ellipse

$$4x^2 + 6y^2 = 12$$

$$\Rightarrow \quad \frac{x^2}{3} + \frac{y^2}{2} = 1$$

Here, $a^2 = 3$ and $b^2 = 2$

$$\because \quad e = \sqrt{1-\frac{b^2}{a^2}} = \sqrt{1-\frac{2}{3}} = \frac{1}{\sqrt{3}}$$

Hence, the focus of ellipse is $(\pm 1, 0)$.

As the hyperbola is confocal with the ellipse, the focus of hyperbola is (1, 0).

Now, $a'e' = 1$

$\Rightarrow \sin\theta\cdot e' = 1$

$$\Rightarrow \quad (e')^2 = \operatorname{cosec}^2\theta$$

$$\therefore \quad b'^2 = a'^2(e'^2 - 1) = \sin^2\theta\,(\operatorname{cosec}^2\theta - 1)$$

$$\Rightarrow \quad b'^2 = 1 - \sin^2\theta = \cos^2\theta$$

Therefore, the equation of hyperbola is

$$\frac{x^2}{\sin^2\theta} - \frac{y^2}{\cos^2\theta} = 1$$

63. *(c)* Total number of balls $= x + y$

$$\text{Required probability} = \frac{{}^xC_1 \times {}^yC_1}{{}^{x+y}C_2} = \frac{1}{2}$$

$$\Rightarrow \frac{2xy}{(x+y)(x+y-1)} = \frac{1}{2}$$

$$4xy = (x+y)(x+y-1)$$

$$4xy = x^2 + xy - x + xy + y^2 - y$$

$$y^2 - (2x+1)y + x^2 - x = 0$$

For real solution,

$$\Delta = (2x+1)^2 - 4(1)(x^2 - x)$$

$$= 4x^2 + 1 + 4x - 4x^2 + 4x = 8x + 1$$

must be a perfect square

$\Rightarrow \quad x = 1, 3, 6, 10$

$\therefore$ Statement I and Statement II both are false.

64. *(d)* Let the source of light be situated at $A(a, 0, 0)$, where $a \neq 0$. Let OA be the incident ray, OB be the reflected ray and ON be the normal to the mirror at O.

$\therefore \quad \angle AON = \angle NOB = \dfrac{\theta}{2}$

Direction ratios of OA are proportional to a, 0, 0 and so its direction cosines are 1, 0, 0.

Direction cosines of ON are $\dfrac{1}{\sqrt{3}}, -\dfrac{1}{\sqrt{3}}, \dfrac{1}{\sqrt{3}}$

$\therefore \cos\dfrac{\theta}{2} = \dfrac{1}{\sqrt{3}}$

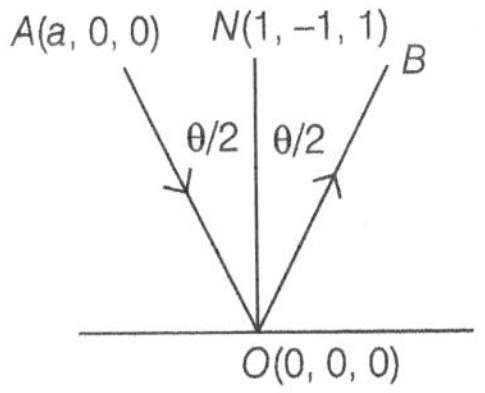

Let l, m, n be the direction cosines of the reflected ray OB.

Thus, $\dfrac{l+1}{2\cos\dfrac{\theta}{2}} = \dfrac{1}{\sqrt{3}}, \dfrac{m+0}{2\cos\dfrac{\theta}{2}} = -\dfrac{1}{\sqrt{3}}$

and $\dfrac{n+0}{2\cos\dfrac{\theta}{2}} = \dfrac{1}{\sqrt{3}}$

$\Rightarrow \quad l = \dfrac{2}{3} - 1, m = -\dfrac{2}{3}, n = \dfrac{2}{3}$

$\Rightarrow \quad l = -\dfrac{1}{3}, m = -\dfrac{2}{3}, n = \dfrac{2}{3}$

Hence, direction cosines of the reflected ray are $-\dfrac{1}{3}, -\dfrac{2}{3}, \dfrac{2}{3}$.

65. *(a)* We have, $\mathbf{b} - 3\mathbf{c} = \lambda\mathbf{a}$

On taking scalar product with $\mathbf{c}$, we get

$(\mathbf{b} - 3\mathbf{c}) \cdot \mathbf{c} = \lambda(\mathbf{a} \cdot \mathbf{c})$

$\Rightarrow \quad \mathbf{b} \cdot \mathbf{c} - 3(\mathbf{c} \cdot \mathbf{c}) = \lambda$

$\Rightarrow \quad \mathbf{b} \cdot \mathbf{c} - 3 = \lambda$

$\Rightarrow \quad \mathbf{b} \cdot \mathbf{c} = 3 + \lambda \quad \ldots(i)$

[$\because |\mathbf{a}| = |\mathbf{c}| = 1$ and $\mathbf{a}, \mathbf{c}$ are collinear vectors]

Again, $\quad \mathbf{b} - 3\mathbf{c} = \lambda\mathbf{a}$

$\Rightarrow \quad |\mathbf{b} - 3\mathbf{c}| = |\lambda\mathbf{a}| \Rightarrow |\mathbf{b} - 3\mathbf{c}|^2 = \lambda^2|\mathbf{a}|^2$

$\Rightarrow |\mathbf{b}|^2 + 9|\mathbf{c}|^2 - 6(\mathbf{b} \cdot \mathbf{c}) = \lambda^2|\mathbf{a}|^2$

$\Rightarrow \quad 36 + 9 - 6(3 + \lambda) = \lambda^2 \quad$ [from Eq. (i)]

$\Rightarrow \quad 27 - 6\lambda = \lambda^2$

$\Rightarrow \quad \lambda^2 + 6\lambda - 27 = 0$

$\Rightarrow \quad \lambda = -9, 3$

Time Saver Tip

We have, $\bar{b} - 3\bar{c} = \lambda\bar{a}$

$\because$ Angle between any two collinear vectors is 0° or 180°.

So, $\quad \bar{a} \cdot \bar{c} = \pm 1$

$\therefore \quad \bar{b} = \lambda\bar{a} + 3\bar{c} \Rightarrow |\bar{b}|^2 = |\lambda\bar{a} + 3\bar{c}|^2$

$\Rightarrow 36 = 9 + \lambda^2 + 6\lambda \Rightarrow \lambda^2 + 6\lambda - 27 = 0 \Rightarrow \lambda = -9, 3$

66. *(c)* $\because \dfrac{X-2}{1} = \dfrac{Y-9}{2} = \dfrac{Z-13}{3} = \mu$

$\Rightarrow \quad X = \mu + 2, Y = 2\mu + 9, Z = 3\mu + 13$

$\because \quad \dfrac{X-\lambda}{-1} = \dfrac{Y-7}{2} = \dfrac{Z+2}{-3} = \delta$

$\Rightarrow \quad X = -\delta + \lambda,$

$Y = 2\delta + 7$

and $\quad Z = -3\delta - 2$

Now, $\mu + 2 = -\delta + \lambda$

$\mu + \delta = -2 + \lambda \quad \ldots(i)$

$\Rightarrow \quad 2\mu + 9 = 2\delta + 7$

$\Rightarrow \quad 2\mu - 2\delta = -2$

$\Rightarrow \quad \mu - \delta = -1 \quad \ldots(ii)$

$\Rightarrow \quad 3\mu + 13 = -3\delta - 2$

$\Rightarrow \quad 3\mu + 3\delta = -15$

$\Rightarrow \quad \mu + \delta = -5 \quad \ldots(iii)$

On solving Eqs. (i), (ii) and (iii), we get

$\lambda = -3,$

$\mu = -3,$

$\delta = -2$

67. *(c)* Equation of line passing through $(5, 1, a)$ and $(3, b, 1)$ is

$$\frac{x-5}{5-3} = \frac{y-1}{1-b} = \frac{z-a}{a-1} = \lambda$$

i.e. $x = 0$

$x = 2\lambda + 5 = 0$

$\Rightarrow \quad \lambda = -\dfrac{5}{2}$

Since, $\quad y = \lambda(1-b) + 1 = \dfrac{17}{2}$

$= -\dfrac{5}{2}(1-b) + 1 = \dfrac{17}{2}$

$\Rightarrow \quad b = 4$

Also, $z = \lambda(a-1) + a = \dfrac{-13}{2}$

$= \dfrac{-5}{2}(a-1) + a = \dfrac{-13}{2}$

$\Rightarrow \quad a = 6$

68. *(c)* $(1-x-x^2+x^3)^6$

$\Rightarrow [(1-x)-x^2(1-x)]^6 \Rightarrow [(1-x)^6(1-x^2)^6]$

$\Rightarrow (1-{}^6C_1x+{}^6C_2x^2-{}^6C_3x^3+{}^6C_4x^4$
$-{}^6C_5x^5+{}^6C_6x^6)$
$\cdot(1-{}^6C_1x^2+{}^6C_2x^4-{}^6C_3x^6+\ldots)$

Coefficient of x^7 is

$${}^6C_1{}^6C_3-{}^6C_3{}^6C_2+{}^6C_5{}^6C_1$$

$$6\cdot\frac{6\cdot5\cdot4}{3\cdot2}-\frac{6\cdot5\cdot4}{3\cdot2\cdot1}\cdot\frac{6\cdot5}{2}+6\cdot6=120-300+36$$

$$=-144$$

69. *(b)* $\therefore \Delta=\sin\theta\cos\phi(0+\sin^2\theta\cos\phi)$
$-\sin\theta\sin\phi(0-\sin^2\theta\sin\phi)$
$+\cos\theta(\cos^2\phi\sin\theta\cos\theta+\cos\theta\sin\theta\sin^2\phi)$

$=\sin^3\theta(\cos^2\phi+\sin^2\phi)+\sin\theta\cos^2\theta(\sin^2\phi+\cos^2\phi)$

$=\sin^3\theta+\sin\theta\cos^2\theta$

$=\sin\theta(\sin^2\theta+\cos^2\theta)$

$=\sin\theta$ $\quad[\because \sin^2\theta+\cos^2\theta=1]$

Which is independent of ϕ.

70. *(b)* The given curves are

$y=\ln(x+e)$

and $x=\ln(1/y)$

$\Rightarrow \dfrac{1}{y}=e^x$

$\Rightarrow y=e^{-x}$

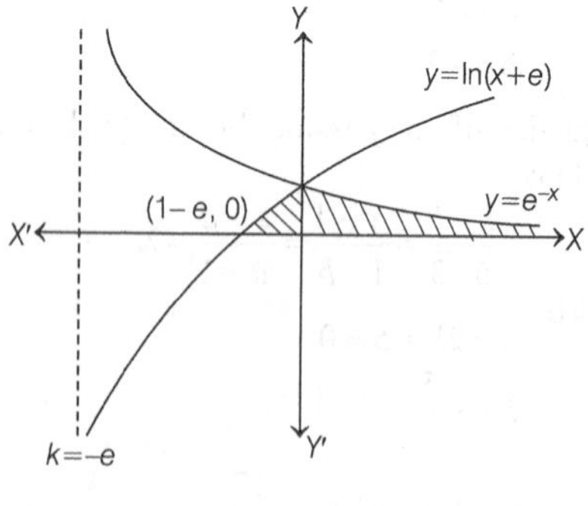

$\therefore$ Required area $=\int_{1-e}^{0}\ln_e(x+e)dx+\int_0^\infty e^{-x}dx$

$=\int_1^e \ln t\,dt+\int_0^\infty e^{-x}dx \quad [\because x+e=t]$

$=[t\ln t-t]_1^e-(e^{-x})_0^\infty$

$=1+1$

$=2$ sq units

71. *(240)* To form any number, we have 10 digits i.e. 0, 1, 2, 3, 4, 5, 6, 7, 8, 9 out of which middle digit can be x, (where $2\le x\le 9$), then the hundredth can be chosen in $(x-1)$ ways (as this place can not be zero) and units place can be chosen in x ways.

Thus, number of required numbers with middle digit is $(x-1)x$.

As $2\le x\le 9$, the number of required numbers is

$$\sum_{x=2}^{9}(x-1)x=\sum_{x=2}^{9}(x^2-x)$$

$$=240$$

72. *(–1)* $\tan^{-1}(x+1)+\tan^{-1}(x-1)=\tan^{-1}3-\tan^{-1}x$

$\Rightarrow \tan^{-1}\left(\dfrac{x+1+x-1}{1-(x+1)(x-1)}\right)=\tan^{-1}\left(\dfrac{3-x}{1+3x}\right),$

when $-\sqrt{2}<x<\sqrt{2}$ and $3x<1$

$\Rightarrow \tan^{-1}\left(\dfrac{2x}{2-x^2}\right)=\tan^{-1}\left(\dfrac{3-x}{1+3x}\right),$

when $-\sqrt{2}<x<\sqrt{2}$ and $x<\dfrac{1}{3}$

$\Rightarrow \dfrac{2x}{2-x^2}=\dfrac{3-x}{1+3x}$, when $-\sqrt{2}<x<\dfrac{1}{3}$

$\Rightarrow 2x(1+3x)=(2-x^2)(3-x)$, when $-\sqrt{2}<x<\dfrac{1}{3}$

$\Rightarrow 2x+6x^2=6-2x-3x^2+x^3$

$\Rightarrow x^3-9x^2+4x+6=0$

$\Rightarrow (x+1)(x^2-10x+6)=0$, when $-\sqrt{2}<x<\dfrac{1}{3}$

$\Rightarrow x=-1$ and neglecting $x^2-10x+6=0$ as its roots does not $\in\left(-\sqrt{2},\dfrac{1}{3}\right)$

Hence, $|x|=1$

73. *(2)* Let $f(x)=x^2-x\sin x-\cos x$

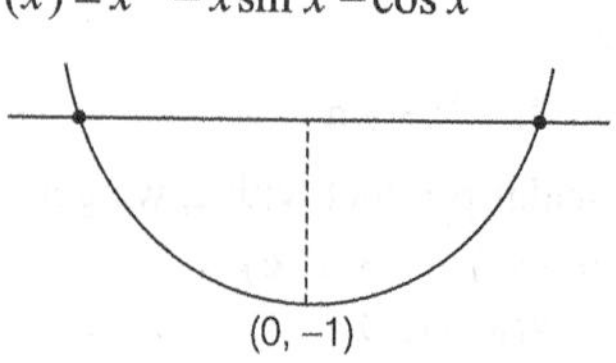

$\therefore f'(x)=2x-x\cos x-\sin x+\sin x$

$=2x-x\cos x$

$f'(x)=0$

$\Rightarrow x(2-\cos x)=0$

or $x=0$ $\quad[\because 2-\cos x>0$ for all real $x]$

Also, $x=0$ is point of minima $f(0)=-1<0$

and $\lim\limits_{x\to\infty} f(x)\to\infty$,

$$\lim_{x \to -\infty} f(x) \to \infty$$

Hence, it meets X-axis at two points and two solutions.

74. *(8)* Median $= \dfrac{\text{25th term} + \text{26th term}}{2}$

$$= \frac{25k + 26k}{2} = 25.5k$$

Mean deviation about median $= 100$

$$\Rightarrow \quad \frac{\Sigma |x_i - 25.5k|}{50} = 100$$

$$\Rightarrow 24.5k + 23.5k + \ldots + 1.5k + 0.5k + 0.5k + 1.5k + \ldots + 23.5k + 24.5k = 5000$$

$$\Rightarrow \quad k + 3k + 5k + \ldots + 49k = 5000$$

$$\Rightarrow \quad k(1 + 3 + 5 + \ldots + 49) = 5000$$

$$\Rightarrow \quad k \cdot \frac{25}{2}(1 + 49) = 5000$$

$$\Rightarrow \quad k = 8$$

75. *(13)* Let ellipse $\dfrac{x^2}{a^2} + \dfrac{y^2}{b^2} = 1$

and circle $x^2 + y^2 = a^2e^2$

Radius of circle $= ae$

Point of intersection of circle and ellipse is

$$\left[\frac{a}{e}\sqrt{2e^2 - 1}, \frac{a}{e}(1 - e^2)\right].$$

Now, area of ΔPF_1F_2

$$= \frac{1}{2}\begin{vmatrix} \frac{a}{e}\sqrt{2e^2 - 1} & \frac{a}{e}(1 - e^2) & 1 \\ ae & 0 & 1 \\ -ae & 0 & 1 \end{vmatrix}$$

$$= \frac{1}{2}\left|\frac{a}{e}(1 - e^2)(2ae)\right| = 30$$

$$\Rightarrow \quad a^2(1 - e^2) = 30 \qquad \text{[given]}$$

$$\Rightarrow \quad a^2e^2 = a^2 - 30 = \left(\frac{17}{2}\right)^2 - 30 = \frac{169}{4}$$

$$\Rightarrow \quad 2ae = 13$$

SECTION I

Foundation Builders

PRACTICE SET 3

Total Marks : 300 | Total Time : 3 hrs | No. of Questions : 75

Instructions

1. There are **three** parts in the questions paper consisting of **Physics**, **Chemistry** and **Mathematics** having 25 questions in each part of equal weightage. Each part (subject) has two sections.
2. **Section A** This section contains 20 multiple choice questions which have only one correct answer. Each question carries **4 marks** for correct answer and **–1 mark** for wrong answer.
3. **Section B** This section contains 5 questions. The answer to each of the questions is a numerical value. Each question carries **4 marks** for correct answer and **–1 mark** for wrong answer. For Section B, the answer should be rounded off to the nearest integer.

PHYSICS

Section A : Objective Type Questions

1. If kinetic energy of a particle remains constant, then

(a) the particle moves in a straight line
(b) the force acted on the particle is perpendicular to its velocity
(c) the force acted on the particle is perpendicular to its acceleration
(d) the object is stationary

2. In a given L-C-R circuit, resistance of the circuit is 40Ω as given, then the voltage across terminals of resistance and current through ammeter will be

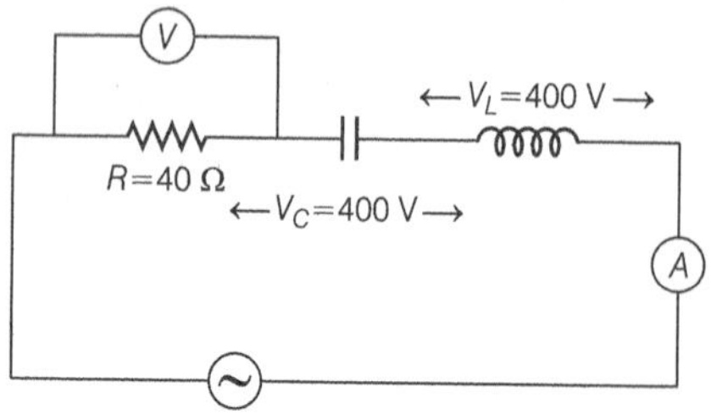

(a) 200 V, 5 A (b) 400 V, 5 A
(c) 400 V, $5\sqrt{2}$ A (d) 200 V, $5\sqrt{2}$ A

3. The approximate depth of an ocean is 5400 m. The compressibility of water is $45.4\times10^{-11}\ \text{Pa}^{-1}$ and density of water is $10^3\ \text{kg/m}^3$. What fractional compression of water will be obtained at the bottom of the ocean?

(a) 2.4×10^{-2} (b) 1.2×10^{-2}
(c) 3.6×10^{-2} (d) 2.4×10^{-5}

4. A horizontal force of 12 N acted on the block placed on a rough floor ($\mu = 0.3$) as shown in the figure

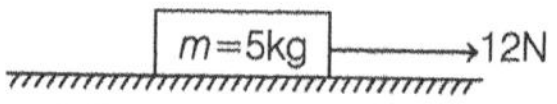

The frictional force between the block and the surface is

(a) 15 N (b) 18 N
(c) 12 N (d) 20 N

5. **Assertion** The rms speed of hydrogen (H_2) is greater than the rms speed of oxygen (O_2).
Reason The rms speed of a gas is inversely proportional to its molecular weight.
In the light of above statements, choose the most appropriate answer from the options given below.

(a) Both Assertion and Reason are correct and Reason is the correct explanation of Assertion.
(b) Both Assertion and Reason are correct, but Reason is not the correct explanation of Assertion.
(c) Assertion is correct, but Reason is incorrect.
(d) Assertion is incorrect, but Reason is correct.

6. The pitch of a screw gauge is 1 mm and there are 100 divisions on the circular scale. While measuring the diameter of a wire, the linear scale reads 1 mm and 47th division on the circular scale coincides with reference line. The length of the wire is 5.6 cm. Find the curved surface area (in cm^2) of the wire in appropriate number of significant figures.

(a) 2 (b) 3 (c) 2.6 (d) 3.6

7. Two semi-circular ring charges with linear charge densities $+\lambda$ and $-\lambda$ are placed to form a ring of radius R. The electric field at the centre of the charged ring will be

(a) $\frac{\lambda}{2\pi\varepsilon_0 R}$ (b) zero (c) $\frac{2\lambda}{\pi\varepsilon_0 R}$ (d) $\frac{\lambda}{\pi\varepsilon_0 R}$

8. The wave nature of light was established by

(i) Maxwell's equations
(ii) de-Broglie relation
(iii) Hertz experiment
(iv) Einstein's theory

Choose the appropriate answer from the given options.

(a) (i) and (ii) only (b) (ii) and (iv) only
(c) (i) and (iii) only (d) (iii) and (iv) only

9. The coulomb force **F** between outermost electron and nucleus of Li^{2+} atom of radius r is

(a) $3K\frac{e^2}{r^2}\hat{\mathbf{r}}$ (b) $-K\frac{e^2}{r^3}\hat{\mathbf{r}}$
(c) $-3K\frac{e^2}{r^3}\mathbf{r}$ (d) $3K\frac{e^2}{r^3}\mathbf{r}$

10. Study the following statement regarding logic gates.

I. NOT gate produces an inverted version of input at its output.
II. If inputs of a NOR gate is 0 and 1, then its output will be 1.
III. AND gate produces same result, when both its input having same value.

Which of the following statement(s) is/are correct ?

(a) Both I and III (b) Only I
(c) Both I and II (d) All I, II, III

11. **Statement I** The equivalent emf of parallel combination of the cells is just the sum of their individual emf.

Statement II In a series combination of two cells, if the polarity of one cell is reversed, then their equivalent emf will be changed, while equivalent resistance remains unchanged.

In the light of above statements, choose the most appropriate answer from the options given below.

(a) Both Statement I and Statement II are true and Statement II is the correct explanation of Statement I
(b) Both Statement I and Statement II are true but Statement II is not the correct explanation of Statement I
(c) Statement I is true but Statement II is false.
(d) Statement I is false but Statement II is true.

12. Unpolarised light of intensity 40 Wm^{-2} passes through three polarisers such that transmission axis of the first and second polariser makes an angle 30° with each other and the transmission axis of the last polariser is crossed with that of the first. The intensity of final emerging light will be

(a) 2.25 W/m^2 (b) 8.35 W/m^2
(c) 3.75 W/m^2 (d) 15 W/m^2

13. Which of the following statement is correct ?

(a) The acceleration due to gravity depends on the mass of the body.
(b) Escape velocity of a body on the surface of the earth is independent of radius of the earth.
(c) Inside a uniform spherical shell, the gravitational potential is same everywhere.
(d) Speed remains constant in a planetary motion.

14.

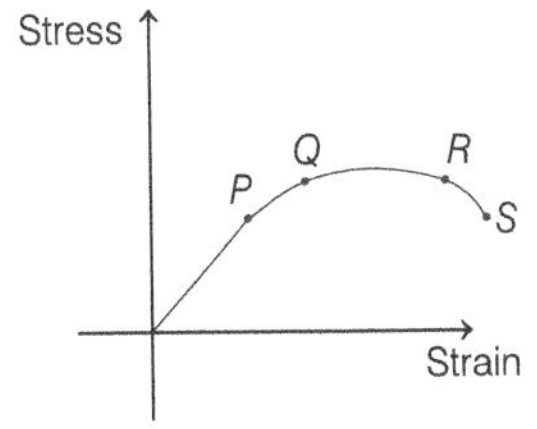

Now, match the terms in Column I with corresponding terms in Column II.

Column I	Column II
A. Yield point	1. P
B. Breaking point	2. Q
C. Ultimate strength point	3. R
D. Proportionally limit point	4. S

Codes

	A	B	C	D
(a)	1	2	3	4
(b)	2	1	3	4
(c)	2	4	3	1
(d)	1	4	3	2

15. The figure shows a current loop having two circular arcs joined by two radial lines. The magnetic field at O is

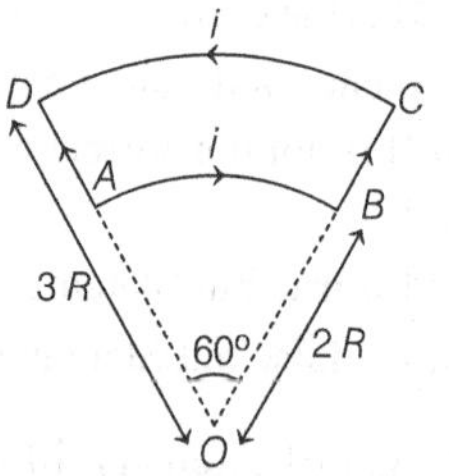

(a) $\frac{\mu_0 i}{36R}$ (b) $\frac{\mu_0 i}{24R}$ (c) $\frac{\mu_0 i}{72R}$ (d) $\frac{\mu_0 i}{48R}$

16. A rectangular loop has a sliding connector AB of length 1m and resistance 3Ω and it is moving with speed 2 m/s as shown. The set-up is placed in uniform magnetic field 5T.

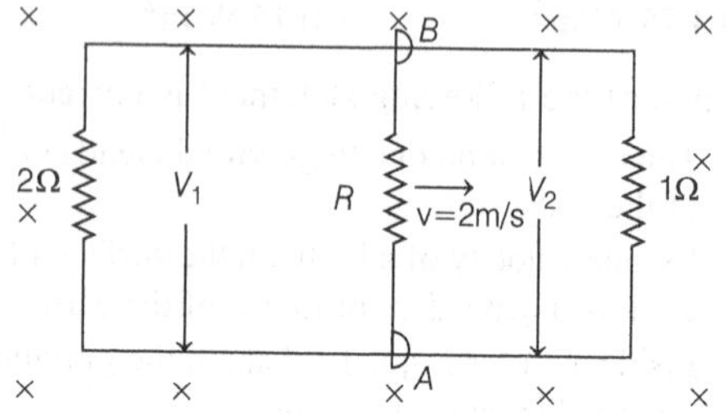

Then, the values of V_1 and V_2 are

(a) $V_1 = \frac{20}{11}$, $V_2 = \frac{90}{11}$ V

(b) $V_1 = \frac{20}{13}$, $V_2 = \frac{21}{13}$ V

(c) $V_1 = V_2 = \frac{20}{11}$ V

(d) $V_1 = V_2 = \frac{21}{11}$ V

17. For the density measurement of a sphere, the mass and radius are measured as (15 ± 0.30) kg and (0.20 ± 0.01) m. The error in the measurement of density is

(a) 0.31 kgm^{-3} (b) 0.17 kgm^{-3}

(c) 0.28 kgm^{-3} (d) 0.27 kgm^{-3}

18. A plane electromagnetic wave of frequency 60 MHz travels in free space along positive x-direction. At a particular time in space, $\mathbf{B} = 2.1 \times 10^{-8}\,\hat{\mathbf{k}}$T, then the corresponding electric field at that point is

(a) $6.3\hat{\mathbf{j}}$ V/m (b) $-6.5\hat{\mathbf{j}}$ V/m

(c) $-6.3\hat{\mathbf{j}}$ V/m (d) $6.5\hat{\mathbf{j}}$ V/m

19. As the electron in Bohr orbit of hydrogen atom passes from state $n = 2$ to $n = 1$, the kinetic energy K and potential energy U changes as

(a) K becomes one-fourth and U becomes four times

(b) K becomes four times, U becomes twice

(c) both K and U become four times

(d) both K and U become twice

20. The null point is found to be 60 cm away from the end A with resistances X and Y in position of a meter bridge as shown. When a resistance of 15Ω is connected in series with Y, then the null point is found to shift by 10 cm towards the end A of the wire. If a resistance of 30Ω wire connected in parallel with Y, then the position of null point is

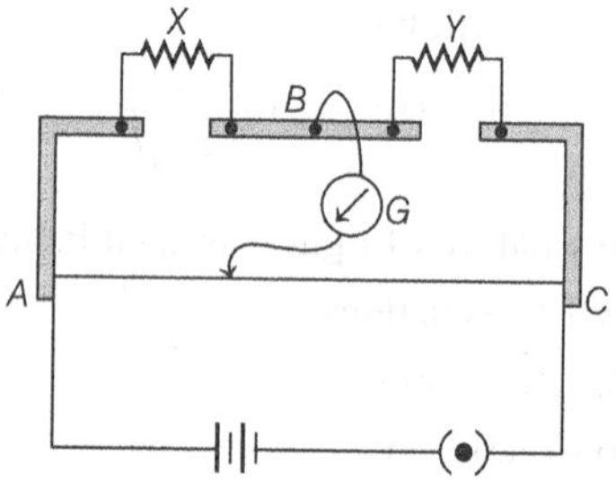

(a) 25 cm from end A (b) 75 cm from end A

(c) 15 cm from end C (d) 45 cm from end C

Section B : Numerical Value Type Questions

21. A body is executing SHM, when its displacement from the mean position is 4 cm and 5cm, the corresponding velocities of the body are 10 cm/s and 8 cm/s, respectively. Then, the time period of the body is $\frac{2\pi}{x}$ s. The value of x is ....

22. Let B_P and B_Q be the magnetic fields produced by the wires P and Q which are placed symmetrically in a rectangular loop $ABCD$ as shown in figure. Current in wire P is I directed inward and in Q is $2I$ directed outwards. If $\int_A^B \mathbf{B}_Q \cdot d\mathbf{l} = +2\mu_0$ T-m, $\int_D^A \mathbf{B}_P \cdot d\mathbf{l} = -2\mu_0$ T-m and $\int_A^B \mathbf{B}_P \cdot d\mathbf{l} = -\mu_0$T-m, then value of I will be ...... A.

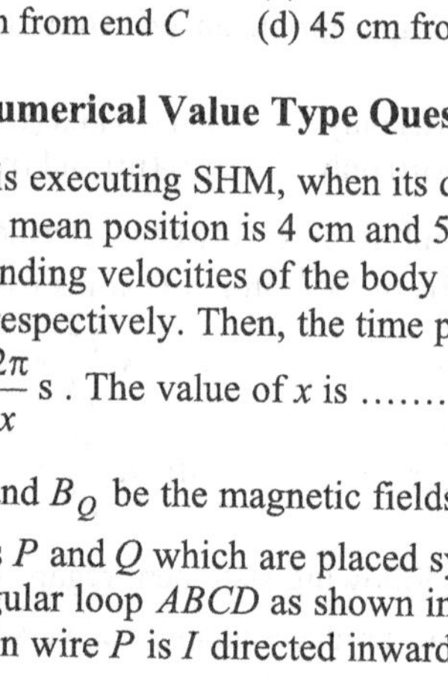

23. The refracting angle of a glass prism is 30°. A ray is incident onto one of the faces perpendicular to it. If the refractive index of glass is 1.5, then the angle between incident ray and emergent ray is (in degree) ....

24. A disc of mass 2 kg and radius 5 cm is rolling with angular speed of 8 rad/s on a horizontal plane as shown in figure, then the magnitude of angular momentum of the disc about the origin O is $x \times 10^{-2}$ kg-m^2/s. The value of x is ....

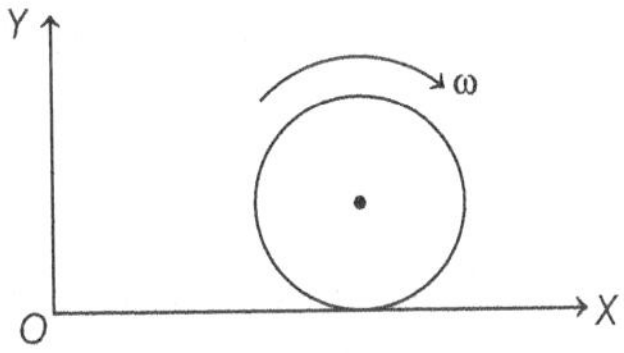

25. An ideal gas ($\gamma = 1.5$) is contained in equal amounts and in same state in two compartments of an adiabatic vessel separated by an adiabatic frictionless separator.

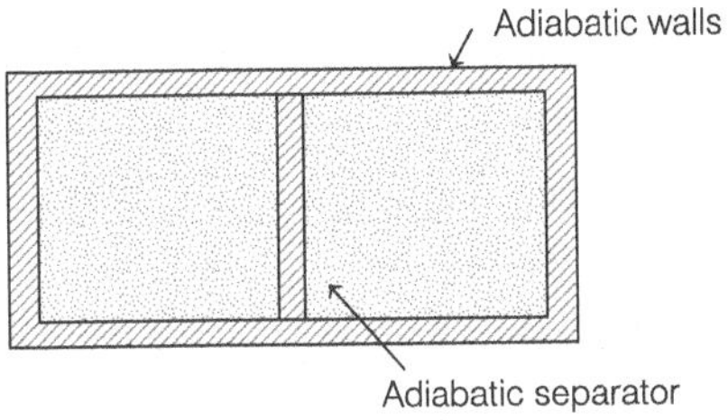

The separator is now sliding slowly towards right such that pressure in right compartment becomes 8 times that in left compartment. Ratio of internal energies of gas in left compartment to that in right compartment is $\frac{x}{2}$. The value of x is......... .

CHEMISTRY

Section A : Objective Type Questions

26. What will be the product formed when most acidic species among the following will react with 3-chloroprop-1-ene?

(a) 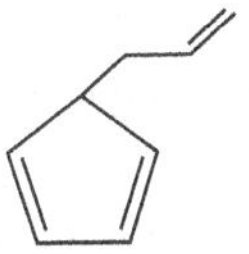 (b) 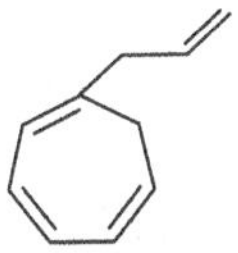

(c) 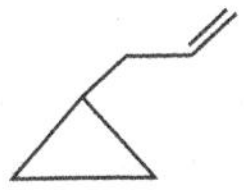 (d) 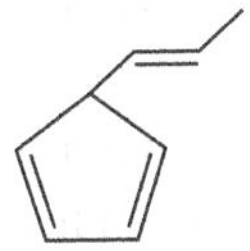

27. Nitrogen exist as N_2 but phosphorus exist as P_4 this is due to

(a) triple bond exist between phosphorus atom
(b) $p\pi$-$p\pi$ bond is weak in P_4
(c) $p\pi$-$p\pi$ bond is weak in N_2
(d) multiple bonds form easily

28. The total number of complexes having same type of geometry is

$[Co(NH_3)_4Cl_2]^+$, $[Pt(NH_3)_2(H_2O)Cl]^+$, $[Pt(NH_3)_3(NO)_3]Cl$, $[Pt(NH_3)_3Cl]Br$, $[Ni(CO)_4]$

(a) 2 (b) 3
(c) 4 (d) 5

29. Match List-I and List-II.

	List-I		List-II
A.	Alcoholic potassium hydroxide	I.	Strong reducing agent.
B.	$Pd/BaSO_4$	II.	Obtained by addition reaction
C.	BHC (Benzene hexachloride)	III.	Used for β-elimination reaction
D.	$LiAlH_4$	IV.	Lindlar's Catalyst

Choose the correct answer from the options given below.

(a) A-III, B-IV, C-II, D-I (b) A-III, B-I, C-IV, D-II
(c) A-II, B-III, C-I, D-IV (d) A-II, B-IV, C-III, D-I

30. Which carbonyl compound gives tertiary alcohol with Grignard reagent followed by acidic hydrolysis?

(a) Formaldehyde (b) Acetaldehyde
(c) Acetone (d) Methyl formate

31. Match List-I with List-II.

	List-I		List-II
A.	$S_2O_3^{2-} + FeCl_3$ solution	I.	Green colouration
B.	$SO_3^{2-} + K_2Cr_2O_7/H^+$	II.	Green colour due to $[Cr(H_2O)_6]^{3+}$ ion

List-I		List-II	
C.	Copper wire test for halogens	III.	Beilstein's test
D.	To test the reducing nature of sugar	IV.	Barfoed test

Choose the correct answer from the options given below.

	A	B	C	D		A	B	C	D
(a)	IV	III	II	I	(b)	I	II	III	IV
(c)	I	IV	II	III	(d)	IV	II	I	III

32. Consider the following reaction,

$$\begin{array}{c} CHO \\ | \\ (CHOH)_4 \\ | \\ CH_2OH \end{array} \xrightarrow{HIO_4} \text{Products}$$

In the above reaction the products are

(a) $5HCOOH + HCHO$

(b) $4HCOOH + HCHO + CO_2$

(c) $3HCOOH + HCHO + \begin{array}{c} COOH \\ | \\ CHO \end{array}$

(d) $4HCOOH + HCHO + CO_2 + H_2O$

33. Λ°_m for NaCl, HCl and Na*A* are 126.4, 425.9 and 100.5 S cm^2 mol^{-1}, respectively. If the conductivity of 0.001 M H*A* is 5×10^{-5} S cm^{-1}. Then, what will be the degree of dissociation of H*A*?

(a) 125×10^{-3} (b) 220×10^{-2}

(c) 225×10^{2} (d) 120×10^{-4}

34. Which of the following set of compounds represents isostructural characteristics?

(a) SF_4, CH_4, NH_3 (b) NF_3, BCl_3, NH_3

(c) BF_3, NF_3, $AlCl_3$ (d) BF_3, BCl_3, BBr_3

35. An organic diketone (*A*) having molecular formula $C_{14}H_{10}O_2$ when undergo reaction with hydroxyl ion (OH^-) converted into benzilic acid (*B*) after acidification. The molecular structure of (*A*) and (*B*) are

(a) 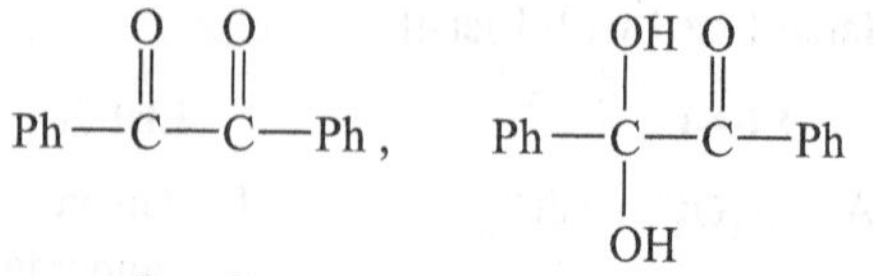

(b) 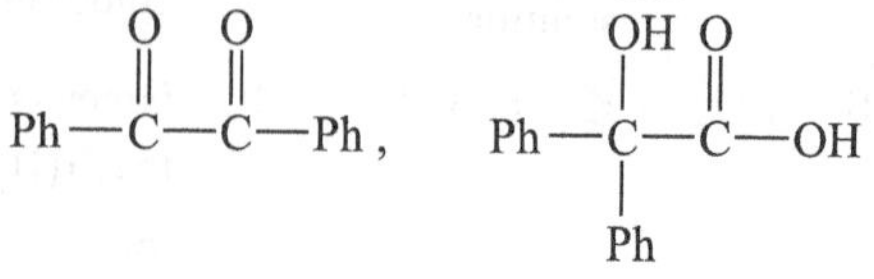

(c) 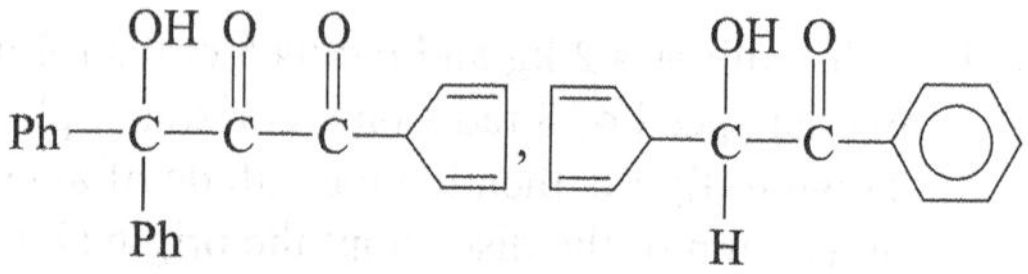

(d) 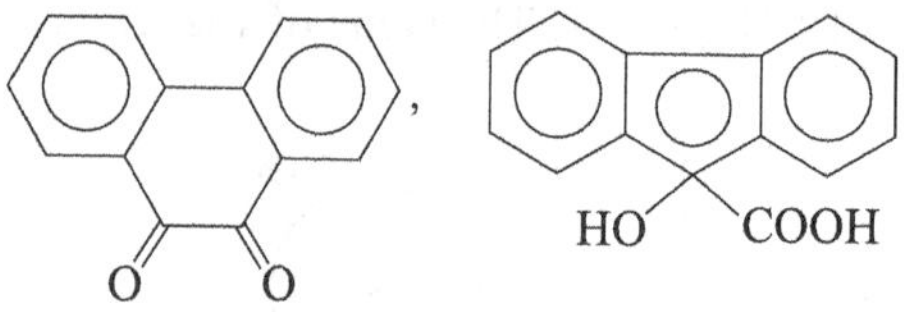

36. Which of the following statements is incorrect?

(a) The electronic configuration of Cr is [Ar] $3d^5 4s^1$ [Atomic number of Cr = 24]

(b) The magnetic quantum number may have a negative value.

(c) In silver atom, 23 electrons have a spin of one type and 24 of the opposite type. (Atomic number of Ag = 47)

(d) The oxidation state of nitrogen in HN_3 is –3.

37. When aqueous NaOH is added to an aqueous solution of chromium (III) ions, a green blue precipitate is formed first which re-dissolves to give a green solution. The green colour of the solution is due to the formation of

(a) $[Cr(H_2O)_6]^{3+}$ (b) $[Cr(OH)_4]^-$

(c) CrO_4^{2-} (d) $[Cr(OH)_3(H_2O)_3]$

38. Given below are two statements one is labelled as Assertion (A) and the other is labelled as Reason (R).

Assertion (A) BF_3 molecules has zero dipole moment.

Reason (R) BF_3 has sp^2-hybridisation and regular trigonal planar geometry.

In the light of the above statements, choose the correct answer from the options given below.

(a) Both A and R are correct and R is the correct explanation of A.

(b) Both A and R are correct but R is not the correct explanation of A.

(c) A is correct but R is incorrect.

(d) A is incorrect but R is correct.

39. Consider the following reaction,

$$C_2H_5NH_2 \begin{cases} \xrightarrow{C_6H_5CHO} [A] \\ \xrightarrow{NOCl} [B] \\ \xrightarrow{C_6H_5SO_2Cl} [C] \end{cases}$$

Which product is identified as a Schiff's base?

(a) *A* (b) *B*

(c) *C* (d) Both *A* and *C*

40. Given below are the two statements.

Statement I Alkyl benzene is not prepared by Friedel-Crafts alkylation of benzene.

Statement II Alkyl halides are less reactive than acyl halides.

In the light of the above statements, choose the correct answer from the option given below.

(a) Both Statement I and Statement II are correct.
(b) Statement I is correct but Statement II is incorrect.
(c) Statement I is incorrect but Statement II is correct.
(d) Both Statement I and Statement II are incorrect.

41. Which of the following species does not exist?

(a) Be_2^+ (b) Be_2 (c) B_2 (d) N_2

42. Which of the following functional groups present in the structure of maltose?

(a) One ketal and hemiketal
(b) Two acetals
(c) One acetal and one hemiacetal
(d) One acetal and one ketal

43. Consider the following reaction,

$$3B_2H_6 + 6NH_3 \xrightarrow[\text{temp.}]{\text{High}} 12H_2 + X,$$

the chemical formula of X is

(a) $B_3N_3H_6$ (b) $(BN)_X$ (c) B_2N (d) B_3H_4

44. Which of the following is the correct order of acidity of the given compounds?

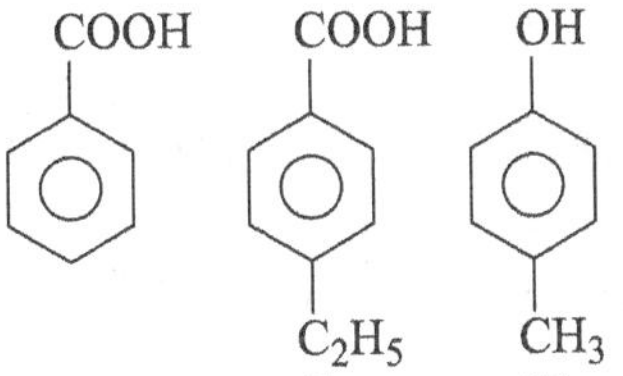

I II III

(a) III > II > I (b) II > I > III
(c) I > II > III (d) III > I > II

45. There are some statements regarding the periodic classification of elements.

A. Non-metallic elements are less in number than metallic elements.

B. The first ionisation enthalpies of elements generally increases with an increase in atomic number as we go along a period.

C. For transition elements, 3d-orbitals are filled with electrons after 3p-orbitals and before 4s-orbitals.

Choose the correct statement from the options given below:

(a) A and B
(b) A and C
(c) B and C
(d) A, B and C

Section B : Numerical Value Type Questions

46. A 200 mL solution of I_2 is divided into two unequal parts, 'A' and 'B'. 'A' part reacts with hypo solution in acidic medium, 30 mL of 0.2 M hypo was consumed. 'B' part was added with 50 mL of 0.6 M NaOH solution. The initial concentration of I_2 was ......$\times 10^{-3}$ M.

47. Addition of 0.643 g of a compound to 50 mL of benzene (density = 0.879 g/mL) lowers the freezing point from 5.51°C to 5.03°C. If K_f for benzene is 5.12 K kg mol^{-1}. The molecular weight of the compound is ...... kg mol^{-1}.

48. Following equilibrium is studied by taking 1 mole of N_2 and 3 moles of H_2 in 1 L flask at a given temperature.

$$N_2(g) + 3H_2(g) \rightleftharpoons 2NH_3(g)$$

$NH_3(g)$ formed at equilibrium is neutralised by 100 mL of 1 M HCl. The value of equilibrium constant, K_C will be ......$\times 10^{-6}$.

49. The gaseous decomposition reaction, $A(g) \longrightarrow 2B(g) + C(g)$ is observed to be first order over the excess of liquid water at 28°C. It is found that after 20 min the total pressure of the system is 200 torr and after a long time (suppose, all reactant is consumed) it is 380 torr. The rate constant of the reaction is ...... $\times 10^{-3}$ min^{-1}.
(Given, vapour pressure of H_2O = 20 torr, ln 2 = 0.3010 and ln 3 = 0.4771)

50. A piston filled with 0.04 mole of an ideal gas expands reversibly from 50 mL to 350 mL at a constant temperature of 310 K. It absorbs 208 J of heat during this process. The division of q and W for this process will be $-x$. The value of x is ....

MATHEMATICS

Section A : Objective Type Questions

51. The radius of the circle passing through foci of the ellipse $\frac{x^2}{4}+\frac{4}{7}y^2=1$ and having its centre at $\left(0, \frac{\sqrt{11}}{2}\right)$ is

(a) $\sqrt{5}$
(b) 3
(c) $\sqrt{12}$
(d) $\frac{7}{2}$

52. $\lim_{x\to 0}\frac{\int_0^x \sin t\,dt}{x^2}$ is equal to

(a) $\frac{1}{2}$
(b) $\frac{2}{3}$
(c) 1
(d) $\frac{1}{4}$

53. Consider the function $f(x)=\begin{cases} x\sin(\log x^2), & x\neq 0 \\ 0, & x=0 \end{cases}$, then

(a) f is continuous and differentiable at $x=0$.
(b) f is not continuous but differentiable at $x=0$.
(c) f is neither continuous nor differentiable at $x=0$.
(d) f is continuous but not differentiable at $x=0$.

54. $\int x^{2/3}(1+x^{1/2})^{-13/3}\,dx$

(a) $\frac{3}{5}(1+x^{-1/2})^{-10/3}+C$
(b) $\frac{3}{5}(1+x^{1/2})^{-10/3}+C$
(c) $\frac{1+x^{1/2}}{13}+C$
(d) None of the above

55. $\int \frac{dx}{x^2(x^n+1)^{\frac{n-1}{n}}}=-[f(x)]^{1/n}+C$, then

(a) $f(x)=1+x^n$
(b) $f(x)=1+x^{-n}$
(c) $f(x)=x^n+x^{-n}$
(d) None of the above

56. Let A be a square matrix of order 3 and matrices B, C and D are such that $B=\text{adj}(A), C=\text{adj}(\text{adj}\,A)$, $D=\text{adj}[\text{adj}(\text{adj}A)]$.
If $|\text{adj}[\text{adj}(\text{adj}(\text{adj}ABCD]\}]|=|A|^k$, then k is equal to

(a) 128
(b) 265
(c) 256
(d) 240

57. The solution of differential equation $\frac{dy}{dx}=\frac{x(2\log x+1)}{\sin y+y\cos y}$ is

(a) $y\sin y=x\log x+C$
(b) $y^2\sin y=x^2\log x+C$
(c) $y\sin y=x^2\log x+\frac{x^2}{2}+C$
(d) $y\sin y=x^2\log x+C$

58. The line $\frac{x-2}{3}=\frac{y+1}{2}=\frac{z-1}{-1}$ intersect the curve $xy=c^2, z=0$, if c is equal to

(a) $\pm\sqrt{5}$
(b) ± 4
(c) $\pm\frac{1}{2}$
(d) $\pm\frac{1}{\sqrt{3}}$

59. If **a**, **b** and **c** are non-zero, non-coplanar vectors and

$\mathbf{b}_1=\mathbf{b}-\frac{\mathbf{b}\cdot\mathbf{a}}{|\mathbf{a}|^2}\mathbf{a},\ \mathbf{b}_2=\mathbf{b}+\frac{\mathbf{b}\cdot\mathbf{a}}{|\mathbf{a}|^2}\mathbf{a}$

$\mathbf{c}_1=\mathbf{c}-\frac{\mathbf{c}\cdot\mathbf{a}}{|\mathbf{a}|^2}\mathbf{a}-\frac{\mathbf{c}\cdot\mathbf{b}}{|\mathbf{b}|^2}\mathbf{b},$

$\mathbf{c}_2=\mathbf{c}-\frac{\mathbf{c}\cdot\mathbf{a}}{|\mathbf{a}|^2}\mathbf{a}-\frac{\mathbf{c}\cdot\mathbf{b}_1}{|\mathbf{b}|^2}\mathbf{b}_1,$

$\mathbf{c}_3=\mathbf{c}-\frac{\mathbf{c}\cdot\mathbf{a}}{|\mathbf{a}|^2}\mathbf{a}-\frac{\mathbf{a}\cdot\mathbf{b}_2}{|\mathbf{b}_2|^2}\mathbf{b}_2,\ \mathbf{c}_4=\mathbf{a}-\frac{\mathbf{c}\cdot\mathbf{a}}{|\mathbf{a}|^2}\mathbf{a}$

Then, which of the following is a set of mutually orthogonal vectors?

(a) $\{\mathbf{a}, \mathbf{b}_1, \mathbf{c}_1\}$
(b) $\{\mathbf{a}, \mathbf{b}_1, \mathbf{c}_2\}$
(c) $\{\mathbf{a}, \mathbf{b}_2, \mathbf{a}_3\}$
(d) $\{\mathbf{a}, \mathbf{b}_2, \mathbf{c}_4\}$

60. If $A=\begin{bmatrix} i & -i \\ -i & i\end{bmatrix}, B=\begin{bmatrix} 1 & -1 \\ -1 & 1\end{bmatrix}$ and $A^8=\lambda B$, then λ is equal to

(a) 128
(b) 64
(c) 32
(d) 16

61. The sides of a rhombus $ABCD$ are parallel to the lines $x-y+2=0$ and $7x-y+3=0$. If the diagonals of the rhombus intersect at $P(1,2)$ and the vertex A (different from the origin) is on the Y-axis, then the ordinate of A is p/q. The value of $p+q$, is

(a) 7
(b) 9
(c) 5
(d) 3

62. **Statement I** If $ax^2+bx+c=0$ is a quadratic equation $(a,b,c\in R)$ such that its roots are α,β and $a+b+c<0,\ a-b+c<0$ and $c>0$, then $[\alpha]+[\beta]=-1$, (where $[\cdot]$ denotes greater integer function).

Statement II If for any two real numbers p and q continuous function $f(x)$ is such that $f(p)f(q)<0$ $\Rightarrow f(x)$ has at least one root lying in (p, q).

(a) Statement I is true, Statement II is true, Statement II is a correct explanation of Statement I.
(b) Statement I is true, Statement II is true, Statement II is not a correct explanation of Statement I.
(c) Statement I is true, Statement II is false.
(d) Statement I is false, Statement II is true.

63. The largest term of the sequence $\frac{1}{503}, \frac{4}{524}, \frac{9}{581}, \frac{16}{692}, \ldots$ is

(a) 5th (b) 6th
(c) 7th (d) 8th

64. If n is a positive integer, then $(\sqrt{5}+1)^{4n}-(\sqrt{5}-1)^{4n}$ is

(a) an odd positive integer
(b) an even positive integer
(c) a rational number other than a positive integer
(d) an irrational number

65. If $[\sqrt{2}\cos x]+[\sin x]=-3$, $x \in[0, 2\pi]$ (where $[\cdot]$ denotes the greatest integer function), then x belongs to

(a) $\left(\pi, \frac{5\pi}{4}\right)$ (b) $\left[\pi, \frac{5\pi}{4}\right]$
(c) $\left(\frac{5\pi}{4}, 2\pi\right)$ (d) $\left[\frac{5\pi}{4}, 2\pi\right]$

66. The variance of first 25 even natural numbers is

(a) $\frac{202}{3}$ (b) 202
(c) $\frac{208}{3}$ (d) 208

67. If ${}^nP_r={}^nP_{r+1}$ and ${}^nC_r={}^nC_{r-1}$, then the value of r is equal to

(a) 1 (b) 2
(c) 3 (d) 4

68. Four unbiased dice A_1, A_2, A_3, A_4 each having six faces numbered 1, 2, 3, 4, 5 and 6 are rolled simultaneously. The probability that A_4 shows a number appearing on one of A_1, A_2 and A_3 is

(a) $\frac{108}{216}$ (b) $\frac{91}{216}$ (c) $\frac{125}{216}$ (d) $\frac{127}{216}$

69. L_1 and L_2 are two lines whose vector equations are

$$L_1: r=\lambda[(\cos\theta+\sqrt{3})\hat{\mathbf{i}}+(\sqrt{2}\sin\theta)\hat{\mathbf{j}}+(\cos\theta-\sqrt{3})\hat{\mathbf{k}}]$$

and $L_2: r=\mu(a\hat{\mathbf{i}}+b\hat{\mathbf{j}}+c\hat{\mathbf{k}})$,

where λ and μ are scalars and α is the acute angle between L_1 and L_2. If the angle α is independent of θ, then the value of α is

(a) $\frac{\pi}{6}$ (b) $\frac{\pi}{4}$ (c) $\frac{\pi}{3}$ (d) $\frac{\pi}{2}$

70. Let $\mathbf{a}$ and $\mathbf{b}$ be two unit vectors such that $|\mathbf{a}+\mathbf{b}|=\sqrt{2}$. If $\mathbf{c}=\mathbf{a}+2\mathbf{b}+3(\mathbf{a}\times\mathbf{b})$ and $2|\mathbf{c}|$ is equal to K, then $\sqrt{14}\,K$ is ....

(a) 14 (b) 28 (c) 42 (d) 56

Section B : Numerical Value Type Questions

71. Let $\cos^{-1}x+\cos^{-1}2x+\cos^{-1}3x=\pi$. If x satisfies the equation $ax^3+bx^2+c=0$, then the value of $a+b+c$ is ....

72. Area bounded by the line $y=x$, curve $y=f(x), f(x)>x, \forall x>1$ and the lines $x=1, x=t$ is $[t+\sqrt{1+t^2}-(1+\sqrt{2})]\ \forall\ t>1$. Then, $f(0)$ is equal to ....

73. The number of integral values of parameter 'a' for which three chords of the ellipse $\frac{x^2}{2a^2}+\frac{y^2}{a^2}=1$ (other than its diameter) passing through the point $P\left(11a, \frac{-a^2}{4}\right)$ are bisected by the parabola $y^2=4ax$ is equal to ....

74. Let S_K denote the sum of infinite geometric series, where $K=1, 2, \ldots, 100$ and first term is $\frac{K-1}{K!}$, common ratio is $\frac{1}{K}$, then the value of $\sum_{K=2}^{100}|(K^2-3K+1)S_K|$ is $3-\frac{\lambda^2}{99!}$, then λ is equal to ....

75. On the curve $y^3=27x$. The total possible number of integral values of y for which the ordinate changes as faster rate than abscissa is equal to ....

DETAILED SOLUTIONS

Physics

1. *(b)* As, kinetic energy (K) = constant

$\therefore \quad \Delta K = K_f - K_i = 0$

$\therefore$From work-energy theorem,

$$W_{\text{net}} = \Delta K = 0$$

$$\mathbf{F} \cdot d\mathbf{r} = 0$$

$\therefore \quad \theta = 90°$

$\therefore$Force is perpendicular to displacement or the velocity of the particle.

2. *(a)* Here, $V_L = V_C = 400\text{V}$

$\therefore$Current passing through inductor and capacitor is same, then $X_L = X_C$

The impedance of the circuit,

$$Z = \sqrt{R^2 + (X_L - X_C)^2}$$

$\because \quad X_L - X_C = 0$

$\therefore \quad Z = R$

and $V_R = V = 200\text{V}$

Now, voltage across resistor is given by

$$V = IZ$$

$$200 = I \times 40 \qquad [\because Z = R = 40\Omega]$$

$$\therefore \quad I = \frac{200}{40} = 5\text{A}$$

Therefore, the voltage across terminals of resistance and current through ammeter are 200 V and 5A, respectively.

3. *(a)* Given, $d = 5400\,\text{m}$

$$\rho = 10^3\,\text{kg/m}^3$$

Compressibility $= 45.4 \times 10^{-11}\ \text{Pa}^{-1}$

The pressure at the bottom of the ocean is given by

$$p = \rho g d = 10^3 \times 10 \times 5400 = 54 \times 10^6\,\text{Pa}$$

So, fractional compression = compressibility × pressure

$$= 45.4 \times 10^{-11} \times 54 \times 10^6$$

$$= 2451.6 \times 10^{-5}$$

$$= 2.4516 \times 10^{-2} = 2.4 \times 10^{-2}$$

4. *(c)* As, $f_{\text{lim}} = \mu mg = 0.3 \times 5 \times 10 = 15\,\text{N}$

Applied force $(F) = 12\,\text{N} < f_{\text{lim}}$

$\therefore$The friction between the surfaces of block and floor is static friction which is equal to applied force.

$\therefore$Frictional force $(f) = 12\,\text{N}$

5. *(a)* The rms speed of a gas, $v_{\text{rms}} = \sqrt{\dfrac{3RT}{M}}$

$$\Rightarrow \quad v_{\text{rms}} \propto \frac{1}{\sqrt{M}}$$

As $\quad M_{H_2} < M_{O_2}$

$\Rightarrow \quad v_{\text{rms}, H_2} > v_{\text{rms}, O_2}$

6. *(c)* Least count of the screw gauge,

$$= \frac{1\,\text{mm}}{100} = 0.01\,\text{mm}$$

Diameter of wire $= 1 \times 1 + 47 \times 0.01$

$$= 1.47\,\text{mm} = 0.147\,\text{cm}$$

Curved surface area of wire $(A) = 2\pi rL$

$$= 2\pi \frac{D}{2} \times L = 3.14 \times 0.147 \times 5.6$$

$$= 2.5848\,\text{cm}^2$$

As length of wire is upto one decimal place,

$\therefore \quad A = 2.6\,\text{cm}^2$

7. *(d)* The given arrangement of charged ring can be shown as

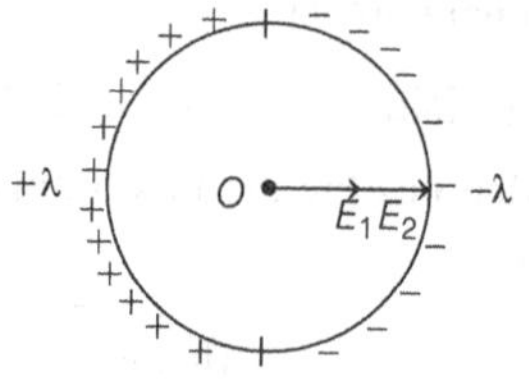

The electric field due to a charged semi-circular ring $(+\lambda)$ is given by

$$E = \frac{\lambda}{2\pi\varepsilon_0 R}$$

$\therefore$Net electric field at the centre O is

$$E_{\text{net}} = E_1 + E_2 = 2E_1 \qquad \left[\because E_1 = E_2 = \frac{\lambda}{2\pi\varepsilon_0 R}\right]$$

$$= 2 \times \frac{\lambda}{2\pi\varepsilon_0 R} = \frac{\lambda}{\pi\varepsilon_0 R}$$

8. *(c)* The Maxwell's equations of electromagnetism and Hertz experiment on the generation and detection of electromagnetic waves in 1887, strongly established the wave nature of light.

9. *(c)* Li^{2+} atom is shown as

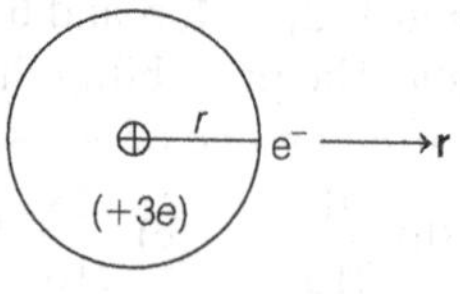

The coulomb force $\mathbf{F}$ between nucleus and outermost e^-,

$$\mathbf{F} = K\frac{(3e)(e)}{r^2}(-\hat{\mathbf{r}})$$

$$\Rightarrow \quad \mathbf{F} = -3K\frac{e^2}{r^3} \cdot \mathbf{r} \qquad \left[\because \hat{\mathbf{r}} = \frac{\mathbf{r}}{r}\right]$$

10. *(a)* For NOT gate, $Y = \overline{A}$

At $A = 0, Y = 1$ (inversion of input A)

At $A = 1, Y = 0$ (inversion of input A)

For AND gate, $Y = A.B$

At $A = 0, B = 0, Y = 0$ (same as that of A, B)

At $A = 1, B = 1, Y = 1$ (same as that of A, B)

11. *(d)* The equivalent emf of series combination of cells is the sum of their individual emf.

Statement I is false.

For series combination of two cells of emf E_1 and E_2, internal resistance r_1 and r_2, if polarity of E_1 is reversed, then

Equivalent emf $= |E_1 - E_2|$ (Affected)

Equivalent resistance $= r_1 + r_2$ (Unaffected)

$\therefore$ Statement II is true.

12. *(c)* Given, angle between P_1 and $P_2 = 30°$

Angle between P_2 and $P_3 = \theta = 90° - 30° = 60°$

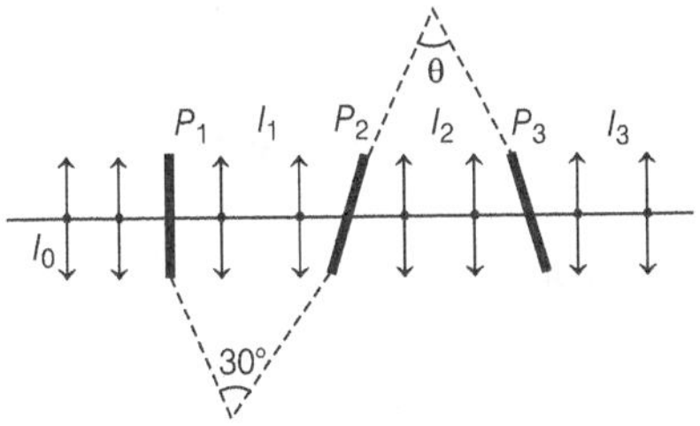

The intensity of light transmitted by P_1,

$$I_1 = \frac{I_0}{2} = \frac{40}{2} = 20 \frac{\text{W}}{\text{m}^2}$$

According to Malus law, the intensity of light transmitted by P_2 is

$$I_2 = I_1 \cos^2 30° = 20\left(\frac{\sqrt{3}}{2}\right)^2 = 15 \frac{\text{W}}{\text{m}^2}$$

Similarly, intensity of light transmitted by P_3,

$$I_3 = I_2 \cos^2 \theta = 15 \cos^2 60°$$

$$= 15 \times \left(\frac{1}{2}\right)^2 = 3.75 \text{ W/m}^2$$

13. *(c)* $g = \frac{GM}{R^2}$, $M =$ Mass of the earth.

The acceleration due to gravity depends on the mass of the Earth and is independent of mass of the body.

The escape velocity of a body on the Earth surface is

$$v_e = \sqrt{\frac{2GM_e}{R_e}} = \sqrt{2gR_e}$$

which is dependent on the radius of the Earth.

Inside a uniform spherical shell, the gravitational field is zero everywhere, i.e. gravitational potential is same everywhere.

The speed of a planet in planetary motion varies due to elliptical shape of the planet's orbit.

14. *(c)* Typical graph of metal under tension is

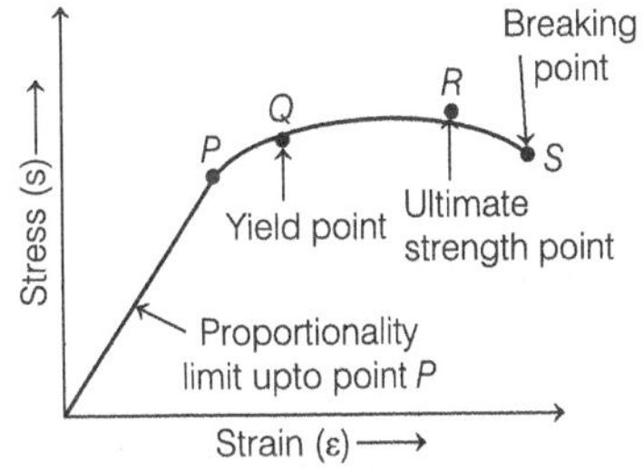

15. *(c)* Magnetic field at point O due to wires CB and DA will be zero.

So, magnetic field due to wire AB will be

$$B_1 = \left(\frac{\theta}{2\pi}\right)\left[\frac{\mu_0 i}{2(2R)}\right]$$

$$= \frac{\frac{\pi}{3}}{2\pi}\left[\frac{\mu_0 i}{4R}\right] = \frac{\mu_0 i}{24R}$$

Direction of field B_1 is going into the plane of paper.

Similarly, magnetic field due to wire CD will be

$$B_2 = \frac{\mu_0 i}{36R}$$

Direction of field B_2 is coming out of the plane of paper.

The resultant field at point O,

$$B = B_1 - B_2$$

$$= \frac{\mu_0 i}{24R} - \frac{\mu_0 i}{36R} = \frac{3\mu_0 i - 2\mu_0 i}{72R} = \frac{\mu_0}{72R}$$

16. *(c)* The circuit diagram is shown as

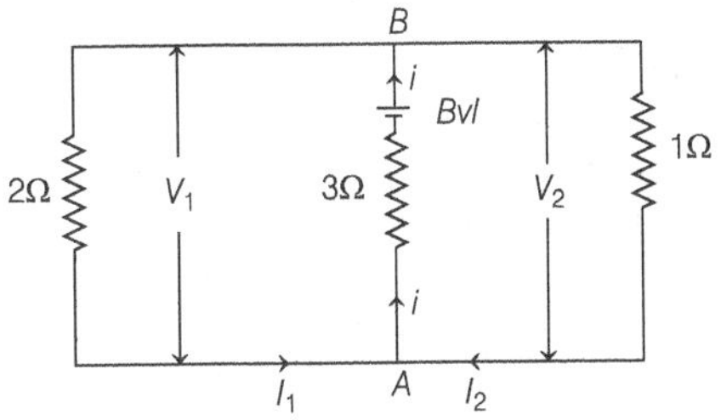

Equivalent resistance between points A and B,.

$$R_{AB} = \frac{2 \times 1}{2+1} + 3 = \frac{11}{3}\Omega$$

$$\therefore \quad i = \frac{Bvl}{R_{AB}} = \frac{5 \times 2 \times 1}{\frac{11}{3}} = \frac{30}{11}\text{ A}$$

$$\therefore \quad V_1 = V_2 = -i \times 3 + Bvl$$

$$= -\frac{30}{11} \times 3 + 5 \times 2 \times 1 = \frac{20}{11}\text{ V}$$

17. *(b)* Volume of sphere, $V=\frac{4}{3}\pi r^3$

$\therefore$ Density of sphere, $d=\frac{m}{V}=\frac{m}{\frac{4}{3}\pi r^3}$

$$\therefore \quad \frac{\Delta d}{d}=\frac{\Delta m}{m}+3\left(\frac{\Delta r}{r}\right)$$

$$=\frac{0.30}{15}+3\times\frac{0.01}{0.20}=0.17\,\text{kgm}^{-3}$$

18. *(a)* $\mathbf{B}=2.1\times10^{-8}\,\hat{\mathbf{k}}$

$\therefore E=cB=3\times10^{8}\times2.1\times10^{-8}=6.3$

As EM wave is in $+x$-direction, **B** is in $+z$-direction. Hence, **E** will be in $+y$-direction.

Here, $\mathbf{E}=6.3\hat{\mathbf{j}}\,\text{V/m}$

19. *(c)* As, we know that, potential energy, $U=2E$, kinetic energy, $K=-E$

and total energy of electron, $E=-\frac{13.6}{n^2}$

For state $n=1$, $E_1=\frac{-13.6}{(1)^2}=-13.6$

For state $n=2$, $E_2=\frac{-13.6}{(2)^2}=\frac{E_1}{4}\Rightarrow E_1=4E_2$

But $\quad K=-E$, so $K_1=4K_2$

Similarly, for state $n=1$, potential energy,

$$U_1=2E_1=2\times\frac{-13.6}{(1)^2}=-2\times13.6$$

For state $n=2$, potential energy,

$$U_2=2E_2=2\times\frac{-13.6}{(2)^2}=\frac{-2\times13.6}{4}$$

$$U_2=\frac{U_1}{4}\Rightarrow U_1=4U_2$$

Thus, both kinetic energy K and potential energy U become four times.

20. *(b)* In first case, $\frac{X}{Y}=\frac{60}{40}$ or $\frac{X}{Y}=\frac{3}{2}$ ...(i)

In second case,

$$\frac{X}{Y+15}=\frac{60-10}{40+10}=\frac{50}{50}=1 \quad ...(ii)$$

Dividing Eq. (i) by Eq. (ii), we get

$$\frac{X}{Y}\times\frac{Y+15}{X}=\frac{3}{2}\times1$$

or $\quad 1+\frac{15}{Y}=\frac{3}{2}$

$$Y=30\Omega$$

or $\quad X=\frac{3}{2}Y=\frac{3}{2}\times30=45\Omega$

When a resistance of 30Ω is connected in parallel with Y, then the resistance in the right gap becomes

$$Y'=\frac{30Y}{30+Y}=\frac{30\times30}{30+30}=15\Omega$$

Suppose, the null point occurs at l cm from end A.

Then, $\frac{X}{Y'}=\frac{l}{100-l}$

$$\frac{45}{15}=\frac{l}{100-l}$$

$$300-3l=l\Rightarrow 4l=300$$

$$l=75\,\text{cm}$$

21. *(2)* As we know,

$$v=A\omega\cos\omega t$$

$$v=A\omega\sqrt{1-\sin^2\omega t}$$

$$\Rightarrow \quad v=A\omega\sqrt{1-\frac{x^2}{A^2}}$$

$$\Rightarrow \quad v=A\omega\frac{1}{A}\sqrt{A^2-x^2}$$

$$\Rightarrow \quad v=\omega\sqrt{A^2-x^2}$$

Here, $v_1=10$ cm/s, $x_1=4$ cm and $v_2=8$ cm/s, $x_2=5$ cm. Then,

$$v_1=\omega\sqrt{A^2-x_1^2} \quad ...(i)$$

and $v_2=\omega\sqrt{A^2-x_2^2}$ ...(ii)

From Eqs. (i) and (ii), we get

$$\frac{v_1}{v_2}=\sqrt{\frac{A^2-x_1^2}{A^2-x_2^2}}$$

$$\left(\frac{v_1}{v_2}\right)^2=\frac{A^2-x_1^2}{A^2-x_2^2}$$

$$\frac{100}{64}=\frac{A^2-16}{A^2-25}$$

$$\frac{25}{16}=\frac{A^2-16}{A^2-25}$$

$$\Rightarrow \quad A^2=41\,\text{cm}$$

From Eq. (i),

$$\omega=\frac{v_1}{\sqrt{A^2-x_1^2}}=\frac{10}{\sqrt{41-16}}$$

$$\omega=\frac{10}{5}=2$$

$$\frac{2\pi}{T}=2\Rightarrow T=\frac{2\pi}{2}$$

$$T=\frac{2\pi}{x}\text{ (given)}$$

$$\therefore \quad x=2$$

22. (*6*) From given diagram in question,

$$\int_A^B \mathbf{B}_P d\mathbf{l} + \int_B^C \mathbf{B}_P d\mathbf{l} + \int_C^D \mathbf{B}_P d\mathbf{l} + \int_D^A \mathbf{B}_P d\mathbf{l} + \int_A^B \mathbf{B}_Q d\mathbf{l} + \int_B^C \mathbf{B}_Q d\mathbf{l} + \int_C^D \mathbf{B}_Q d\mathbf{l} + \int_D^A \mathbf{B}_Q d\mathbf{l} = \mu_0 (2I - I)$$

$$\Rightarrow \quad (-\mu_0 - 2\mu_0 - \mu_0 - 2\mu_0) + (2\mu_0 + 4\mu_0 + 2\mu_0 + 4\mu_0) = \mu_0 I$$

$$\Rightarrow \quad 6\mu_0 = \mu_0 I$$

$$\Rightarrow \quad I = 6\,\text{A}$$

23. (19) Given, $A = 30°, \mu = 1.5$ and $i_1 = 0°$

Since, $i_1 = 0°$, therefore, r_1 is also equal to 0°.

Further, since, $r_1 + r_2 = A$

$$\therefore \quad r_2 = A = 30°$$

Using, $\mu = \dfrac{\sin i_2}{\sin r_2}$

we have, $1.5 = \dfrac{\sin i_2}{\sin 30°}$

or $\quad \sin i_2 = 1.5 \sin 30° = 1.5 \times \dfrac{1}{2} = 0.75$

$$\therefore \quad i_2 = \sin^{-1}(0.75) = 48.6°$$

Now, the deviation,

$$\delta = (i_1 + i_2) - A = (0° + 48.6°) - 30°$$

or $\quad \delta = 18.6°$

$$\delta \approx 19°$$

24. (6)

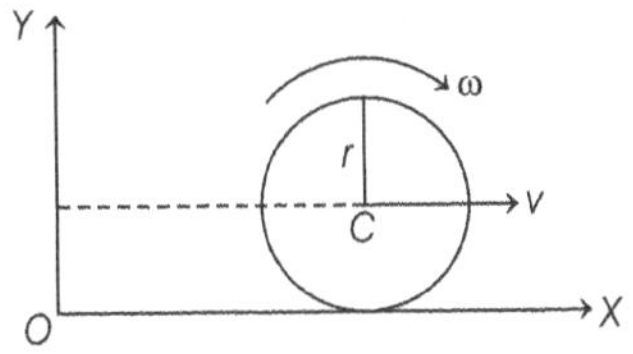

The total angular momentum of disc about O is given by

$$L_{\text{net}} = L_T + L_R = mv \times r + I_{\text{CM}}\omega$$

$$= m(r\omega)\cdot r + \frac{1}{2}mr^2 \cdot \omega$$

$$= \frac{3}{2}mr^2\omega$$

$$= \frac{3}{2} \times 2 \times (5 \times 10^{-2})^2 \times 8 = 0.06\ \text{kg-m}^2/\text{s}$$

$$= 6 \times 10^{-2}\ \text{kg-m}^2/\text{s}$$

$$= x \times 10^{-2}\ \text{kg-m}^2/\text{s}$$

$$\therefore \quad x = 6$$

25. (1) In an adiabatic process, $T \propto p^{1-\frac{1}{\gamma}}$

$$\Rightarrow \quad T \propto p^{\frac{1}{3}} \qquad [\because \gamma = 1.5]$$

If T_L and T_R are final temperatures of gases in left and right compartments respectively, p_L and p_R are respective pressures, then the ratio of internal energy U_L and U_R is

$$\frac{U_L}{U_R} = \frac{\frac{f}{2}nRT_L}{\frac{f}{2}nRT_R} = \frac{T_L}{T_R}$$

Using Eq. (i), then

$$\Rightarrow \quad \frac{U_L}{U_R} = \left(\frac{p_L}{p_R}\right)^{\frac{1}{3}} = \left(\frac{1}{8}\right)^{\frac{1}{3}} = \frac{1}{2} = \frac{x}{2} \qquad \text{(given)}$$

$$\therefore \quad x = 1$$

Chemistry

26. (*a*) **Acidic character** The species which easily donate its hydrogen and produces stable conjugate base is acid. The species which produces more stable conjugate base is more stronger acid.

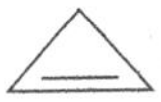 and  do not lose H^+.

Hence, these are not acidic.

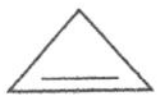 loses the H^+ ion but does not produces a stable aromatic ion. i.e. 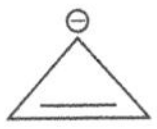

(Non-aromatic)

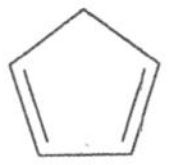 loses the H^+ ion easily and produces stable aromatic cyclopentadienyl anion.

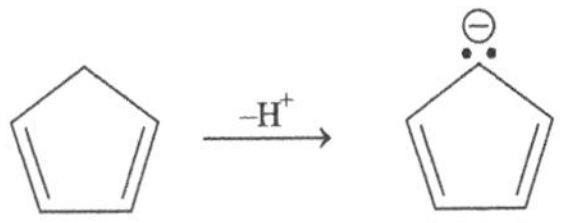

Non-aromatic

Aromatic (Cyclopentadienyl anion)

Now, cyclopentadienyl anion on reaction with 3-chloroprop-1-ene produces the product *via* nucleophilic substitution reaction.

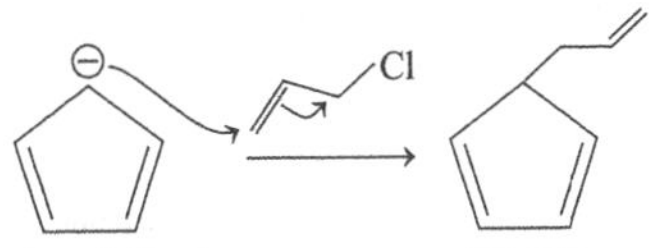

Cyclopentadienyl anion

Product

27. (*b*) Existence of nitrogen as N_2 is due to strong $p\pi$- $p\pi$ bonding between small sized *p*-orbitals of N.

Existence of phosphorus as P_4 is due to existence of weak $p\pi$- $p\pi$ bonding and large size of *p*-orbital of phosphorous atom.

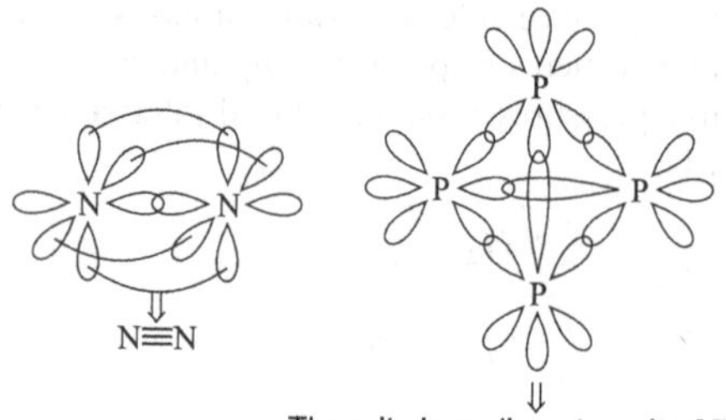

Thus, it show discrete unit of P_4, (due to large size of *p*-orbital or phosphorus)

28. (*b*) $[Pt(NH_3)_2(H_2O)Cl]^+$, $[Pt(NH_3)_3(NO_3)]Cl$ and $[Pt(NH_3)_3Cl]Br$ have square planar geometry. $[Co(NH_3)_4Cl_2]^+$ has octahedral geometry and $[Ni(CO)_4]$ has tetrahedral geometry.

Hence option (b) is correct.

29. (*a*) The correct match is

$$A \rightarrow (III), B \rightarrow (IV), C \rightarrow (II), D \rightarrow (I).$$

- Alcoholic KOH is used for β-elimination reaction.

$$\overset{+\,-}{KOH} + CH_3CH_2OH \longrightarrow CH_3CH_2\bar{O} + H_2O$$

Acts as a base to abstract β-hydrogen

- Pd / $BaSO_4$ known as Lindlar's catalyst.
- Benzene hexachloride (BHC) is obtained by addition reaction.

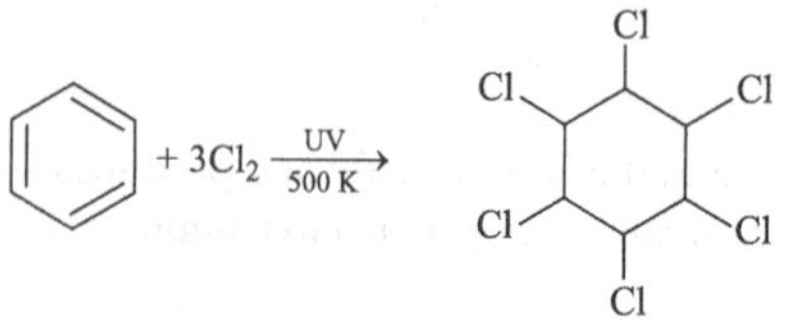

Benzene hexachloride (BHC)

- $LiAlH_4$ is a strong reducing agent.

30. (*c*) The reaction between Grignard reagent and ketone produces tertiary alcohol as a product.

The reaction is carried out in a single step and involves the formation of a six-membered transition state, which is formed when the Grignard reagent attacks the carbonyl centre of the ketone.

31. (*b*) The correct match is A-I, B-II, C-III, D-IV

A. $FeCl_3 + 2S_2O_3^{2-} \longrightarrow \underset{\text{Violet solution}}{[Fe(S_2O_3)_2]^-} + 3Cl^-$

$$[Fe(S_2O_3)_2]^- + Fe^{3+} \longrightarrow \underset{\text{(Green)}}{2Fe^{2+}} + S_4O_6^{2-}$$

B. $Cr_2O_7^{2-} + 8H^+ + 3SO_3^{2-} \longrightarrow \underset{\text{(Green)}}{2[Cr(H_2O)_6]^{3+}} + 3SO_4^{2-} + 4H_2O$

C. The Beilstein test is a simple qualitative chemical test for organic halides.

D. Barfoed's test is a chemical test used for detecting the presence of monosaccharides. It is based on the reduction of copper (II) acetate to copper (I) oxide (Cu_2O), which forms a brick red precipitate.

32. (*a*) Glucose reacts with excess of HIO_4 to form 5 moles of formic acid and 1 mole of formaldehyde.

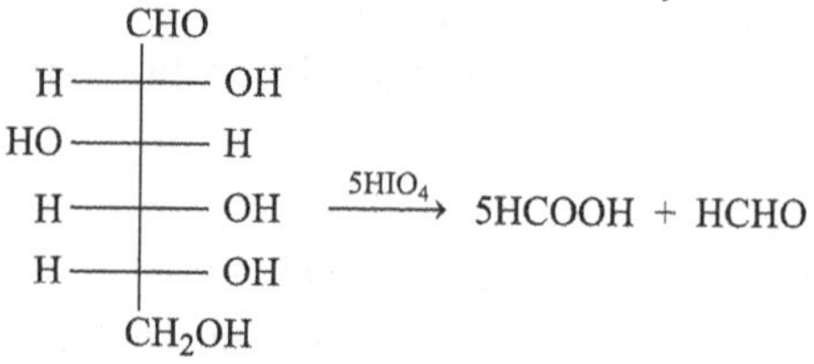

33. (*a*) $\Lambda^\circ_m(NaCl) = 126.4\ S\ cm^2\ mol^{-1}$

$$\Lambda^\circ_m(HCl) = 425.9\ S\ cm^2\ mol^{-1}$$

$$\Lambda^\circ_m(NaA) = 100.5\ S\ cm^2\ mol^{-1}$$

$$\Lambda^\circ_m(HA) = \Lambda^\circ_m(HCl) + \Lambda^\circ_m(NaA) - \Lambda^\circ_m(NaCl)$$

$$= 425.9 + 100.5 - 126.4$$

$$= 400\ S\ cm^2\ mol^{-1}$$

$$\Lambda_m = \frac{\kappa(HA) \times 1000}{\text{Molarity of } HA} = \frac{5 \times 10^{-5} \times 1000}{0.001}$$

$$= 50\ S\ cm^2\ mol^{-1}$$

$$\alpha = \frac{\Lambda_m(HA)}{\Lambda^\circ_m(HA)} = \frac{50}{400} = 0.125$$

$$= 125 \times 10^{-3}$$

34. (*d*) As, trihalides of boron have same structure due to same hybridisation

$$H = \frac{V + M - C + A}{2} = \frac{3+3}{2} = 3$$

Hybridisation $= sp^2$

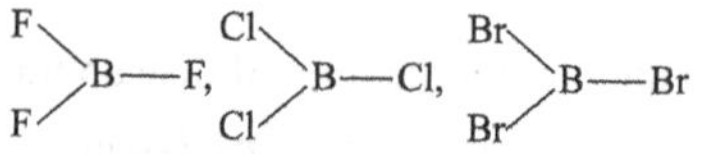

35. (*b*) **Molecular structure determination** Molecular structure of compound having molecular formula $C_{14}H_{10}O_2$ is determined by calculating degree of unsaturation.

$$u = (C + 1) - \frac{H}{2} - \frac{X}{2} + \frac{N}{2} = (14 + 1) - \frac{10}{2}$$

$$= 15 - 5 = 10$$

Degree of unsaturation is 10 in which 2 units are already considered to be used as diketo group $\left(\overset{O}{\overset{\|}{-C}}-\overset{O}{\overset{\|}{C}}- \right)$.

Rest 8 unit of unsaturation is satisfied by two phenyl ring each having, $u = 4$.

Hence, correct structure is

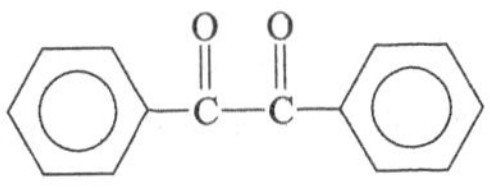

Benzilic acid rearrangement

Conversion of benzil to benzilic acid in presence of base is known as benzilic acid rearrangement. In general benzilic acid is α-hydroxycarboxylic acid. The reaction is believed to occur as

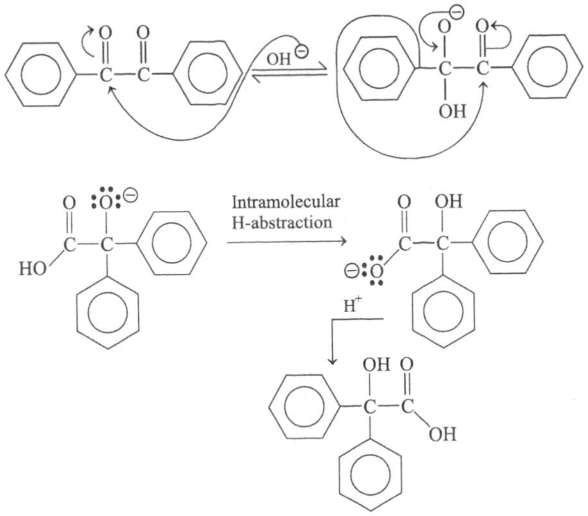

Benzilic acid

Study Tactics

This problem include conceptual mixing of molecular structure determination and benzilic acid rearrangement. This problem can be solved by using following sequential step.

I. Calculate the degree of unsaturation and then determine the appropriate molecular structure.

II. Complete the reaction using the concept of benzilic acid rearrangement in which diketone undergo benzilic acid rearrangement in presence of base to produce corresponding benzilic acid.

36. (*d*) • $_{24}Cr = 1s^2, 2s^2 2p^6, 3s^2 3p^6 3d^5, 4s^1$

$= [Ar] 3d^5, 4s^1$

• For magnetic quantum number (m), negative values are possible.

For, s-subshell, $l = 0$, hence $m = 0$.

For p-subshell, $l = 1$, hence $m = -1, 0, +1$

• $_{47}Ag = 1s^2, 2s^2 2p^6, 3s^2 3p^6 3d^{10}, 4s^2 4p^6 4d^{10} 5s^1$

Hence, 23 electrons have a spin of one type and 24 of the opposite type.

• Oxidation state of N in N_3H is $-1/3$.

37. (*b*) When aqueous solution of NaOH is added to aqueous solution of Cr (III) ion it produces $[Cr(H_2O)_3(OH)_3]$.

$$[Cr(H_2O)_6]^{3+} + NaOH \longrightarrow \underset{\text{Green blue ppt.}}{[Cr(OH)_3(H_2O)_3]} + H_2O$$

Which on further redissolves in *aq.* NaOH (in excess) to produce dark green solution due to formation of $[Cr(OH)_4]^-$.

$$[Cr(OH)_3(H_2O)_3] + \underset{\text{Excess}}{OH^-} \longrightarrow \underset{\text{Dark green solution}}{[Cr(OH)_4]^-}$$

38. (*b*) Both A and R are correct but R is not the correct explanation of A.

39. (*a*) The reactions of $C_2H_5NH_2$ with given reagents are given below.

$$C_2H_5NH_2 \xrightarrow{C_6H_5CHO} \underset{(A)}{C_2H_5{-}N{=}CHC_6H_5}$$

$$C_2H_5NH_2 \xrightarrow{NOCl} \underset{(B)}{C_2H_5Cl} + N_2 + H_2O$$

$$C_2H_5NH_2 \xrightarrow{C_6H_5SO_2Cl} \underset{(C)}{C_6H_5SO_2NHC_2H_5} + HCl$$

Here, in the first reaction, (A) is Schiff's base because it resembles with general structure of Schiff's base, i.e. $R_2C{=}NR'$.

40. (*a*) Due to $+I$-effect of the alkyl group, electron density increases in alkylbenzene relative to benzene and hence further substitution occurs to form a poly alkyl substituted benzene. Thus, Statement I is correct.

Now, the C-atoms of alkyl halides are less electrophilic than the C-atom of carbonyl.

So, alkyl halides are less reactive than acyl halides.

41. (*b*) $Be_2(8) = \sigma 1s^2, \sigma^* 1s^2, \sigma 2s^2, \sigma^* 2s^2$

Bond order $= \dfrac{4-4}{2} = 0$

So, no bond is formed between two atoms of Be. Hence, Be_2 does not exist.

42. (*c*)

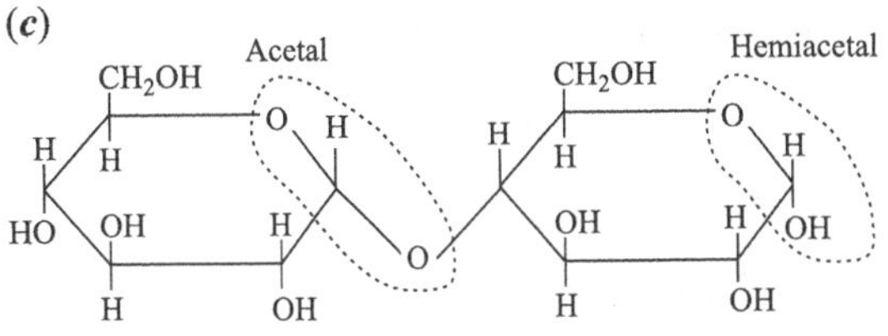

One acetal and one hemiacetal group is present in maltose.

43. (*a*) The reaction is represented as

$$3B_2H_6 + 6NH_3 \longrightarrow 3[BH_2(NH_3)_2]^+[BH_4]^- \longrightarrow \underset{(X)}{2B_3N_3H_6} + 12H_2$$

The chemical formula of X is $B_3N_3H_6$.

44. (*c*) The decreasing order of acidic strength of given compound is I > II > III.

The benzoate ion is more stabilised because the negative charge is on the more electronegative oxygen atom, whereas in phenoxide ion, it is on the less electronegative carbon atoms. This causes benzoic acid to be a stronger acid than phenol.

The alkyl group attached to *para* position to carboxylic group in aromatic carboxylic acid reduces the acidity. Similarly, electron releasing group, i.e. $—CH_3$ present in the para position to $—OH$ group decreases its acidity.

Study Tactics

The more stable is the conjugate base the stronger is the acid. Also, ERG decreases the acidity while EWG increases the acidity.

45. (*a*) In case of transition elements (or any other elements), the order of filling of electrons in various orbital is $3p > 4s > 3d$.

So, according to Aufbau principle, 3*d*-orbital is filled when 4*s*-orbital gets completely filled.

Hence, among the given statements, only (C) is incorrect while (A) and (B) are correct.

46. (*90*) $I_2 + 2Na_2S_2O_3 \longrightarrow 2NaI + Na_2S_4O_6$ ...(i)

Moles of $Na_2S_2O_3$ consumed

$$= 0.03 \times 0.2 = 0.006 \text{ moles}$$

$$\text{Moles of } I_2 \text{ consumed} = \frac{0.006}{2} = 0.003 \text{ moles}$$

$3I_2 + 6NaOH \longrightarrow 5NaI + NaIO_3 + 3H_2O$ ...(ii)

Moles of I_2 reacted with NaOH,

$$\frac{3 \times 0.05 \times 0.6}{6} = 0.015 \text{ moles}$$

Total moles of I_2 consumed in reactions (i) and (ii)

$$= 0.003 + 0.015 = 0.018 \text{ moles}$$

$$\text{Molarity of } I_2 = \frac{0.018}{0.2} = 0.09 \text{ M}$$

$$= 90 \times 10^{-3} \text{ M}$$

47. (*156*) $K_f = 5.12 \text{ K kg mol}^{-1}$

$W_{\text{solute}} = 0.643 \text{ g}$

$\Delta T_f = T_0 - T_f = 5.51 - 5.03 = 0.48 \text{ K}$

$W_{\text{solvent}} = 43.95 \text{ g}$

$$\therefore \left[d = \frac{M}{V}\right] = (V \times d = 50 \text{ mL} \times 0.879)$$

$$M_{\text{solute}} = \frac{K_f \times w_{\text{solute}} \times 1000}{\Delta T_f \times W_{\text{solvent}}}$$

$$= \frac{5.12 \times 0.643 \times 1000}{0.48 \times 43.95} = 156 \text{ kg mol}^{-1}$$

48. (*455*) $NH_3(g) + HCl(aq) \rightleftharpoons NH_4Cl(aq)$

$\Rightarrow$ 1 mole of NH_3 = 1 mole of HCl

100 mL of 1 M HCl $\equiv$ 100 millimole of HCl

= 0.1 mole of HCl

= 0.1 mole of NH_3

	$N_2(g)$ +	$3H_2(g)$ $\rightleftharpoons$	$2NH_3(g)$
Initially	1	3	0
At equilibrium	$1-x$	$3-3x$	$2x$

$\Rightarrow$ $2x = 0.1$,

i.e. $x = \frac{0.1}{2} = 0.05$ mol

and $V = 1$ L

$$\text{Therefore, } K_C = \frac{[NH_3]^2}{[N_2][H_2]^3}$$

$$= \frac{\left(\frac{2x}{V}\right)^2}{\left(\frac{1-x}{V}\right)\left(\frac{3-3x}{V}\right)^3} = \frac{4x^2V^2}{27(1-x)^4}$$

$$= \frac{4 \times (0.05)^2 \times (1)^2}{27(1-0.05)^4}$$

$$= 4.55 \times 10^{-4}$$

$$= 455 \times 10^{-6}$$

49. (*14*) $A(g) \longrightarrow 2B(g) + C(g)$

	$A(g)$	$2B(g)$	$C(g)$
After 20 mins	$p_0 - x$	$2x$	x
After long time, $t \to \infty$	0	$2p_0$	p_0

As given, $p_0 - x + 2x + x$ + vapour pressure of water $= 200$

$p_0 + 2x + 20 = 200$

$p_0 + 2x = 180$ ...(i)

and $2p_0 + p_0 + 20 = 380$

$3p_0 = 360$

$p_0 = 120$ torr ...(ii)

On substituting in Eq, (i) we get

$120 + 2x = 180$

$2x = 60$

$x = 30$ torr

$$k = \frac{1}{t}\ln\left(\frac{p_0}{p_0 - x}\right) = \frac{1}{20}\ln\left(\frac{120}{90}\right)$$

$$k = \frac{1}{20}(\ln 12 - \ln 9)$$

$$= \frac{1}{20}(\ln 4 + \ln 3 - 2\ln 3)$$

$$= 0.014 \text{ min}^{-1}$$

$$= 14 \times 10^{-3} \text{ min}^{-1}$$

50. (1) As we know, $\Delta U = q + W$, and the process is reversibly isothermal and expansion occurs, here.

Therefore,

$$q = -W$$

and $\Delta U = 0$

$$W = -2.303\, nRT \log \frac{V_2}{V_1}$$

$$= -2.303 \times 0.04 \times 8.314 \times 310 \times \log \frac{350}{50}$$

$$= -237.42 \times \log 7$$

$$= -200.64$$

So, $q = +200.64$

The division of q and $W = \dfrac{-200.64}{200.64} = -1$

Mathematics

51. *(a)* Given equation of ellipse is $\dfrac{x^2}{4} + \dfrac{y^2}{7/4} = 1$

Here, $a^2 = 4$ and $b^2 = \dfrac{7}{4}$

$\therefore \quad b^2 = a^2(1-e^2)$

$\Rightarrow \quad \dfrac{7}{4} = 4(1-e^2)$

$\Rightarrow \quad e^2 = 1 - \dfrac{7}{16} = \dfrac{9}{16} \Rightarrow e = \dfrac{3}{4}$

Thus, foci are $\left(\pm\dfrac{3}{2}, 0\right)$.

The radius of required circle

$$= \sqrt{\left(\pm\frac{3}{2} - 0\right)^2 + \left(0 - \frac{\sqrt{11}}{2}\right)^2} = \sqrt{5}$$

52. *(a)* $\lim\limits_{x \to 0} \dfrac{\int_0^x \sin t\, dt}{x^2}$ $\quad \left[\dfrac{0}{0} \text{ form}\right]$

$= \lim\limits_{x \to 0} \dfrac{\sin x}{2x}$ [using L-Hospital's rule]

$= \dfrac{1}{2} \lim\limits_{x \to 0} \left(\dfrac{\sin x}{x}\right) = \dfrac{1}{2} \times 1 = \dfrac{1}{2}$

53. *(d)* Given, $f(x) = \begin{cases} x \sin(\log x^2), & x \neq 0 \\ 0, & x = 0 \end{cases}$

Test for continuity

LHL $= \lim\limits_{h \to 0} f(0-h)$

$= \lim\limits_{h \to 0} (-h)\sin(\log(-h)^2) = -\lim\limits_{h \to 0} h \sin(\log h^2)$

As $h \to 0$, $\log h^2 \to -\infty$

$\Rightarrow \sin(\log h^2)$ oscillates between -1 and 1.

So, LHL $= 0 \times$ a number between -1 and 1

$= 0$

RHL $= \lim\limits_{h \to 0} f(0+h)$

$= \lim\limits_{h \to 0} h \sin(\log h^2)$

$= 0 \times$ a number between -1 and $1 = 0$

Also, $f(0) = 0$ [given]

$\Rightarrow \quad$ LHL = RHL $= f(0)$

$f(x)$ is continuous at $x = 0$

Test of differentiability

$$Lf'(0) = \lim_{h \to 0} \frac{f(0-h) - f(0)}{-h}$$

$$= \lim_{h \to 0} \frac{-h \sin[\log(-h)^2] - 0}{-h}$$

$$= \lim_{h \to 0} [\sin(\log h^2)]$$

= a number between -1 and 1

$\Rightarrow$ Limit does not exist.

$\therefore$ Left hand derivation is not defined.

$f(x)$ is not differentiable at $x = 0$.

54. *(a)* Let $I = \int x^{2/3}(1 + x^{1/2})^{-13/3}\, dx$

$$= \int x^{2/3} x^{-13/6} (x^{-1/2} + 1)^{-13/3}\, dx$$

$$= \int x^{-3/2} (x^{-1/2} + 1)^{-13/3}\, dx$$

Put $x^{-1/2} + 1 = t \Rightarrow x^{-3/2} dx = -2dt$

$\therefore \quad I = -2\int t^{-13/3}\, dt$

$$= -2 \frac{t^{-10/3}}{\left(\frac{-10}{3}\right)} = \frac{3}{5}(x^{-1/2} + 1)^{-10/3} + C$$

55. *(b)* Let $I = \int \dfrac{dx}{x^2 (x^n + 1)^{\frac{n-1}{n}}}$

$$= \int \frac{dx}{x^2 x^{n\frac{(n-1)}{n}} (1 + x^{-n})^{\frac{n-1}{n}}}$$

$$= \int \frac{dx}{x^{n+1} (1 + x^{-n})^{\frac{n-1}{n}}}$$

Put $1 + x^{-n} = t \Rightarrow x^{-(n+1)} dx = \dfrac{dt}{-n}$

$\Rightarrow \quad \dfrac{dx}{x^{n+1}} = \dfrac{-dt}{n}$

$\therefore \quad I = \int \dfrac{-dt}{nt^{\left(\frac{n-1}{n}\right)}} = -\dfrac{1}{n} \int t^{\left(\frac{1}{n} - 1\right)} dt$

$$= -\frac{1}{n}\frac{t^{1/n}}{\frac{1}{n}} + C = -t^{1/n} + C$$

$$= -(1+x^{-n})^{1/n} + C$$

Thus, $f(x) = 1 + x^{-n}$

56. *(d)* Here, A is a matrix of order 3 and $B = \text{adj}(A)$, $C = \text{adj}(\text{adj} A)$, $D = \text{adj}[\text{adj}(\text{adj} A)]$

Let $ABCD = E$, then

$|\text{adj}[\text{adj}\{\text{adj}(\text{adj} E)\}]| = |E|^{(3-1)^4} = |E|^{16}$

$|ABCD|^{16} = |A|^{16}|B|^{16}|C|^{16}|D|^{16}$

$= |A|^{16} \cdot (|A|^2)^{16} \cdot (|A|^4)^{16} \cdot |(A)^8|^{16}$

$= |A|^{16+32+64+128} = |A|^{240}$

$\Rightarrow \quad |A|^k = |A|^{240} \Rightarrow k = 240$

57. *(d)* We have, $\frac{dy}{dx} = \frac{x(2\log x + 1)}{\sin y + y\cos y}$

$\Rightarrow (\sin y + y\cos y)\, dy = x(2\log x + 1)\, dx$

$\Rightarrow -\cos y + \{(y\sin y) + \cos y\}$

$$= 2\left\{\frac{x^2}{2}\log x - \frac{1}{2}\int x\, dx\right\} + \frac{x^2}{2}$$

$\Rightarrow \quad y\sin y = x^2 \log x + C$

58. *(a)* For the points, where the line $\frac{x-2}{3} = \frac{y+1}{2} = \frac{z-1}{-1}$ intersects the curve $xy = c^2$, $z = 0$, we have

$\therefore \quad \frac{x-2}{3} = \frac{y+1}{2} = \frac{0-1}{-1}$

$\Rightarrow \quad \frac{x-2}{3} = 1$ and $\frac{y+1}{2} = 1$

$\Rightarrow \quad x = 5, y = 1$

On putting the values of x and y in $xy = c^2$, we get

$$c = \pm\sqrt{5}$$

59. *(b)* Since, $\mathbf{b}_1 = \mathbf{b} - \frac{\mathbf{b}\cdot\mathbf{a}}{|\mathbf{a}|^2}\mathbf{a}$, $\mathbf{b}_2 = \mathbf{b} + \frac{\mathbf{b}\cdot\mathbf{a}}{|\mathbf{a}|^2}\mathbf{a}$

and $\quad \mathbf{c}_1 = \mathbf{c} - \frac{\mathbf{c}\cdot\mathbf{a}}{|\mathbf{a}|^2}\mathbf{a} - \frac{\mathbf{c}\cdot\mathbf{b}}{|\mathbf{b}|^2}\mathbf{b}$,

$\mathbf{c}_2 = \mathbf{c} - \frac{\mathbf{c}\cdot\mathbf{a}}{|\mathbf{a}|^2}\mathbf{a} - \frac{\mathbf{c}\cdot\mathbf{b}_1}{|\mathbf{b}|^2}\mathbf{b}_1$,

$\mathbf{c}_3 = \mathbf{c} - \frac{\mathbf{c}\cdot\mathbf{a}}{|\mathbf{a}|^2}\mathbf{a} - \frac{\mathbf{a}\cdot\mathbf{b}_2}{|\mathbf{b}_2|^2}\mathbf{b}_2$, $\mathbf{c}_4 = \mathbf{a} - \frac{\mathbf{c}\cdot\mathbf{a}}{|\mathbf{a}|^2}\mathbf{a}$

which shows $\mathbf{a}\cdot\mathbf{b}_1 = 0 = \mathbf{a}\cdot\mathbf{c}_2 = \mathbf{b}_1\cdot\mathbf{c}_2$

So, $\{\mathbf{a}, \mathbf{b}_1, \mathbf{c}_2\}$ are mutually orhtogonal vectors.

60. *(a)* Given, $A = \begin{bmatrix} i & -i \\ -i & i \end{bmatrix}$ and $B = \begin{bmatrix} 1 & -1 \\ -1 & 1 \end{bmatrix}$

Now, $\quad A = i\begin{bmatrix} 1 & -1 \\ -1 & 1 \end{bmatrix} = iB$

$A^2 = (iB)^2 = i^2B^2 = -B^2$

$= -\begin{bmatrix} 1 & -1 \\ -1 & 1 \end{bmatrix}\begin{bmatrix} 1 & -1 \\ -1 & 1 \end{bmatrix} = -\begin{bmatrix} 2 & -2 \\ -2 & 2 \end{bmatrix} = -2B$

$\Rightarrow \quad A^2 = -2B$

$\Rightarrow \quad A^4 = (-2B)^2 = 4B^2 = 4(2B) = 8B$

$\therefore \quad A^8 = (8B)^2 = 64B^2 = 64(2B) = 128B = \lambda B$

$\Rightarrow \quad \lambda = 128$

61. *(a)* Let $A(0, a)$ on the Y-axis and opposite vertex $C(\alpha, \beta)$, then mid-point of AC is $P(1, 2)$.

$\Rightarrow \frac{\alpha + 0}{2} = 1$ and $\frac{a + \beta}{2} = 2$

$\Rightarrow \quad \alpha = 2$ and $a + \beta = 4$

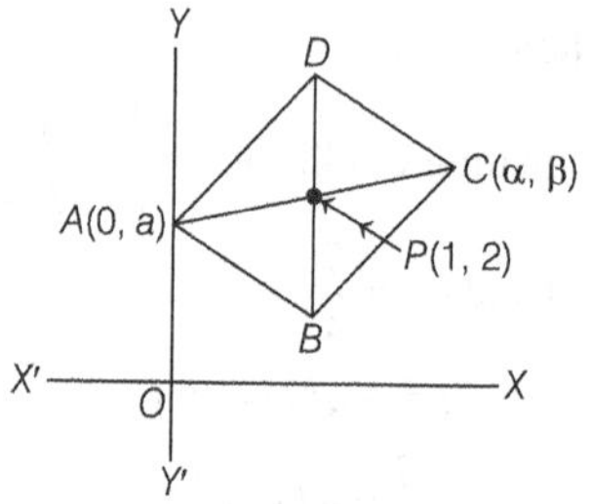

Sides of rhombus are parallel to the lines

$$x - y + 2 = 0 \text{ and } 7x - y + 3 = 0$$

Equation of lines parallel to diagonal is given by

$$\frac{x - y + 2}{\sqrt{2}} = \pm\frac{7x - y + 3}{5\sqrt{2}}$$

$\Rightarrow \quad 2x + 4y - 7 = 0$ and $12x - 6y + 13 = 0$

So, the slope of AC is $-\frac{1}{2}$ or 2.

$\therefore \quad$ Slope of AP = Slope of AC

$\Rightarrow \quad \frac{2-a}{1-0} = \frac{-1}{2}$ or 2

$\Rightarrow \quad 2 - a = \frac{-1}{2}$ or 2

$\Rightarrow \quad a = \frac{5}{2}$ or 0

$\therefore A\left(0, \frac{5}{2}\right)$

Ordinate of $A = \frac{5}{2} = \frac{p}{q}$

Hence, $p + q = 5 + 2 = 7$

62. *(a)* Let $f(x) \equiv ax^2 + bx + c$

$f(0) = c > 0$

$f(1) = a + b + c < 0$

$f(-1) = a - b + c < 0$

Hence, one root lies in $(-1, 0)$ and the other lies in $(0, 1)$.

$\therefore \quad [\alpha] = -1$ and $[\beta] = 0 \Rightarrow [\alpha] + [\beta] = -1$

Therefore, Statement I is true.

Statement II is true and it is a correct explanation of Statement I.

63. *(c)* Given sequence $\frac{1}{503}, \frac{4}{524}, \frac{9}{581}, \frac{16}{692}, \ldots$

General term is

$$T_n = \frac{n^2}{500 + 3n^3}$$

For largest term, $\frac{dT_n}{dn} = 0$

$$\Rightarrow \frac{n(1000 - 3n^3)}{(500 + 3n^3)^2} = 0$$

$$\Rightarrow n(1000 - 3n^3) = 0$$

$$\Rightarrow n = 0 \text{ or } 1000 - 3n^3 = 0$$

$$\Rightarrow n = \left(\frac{1000}{3}\right)^{1/3} \Rightarrow 6 < n < 7$$

$\Rightarrow$ 7th term is largest.

64. *(d)* Given, $(\sqrt{5} + 1)^{4n} - (\sqrt{5} - 1)^{4n}$

Using Binomial Expansion,

$$[(x + y)^n - (x - y)^n = 2[{}^nC_1 x^{n-1} y + {}^nC_3 x^{n-3} y^3 + {}^nC_5 x^{n-5} y^5 + \ldots]]$$

$$= 2[{}^{4n}C_1 (\sqrt{5})^{4n-1} + {}^{4n}C_3 (\sqrt{5})^{4n-3} + {}^{4n}C_5 (\sqrt{5})^{4n-5} + \ldots + {}^{4n}C_{4n-1} (\sqrt{5})]$$

$$= 2\sqrt{5}\,[{}^{4n}C_1 (5)^{2n-1} + {}^{4n}C_3 (5)^{2n-2} + \ldots + {}^{4n}C_{4n-1}]$$

This is an irrational number.

65. *(a)* Given, $[\sqrt{2}\cos x] + [\sin x] = -3$

$$\Rightarrow [\sqrt{2}\cos x] = -2 \text{ and } [\sin x] = -1$$

$$\Rightarrow -2 \le \sqrt{2}\cos x < -1 \text{ and } -1 \le \sin x < 0$$

$$\Rightarrow -\sqrt{2} \le \cos x < \frac{-1}{\sqrt{2}} \text{ and } -1 \le \sin x < 0$$

$$\Rightarrow -1 \le \cos x < -\frac{1}{\sqrt{2}} \text{ and } -1 \le \sin x < 0$$

$$\Rightarrow x \in \left(\frac{3\pi}{4}, \frac{5\pi}{4}\right) \text{ and } x \in (\pi, 2\pi)$$

$$\Rightarrow x \in \left(\frac{3\pi}{4}, \frac{5\pi}{4}\right) \cap (\pi, 2\pi) \Rightarrow x \in \left(\pi, \frac{5\pi}{4}\right)$$

66. *(d)* First 25 even natural numbers are 2, 4, 6, ... 50.

$$\text{Variance} = \frac{\Sigma x^2}{n} - (\bar{x})^2$$

$$= \frac{2^2 + 4^2 + 6^2 + \ldots + 50^2}{25} - \left(\frac{2 + 4 + 6 + \ldots + 50}{25}\right)^2$$

$$= \frac{22100}{25} - (26)^2 = 884 - 676 = 208$$

Time Saver Tip

$\because$ Variance of first n even natural numbers $= \frac{n^2 - 1}{3}$

$\therefore$ Variance of first 25 even natural numbers

$$= \frac{25^2 - 1}{3} = \frac{624}{3} = 208$$

67. *(b)* Since, ${}^nP_r = {}^nP_{r+1}$

$$\Rightarrow \frac{n!}{(n-r)!} = \frac{n!}{(n-r-1)!}$$

$$\Rightarrow \frac{n!}{(n-r)(n-r-1)!} = \frac{n!}{(n-r-1)!}$$

$$\Rightarrow n - r = 1 \quad \ldots \text{(i)}$$

and ${}^nC_r = {}^nC_{r-1}$

$$\Rightarrow \frac{n!}{r!(n-r)!} = \frac{n!}{(r-1)!(n-r+1)!}$$

$$\Rightarrow \frac{n!}{r(r-1)!(n-r)!} = \frac{n!}{(r-1)!(n-r+1)(n-r)!}$$

$$\Rightarrow n - r + 1 = r$$

$$\Rightarrow 1 + 1 = r \quad [\because n - r = 1]$$

$$\Rightarrow r = 2$$

68. *(b)* For given condition, we have three cases

I. All dice shows same number.

II. Number appearing on A_4 appears on any one of A_1, A_2, A_3.

III. Number appearing on A_4 appears on any two A_1, A_2, A_3.

Now, sample space $|S| = 6 \times 6 \times 6 \times 6 = 6^4$

For Case I

Select one number on A_4 which appears on all number of favourable cases $= {}^6C_1 \times 1 = 6$

For Case II

For A_4 number of ways ${}^6C_1$, for other two there are 5×5 ways.

Number of favourable ways $= {}^6C_1 \times {}^3C_1 \times 5 \times 5$

For Case III

For A_4 there are ${}^6C_1$ ways.

It appears on any two of A_1, A_2, A_3 is ${}^3C_2 \times 1^2$ ways for remaining one there are 5 ways.

Number of favourable ways $= {}^6C_1 \times {}^3C_2 \times 1^2 \times 5$

Required probability

$$= \frac{{}^6C_1 + {}^6C_1 \times {}^3C_1 \times 5^2 + {}^6C_1 \times {}^3C_2 \times 5}{6^4}$$

$$= \frac{6 + 6\times 75 + 6\times 15}{6^4}$$

$$= \frac{91}{216}$$

69. *(a)* Both the lines pass through origin. Line L_1 is parallel to the vector

$\mathbf{V}_1 = (\cos\theta + \sqrt{3}\,\hat{\mathbf{i}}) + (\sqrt{2}\sin\theta)\hat{\mathbf{j}} + (\cos\theta - \sqrt{3})\hat{\mathbf{k}}$

and L_2 is parallel to the vector

$\mathbf{V}_2 = a\hat{\mathbf{i}} + b\hat{\mathbf{j}} + c\hat{\mathbf{k}}$

$$\therefore \cos\alpha = \frac{\mathbf{V}_1 \cdot \mathbf{V}_2}{|\mathbf{V}_1|\,|\mathbf{V}_2|}$$

$$= \frac{a(\cos\theta + \sqrt{3}) + (b\sqrt{2})\sin\theta + c(\cos\theta - \sqrt{3})}{\sqrt{a^2+b^2+c^2}\sqrt{(\cos\theta+\sqrt{3})^2 + 2\sin^2\theta + (\cos\theta - \sqrt{3})^2}}$$

$$= \frac{(a+c)\cos\theta + b\sqrt{2}\sin\theta + (a-c)\sqrt{3}}{\sqrt{a^2+b^2+c^2}\sqrt{2+6}}$$

In order that $\cos\alpha$ is the independent of θ.

$$a + c = 0 \text{ and } b = 0$$

$$\therefore \cos\alpha = \frac{2a\sqrt{3}}{a\sqrt{2}\cdot 2\sqrt{2}} = \frac{\sqrt{3}}{2}$$

$$\alpha = \frac{\pi}{6}$$

70. *(b)* Given that $|\mathbf{a} + \mathbf{b}| = \sqrt{2}$ ...(i)

On squaring Eq. (i) on both sides, we get

$$|\mathbf{a} + \mathbf{b}|^2 = 2$$

$$|\mathbf{a}|^2 + |\mathbf{b}|^2 - 2\mathbf{a}\cdot\mathbf{b} = 2$$

$$1 + 1 + 2\cos\theta = 2$$

$$\cos\theta = 0$$

$$\theta = 90^\circ$$

$\mathbf{a}\times\mathbf{b}$ is perpendicular to plane containing $\mathbf{a}$ and $\mathbf{b}$.

$$\mathbf{c} = \mathbf{a} + 2\mathbf{b} + 3(\mathbf{a}\times\mathbf{b}) \quad \text{...(ii)}$$

On squaring both sides, we get

$$|\mathbf{c}|^2 = |\mathbf{a}|^2 + 4|\mathbf{b}|^2 + 9|\mathbf{a}\times\mathbf{b}|^2 + 4(\mathbf{a}\cdot\mathbf{b}) + 6\mathbf{a}\cdot(\mathbf{a}\times\mathbf{b}) + |12\mathbf{b}\cdot(\mathbf{a}\times\mathbf{b})|$$

$$= 1 + 4 + 9\sin^2\theta + 4\cos\theta + 0 + 0$$

$$= 1 + 4 + 9(1)^2 + 4(0)$$

$$\Rightarrow \quad |\mathbf{c}|^2 = 14$$

$$\Rightarrow \quad 2|\mathbf{c}| = 2\sqrt{14} = K \quad \text{(given)}$$

So, $\sqrt{14}\,K = 28$

71. *(25)* $\cos^{-1}x + \cos^{-1}2x + \cos^{-1}3x = \pi$

or $\quad \cos^{-1}2x + \cos^{-1}3x = \pi - \cos^{-1}x$

$$= \cos^{-1}(-x)$$

$$\Rightarrow \cos^{-1}[(2x)(3x) - \sqrt{1-4x^2}\sqrt{1-9x^2}] = \cos^{-1}(-x)$$

$$\Rightarrow \quad 6x^2 - \sqrt{1-4x^2}\sqrt{1-9x^2} = -x$$

$$\Rightarrow \quad (6x^2 + x)^2 = (1-4x^2)(1-9x^2)$$

$$\Rightarrow \quad x^2 + 12x^3 = 1 - 13x^2$$

or $\quad 12x^3 + 14x^2 - 1 = 0$ ...(i)

On comparing Eqs. (i) with $ax^3 + bx^2 + c = 0$, we get

$$a = 12,\ b = 14 \text{ and } c = -1$$

So, $a + b + c = 12 + 14 - 1 = 25$

72. *(1)* The area bounded by $y = f(x)$ and $y = x$ between the lines $x = 1$ and $x = t$ is $\int_1^t [f(x) - x]\,dx$.

But it is equal to $[t + \sqrt{1+t^2} - (1+\sqrt{2})]$

$$\therefore \int_1^t [f(x) - x]\,dx = (t + \sqrt{1+t^2}) - (1+\sqrt{2})$$

On differentiating both sides w.r.t. t, we get

$$f(t) - t = 1 + \frac{t}{\sqrt{1+t^2}}$$

$$\Rightarrow \quad f(t) = 1 + t + \frac{t}{\sqrt{1+t^2}}$$

$$\therefore \quad f(0) = 1 + 0 + \frac{0}{\sqrt{1+0}} \Rightarrow f(0) = 1$$

73. *(2)* Let point on parabola be $(at^2, 2at)$.

It is also mid-point of chord.

$\therefore$ Equation of chord : $T = S_1$

$$\frac{at^2\cdot x}{2a^2} + \frac{2aty}{a^2} - 1 = \frac{a^2t^4}{2a^2} + \frac{4a^2t^2}{a^2} - 1$$

$$xt + 4y = at^3 + 8at$$

It passes $\left(11a, \dfrac{-a^2}{4}\right)$.

$$11at - a^2 = at^3 + 8at$$

$$t^3 - 3t + a = 0 \quad \text{...(i)}$$

From Eq. (i), we must have three values of t.

Let $f(t) = t^3 - 3t + a$

$$f'(t) = 3t^2 - 3 = 3(t^2 - 1)$$

$$f(1)\,f(-1) < 0$$

$$(a-2)(a+2) < 0$$

$$a \in (-2, 2) - \{0\} \quad [\because a \text{ cannot be zero}]$$

Hence, $a = -1, 1$ two values.

74. *(10)* First term $(a) = \dfrac{K-1}{K!}$

Common ratio $(r) = \dfrac{1}{K}$

Sum of infinite geometric series $= \dfrac{a}{1-r}$

$$\Rightarrow \quad S_K = \frac{\dfrac{K-1}{K!}}{1-\dfrac{1}{K}} = \frac{1}{(K-1)!}$$

Now, $(K^2-3K+1)S_K = \dfrac{(K^2-2K+1)-K}{(K-1)!}$

$$= \frac{(K-1)^2-K}{(K-1)!} = \frac{K-1}{(K-2)!} - \frac{K}{(K-1)!}$$

$$\therefore \sum_{K=2}^{100} |(K^2-3K+1)S_K| = \sum_{K=2}^{100} \left|\frac{K-1}{(K-2)!} - \frac{K}{(K-1)!}\right|$$

$$\Rightarrow 3 - \frac{\lambda^2}{99!} = \left|\frac{1}{0!} - \frac{2}{1!}\right| + \left|\frac{2}{1!} - \frac{3}{2!}\right| + \left|\frac{3}{2!} - \frac{4}{3!}\right| + \ldots + \left|\frac{99}{98!} - \frac{100}{99!}\right|$$

$$= \frac{2}{1!} - \frac{1}{0!} + \frac{2}{1!} - \frac{3}{2!} + \frac{3}{2!} - \frac{4}{3!} + \ldots + \frac{99}{98!} - \frac{100}{99!}$$

$$= 1 + \frac{2}{1!} - \frac{100}{99!}$$

$$\Rightarrow 3 - \frac{\lambda^2}{99!} = 3 - \frac{100}{99!} \Rightarrow \lambda^2 = 100 \Rightarrow \lambda = 10$$

75. *(4)* Given, $y^3 = 27x$

On differentiating w.r.t. x, we get

$$3y^2 \frac{dy}{dx} = 27$$

$$\Rightarrow \quad \frac{dy}{dx} = \frac{9}{y^2} \quad \ldots\text{(i)}$$

Now, ordinate changes as a faster rate than the abscissa, then

$$\left|\frac{dy}{dt}\right| > \left|\frac{dx}{dt}\right|$$

$$\Rightarrow \quad \left|\frac{\dfrac{dy}{dt}}{\dfrac{dx}{dt}}\right| > 1 \Rightarrow \left|\frac{dy}{dx}\right| > 1$$

$$\Rightarrow \quad \left|\frac{9}{y^2}\right| > 1 \Rightarrow \frac{9}{y^2} > 1$$

$$\Rightarrow \quad 1 - \frac{9}{y^2} < 0$$

$$\Rightarrow \quad \frac{(y+3)(y-3)}{y^2} < 0$$

$$\therefore \quad y \in (-3, 3) - \{0\}$$

Hence, the total possible number of integral values of y for which the ordinates changes as faster rate than abscissa $= 4$.

JEE Main

TEST DRIVE

SECTION II

Mastery Builders

This section contains **10 Practice Sets** designed to match the exact difficulty level of JEE Main. These practice papers will simulate the real exam experience, preparing you to tackle any question that comes your way.

SECTION II

Mastery Builders

PRACTICE SET 4

Total Marks : 300	Total Time : 3 hrs	No. of Questions : 75

Instructions

1. There are **three** parts in the questions paper consisting of **Physics**, **Chemistry** and **Mathematics** having 25 questions in each part of equal weightage. Each part (subject) has two sections.
2. **Section A** This section contains 20 multiple choice questions which have only one correct answer. Each question carries **4 marks** for correct answer and **– 1 mark** for wrong answer.
3. **Section B** This section contains 5 questions. The answer to each of the questions is a numerical value. Each question carries **4 marks** for correct answer and **– 1 mark** for wrong answer. For Section B, the answer should be rounded off to the nearest integer.

PHYSICS

Section A : Objective Type Questions

1. In a Wheatstone bridge, $P = 90\,\Omega, Q = 110\,\Omega$, $R = 40\,\Omega$ and $S = 60\Omega$ and a cell of emf 4V. Then, the potential difference between the diagonal along which a galvanometer is connected, is

(a) -0.2V (b) $+\ 0.2$V
(c) -1V (d) $+1$V

2. A wall is made of equally thick layers A and B of different materials. Thermal conductivity of A is twice that of B. In the steady state, the temperature difference across the wall is 72°C. The temperature difference across the layer A is

(a) 12° C (b) 36°C
(c) 24°C (d) 0° C

3. Identify the graph (s) which correctly represents an isotherm at temperatures T_1 and $T_2 (T_2 > T_1)$

(a)
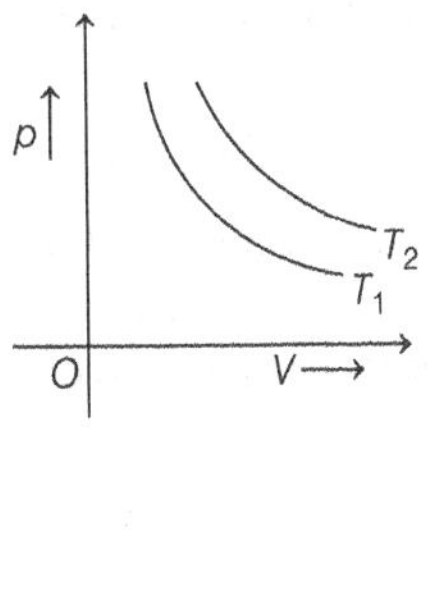

(b)
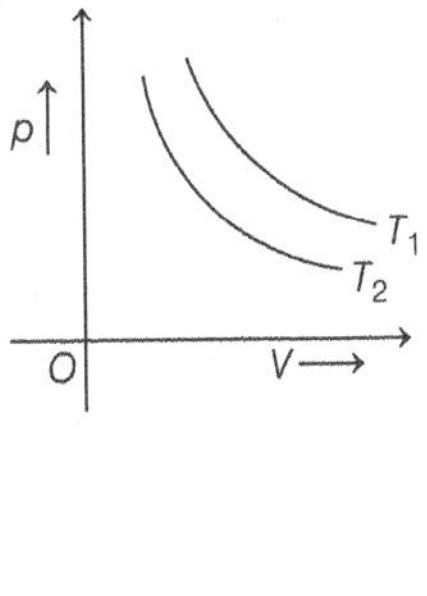

(c)
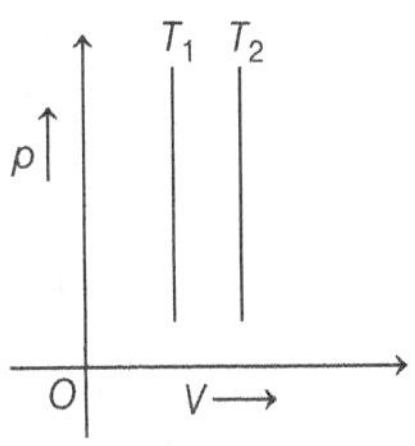

(d)
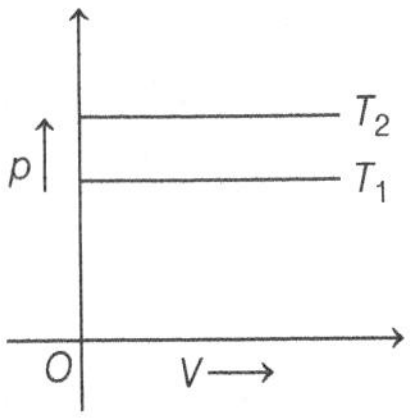

4. Binding energy per nucleon of helium and deuterium nuclei are 7 MeV and 1.1 MeV, respectively. When two deuterium nuclei undergo fusion to form helium nucleus, energy released in the process is

(a) 23.6 J (b) 37.76×10^{-13} MeV
(c) 23.6 eV (d) 37.76×10^{-13} J

5. A beam consisting of 4 different types of ions named α, β, γ and δ enters a region of magnetic field as shown in the figure below.

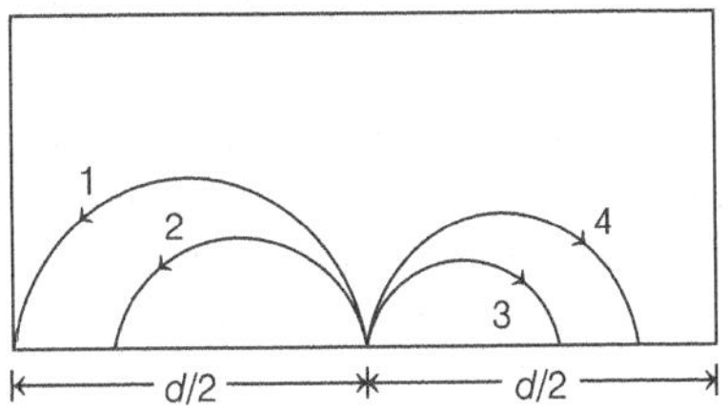

Magnetic field is perpendicular to the velocity of beam. All ions in the beam travels with same speed. Magnetic field is directed out of the plane of the figure and velocity of particles are in + y - direction. Mass and charge of ions are given as

Ion	Mass	Charge
α	$2m$	e
β	$4m$	e^-
γ	$2m$	e^-
δ	$3m$	e

Now, match the trajectories with its ions.

	Column I (Ion)		Column II (Path)
i.	α	p.	1
ii.	β	q.	2
iii.	γ	r.	3
iv.	δ	s.	4

Codes

	i	ii	iii	iv		i	ii	iii	iv
(a)	p	r	s	q	(b)	q	s	r	p
(c)	r	p	q	s	(d)	s	q	p	r

6. 5.6 L of helium gas at STP is compressed adiabatically to 0.7 L. If the initial temperature of the gas be T_1, the work done in the process is found to be $-\frac{m}{n}RT_1$, then the value of $m-n$ is

(a) 1 (b) 1.5 (c) 2 (d) 2.5

7. When an AC voltage of variable frequency is applied across the L-C-R circuit, the current in the circuit was found to be same at 8 MHz and 18 MHz. Then, the frequency at which current will be maximum is

(a) 24 MHz (b) 13 MHz
(c) 10 MHz (d) 12 MHz

8. Two charges are placed at a certain distance in air experience a force of F Newton. If both charges were placed in a liquid of dielectric constant K, at some distance, then the force between them will be

(a) FK (b) $F\sqrt{K}$ (c) $\frac{F}{K}$ (d) $\sqrt{FK}$

9. A cubical block of mass m and edge a slides down a rough inclined plane of inclination θ with a uniform velocity. The torque of the normal force on the block about its centre has a magnitude

(a) zero (b) mga (c) $mga\sin\theta$ (d) $\frac{mga\sin\theta}{2}$

10. Two particles execute SHM of the same amplitude and frequency along the same straight line. If they pass one another when going in opposite directions, each time their displacement is half of their amplitude, the phase difference between them is

(a) $\frac{\pi}{6}$ (b) $\frac{\pi}{4}$ (c) $\frac{\pi}{3}$ (d) $\frac{2\pi}{3}$

11. Heat is supplied to a certain homogeneous sample of matter, at a uniform rate. Its temperature is plotted against time, as shown. Which of the following conclusions can be drawn?

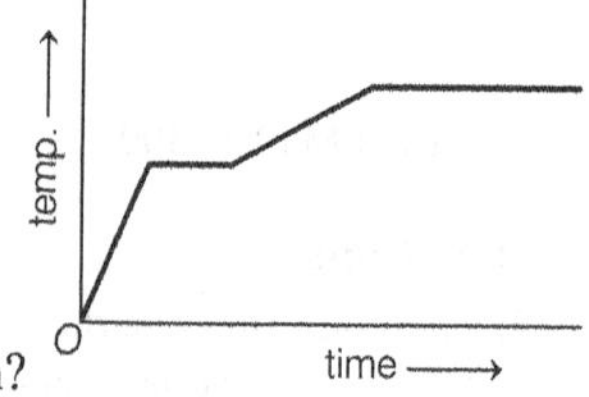

(a) Its specific heat capacity is greater in the solid state than in the liquid state
(b) Its specific heat capacity is greater in the liquid state than in the solid state
(c) Its latent heat of vaporisation is equal to its latent heat of fusion
(d) Its latent heat of vaporisation is smaller than its latent heat of fusion

12. The figure given below shows two capacitors $C_1 = 4\mu F$ and $C_2 = 8\mu F$ connected to a 12 V battery and two switches S_1 and S_2. Switch S_2 is initially kept open and S_1 is kept closed until the capacitance C_1 is fully charged. Switch S_1 is now opened and S_2 is closed until electrostatic conditions are restored. In this process, the percentage loss of energy stored in C_1 is

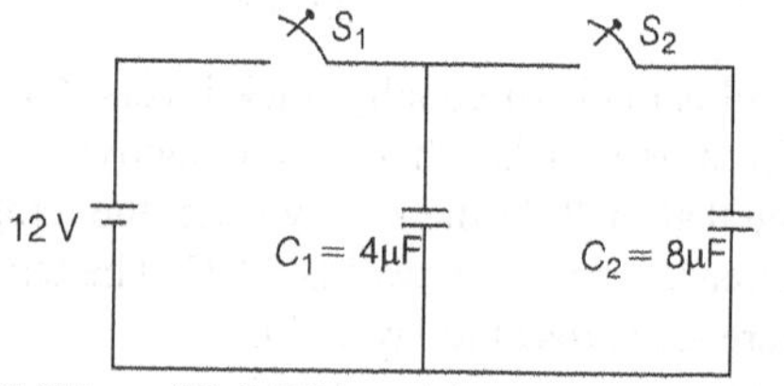

(a) 88.9% (b) 76.5% (c) 63.4% (d) 55.0%

13. The maximum height attained by a projectile is increased by 10% by increasing its speed of projection without changing the angle of projection. The percentage increase in the horizontal range will be

(a) 10% (b) 20% (c) 15% (d) 25%

14. If proton and α-particle are accelerated by same potential V, having kinetic energy 1 eV and 2 eV respectively, then the ratio of their wavelengths $\lambda_p : \lambda_\alpha$ would be

(a) $2\sqrt{2}:1$ (b) $1:2\sqrt{2}$ (c) $2:1$ (d) $1:2$

15. Read the following statements

I. If the stone tied at the end of a string is whirled and the string breaks, the stone flies off tangentially.

II. The necessary centripetal force required to keep the stone moving in a circle is provided by tension in the string.

III. Centripetal force is a real force.

Choose the correct statements

(a) I only (b) II and III only
(c) I and III only (d) I, II and III

16. In a reverse biased *p-n* junction, when the applied bias voltage is equal to the breakdown voltage, then

(a) current remains constant, while voltage increases sharply
(b) voltage remains constant, while current increases sharply
(c) current and voltage increase
(d) current and voltage decrease

17. The dimensional formula of Young's modulus is $[ML^{-1}T^{-2}]$. Find its new dimensional formula in the form of velocity (v), acceleration (A) and Force (F), if these all are taken as fundamental quantities instead of mass (M), length (L) and time (T).

(a) $[F^2A^2v^{-2}]$ (b) $[FA^2v^{-4}]$
(c) $[FA^3v^{-2}]$ (d) $[F^2Av^{-5}]$

18. **Assertion** If the velocity of a revolving satellite is increased to $\sqrt{2}$ times, then it escapes out from the earth.

Reason Escape velocity does not depend on the direction of projection.

In the light of above statements, choose the most appropriate answer from the options given below.

(a) Both Assertion and Reason are true and Reason is the correct explanation of Assertion.
(b) Both Assertion and Reason are correct but Reason is not the correct explanation of Assertion.
(c) Assertion is true and Reason is false.
(d) Assertion is false and Reason is true.

19. Diameter of a cylindrical thin wire measured by a screw gauge of least count 0.001 cm is 0.05 cm. The length measured by a scale of least count 0.1 cm is 110.0 cm. When a weight of 50 N is suspended from the wire, the extension is measured to be 0.125 cm by a micrometer of least count 0.001 cm. The maximum error in the measurement of Young's modulus of the material of the wire is

(a) $2.05\times10^9\,Nm^{-2}$ (b) $1.09\times10^{10}\,Nm^{-2}$
(c) $3.07\times10^9\,Nm^{-2}$ (d) $7.03\times10^{10}\,Nm^{-2}$

20. The two statements are given below.

Statement I Electromagnetic waves consist of oscillating electric and magnetic fields, that are parallel to each other and perpendicular to the direction of propagation of EM waves.

Statement II Electromagnetic waves are transverse in nature and travel through vacuum with same speed of 3×10^8 m/s.

In the light of above statements, choose the most appropriate answer from the options given below.

(a) Both Statement I and Statement II are true.
(b) Both Statement I and Statement II are false
(c) Statement I is true but Statement II is false.
(d) Statement I is false but Statement II is true.

Section B : Numerical Value Type Questions

21. A solid sphere of radius *r* made of a material of bulk modulus *K* is surrounded by a liquid in a cylindrical container. A massless piston of area *a* floats on the surface of the liquid. When a mass *m* is placed on the piston to compress the liquid, the fractional change in the radius of the sphere $(\Delta r/r)$ is found to be $\frac{mg}{xKa}$, then the value of *x* is ....

22. A cell of constant emf first connected to a resistance $18\,\Omega$ and then connected to a resistance $8\,\Omega$. If the power delivered in both the cases is same, then the internal resistance (in Ω) of cell will be ....

23. A block *A* is moving with a speed 10 m/s towards an another identical block, *B*, which was initially at rest and having mass 4 kg, then the velocity of block *B* after the collision will be ...... m/s.
(Assume collision is perfectly inelastic)

24. A juggler throws balls into air. He throws one whenever previous one is at its highest point. If he throws *n* balls per second, then the height to which each ball will rise is found to be $\frac{g}{2n^p}$, then the value of *p* will be ....

25. A parallel sided block of glass of refractive index 1.5, which is 36 mm thick rests on the bottom of a tank, filled with water (refractive index = 4/3). The difference between the apparent depth of bottom at *A* and *B*, when seen from vertically above is found to be *x* mm, then the value of *x* will be ....

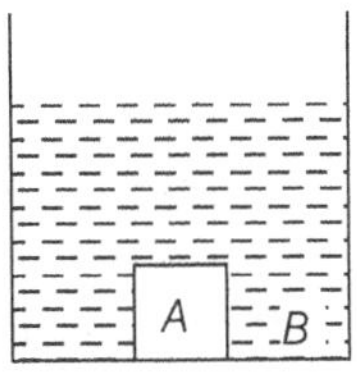

CHEMISTRY

Section A : Objective Type Questions

26. Glucose on treatment with conc. HCl produces

(a) formic acid (b) maltose + fructose
(c) fructose (d) levulinic acid

27. For the chemical equilibrium,

$$CaCO_3(s) \rightleftharpoons CaO(s) + CO_2(g)$$

ΔH_r° can be determined from which one of the following plots?

(a) 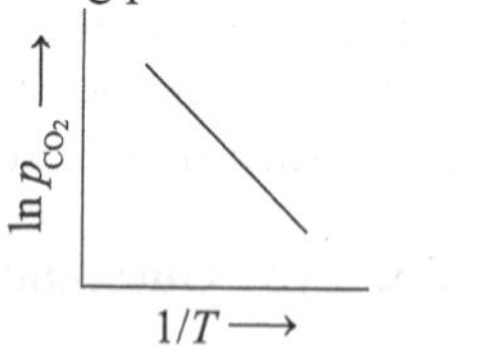

(b) 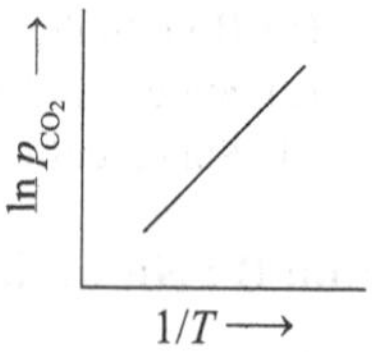

(c) 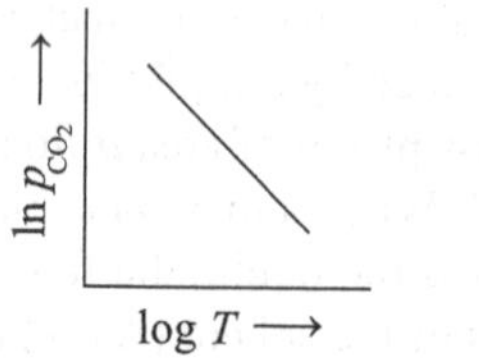

(d) 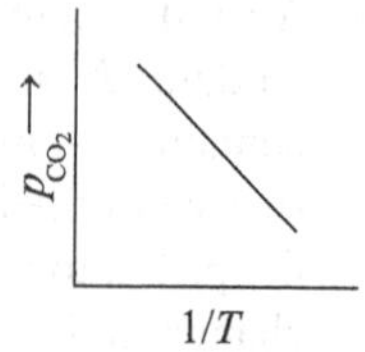

28. Among the following compounds, the total number of compounds which give positive Lassaigne's test for nitrogen is

(i) $H_2N—CH_2—COOH$
(ii) $NH_2—NH_2$
(iii) PhN_2Cl
(iv) $NH_2—C(=O)—NH_2$
(v) NH_2OH
(vi) $PhNHC(=O)—Ph$
(vii) $C_6H_5NHNH_2$

(a) 4 (b) 5 (c) 3 (d) 6

29. The correct order of ionic radii of Ce, La, Pm and Yb in +3 oxidation state is

(a) $Yb^{3+} < Pm^{3+} < Ce^{3+} < La^{3+}$
(b) $La^{3+} < Pm^{3+} < Ce^{3+} < Yb^{3+}$
(c) $La^{3+} < Ce^{3+} < Pm^{3+} < Yb^{3+}$
(d) $Yb^{3+} < Ce^{3+} < Pm^{3+} < La^{3+}$

30. An organic compound gave positive iodoform and Tollen's test. The compound is

(a) CH_3CH_2OH (b) $CH_3CH_2CH_2CHO$
(c) CH_3CHO (d) CH_3COCH_3

31. *P* and *Q* are isomers of a dicarboxylic acid $C_4H_4O_4$. Both decolourise Br_2/H_2O. On heating *P* forms a cyclic anhydride. Upon treatment with dilute alkaline $KMnO_4$, *P* as well as *Q* could produce one or more than one form *S*, *T* and *U*.

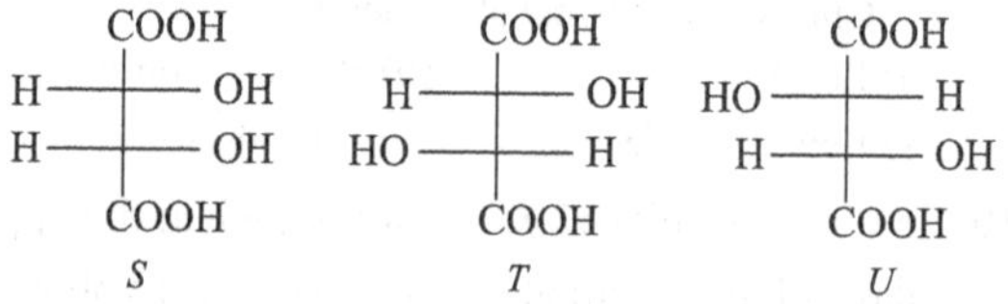

(a) Optically active *S* and optically active pair (*T*, *U*).
(b) Optically inactive *S* and optically active pair (*T*, *U*).
(c) Optically active pair (*T*, *U*) and optically active *S*.
(d) Optically inactive pair (*T*, *U*) and optically inactive *S*.

32. Consider the following series of reactions,

$(A) \xrightarrow{KOH} (B)$ (gas turns red litmus blue) + (C)

$(C) \xrightarrow{Zn + KOH} (B)$ (gas)

$(A) \xrightarrow{Heat}$ Gas (does not support combustion)

Identify *A*, *B* and *C*.

(a) $A = NH_4NO_3, B = N_2, C = KNO_3$
(b) $A = NH_4NO_2, B = NH_3, C = KNO_2$
(c) $A = (NH_4)_2SO_4, B = NH_3, C = KNO_3$
(d) $A = NH_4Cl, B = N_2, C = KNO_3$

33. An organic compound (*A*) having molecular formula, $C_9H_{10}O$ is supplied to a student to determine the structural formula of compound. He observed that compound forms an orange red precipitate with 2,4-DNP reagent and yellow precipitate on heating with I_2 / NaOH. It neither reduces Tollen's reagent or Fehling solution nor it decolourises bromine water solution. On drastic oxidation with CrO_3, it produces a carboxylic acid having molecular formula $C_7H_6O_2$. Identify the correct compound *A*.

(a) 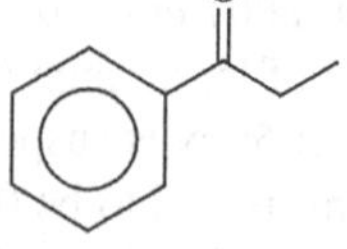

(b) 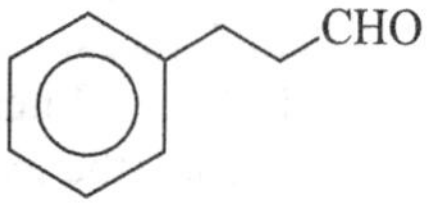

(c) 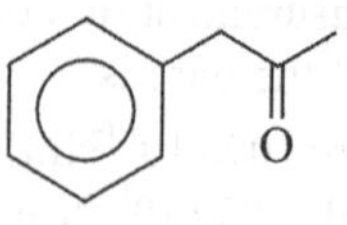

(d) 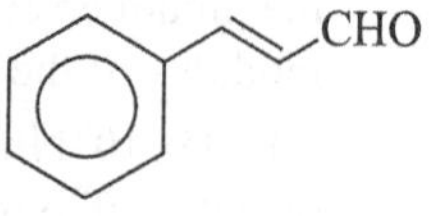

34. When an inorganic compound X having 3 centre-2-electron as well as $2C—2e^-$ bonds react with NH_3 gas at a certain temperature, gives a compound Y, isostructural with benzene. Compound X with ammonia at a high temperature produces a substance Z, then

(a) X is BH_3, Y is $B_2N_2H_3$, Z is inorganic benzene.
(b) X is B_2H_6, Y is $B_3N_3H_6$, Z is inorganic graphite.
(c) X is borax, Y is B_2O_3, Z is inorganic benzene.
(d) X is inorganic benzene, Y is diborane, Z is borax

35. Consider the following sequence of reactions,

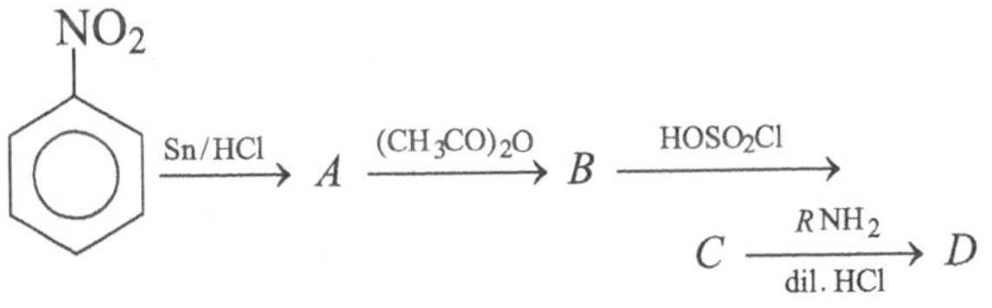

The correct structure of product (D) is

(a) 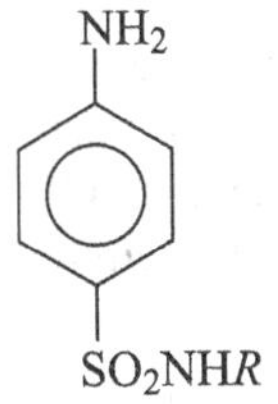

(b) 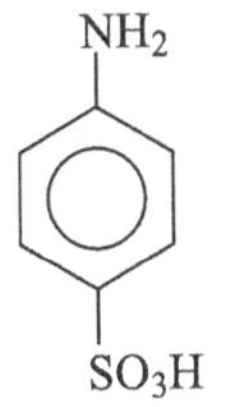

(c) 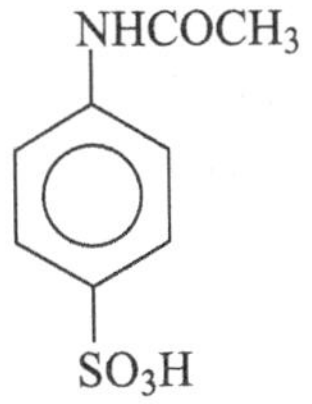

(d) 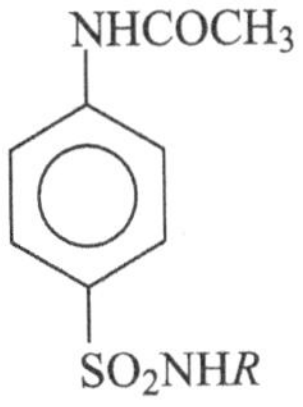

36. Match List-I with List-II.

List-I		List-I	
A.	BrF_5	I.	bent
B.	$[CrF_6]^{3-}$	II.	square pyramidal
C.	O_3	III.	trigonal bipyramidal
D.	PCl_5	IV.	octahedral

Choose the correct answer from the options given below.

(a) A-II, B-IV, C-I, D-III (b) A-III, B-I, C-II, D-IV
(c) A-II, B-IV, C-III, D-I (d) A-I, B-III, C-II, D-IV

37. Consider the following reactions

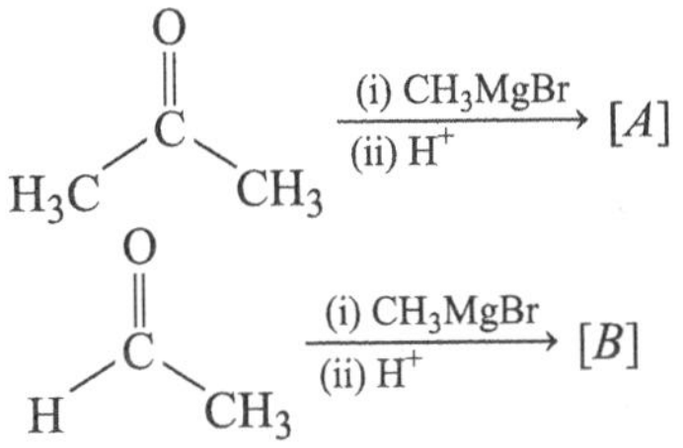

The reagent used to differentiate between A and B is

(a) $FeCl_3$ (b) active metal Na
(c) Lucas reagent (d) Ninhydrin Test

38. The coordination number of Ni^{2+} is 4.

$NiCl_2$ + KCN (excess) ⟶ A (Cyano complex)
$NiCl_2$ + Conc. HCl (excess) ⟶ B (Chloro complex)

The IUPAC name of A and B are

(a) potassiumtetracyanonickelate(II), potassiumtetrachloronickelate(II)
(b) tetracyanopotassiumnickelate (II), tetrachloropotasium-nickelate(II)
(c) tetracyanonickel(II), tetrachloronickel(II)
(d) potassiumtetracyanonickel(II), potassium tetrachloronickel(II)

39. Given below are two statements.

Statement I The graph of molar conductivity for KI increases steeply with dilution.

Statement II The graph of molar conductivity for carbonic acid increases slowly with dilution.

In the light of the above statements, choose the correct answer from the options given below.

(a) Both Statement I and Statement II are correct.
(b) Both Statement I and Statement II are incorrect.
(c) Statement I is correct and Statement II is incorrect.
(d) Statement I is incorrect and Statement II is correct.

40. Given below are two statements one is labelled as Assertion (A) and other is labelled as Reason (R).

Assertion (A) Following structure is acceptable.

$$:N\equiv N-\ddot{\underset{\cdot\cdot}{O}}:$$

Reason (R) Octet rule is followed for each atom, central N has positive charge and oxygen has negative charge.

In the light of the above statements. Choose the correct answer from the options given below.

(a) Both (A) and (R) are true, but (R) is the correct explanation of (A).

(b) Both (A) and (R) are true but (R) is not the correct explanation of (A).
(c) (A) is true but (R) is false.
(d) (A) is false but (R) is true.

41. The IUPAC name of $[Fe(O_2)(CN)_4Cl]^{4-}$ is
(a) chlorotetracyanodioxoferrate (II) ion
(b) chlorotetracyanoperoxoferrate (II) ion
(c) chlorotetracyanosuperoxoferrate (II) ion
(d) chlorotetracyanooxoferrate (II) ion

42. The chlorobenzene is generally obtained from a corresponding diazonium salt by reacting it with
(a) Cu_2Cl_2 (b) $CuSO_4$
(c) Cu (d) $[Cu(NH_3)_4]^{2+}$

43. Consider the following compounds.

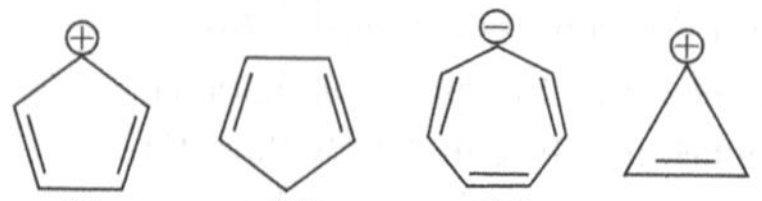

(i) (ii) (iii) (iv)

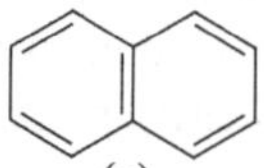

(v)

The number of compounds which obey Huckel's rule is
(a) 2 (b) 1 (c) 3 (d) 4

44. When N_2 is converted into N_2^+ and O_2 is converted into O_2^+, then the bond strength, respectively
(a) increases, increases (b) decreases, decreases
(c) increases, decreases (d) decreases, increases

45. Match the List-I with List-II.

List-I	List-II
A. $[B_4O_7]^{2-} \longrightarrow [B(OH)_3]$	I. Heat
B. $[AlO_2]^- \longrightarrow Al(OH)_3$	II. Hydrolysis
C. $[SiO_4]^{4-} \longrightarrow [Si_2O_7]^{6-}$	III. Acidification
D. $Bi^{3+} \longrightarrow (BiO)^+$	IV. Dilution by water

Choose the correct answer from the options given below.

	A	B	C	D
(a)	I	III	IV	II
(b)	IV	II	III	I
(c)	III	IV	I	II
(d)	II	IV	I	III

Section B : Numerical Value Type Questions

46. The number of structural isomers for C_6H_{14} is ....

47. 1 mole of carbon dioxide gas at 300 K is expanded under adiabatic condition such that its volume becomes 27 times. The work done during the process is + ... cal
(Given, $\gamma = 1.33$ and $C_V = 6$ cal mol^{-1} for CO_2)

48. Photoelectrons are liberated by ultraviolet light of wavelength 3000 Å from a metallic surface for which the photoelectric threshold is 4000 Å. The de-Broglie wavelength of electron emitted with maximum kinetic energy is ...... 10^{-9} m.
[Nearest integer] [Given: $h = 6.626 \times 10^{-34}$ Js]

49. Reaction, $A + B \longrightarrow C + D$, follows rate law; $R = k[A]^{1/2}[B]^{1/2}$ starting with 1 M of each A and B. The time taken for concentration of A become 0.1 M will be ....s.
[Given, $k = 4.606 \times 10^{-4}\ s^{-1}$]

50. The standard reduction potentials are 1.51 V for MnO_4^- / Mn^{2+}, 1.36 V for Cl_2/Cl^-, 1.07 V for Br_2/Br^-, and 0.54 V for I_2/I^- at 298 K.
At pH = 3, the number of species expected to oxidise by permanganate is/are ....

$$\left(\frac{RT}{F} = 0.059\text{ V}\right)$$

MATHEMATICS

Section A : Objective Type Questions

51. If $\int \frac{2\cos x - \sin x + \lambda}{\cos x + \sin x - 2} dx$

$= A \ln|\cos x + \sin x - 2| + Bx + C,$

then ordered triplet A, B and λ is

(a) $\left(\frac{1}{2}, \frac{3}{2}, -1\right)$ (b) $\left(\frac{3}{2}, \frac{1}{2}, -1\right)$

(c) $\left(\frac{1}{2}, -1, -\frac{3}{2}\right)$ (d) $\left(\frac{3}{2}, -1, \frac{1}{2}\right)$

52. The area bounded by the curve $y = \log x$, $y = \log|x|$, $y = |\log x|$ and $y = |\log|x||$ is

(a) 4 sq units (b) 6 sq units

(c) 10 sq units (d) None of these

53. The value of (a, b), so that the function

$$f(x) = \begin{cases} x + a\sqrt{2}\sin x & , \quad 0 \le x < \frac{\pi}{4} \\ 2x\cot x + b & , \quad \frac{\pi}{4} \le x \le \frac{\pi}{2} \\ a\cos 2x - b\sin x & , \quad \frac{\pi}{2} < x \le \pi \end{cases}$$

is continuous for $x \in [0, \pi]$

(a) $\left(\frac{\pi}{6}, \frac{\pi}{12}\right)$ (b) $\left(\frac{\pi}{6}, -\frac{\pi}{12}\right)$

(c) $\left(-\frac{\pi}{4}, \frac{\pi}{12}\right)$ (d) $\left(\frac{\pi}{4}, \frac{\pi}{6}\right)$

54. The value of k for which

$$\cos^3 2\theta + 3\cos 2\theta = k(\cos^6\theta - \sin^6\theta)$$

(a) 1 (b) 0

(c) 2 (d) 4

55. All the three vertices of an equilateral triangle lie on the parabola $y = x^2$ and one of its sides has a slope of 2. Then, sum of the x-coordinates of three vertices is

(a) $\frac{3}{11}$ (b) $\frac{2}{7}$

(c) $\frac{13}{17}$ (d) $\frac{4}{7}$

56. If $f(x) = \min\{2\sin x, 1 - \cos x, 1\}$, then $\int_0^\pi f(x)\,dx$ is equal to

(a) $\sqrt{3} - 1 + \frac{5\pi}{6}$ (b) $\sqrt{3} - 1 + \frac{2\pi}{3}$

(c) $1 - \sqrt{3} + \frac{2\pi}{3}$ (d) $1 - \sqrt{3} + \frac{5\pi}{6}$

57. Sum of the given series,

$$\cot^{-1}\left(1^2 + \frac{3}{4}\right) + \cot^{-1}\left(2^2 + \frac{3}{4}\right) + \cot^{-1}\left(3^2 + \frac{3}{4}\right) + \ldots \infty \text{ is}$$

(a) $\frac{\pi}{4}$ (b) $\tan^{-1} 2$

(c) $\tan^{-1} 3$ (d) None of these

58. Let $P(\alpha, \beta)$ lies in the first quadrant. If the two circles passes through P touches the coordinate axes cut each other orthogonally, then

(a) $\alpha^2 + \alpha\beta + \beta^2 = 0$ (b) $\alpha^2 - 4\alpha\beta - \beta^2 = 0$

(c) $\alpha^2 - 4\alpha\beta + \beta^2 = 0$ (d) $\alpha^2 + 4\alpha\beta + \beta^2 = 0$

59. If the point $P(z)$, $Q(-z)$ and $R(1 - z)$ are the vertices of an equilateral ΔABC, then Re (z) is equal to

(a) $\frac{1}{2}$ (b) 2 (c) 1 (d) $\frac{1}{4}$

60. Let $f_1 = \sum_{i=1}^{10} i(i-1)\,{}^{10}C_i$,

$$f_2 = \sum_{i=1}^{10} i\,{}^{10}C_i \text{ and } f_3 = \sum_{i=1}^{10} i^2\,{}^{10}C_i$$

Statement I $f_3 = 55 \times 2^9$

Statement II $f_1 = 90 \times 2^8$ and $f_2 = 10 \times 2^8$,

which of the following is correct?

(a) Statement I is false and Statement II is true.

(b) Statement I is true and Statement II is false.

(c) Statement I and Statement II both are true.

(d) Statement I and Statement II both are false.

61. Let $f(x) = \frac{x-3}{x+1}, x \ne -1$, then $f^{2100}(2100)$ is equal to, where $f^n(x) = (f_o f_o \ldots\ldots o f)(x)$ (n times) is

(a) 2100 (b) 4200

(c) 1050 (d) 1

62. The radius of the base of a cone is increasing at the rate of 3 cm/min and the altitude is decreasing at the rate of 4 cm/min. The rate of change of lateral surface area when the radius is 7 cm and altitude is 24 cm is

(a) 50 cm^2/min (b) 54 π cm^2/min

(c) 62π cm^2/min (d) 66 π cm^2/min

63. If p, q and r are three complex numbers such that $p^2+q^2+r^2=0$ and

$$\Delta = \begin{vmatrix} q^2+r^2 & pq & pr \\ pq & r^2+p^2 & qr \\ pr & qr & p^2+q^2 \end{vmatrix} = \left(\frac{K}{2}\right)^2 p^2q^2r^2,$$

then K is equal to

(a) ± 1 (b) ± 2
(c) ± 4 (d) ± 6

64. If t is any fixed real number such that $\frac{t-p}{ap}=\frac{t-q}{bq}=\frac{t-r}{cr}$, if a, b and c are in AP, then p, q and r are in

(a) arithmetic progression.
(b) geometric progression.
(c) harmonic progression.
(d) Both (a) and (b)

65. If ${}^mC_{p-1}=(\lambda^2-8)\,{}^{m+1}C_p$, then λ belongs to

(a) $[-3,-2\sqrt{2})$ (b) $[-3,-2\sqrt{2}]$
(c) $(-3,-2\sqrt{2})$ (d) $(-3, 3)$

66. There are exactly two points on the ellipse $\frac{x^2}{a^2}+\frac{y^2}{b^2}=1$, whose distances from its centre are same and are equal to $\sqrt{\frac{(2a^2+3b^2)}{3}}$. Then, the eccentricity of the ellipse is

(a) $\frac{1}{3}$ (b) $\sqrt{\frac{2}{3}}$ (c) $\frac{1}{\sqrt{3}}$ (d) $\frac{1}{2\sqrt{3}}$

67. The straight lines whose direction cosines are given by $al+bm+cn=0$ and $fmn+gnl+hlm=0$ are perpendicular to each other, then the value of $\left(\frac{f}{a}+\frac{g}{b}+\frac{h}{c}\right)!+2$ is equal to

(a) 2 (b) 3 (c) 9 (d) 8

68. If **r** is the position vector of a point which is equidistant form the points **a** and **b**, then $\left|\mathbf{r}-\frac{\mathbf{a}+\mathbf{b}}{2}\right|\cdot(\mathbf{a}-\mathbf{b})$ is equal to

(a) $|\mathbf{a}|$ (b) $|\mathbf{b}|$ (c) 1 (d) 0

69. A set X has 5 elements. A_1 and A_2 are subsets of X. The probability that A_1 and A_2 are not disjoint, is

(a) $\frac{781}{1024}$ (b) $\frac{243}{1024}$
(c) $\frac{32}{1024}$ (d) $\frac{275}{1024}$

70. If Z_n, $n=1, 2, 3, \ldots$ are the solutions of the system of equations $\left|\frac{z-12}{z-8i}\right|=\frac{5}{3}$; $\left|\frac{z-4}{z-8}\right|=1$, then $\Sigma|z_n|$ is equal to

(a) $10-\sqrt{325}$ (b) $5+\sqrt{325}$
(c) $10-5\sqrt{13}$ (d) $10+5\sqrt{13}$

Section B : Numerical Value Type Questions

71. If $x\sin x\frac{dy}{dx}+(x+x\cos x+\sin x)y=\sin x$, $y\left(\frac{\pi}{2}\right)=1-\frac{2}{\pi}$, then $3\lim_{x\to 0} y(x)$ is equal to ....

72. If for $0<x<\frac{\pi}{2}$, exp $[(\sin^2 x+\sin^4 x+\sin^6 x+\ldots\infty)$ $\log_e 2]$ satisfies the quadratic equation $x^2-9x+8=0$. If the value of $\frac{\sin x-\cos x}{\sin x+\cos x}$ is $a-\sqrt{b}$, then $a+b$ equals ....

73. If $\sum_{i=1}^{n}(x_i+1)^2=9n$ and $\sum_{i=1}^{n}(x_i-1)^2=5n$, then the variance of $x_1, x_2, \ldots, x_n$ is ....

74. If sum of the last three digits of 17^{256} is x, then x is equal to ....

75. The interior angles of a regular polygon measure 150° each. The number of diagonals of the polygon are ....

DETAILED SOLUTIONS

Physics

1. *(a)* Current through resistance P and Q,

$$i_1 = \frac{4}{90+110} = \frac{1}{50}\text{ A}$$

$$V_A - V_B = Pi_1 = 90 \times \frac{1}{50} = 1.8\text{ V}$$

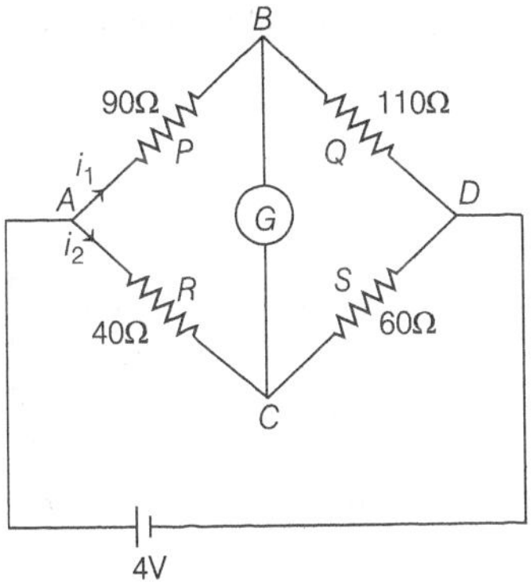

Current through resistance R and S,

$$i_2 = \frac{4}{40+60} = \frac{1}{25}\text{ A}$$

$$V_A - V_C = Ri_2 = 40 \times \frac{1}{25} = 1.6\text{ V}$$

$$V_B - V_C = (V_A - V_C) - (V_A - V_B)$$
$$= 1.6 - 1.8 = -0.2\text{ V}$$

2. *(c)* Given, $K_A = 2K_B$

$$T_A - T_B = 72°\text{C}$$

Let T be the temperature at the junction.

As, $\left(\frac{\Delta Q}{\Delta t}\right)_A = \left(\frac{\Delta Q}{\Delta t}\right)_B$

$\therefore \quad \frac{K_A A(T_A - T)}{x} = \frac{K_B A(T - T_B)}{x}$

or $\quad 2K_B(T_A - T) = K_B(T - T_B)$

or $\quad 2(T_A - T) = T - T_B$

Add $(T_A - T)$ on both sides, we get

$3(T_A - T) = T_A - T + T - T_B$

or $\quad 3(T_A - T) = T_A - T_B$

or $\quad T_A - T = \frac{T_A - T_B}{3}$

$= \frac{72}{3} = 24°\text{C}$

$\therefore$Temperature difference across the layer A

$= T_A - T = 24°\text{C}$

3. *(a)* For isothermal process, $pV = \text{constant}$.

Therefore, p-V graph is a rectangular hyperbola.

Since, $T_2 > T_1 \Rightarrow (pV)_{\text{at } T_2} > (pV)_{\text{at } T_1}$

Thus, the graph given in option (a) correctly represents the given situation.

4. *(d)* Reaction under consideration is

$${}_1^2\text{H} + {}_1^2\text{H} \longrightarrow {}_2^4\text{He} + Q$$

Here, $BE_{\text{H}} = 1.1\text{ MeV}$ and $BE_{\text{He}} = 7\text{ MeV}$

As Q value $= \Sigma BE_{\text{products}} - \Sigma BE_{\text{reactants}}$

$$Q = 4 \times 7 - (2 \times 1.1 + 2 \times 1.1)$$
$$= 23.6\text{ MeV} = 23.6 \times 10^6 \times 1.6 \times 10^{-19}\text{ J}$$
$$= 37.76 \times 10^{-13}\text{ J}$$

5. *(c)* $\mathbf{F} = q(\mathbf{v} \times \mathbf{B})$

For positive charges, i.e. α and δ, force experienced will be in the direction of $\mathbf{v} \times \mathbf{B}$, i.e. towards $+x$-direction.

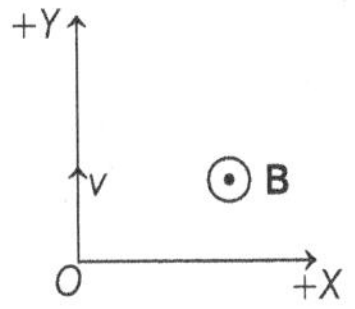

For negative charge, i.e. β and γ, force will be along $-x$-direction.

Now, $r = \frac{mv}{Bq} = \frac{v}{B\left(\frac{q}{m}\right)}$

$\Rightarrow r \propto \frac{1}{s}$ where, $s = \frac{q}{m}$

Since, magnitude of charge is same

$\Rightarrow \quad r \propto m$

Hence, (i) - (r), (ii)-(p), (iii)-(q), (iv)-(s)

6. *(a)* At STP, 22.4 L of any gas is 1 mole

$\therefore \quad 5.6\text{ L} = \frac{5.6}{22.4} = \frac{1}{4}$ moles $= n$

In adiabatic process, $TV^{\gamma-1} = \text{constant}$

$\therefore \; T_2V_2^{\gamma-1} = T_1V_1^{\gamma-1}$

or $\quad T_2 = T_1\left(\frac{V_1}{V_2}\right)^{\gamma-1}$

$= T_1\left(\frac{5.6}{0.7}\right)^{5/3-1} \quad \left[\because \gamma = \frac{C_p}{C_V} = \frac{5}{3}\right]$

$= 4T_1$

Further, in adiabatic process, $Q = 0$

$\therefore W + \Delta U = 0$

$\therefore \quad W = -\Delta U = -nC_V \Delta T$

$$= -n\left(\frac{R}{\gamma - 1}\right)(T_2 - T_1)$$

$$= -\frac{1}{4}\left(\frac{R}{\frac{5}{3} - 1}\right)(4T_1 - T_1)$$

$$= -\frac{9}{8}RT_1$$

$$= -\frac{m}{n}RT_1 \text{ (given)}$$

$\therefore \quad m = 9, n = 8$

Hence, $m - n = 9 - 8 = 1$

7. *(d)* In a given *L-C-R* circuit, current will be same at different frequencies if impedance Z is same. Current in the circuit will be maximum, if circuit is in resonance. Typical graph of current *versus* frequency of *L-C-R* circuit is

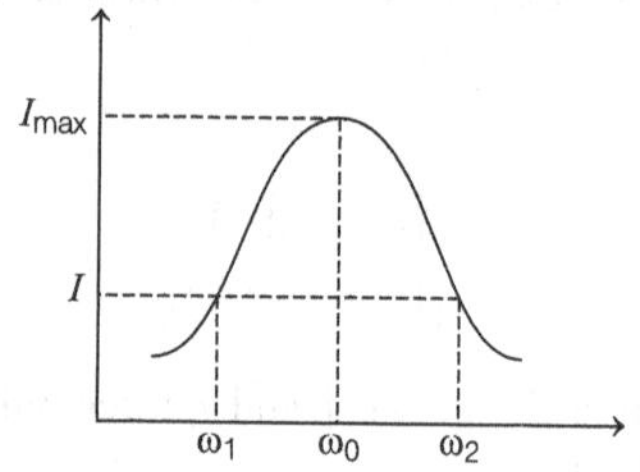

Now, as resistance is independent of frequency. So, current will same when impedance Z at two frequencies is same. At lower frequency, capacitive reactance is more but at higher frequencies, inductive reactance dominates.

So, $\quad Z_{1 \text{ at } \omega_1} = \frac{1}{\omega_1 C} - \omega_1 L$

$$Z_{2 \text{ at } \omega_2} = \omega_2 L - \frac{1}{\omega_2 C}$$

For current to be same

$$Z_1 = Z_2$$

$$\Rightarrow \quad \frac{1}{\omega_1 C} - \omega_1 L = \omega_2 L - \frac{1}{\omega_2 C}$$

$$\Rightarrow \quad \left(\frac{1}{\omega_1} + \frac{1}{\omega_2}\right)\frac{1}{C} = (\omega_1 + \omega_2)L$$

$$\Rightarrow \quad \frac{\omega_1 + \omega_2}{\omega_1 \omega_2} = (\omega_1 + \omega_2)LC$$

Now, current is maximum at $\omega_0 = \frac{1}{\sqrt{LC}}$

$\therefore$Resonant frequency, $f_r = \frac{\sqrt{2\pi f_1 \times 2\pi f_2}}{2\pi}$

For given problem,

$$f_r = \sqrt{8 \times 18} = 12\,\text{MHz}$$

8. *(c)* $F_{\text{air}} = \frac{1}{4\pi\varepsilon_0} \cdot \frac{q_1 q_2}{r^2}$

$$F_{\text{liquid}} = \frac{1}{4\pi\varepsilon_0 K} \cdot \frac{q_1 q_2}{r^2} = \frac{F_{\text{air}}}{K} = \frac{F}{K}$$

9. *(d)* Block is moving with uniform velocity, so there will be rotational as well as translational equilibrium.

$$mg\cos\theta = N \qquad \text{...(i)}$$

To avoid toppling, normal force will shift towards left by distance x.

$$f = mg\sin\theta \qquad \text{...(ii)}$$

$$f \times \frac{a}{2} = N \times x$$

$$mg\sin\theta \times \frac{a}{2} = mg\cos\theta \times x$$

$$x = \frac{a}{2}\tan\theta$$

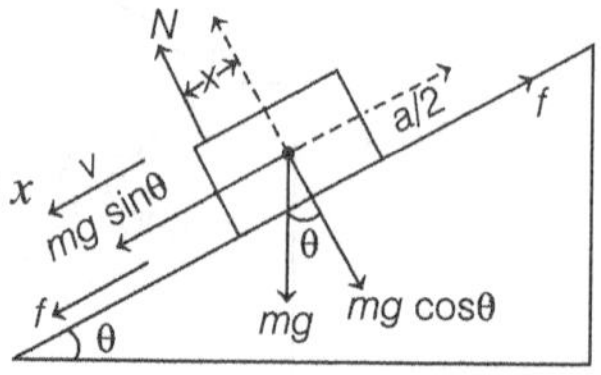

Torque of the normal force on the block about its centre

$$= N \times x$$

$$= mg\cos\theta \times \frac{a}{2}\tan\theta$$

$$= \frac{mg\,a\sin\theta}{2}$$

10. *(d)* Equation of simple harmonic wave is given as

$$y = A\sin(\omega t + \phi)$$

Here, $y = \frac{A}{2} \Rightarrow A\sin(\omega t + \phi) = \frac{A}{2}$

So, $\quad \delta = \omega t + \phi = \frac{\pi}{6} \text{ or } \frac{5\pi}{6}$

So, the phase difference of the two particles when they are crossing each other at $y = A/2$ in opposite direction are

$$\delta = \delta_1 - \delta_2 = \frac{5\pi}{6} - \frac{\pi}{6} = \frac{2\pi}{3}$$

11. *(b)* From graph, we can say temperature is increasing then become constant which suggests phase transition and once its done, its temperature further increases and we see another phase transition.

Here, we can say initially substance was in solid state then liquid and gas at the end. As we see, the slope of solid state is more than liquid state, so specific heat is greater in liquid state. The latent heat of vaporisation is more that its latent heat of fusion.

12. *(a)* When the capacitor is fully charged, the voltage V_1 across it = voltage of the battery = 12 V

The charge on C_1, $Q = C_1 V_1 = 4 \times 10^{-6} \times 12$

$$= 48 \times 10^{-6} = 48\,\mu C$$

When S_1 is opened and S_2 is closed, charge will flow from C_1 and C_2 until they have the same potential V which is given by $V = \dfrac{Q}{C_1 + C_2} = \dfrac{48}{4+8} = \dfrac{48}{12} = 4\text{ V}$

Since, C_1 and C_2 are parallel, therefore they have the same, i.e. common potential.

Now, $U_i = \frac{1}{2}C_1V_1^2$

$$= \frac{1}{2} \times 4 \times 10^{-6} \times (12)^2$$
$$= 288 \times 10^{-6}\text{ J}$$
$$U_f = \frac{1}{2}C_1V^2 = \frac{1}{2} \times 4 \times 10^{-6} \times (4)^2$$
$$= \frac{1}{2} \times 4 \times 10^{-6} \times 4 \times 4 = 32 \times 10^{-6}\text{ J}$$

$\therefore$ Loss of energy stored in C_1

$$= 288 \times 10^{-6} - 32 \times 10^{-6} = 256 \times 10^{-6}\text{ J}$$

The percentage loss of energy is $\dfrac{256 \times 10^{-6}}{288 \times 10^{-6}} \times 100$

$$= 88.88\% \approx 88.9\%$$

13. *(a)* The maximum height,

$$h = \frac{u^2 \sin^2 \theta}{2g} \quad \ldots\text{(i)}$$

The change in height is given by differentiating this expression. Thus,

$$dh = \frac{2udu \sin^2 \theta}{2g} \quad \ldots\text{(ii)}$$

On dividing Eq. (ii) by Eq. (i), we get

$$\frac{dh}{h} = \frac{2du}{u}$$

$$\Rightarrow \quad \frac{2du}{u} = 0.1 \quad \left[\because \frac{dh}{h} = 10\%\right]$$

Now, the range, $R = \dfrac{u^2 \sin 2\theta}{g}$ ...(iii)

$$\therefore \quad dR = \frac{2u\,du \sin 2\theta}{g} \quad \ldots\text{(iv)}$$

On dividing Eq. (iv) by Eq. (iii), we get

$$\Rightarrow \quad \frac{dR}{R} = \frac{2du}{u} = 0.1$$

Hence, range is also increased by 10%.

14. *(a)* KE of proton and α-particle is given as 1 eV and 2 eV, respectively.

Now, momentum, $p = \sqrt{2m\,\text{KE}}$

$\therefore$ Momentum of proton, $p_p = \sqrt{2m_p \times 1}$

$\therefore$ Momentum of α-particle whose mass $m_\alpha = 4m_p$ is

$$p_\alpha = \sqrt{8m_p \times (2)}$$
$$= \sqrt{16m_p}$$

$\therefore$ Now, $\lambda = \dfrac{h}{p}$ [de-Broglie relation]

$$\Rightarrow \quad \frac{\lambda_p}{\lambda_\alpha} = \frac{p_\alpha}{p_p} = \frac{\sqrt{16m_p}}{\sqrt{2m_p}} = 2\sqrt{2} : 1$$

15. *(d)* All the statements I, II and III are correct. When a stone is moving around a circular path, its velocity acts tangentially to the circle. When the string breaks, the centripetal force will not act. Due to inertia, the stone continues to move along the tangent to the circular path and stone flies off tangentially. The necessary centripetal force required to keep the stone moving in a circle is provided by tension in the string. Hence, the centripetal force is a real force.

16. *(b)* In a reverse biased *p-n* junction, when the applied bias voltage is equal to the breakdown voltage, then voltage remains constant, while current increases sharply. Typical voltage-current graph of a reversed biased *p-n* diode is

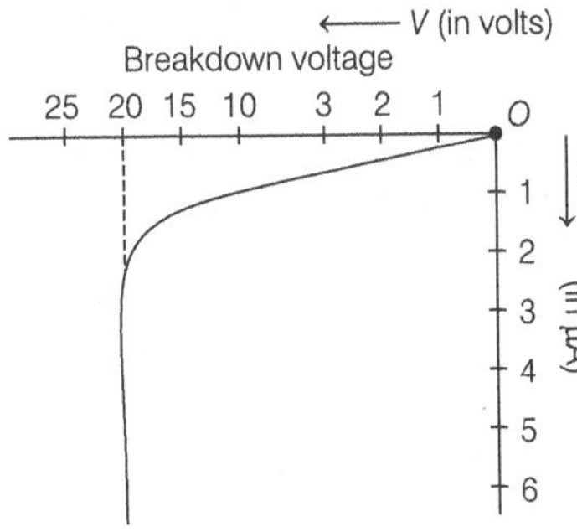

17. *(b)* The dimensional formula of Young's modulus is $[ML^{-1}T^{-2}]$.

Now, the dimensions of v, A and F in terms of M, L and T are

$$[v] = [LT^{-1}], [A] = [LT^{-2}]$$

and $[F] = [MLT^{-2}]$

Let $Y = [v^a A^b F^c]$

$$[M^1L^{-1}T^{-2}] = [LT^{-1}]^a \times [LT^{-2}]^b \times [MLT^{-2}]^c$$
$$[M^1L^{-1}T^{-2}] = [M^cL^{a+b+c}T^{-a-2b-2c}]$$

On equating powers of M, L and T, we have $c = 1$,

$$a + b + c = -1, -a - 2b - 2c = -2$$

On solving above equations, we get

$$a = -4, b = 2$$

Thus, new dimensional formula of

$$[Y] = [FA^2v^{-4}]$$

18. *(b)* Escape velocity does not depend on the direction of projection of the body from the surface of the earth.

If the velocity of a revolving satellite around the earth is increased to $\sqrt{2}$ times or 41.4%, then it escapes out from the earth because in this condition, its velocity is equal to escape velocity.

19. *(b)* We know that Young's modulus of elasticity is given by

$$Y = \frac{FL}{lA} = \frac{FL}{l\left(\frac{\pi d^2}{4}\right)}$$

Substituting the values given, we get

$$Y = \frac{50 \times 1.1 \times 4}{(1.25 \times 10^{-3}) \times \pi \times (5.0 \times 10^{-4})^2}$$

$$= 2.24 \times 10^{11}\ \text{Nm}^{-2}$$

Again, $\frac{\Delta Y}{Y} = \frac{\Delta L}{L} + \frac{\Delta l}{l} + 2\frac{\Delta d}{d}$

$$= \frac{0.1}{110} + \frac{0.001}{0.125} + 2 \times \frac{(0.001)}{0.05} = 0.0489$$

$\Rightarrow \quad \Delta Y = 0.0489\, Y$

$$= (0.0489)(2.24 \times 10^{11})\text{Nm}^{-2}$$

$$= 1.09 \times 10^{10}\ \text{Nm}^{-2}$$

20. *(d)* Electromagnetic waves consist of oscillating electric and magnetic field that are perpendicular to each other and also perpendicular to the direction of propagation of EM waves. Thus, electromagnetic waves are transverse in nature and travel through vacuum with same speed of 3×10^8 m/s.

21. (3) The volume of sphere in liquid,

$$V = \frac{4}{3}\pi r^3$$

When mass m is placed on the piston, the increased pressure, $\quad p = \frac{mg}{a}$

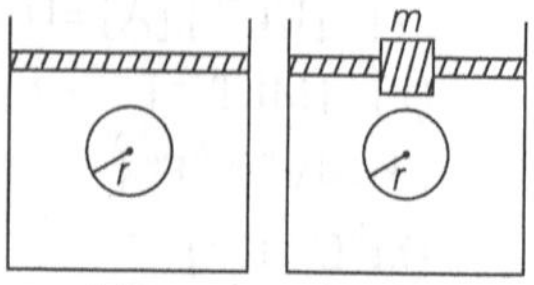

(a) (b)

Since, this increased pressure is equally applicable to all directions on the sphere, so there will be decrease in volume of sphere, due to decrease in its radius.

$$\Delta V = \frac{4\pi}{3} \times 3r^2 \Delta r = 4\pi r^2 \Delta r$$

$$\Rightarrow \quad \frac{\Delta V}{V} = \frac{4\pi r^2 \Delta r}{\frac{4}{3}\pi r^3} = \frac{3\Delta r}{r}$$

$\therefore$ Bulk modulus, $K = \frac{pV}{\Delta V} = \frac{mg}{a} \times \frac{r}{3\Delta r}$

$$\therefore \quad \frac{\Delta r}{r} = \frac{mg}{3Ka}$$

On comparing with the given equation,

$$= \frac{mg}{xKa}$$

$$\therefore \quad x = 3$$

22. (12) Consider two cases,

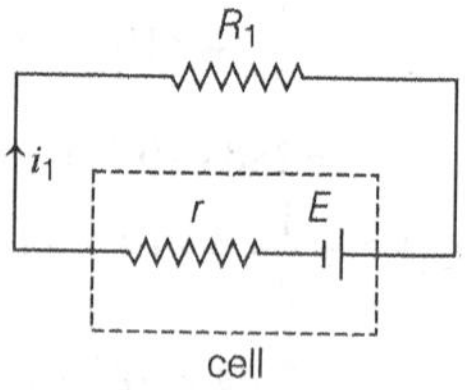

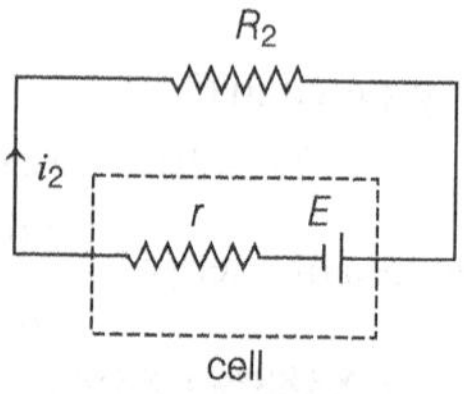

Power delivered in both the cases is same.

Thus, $\quad i_i^2 R_1 = i_2^2 \times R_2$

$$\Rightarrow \quad \left(\frac{E}{R_1 + r}\right)^2 R_1 = \left(\frac{E}{R_2 + r}\right)^2 R_2$$

$$\Rightarrow \quad (R_2 + r)^2 \cdot R_1 = (R_1 + r)^2 \cdot R_2$$

$$\Rightarrow \quad (R_1 - R_2) r^2 = (R_1 - R_2)(R_1 R_2)$$

$$\Rightarrow \quad r = \sqrt{R_1 R_2}$$

Put, $R_1 = 18\Omega$ and $R_2 = 8\Omega$

Then, $r = \sqrt{144} = 12\Omega$

23. (5) In perfectly inelastic collision, blocks stick togehter after impact.

From conservation of momentum theorem,

$$4 \times 10 = (4 + 4) \times v' \text{ or } v' = 5 \text{ m/s}$$

24. (2) Time taken by each ball to reach the highest point,

$t = \frac{1}{n}$ s

As the juggler throws second ball, when the first ball is at its highest point, so $v = 0$

Using $\quad v = u - gt$, we have

$$0 = u - g \times 1/n \Rightarrow u = g/n$$

Also, $\quad v^2 = u^2 + 2(-g)h$

$$0 = (g/n)^2 + 2(-g) \times h \text{ or } h = \frac{g}{2n^2}$$

Comparing with given equation

$$h = \frac{g}{2n^2} = \frac{g}{2n^p}$$

$\therefore \quad p = 2$

25. (3) Apparent depth of A,

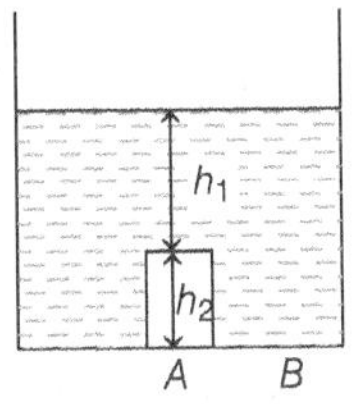

$$d_1 = \frac{h_1}{4/3} + \frac{h_2}{3/2} = \frac{3}{4}h_1 + \frac{2}{3}h_2 \quad \text{...(i)}$$

Apparent depth of B,

$$d_2 = \frac{(h_1 + h_2)}{4/3} = \frac{3}{4}(h_1 + h_2) \quad \text{...(ii)}$$

From Eqs. (i) and (ii), we get

$$\therefore \quad d_2 - d_1 = \left(\frac{3}{4} - \frac{2}{3}\right)h_2 = 3\,\text{mm} \quad (\text{where, } h_2 = 36\,\text{mm})$$

$$\Rightarrow \quad x = 3$$

Chemistry

26. (*d*) Glucose on reacting with conc. HCl forms levulinic acid.

$$C_6H_{12}O_6 \xrightarrow{\text{Conc. HCl}} \underset{\text{Levulinic acid}}{CH_3COCH_2CH_2COOH} + HCOOH + H_2O$$

27. (*a*) It is plot of $\ln p_{CO_2}$ v/s $\frac{1}{T}$

$$K_p = Ae^{-\Delta H_r^\circ/RT}$$

$$\ln K_p = \ln A - \Delta H_r^\circ / RT$$

$$K_p = [p_{CO_2}]$$

$$\ln [p_{CO_2}] = \ln A - \Delta H_r^\circ / RT$$

$$\text{Slope} = -\Delta H_r^\circ / R$$

Therefore, option (a) is correct.

28. (*a*) Lassaigne's test is given by the compound that contain C and N element and forms NaCN with Lassaigne's extract. Among the given compounds, only (i), (iv), (vi) and (vii) gives positive Lassaigne's test while (ii), (iii) and (v) not.

29. (*a*) Ionic radii decrease as we move across lanthanide series due to lanthanoid contraction.

Hence, the correct order is $Yb^{3+} < Pm^{3+} < Ce^{3+} < La^{3+}$.

30. (*c*) The compound which contains —CHO group, gives positive Tollen's test and the compound with $CH_3-\underset{\underset{O}{\|}}{C}-$ group gives iodoform test.

Thus, the organic compound is $CH_3-\underset{\underset{H}{|}}{C}=O$

31. (*d*) Formation of anhydride from dicarboxylic acid indicates *cis* isomers.

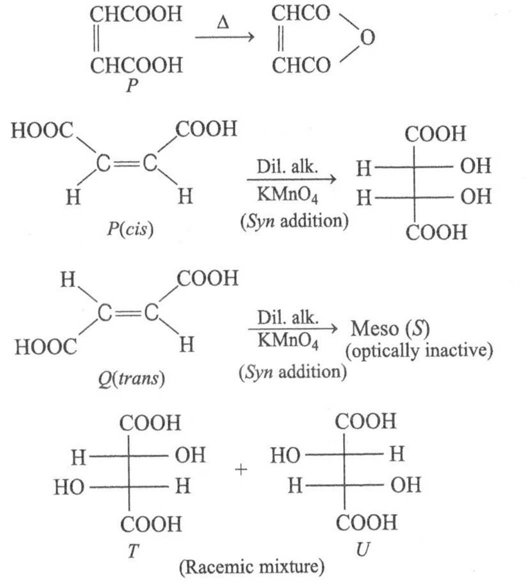

Racemic mixture is optically inactive.

32. (*b*) The given reaction are represented as

$$\underset{(A)}{NH_4NO_2} + KOH \longrightarrow \underset{(B)}{NH_3\uparrow} + \underset{(C)}{KNO_2} + H_2O$$

$$\underset{(C)}{KNO_2} + 3\,Zn + 5\,KOH + 5H_2O \longrightarrow 3K_2[Zn(OH)_4] + \underset{(B)}{NH_3\uparrow}$$

$$\underset{(A)}{NH_4NO_2} \xrightarrow{\text{Heat}} \underset{\text{(Does not support combustion)}}{N_2\uparrow + 2\,H_2O}$$

33. (*c*)

- **2, 4-DNP test** Compound that gives positive 2,4-DNP test means the compound contain carbonyl group.
- **Iodoform test** A compound show iodoform test means it must contain keto group (not aldehydic) $-\overset{\overset{O}{\|}}{C}-CH_3$.
  This is also confirmed by negative Tollen's test and Fehling test.
- **Baeyer's test** If compound does not decolourises bromine water mean there is no any free double bond or triple bond.

Molecular formula $= C_9H_{10}O$

$$u = (9 + 1) - \frac{10}{2} = 5$$

Degree of unsaturation is 5 and no any double or triple bond thus there must be a benzene ring and one C=O group.

Out of given four choices only option (c) is satisfying all criterion.

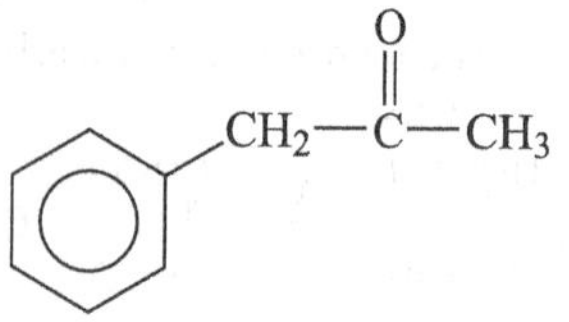

Hence, (c) is the correct choice.

34. (*b*) The given reaction are represented as

$$\underset{\substack{(X)\\(3C-2e^-,\\2C-2e^-\text{ bonds})}}{B_2H_6} + 2NH_3 \xrightarrow{\text{Low temperature}} B_2H_6 \cdot 2NH_3 \xrightarrow{200^\circ C} \underset{(Y)}{B_3N_3H_6}$$

Y has structure similar to benzene (inorganic benzene).

$$nB_2H_6 \xrightarrow[\text{High temperature}]{\text{Excess } NH_3} \underset{\substack{(Z)\\\text{Inorganic}\\\text{graphite}}}{(BN)_n} + 6H_2$$

35. (*a*) The reaction sequence involves various processes are shown as

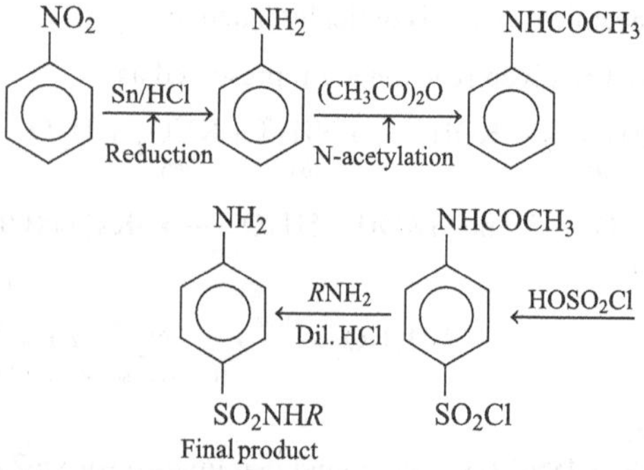

36. (*a*) The correct match is A-II, B-IV, C-I, D-III

A. $BrF_5 \equiv$

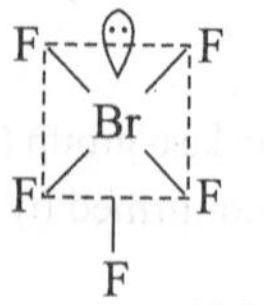

Square pyramidal

B. $[CrF_6]^{3-} \equiv$

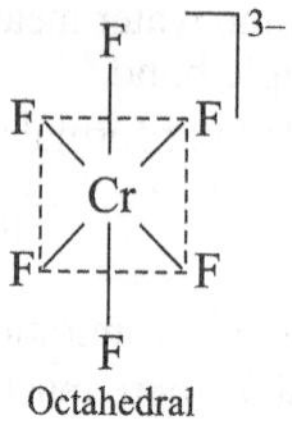

Octahedral

C. $O_3 \equiv$

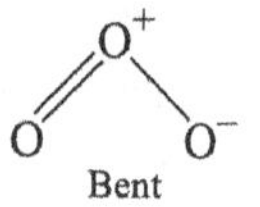

Bent

D. $PCl_5 \equiv$

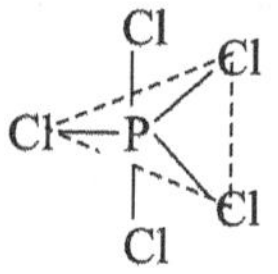

Trigonal bipyramidal

37. (*c*) Secondary and tertiary alcohols are differentiated by Lucas reagent. The reactions are represented as

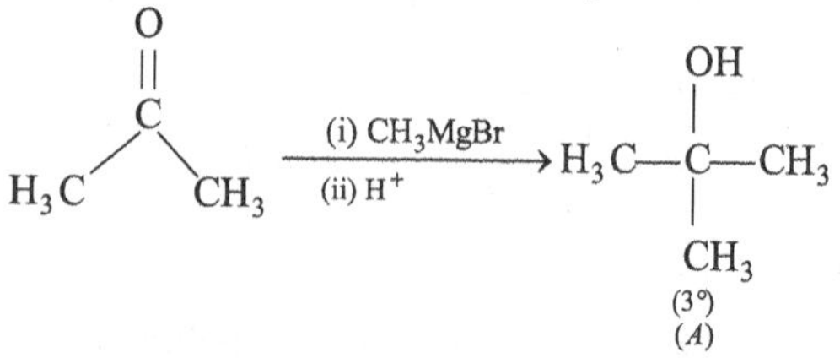

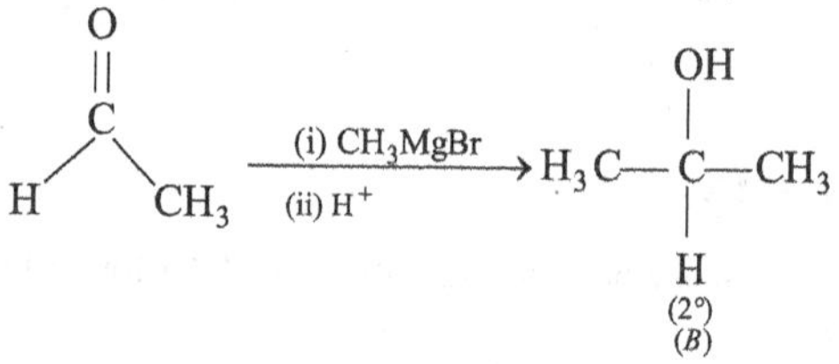

38. (*a*) Potassiumtetracyanonickelate(II), potassiumtetrachloronickelate(II).

The reactions are as follows:

$$NiCl_2 + \underset{(\text{excess})}{4KCN} \longrightarrow K_2[Ni(CN)_4] + 2KCl$$

A is $K_2[Ni(CN)_4]$ also called potassiumtetracyanonickelate (II).

$$NiCl_2 + 4HCl(\text{conc.}) \longrightarrow K_2[Ni(Cl)_4] + 4HCN$$

B is $K_2[NiCl_4]$ also called potassiumtetrachloronickelate(II).

39. (*b*) Both Statement I and Statement II are incorrect.

For KI, molar conductivity increases slowly with dilution. For carbonic acid, molar conductivity increases steeply with dilution.

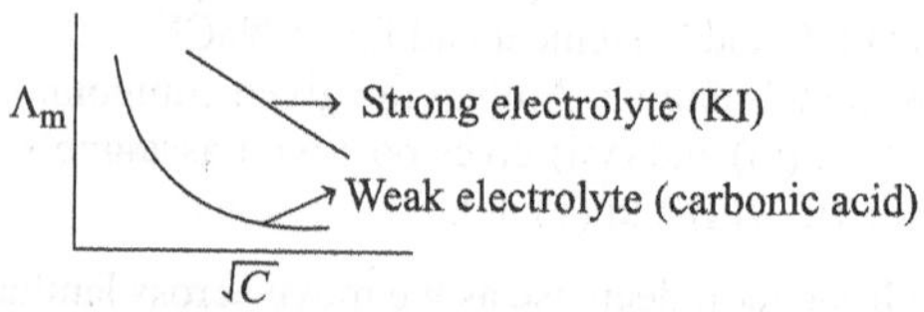

40. (*a*) Both (A) and (R) are true and (R) is the correct explanation of (A).

41. (*c*) The correct IUPAC name of $[Fe(O_2)(CN)_4Cl]^{4-}$ is chlorotetracyanosuperoxoferrate (II) ion.

42. *(a)* Chlorobenzene is formed by reacting diazonium salt with Cu_2Cl_2. The reaction is represented as

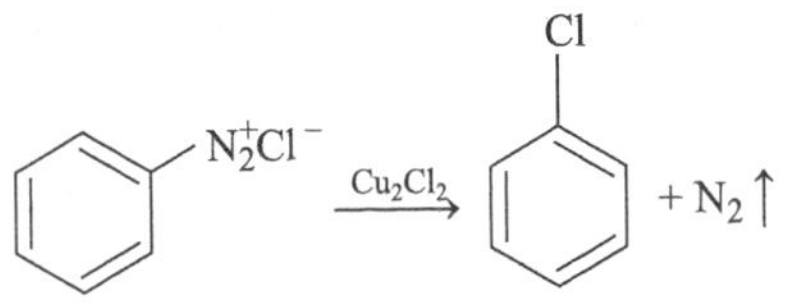

43. (*a*) Huckel rule states that a planar and cyclic molecules and aromatic, if it has $(4n+2)$ π electrons. Among the given compounds only (iv) and (v) obeys Huckel's rule while other's not.

44. (*d*) $$BO = \frac{N_b - N_a}{2}$$

Bond strength $\propto$ bond order

M.O.E.C of $N_2 = \sigma 1s^2, \sigma^* 1s^2, \sigma 2s^2, \sigma^* 2s^2, \pi 2p_x^2$
$= \pi 2p_y^2, \sigma 2p_z^2$

$$\text{Bond order} = \frac{10-4}{2} = 3$$

M.O.E.C of $N_2^+ = \sigma 1s^2, \sigma^* 1s^2, \sigma 2s^2, \sigma^* 2s^2, \pi 2p_x^2$
$= \pi 2p_y^2, \sigma 2p_z^1$

$$\text{Bond order} = \frac{9-4}{2} = 2.5$$

Here, on converting N_2 to N_2^+ bond order decreases, so bond strength decreases.

M.O.E.C of $O_2 = \sigma 1s^2, \sigma^* 1s^2, \sigma 2s^2, \sigma^* 2s^2, \sigma 2p_z^2, \pi 2p_x^2$
$\equiv \pi 2p_y^2, \pi^* 2p_x^1 \equiv \pi^* 2p_y^1$

$$\text{Bond order} = \frac{10-6}{2} = 2$$

M.O.E.C of $O_2^+ = \sigma 1s^2\ \sigma^* 1s^2, \sigma 2s^2\ \sigma^* 2s^2, \sigma 2p_z^2$
$\pi 2p_x^2 \equiv \pi 2p_y^2, \pi^* 2p_x^1 \equiv \pi^* 2p_y^0$

$$\text{Bond order} = \frac{10-5}{2} = 2.5$$

On conversion of O_2 to O_2^+ bond order increases and hence bond strength also increases.

Study Tactics

To solve this question, remember that molecular orbital theory consider the interaction of atomic orbitals to form molecular orbitals and more electrons in bonding molecular orbitals lead to stronger bonds.

45. *(c)* A. III. On acidification $(B_4O_7)^{2-}$ gives $B(OH)_3$ or (H_3BO_3)

B. IV. $(AlO)_2^-$ on dilution yields a white ppt. of $Al(OH)_3$.

C. I. When heated, $(SiO_4)^{4-}$ changes to $(Si_2O_7)^{6-}$.

D. II. Bi^{3+} hydrolyses to yield $(BiO)^+$ ion.

46. (*5*) The structural isomers are represented as

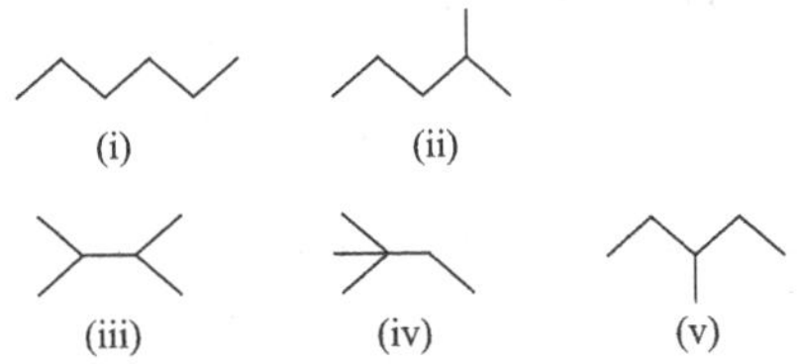

47. (*1200*) For adiabatic condition,

$$\frac{T_2}{T_1} = \left(\frac{V_1}{V_2}\right)^{\gamma-1}$$

or $$\frac{T_2}{T_1} = \left(\frac{1}{27}\right)^{1.33-1} = \left(\frac{1}{27}\right)^{0.33} = \left(\frac{1}{27}\right)^{1/3} = \frac{1}{3}$$

or $$T_2 = \frac{1}{3} \times T_1$$

$$T_2 = \frac{1}{3} \times 300 = 100\text{ K}$$

From first law of thermodynamics,

$$\Delta E = q + W$$
$$W = -\Delta E \qquad (\because q = 0 \text{ for adiabatic expansion})$$
$$= -C_V(T_2 - T_1)$$
$$= -6(100-300) = +1200\text{ cal}$$

48. (*1*) KE = quantum energy – threshold energy

$$= \frac{6.626\times10^{-34}\times3\times10^8}{3000\times10^{-10}} - \frac{6.626\times10^{-34}\times3\times10^8}{4000\times10^{-10}}$$
$$= 6.626\times10^{-19} - 4.9695\times10^{-19}$$
$$= 1.6565\times10^{-19}\text{ J}$$

$$KE = 1.6565\times10^{-19} = \frac{1}{2}mv^2$$

$$\therefore m^2v^2 = 2\times1.6565\times10^{-19}\times9.1\times10^{-31}$$
$$mv = 5.49\times10^{-25}$$

de-Broglie wavelength, $$\lambda = \frac{h}{mv} = \frac{6.626\times10^{-34}}{5.49\times10^{-25}}$$
$$= 1.2\times10^{-9}\text{ m} \approx 1\times10^{-9}\text{ m}$$

49. (*5000*) For the reaction,

	A	+ B	$\longrightarrow$ C	+ D
At $t=0$	1	1	0	0
At $t=t$	$1-x$	$1-x$	x	x

$$R = k[A]^{1/2}[B]^{1/2}$$
$$\frac{dx}{dt} = k(1-x)^{1/2}(1-x)^{1/2}$$
$$\frac{dx}{dt} = k(1-x)$$

Integrating on both sides,

$$k = \frac{2.303}{t}\log\frac{1}{1-x}$$

$$4.606\times 10^{-4}=\frac{2.303}{t}\log\frac{1}{0.1}$$

$$2\times 10^{-4}=\frac{1}{t} \text{ or } t=\frac{10^4}{2}=\frac{10000}{2}=5000\text{ s}$$

50. *(2)* $MnO_4^- + 8H^+ + 5e^- \longrightarrow Mn^{2+} + 4H_2O$

$$E=1.51-\frac{0.059}{5}\log\frac{[Mn^{2+}]}{[MnO_4^-][H^+]^8}$$

Taking Mn^{2+} and MnO_4^- in standard state, i.e. 1M,

$$E=1.51-\frac{0.059}{5}\times 8\log\frac{1}{[H^+]}$$

$$E=1.51-\frac{0.059}{5}\times 8\times 3=1.2268\text{ V}$$

Hence, at this pH, MnO_4^- will oxidise only Br^- and I^- as standard reduction potential of Cl_2/Cl^- is 1.36 V which is greater than that for MnO_4^-/Mn^{2+}.

Mathematics

51. *(b)* Given, $A\ln|\cos x+\sin x-2|+Bx+C$

$$=\int\frac{2\cos x-\sin x+\lambda}{\cos x+\sin x-2}dx$$

On differentiating both sides, we get

$$\frac{d}{dx}[A\ln|\cos x+\sin x-2|+Bx+C]=\frac{2\cos x-\sin x+\lambda}{\cos x+\sin x-2}$$

$$\Rightarrow \frac{A(\cos x-\sin x)}{\cos x+\sin x-2}+B=\frac{2\cos x-\sin x+\lambda}{\cos x+\sin x-2}$$

$$\Rightarrow \frac{(A+B)\cos x+(B-A)\sin x-2B}{\cos x+\sin x-2}=\frac{2\cos x-\sin x+\lambda}{\cos x+\sin x-2}$$

On comparing both sides, we get

$$A+B=2, B-A=-1 \text{ and } \lambda=-2B$$

On solving, we get

$$A=\frac{3}{2},\ B=\frac{1}{2} \text{ and } \lambda=-1$$

52. *(a)* $\log x$ is defined for $x>0$,

$|\log x|\geq 0$ and $|\log|x||\geq 0$

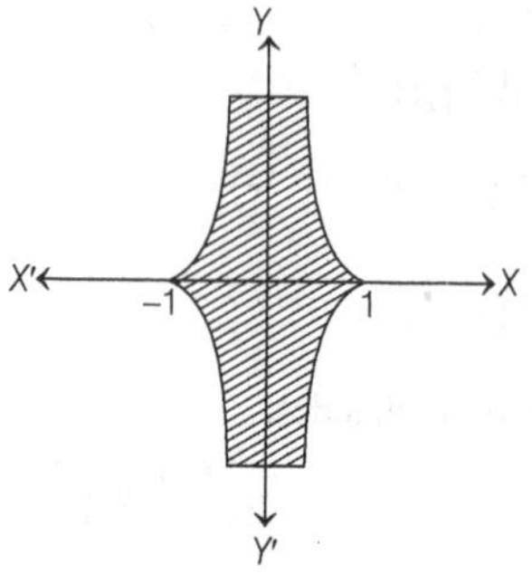

$$\therefore\quad A=4\int_0^1|\log x|dx=-4\int_0^1\log x\,dx$$

$$=-4[x\log x-x]_0^1=-4(-1)=4\text{ sq units}$$

53. *(b)* $f(x)=\begin{cases} x+a\sqrt{2}\sin x, & 0\leq x<\frac{\pi}{4} \\ 2x\cot x+b, & \frac{\pi}{4}\leq x\leq\frac{\pi}{2} \\ a\cos 2x-b\sin x, & \frac{\pi}{2}<x\leq\pi \end{cases}$

At $x=\frac{\pi}{4}$, LHL $=\lim_{x\to\frac{\pi}{4}^-}f(x)$

$$=\lim_{x\to\frac{\pi}{4}^-}(x+a\sqrt{2}\sin x)$$

$$=\frac{\pi}{4}+a$$

RHL $=\lim_{x\to\frac{\pi}{4}^+}f(x)$

$$=\lim_{x\to\frac{\pi}{4}^+}(2x\cot x+b)=\frac{\pi}{2}+b$$

and $f\left(\frac{\pi}{4}\right)=2\times\frac{\pi}{4}\cot\left(\frac{\pi}{4}\right)+b$

$$=\frac{\pi}{2}+b$$

Since, $f(x)$ is continuous.

$$\therefore\quad \text{LHL}=\text{RHL}=f\left(\frac{\pi}{4}\right)$$

$$\Rightarrow\quad \frac{\pi}{4}+a=\frac{\pi}{2}+b\Rightarrow a-b=\frac{\pi}{4} \qquad \ldots(i)$$

At $x=\frac{\pi}{2}$, LHL $=\lim_{x\to\frac{\pi}{2}^-}f(x)$

$$=\lim_{x\to\frac{\pi}{2}^-}(2x\cot x+b)=b$$

RHL $=\lim_{x\to\frac{\pi}{2}^+}f(x)$

$$=\lim_{x\to\frac{\pi}{2}^+}(a\cos 2x-b\sin x)=-a-b$$

and $f\left(\frac{\pi}{2}\right)=2\frac{\pi}{2}\cot\frac{\pi}{2}+b=b$

$\because f(x)$ is continuous.

$$\therefore\quad \text{LHL}=\text{RHL}=f\left(\frac{\pi}{2}\right)$$

$$\Rightarrow\quad b=-a-b$$

$\Rightarrow \quad a + 2b = 0 \quad \ldots(ii)$

On solving Eqs. (i) and (ii), we get

$$a = \frac{\pi}{6}, \; b = -\frac{\pi}{12}$$

54. *(d)* $(\cos^2\theta - \sin^2\theta)^3 + 3(\cos^2\theta - \sin^2\theta)$

$= \cos^6\theta - \sin^6\theta - 3\cos^4\theta\sin^2\theta + 3\cos^2\theta\sin^4\theta + 3\cos^2\theta - 3\sin^2\theta$

$= \cos^6\theta - \sin^6\theta - 3\cos^4\theta(1-\cos^2\theta) + 3(1-\sin^2\theta)\sin^4\theta + 3\cos^2\theta - 3\sin^2\theta$

$= \cos^6\theta - \sin^6\theta + 3\cos^6\theta + 3\sin^4\theta - 3\cos^4\theta - 3\sin^6\theta + 3\cos^2\theta - 3\sin^2\theta$

$= 4(\cos^6\theta - \sin^6\theta) + 3\cos^2\theta(1-\cos^2\theta) - 3\sin^2\theta(1-\sin^2\theta)$

$= 4(\cos^6\theta - \sin^6\theta)$

Hence, $k = 4$

55. *(a)* $m_1 = 2 = \dfrac{t_2^2 - t_1^2}{t_2 - t_1} \Rightarrow t_1 + t_2 = 2 \quad \ldots(i)$

Now, $\dfrac{2 - m_2}{1 + 2m_2} = \sqrt{3}$

[$\because$ given triangle is an equilateral triangle]

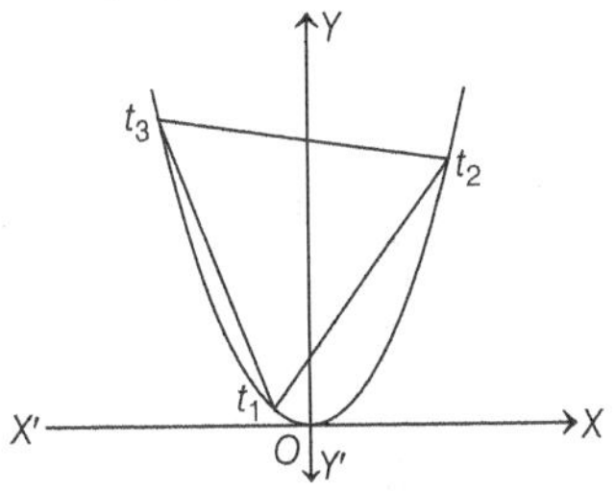

$\therefore \quad 2 - m_2 = \sqrt{3} + 2\sqrt{3}m_2$

$\therefore \quad m_2(2\sqrt{3}+1) = 2 - \sqrt{3}$

$\therefore \quad t_1 + t_3 = m_2 = \dfrac{2-\sqrt{3}}{1+2\sqrt{3}} \quad \ldots(ii)$

Again, $\dfrac{m_3 - 2}{1 + 2m_3} = \sqrt{3}$

$\therefore \quad m_3 - 2 = \sqrt{3} + 2\sqrt{3}m_3$

$\therefore \quad m_3(1 - 2\sqrt{3}) = 2 + \sqrt{3}$

$\Rightarrow \quad m_3 = \dfrac{2+\sqrt{3}}{1-2\sqrt{3}} = t_2 + t_3 \quad \ldots(iii)$

On Adding Eqs. (i), (ii) and (iii), we get

$$2(t_1 + t_2 + t_3) = 2 + \frac{2-\sqrt{3}}{1+2\sqrt{3}} + \frac{2+\sqrt{3}}{1-2\sqrt{3}}$$

$$= 2 + \frac{(2-\sqrt{3})(2\sqrt{3}-1) - (2+\sqrt{3})(2\sqrt{3}+1)}{11}$$

$$= 2 + \frac{(5\sqrt{3}-8) - (5\sqrt{3}+8)}{11} = 2 - \frac{16}{11} = \frac{6}{11}$$

Hence, $t_1 + t_2 + t_3 = \dfrac{3}{11}$

Study Tactics

Using formula, $\dfrac{m_1 - m_2}{1 + m_1 m_2}$

$\because m_1$ is given, we get m_2 and m_3 in the form of t, then add m_1, m_2 and m_3, we get the required result.

56. *(d)* $f(x) = \text{Minimum}\{2\sin x, 1 - \cos x, 1\}$

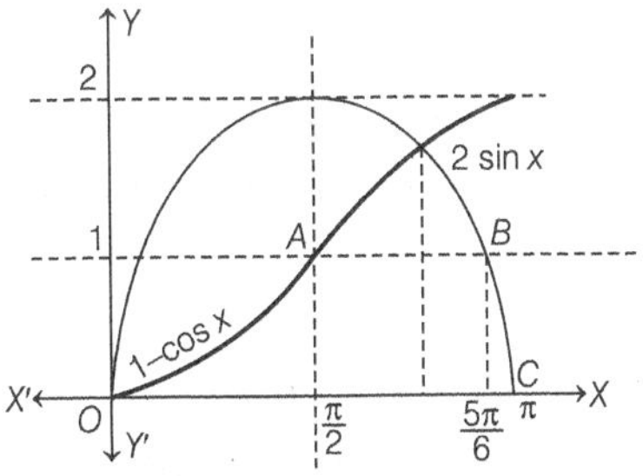

$$\int_0^\pi f(x)\,dx = \int_{OA} f(x)\,dx + \int_{AB} f(x)\,dx + \int_{BC} f(x)\,dx$$

$$= \int_0^{\pi/2} (1 - \cos x)\,dx + \int_{\pi/2}^{5\pi/6} dx + \int_{5\pi/6}^{\pi} 2\sin x\,dx$$

$$= [x - \sin x]_0^{\pi/2} + [x]_{\pi/2}^{5\pi/6} - 2[\cos x]_{5\pi/6}^{\pi}$$

$$= \frac{\pi}{2} - 1 + \frac{5\pi}{6} - \frac{\pi}{2} - 2\left(-1 + \frac{\sqrt{3}}{2}\right)$$

$$= 1 - \sqrt{3} + \frac{5\pi}{6}$$

57. *(b)* Here, $T_n = \cot^{-1}\left(n^2 + \dfrac{3}{4}\right)$

$$= \tan^{-1}\left(\frac{4}{4n^2 + 3}\right)$$

$$= \tan^{-1}\left[\frac{1}{1 + \left(n + \frac{1}{2}\right)\left(n - \frac{1}{2}\right)}\right]$$

$$= \tan^{-1}\left[\frac{\left(n + \frac{1}{2}\right) - \left(n - \frac{1}{2}\right)}{1 + \left(n + \frac{1}{2}\right)\left(n - \frac{1}{2}\right)}\right]$$

$$= \tan^{-1}\left(n+\frac{1}{2}\right)-\tan^{-1}\left(n-\frac{1}{2}\right)$$

$$\therefore \quad S_{\infty} = \tan^{-1}\left(\infty+\frac{1}{2}\right)-\tan^{-1}\left(\frac{1}{2}\right)$$

$$= \frac{\pi}{2}-\tan^{-1}\left(\frac{1}{2}\right)$$

$$\Rightarrow \quad S_{\infty} = \cot^{-1}\left(\frac{1}{2}\right) \Rightarrow S_{\infty} = \tan^{-1} 2$$

58. *(c)* Since, circles touches coordinate axes.

Let radius $= r$, then centre (r, r)

$\therefore$ Equation of circle will be

$$(x-r)^2 + (y-r)^2 = r^2$$

$$\Rightarrow \quad x^2 + y^2 - 2rx - 2ry + r^2 = 0$$

This represents two circles having radius r_1 and r_2, then equation of circles is

$$x^2 + y^2 - 2r_1x - 2r_1y + r_1^2 = 0 \quad \ldots(i)$$

and $$x^2 + y^2 - 2r_2x - 2r_2y + r_2^2 = 0 \quad \ldots(ii)$$

Circles Eqs. (i) and (ii) cuts each other orthogonally.

$$\therefore \quad 2r_1r_2 + 2r_1r_2 = r_1^2 + r_2^2$$

$$\Rightarrow \quad r_1^2 + r_2^2 = 4r_1r_2$$

$$\Rightarrow \quad (r_1 + r_2)^2 = 6r_1r_2 \quad \ldots(iii)$$

Since, circles passes through $P(\alpha, \beta)$.

$$\therefore \quad r^2 - 2(\alpha+\beta)r + \alpha^2 + \beta^2 = 0$$

Sum and product of roots

$$r_1 + r_2 = 2(\alpha+\beta)$$

$$r_1r_2 = \alpha^2 + \beta^2$$

From Eq. (iii),

$$4(\alpha+\beta)^2 = 6(\alpha^2+\beta^2)$$

$$\Rightarrow \quad \alpha^2 - 4\alpha\beta + \beta^2 = 0$$

59. *(d)* $|PQ| = |z + z| = 2|z|$

$|QR| = |1 - z + z| = 1$

$|PR| = |1 - z - z| = |1 - 2z|$

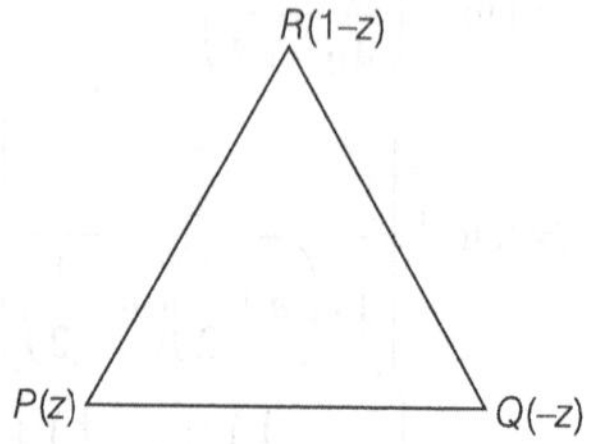

Since, triangle is equilateral.

$|PQ| = |QR| = |PR|$

$$\Rightarrow \quad 2|z| = 1 = |1-2z|$$

$$\Rightarrow \quad |z| = \frac{1}{2} \text{ and } |1-2z| = 1$$

$$\because \quad |1-2z| = 1$$

$$\Rightarrow \quad |1-2z|^2 = 1$$

$$\Rightarrow \quad (1-2z)(1-2\bar{z}) = 1$$

$$\Rightarrow \quad 1 - 2\bar{z} - 2z + 4z\bar{z} = 1$$

$$\Rightarrow \quad 1 + 4|z|^2 - 2(z+\bar{z}) = 1$$

$$\Rightarrow \quad 1 + 4\times\frac{1}{4} - 2(z+\bar{z}) = 1$$

$$\Rightarrow \quad z + \bar{z} = \frac{1}{2}$$

$$\because \quad z + \bar{z} = 2\,\text{Re}(z) = \frac{1}{2}$$

$$\Rightarrow \quad \text{Re}(z) = \frac{1}{4}$$

60. *(b)* Let $(1+x)^{10} = \sum_{i=0}^{10} {}^{10}C_i x^i$

On differentiating both sides w.r.t. x, we get

$$10(1+x)^9 = \sum_{i=1}^{10} i\,{}^{10}C_i x^{i-1} \quad \ldots(i)$$

Again, on differentiating both sides w.r.t. x, we get

$$10\times 9(1+x)^8 = \sum_{i=1}^{10} i(i-1)\,{}^{10}C_i x^{i-2} \quad \ldots(ii)$$

On substituting $x = 1$ in Eqs. (i) and (ii), we get

$$10\times 2^9 = \sum_{i=1}^{10} i\,{}^{10}C_i = f_2 \quad \ldots(iii)$$

and $$90\times 2^8 = \sum_{i=1}^{10} i(i-1)\,{}^{10}C_i = f_1 \quad \ldots(iv)$$

On adding Eqs. (iii) and (iv), we get

$$10\times 2^9 + 90\times 2^8 = \sum_{i=1}^{10} [i + i(i-1)]\,{}^{10}C_i$$

$$10\times 2^8(2+9) = \sum_{i=1}^{10} i^2\,{}^{10}C_i = f_3$$

$$\Rightarrow \quad f_3 = 55\times 2^9$$

Thus, Statement I is true but Statement II is false.

Study Tactics

Differentiate the given expression twice with respect to x, then put $x = 1$ in both the equation and add both equation.

61. *(a)* $f(x)=\dfrac{x-3}{x+1}$

$$f^2(x)=(fof)(x)=\dfrac{\dfrac{x-3}{x+1}-3}{\dfrac{x-3}{x+1}+1}$$

$$=\dfrac{x-3-3x-3}{x-3+x+1}=\dfrac{-2x-6}{2x-2}$$

$$=-\dfrac{(x+3)}{x-1}$$

$$f^3(x)=(fofof)(x)=-\dfrac{\left(\dfrac{x-3}{x+1}+3\right)}{\dfrac{x-3}{x+1}-1}=x$$

$\therefore f^{3n}(x)=x,\ n\in N$

$\therefore f^{2100}(x)=x$

$\Rightarrow f^{2100}(2100)=2100$

62. *(b)* Let r, l and h be the radius, slant height and height of cone respectively, at any time t.

Then, $l^2=r^2+h^2$

$\Rightarrow\ 2l\dfrac{dl}{dt}=2r\dfrac{dr}{dt}+2h\dfrac{dh}{dt}$

$\Rightarrow\ l\dfrac{dl}{dt}=r\dfrac{dr}{dt}+h\dfrac{dh}{dt}$

$=7(3)+24(-4)$ $\quad\left[\because \dfrac{dh}{dt}=-4 \text{ and } \dfrac{dr}{dt}=3\right]$

$=-75$

Now, $l^2=7^2+24^2$

$\Rightarrow\ l^2=625$

$\Rightarrow\ l=25$

$\therefore\ \dfrac{dl}{dt}=-3$

$\therefore\ \dfrac{ds}{dt}=\dfrac{d}{dt}(\pi rl)$

$=\pi\left[l\dfrac{dr}{dt}+r\dfrac{dl}{dt}\right]$

$=\pi[25\times 3+7\times(-3)]$

$=\pi\times 54=54\,\pi\ \text{cm}^2/\text{min}$

63. *(c)* $\Delta=\begin{vmatrix} q^2+r^2 & pq & pr\\ pq & r^2+p^2 & qr\\ pr & qr & p^2+q^2\end{vmatrix}$

$=(q^2+r^2)[(r^2+p^2)(p^2+q^2)-q^2r^2]$
$\quad -pq[pq(p^2+q^2)-pqr^2]+pr[pq^2r-(r^2+p^2)pr]$

$=(q^2+r^2)[r^2p^2+p^4+q^2r^2+p^2q^2-q^2r^2]$
$\quad -pq[p^3q+pq^3-pqr^2]+[p^2r^2q^2-p^2r^2(r^2+p^2)]$

$=p^2q^2r^2+p^4q^2+p^2q^4+r^4p^2+r^2p^4+p^2q^2r^2$
$\quad -p^4q^2-p^2q^4+p^2q^2r^2+p^2r^2q^2-p^2r^4-p^4r^2$

$=p^2q^2r^2(4)$

$\Rightarrow\ \left(\dfrac{K}{2}\right)^2=4$

$\Rightarrow\ \dfrac{K}{2}=\pm 2 \Rightarrow K=\pm 4$

64. *(c)* Given, a, b and c are in AP.

$\Rightarrow\ b-a=c-b=\lambda$(let)

$\because\ \dfrac{t-p}{ap}=\dfrac{t-q}{bq}=\dfrac{t-r}{cr}$

$\Rightarrow\ \dfrac{\dfrac{t}{p}-1}{a}=\dfrac{\dfrac{t}{q}-1}{b}=\dfrac{\dfrac{t}{r}-1}{c}$

$\Rightarrow\ \dfrac{\left(\dfrac{t}{p}-1\right)-\left(\dfrac{t}{q}-1\right)}{a-b}=\dfrac{\left(\dfrac{t}{q}-1\right)-\left(\dfrac{t}{r}-1\right)}{b-c}$

$\Rightarrow\ t\left(\dfrac{1}{p}-\dfrac{1}{q}\right)=t\left(\dfrac{1}{q}-\dfrac{1}{r}\right)$

$\Rightarrow\ \dfrac{1}{p}-\dfrac{1}{q}=\dfrac{1}{q}-\dfrac{1}{r}\Rightarrow\ \dfrac{2}{q}=\dfrac{1}{p}+\dfrac{1}{r}$

Hence, p, q and r are in HP. [$\because$ HP is inverse of AP]

65. *(a)* Here, $p-1\ge 0,\ p\le m+1$

$\Rightarrow\ 1\le p\le m+1$

$\Rightarrow\ \dfrac{1}{m+1}\le\dfrac{p}{m+1}\le 1$

Now, $\lambda^2-8=\dfrac{{}^mC_{p-1}}{{}^{m+1}C_p}$

$=\dfrac{m!}{(p-1)!(m-p+1)!}\times\dfrac{p!(m+1-p)!}{(m+1)!}$

$=\dfrac{p}{m+1}$

$\therefore\ \dfrac{1}{m+1}\le\lambda^2-8\le 1$

$\Rightarrow\ 8<\dfrac{1}{m+1}+8\le\lambda^2\le 9$

$\Rightarrow\ 8<\lambda^2\le 9$

$\Rightarrow\ -3\le\lambda<-2\sqrt{2}$ or $2\sqrt{2}<\lambda\le 3$

Hence, $\lambda\in[-3,-2\sqrt{2})$

66. *(b)* Since, there are exactly two points on the ellipse $\frac{x^2}{a^2}+\frac{y^2}{b^2}=1$, whose distance from the centre is the same, the points would either be the endpoints of the major axis or of the minor axis.

But $\sqrt{\frac{(2a^2+3b^2)}{3}} > b.$

So, the points are the vertices of the major axis.

Hence, $a=\sqrt{\frac{2a^2+3b^2}{3}}$

On squaring both sides, we get

$$a^2=\frac{2a^2+3b^2}{3}$$

$$\Rightarrow 3a^2-2a^2=3b^2$$

$$\Rightarrow \quad b^2=\frac{1}{3}a^2$$

$$\because \quad e=\sqrt{1-\frac{b^2}{a^2}}$$

$$\therefore \quad e=\sqrt{1-\frac{a^2}{3a^2}}=\sqrt{1-\frac{1}{3}}=\sqrt{\frac{2}{3}}$$

67. *(b)* We have,

$$al+bm+cn=0 \quad \text{...(i)}$$

$$fmn+gnl+hlm=0 \quad \text{...(ii)}$$

From Eq. (i),

$$n=-\left(\frac{al+bm}{c}\right)$$

Put the value of n in Eq. (ii), we get

$$fm\left(-\frac{(al+bm)}{c}\right)+gl\left(-\frac{(al+bm)}{c}\right)+hlm=0$$

or $$ag\frac{l^2}{m^2}+\frac{l}{m}(af+bg-ch)+bf=0 \quad \text{...(iii)}$$

Let $X=\frac{l}{m}$

$$agX^2+(af+bg-ch)X+bf=0$$

Let two values of $X\left(=\frac{l}{m}\right)$ be $\frac{l_1}{m_1}$ and $\frac{l_2}{m_2}$.

Product of roots $=\frac{l_1l_2}{m_1m_2}=\frac{bf}{ag}$

$$\therefore \quad \frac{l_1l_2}{\frac{f}{a}}=\frac{m_1m_2}{\frac{g}{b}}=\frac{n_1n_2}{\frac{h}{c}}$$

Now, lines are perpendicular, if

$$l_1l_2+m_1m_2+n_1n_2=0$$

$$\Rightarrow \quad k\left(\frac{f}{a}+\frac{g}{b}+\frac{h}{c}\right)=0$$

$$\Rightarrow \quad \frac{f}{a}+\frac{g}{b}+\frac{h}{c}=0$$

Hence, $\left(\frac{f}{a}+\frac{g}{b}+\frac{h}{c}\right)!+2=0!+2=3 \quad [\because 0!=1]$

68. *(d)* $\because |\mathbf{r}-\mathbf{a}|=|\mathbf{r}-\mathbf{b}|$

$$\Rightarrow |\mathbf{r}-\mathbf{a}|^2=|\mathbf{r}-\mathbf{b}|^2$$

$$\Rightarrow (\mathbf{r}-\mathbf{a})\cdot(\mathbf{r}-\mathbf{a})=(\mathbf{r}-\mathbf{b})\cdot(\mathbf{r}-\mathbf{b})$$

$$\Rightarrow |\mathbf{r}|^2+|\mathbf{a}|^2-2\mathbf{r}\cdot\mathbf{a}=|\mathbf{r}|^2+|\mathbf{b}|^2-2\mathbf{r}\cdot\mathbf{b}$$

$$\Rightarrow 2\mathbf{r}\cdot(\mathbf{a}-\mathbf{b})-|\mathbf{a}|^2+|\mathbf{b}|^2=0$$

$$\Rightarrow 2\mathbf{r}\cdot(\mathbf{a}-\mathbf{b})-(\mathbf{a}-\mathbf{b})\cdot(\mathbf{b}+\mathbf{a})=0$$

$$\Rightarrow \left|\mathbf{r}-\frac{\mathbf{a}+\mathbf{b}}{2}\right|\cdot(\mathbf{a}-\mathbf{b})=0$$

69. *(a)* Number of elements in $X=5$

Number of subsets of $X=2^5=32$

Total combination of A_1 and $A_2=32\times 32=1024$

For disjoint combination there are three possibilities for $x\in X$

(i) $x\in A_1$, then $x\notin A_2$

(ii) $x\notin A_1$, then $x\in A_2$

(iii) $x\notin A_1$ and $x\notin A_2$

Number of disjoint combination $=3^5=243$

Probability that A_1 and A_2 are not disjoint

$$=\frac{1024-243}{1024}=\frac{781}{1024}$$

70. *(d)* $\because \quad \left|\frac{z-12}{z-8i}\right|=\frac{5}{3}$

$$\Rightarrow \quad 3|x+iy-12|=5|x+iy-8i|$$

$$\Rightarrow 9[(x-12)^2+y^2]=25[x^2+(y-8)^2]$$

$$\Rightarrow 2x^2+2y^2+27x-50y+38=0 \quad \text{...(i)}$$

and $\left|\frac{z-4}{z-8}\right|=1\Rightarrow x=6 \quad \text{....(ii)}$

On solving Eqs. (i) and (ii), we get

Point of intersection are (6, 8), (6, 17)

$\therefore \quad z_1=6+8i,\ z_2=6+17i$

$$\therefore \Sigma|z_n|=|z_1|+|z_2|=10+\sqrt{325}=10+5\sqrt{13}$$

71. *(1)* We have,

$$x\sin x\frac{dy}{dx}+(x+x\cos x+\sin x)y=\sin x$$

$$\Rightarrow \quad \frac{dy}{dx}+\left(\frac{1}{\sin x}+\cot x+\frac{1}{x}\right)y=\frac{1}{x}$$

Here, $P=\frac{1}{\sin x}+\cot x+\frac{1}{x}$ and $Q=\frac{1}{x}$

$$\text{IF} = e^{\int \left(\frac{1}{\sin x} + \cot x + \frac{1}{x}\right) dx}$$

$$= e^{\log\left(x \tan\frac{x}{2} \sin x\right)}$$

$$= x \tan\frac{x}{2} \sin x$$

$$= x \tan\frac{x}{2} 2\sin\frac{x}{2} \cos\frac{x}{2}$$

$$= x(1 - \cos x)$$

Now, solution is given by

$$y(\text{IF}) = \int Q(\text{IF})\, dx + C$$

$$xy(1-\cos x) = \int \frac{1}{x} \times x(1-\cos x) dx + C = x - \sin x + C$$

Now, $y\left(\frac{\pi}{2}\right) = 1 - \frac{2}{\pi}$

$\Rightarrow \quad C = 0$

$\therefore \quad y(x) = \dfrac{x - \sin x}{x(1-\cos x)}$

$$= \frac{x - \left(x - \frac{x^3}{3!} + \dots\right)}{x\left[1 - \left(1 - \frac{x^2}{2!} + \dots\right)\right]}$$

Hence, $\lim\limits_{x \to 0} y(x) = 3 \lim\limits_{x \to 0} \dfrac{x^3}{6} \times \dfrac{1}{x^3/2}$

$$= 3 \times \frac{1}{3} = 1 \qquad [\text{as } x \to 0]$$

Study Tactics

The given equation is in the form of linear differential equation. So, first we solve for integrating factor in which we use the property of logarithm, then we get the required solution of differential equation. At last, use expansion series of sin x and cos x.

72. *(5)* $\because 0 < x < \dfrac{\pi}{2}$

$\therefore 0 < \sin^2 x < 1$

Then, $\sin^2 x + \sin^4 x + \sin^6 x + \dots \infty = \dfrac{\sin^2 x}{1 - \sin^2 x}$

$$= \tan^2 x$$

$\therefore \exp[(\sin^2 x + \sin^4 x + \sin^6 x + \dots \infty) \log_e 2]$

$$= \exp[\tan^2 x \log_e 2]$$

$$= \exp[\log_e 2^{\tan^2 x}] = e^{\log_e 2^{\tan^2 x}}$$

$$= 2^{\tan^2 x}$$

Let $y = 2^{\tan^2 x}$

$\because$ y satisfies the quadratic equation $y^2 - 9y + 8 = 0$.

$\Rightarrow \quad y = 1, 8$

If $y = 1 = 2^{\tan^2 x}$

$\therefore \quad 2^{\tan^2 x} = 2^0$

$\Rightarrow \quad \tan^2 x = 0$

$x = 0$ (impossible) $\quad [\because x > 0]$

Now, if $\quad y = 8 = 2^{\tan^2 x}$

$\therefore \quad 2^{\tan^2 x} = 2^3$

$\Rightarrow \quad \tan^2 x = 3$

$\Rightarrow \quad \tan x = \sqrt{3}$

So, $\dfrac{\sin x - \cos x}{\sin x + \cos x} = \dfrac{\tan x - 1}{\tan x + 1}$

$$= \frac{\sqrt{3}-1}{\sqrt{3}+1} \times \frac{\sqrt{3}-1}{\sqrt{3}-1}$$

$$= \frac{(\sqrt{3}-1)^2}{3-1}$$

$$= \frac{3 + 1 - 2\sqrt{3}}{2} = 2 - \sqrt{3}$$

On comparing it with $a - \sqrt{b}$, we get

$$a = 2 \text{ and } b = 3$$

Hence, $a + b = 5$

73. *(5)* $\sum\limits_{i=1}^{n} (x_i + 1)^2 = 9n \quad \dots(\text{i})$

$\sum\limits_{i=1}^{n} (x_i - 1)^2 = 5n \quad \dots(\text{ii})$

On adding Eqs. (i) and (ii), we get

$$\sum_{i=1}^{n} \{(x_i+1)^2 + (x_i-1)^2\} = 14n$$

$\Rightarrow \quad \sum\limits_{i=1}^{n} 2(x_i^2 + 1) = 14n$

$\Rightarrow \quad \sum\limits_{i=1}^{n} x_i^2 + \sum\limits_{i=1}^{n} 1 = \dfrac{14n}{2}$

$\Rightarrow \quad \sum\limits_{i=1}^{n} x_i^2 = 6n$

$\Rightarrow \quad \dfrac{\Sigma x_i^2}{n} = 6 \quad \dots(\text{iii})$

On subtracting Eqs. (i) and (ii), we get

$$\sum_{i=1}^{n}\{(x_i+1)^2-(x_i-1)^2\}=4n$$

$$\Rightarrow \quad 4\sum_{i=1}^{n} x_i = 4n$$

$$\Rightarrow \quad \frac{\Sigma x_i}{n}=1$$

$$\therefore \text{ Variance} = \frac{\Sigma x_i^2}{n}-\left(\frac{\Sigma x_i}{n}\right)^2$$

$$=6-1=5$$

74. *(15)* $\because 17^2 = 289 = 290-1$

Now, $17^{256}=(290-1)^{128}$

$$=(1-290)^{128}$$

$$=(290)^{128}-{}^{128}C_1(290)^{127}+{}^{128}C_2(290)^{126}-\dots$$

$$\dots-{}^{128}C_{125}(290)^3+{}^{128}C_{126}(290)^2$$

$$-{}^{128}C_{127}(290)+1$$

$$=1000m+{}^{128}C_{126}(290)^2-{}^{128}C_{127}(290)+1$$

where, m is a positive integer.

$$=1000m+\frac{128\times 127}{2}\times(290)^2-128(290)+1$$

$$=1000m+683527680+1$$

Last three digits are 6, 8 and 1.

$$\therefore \quad x=6+8+1$$

$$=15$$

75. *(54)* Let n be the number of sides of the polygon.

$$n150=(n-2)180$$

$$\Rightarrow \quad 150n-180n=-360$$

$$\Rightarrow \quad 30n=360$$

$$\Rightarrow \quad n=12$$

Then, the number of diagonals $={}^{12}C_2-12$

$$=66-12$$

$$=54$$

Time Saver Tip

We know that

$$\text{Number of sides}=\frac{360^\circ}{180^\circ-\text{Interior angle}}$$

$$=\frac{360^\circ}{30^\circ}=12$$

Hence, the number of diagonals $={}^nC_2-n$

$$=54$$

SECTION II

Mastery Builders

PRACTICE SET 5

Total Marks : 300	Total Time : 3 hrs	No. of Questions : 75

Instructions

1. There are **three** parts in the questions paper consisting of **Physics**, **Chemistry** and **Mathematics** having 25 questions in each part of equal weightage. Each part (subject) has two sections.
2. **Section A** This section contains 20 multiple choice questions which have only one correct answer. Each question carries **4 marks** for correct answer and **– 1 mark** for wrong answer.
3. **Section B** This section contains 5 questions. The answer to each of the questions is a numerical value. Each question carries **4 marks** for correct answer and **– 1 mark** for wrong answer. For Section B, the answer should be rounded off to the nearest integer.

PHYSICS

Section A : Objective Type Questions

1. Match Column I (activity) with Column II (EM Waves) and select the correct answer from the codes given below.

Column I	Column II
A. Investigation of solid structure	1. Infrared rays
B. Used in remote switches of household electric system	2. Ultraviolet rays
C. Ozone layer stops	3. Gamma rays
D. Kill pathogenic microorganisms	4. X-rays

Codes

	A	B	C	D		A	B	C	D
(a)	1	4	2	3	(b)	4	2	1	3
(c)	4	1	2	3	(d)	3	2	4	1

2. A rolling body was initially at rest on a rough floor and a force F (parallel to floor) is applied on the top of the rolling body. Choose the correct option for the pure accelerated rolling.

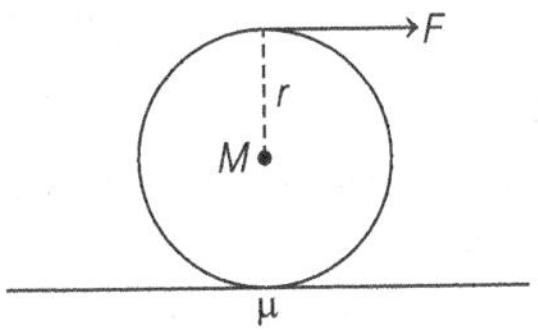

(a) The kinetic friction (f_k) will act in opposite direction of force
(b) The static friction must act
(c) The friction will not act in the backward direction
(d) If rolling body is a disc, the static friction will act in the backward direction

3. **Assertion** The application of brakes on the driving wheels of a bus means the reversal of the direction of friction between the wheels and the ground.
Reason The application of brakes means the application of force on the axle in the opposite direction. Consequently, the frictional force is also reversed.
In the light of above statements, choose the most appropriate answer from the options given below.
(a) Both Assertion and Reason are true and Reason is the correct explanation of Assertion
(b) Both Assertion and Reason are true Reason is not the correct explanation of Assertion
(c) Assertion is true but Reason is false
(d) Assertion is false but Reason is true

4. A particle is moving with velocity $\mathbf{v} = k(y\hat{\mathbf{i}} + x\hat{\mathbf{j}})$, where k is a constant. The general equation for its path is

(a) $xy = \text{constant}$ (b) $y^2 = x^2 + \text{constant}$
(c) $y^2 = x + \text{constant}$ (d) $y = x^2 + \text{constant}$

5. Distance travelled by a particle as a function of time is given below as

$$s = \frac{t^3}{3} - 9t + 8$$

Then, find the time t in second when power delivered to the body is minimum.

(a) $\sqrt{2}$ s (b) $\sqrt{3}$ s
(c) 2 s (d) $\sqrt{5}$ s

6. An object is at distance $u = (60 \pm 0.6)$ cm and image formed by a mirror on the same side at a distance $v = (30 \pm 0.3)$ cm, then focal length is

(a) (20 ± 0.4) cm (b) (20 ± 0.2) cm
(c) (20 ± 0.1) cm (d) (20 ± 0.6) cm

7. A length is measured by a vernier calliper which has positive zero error of 0.3 mm. While the measurement of length, it is found that 0 on the vernier scale lies between 6.5 cm and 7.5 cm. The vernier coincide is 4, then the correct value of measurement of length is (Take least count = 0.01 cm)

(a) 6.50 cm (b) 6.51 cm
(c) 6.52 cm (d) 6.53 cm

8. Choose the incorrect statements in the option given below

I. To form a *p-n* junction, as *n*-type silicon crystal grown the process is abruptly altered to grow a *p*-type silicon crystal.

II. Depletion zone appears in a *p-n* junction, when *p*-type silicon is given a negative potential and *n*-type silicon is given a positive potential

III. Diffusion of electrons in a newly formed *p-n* junction creates a junction field that stops further diffusion of electrons from *n* to *p*-side.

(a) I and II only (b) II and III only
(c) III and I only (d) I, II and III

9. Energy levels A, B and C of a certain atom corresponding to increasing values of energy, i.e., $E_A < E_B < E_C$. If λ_1, λ_2 and λ_3 are the wavelengths of radiations corresponding to transition C to B, B to A and C to A, respectively. Then, which of the following statements relations is correct ?

(a) $\lambda_3 = \lambda_1 + \lambda_2$ (b) $\lambda_3 = \dfrac{\lambda_1 \cdot \lambda_2}{\lambda_1 + \lambda_2}$
(c) $\lambda_1 + \lambda_2 + \lambda_3 = 0$ (d) $\lambda_3^2 = \lambda_1^2 + \lambda_2^2$

10. Maximum kinetic energy of a photoelectron is E, when the wavelength of incident light is λ. Its energy becomes 4 times, when wavelength is reduced to $\frac{1}{3}\lambda$, then work-function is

(a) $\dfrac{3hc}{\lambda}$ (b) $\dfrac{hc}{\lambda}$ (c) $\dfrac{hc}{2\lambda}$ (d) $\dfrac{hc}{3\lambda}$

11. Two mirrors are inclined at an angle θ as shown in figure. Light ray is incident parallel to one of the mirrors. Light will start retracing its path after third reflection, if

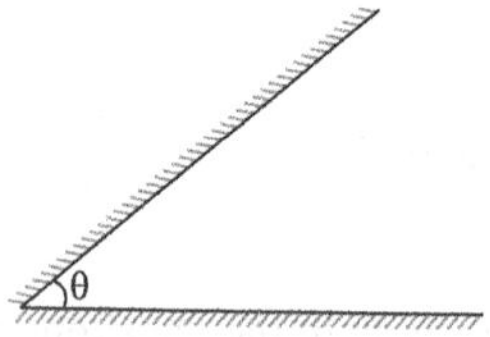

(a) $\theta = 45°$ (b) $\theta = 60°$ (c) $\theta = 30°$ (d) All of these

12. **Statement I** The peak voltage in 220 V AC source is about 311 V.

Statement II A capacitor acts as an infinite resistance for DC.

Choose the correct options

(a) Statement I is correct and Statement II is incorrect
(b) Statement I is incorrect and Statement II is correct
(c) Both Statement I and Statement II are correct
(d) Both Statement I and Statement II are incorrect

13. Two circular loops 1 and 2 are made by the same copper wire but the radius of the first loop is twice to that of the second loop. The ratio of the potential difference applied across the loops, if the magnetic field produced at their centres is equal to

(a) 3 (b) 4 (c) 6 (d) 2

14. A particle of charge $-q$ and mass m moves in a circle of radius r around an infinitely long line charge of linear density $+\lambda$. Then, the time period will be

(a) $T = 2\pi r\sqrt{\dfrac{m}{2K\lambda q}}$ (b) $T^2 = \dfrac{4\pi^2 m}{2K\lambda q} \cdot r^3$
(c) $T = \dfrac{1}{2\pi r}\sqrt{\dfrac{2K\lambda q}{m}}$ (d) $T = \dfrac{1}{2\pi r}\sqrt{\dfrac{m}{2K\lambda q}}$

15. A transverse wave travels on a steel wire with a velocity v when tension in it is 2.06×10^6 N. When the tension is changed to T, the velocity changed to $v/3$. The value of T is close to

(a) 5.28×10^4 N (b) 2.28×10^5 N
(c) 10.50×10^5 N (d) 15.30×10^5 N

16. The average translational energy and the rms speed of molecules in a sample of oxygen at 300 K are 6.21×10^{-21} J and 484 m/s respectively.

The corresponding values at 600 K are nearly (assuming ideal gas behaviour).

(a) 12.42×10^{-21} J, 968 m/s
(b) 6.21×10^{-21} J, 968 m/s
(c) 8.78×10^{-21} J, 684 m/s
(d) 12.42×10^{-21} J, 684 m/s

17. In a thermodynamic process, pressure of fixed mass of a gas is changed in such a manner that the gas releases 20 J of heat and 8 J of work is done on the gas. If initial internal energy of the gas was 40 J, what will be final internal energy?

(a) 18 J (b) 28 J
(c) 52 J (d) 32 J

18. The radius of metal sphere at temperature T is R. If the sphere is heated by a little temperature ΔT, then the increase in volume of sphere (if α is its coefficient of linear expansion) will be

(a) $2\pi R\,\alpha\,\Delta T$ (b) $4\pi R^2\,\alpha\,\Delta T$
(c) $\frac{4\pi}{3}R^3\,\alpha\,\Delta T$ (d) $4\pi R^3\,\alpha\,\Delta T$

19. A bigger drop of liquid is converted into a smaller drop, then (S = surface tension)

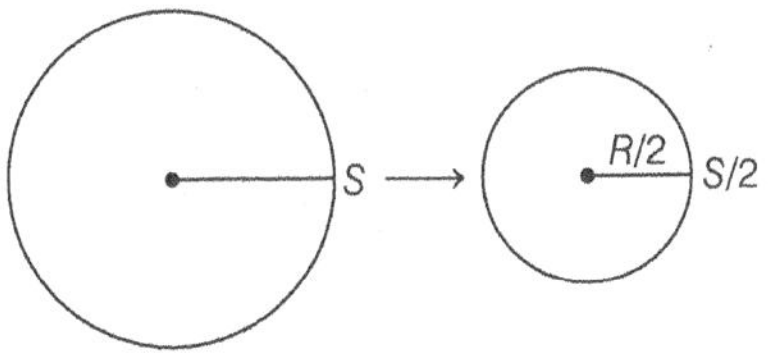

(a) energy released during process = 0
(b) energy released during process $=\frac{\pi R^2 S\times7}{2}$
(c) the pressure inside the bigger drop > pressure inside the smaller drop
(d) if two drops connected by a pipe, the liquid will flow from smaller drop to bigger drop

20. A satellite is revolving around the earth with -200 J of total mechanical energy. The energy needed to throw it out of the Earth's gravitational field is

(a) -200 J (b) $+200$ J
(c) $+100$ J (d) -100 J

Section B : Numerical Value Type Questions

21. A pure germanium plate of area $3.5\times10^{-4}\,m^2$ and of thickness 1.5×10^{-3} m is connected across a battery of potential 5V. The amount of heat generated in plate in 120 s is $x\times10^{-11}$ J. Given that, the concentration of carriers in germanium at room temperature is 1.6×10^{6} per cubic metre. The mobilities of electron and holes are $0.42\,m^2V^{-1}s^{-1}$ and $0.2\,m^2V^{-1}s^{-1}$, respectively. The value of x is ....

22. When switch S is closed, then the reading of ammeter is ....A.

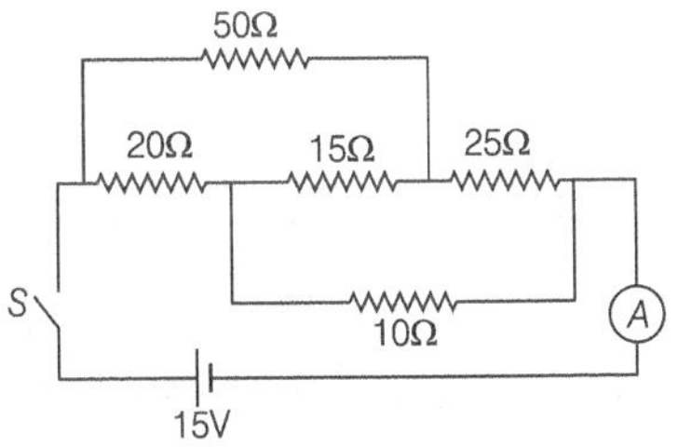

23. A hydrogen atom initially in the ground level absorbs a photon, which excites it to the $n=4$ level. The wavelength of photon is .... Å.

24. The horizontal range of a projectile fired at an angle of 15° is 50 m. If it is fired with the same speed at an angle of 45°, its range will be .... m

25. A pendulum of length $l=1$ m is released from $\theta_0=60°$. The rate of change of speed of the bob at $\theta=30°$ is ... ms^{-2}.

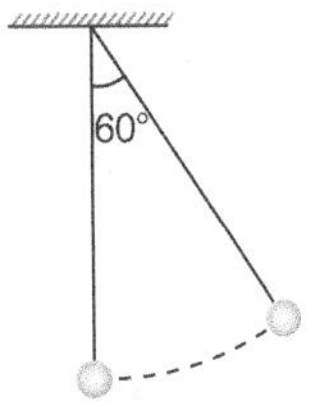

CHEMISTRY

Section A : Objective Type Questions

26. Which of the following statement is correct regarding following process?

(i) $Cl \xrightarrow{EA} Cl^-$ (ii) $Cl^- \xrightarrow{IE} Cl$

(iii) $Cl \xrightarrow{IE} Cl^+$ (iv) $Cl \xrightarrow{IE} Cl^{2+}$

(a) |IE of Process (iv)| = |IE of Process (iii)|
(b) |IE of Process (iii)| = |IE of Process (ii)|
(c) |IE of Process (ii)| = |EA of Process (i)|
(d) |IE of Process (iv)| = |EA of Process (i)|

27. Match List-I with List-II.

	List-I		List-I
A.	Silica gel	I.	Transformer plates
B.	Silicon	II.	Ion-exchanger
C.	Silicone	III.	Drying agent
D.	Silicate	IV.	Sealant

Choose the correct answer from the options given below.

(a) A-II, B-I, C-III, D-IV (b) A-III, B-I, C-II, D-IV
(c) A-III, B-I, C-IV, D-II (d) A-IV, B-II, C-III, D-I

28. $NiCl_2[P(C_2H_5)_2(C_6H_5)]_2$ exhibits temperature dependent magnetic behaviour (paramagnetic/diamagnetic). The coordination geometries of Ni^{2+} in the paramagnetic and diamagnetic states respectively are

(a) tetrahedral and tetrahedral
(b) square planar and square planar
(c) tetrahedral and square planar
(d) square planar and tetrahedral

29. The oxidation number of nitrogen atoms in ammonium nitrate are

(a) $-3, +5$ (b) $-3, -3$ (c) $+5, +5$ (d) $+2, -3$

30. Consider the following reaction,

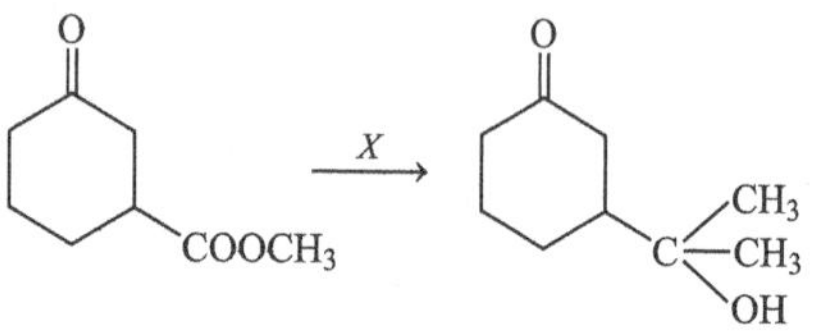

The reagent 'X' in the reaction is

(a) $HOCH_2—CH_2OH \mid H^+$, $LiAlH_4$ |ether, $2CH_3MgBr$, H_3O^+
(b) $2CH_3MgBr$ and H_3O^+
(c) $HOCH_2—CH_2OH \mid H^+$, $2CH_3MgBr$, H_3O^+
(d) $HOCH_2—CH_2OH \mid H^+$, H_2 |Pt, CH_3OH, H^+

31. Which of the following species has least bond angle?

(a) NH_3 (b) AsH_3 (c) PH_3 (d) O_2

32. The maximum probability of finding an electron in the d_{xz}-orbital is along

(a) x-axis (b) z-axis
(c) 45° from x and z-axis (d) 90° from x and z-axis

33. For a given reaction energy of activation for forward reaction $(E_a)_f$ is 80 kJ mol^{-1}, $\Delta H = -40$ kJ mol^{-1} for the reaction. A catalyst lowers $(E_a)_f$ to 20 kJ mol^{-1}. The ratio of energy of activation for reverse reaction before and after addition of catalyst is

(a) 1 : 2 (b) 2 : 5 (c) 5 : 2 (d) 2 : 1

34. Which of the following carbocation is most stable?

(a) 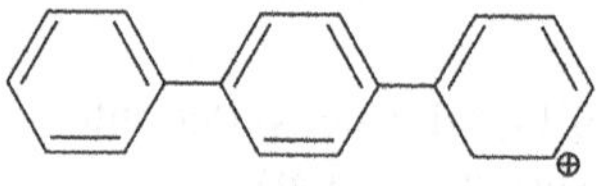

(b) 

(c) 

(d) 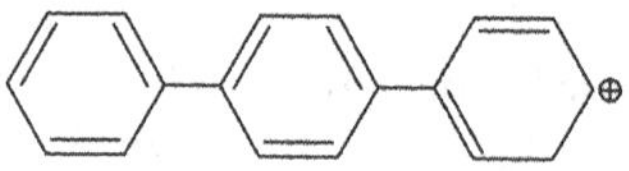

35. Consider the plots for the types of reaction,

$$nA \longrightarrow B + C$$

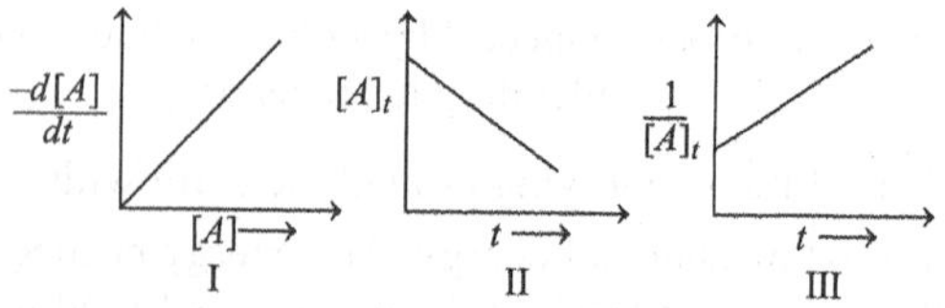

These plots respectively correspond to the reaction orders

(a) 0, 2, 1 (b) 0, 1, 2 (c) 1, 1, 2 (d) 1, 0, 2

36. Which of the following is the correct order of stability of conformational isomers of 2-amino ethan-l-ol?

(a) *Gauche* > eclipsed > *anti*
(b) *Gauche* > *anti* > eclipsed
(c) Eclipsed > *gauche* > *anti*
(d) *Anti* > eclipsed > *gauche*

37. With respect to the following reaction, consider the given statements,

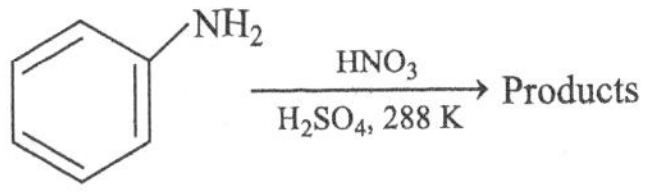

A. *o*-nitroaniline and *p*-nitroaniline are the predominant products.
B. *p*-nitroaniline and *m*-nitroaniline are the predominant products.
C. HNO_3 acts as an acid.
D. H_2SO_4 acts as an acid.

Choose the most appropriate answer from the options given below.

(a) (A) and (C) are correct
(b) (A) and (D) are correct
(c) (B) and (D) are correct
(d) (B) and (C) are correct

38. Which of the following given molecule has $p\pi$-$d\pi$ bonding?

(a) SO_3^{2-} (b) NO_3^- (c) CO_3^{2-} (d) BO_3^{3-}

39. Identify the correct order for the given property for following compounds.

A. Unpaired electrons $Mn^{2+} > Cr^{2+} > V^{2+} > Cu^{2+}$
B. Bond angle $Cr_2O_7^{2-} > CrO_4^{2-} > MnO_4^-$
C. Enthalpy of atomisation, $\Delta aH^\ominus$ (kJ/mol) Zn > Ni > Co > V
D. Number of (Cr—O) bonds $Cr_2O_7^{2-} > CrO_4^{2-}$
E. Density(g cm^{-3}) Ni > Mn > Cr > V

Choose the correct answer from the options given below.

(a) B, A and C (b) Only D
(c) C, D and E (d) A, D and E

40. Match the List-I with List-II.

List-I		List-II
A. Proteins	I.	Monosaccharides
B. Carbohydrates	II.	Triglycerides
C. Lipids	III.	Nucleotides
D. Nucleic acids	IV.	Amino acids

Choose the correct option given below.

(a) A-I, B-II, C-IV, D-III
(b) A-IV, B-I, C-II, D-III
(c) A-II, B-III, C-I, D-IV
(d) A-IV, B-III, C-II, D-I

41. Consider the following statements.

Statement I Thermal decomposition of $Ba(N_3)_2$ produces ammonia (NH_3) gas.

Statement II The pentavalent oxide of group-15 element E_2O_5 is less acidic than trivalent oxide E_2O_3 of the same element.

In light of the above statements choose the correct answer from the options given below.

(a) Both Statement I and Statement II are correct.
(b) Both Statement I and Statement II are incorrect.
(c) Statement I is correct but Statement II is incorrect.
(d) Statement I is incorrect but Statement II is correct.

42. Given below are two statements one is labelled as Assertion (A) and other is labelled as Reason (R).

Assertion (A) Phenolphthalein is a pH dependent indicator, remains colourless in basic medium and gives pink colour in acidic solution.

Reason (R) Phenolphthalein is a weak acid. It dissociates in basic medium.

In light of the above statements choose the correct answer from the options given below.

(a) Both A and R are true but R is the correct explanation of A.
(b) Both A and R are true but R is not the correct explanation of A.
(c) A is true but R is false.
(d) A is false but R is true.

43. For independent process at 350 K.

Process	ΔH (kJ mol^{-1})	ΔS (JK^{-1})
A	35	15
B	25	– 16
C	–15	20
D	– 40	–15
E	10	15

The number of spontaneous processes from the given data is

(a) 2 (b) 3 (c) 4 (d) 1

44. Identify the name of the given reaction,

$$\text{Phenol} \xrightarrow{NaNO_2/H_2SO_4} B \xrightarrow{H_2O} C \xrightarrow{NaOH} \text{Indophenol salt}$$

(a) Phthalein fusion test
(b) Liebermann's reaction
(c) Schotten-Baumann reaction
(d) Reimer-Tiemann reaction

45. The final major product B of the following reaction is

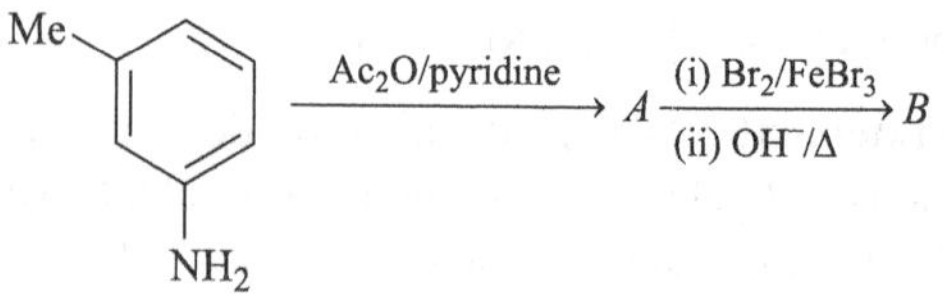

(a) 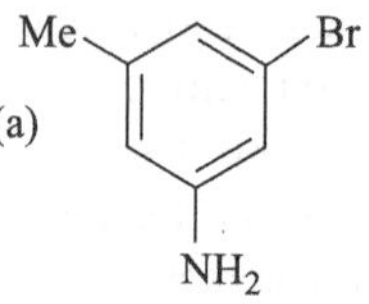

(b) 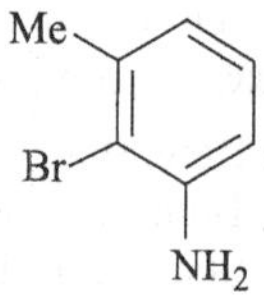

(c) 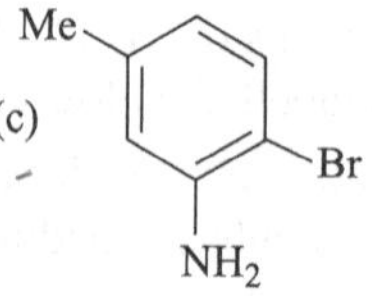

(d) 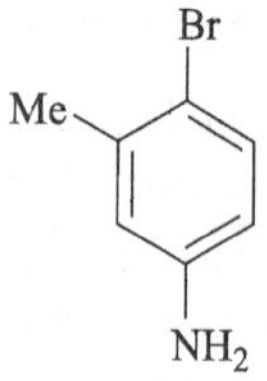

Section B : Numerical Value Type Questions

46. Given, Pt | H_2 (g, 0.1 atm) | H^+ (*aq*, 10^{-2} M)|| MnO_4^- (*aq*, 0.1 M)| Mn^{2+} (*aq*, 0.01M) M), | Pt

The potential of the above cell will be ....$\times 10^{-2}$ V. $[E^\circ_{MnO_4^-/Mn^{2+}} = 1.50 \text{ V}]$

47. The vapour pressure of water at 20°C is 17.5 mm Hg. If 18 g glucose ($C_6H_{12}O_6$) is added to 178.2 g of water at 20°C, the vapour pressure of the resulting solution will be ......mm Hg.

48. 5.75 g of a dibasic acid was dissolved in water and diluted to 250 mL. 35 mL of this solution was neutralised by 15 mL of 1 N NaOH solution. The molecular weight of the acid will be ...... g/mol.

49. A solution is prepared by mixing 0.2 mol each of H_2CO_3, $NaHCO_3$, Na_2CO_3 and NaOH in 200 mL of water. pH of the solution is ....

Given, pKa_1 and pKa_2 of H_2CO_3 are 6.37 and 10.32

50. In Duma's method for estimation of nitrogen, 0.40 g of organic compound gave 50 cm^3 of nitrogen gas collected at 300 K and 715 mm Hg pressure. If vapour pressure of water at 300 K is 15 mm Hg, the value of percentage of nitrogen in the compound is .... %.

MATHEMATICS

Section A : Objective Type Questions

51. The number of permutations of the letters *a*, *b*, *c* and *d* such that *b* does not follow *a* and *c* does not follow *b* and *d* does not follow *c*, is

(a) 9 (b) 11 (c) 13 (d) 14

52. The sum to m term of the series

$$\frac{3}{1^2 \times 2^2} + \frac{5}{2^2 \times 3^2} + \frac{7}{3^2 \times 4^2} + \dots$$

(a) $\frac{2m+m^2}{(m+1)^2}$ (b) $\frac{2m+1}{m^2(m+1)^2}$

(c) $\left[\frac{1}{m(m+1)}\right]^2$ (d) $\frac{1}{(m+1)^2}$

53. If $(x+y)^2 \frac{dy}{dx} = a^2$, $y(0) = 0$ at $y = a$, then $\frac{x}{a}$ is equal to

(a) $2 - \tan 2$ (b) $\tan 2$

(c) $\tan 2 + 1$ (d) $\tan 1 - 1$

54. If $A(z_1)$, $B(z_2)$ and $C(z_3)$ are the vertices of an equilateral ΔABC, then the value of

$$\frac{1}{2}\left[\arg\left(\frac{z_2+z_3-2z_1}{z_3-z_2}\right)\right]$$ is equal to (where, [·] is greatest integer function)

(a) 1 (b) $\frac{1}{2}$ (c) 2 (d) $\frac{3}{2}$

55. A fair coin is tossed until two consecutive tails is obtained. The probability that two consecutive tails occur on the seventh and eighth flips is equal to

(a) $\frac{11}{256}$ (b) $\frac{15}{256}$ (c) $\frac{13}{256}$ (d) $\frac{17}{256}$

56. Let $g(x) = \begin{vmatrix} a & a^2 & a^3 \\ 0 & -1 & 6 \\ \cos x & \sin x & x^3 \end{vmatrix}$, where a is a constant.

Then,

Statement I $\frac{d^3}{dx^3}[g(x)]$ at $x = 0$ is independent of a.

Statement II $\frac{d^2}{dx^2}[g(x)]$ at $x = 0$ is $a + a^3$.

(a) Statement I and Statement II both are true.
(b) Statement I is true and Statement II is false.
(c) Statement I is false and Statement II is true.
(d) Statement I and Statement II both are false.

57. $\int \frac{x+\sqrt[3]{x^2}+\sqrt[6]{x}}{x(1+\sqrt[3]{x})}dx$ is equal to

(a) $\frac{3}{2}x^{2/3}+\tan^{-1}x^6+C$

(b) $\frac{3}{2}x^{2/3}+6\tan^{-1}x^6+C$

(c) $\frac{3}{2}x^{2/3}+6\tan^{-1}x^{\frac{1}{6}}+C$

(d) $\frac{3}{2}x+6\tan^{-1}x$

58. The value of p, for which the equation $4\operatorname{cosec}^2\{\pi(p+x)\}+p^2-4p=0$ has a real solution, is

(a) 0 (b) 1
(c) 2 (d) 3

59. If the length of the perpendicular from the point $(\alpha, 0, \alpha)$ $(\alpha \neq 0)$ to the line $\frac{x}{1}=\frac{y-1}{0}=\frac{z+1}{-1}$ is $\sqrt{\frac{3}{2}}$, then α is equal to

(a) 1 (b) 2
(c) −1 (d) −2

60. The jump of the function at the point of the discontinuity of the function

$f(x)=\frac{1-k^{1/x}}{1+k^{1/x}}$ $(0<k<1)$, is

(a) 3 (b) 1 (c) 0 (d) 2

61. The vectors **a** and **b** are non-collinear. The value of $(-2x)$ for which the vectors $\mathbf{c}=(2x+3)\mathbf{a}+\mathbf{b}$ and $\mathbf{d}=(2x+3)\mathbf{a}-\mathbf{b}$ are collinear, is

(a) 1 (b) 2
(c) 3 (d) 4

62. A line segment has length 56 units and direction ratios are 3, –2, 6. If the line makes an obtuse angle with X-axis, the sum of components of the line vector are

(a) – 24, 16, – 48
(b) – 56, – 22, 84
(c) – 42, 42, – 84
(d) – 62, 52, 89

63. If $f(x)=x\sin x-\frac{1}{2}\sin^2 x$, $x\in\left(0,\frac{\pi}{2}\right)$ and the range of $f(x)$ is $\left(0,\frac{\pi-k}{k+1}\right)$, then k is equal to

(a) 1 (b) 2 (c) 3 (d) 4

64. Let $g(x)=\frac{1}{x}f(x)$, where

$f(x)=\begin{cases} x\sin\frac{1}{x}, & x\neq 0 \\ 0, & x=0\end{cases}$, then at $x=0$ is

(a) g is differentiable but $f(x)$ is not differentiable.
(b) Both g and f are differentiable.
(c) g is not continuous and f is differentiable.
(d) g is not continuous and f is not differentiable.

65. The domain of definition of the function $f(x)=\log_{10}\{1-\log_{10}(x^2-5x+16)\}$, is

(a) [2, 3] (b) [2, 3) (c) (2, 3] (d) (2, 3)

66. If z is a complex number satisfying $|z|^2+2(z+\bar{z})+3i(z-\bar{z})+4=0$, then complex number $z+3+2i$ will lie on

(a) circle with centre $(1-5i)$, radius 4
(b) circle with centre $(1+5i)$, radius 4
(c) circle with centre $(1+5i)$, radius 3
(d) circle with centre $(1-5i)$, radius 3

67. If the median and the range of four numbers $\{x, y, 2x+y, x-y\}$, where $0<y<x<2y$ are 10 and 28 respectively, then $\frac{y+x}{\text{Mean}}$ is equal to

(a) $\frac{7}{10}$ (b) $\frac{4}{7}$ (c) $\frac{6}{7}$ (d) $\frac{10}{7}$

68. If $ax+by+c=0$ is an equation of the line joining the extremities of a pair of semi-conjugate diameters of an ellipse $\frac{x^2}{9}+\frac{y^2}{4}=1$, then $\sqrt{\frac{9a^2+4b^2}{c^2}}$ is equal to

(a) 1 (b) $2\sqrt{2}$ (c) $\sqrt{2}$ (d) $2\sqrt{3}$

69. $P(a\sec\theta, b\tan\theta)$ and $Q(a\sec\phi, b\tan\phi)$ are the ends of a focal chords of $\frac{x^2}{a^2}-\frac{y^2}{b^2}=1$, then $\tan\frac{\theta}{2}\tan\frac{\phi}{2}$ is equal to

(a) $\frac{e-1}{e+1}$ (b) $\frac{1-e}{1+e}$ (c) $\frac{1+e}{1-e}$ (d) $\frac{e+1}{e-1}$

70. If $A = \begin{bmatrix} 1 & -\tan\theta \\ \tan\theta & 1 \end{bmatrix} \begin{bmatrix} 1 & \tan\theta \\ -\tan\theta & 1 \end{bmatrix}^{-1} = \begin{bmatrix} p & -q \\ q & p \end{bmatrix}$,

then $p^2 + q^2$ is equal to

(a) 1 (b) 0 (c) -1 (d) 2

Section B : Numerical Value Type Questions

71. Suppose $f(x) = ax + b$ and $g(x) = bx + a$, where $a, b \in N$. If $f(g(50)) - g(f(50)) = 28$, then number of all possible values of the product ab is equal to …… .

72. If m_1 and m_2 are the roots of the equation $x^2 + (\sqrt{3}+2)x + (\sqrt{3}-1) = 0$ and if area of the triangle formed by the lines $y = m_1 x$, $y = m_2 x$ and $y = c$ is $(a+b)c^2$, then the value of 2008 $(a^2 + b^2)$ must be ……… .

73. If area between the curves $y = xe^x$ and $y = xe^{-x}$ and the line $x = 1$ is λ sq unit, then the value of 128 $e\lambda$ must be ……… .

74. Suppose p and q are 5-digit numbers. If the numbers of ways of forming p and q, so that these numbers can be added without carrying at any stage is $36(55)^k$, then k is equal to ……… .

75. If $1 + x^3 + x^4 = \sum_{i=0}^{4} a_i (1+x)^i$, $\forall x \in R$, then $|a_3|$ is …

DETAILED SOLUTIONS

Physics

1. *(c)* The structure of solids is investigated by X-rays. Infrared rays are widely used in remote switches of household electronics system such as TV sets, video recorders etc. Ozone layer of atmosphere blocks ultraviolet rays.

Gamma rays are used in food industry to kill pathogenic microorganism.

2. *(c)* This question is based on pure accelerated rolling. For pure accelerated rolling, $a_{CM} = R\alpha$. In this case, the friction will be zero if the rolling body is ring and for any other rolling body, static friction will be in the forward direction.

The rolling body may be of different types, so moment of inertia of different rolling bodies will be different.

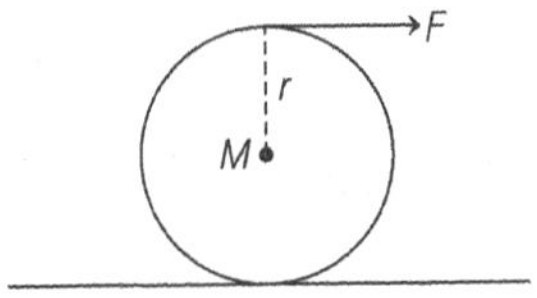

For ring,

$$\frac{F}{M} = a_{CM}$$

$\Rightarrow$ $$\tau = I\alpha \quad \Rightarrow Fr = I\alpha$$

$\Rightarrow$ $$\alpha = \frac{Fr}{I} = \frac{Fr}{Mr^2} = \frac{F}{Mr}$$

For pure accelerated rolling,

$$a_{CM} = r\alpha$$

$$\frac{F}{M} = r\frac{F}{Mr}$$

So, for ring, there is no need of friction.

Other than ring

$$I_{\text{other rolling body}} < I_{\text{ring}}$$

3. *(a)* The application of brakes means the application of force on the axle in the opposite direction. Consequently, the frictional force is also reversed. Hence, the application of brakes on the driving wheels of a bus means the reversal of the direction of friction between the wheels and the ground.

4. *(b)* The velocity of the particle, $\mathbf{v} = ky\hat{\mathbf{i}} + kx\hat{\mathbf{j}}$

$\Rightarrow$ $$\frac{dx}{dt} = ky, \frac{dy}{dt} = kx$$

$$\frac{dy}{dx} = \frac{dy}{dt} \times \frac{dt}{dx} = \frac{kx}{ky}$$

$$ydy = xdx$$

$\Rightarrow$ $$y^2 = x^2 + c$$

This is the required general equation.

5. *(b)* Given, $s = \frac{t^3}{3} - 9t + 8$

Velocity, $v = \frac{ds}{dt} = t^2 - 9$

and acceleration is, $a = \frac{d^2s}{dt^2} = 2t$

So, force acting on it is

$$F = m \times a = 2mt$$

where, m is the mass of particle.

Now, power is $P = F \cdot v = 2mt\,(t^2 - 9)$

When power is minimum, then

$$\frac{dP}{dt} = 0$$

$\Rightarrow$ $$2m[3t^2 - 9] = 0$$

$\Rightarrow$ $$3t^2 - 9 = 0$$

$\Rightarrow$ $$t^2 = 3$$

or $$t = \sqrt{3}\text{ s}$$

also $$\left.\frac{d^2P}{dt^2}\right]_{\text{at } t = \sqrt{3}}$$ is positive.

So, P is minimum at $t = \sqrt{3}$ s.

Study Tactics

The formula of power in terms of force and velocity is

$$P = F \cdot v = mav = m\frac{d^2s}{dt^2} \cdot \frac{ds}{dt}$$

$\Rightarrow$ $$\frac{dP}{dt} = m\left(\frac{d^3s}{dt^3}\frac{ds}{dt} + \frac{d^2s}{dt^2}\frac{d^2s}{dt^2}\right)$$

$$0 = m\left[\frac{d^3s}{dt^3}\frac{ds}{dt} + \left(\frac{d^2s}{dt^2}\right)^2\right]$$

Use this method to find the value of t.

6. *(b)* Focal length is given by

$$\frac{1}{f} = \frac{1}{u} + \frac{1}{v} \quad \ldots\text{(i)}$$

So, $$\frac{1}{f} = \frac{1}{-60} + \frac{1}{-30}$$

$\Rightarrow$ $$f = -\frac{60 \times 30}{60 + 30} = -20\text{ cm}$$

To get the error in focal length, differentiate Eq. (i)

$$\therefore \quad -\frac{df}{f^2} = \frac{-1}{u^2} \cdot du - \frac{1}{v^2} \cdot dv$$

Cancelling negative signs and approximating differential change with small Δ change, we get

$$\frac{\Delta f}{f^2} = \frac{\Delta u}{u^2} + \frac{\Delta v}{v^2}$$

or
$$\Delta f = \left[\frac{\Delta u}{u^2} + \frac{\Delta v}{v^2}\right] \cdot f^2$$

$$\Rightarrow \quad \Delta f = \left[\frac{0.6}{60 \times 60} + \frac{0.3}{30 \times 30}\right] \times (-20) \times (-20) = 0.2\ \text{cm}$$

Hence, focal length is (20 ± 0.2) cm taking only magnitude.

7. *(b)* Positive error = 0.3 mm

$\therefore$Main scale reading = 6.5 cm

Vernier scale reading = LC × vernier coincide

$$= 0.01 \times 4 = 0.04\ \text{cm}$$

$\therefore$Final value of measured length is

$$L = \text{MSR} + \text{VSR} - \text{Zero error}$$
$$= 6.5 + 0.04 - 0.03\ \text{cm} = 6.51\ \text{cm}$$

8. *(b)* Statement I is correct and Statement II and III are incorrect. Statement II is incorrect as depletion zone exists even in a *p-n* junction with zero bias. Statement III is incorrect as junction field is created by ions formed due to diffusion.

9. *(b)* Consider the energy levels of an atom

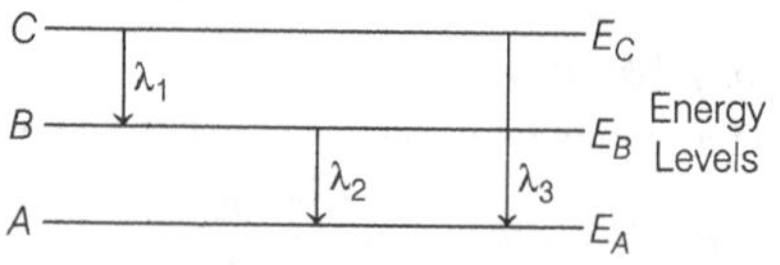

From the figure,

$$(E_B - E_A) + (E_C - E_B) = (E_C - E_A)$$
$$\frac{hc}{\lambda_1} + \frac{hc}{\lambda_2} = \frac{hc}{\lambda_3}$$
$$\lambda_3 = \frac{\lambda_1 \lambda_2}{\lambda_1 + \lambda_2}$$

10. *(d)* Let W be the work-function which corresponds to the energy required to pick out an electron from a material then,

$$\Rightarrow \quad \text{KE} = E = \frac{hc}{\lambda} - W \quad \ldots\text{(i)}$$

$$\Rightarrow \quad 4E = \frac{3hc}{\lambda} - W \quad \ldots\text{(ii)}$$

On dividing Eqs. (i) and (ii), we get

$$\Rightarrow \quad \frac{1}{4} = \frac{\frac{hc}{\lambda} - W}{\frac{3hc}{\lambda} - W}$$

$$\Rightarrow \quad W = \frac{hc}{3\lambda}$$

11. *(c)*

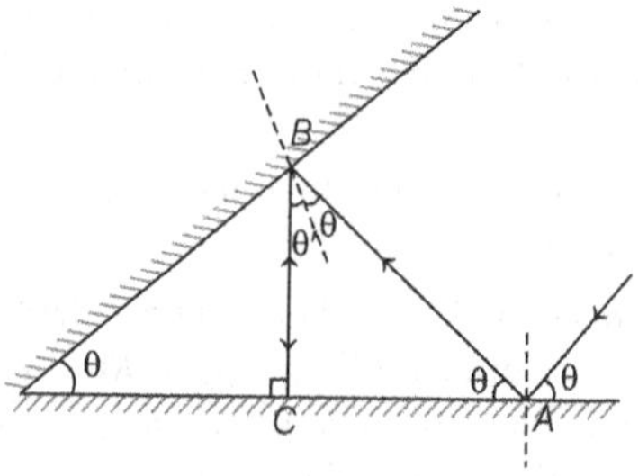

In ΔABC, $90° + 3\theta = 180°$, $\theta = 30°$

12. *(c)* The peak voltage can be calculated by formula

$$V_0 = \sqrt{2} V_{\text{rms}} = \sqrt{2} \times 220 \approx 311\ \text{V}$$

Capacitive resistance can be given by

$$X_C = \frac{1}{\omega C}, \text{ for DC, } \omega = 0$$

So, $X_C = \infty$

13. *(b)* If r_1 and r_2 are the radii of the circular loops 1 and 2 and B_1 and B_2 are magnetic fields at their centres, then

$$B_1 = \frac{\mu_0 I_1}{2r_1} \quad \ldots\text{(i)}$$

$$B_2 = \frac{\mu_0 I_2}{2r_2} \quad \ldots\text{(ii)}$$

Given, $B_1 = B_2$, $r_1 = 2r_2$

From Eqs. (i) and (ii), we get

$$\frac{B_1}{B_2} = \frac{\mu_0 I_1}{2r_1} \times \frac{r_1}{\mu_0 I_2} \quad (\because B_1 = B_2)$$

$$I_1 = 2I_2$$

Suppose R_1 and R_2 are the resistances of the coil 1 and 2, then $R_1 = 2R_2$

Because, resistance of coil $\propto$ length,

Length of first loop, $l_1 = 2\pi r_1$ and

Length of second loop, $l_2 = 2\pi r_2$.

If V_1, V_2 are the potential difference across them, then

$$\frac{V_1}{V_2} = \frac{I_1 R_1}{I_2 R_2} = \frac{(2I_2)(2R_2)}{I_2 R_2} = 4$$

14. *(a)* The electric force on charge due to infinitely long wire provides a centripetal force.

Electric field around an infinite line charge is calculated using Gauss' law

$$E = \frac{\lambda}{2\pi\varepsilon_0 r}$$

Now, particle is moving in a circle. So, centripetal force is provided by electric force, $\frac{mv^2}{r} = \frac{\lambda}{2\pi\varepsilon_0 r}\cdot q$

$$v = \left(\frac{2K\lambda q}{m}\right)^{1/2} \quad \left[\text{Using } K = \frac{1}{4\pi\varepsilon_0}\right]$$

Now the time period, $T = \frac{2\pi r}{v} = 2\pi r \times \sqrt{\frac{m}{2k\lambda q}}$

Study Tactics

Use the electric field due to infinite line charge and obtain the expression of electric force on charge q.

Equate this force with $\frac{mv^2}{r}$ and obtain v. Finally, obtain time period, T using $T = \frac{2\pi r}{v}$.

15. *(b)* Speed of transverse wave over a string is given by

$$v = \sqrt{\frac{T}{\mu}} \quad \ldots\text{(i)}$$

where, T = tension in string

μ = mass per unit length of string.

Here, when velocity is v, then tension, $T_1 = 2.06 \times 10^6$ N

Let when velocity is $v/3$, then tension is T, hence from Eq. (i), we get

$$\frac{v}{v/3} = \sqrt{\frac{T_1}{T}}$$

$$T = \frac{T_1}{9} = \frac{2.06\times 10^6}{9} = 2.28\times 10^5 \text{ N}$$

16. *(d)* As $\text{KE} = \frac{3}{2}KT$ and $v_{\text{rms}} = \sqrt{\frac{3RT}{M}}$

$\text{KE} \propto T$

$\therefore \quad \text{KE}_2 = 2\text{KE}_1 = 2\times 6.21\times 10^{-21}$ J

$\text{KE}_2 = 12.42\times 10^{-21}$ J

$$\frac{v_{\text{rms}2}}{v_{\text{rms}1}} = \sqrt{\frac{T_2}{T_1}} = \sqrt{2}$$

$\therefore \quad v_{\text{rms}2} = \sqrt{2}\times v_{\text{rms}1}$ (where, $v_{\text{rms}1} = 484$ m/s)

$= 684$ m/s

17. *(b)* Here, heat release by gas, $\Delta Q = -20$ J

Work done on gas, $\Delta W = -8$ J

Here, heat released by gas and work done on gas are taken negative.

We know that, first law of thermodynamics,

$$\Delta Q = \Delta U + \Delta W$$

$$\Rightarrow \Delta U = \Delta Q - \Delta W$$

$$= -20 + 8 = -12 \text{ J}$$

Now, given that, initial internal energy of gas,

$$U_{\text{initial}} = 40 \text{ J}$$

$\because \quad \Delta U = U_{\text{final}} - U_{\text{initial}}$

$\therefore \quad U_{\text{final}} = \Delta U + U_{\text{initial}}$

$= (-12 + 40)$ J $= 28$ J

18. *(d)* The volume of a sphere is given by $V = \frac{4}{3}\pi R^3$

$$\therefore \quad \frac{\Delta V}{V} = 3\left(\frac{\Delta R}{R}\right) \quad \ldots\text{(i)}$$

As, $R' = R(1 + \alpha\,\Delta T)$

$\Rightarrow \quad R' - R = \Delta R = R\,\alpha\,\Delta T$

$$\Rightarrow \quad \frac{\Delta R}{R} = \alpha\,\Delta T \quad \ldots\text{(ii)}$$

From Eqs. (i) and (ii), we get

$$\Delta V = (3\alpha\,\Delta T)V$$

$$= 3\alpha\,\Delta T\cdot\frac{4}{3}\pi R^3$$

$$= 4\pi R^3\alpha\,\Delta T$$

19. *(b)* The free surface of a liquid possesses some potential energy which is known as surface energy. As the surface area increases, the surface energy will increase. That means, to increase the surface area, some external work is required and *vice-versa*.

Energy released, $\Delta U = U_{\text{bigger}} - U_{\text{smaller}}$

$$= -\frac{4\pi R^2}{4}\times\frac{S}{2} + 4\pi R^2\times S$$

$$\Rightarrow \quad \Delta U = \pi R^2 S\left[\frac{7}{2}\right]$$

$$(p_{\text{excess}})_{\text{bigger}} = \frac{2S}{R}$$

$$(p_{\text{excess}})_{\text{smaller}} = \frac{2\times S/2}{R/2} = \frac{2S}{R}$$

20. *(b)* An object bound to Earth's gravitational field must have negative potential energy (here PE at infinity to be taken zero). If the total mechanical energy of an object is positive, then it will definitely leave the Earth's gravitational field.

21. (11) Conductivity of semiconductor is given by

$$\sigma = n_e e\mu_e + n_h e\mu_h$$

Since, it is an intrinsic semiconductor.

$\Rightarrow \quad n_e = n_h = n_i$

$$\sigma = n_i e(\mu_e + \mu_h)$$

$$= 1.6\times 10^6 \times 1.6\times 10^{-19}[(0.42) + 0.2]$$

$$= 1.5872\times 10^{-13}\ \Omega^{-1}\text{m}^{-1}$$

Current flowing,

$$I = JA \quad \ldots\text{(i)}$$

where, $J = \sigma E = \sigma \frac{V}{d}$

From Eq. (i), we get

$$I = 1.5872 \times 10^{-13} \left(\frac{5}{1.5 \times 10^{-3}} \right) \times 3.5 \times 10^{-4}$$

$$= 1.852 \times 10^{-13} \text{ A}$$

Now, heat produced,

$$H = VIt = 5 \times 1.852 \times 10^{-13} \times 120$$

$$= 11.11 \times 10^{-11} \text{ J}$$

$$\simeq 11 \times 10^{-11} \text{J}$$

22. (1) The given problem is the application of balanced Wheatstone bridge, i.e.,

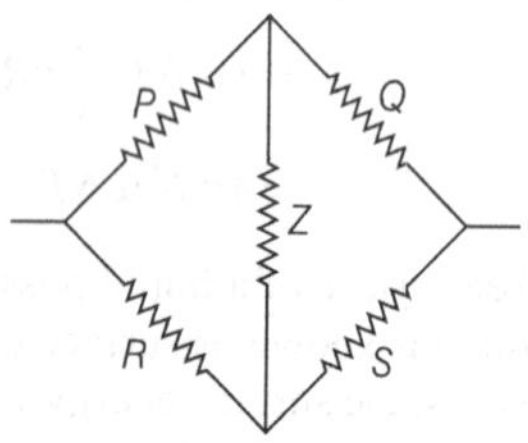

If $\frac{P}{Q} = \frac{R}{S}$, then Z can be removed.

Given, arrangement is a balanced Wheatstone bridge

Clearly, $\frac{50}{20} = \frac{25}{10}$

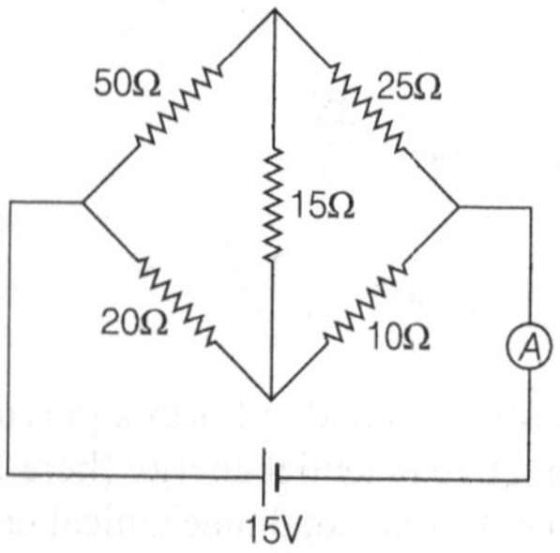

If total resistance is R_T, then

$$\frac{1}{R_T} = \frac{1}{(50+25)} + \frac{1}{(20+10)} = \frac{1}{75} + \frac{1}{30}$$

$$\Rightarrow \quad R_T = \frac{150}{7} \Omega$$

Reading of ammeter,

$$I = \frac{15}{150/7} = 0.7 \text{A}$$

$$\simeq 1 \text{A}$$

23. (974) If $E = h\nu$ be the energy absorbed, then

$$h\nu = \Delta E \Rightarrow h\nu = E_4 - E_1$$

Now, using the relation, $E_n = \frac{-13.6}{n^2}$

For $n = 4$, $E_4 = \frac{-13.6}{4^2}$

$$= \frac{-13.6}{16} = -0.85 \text{ eV} \quad \ldots \text{(i)}$$

As, H-atom initially at ground level, for $n = 1$,

[$\because$ 1eV $= 1.6 \times 10^{-19}$ J and $h = 6.62 \times 10^{-34}$ J]

$$E_1 = -13.6 \text{ eV}$$

$\therefore$ From Eqs. (i) and (ii), we get

$$h\nu = -0.85 - (-13.6)$$

$$= 12.75 \text{ eV} = 12.75 \times 1.6 \times 10^{-19} \text{ J}$$

$$\nu = \frac{12.75 \times 1.6 \times 10^{-19}}{6.62 \times 10^{-34}}$$

$$= 3.08 \times 10^{15} \text{ Hz}$$

Using the relation, $c = \nu\lambda$, we get

$$\lambda = \frac{c}{\nu} = \frac{3 \times 10^8}{3.08 \times 10^{15}}$$

$$= 974.02 \times 10^{-10} \text{ m}$$

$$\approx 974 \text{ Å}$$

24. (100)

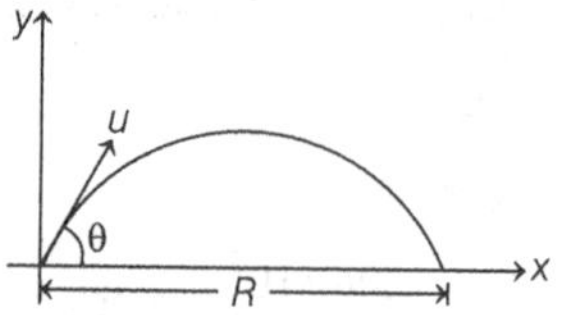

Given, $\theta = 15°$ and $R = 50$ m

Range, $R = \frac{u^2 \sin 2\theta}{g}$

Putting all the given values in the formula, we get

$$\Rightarrow R = 50 \text{m} = \frac{u^2 \sin (2 \times 15°)}{g}$$

$$\Rightarrow 50 \times g = u^2 \sin 30° = u^2 \times \frac{1}{2}$$

$$\Rightarrow 50 \times g \times 2 = u^2$$

$$\Rightarrow u^2 = 50 \times 9.8 \times 2 = 100 \times 9.8 = 980$$

$$\Rightarrow \quad u = \sqrt{980} = \sqrt{49 \times 20} = 7 \times 2 \times \sqrt{5}$$

$$= 14 \times 2.23 \text{ ms}^{-1} = 31.304 \text{ ms}^{-1}$$

For $\theta = 45°$, $R = \frac{u^2 \sin (2 \times 45°)}{g} = \frac{u^2}{g}$ $\quad (\because \sin 90° = 1)$

$$\Rightarrow R = \frac{(14\sqrt{5})^2}{g} = \frac{14 \times 14 \times 5}{9.8} = 100 \text{m}$$

25. (5) Rate of change of speed $= \dfrac{dv}{dt}$

$= \text{tangential acceleration}$

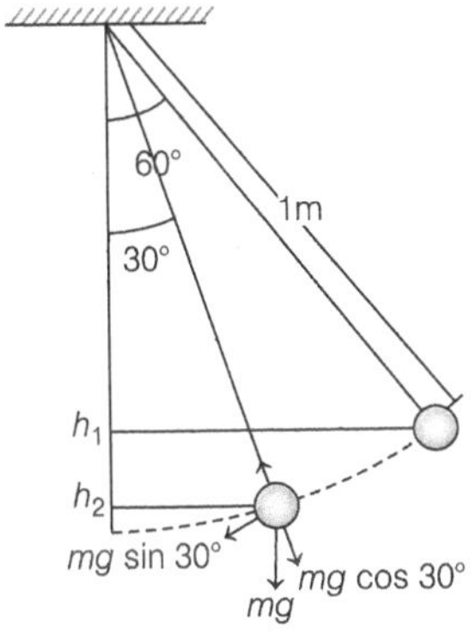

$= \dfrac{\text{tangential forces}}{\text{mass}}$

$= \dfrac{mg \sin 30°}{m}$

$= g \sin 30°$

$= 10 \times \dfrac{1}{2} \text{ ms}^{-2} = 5 \text{ ms}^{-2}$

Chemistry

26. (*c*) Amount of energy involved in losing electron from outermost shell is known as IE while amount of energy involved in accepting one extra electron is known as EA and we know that

$$|\text{IE}| = |\text{EA}|$$

In process (ii), IE is involved.

$$\text{Cl}^- \longrightarrow \text{Cl} + e^-$$

While in process (i), EA is involved.

$$\text{Cl} + e^- \longrightarrow \text{Cl}^-$$

So, same amount of energy is involved in losing or gaining of an electron in both processess.

So, |IE of process (ii)| = |EA of process (i)|

Study Tactics

This problem includes conceptual mixing of IE and EA. It can be solved by knowing the exact concept involved in IE and EA.

27. (*c*) The correct match is A-III, B-I, C-IV, D-II

Silica gel absorbs moisture from the surrounding, thus used as drying agent. Silicon being a semiconductor is used in transformer plates. Silicone is a sealant while silicates including zeolites are ion-exchangers.

28. (*c*) $NiCl_2 [P(Et)_2 (Ph)]_2$ exhibits temperature dependent magnetic behaviour.

Oxidation state of Ni = + 2

Electronic configuration of $Ni^{2+} = [Ar]\, 3d^8\, 4s^0$

When paramagnetic i.e. presence of unpaired electrons then,

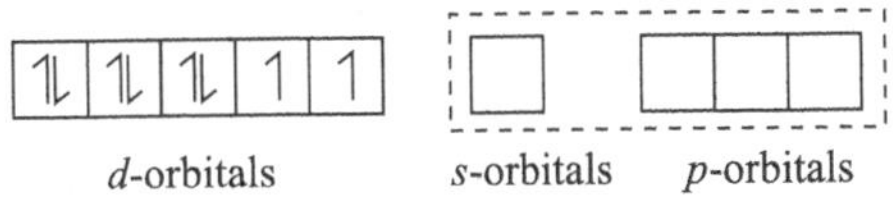

sp^3 hybridisation i.e. tetrahedral geometry.

When diamagnetic i.e. all the electrons are paired.

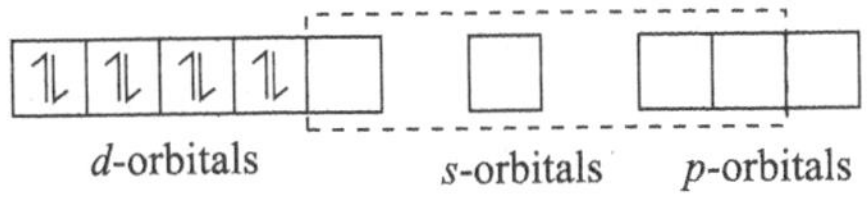

dsp^2 hybridisation i.e. square planar geometry.

Hence, coordination geometries of Ni^{2+} in paramagnetic and diamagnetic states are tetrahedral and square planar respectively.

29. (***a***) Ammonium nitrate (NH_4NO_3) exists as $NH_4^+ \; NO_3^-$.

In NH_4^+, N have oxidation state.

$$x + 4 = +1 \Rightarrow x = -3$$

In NO_3^-, N have oxidation state.

$$x - 6 = -1 \Rightarrow x = +5$$

30. (*c*) Aldehyde and ketone are protected by formation of acetal. The reaction is represented as

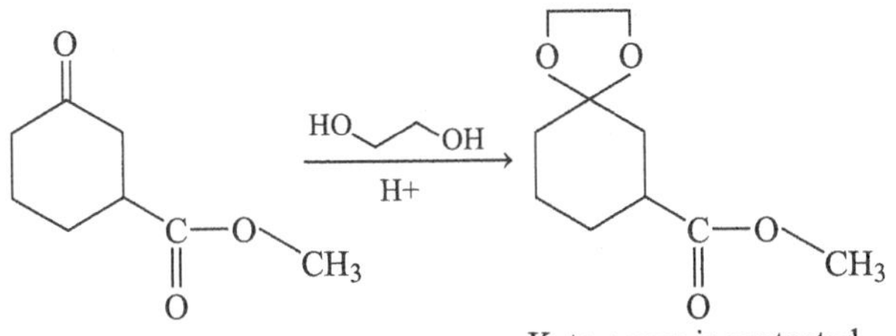

Keto group is protected

Addition of Grignard reagent to ester group leads to formation of ketone and then alcohol after hydrolysis.

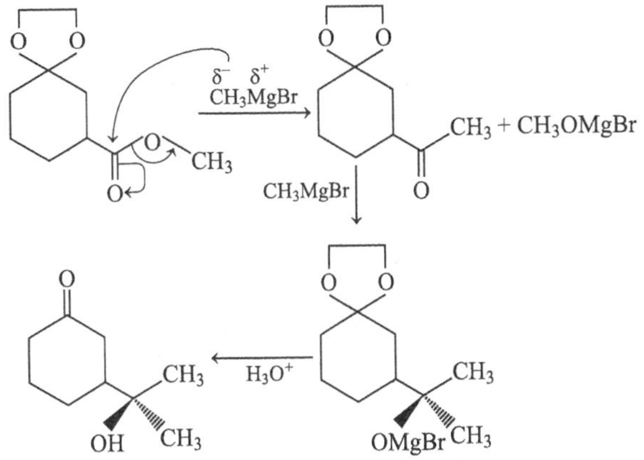

Study Tactics

This problem includes conceptual mixing of protecting carbonyl group and addition reaction to the carbonyl group. It is solved by protecting the keto group first followed by nucleophilic addition reaction to carbonyl centre and subsequent hydrolysis to get the desired product.

31. (*b*) According to Drago's rule, bond angle of NH_3, PH_3 and AsH_3 decreases from top to bottom

$$NH_3 > PH_3 > AsH_3$$

Here, AsH_3 has smallest bond angle. Bond angle of O_2 is maximum because two oxygen atoms are connected together with a double bond. The bond angle is 180°.

32. (*c*) Diagram of d_{xz}-orbital is shown below.

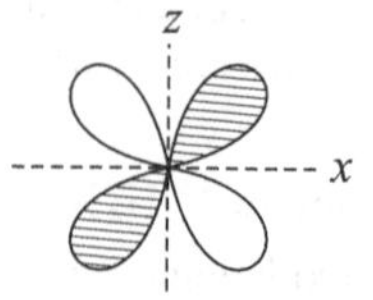

So, in this d_{xz}-orbital, maximum probability of finding an electron is at an angle of 45° from x and z-axis as the lobes of d_{xz}-orbital are oriented in between the x and z axis of 45°.

Study Tactics

To determine the maximum probability of finding an electron in the d_{xz}-orbital first visualize its shape. Remember that the electron is most likely to be found at the place where the electron density is highest.

33. (*d*) Given, $\Delta H = -40 \text{ kJ mol}^{-1}$

$$(E_a)_f = 80 \text{ kJ mol}^{-1}$$

For a reaction,

$$\Delta H = (E_a)_f - (E_a)_b$$

$$(E_a)_b = (E_a)_f - \Delta H = 80 - (-40)$$

$$= 120 \text{ kJ mol}^{-1} \quad \ldots\text{(i)}$$

$\because$ Catalyst lowers $(E_a)_f$ by 20 kJ/mol hence,

$$E_a{'}_f = 20 \text{ kJ mol}^{-1}$$

Now, $\Delta H = (E_a)_f{'} - (E_a)_b{'}$

$$-40 = 20 - (E_a)_b{'}$$

$$(E_a)_b{'} = 60 \text{ kJ mol}^{-1} \quad \ldots\text{(ii)}$$

Now, ratio of energy of activation for reverse reaction before and after addition of catalyst is

Divide equation (i) with equation (ii), then,

$$\frac{(E_a)_b}{(E_a)_b{'}} = \frac{120}{60} = 2:1$$

34. (*a*) In option (a) charge will disperse over maximum number of carbon atoms.

35. (*d*) • In first order reactions,

$$-\frac{d[A]}{dt} = k[A].$$

• In zero order reactions,

$$[A]_t = [A]_0 - kt$$

• In second order reactions,

$$\frac{t-1}{[A]_t} = \frac{1}{[A]_0} + kt$$

On comparing the above equations with $y = mx + c$, we get the given plots (I, II, III) corresponding to reaction order 1, 0 and 2 respectively.

36. (*b*) **Hydrogen bonding** The presence of hydrogen bonding stabilises the structure.

The conformations of 2-amino-ethan-1-ol are

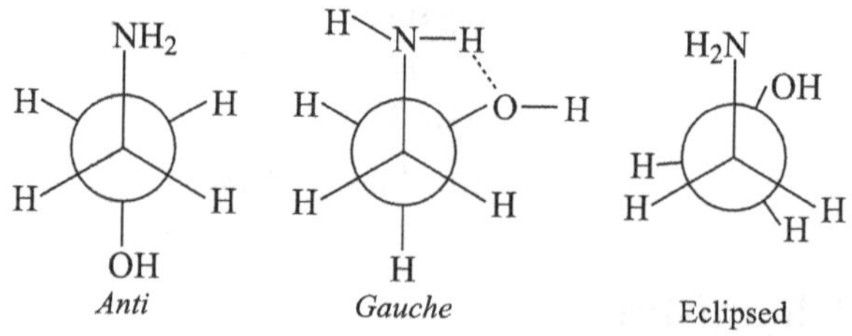

Out of above, gauche form have more stability due to hydrogen bonding.

So, the correct order is *gauche* > *anti* > eclipsed.

Study Tactics

This problem includes conceptual mixing of conformational structure of molecule and H-bonding. Students are advised to draw the conformational structure of organic molecule first followed by analysing the effect of H-bonding on molecule.

37. (*c*) Among the given statement B and D are correct. The reaction is represented as

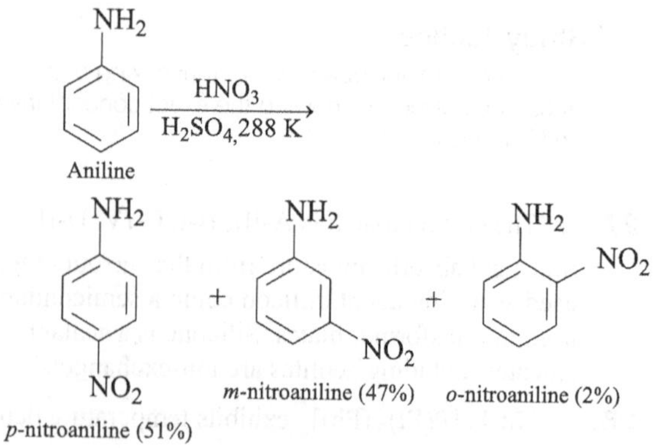

Nitration involves an electrophilic substitution reaction. The addition of HNO_3 and H_2SO_4 will lead to the nitration of compound. H_2SO_4 acts as an acid by donating hydrogen ion.

38. ***(a)*** The sulphur atom is sp^3-hybridised in SO_3^{2-}. In sulphur, three p-orbitals forms a bond with three oxygen atoms and unhybridised orbital is involved in bond formation.

$$_8O = 1s^2, 2s^2, 2p_x^2, 2p_y^1, 2p_z^1$$

In oxygen, two unpaired p-orbitals are present. One is involved in σ bond formation while other is used in π-bond formation.

Thus, in SO_3^{2-} $p\pi$-orbitals are involved for $p\pi$-$d\pi$ bonding.

39. ***(d)*** Among the given order, A, D and E are correct while B and C are incorrect.

- 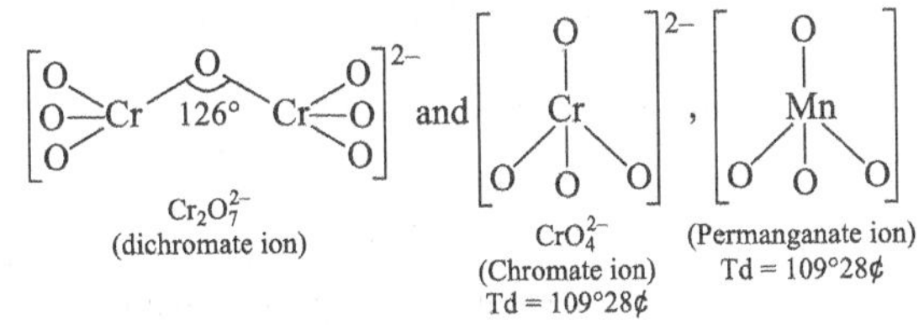

- Increasing order of enthalpy of atomisation is

V	> Ni	> Co	> Zn
515	430	425	126

40. ***(b)*** The correct match is A-(IV), B-(I), C-(II), D-(III).
The monomer unit of protein is amino acid.
The simplest form of a carbohydrate is monosaccharide.
The simplest form of lipid is triglyceride.
The monomeric unit of nucleic acid is nucleotide.

41. ***(b)*** Both Statement I and Statement II are incorrect.
Thermal decomposition of $Ba(N_3)_2$ produces nitrogen gas.

$$Ba(N_3)_2 \xrightarrow[\text{decomposition}]{\text{Thermal}} Ba + 3N_2\uparrow$$

As, positive oxidation state increase, acidic character increases.
∵ Pentavalent oxide, E_2O_5 has + 5 oxidation state.
Trivalent oxide, E_2O_3 has + 3 oxidation state.
So, acidic character, $E_2O_5 > E_2O_3$.

42. ***(d)*** (A) is false but (R) is true.
Phenolphthalein is a pH dependent indicator. It is a weak acid which is colourless in the acidic solution but gives pink colour in basic medium. It dissociates in basic medium and the pink colour is due to its conjugate form.

$$\underset{\text{(Colourless)}}{HPh(aq)} \rightleftharpoons H^+ + \underset{\text{(Pink)}}{Ph^-}$$

43. ***(a)*** If the value of ΔG is negative, then the process is spontaneous.
The value of ΔG can be calculated by using the formula
$\Delta G = \Delta H - T\Delta S$
Hence, there are two spontaneous processes.

44. ***(b)*** $\text{Phenol} \xrightarrow[\text{In excess of } H_2O]{NaNO_2} \text{Red coloured compound}$

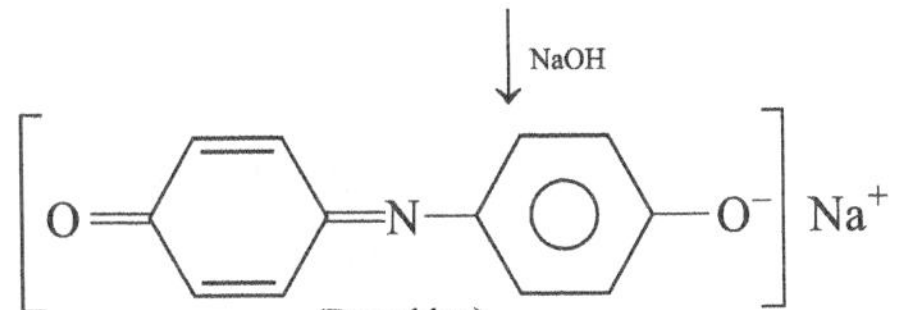

(Deep blue)
Sodium salt of indophenol

The name of the reaction is Liebermann's reaction.
It is a test for phenol.

45. ***(d)*** The mechanism of the given reaction is follow as

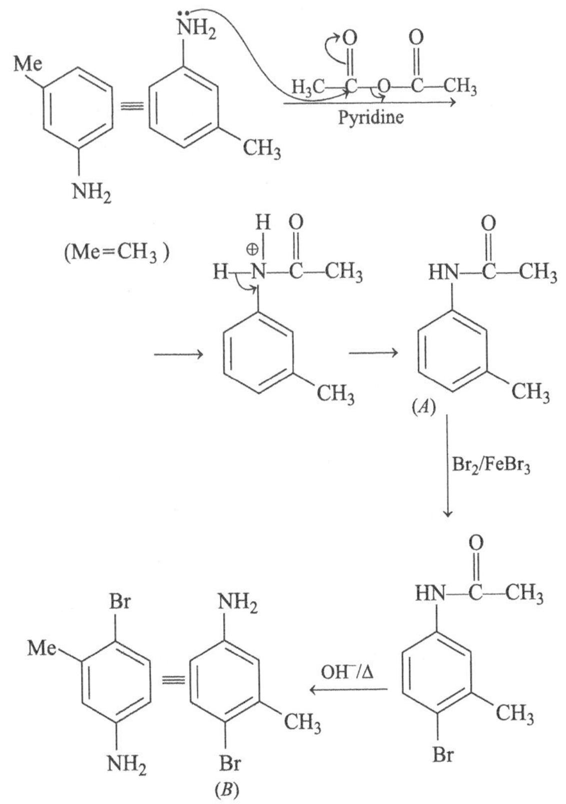

46. ***(141)*** **Anode**

$$[H_2(g) \longrightarrow 2H^+(aq) + 2e^-, E° = 0\,V] \times 5$$

Cathode $[MnO_4^-(aq) + 8H^+(aq) + 5e^- \longrightarrow$
$Mn^{2+}(aq) + 4H_2O(aq)] \times 2$

Net reaction

$$2MnO_4^-(aq) + 16H^+(aq) + 5H_2(g) \longrightarrow 10H^+(aq) + 2Mn^{2+}(aq) + 8H_2O(aq)$$

According of nernst equation

$$E_{\text{cell}} = E^\circ_{\text{cell}} - \frac{0.0591}{n}\log Q$$

$$E_{\text{cell}} = E^\circ_{\text{cell}} - \frac{0.0591}{n}\log\frac{[H^+]^{10}[Mn^{2+}]^2}{[MnO_4^-]^2[H^+]^{16}[H_2]^5}$$

$n = 10$ (no. of electrons transferred)

$$E_{\text{cell}} = 1.50 - \frac{0.0591}{10}\log\left[\frac{(10^{-2})^{10}[0.01]^2}{(0.1)^2[10^{-2}]^{16}[0.1]^5}\right]$$

$$= 1.50 - 0.006\log 10^{15}$$

$$= 1.50 - 0.006\times 15 = 1.50 - 0.09$$

$$= 1.41\text{ V} = 141\times 10^{-2}\text{ V}$$

47. (17) $\dfrac{p^\circ - p_s}{p^\circ} = \dfrac{n_2}{n_1+n_2} = \dfrac{w_2/M_2}{w_1/M_1 + w_2/M_2}$

or $\dfrac{17.5 - p_s}{p^\circ} = \dfrac{18/180}{178.2/18 + 18/180}$

or $\dfrac{17.5 - p_s}{17.5} = \dfrac{1/10}{9.9 + 0.1}$

or $p_s = 17.325\text{ mm Hg} \approx 17\text{ mm Hg}$

48. (*107*) Let the equivalent weight of the acid be x.

Eq. weight of acid $= \dfrac{5.75}{x}$

Millieq. of the acid $= \dfrac{5.75\times 1000}{x} = \dfrac{5750}{x}$

Now, 250 mL of acid contains $= \dfrac{5750}{x}$ meq.

1 mL of acid contains $= \dfrac{5750}{x}\times\dfrac{1}{250}$

35 mL of acid contains $= \dfrac{5750}{x}\times\dfrac{35}{250} = \dfrac{805}{x}$ meq.

Millieq. of 35 mL of acid = Millieq. of NaOH

$\Rightarrow \dfrac{805}{x} = 1\times 15 \Rightarrow x = 53.6$

Molecular weight $= 2\times 53.6 = 107.2$ g/mol ≈ 107 g/mol

49. (*10*) Reaction between H_2CO_3 and NaOH take place as such.

$$\underset{0.2}{H_2CO_3} + \underset{0.2}{NaOH} \longrightarrow \underset{0.2}{NaHCO_3} + H_2O$$

In the final solution, we have 0.2 mol.

Na_2CO_3 and 0.4 mol of $NaHCO_3$.

It is a buffer of $NaHCO_3$ and Na_2CO_3.

$$\therefore\ pH = pKa_2 + \log\frac{(\text{Salt})}{(\text{Acid})} = 10.32 + \log\frac{\left(\frac{0.2}{0.2}\right)}{\left(\frac{0.4}{0.2}\right)}$$

$$= 10.32 + \log 1 - \log 2$$

$$= 10.32 - 0.3010 = 10.02 = 10$$

50. (*13*) **Determination of volume of nitrogen**

Given, mass of substance = 0.40 g

V_{N_2} (volume of N_2) = 50 cm^3

Atmospheric pressure = 715 mm Hg

Room temperature = 300 K

Vapour pressure of water at 300 K = 15 mm Hg

Actual vapour pressure of dry gas

$= 715 - 15 = 700$ mm Hg

Now to determine the volume of N_2, we use

$$\frac{p_1V_1}{T_1} = \frac{p_2V_2}{T_2} \Rightarrow \frac{700\times 50}{300} = \frac{760\times V_2}{273}$$

$$V_2 = \frac{273\times 700\times 50}{300\times 760} = 41.9\text{ cm}^3$$

The mass of nitrogen in 41.9 cm^3 gas is calculated by using concept of gram molar volume.

22400 cm^3 of N_2 at STP weighs = 28 g

41.9 cm^3 of N_2 at STP weigh $= \dfrac{28\times 41.9}{22400}$ g

Percentage of N_2

$$= \frac{\text{Mass of } N_2}{\text{Total mass of organic compound}}\times 100$$

$$= \frac{28\times 41.9}{22400}\times\frac{100}{0.4} = 13.09\,\% \approx 13\%$$

Mathematics

51. (*b*) **Case I**

a	x		

Here, $x = c$ or d

Number of ways = 2

Case II

b	y	z	

Here, $y = a$ or d

If $y = a$, then $z = c$ or d

$\therefore$ Number of ways = 3

Case III

c	u	p	

Here, $u = a$ or b

If $u = a$, then $p = d$

and if $u = b$, then $p = a$ or d

$\therefore$ Number of ways = 3

Case IV

d	q	r	s

Here, $q = a$ or b or c

If $q = a$, then $r = c$ and $s = b$

If $q = b$, then $r = a$ and $s = c$

If $q = c$, then $r = a$ or b

If $r = a$, then $s = b$ (impossible)

and if $r = b$, then $s = a$ (possible)

$\therefore$ Number of ways $= 3$

$\therefore$ Total ways $= 2+3+3+3 = 11$ ways

52. *(a)* Let T_r be the rth term of the given series $\frac{3}{1^2\times 2^2}+\frac{5}{2^2\times 3^2}+\frac{7}{3^2\times 4^2}+\ldots$, then

$$T_r = \frac{2r+1}{r^2(r+1)^2},\ r=1,2,3,\ldots$$
$$= \frac{1}{r^2}-\frac{1}{(r+1)^2},\ r=1,2,3,\ldots$$

Required sum, $S_m = \sum_{r=1}^{m} T_r = \sum_{r=1}^{m}\left[\frac{1}{r^2}-\frac{1}{(r+1)^2}\right]$

$$= \left(\frac{1}{1^2}-\frac{1}{2^2}\right)+\left(\frac{1}{2^2}-\frac{1}{3^2}\right)+\left(\frac{1}{3^2}-\frac{1}{4^2}\right)+\ldots+\left(\frac{1}{m^2}-\frac{1}{(m+1)^2}\right)$$
$$= 1-\frac{1}{(m+1)^2} = \frac{(m+1)^2-1}{(m+1)^2}$$
$$= \frac{m^2+2m}{(m+1)^2}$$

53. *(d)* Given that $(x+y)^2\frac{dy}{dx} = a^2$ ...(i)

Let $x+y=t$

On differentiating $x+y=t$ w.r.t. x, we get

$$1+\frac{dy}{dx} = \frac{dt}{dx}$$
$$\Rightarrow \quad \frac{dy}{dx} = \frac{dt}{dx}-1$$

$\therefore$ Eq. (i) becomes, $t^2\left(\frac{dt}{dx}-1\right) = a^2$

$$\Rightarrow \quad \frac{dt}{dx}-1 = \frac{a^2}{t^2}$$
$$\Rightarrow \quad \frac{dt}{dx} = \frac{a^2}{t^2}+1 = \frac{a^2+t^2}{t^2}$$
$$\Rightarrow \quad \frac{t^2}{a^2+t^2}dt = dx$$

On integrating both sides, we get

$$\int \frac{t^2}{a^2+t^2}dt = \int dx$$
$$\Rightarrow \quad \int dt - \int \frac{a^2}{a^2+t^2}dt = \int dx$$
$$\Rightarrow \quad t - a\tan^{-1}\left(\frac{t}{a}\right) = x+C$$
$$\Rightarrow \quad x+y-a\tan^{-1}\left(\frac{x+y}{a}\right) = x+C \qquad [\because t = x+y]$$
$$\Rightarrow \quad y-C = a\tan^{-1}\left(\frac{x+y}{a}\right)$$

At $x=0,\ y=0$

Then, $0-C = a\tan^{-1}0$

$$C = 0$$
$$\therefore \quad y = a\tan^{-1}\left(\frac{x+y}{a}\right)$$
$$\Rightarrow \quad \frac{y}{a} = \tan^{-1}\left(\frac{x+y}{a}\right)$$
$$\Rightarrow \quad \tan\frac{y}{a} = \frac{x+y}{a}$$

At $y=a$,

$$\tan\frac{y}{a} = \tan 1$$
$$= \frac{x+a}{a} = \frac{x}{a}+1$$
$$\Rightarrow \quad \frac{x}{a} = \tan 1 - 1$$

54. *(b)* ΔABC is an equilateral triangle.

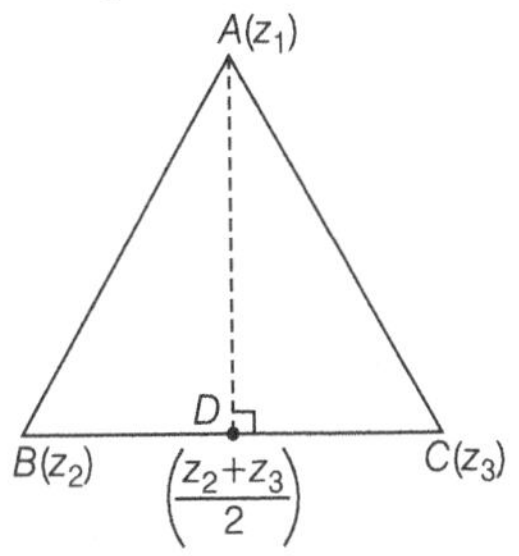

$$\arg\left(\frac{z_2+z_3-2z_1}{z_3-z_2}\right)$$
$$= \arg\left[\frac{2\left(\frac{z_2+z_3}{2}-z_1\right)}{z_3-z_2}\right]$$
$$= \arg(2)+\arg\left(\frac{\frac{z_2+z_3}{2}-z_1}{z_3-z_2}\right)$$
$$= \arg\left(\frac{\frac{z_2+z_3}{2}-z_1}{z_3-z_2}\right)$$

Here, $\frac{z_2+z_3}{2}$ is the mid-point D of BC and ΔABC is equilateral.

$\therefore AD \perp BC$

Thus, $$\arg\left(\frac{\frac{z_2+z_3}{2}-z_1}{z_3-z_2}\right)=\frac{\pi}{2}$$

$$\Rightarrow \left[\arg\left(\frac{\frac{z_2+z_3}{2}-z_1}{z_3-z_2}\right)\right]=\left[\frac{\pi}{2}\right]=1$$

$$\Rightarrow \frac{1}{2}\left[\arg\left(\frac{\frac{z_2+z_3}{2}-z_1}{z_3-z_2}\right)\right]=\frac{1}{2}$$

Study Tactics

$\therefore A(z_1), B(z_2)$ and $C(z_3)$ are the vertices of an equilateral ΔABC.

Then, mid-point D of BC will be $\frac{z_2+z_3}{2}$ and also $AD \perp BC$.

So, $$\arg\left(\frac{\frac{z_2+z_3}{2}-z_1}{z_3-z_2}\right)=\frac{\pi}{2}$$

55. *(c)* $\because$ Tossing coin eight times

$$\times\times\times\times\times HTT$$

Head must be at sixth place as seventh and eighth is for tails.

Possible situation for first 5 places.

(i) $HHHHH$ all heads $=1$ way

(ii) 4 H and 1T

i.e. $HHHHT=\frac{5!}{4!}=5$ ways

(iii) 3H and 2T but not consecutive tails HHH (4 gaps) and place 2 tails in gaps

$\Rightarrow \quad {}^4C_2=6$ ways

(iv) 2H and 3T but not consecutive tails HH (3 gaps) and places 3 tails in gaps

$\Rightarrow \quad {}^3C_3=1$ way

Total possible ways $=13$

Total number of ways (sample space)

$$=2^8=256$$

$\therefore$ Required probability $=\frac{13}{256}$

56. *(b)* Given, $$g(x)=\begin{vmatrix} a & a^2 & a^3 \\ 0 & -1 & 6 \\ \cos x & \sin x & x^3 \end{vmatrix}$$

$$g'(x)=\begin{vmatrix} a & a^2 & a^3 \\ 0 & -1 & 6 \\ -\sin x & \cos x & 3x^2 \end{vmatrix}$$

$$g''(x)=\begin{vmatrix} a & a^2 & a^3 \\ 0 & -1 & 6 \\ -\cos x & -\sin x & 6x \end{vmatrix}$$

$$g'''(x)=\begin{vmatrix} a & a^2 & a^3 \\ 0 & -1 & 6 \\ \sin x & -\cos x & 6 \end{vmatrix}$$

Now, $$g''(0)=\begin{vmatrix} a & a^2 & a^3 \\ 0 & -1 & 6 \\ -1 & 0 & 0 \end{vmatrix}=-6a^2-a^3$$

Statement II is false.

and $$g'''(0)=\begin{vmatrix} a & a^2 & a^3 \\ 0 & -1 & 6 \\ 0 & -1 & 6 \end{vmatrix}=0$$

Statement I is true.

57. *(c)* Here, LCM $(3, 6)=6$

Now, put $x=t^6$

$\Rightarrow \quad dx=6t^5\,dt$

$$\therefore \int \frac{x+\sqrt[3]{x^2}+\sqrt[6]{x}}{x(1+\sqrt[3]{x})}dx=\int \frac{t^6+\sqrt[3]{t^{12}}+\sqrt[6]{t^6}}{t^6(1+\sqrt[3]{t^6})}\cdot 6t^5\,dt$$

$$=\int \frac{t^6+t^4+t}{t^6(1+t^2)}\cdot 6t^5\,dt$$

$$=6\int \frac{t^5+t^3+1}{1+t^2}dt$$

$$=6\int\left(\frac{t^3(1+t^2)}{1+t^2}+\frac{1}{1+t^2}\right)dt$$

$$=6\int t^3\,dt+6\int\frac{1}{1+t^2}dt$$

$$=\frac{3}{2}t^4+6\tan^{-1}t+C$$

$$=\frac{3}{2}x^{2/3}+6\tan^{-1}x^{1/6}+C$$

58. *(c)* Given, $4\ \text{cosec}^2\{\pi(p+x)\}+p^2-4p=0$

$$\Rightarrow \quad 4\text{cosec}^2\{\pi(p+x)\}=4p-p^2$$

$$\Rightarrow \quad \sin^2\{\pi(p+x)\}=\frac{4}{4p-p^2}$$

For solution $0\le\frac{4}{4p-p^2}\le 1$

Now, $\frac{4}{4p-p^2}\ge 0$

$$\Rightarrow \quad 4p-p^2>0$$

$$\Rightarrow \quad 0<p<4 \quad \ldots\text{(i)}$$

Also, $p\in(-\infty,0)\cup(4,\infty)\cup\{2\}$ ...(ii)

From Eqs. (i) and (ii), we get

$$p=2$$

59. *(c)* Given, $\frac{x}{1}=\frac{y-1}{0}=\frac{z+1}{-1}=\lambda$ (let)

and any point $P(\alpha,0,\alpha)$.

Any point on line is represented by $A(\lambda,1,-\lambda-1)$.

Direction ratios of AP are

$$\lambda-\alpha,1,-\lambda-1-\alpha$$

Which is perpendicular to the line.

$$\therefore \quad (\lambda-\alpha)\cdot 1+0\cdot 1-1(-\lambda-1-\alpha)=0$$

$$\Rightarrow \quad \lambda-\alpha+\lambda+1+\alpha=0$$

$$\Rightarrow \quad \lambda=-\frac{1}{2}$$

$\therefore$ Coordinates of point A are $\left(-\frac{1}{2},1,-\frac{1}{2}\right)$.

Given, $AP=\sqrt{\frac{3}{2}}$

$$\Rightarrow \quad AP^2=\frac{3}{2}$$

$$\Rightarrow \quad \left(\alpha+\frac{1}{2}\right)^2+1+\left(\alpha+\frac{1}{2}\right)^2=\frac{3}{2}$$

$$\Rightarrow \quad 2\left(\alpha+\frac{1}{2}\right)^2=\frac{3}{2}-1$$

$$\Rightarrow \quad \left(\alpha+\frac{1}{2}\right)^2=\frac{1}{4}$$

$$\Rightarrow \quad \alpha=0,-1\ (\alpha\ne 0)\Rightarrow \alpha=-1$$

60. *(d)* Check at $x=0$

$$\text{LHL}=\lim_{x\to 0^-} f(x)$$

$$=\lim_{h\to 0} f(0-h)$$

$$=\lim_{h\to 0}\frac{1-k^{-1/h}}{1+k^{-1/h}}$$

$$=\frac{1-0}{1+0} \quad [\text{for } 0<k<1]$$

and $\text{RHL}=\lim_{x\to 0^+} f(x)$

$$=\lim_{h\to 0} f(0+h)$$

$$=\lim_{h\to 0}\frac{1-k^{\frac{1}{h}}}{1+k^{\frac{1}{h}}}$$

$$=\lim_{h\to 0}\frac{k^{-1/h}-1}{k^{-1/h}+1}$$

$$=\frac{0-1}{0+1}=-1$$

$$\therefore \quad \text{Jump}=\text{LHL}-\text{RHL}=1-(-1)=2$$

61. *(c)* Since, **d** is collinear to vector **c**.

$$\therefore \quad \mathbf{c}=\lambda\mathbf{d}$$

$$\Rightarrow \quad (2x+3)\mathbf{a}+\mathbf{b}=\lambda[(2x+3)\,\mathbf{a}-\mathbf{b}]$$

$$\Rightarrow \quad 2x\mathbf{a}+3\mathbf{a}+\mathbf{b}=2\lambda x\mathbf{a}+3\lambda\mathbf{a}-\mathbf{b}\lambda$$

$$\Rightarrow \quad (2x\mathbf{a}-2\lambda x\mathbf{a}+3\mathbf{a}-3\lambda\mathbf{a})+(\mathbf{b}\lambda+\mathbf{b})=0$$

$$\Rightarrow \quad (2x-2\lambda x+3-3\lambda)\,\mathbf{a}+(\lambda+1)\,\mathbf{b}=0$$

$$\therefore \quad (2x-2\lambda x+3-3\lambda)=0$$

and $(\lambda+1)=0$

$$\Rightarrow \quad (2x+3)-\lambda(2x+3)=0 \text{ and } \lambda=-1$$

$$\Rightarrow \quad (2x+3)(1-\lambda)=0$$

$$\therefore \quad x=-\frac{3}{2} \text{ and } \lambda=1$$

$$\Rightarrow \quad -2x=3$$

62. *(a)* Let the components of the line segment vector be a, b and c, then

$$a^2+b^2+c^2=(56)^2 \quad \ldots\text{(i)}$$

Also, $\frac{a}{3}=\frac{b}{-2}=\frac{c}{6}=\lambda$ (say)

Then, $a=3\lambda$, $b=-2\lambda$ and $c=6\lambda$

and from Eq. (i), we get

$$9\lambda^2+4\lambda^2+36\lambda^2=(56)^2$$

$$\Rightarrow \quad 49\lambda^2=(56)^2\Rightarrow\lambda=\pm 8$$

Since, $a=3\lambda<0$ as the line makes an obtuse angle with X-axis, $\lambda=-8$ and required components are $-24, 16, -48$.

63. *(a)* We have, $f(x)=x\sin x-\frac{1}{2}\sin^2 x$

$$f'(x)=x\cos x+\sin x-\sin x\cos x$$

$$=\sin x(1-\cos x)+x\cos x$$

For $x\in\left(0,\frac{\pi}{2}\right)$, $\sin x>0$, $(1-\cos x)>0$, $\cos x>0$

$\Rightarrow \qquad f'(x) > 0, \forall x \in \left(0, \frac{\pi}{2}\right)$

$\Rightarrow$ $f(x)$ is strictly increasing in $\left(0, \frac{\pi}{2}\right)$.

Now, $\lim_{x \to 0} f(x) = 0$ and $\lim_{x \to \frac{\pi}{2}} f(x) = \frac{\pi - 1}{2}$

$\therefore$ Range of $f(x) = \left(0, \frac{\pi - 1}{2}\right)$

But range of $f(x) = \left(0, \frac{\pi - k}{k+1}\right)$

Hence, $k = 1$

64. *(d)* $f(x) = \begin{cases} x \sin \frac{1}{x}, & x \neq 0 \\ 0, & x = 0 \end{cases}$

$$\text{LHL} = \lim_{h \to 0} f(0-h) = \lim_{h \to 0} (0-h) \sin \frac{1}{(0-h)}$$

$$= \lim_{h \to 0} h \sin \frac{1}{h}$$

$= 0 \times$ a finite value between -1 and $1 = 0$

$$\text{RHL} = \lim_{h \to 0} f(0+h) = \lim_{h \to 0} h \sin \frac{1}{h}$$

$= 0 \times$ a finite value between -1 and $1 = 0$

Also, $f(0) = 0$

$\Rightarrow \qquad \text{LHL} = \text{RHL} = f(0)$

$\therefore f$ is continuous at $x = 0$

Differentiability of $f(x)$ at $x = 0$

$$Lf'(0) = \lim_{h \to 0} \frac{f(0-h) - f(0)}{-h}$$

$$= \lim_{h \to 0} \frac{(0-h)\sin \frac{1}{(0-h)} - 0}{-h}$$

$= \lim_{h \to 0} -\sin \frac{1}{h} =$ a finite value between -1 and 1

$\Rightarrow$ limit does not exist.

$\therefore f(x)$ is not differentiable at $x = 0$

Now, $g(x) = \frac{1}{x} f(x) = \begin{cases} \sin \frac{1}{x}, & x \neq 0 \\ 0, & x = 0 \end{cases}$

$$\lim_{x \to 0} g(x) = \lim_{x \to 0} \sin \frac{1}{x}$$

$=$ a finite value between -1 and 1

$=$ limit does not exist

$\therefore g(x)$ is not continuous at $x = 0$.

$\Rightarrow g(x)$ is not differentiable at $x = 0$.

65. *(d)* We have, $f(x) = \log_{10}\{1 - \log_{10}(x^2 - 5x + 16)\}$

$f(x)$ is defined, when

$1 - \log_{10}(x^2 - 5x + 16) > 0$ and $x^2 - 5x + 16 > 0$

$\Rightarrow \quad \log_{10}(x^2 - 5x + 16) < 1$ and $x^2 - 5x + 16 > 0$

$\Rightarrow \quad x^2 - 5x + 16 < 10$ and $\left(x - \frac{5}{2}\right)^2 + \frac{39}{4} > 0$

$\Rightarrow \quad x^2 - 5x + 6 < 0$ and $\left(x - \frac{5}{2}\right)^2 + \frac{39}{4} > 0$

$\Rightarrow \quad x - 3x - 2x + 6 < 0$ and $x \in R$

$\Rightarrow \quad (x-2)(x-3) < 0$ and $x \in R$

$\Rightarrow \quad x \in (2, 3)$ and $x \in R$

$\therefore \quad x \in (2, 3)$

Hence, $D_f = (2, 3)$

66. *(c)* Given, equation of circle

$$|z|^2 + z(2+3i) + \bar{z}(2-3i) + 4 = 0$$

Centre $= (-2+3i)$ and radius $= \sqrt{13-4} = 3$

$\therefore \quad |z + (2-3i)| = 3$

Let $\omega = z + 3 + 2i = z + 2 - 3i + 1 + 5i$

$\therefore \quad \omega - 1 - 5i = z + 2 - 3i$

$\Rightarrow \quad |\omega - 1 - 5i| = |z + 2 - 3i| = 3$ or $|\omega - 1 - 5i| = 3$

So, ω lies on a circle whose centre is $(1+5i)$ and radius is 3.

67. *(d)* Given, $0 < y < x < 2y$

Arranging given four numbers in ascending order

$$x - y, y, x, 2x + y$$

$$\text{Median} = \frac{\text{2nd term} + \text{3rd term}}{2} = 10$$

$\Rightarrow \qquad y + x = 20 \qquad \ldots(\text{i})$

Range $= (2x + y) - (x - y) = 28$

$x + 2y = 28 \qquad \ldots(\text{ii})$

On solving Eqs. (i) and (ii), we get

$$x = 12 \text{ and } y = 8$$

So, numbers are 4, 8, 12 and 32.

$$\text{Mean} = \frac{4 + 8 + 12 + 32}{4} = 14$$

$$\therefore \quad \frac{x+y}{\text{Mean}} = \frac{12+8}{14} = \frac{20}{14} = \frac{10}{7}$$

68. *(c)* Let the extremities of a pair of semi-conjugate diameters of the ellipse be $P(3\cos\theta, 2\sin\theta)$ and $Q(3\cos\alpha, 2\sin\alpha)$, C being the centre.

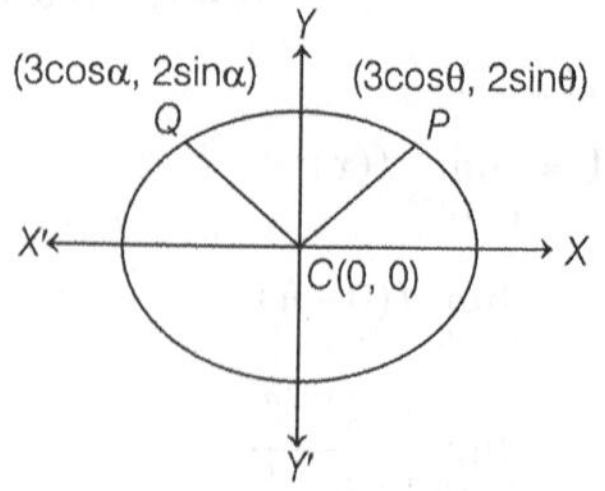

$\therefore$ Slope of CP × Slope of $CQ = \frac{-4}{9}$

$$\Rightarrow \quad \frac{2}{3}\tan\theta \times \frac{2}{3}\tan\alpha = -\frac{4}{9}$$

$$\Rightarrow \quad 1 + \tan\theta \cdot \tan\alpha = 0 \Rightarrow \alpha - \theta = \frac{\pi}{2}$$

$\Rightarrow$ Coordinates of Q are $(-3\sin\theta, 2\cos\theta)$ and the equation of PQ is

$$2(\sin\theta - \cos\theta)x - 3(\cos\theta + \sin\theta)y + 6 = 0$$

which represents $ax + by + c = 0$, then

$$\frac{a}{2(\sin\theta - \cos\theta)} = \frac{b}{-3(\cos\theta + \sin\theta)} = \frac{c}{6}$$

So, $\sqrt{\frac{9a^2 + 4b^2}{c^2}} = \sqrt{(\sin\theta - \cos\theta)^2 + (\cos\theta + \sin\theta)^2}$

$$= \sqrt{1 - 2\sin\theta \cdot \cos\theta + 1 + 2\sin\theta\cos\theta} = \sqrt{2}$$

69. *(b)* Equation of PQ is

$$\frac{x}{a}\cos\left(\frac{\theta-\phi}{2}\right) - \frac{y}{b}\sin\left(\frac{\theta+\phi}{2}\right) = \cos\left(\frac{\theta+\phi}{2}\right)$$

It passes through $(ae, 0)$

$$\therefore \quad e\cos\left(\frac{\theta-\phi}{2}\right) - 0 = \cos\left(\frac{\theta+\phi}{2}\right) \Rightarrow \frac{\cos\left(\frac{\theta-\phi}{2}\right)}{\cos\left(\frac{\theta+\phi}{2}\right)} = \frac{1}{e}$$

$$\Rightarrow \quad \frac{\cos\left(\frac{\theta-\phi}{2}\right) - \cos\left(\frac{\theta+\phi}{2}\right)}{\cos\left(\frac{\theta-\phi}{2}\right) + \cos\left(\frac{\theta+\phi}{2}\right)} = \frac{1-e}{1+e}$$

[by componendo and dividendo method]

$$\Rightarrow \quad \tan\frac{\theta}{2}\tan\frac{\phi}{2} = \frac{1-e}{1+e}$$

70. *(a)* We have,

$$A = \begin{bmatrix} 1 & -\tan\theta \\ \tan\theta & 1 \end{bmatrix}\begin{bmatrix} 1 & \tan\theta \\ -\tan\theta & 1 \end{bmatrix}^{-1} = \begin{bmatrix} p & -q \\ q & p \end{bmatrix}$$

$$\Rightarrow A = \begin{bmatrix} 1 & -\tan\theta \\ \tan\theta & 1 \end{bmatrix} \cdot \frac{1}{1+\tan^2\theta}\begin{bmatrix} 1 & -\tan\theta \\ \tan\theta & 1 \end{bmatrix}$$

$$= \frac{1}{1+\tan^2\theta}\begin{bmatrix} 1-\tan^2\theta & -2\tan\theta \\ 2\tan\theta & 1-\tan^2\theta \end{bmatrix}$$

$$= \begin{bmatrix} \frac{1-\tan^2\theta}{1+\tan^2\theta} & \frac{-2\tan\theta}{1+\tan^2\theta} \\ \frac{2\tan\theta}{1+\tan^2\theta} & \frac{1-\tan^2\theta}{1+\tan^2\theta} \end{bmatrix}$$

$$= \begin{bmatrix} \cos 2\theta & -\sin 2\theta \\ \sin 2\theta & \cos 2\theta \end{bmatrix} = \begin{bmatrix} p & -q \\ q & p \end{bmatrix}$$

On comparing, we get

$$p = \cos 2\theta \text{ and } q = \sin 2\theta$$

$$\therefore \quad p^2 + q^2 = (\cos 2\theta)^2 + (\sin 2\theta)^2 = 1$$

71. *(2)* $\because f[g(50)] = f(50b + a) = a(50b + a) + b$

$$= 50ab + a^2 + b$$

and $g[f(50)] = g(50a + b)$

$$= b(50a + b) + a$$
$$= 50ab + b^2 + a$$

$\therefore \quad f[g(50)] - g[f(50)]$

$$= (50ab + a^2 + b) - (50ab + b^2 + a)$$
$$= (a^2 - b^2) - (a - b) = (a - b)(a + b - 1)$$

But given $f[g(50)] - g[f(50)] = 28$

or $\quad (a-b)(a+b-1) = 28 = 1\times 28, 2\times 14, 4\times 7$

$$\Rightarrow \quad a - b = 1, a + b - 1 = 28$$

$$\Rightarrow \quad a = 15, b = 14 \Rightarrow ab = 210$$

and $a - b = 2, a + b - 1 = 14 \Rightarrow a = \frac{17}{2}, b = \frac{13}{2}$

and $a - b = 4, a + b - 1 = 7 \Rightarrow a = 6, b = 2$ and $ab = 12$

Hence, required numbers = 2

72. *(5522)* Given, m_1 and m_2 are the roots of the equation $x^2 + (\sqrt{3}+2)x + (\sqrt{3}-1) = 0$.

Then, $m_1 + m_2 = -(\sqrt{3}+2), m_1 m_2 = (\sqrt{3}-1)$

$$\therefore \quad m_1 - m_2 = \sqrt{(m_1 + m_2)^2 - 4m_1 m_2}$$

$$= \sqrt{3 + 4 + 4\sqrt{3} - 4\sqrt{3} + 4} = \sqrt{11}$$

and coordinates of the vertices of the given triangle are $(0, 0), \left(\frac{c}{m_1}, c\right)$ and $\left(\frac{c}{m_2}, c\right)$.

Hence, the required area of triangle

$$= \frac{1}{2}\left|\frac{c}{m_1} \times c - \frac{c}{m_2} \times c\right|$$

$$= \frac{1}{2}c^2\left|\left(\frac{1}{m_1} - \frac{1}{m_2}\right)\right| = \frac{1}{2}c^2\frac{|m_2 - m_1|}{|m_1 m_2|}$$

$$= \frac{1}{2}c^2\frac{\sqrt{11}}{\sqrt{3}-1} = \frac{1}{2}c^2\frac{\sqrt{11}(\sqrt{3}+1)}{(\sqrt{3}-1)(\sqrt{3}+1)}$$

$$= \left(\frac{\sqrt{33}+\sqrt{11}}{4}\right)c^2$$

On comparing, $a = \frac{\sqrt{33}}{4}$ and $b = \frac{\sqrt{11}}{4}$

or $\quad a = \frac{\sqrt{11}}{4}$ and $b = \frac{\sqrt{33}}{4}$

$$\therefore \quad a^2 + b^2 = \frac{11}{16} + \frac{33}{16} = \frac{44}{16} = \frac{11}{4}$$

$$\Rightarrow 2008(a^2+b^2)=2008\times\frac{11}{4}=502\times 11=5522$$

Study Tactics

$\therefore$ Equations of lines of triangle are

$$y=m_1x,\ y=m_2x \text{ and } y=c$$

From above equations, the coordinates of vertices of triangle are (0, 0), $\left(\frac{c}{m_1}, c\right)$ and $\left(\frac{c}{m_2}, c\right)$.

Now, use formula,

Area of triangle $=\frac{1}{2}[x_1(y_2-y_1)+x_2(y_3-y_1)+x_3(y_1-y_2)]$

where, (x_1, y_1), (x_2, y_2) and (x_3, y_3) are the vertices of triangle.

73. *(256)* Required area

$$=\int_0^1 (xe^x-xe^{-x})\,dx$$

$$=[x(e^x+e^{-x})-(e^x-e^{-x})]_0^1$$

$$=[(e+e^{-1})-(e-e^{-1})]-[0-(1-1)]=\frac{2}{e}$$

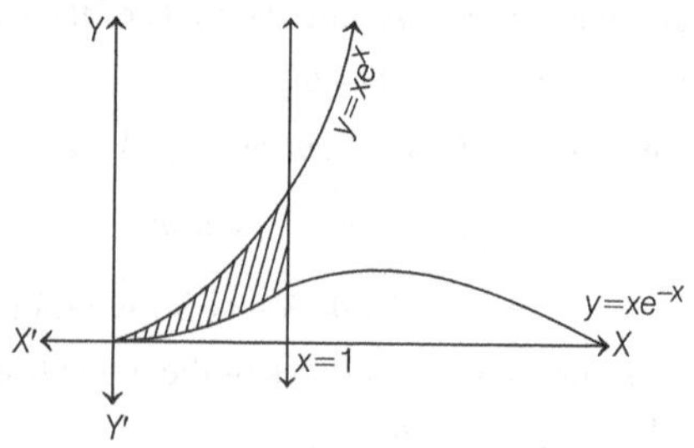

$$\therefore \quad 128e\lambda=128e\times\frac{2}{e}=256$$

74. *(4)* Let $p=x_1x_2x_3x_4x_5$

$$q=y_1y_2y_3y_4y_5$$

p and q can be added without carrying at any stage i.e. $x_i+y_i\le 9$

Thus, the number of ways to select x_5 and y_5 can be $10+9+8+\ldots+1=55$ ways

Similarly, each pair (x_4, y_4), (x_3, y_3), (x_2, y_2) can be selected in 55 ways.

But (x_1, y_1) can be selected in $1+2+3+\ldots+8=36$ ways as this pair cannot have 0 or 9.

Thus, the number of ways $=36(55)^4$

$$\therefore \quad k=4$$

75. *(3)* Given, $(1+x^3+x^4)=\sum_{i=0}^{4} a_i(1+x)^i,\ \forall\, x\in R$

Replacing $x\to(x-1)$, we get

$$\therefore \sum_{i=0}^{4} a_i(1+x-1)^i=1+(x-1)^3+(x-1)^4$$

$$\sum_{i=0}^{4} a_ix^i=1-(1-x)^3+(1-x)^4$$

a_3 is the coefficient of x^3 in RHS $={}^3C_3-{}^4C_3$

$$=1-4=-3$$

SECTION II

Mastery Builders

PRACTICE SET 6

Total Marks : 300 | Total Time : 3 hrs | No. of Questions : 75

Instructions

1. There are **three** parts in the questions paper consisting of **Physics**, **Chemistry** and **Mathematics** having 25 questions in each part of equal weightage. Each part (subject) has two sections.
2. **Section A** This section contains 20 multiple choice questions which have only one correct answer. Each question carries **4 marks** for correct answer and **–1 mark** for wrong answer.
3. **Section B** This section contains 5 questions. The answer to each of the questions is a numerical value. Each question carries **4 marks** for correct answer and **–1 mark** for wrong answer. For Section B, the answer should be rounded off to the nearest integer.

PHYSICS

Section A : Objective Type Questions

1. The hydrogen atom emits a line spectrum consisting of various series. Which series of hydrogen atomic spectra is lying in U-V region ?

(a) Lyman series (b) Brackett series
(c) Balmer series (d) Paschen series

2. A car moving with a speed of 4 m/s, is decelerated at a rate given by $\frac{dv}{dt} = -2\sqrt{v}$, where v is the instantaneous speed. The time taken by the object to come to rest would be

(a) 1 s (b) 2 s (c) 4 s (d) 8 s

3. At what temperature does a lamp (100 W, 200 V) work, if it is given that resistance of cold tungsten filament is 40Ω ?
(Take, $\alpha = 3 \times 10^{-3}$ /°C and 25°C as normal room temperature)

(a) 3000°C (b) 325°C (c) 3025°C (d) 300°C

4. Current through an AC series circuit is 4 A. If operated at resonant frequency $\frac{\omega_0}{2}$, current reduces to 2A. Then at $2\omega_0$, current in the circuit will be

(a) $\sqrt{2}$A (b) 3 A (c) 1 A (d) 2 A

5. For a particle performing uniform circular motion, choose the correct options from the following.

I. Magnitude of particle velocity (speed) remains constant.
II. Direction of acceleration keeps changing as particle moves.
III. Angular momentum is constant in magnitude but its direction keeps changing.
IV. Particle velocity remains directed perpendicular to radius vector.

(a) Both II, III (b) I, II and III
(c) I, II and IV (d) All I, II, III and IV

6. The electric field of an electromagnetic wave is given by $\mathbf{E} = E_0 \sin(kx - \omega t + \phi)\,\hat{\mathbf{i}}$, then the range of magnitude of **E**, when E_0 is a positive constant, is

(a) $-E_0 \le E \le +E_0$ (b) $-\frac{E_0}{2} \le E \le +\frac{E}{2}$
(c) $-1 \le E \le +1$ (d) $0 \le E \le E_0$

7. **Statement I** When a magnet is brought near iron nails, only translatory force act on it.

Statement II The field due to a magnet is generally uniform.

(a) Only Statement I is correct
(b) Only Statement II is correct
(c) Both Statement I and Statement II are incorrect
(d) Both Statement I and Statement II are correct

8. Assertion Static friction is a self-adjusting force.

Reason The interatomic force at the point of contact give rise to friction between the surfaces.

(a) Both Assertion and Reason are correct but Reason is not the correct explanation for Assertion.
(b) Both Assertion and Reason are correct and Reason is the correct explanation for Assertion.
(c) Assertion is correct but Reason is incorrect.
(d) Both Assertion and Reason are incorrect.

9. A monoatomic gas at a pressure p, having a volume V expands isothermally to a volume $2V$ and then adiabatically to a volume $16V$. The final pressure of the gas is (take, $\gamma = \frac{5}{3}$)

(a) $64\,p$ (b) $32\,p$ (c) $\frac{p}{64}$ (d) $16\,p$

10. One main scale division of a vernier callipers is a cm and nth division of the vernier scale coincide with $(n-1)$ th division of the main scale. The least count of the callipers in mm is

(a) $\frac{10na}{(n-1)}$ (b) $\frac{10a}{(n-1)}$ (c) $\left(\frac{n-1}{10n}\right)$ (d) $\frac{10a}{n}$

11. A metal rod of cross-sectional area A and length L is fixed between two rigid walls. If the rod is heated by temperature ΔT, then what will be the reaction generated by walls given that Young's modulus of rod is Y and coefficient of thermal expansion is α ?

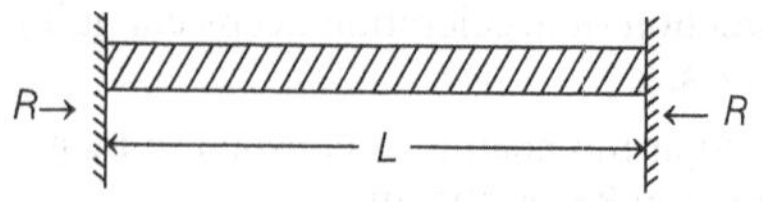

(a) $\frac{\alpha AY(\Delta T)}{2}$ (b) $\frac{2\alpha(\Delta T)}{AY}$
(c) $\alpha AY(\Delta T)$ (d) $\frac{\alpha(\Delta T)}{2AY}$

12. A spherical capacitor of inner radius $a = 1$ cm and outer radius $b = 2$ cm is earthed as shown. It can be connected to an isolated metallic sphere of radius $c = 1$ cm through a switch S and a very long conducting wire. If initial charge on inner sphere is $q = 30\,\mu$C, then charge on the sphere of radius c, when switch S is closed, will be

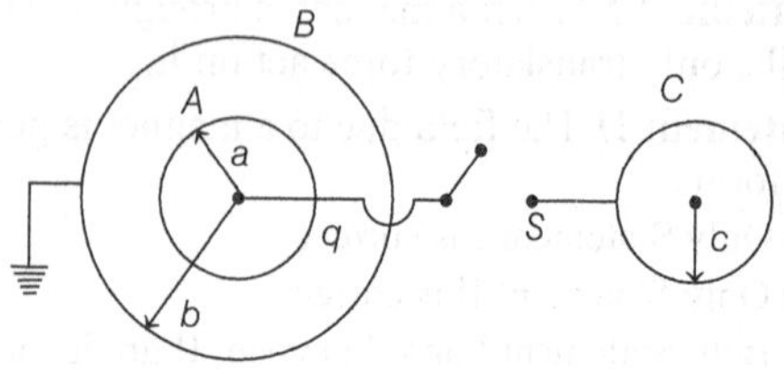

(a) 10μC (b) 20μC
(c) 30μC (d) 25μC

13. Match the Column I (physical quantity) with Column II (dimensions) and select the correct answer from the codes given below.

	Column I		Column II
A.	Electric field	1.	$[ML^3T^{-3}A^{-1}]$
B.	Electric flux	2.	$[ML^2T^{-3}A^{-1}]$
C.	Electric potential	3.	$[MLT^{-3}A^{-1}]$
D.	Permittivity of free space	4.	None of these

Codes

	A	B	C	D
(a)	3	1	2	4
(b)	4	3	1	2
(c)	2	3	1	4
(d)	3	2	1	4

14. In a certain region of space, the gravitational field is given by $\frac{-3}{r}$, where r is the distance. If gravitational potential at a distance 6 m is 12 Jkg^{-1}, then the gravitational potential at a distance 12 m is

(a) 12.079 J kg^{-1} (b) 14.079 J kg^{-1}
(c) 12.79 J kg^{-1} (d) 13.97 J kg^{-1}

15. In the circuit given below, the current through 5Ω resistor is

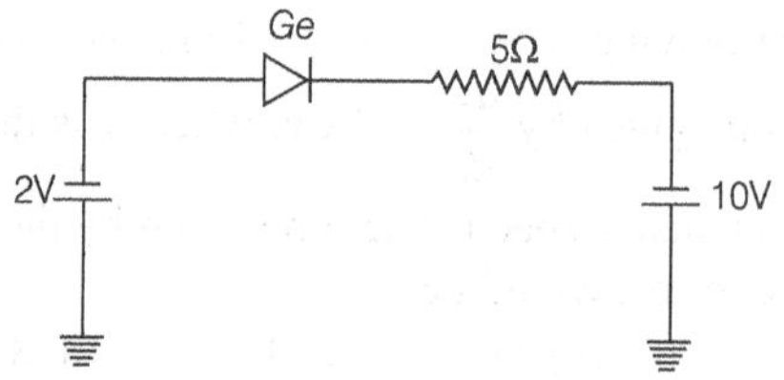

(a) 1.25 A (b) 1.52 A
(c) 1.54 A (d) 1.62 A

16. The coefficient of apparent expansion of a liquid when determined using two different vessels P and Q are γ_1 and γ_2, respectively. If the coefficient of linear expansion of vessel P is α, then the coefficient of linear expansion of vessel Q is

(a) $\frac{\gamma_1 + \gamma_2}{3} + \alpha$ (b) $\frac{\gamma_1 + \gamma_2}{3}$
(c) $\frac{\gamma_1 - \gamma_2}{3} + \alpha$ (d) None of these

17. In an oblique collision, first ball collides with a second ball. First ball starts moving towards $+Y$-axis and second ball starts moving making an angle 60° with the X-axis with a velocity $2v$. The velocity of first ball after collision is

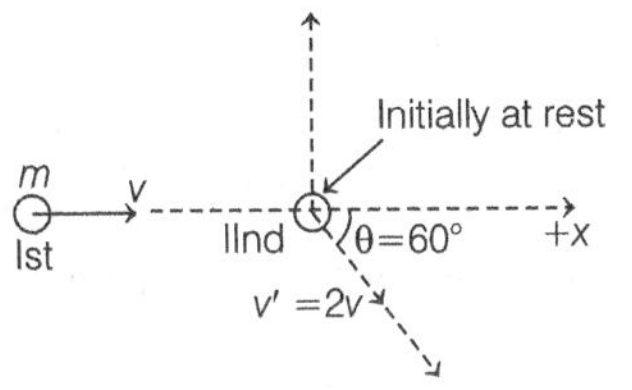

(a) $v/2$ (b) $2v$
(c) $\frac{2v}{\sqrt{3}}$ (d) $\sqrt{3}v$

18. A polyatomic ideal gas has 48 vibrational modes. What is the value of γ?

(a) 1.02 (b) 1.04
(c) 1.08 (d) 10.2

19. A spherical metal shell A of radius R_A and a solid metal sphere B of radius R_B ($< R_A$) are kept far apart and each is given charge $+Q$. Now, they are connected by a thin metal wire. Then,

(a) $E_{A \text{ (inside)}} > 0$
(b) $Q_A > Q_B$
(c) $\frac{\sigma_A}{\sigma_B} > \frac{R_B}{R_A}$
(d) $E_{A \text{ (on surface)}} > E_{B \text{ (on surface)}}$

20. Which of the following statements is correct ?

(a) In closed organ pipe, only odd harmonics are present.
(b) In open organ pipe, only odd harmonics are present.
(c) In open pipe, even harmonics are not present.
(d) In closed organ pipe, only even harmonics are present.

Section B : Numerical Value Type Questions

21. Hydrogen ($_1H^1$), deuterium ($_1H^2$), singly ionised helium $(_2He^4)^+$ and doubly ionised lithium $(_3Li^6)^{++}$ all have one electron around the nucleus. Consider an electron transition from $n=2$ to $n=1$. If the wavelengths of emitted radiation are $\lambda_1, \lambda_2, \lambda_3$ and λ_4 respectively, for four elements. Then, the value of $\frac{\lambda_1\lambda_2}{\lambda_3\lambda_4}$ is ....

22. A particle of mass 20g is released with an initial velocity 10 m/s along the curve from the point A, as shown in the figure. The point A is at a height h from point B. The particle slides along the frictionless surface. When the particle reaches point B, its angular momentum about O will be $x \times 10^{-2}$ kg m^{2}s^{-1}. The value of x is ....

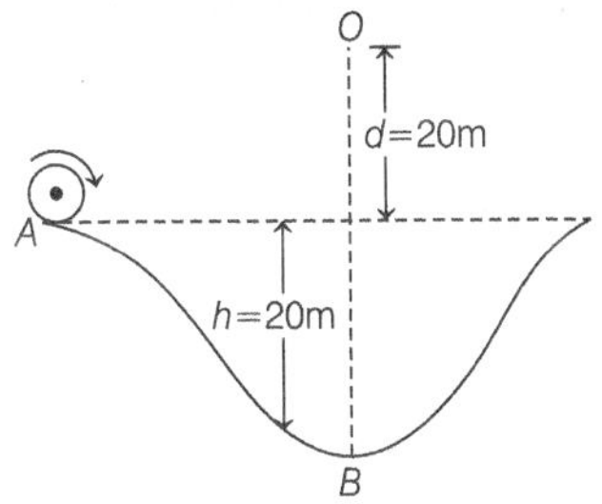

23. Current through wire XY of circuit shown is .... A.

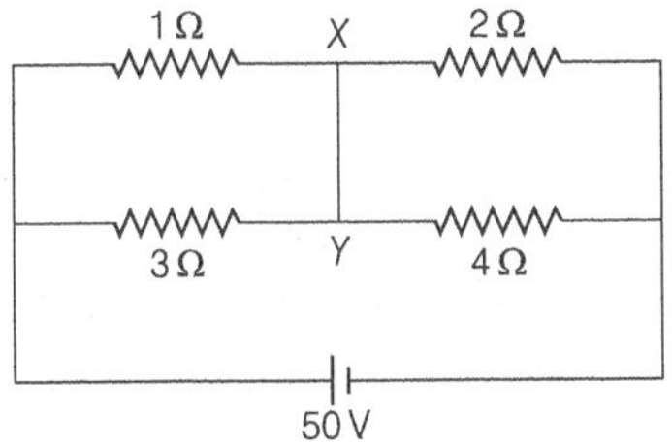

24. Magnetic field of 1T is applied to this arrangement. If connecting wires and rods have negligible resistance, then current through 5Ω resistor is ......mA.

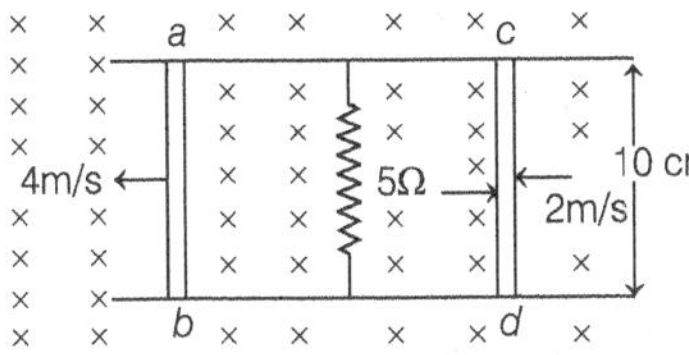

25. If velocity of a particle A is 50% of velocity of particle B and mass of B is 25% of mass of A, then find de-Broglie wavelength of B in angstrom (Å), if wavelength of A is 2Å .

CHEMISTRY

Section A : Objective Type Questions

26. Which of the following pair of molecules/ions have same value of bond order?

(a) O_2^+, NO, $CN^{\ominus}$ (b) $CN^{\ominus}$, N_2^+, N_2
(c) N_2^+, NO, O_2^+ (d) O_2^+, $CN^{\ominus}$, N_2^+

27. Which of the following non-essential amino acid will undergo diazo coupling reaction?

(a) Tyrosine (b) Glycine
(c) Alanine (d) Aspartic acid

28. The correct order of ionic size of $N^{3-}, Na^+, O^{2-}, F^-, Mg^{2+}$ will be

(a) $O^{2-} < Na^+ < F^- < N^{3-} < Mg^{2+}$
(b) $Mg^{2+} < Na^+ < F^- < O^{2-} < N^{3-}$
(c) $F^- < N^{3-} < O^{2-} < Mg^{2+} < Na^+$
(d) $F^- < Mg^{2+} < O^{2-} < Na^+ < N^{3-}$

29. Which type of bonds are present in $[Re_2Cl_8]^{2-}$?

(a) $1\sigma + 2\pi + 1\delta$ (b) $1\sigma + 2\pi$
(c) $1\sigma + 1\pi + 1\delta$ (d) $1\delta + 3\pi$

30. An organic compound contains N, S, C and H. Which of the following method can not be used for quantitative estimation of elements in the given organic compound?

(a) Duma's method
(b) Carius method
(c) Kjeldahl's method
(d) Liebig method

31. Consider the following metal coins.
$Sr^{2+}, Ba^{2+}, Ca^{2+}, Cu^{2+}, Fe^{2+}$
The total number of ions that will be characterised by flame test.

(a) 4 (b) 3 (c) 5 (d) 2

32. The standard state Gibb's free energies of formation of $C_{(graphite)}$ and $C_{(diamond)}$ at $T = 298$ K are
$\Delta_f G°[C_{(graphite)}] = 0$ kJ/mol
$\Delta_r G°[C_{(diamond)}] = 2.9$ kJ/mol
The standard state means that the pressure should be 1 bar, and substance should be pure at a given temperature. The conversion of graphite $[C_{(graphite)}]$ to diamond $[C_{(diamond)}]$ reduces its volume by 2×10^{-6} m^3 mol^{-1}. If $C_{(graphite)}$ is converted to $C_{(diamond)}$ isothermally at $T = 298$ K, the pressure which $C_{(graphite)}$ is in equilibrium with $C_{(diamond)}$ is [useful information : $1 J = 1 kg\ m^2 s^{-2}$, $1 Pa = 1 kg\ m^{-1} s^{-2}$, $1 bar = 10^5 Pa$]

(a) 14501 bar (b) 58001 bar
(c) 1450 bar (d) 29001 bar

33. The major product of the following reaction is

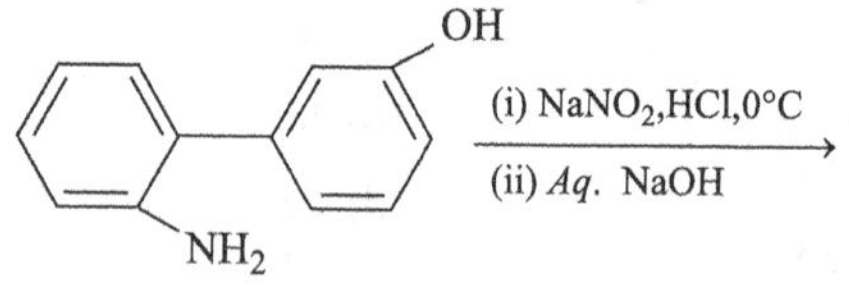

(a)
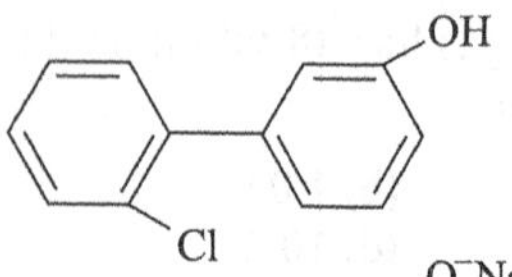

(b)
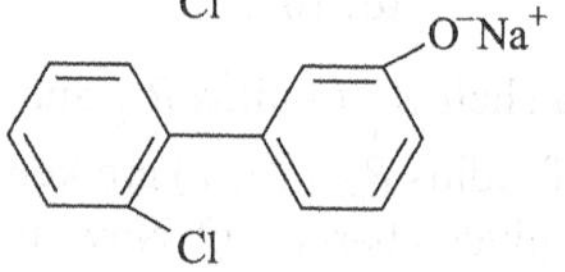

(c)
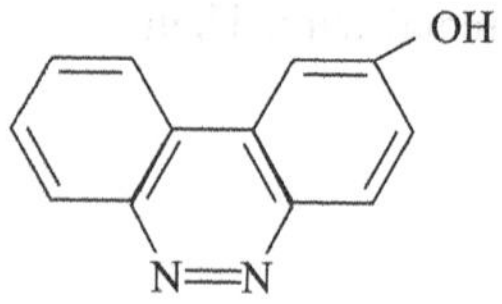

(d)
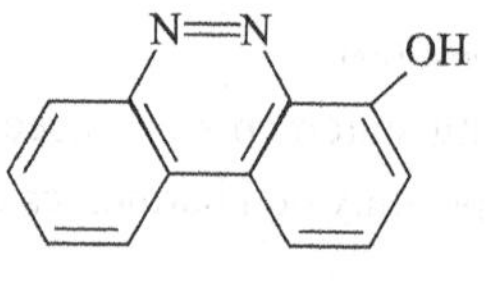

34. The order of basicity among the following compounds is

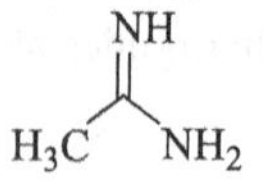

I

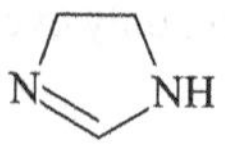

II

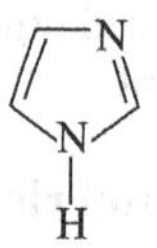

III

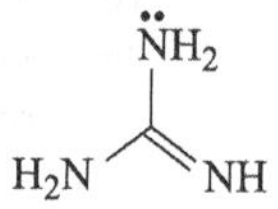

IV

(a) II > I > IV > III (b) IV > II > III > I
(c) IV > I > II > III (d) I > IV > III > II

35. Addition of excess aqueous ammonia to a pink coloured aqueous solution of $MCl_2 \cdot 6H_2O (X)$ and NH_4Cl gives an octahedral complex Y in the presence of air. In aqueous solution, complex Y behaves as 1 : 3 electrolyte. The reaction of X with excess HCl at room temperature results in the formation of a blue coloured complex Z.

The calculated spin only magnetic moment of X and Z is 3.87 BM, whereas it is zero for complex Y. Among the following options, which statement(s) is (are) correct?

I. Addition of silver nitrate to Y gives only two equivalents of silver chloride.

II. The hybridisation of the central metal ion in Y is dsp^3.

III. Z is tetrahedral complex.

IV. When X and Z are in equilibrium at 0°C, the colour of the solution is pink.

Choose the most appropriate answer from the options given below.

(a) I, III and IV
(b) III and IV
(c) I and III
(d) I and IV

36. The standard potential at 298 K for the following half-cell reactions are given against each.

$$Zn^{2+}/Zn = -0.762\ V$$
$$H^+/H_2 = 0\ V$$
$$Cr^{3+}/Cr = -0.740\ V$$
$$Fe^{3+}/Fe^{2+} = 0.770\ V$$

Which among the following is the strongest reducing agent?

(a) $Zn(s)$ (b) $Cr(s)$
(c) $H_2(g)$ (d) $Fe^{2+}(aq)$

37. The major product formed in the following reaction is

$CH_3-C(CH_3)(H)-CH_2Br \xrightarrow[CH_3OH]{CH_3O^-}$

(a) $CH_3-C(CH_3)(H)-CH_2CH_3$ (b) $CH_3-CH(OCH_3)-CH_2CH_3$

(c) $CH_3-C(CH_3)=CH_2$ (d) $CH_3-C(CH_3)(OCH_3)-CH_3$

38. What will be the product formed by reaction between A and B in presence of NaOEt? A is prepared by hydrolysis of CaC_2. B is prepared by reaction of CH_3CH_2OH with PCC?

(a) $CH_3CH_2-COOCH_3$ (b) $H-C(=O)-C\equiv CH$

(c) $CH_3-C(OH)(H)-C\equiv CH$ (d) $CH_3-C(=O)-C\equiv C-H$

39. Match List-I with List-II.

	List-I		List-II
A.	Sucrose	I.	β-D-galactose and β-D-glucose
B.	Lactose	II.	α-D-glucose and β-D-fructose
C.	Maltose	III.	Glucose, fructose + Galactose
D.	Raffinose	IV.	α-D-glucose and α-D-glucose

Choose the correct answer from the options given below.

(a) A-II, B-I, C-IV, D-III (b) A-I, B-II, C-III, D-IV
(c) A-III, B-I, C-II, D-IV (d) A-III, B-II, C-I, D-IV

40.

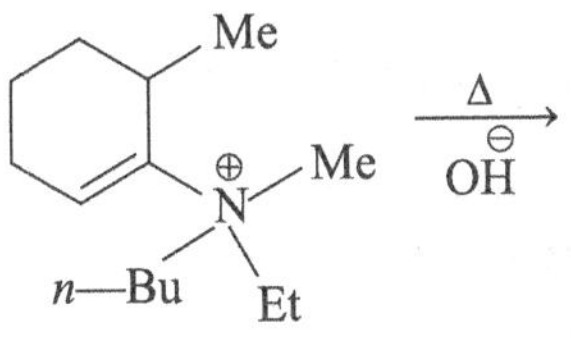

The alkene formed as a major product in the above elimination reaction is

(a) 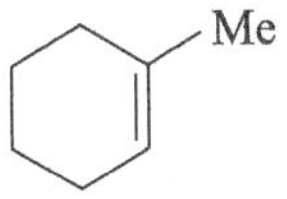

(b) 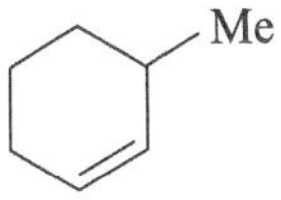

(c) 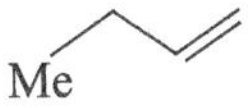

(d) $CH_2=CH_2$

41. Which among the following is incorrect statement?

(a) Organic compound shows electromeric effect in the presence of the reagent only.
(b) Electromeric effect is temporary effect.
(c) Hydrogen shows positive electromeric effect.
(d) The central carbon-carbon bond in n-butane is shorter than that of 1, 3-butadiene.

42. Consider the following statements given below.

Statement I Alkene on reacting with ozone followed by treatment with Zn/H_2O gives carbonyl compounds.

Statement II Carbonyl compounds react with hydroxylamine to give oximes.

In the light of the above statements, choose the correct answer from the options given below.
(a) Both Statement I and Statement II are correct.
(b) Both Statement I and Statement II are incorrect.
(c) Statement I is correct but Statement II is incorrect.
(d) Statement I is incorrect but Statement II is correct.

43. Consider the following ions.
$Eu^{2+}, Tm^{2+}, Dy^{3+}, Yb^{3+}, Tb^{4+}, Lu^{3+}, Ce^{3+}$
The total number of ions in which f-orbitals are half and fully filled.
(a) 4 (b) 3
(c) 5 (d) 2

44. Match List-I with List-II.

	List-I Electronic configuration of element		List-II Ionisation enthalpy/(kJ mol^{-1})
A.	$[Xe]6s^2$	I.	374
B.	$[He]2s^2 2p^1$	II.	520
C.	$[He]2s^1$	III.	503
D.	$[Xe]6s^1$	IV	801

Choose the most appropriate answer from the options given below.
(a) A-IV, B-II, C-III, D-I
(b) A-III, B-IV, C-II, D-I
(c) A-II, B-I, C-IV, D-III
(d) A-I, B-III, C-IV, D-II

45. Given below are two statements, one is labelled as Assertion (A) and other is labelled as Reason (R).
Assertion (A) In the redox process of XeF_2 with I_2, XeF_2 is powerful oxidising agent.
Reason (R) The hydrolysis of XeF_2 is slow in dilute acid but rapid in basic solution.
In the light of the above statements, choose the correct answer from the options given below.
(a) Both (A) and (R) are true and (R) is the correct explanantion of (A).
(b) Both(A) and (R) are true but (R) is not the correct explanantion of (A).
(c) (A) is true but (R) is false.
(d) (A) is false but (R) is true.

Section B : Numerical Value Type Questions

46. A mixture of H_2SO_4, $H_2C_2O_4$ and some inert impurity weighing 4.185 g was dissolved in water and the solution is made upto 1 L. 10 mL of this solution required 1.5 mL of 0.2 N NaOH for complete neutralisation.

In another experiment, 100 mL of the same solution in hot condition required 2 mL of 0.04 M $KMnO_4$ solution for complete reaction. The weight percentage (%) of H_2SO_4 in the mixture was ....

[Nearest integer]

47. The reaction, $X \longrightarrow Y$ is an exothermic reaction. Activation energy of the reaction for conversion of X into Y is 150 kJ mol^{-1}. Enthalpy is 135 kJ mol^{-1}. The activation energy for the reverse reaction, $Y \longrightarrow X$ will be ...... kJ mol^{-1}.

48. A 5.25% solution of a substance is isotonic with a 1.5% solution of urea (molar mass = 60 g mol^{-1}) in the same solvent.

If the densities of both the solutions are assumed to be 1.0 g cm^{-3}, molar mass of the substance will be .... g mol^{-1}.

49. In a saturated solution of the sparingly soluble strong electrolyte $AgIO_3$ (molecular mass = 283), the equilibrium sets as

$$AgIO_3(s) \rightleftharpoons Ag^+(aq) + IO_3^-(aq)$$

If the solubility product constant, K_{sp} of $AgIO_3$ at a given temperature is 1.0×10^{-8}. The mass of $AgIO_3$ contained in 100 mL of its saturated solution is ...... 10^{-5} g.

50. The angular momentum of an electron in a Bohr's orbit of He^+ is 3.16×10^{-34} Js. The wave number in terms of Rydberg's constant (R) of the spectral line when an electron emitted from this state to the first excited state will be $\frac{x}{y}R$.

The sum of x and y will be ....

[Given, $h = 6.62 \times 10^{-34}$ Js]

MATHEMATICS

Section A : Objective Type Questions

51. If $f(x+y)=f(x)f(y)\ \forall x, y \in R$ and $f(0)\neq 0$, then the function $g(x)=\dfrac{f(x)}{1+\{f(x)\}^2}$ is

(a) even function (b) odd function
(c) even if $f(x)>0$ (d) odd if $f(x)>0$

52. If $z = cis\alpha, 0<\alpha<\dfrac{\pi}{6}$, then argument of $1-z^4$ is

(a) $\dfrac{\pi}{2}+\dfrac{\alpha}{2}$ (b) $2\alpha-\dfrac{\pi}{2}$ (c) $\dfrac{\pi}{2}-\dfrac{\alpha}{2}$ (d) $\dfrac{\pi}{2}+2\alpha$

53. Consider a parabola $x^2=4y$ and the point $F(0,1)$. Let $P_1(x_1,y_1), P_2(x_2,y_2), P_3(x_3,y_3),\ldots P_n(x_n,y_n)$ be n points on the parabola such that $x_k>0$ and $\angle OFP_k=\dfrac{k\pi}{2n}$ $(k=1,2,3,\ldots n)$. Then, the value of $\int_0^1 FP_k\,dx\left[\text{where, } \dfrac{k}{n}=x\right]$ is

(a) $\dfrac{2}{\pi}$ (b) $\dfrac{4}{\pi}$ (c) $\dfrac{\pi}{2}$ (d) $\dfrac{\pi}{4}$

54. The function $f(x)=[x]^2-[x^2]$ is discontinuous at (where, $[\cdot]$ is greatest integer function)

(a) all integers (b) all integers except 0 and 1
(c) all integers except 0 (d) all integers except 1

55. If the adjoint of a 3×3 matrix P is $\begin{bmatrix}1&4&4\\2&1&7\\1&1&3\end{bmatrix}$, then the possible values of the determinant of P are

(a) ± 2 (b) ± 1 (c) ± 3 (d) ± 4

56. If α and β be the coefficients of x^6 and x^4 respectively in the expansion of $(x+\sqrt{x^2-1})^8+(x-\sqrt{x^2-1})^8$, then

(a) $\alpha+\beta=30$ (b) $\alpha-\beta=580$
(c) $\alpha-\beta=-132$ (d) $\alpha+\beta=-472$

57. The value of $\int_{\pi/2}^{3\pi/2}[2\sin x]dx$ is equal to (where, $[\cdot]$ is the greatest integer function)

(a) $-\pi$ (b) 0
(c) $-\dfrac{\pi}{2}$ (d) $\dfrac{\pi}{2}$

58. The area bounded by the curves $y^2=x, 2y+3=x$ and X-axis in the first quadrant, is

(a) 36 (b) 18
(c) 9 (d) $\dfrac{27}{4}$

59. The numbers 1, 3, 6, 10, 15, 21, 28,... are called triangular numbers. If a_n denotes the nth triangular number, then a_{100} must be equal to

(a) 5050 (b) 5151
(c) 5252 (d) 5353

60. The positive integers p, q, r and s satisfying $\tan\dfrac{\pi}{24}=(\sqrt{p}-\sqrt{q})(\sqrt{r}-s)$. Then, the value of $p+q+r+s$ is

(a) 5 (b) 8
(c) 7 (d) 4

61. If the length of the perpendicular drawn from the point $P(a,-1,11)$, $a>1$ to the line $\dfrac{x}{2}=\dfrac{y-2}{3}=\dfrac{z-3}{4}$ is $\sqrt{53}$ units and $Q(\alpha_1,\alpha_2,\alpha_3)$ is the image of the point P in this line, then $a+\sum_{i=1}^{3}\alpha_i$ is equal to

(a) 12 (b) 18
(c) 19 (d) 20

62. The value of $\int_{-\pi/4}^{3\pi/4}\dfrac{\sin x+\cos x}{e^{x-\frac{\pi}{4}}+1}dx$ is equal to

(a) 2 (b) $\sqrt{2}$
(c) $\sqrt{3}$ (d) $\sqrt{5}$

63. If we multiply first 10 positive integers by -1 and then add 1 to each number, then the variance of the numbers so obtained is

(a) 8.25 (b) 2.25
(c) 6.5 (d) 3.5

64. $\dfrac{d^2x}{dy^2}$ equals

(a) $\left(\dfrac{d^2y}{dx^2}\right)^{-1}$ (b) $-\left(\dfrac{d^2y}{dx^2}\right)^{-1}\left(\dfrac{dy}{dx}\right)^{-3}$
(c) $\left(\dfrac{d^2y}{dx^2}\right)\left(\dfrac{dy}{dx}\right)^{-2}$ (d) $-\left(\dfrac{d^2y}{dx^2}\right)\left(\dfrac{dy}{dx}\right)^{-3}$

65. Let f be a function defined by $f(x) = \tan 3x + \cos \frac{5x}{2}$ and $g(x) = \cos x + \{x\}$. Consider two statements.

Statement I f is periodic function with period 4π.

Statement II g is also periodic function with period $\frac{3\pi}{2}$

(a) Statement I is true and Statement II is false
(b) Statement I is false and Statement II is true
(c) Statement I and Statement II both are true
(d) Statement I and Statement II both are false

66. If the length of the latusrectum of an ellipse is 4 units and the distance between a focus and its nearest vertex on the major axis is $\frac{3}{2}$ units, then its eccentricity is

(a) $\frac{1}{2}$ (b) $\frac{1}{3}$ (c) $\frac{2}{3}$ (d) $\frac{1}{9}$

67. Total number of solution of $\log_{(2x+3)}(6x^2 + 23x + 21) + \log_{(3x+7)}(4x^2 + 12x + 9) = 4$, is

(a) 2 (b) 4
(c) 1 (d) 0

68. If $|z_1 z_2| = \left|2\cos\frac{\pi}{4} - 2i\sin\frac{\pi}{4}\right|$ and $\arg\left|\frac{z_1}{z_2}\right| = \frac{\pi}{3}$, then $z_1^2 \bar{z}_2^2$ equal to

(a) $4(-1 + i\sqrt{3})$ (b) $4(1 + i\sqrt{3})$
(c) $2(-1 - i\sqrt{3})$ (d) $2(-1 + i\sqrt{3})$

69. Let $T_n = \left(1 + \frac{1}{n}\right)^n$, then for each $n \in N$

(a) $T_n \geq 3$ (b) $2 \leq T_n < 3$
(c) $T_n > 4$ (d) $T_n < 2$

70. Let $p, q, r \in R$ such that $pqr = 5$, if $A = \begin{bmatrix} 3p & q & r \\ q & 3r & p \\ r & p & 3q \end{bmatrix}$ and $AA^T = 4I$ and $|A|$ is a positive real number, then the value of $p^3 + q^3 + r^3$ is

(a) 125 (b) 137
(c) 64 (d) 16

Section B : Numerical Value Type Questions

71. The largest term in the sequence $a_n = \frac{n^2}{n^3 + 100}$ $[\forall n \in [1, \infty)]$ is λ. Then, the value of 79λ is ....

72. The number of words can be formed with the letters of the word PATALIPUTRA without changing the relative positions of vowels and consonants is k^2, then k is equal to ....

73. Consider **r**, **a**, **b** and **c** are non-zero vectors such that $\mathbf{r} \cdot \mathbf{a} = 0$, $|\mathbf{r} \times \mathbf{b}| = |\mathbf{r}||\mathbf{b}|$, $|\mathbf{r} \times \mathbf{c}| = |\mathbf{r}||\mathbf{c}|$, then $(\mathbf{a} \times \mathbf{b}) \cdot \mathbf{c}$ is equal to ....

74. If line $y = 2x + 3$ does not intersect any member of family of parabolas $y^2 = 8ax$, $(a \in R^+)$ at two distinct points, then the maximum length of latusrectum of parabola is ....

75. Players A and B alternatively toss a biased coin with A going first. A wins, if A tosses a tail before B tosses a head, otherwise B wins. If the probability of a head is $\frac{\sqrt{A} + B}{C}$, then the value of $A + B + C$ for which the game is fair to both players is ....

DETAILED SOLUTIONS

Physics

1. *(a)* The wavelength in various series is given as

$$\frac{1}{\lambda} = R\left(\frac{1}{n_1^2} - \frac{1}{n_2^2}\right)$$

For Lyman series, $n_1 = 1, n_2 = 2, 3, 4$
which is lying in U-V region.
Therefore, it must be Lyman series.

2. *(b)* Given, $\frac{dv}{dt} = -2\sqrt{v}$

$$\frac{dv}{\sqrt{v}} = -2dt$$

On integrating within limit $v = 4$ to 0 and $t = 0$ to t, we get

$$\int_4^0 \frac{1}{\sqrt{v}} dv = -2\int_0^t dt$$

$$2[v^{1/2}]_4^0 = -2t$$

$$2[0-(4)^{1/2}] = -2t$$

$$t = 2\text{s}$$

3. *(c)* Power, $P = \frac{V^2}{R}$

Resistance of hot tungsten

$$\frac{V^2}{P} = \frac{200 \times 200}{100}$$

$$R_{\text{hot}} = 400\Omega, R_{\text{cold}} = 40\Omega$$

Now, $R_{\text{hot}} = R_{\text{cold}}(1 + \alpha \cdot \Delta T)$

$$\frac{400}{40} = 1 + \alpha \times \Delta T$$

$$\Delta T = \frac{9}{3 \times 10^{-3}} = 3000°\text{C}$$

$\therefore$ $T_{\text{hot}} = 3000°\text{C} + 25°$

$$T_{\text{hot}} = 3025°\text{C}$$

4. *(d)* At resonance,

$$X_C = X_L$$

At, $\omega = \frac{\omega_0}{2}, X_C' = 2X_C$ and $X_L' = \frac{X_L}{2}$

$$\left[\because X_C = \frac{1}{\omega C}, \text{i.e} X_C \propto \frac{1}{\omega}\right]$$

Then, $Z' = \sqrt{X_C'^2 + X_L'^2}$ $[\because X_L = \omega L, \text{ i.e. } X_L \propto \omega]$

$$= \sqrt{4 \times X_C^2 + \frac{X_L^2}{4}}$$

$$Z' = \frac{\sqrt{17}}{2} \cdot X_C \qquad [\because X_C = X_L]$$

Now, when frequency is $2\omega_0$, then

$$X_C' = \frac{X_C}{2} \text{ and } X_L' = 2X_L$$

So, $Z' = \sqrt{X_C'^2 + X_L'^2} = \frac{\sqrt{17}}{2} X_C = Z'$

Since, impedance remains the same at $\omega = \frac{\omega_0}{2}$ and at $\omega = 2\omega_0$. So, current should be 2 A.

5. *(c)* While a particle is in uniform circular motion. Then, the following statements are true.

(i) Speed will be always constant throughout motion.

(ii) The centripetal acceleration $a = v^2 / r$ and its direction will always towards centre of the circular trajectory.

(iii) Angular momentum (mvr) is constant in magnitude and direction. And its direction is perpendicular to the plane containing r and v.

(iv) Velocity will be always tangential to the direction of motion at every point of the trajectory.

6. *(a)* The graph of $a \sin x$ is given by

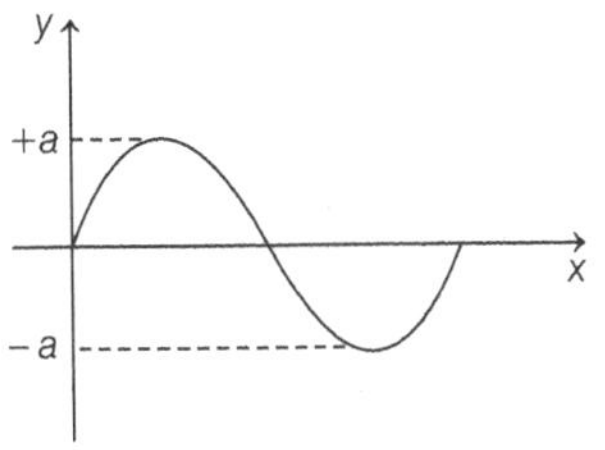

$\therefore$ By comparing equation of electric field with $a \sin x$, we get

$$-E_0 \le E \le +E_0$$

7. *(c)* In general, the field due to a magnet is non-uniform. Therefore, it exerts both, a net force and a torque on this nail which will translate and also rotate the nails before striking to the north pole of the magnet with their induced south poles and *vice-versa*.

8. *(a)* Static friction is the opposing force that comes into play when a body is at rest or tends to move over the surface of another body, but the actual motion has not yet started. Static friction is a self-adjusting force because neither it has fixed magnitude nor direction. It adjusts itself according to the force applied. Higher the force applied, higher the frictional force upto a limit.

9. *(c)* For isothermal expansion,

i.e., $pV = p' \times 2V$ $\qquad [\because V' = 2V]$

$$p' = \frac{p}{2}$$

For adiabatic expansion,

$$pV^{\gamma} = \text{constant}$$

$$p'V'^{\gamma} = p''V''^{\gamma}$$

$$\frac{p}{2}[2V]^{5/3} = p''[16V]^{5/3} \qquad \left[\because \gamma = \frac{5}{3}\right]$$

$$p'' = \frac{p}{2}\left[\frac{2V}{16V}\right]^{5/3}$$

$$= \frac{p}{2}\left[\frac{1}{8}\right]^{5/3}$$

$$p'' = \frac{p}{64}$$

The final pressure of the gas $(p'') = \frac{p}{64}$

10. *(d)* $a = \text{MSD}$

$$a' = \text{VSD}$$

$$(n-1)a = n(a')$$

$$a' = \frac{(n-1)a}{n}$$

$\therefore$ Least count $= 1\,\text{MSD} - 1\,\text{VSD}$

$$= (a - a')\,\text{cm}$$

$$= a - \frac{(n-1)a}{n}$$

$$= \frac{na - na + a}{n} = \frac{a}{n}\,\text{cm}$$

$$= \left(\frac{10a}{n}\right)\text{mm}$$

11. *(c)* Due to heating, there shall be expansion in rod given by

$$\Delta L_1 = \alpha(\Delta T)L \qquad \ldots\text{(i)}$$

But this expansion is resisted by rigid walls. If R is the reaction produced, then stress in rod is $\frac{R}{A}$.

As, $\text{strain} = \frac{\text{stress}}{Y}$

$$\frac{\Delta L_1}{L} = \frac{R/A}{Y}$$

[ΔL_1 is compression by stress due to wall]

$$\Delta L_1 = \frac{R}{AY}\cdot L \qquad \ldots\text{(ii)}$$

From Eqs. (i) and (ii), we have

$$R = \alpha AY(\Delta T)$$

12. *(a)* After connecting two spheres, their potential will be same.

When switch S is closed, then charge on sphere A will be redistributed on A and C in ratio of their capacitances.

$$\frac{q_1}{q - q_1} = \frac{4\pi\varepsilon_0\left(\frac{ab}{b-a}\right)}{4\pi\varepsilon_0 c}$$

$$\frac{q_1}{q - q_1} = \frac{ab}{c(b-a)} = \frac{ab}{bc - ac}$$

$$q_1 = \frac{(q - q_1)ab}{bc - ac} = \frac{q\,ab}{bc - ac} - \frac{q_1 ab}{bc - ac}$$

$$q_1\left(1 + \frac{ab}{bc - ac}\right) = \frac{q\,ab}{bc - ac}$$

$$q_1 = \frac{q\,ab}{ab + bc - ca}$$

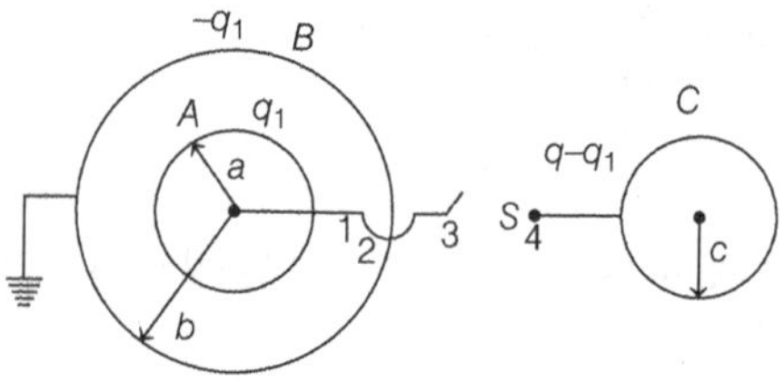

Taking, $q = 30\,\mu\text{C}$

$$a = 1\text{ cm}, b = 2\text{ cm and } c = 1\text{ cm}$$

We get, $q_1 = 20\,\mu\text{C}$

$\therefore \qquad q - q_1 = 30 - 20 = 10\,\mu\text{C}$

13. *(a)* Electric field, $[E]$

$$= \frac{[F]}{[q]} = \frac{[\text{MLT}^{-2}]}{[\text{AT}]} = [\text{MLT}^{-3}\text{A}^{-1}]$$

Electric potential $[V]$

$$= [E]\cdot[d] = [\text{MLT}^{-3}\text{A}^{-1}][\text{L}]$$

$$= [\text{ML}^2\text{T}^{-3}\text{A}^{-1}]$$

Electric flux, $[\phi] = [E]\cdot[A]$

$$= [\text{MLT}^{-3}\text{A}^{-1}][\text{L}^2] = [\text{ML}^3\text{T}^{-3}\text{A}^{-1}]$$

Permittivity of free space,

$$[\varepsilon_0] = \frac{[q]}{[4\pi Er^2]} = \frac{[\text{AT}]}{[\text{MLT}^{-3}\text{A}^{-1}][\text{L}^2]}$$

$$= [\text{M}^{-1}\text{L}^{-3}\text{T}^4\text{A}^2]$$

14. *(b)* Given, $E = -\frac{3}{r}$, $r_1 = 6\,\text{m}$, $r_2 = 12\,\text{m}$, $V_1 = 12\,\text{J kg}^{-1}$

As, $\qquad dV = -E\cdot dr$

$$\Rightarrow \int_{V_1}^{V_2} dV = -\int_{r_1}^{r_2} \frac{-3}{r}\,dr$$

$$\Rightarrow V_2 - V_1 = 3\int_6^{12}\frac{dr}{r} = 3\ln\frac{12}{6}$$

$\therefore \qquad V_2 = V_1 + 3\ln 2 = 12 + 3\times 0.693$

$\therefore \qquad V_2 = 14.079\,\text{J kg}^{-1}$

15. *(c)* As the germanium diode is forward biased and its knee voltage is 0.3 V.

So, the given circuit can be redrawn as

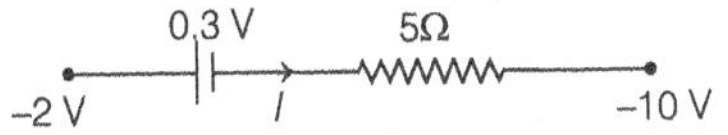

$\therefore \quad I = \frac{V}{R} = \frac{-2-0.3-(-10)}{5} = \frac{7.7}{5} = 1.54\text{ A}$

16. *(c)* Using, $\gamma_{real} = \gamma_{app} + \gamma_{vessel}$

$\therefore \quad \gamma_{vessel} = 3\alpha$

For vessel P, $\gamma_{real} = \gamma_1 + 3\alpha_1$

For vessel Q, $\gamma_{real} = \gamma_2 + 3\alpha_2$

Hence, $\gamma_1 + 3\alpha_1 = \gamma_2 + 3\alpha_2$

$$\gamma_1 - \gamma_2 = 3\alpha_2 - 3\alpha_1$$

$$\frac{\gamma_1 - \gamma_2}{3} = \alpha_2 - \alpha_1$$

$$\alpha_2 = \frac{\gamma_1 - \gamma_2}{3} + \alpha_1 = \frac{\gamma_1 - \gamma_2}{3} + \alpha$$

17. *(d)* As the net external force on the system is zero.

$\therefore \quad p_i = p_f$

Applying conservation of momentum in x-direction,

$$mv = 2m' v\cos 60^\circ$$

$\therefore \quad m' = m$

Applying conservation of momentum in y-direction,

$$0 = m' v' - 2m' v\sin 60^\circ$$

$\Rightarrow \quad v' = 2v \times \frac{\sqrt{3}}{2} = \sqrt{3}v$

$\therefore \quad v' = \sqrt{3}v$

18. *(a)* Since, each vibrational mode has 2 degrees of freedom.

Hence, total vibrational degrees of freedom

$$= 48 \times 2 = 96$$

$\therefore$ Total number of degrees of freedom

$$= 3 + 3 + 96 = 102$$

$\therefore \quad \gamma = 1 + \frac{2}{f} = 1 + \frac{2}{102} = \frac{104}{102} = 1.0196 \approx 1.02$

19. *(b)* We have, $Q_A + Q_B = 2Q$ ...(i)

$$\frac{Q_A}{R_A} = \frac{Q_B}{R_B} \quad \text{...(ii)}$$

From Eqs. (i) and (ii), we get

$$Q_A = Q_B\left(\frac{R_A}{R_B}\right) \text{ and } Q_B\left(1 + \frac{R_A}{R_B}\right) = 2Q$$

$\Rightarrow \quad Q_B = \frac{2Q}{\left(1 + \frac{R_A}{R_B}\right)} = \frac{2QR_B}{R_A + R_B}$

and $\quad Q_A = \frac{2QR_A}{R_A + R_B} \Rightarrow Q_A > Q_B \quad [\because R_B < R_A]$

Ratio of surface charge density of solid sphere and shell is

$$\frac{\sigma_A}{\sigma_B} = \frac{R_B}{R_A}$$

$$E_{A\,(\text{on surface})} < E_{B\,(\text{on surface})}$$

20. *(a)* For closed organ pipe,

Fundamental frequency, $\nu_0 = \frac{v}{4l}$

and frequency of first harmonics,

$$\nu_1 = \left(n + \frac{1}{2}\right)\frac{v}{2l} = \frac{3v}{4l} \text{ and } \nu_2 = \frac{5v}{4l}$$

$\therefore \nu_0 : \nu_1 : \nu_2 = 1:3:5$

21. (36) For hydrogen atom, we get

$$\frac{1}{\lambda_1} = RZ^2\left(\frac{1}{1^2} - \frac{1}{2^2}\right) \Rightarrow \frac{1}{\lambda_1} = R(1)^2\left(\frac{3}{4}\right)$$

$\Rightarrow \quad \lambda_1 = \frac{4}{3R}$ ...(i)

For deuterium $(Z = 1)$, we get

$$\frac{1}{\lambda_2} = R(1)^2\left(\frac{3}{4}\right)$$

$\Rightarrow \quad \lambda_2 = \frac{4}{3R}$ ...(ii)

For singly ionised helium $(Z = 2)$, we get

$$\frac{1}{\lambda_3} = R(2)^2\left(\frac{3}{4}\right) \Rightarrow \lambda_3 = \frac{1}{4}\cdot\frac{4}{3R} = \frac{1}{3R}$$

$\Rightarrow \quad \lambda_3 = \frac{1}{3R}$ ...(iii)

For doubly ionised lithium $(Z = 3)$, we get

$$\frac{1}{\lambda_4} = R(3)^2\left(\frac{3}{4}\right) \Rightarrow \lambda_4 = \frac{1}{9}\cdot\frac{4}{3R} = \frac{4}{27R}$$

$\Rightarrow \quad \lambda_4 = \frac{4}{27R}$ ...(iv)

From Eqs. (i), (ii), (iii) and (iv), we get

$$\because \quad \frac{\lambda_1\lambda_2}{\lambda_3\lambda_4} = \frac{\frac{4}{3R}\cdot\frac{4}{3R}}{\frac{1}{3R}\cdot\frac{4}{27R}} = 36$$

22. (1788) The given figure is shown below as

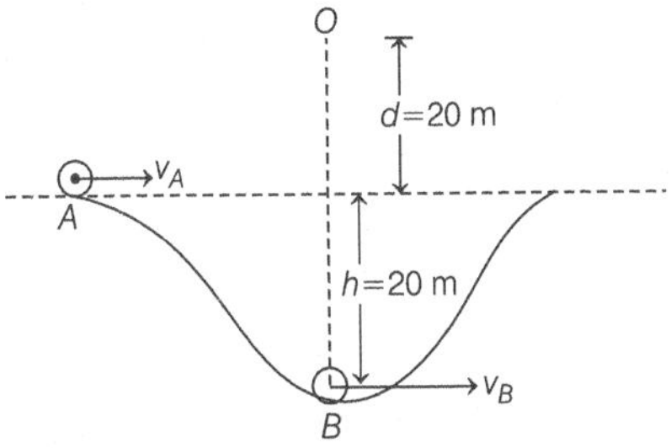

As friction is absent, energy at A = energy at B

$$\Rightarrow \quad \frac{1}{2}mv_A^2 + mgh = \frac{1}{2}mv_B^2$$

$$\Rightarrow \quad v_A^2 + 2gh = v_B^2$$

$$\Rightarrow \quad v_B^2 = (10)^2 + 2\times 10\times 20 = 500$$

$$\Rightarrow \quad v_B = \sqrt{500}\ \text{ms}^{-1} = 22.36\ \text{ms}^{-1}$$

Angular momentum about point $O = mv_B r_B$

$$= 20\times 10^{-3} \times 22.36\times 40$$
$$= 1788\ \text{kg m}^2\text{s}^{-1}$$
$$= x\times 10^{-2}\ \text{kg m}^2\text{s}^{-1}$$

$$\therefore \quad x = 1788$$

Study Tactics

Initially, calculate the velocity with respect to B, then find the angular momentum about point O by using formula mv_Br_B.

23. (2) Let current through XY is i_3.

Applying Kirchhoff's law to loops (1) and (2),

$$i_1 \times 1 + 0\times i_3 - 3i_2 = 0$$

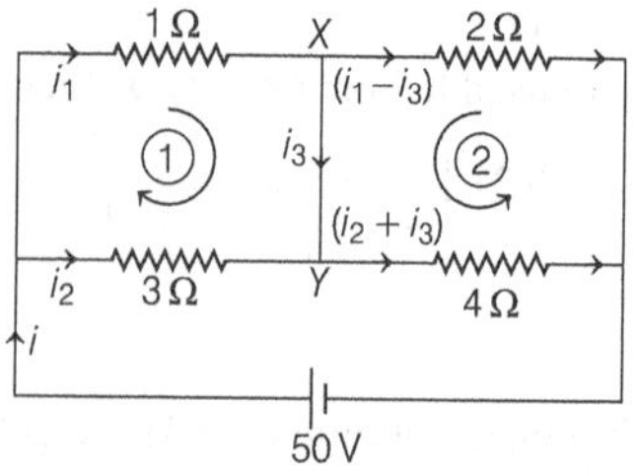

$\therefore\ i_1 = 3i_2$ ...(i)

and $-2(i_1 - i_3) + 4(i_2 + i_3) = 0$

So, $\quad 2i_1 - 4i_2 = 6i_3$ ...(ii)

Also, $50 = 3i_2 + 4(i_2 + i_3)$

$\therefore \quad 7i_2 + 4i_3 = 50$ ...(iii)

From Eqs. (i), (ii) and (iii), we get

$$i_3 = 2\text{A}$$

24. (40) Motional emf will be induced in both the rods.

Potential difference in

$$ab = Blv = 1\times 0.1\times 4 = 0.4\ \text{V}$$

Potential difference in $cd = Blv = 1\times 0.1\times 2 = 0.2$ V

So, the arrangement would act like

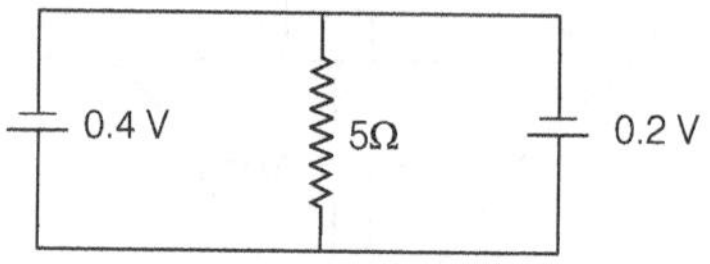

$$\therefore \text{ Current, } I = \frac{0.4-0.2}{5} = \frac{0.2}{5} = 0.04\ \text{A} = 40\ \text{mA}$$

25. (4) m_A and m_B and v_A and v_B be masses and velocities of A and B particles, respectively.

$$v_A = (50\%)\text{ of } v_B = \frac{v_B}{2},$$

$$m_B = (25\%)\text{ of } m_A = \frac{m_A}{4}$$

$$\therefore \quad v_B = 2v_A$$
$$m_B = m_A/4 \quad \text{...(i)}$$

From de-Broglie wavelength, we know that

$$\lambda = \frac{h}{mv}$$

$$\therefore \quad \lambda \propto \frac{1}{mv} \quad (h = \text{Planck's constant})$$

$$\therefore \quad \frac{\lambda_A}{\lambda_B} = \frac{m_B v_B}{m_A v_A}$$

$$\lambda_B = \frac{\lambda_A (m_A v_A)}{(m_B v_B)} = 2\frac{(m_A v_A)}{m_B v_B}\ \text{Å}$$

Putting values of m_B and v_B from Eq. (i) in terms of A

$$\lambda_B = \frac{2(m_A v_A)\cdot 4}{(m_A)(2v_A)} = 4\ \text{Å}$$

Chemistry

26. (c) Order of filling of molecular orbital occurs in two ways

- If total number of electrons present in molecule/ion is less than or equal to 14.

$$\sigma 1s, \sigma^* 1s, \sigma 2s, \sigma^* 2s, \pi 2p_x = \pi 2p_y, \sigma 2p_z, \pi^* 2p_x = \pi^* 2p_y, \sigma^* 2p_z$$

- If total number of electrons present in molecules/ions is greater than 14.

$$\sigma 1s, \sigma^* 1s, \sigma 2s, \sigma^* 2s, \sigma 2p_z, \pi 2p_x = \pi 2p_y, \pi^* 2p_x = \pi^* 2p_y, \sigma^* 2p_z$$

Molecular orbital electronic configuration of N_2^+ is

$$\sigma 1s^2, \sigma^* 1s^2, \sigma 2s^2, \sigma^* 2s^2, \pi 2p_x^2 = \pi 2p_y^2, \sigma 2p_z^1$$

Bond order of $N_2^+ = \frac{9-4}{2} = 2.5$

Molecular orbital electronic configuration of NO

$$\sigma 1s^2, \sigma^* 1s^2, \sigma 2s^2, \sigma^* 2s^2, \sigma 2p_z^2, \pi 2p_x^2 = \pi 2p_y^2, \pi^* 2p_x^1 = \pi^* 2p_y^0$$

Bond order of NO $= \frac{10-5}{2} = 2.5$

Molecular orbital electronic configuration of O_2^+

$$\sigma 1s^2, \sigma^* 1s^2, \sigma 2s^2, \sigma^* 2s^2, \sigma 2p_z^2, \pi 2p_x^2$$
$$\pi 2p_y^2, \pi^* 2p_x^1 \equiv \pi^* 2p_y^0$$

Bond order of $O_2^+ = \frac{10-5}{2} = 2.5$

27. (*a*) **Non-essential amino acids** The amino acids which can be synthesised in the body are known as non-essential amino acids.

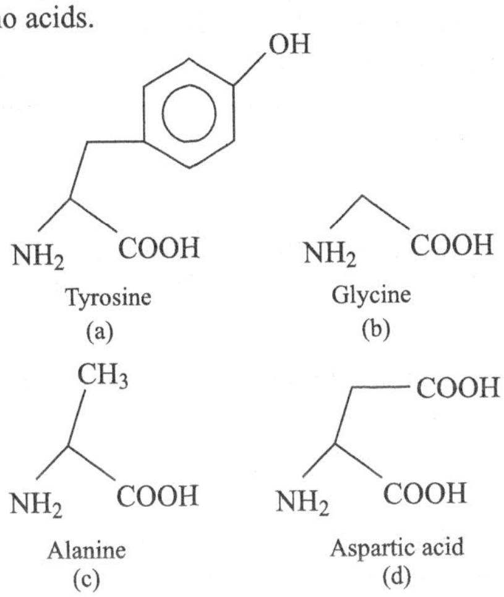

Aromatic amino acid undergo diazo coupling reaction. Tyrosine have aromatic ring it is a non-essential amino acid. Hence, option (a) is correct.

28. (*b*) All given species are isoelectronic having 10 electrons. In this cases more is the negative charge on atom larger will be its radius and more is the positive charge smaller will be radius. Hence, the correct order is

$$Mg^{2+} < Na^{+} < F^{-} < O^{2-} < N^{3-}$$

29. (*a*) **Bonding in $[Re_2Cl_8]^{2-}$** : σ-bond is formed by head on overlapping of orbitals of Re.

π-bond is formed by lateral overlapping of *p*-orbitals of Re.

1 σ and 2π-bonds in $[Re_2Cl_8]^{2-}$ can be represented as

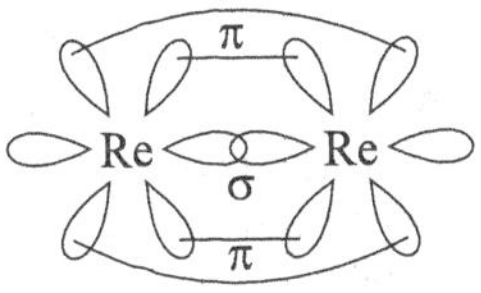

While δ-bonds are formed by overlapping of d_{xy} orbitals as

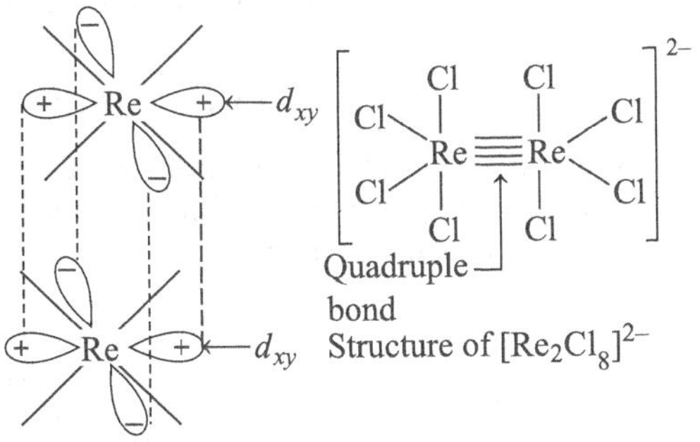

30. (*b*) **Quantitative detection of elements** The list of elements and their methods for quantitative analysis are given below as

Element	Methods of quantitative analysis
C	Liebig method
H	Liebig method
N	Duma's method, Kjeldahl's method
Halogen, P	Carius method

Since, the above compound does not contain any halogen or phosphorus. So, Carius method cannot be used for the given organic compound.

31. (*a*) A flame test is a qualitative analysis used to identify the metal and metalloid ion in the sample.
Among the given metal ions only four, i.e. Sr^{2+}, Ba^{2+}, Ca^{2+} and Cu^{2+} will respond to flame test.

32. (*a*) $C_{(graphite)} \longrightarrow C_{(diamond)}$

$\Delta_r G° = \Delta G°_{(diamond)} - \Delta G°_{(graphite)}$

2.9 kJ/mol $- 0$ kJ/mol $= 2.9$ kJ/mol

$\Delta_r G° = \Delta_r H° - T\Delta_r S°$

$\Delta_r S° \approx 0$, therefore $T\Delta_r S° = 0$

$\Delta_r H° = \Delta U + p\Delta V$

For isothermal process $\Delta T = 0$

$\therefore \quad \Delta U = 0$

$$\Delta_r H° = 2.9 \times 10^3 \text{ J/mol} = p\Delta V$$

At equilibrium pressure is constant

2.9×10^3 J/mol $= p(V_2 - V_1)$

$$p = \frac{2.9 \times 10^3 \text{ J/mol}}{2 \times 10^{-6} \text{ m}^3\text{/mol}}$$

$$p = 1450 \times 10^6 \text{ Pascal} = 14500 \text{ bar}$$

Total pressure = equilibrium pressure + initial pressure
= 14501 bar

33. (*c*) ***Step* 1**

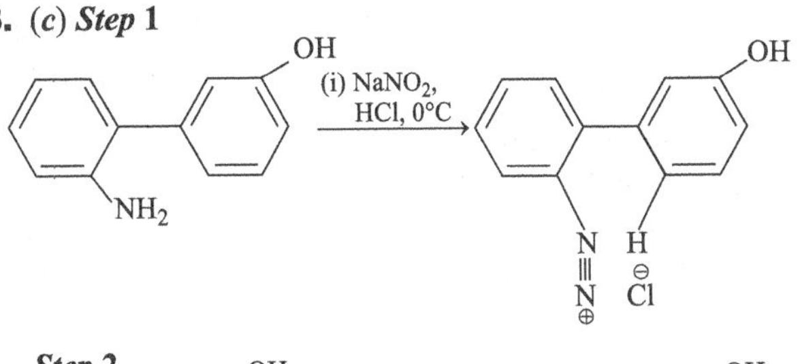

***Step* 2**

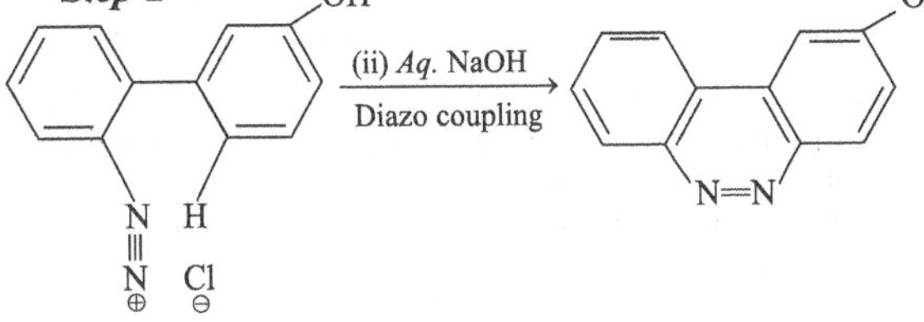

34. (*c*)

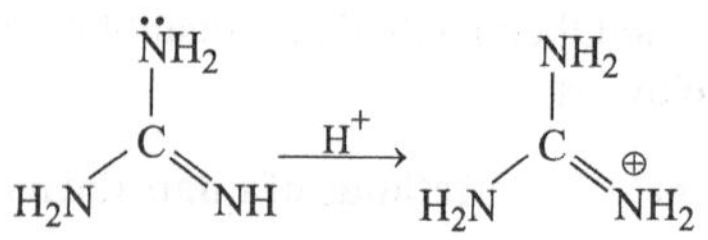

(IV)

The conjugate acid is stabilised by resonance with two different —NH_2 group. Hence, electron density increases on N of =NH.

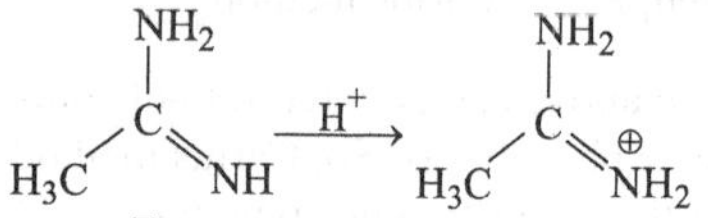

(I)

The conjugate acid is stabilised by resonance with one —NH_2 group.

Hence, as compared to IV lesser increase of electron density on N of =NH.

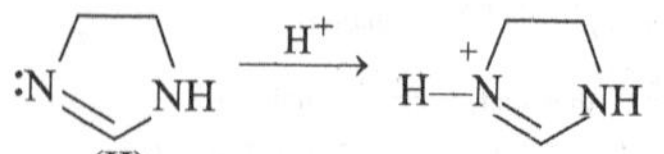

(II)

Lone pair is not involved in aromaticity. Hence, are more available.

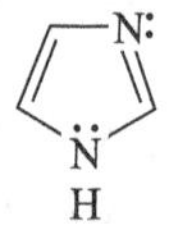

(III)

Lone pair is involved in aromatic sextet. Hence, not available.

Therefore, the correct order of basic strength is

IV > I > II > III.

35. (*b*) $\underset{(X)}{[Co(H_2O)_6]Cl_2} \xrightarrow[NH_4Cl\ +\ Air]{Excess\ NH_3(aq)} \underset{(Y)}{[Co(NH_3)_6]Cl_3}$

Pink solution

$\downarrow$ HCl (excess) room temperature

$\underset{(Z)}{[CoCl_4]^{2-}}$

Blue colour (tetrahedral sp^3 as Cl^- is a weak field ligand)

1:3 electrolyte (d^2sp^3 hybridisation of central metal atom as NH_3 act as a strong field ligand)

$\downarrow$ $3AgNO_3(aq)$

3AgCl(3 eq. of AgCl)

$$\underset{(Blue)}{[CoCl_4]^{2-}} + 6H_2O \rightleftharpoons \underset{Pink}{[Co(H_2O)_6]^{2+}} + 4Cl^-$$

X and *Z* in equilibrium at 0°C ⇒ then equilibrium is shifted towards *X*, making colour of solution pink.

36. (*a*) The more negative the standard potential, the least the reduction tendency of the ion. The corresponding atoms has largest oxidation tendency and thus is a strong reducing agent. In the present case, Zn is the strongest reducing agent.

37. (*c*) The compound $CH_3—\overset{\overset{CH_3}{|}}{C}H—CH_2Br$ on reaction with CH_3O^-/CH_3OH follows mechanism E_{1CB} and the major product is Zaitsev product. The reaction involved is as follows

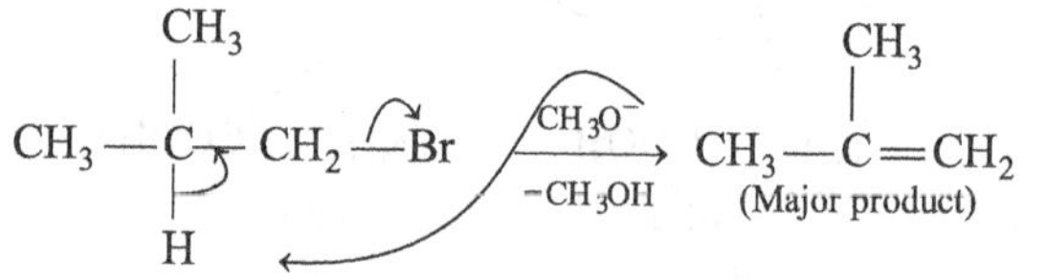

38. (*c*) **Preparation of alkyne** When CaC_2 is hydrolysed, it produces acetylene gas *A*.

$$CaC_2 \xrightarrow{H_3O^+} \underset{(A)}{HC{\equiv}CH}$$

Preparation of aldehyde When alcohol is oxidised by PCC, it produces aldehyde (*B*).

$$CH_3CH_2OH \xrightarrow{PCC} \underset{(B)}{CH_3CHO}$$

Reaction between *A* and *B* in presence of base NaOEt

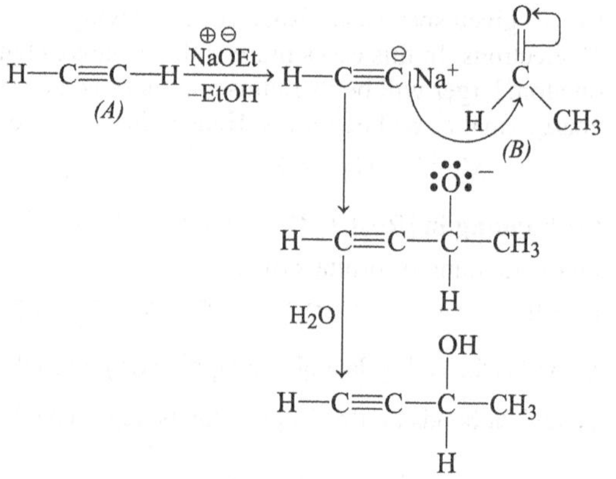

Study Tactics

This problem is based on conceptual mixing of preparation of alkyne and reaction of alkyne with carbonyl compound.

Complete the reaction using information supplied in the question then undergo sequential oxidation and nucleophilic addition reaction to get the final product.

39. (*a*) The correct match is A-II, B-I, C-IV, D-III.

40. (*d*) **Hofmann's rule** When theoretically more than one type of alkenes are possible, the alkene containing least alkylated double bond is formed.

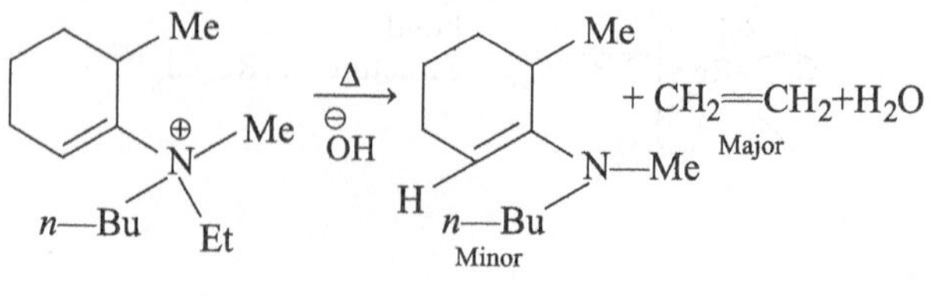

41. (*d*)

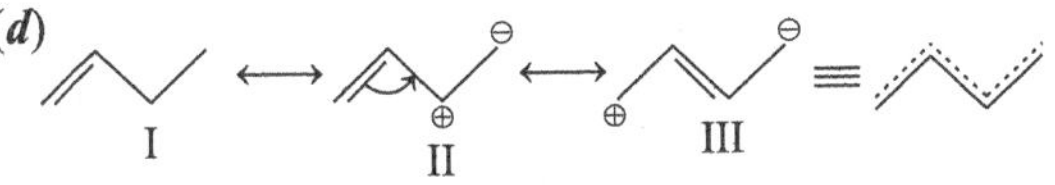

The above shown resonance introduces some double bond character to central C—C bond. Therefore, the central C—C bond in 1, 3-butadiene is shorter than C—C bond in butane.

42. (*a*) Alkene on reacting with ozone (O_3) followed by treatment with Zn/H_2O gives carbonyl compounds.

e.g.

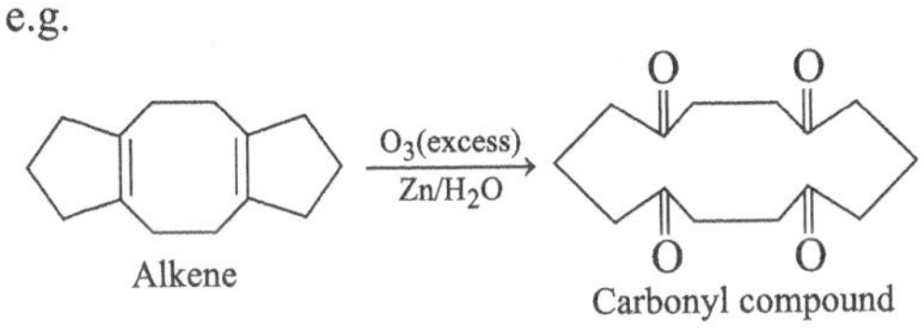

When carbonyl compound reacts with NH_2OH, oxime is obtained.

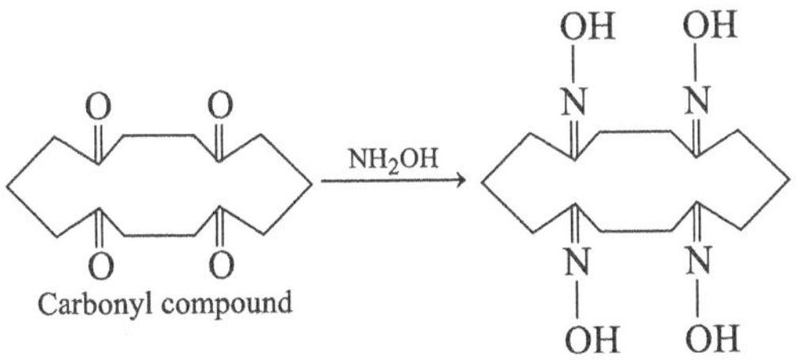

43. (*b*) Among the given ions only three ions in which *f*-orbitals are half and fully filled.

Element	+2	+3	+4
Eu	$4f^7$		
Tm	$4f^{13}$		
Dy		$4f^9$	
Yb		$4f^{13}$	
Tb			$4f^7$
Lu		$4f^{14}$	
Ce		$4f^1$	

44. (*b*)

Electronic configuration	Element
A. $[Xe]6s^2$	Barium (Ba)
B. $[He]2s^2, 2p^1$	Boron (B)
C. $[He]2s^1$	Lithium (Li)
D. $[Xe]6s^1$	Caesium (Cs)

Li and Cs belong to same group, i.e. first group. While, Ba belong to second group elements and boron belong to group 13.

On the other hand, Li and boron are present in period 2 and Cs and Ba are present in period 6.

As we go down the group, the size of the atoms increases due to which the attraction between the nucleus and the electrons in the outermost shell decreases. Therefore, the ionisation enthalpy decreases.

However, as we move from left to right across a period, the ionisation energy of elements increases due to decrease in the size of atoms.

So, according to the above facts, the ionisation enthalpy order is Cs < Ba < Li < B

Hence, $[Xe]6s^2 \longrightarrow 503$ kJ/mol

$[He]2s^2 2p^1 \longrightarrow 801$ kJ/mol

$[He]2s^1 \longrightarrow 520$ kJ/mol

$[Xe]6s^1 \longrightarrow 374$ kJ/mol

45. (*b*) Both (A) and (R) are true but (R) is not the correct explanation of (A).

XeF_2 is powerful oxidising agent because the Xe—F bond has the lowest bond energy.

46. (*30*) Let millimoles of $H_2SO_4 = x$

and millimoles of $H_2C_2O_4 = y$

Milliequivalent of acid = Milliequivalent of base

$$(x + y) \times 2 = 1.5 \times 0.2 \times \frac{1000}{10}$$

So, $x + y = 15$

In IInd case, $KMnO_4$ reacted with $H_2C_2O_4$ in which Mn^{7+} converted into Mn^{2+} and $C_2O_4^{2-}$ gets converted into CO_2 (change in oxidation number = 5)

Milliequivalent of $H_2C_2O_4$ = Milliequivalent of $KMnO_4$

$$y \times 2 = 0.04 \times 2 \times 5 \times \frac{1000}{100}$$

$$y = 2$$

$\therefore \quad x + 2 = 15, \; x = 13$

Weight of $H_2SO_4 = 13 \times 10^{-3} \times 98 = 1.274$ g

$$\text{Weight \% of } H_2SO_4 = \frac{1.274}{4.185} \times 100 = 30.44\% \approx 30\%$$

47. (*285*) An exothermic reaction can be represented as,

$$X \longrightarrow Y + \text{Energy}$$

i.e. energy of $Y << X$.

So, the energy profile diagram is

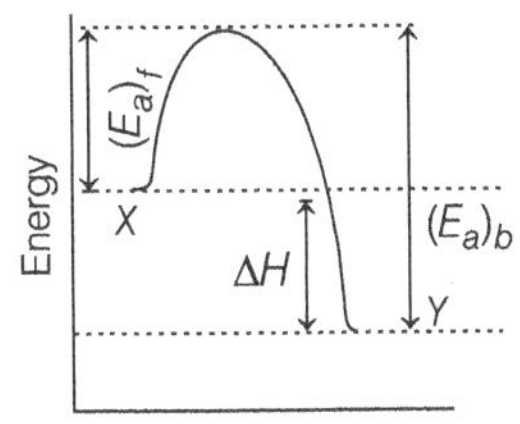

$$\therefore \quad (E_a)_f + \Delta H = (E_a)_b$$
$$150 + 135 = (E_a)_b$$
$$(E_a)_b = 285 \text{ kJ mol}^{-1}$$

48. (*210*) Molar concentration of the substance = Molar concentration of urea

$$\frac{5.25 \text{ g L}^{-1}}{M} = \frac{1.5 \text{ gL}^{-1}}{60 \text{ g mol}^{-1}}$$

($\because$ 100 g solution = 100 mL as $d = 1 \text{ g cm}^{-3}$ and 1L $= 1000 \text{ cm}^3 = 1000$ mL)

$$\therefore \quad M = 210 \text{ g mol}^{-1}$$

Study Tactics

When two solutions are isotonic, then the osmotic pressure of one solution is equal to the osmotic pressure of other solution.

49. (*283*) $AgIO_3(s) \rightleftharpoons Ag^+(aq) + IO_3^-(aq)$

$$K_{sp} = [Ag^+][IO_3^-]$$
$$1.0 \times 10^{-8} = (S)(S)$$
$$1.0 \times 10^{-8} = S^2$$
$$S = \sqrt{1.0 \times 10^{-8}}$$
$$S = 1.0 \times 10^{-4} \text{ mol/L}$$

In 1000 mL, moles of $AgIO_3$ dissolved $= 1 \times 10^{-4}$ moles

In 100 mL, moles of $AgIO_3$ dissolved $= 1 \times 10^{-5}$ moles

Mass of $AgIO_3$ in 100 mL $= 1 \times 10^{-5} \times 283 = 283 \times 10^{-5}$ g

50. (*14*) $\bar{\nu} = RZ^2\left(\frac{1}{n_1^2} - \frac{1}{n_2^2}\right)$

Angular momentum $= \frac{nh}{2\pi}$

$$3.16 \times 10^{-34} = \frac{n \times 6.62 \times 10^{-34}}{2\pi}$$
$$n = 2.99 \simeq 3$$
$$\bar{\nu} = RZ^2\left(\frac{1}{n_1^2} - \frac{1}{n_2^2}\right),$$
$$n_1 = 2, n_2 = 3$$
$$\bar{\nu} = RZ^2\left(\frac{1}{4} - \frac{1}{9}\right), \quad \bar{\nu} = \frac{5R}{9}$$
$$\therefore \quad \frac{5}{9}R = \frac{x}{y}R$$

So, $x + y = 5 + 9 = 14$.

Mathematics

51. (*a*) $\because f(x+y) = f(x) \cdot f(y)$

If $x = y = 0$, then $f(0) = 1$

If $y = -x$, then $f(0) = f(x) f(-x)$

$$\Rightarrow \quad f(-x) = \frac{1}{f(x)}$$

Now, $$g(x) = \frac{f(x)}{1 + \{f(x)\}^2}$$

$$\Rightarrow \quad g(-x) = \frac{f(-x)}{1 + \{f(-x)\}^2}$$
$$= \frac{\frac{1}{f(x)}}{1 + \left\{\frac{1}{f(x)}\right\}^2} = \frac{f(x)}{1 + \{f(x)\}^2} = g(x)$$

$\therefore g(x)$ is an even function.

52. (*b*) Given, $z = cis\,\alpha = \cos\alpha + i\sin\alpha$

$$\therefore \quad 1 - z^4 = 1 - (\cos\alpha + i\sin\alpha)^4$$
$$= 1 - (\cos 4\alpha + i\sin 4\alpha)$$

[using De-moivre's theorem]

$$= (1 - \cos 4\alpha) - i\sin 4\alpha$$
$$\arg(1 - z^4) = -\tan^{-1}\left(\frac{\sin 4\alpha}{1 - \cos 4\alpha}\right)$$
$$= -\tan^{-1}\left(\frac{2\sin 2\alpha \cos 2\alpha}{2\sin^2 2\alpha}\right)$$
$$= -\tan^{-1}(\cot 2\alpha)$$
$$= \tan^{-1}\left[\tan\left(2\alpha - \frac{\pi}{2}\right)\right] = 2\alpha - \frac{\pi}{2}$$

53. (*b*) Given, parabola $x^2 = 4y$

Let parametric coordinate of $P_k = (2t, t^2)$

Slope of $FP_k = \frac{t^2 - 1}{2t - 0} = \tan\left(\frac{\pi}{2} + \theta_k\right)$

$$\Rightarrow \quad \cot\theta_k = -\left(\frac{t^2 - 1}{2t}\right)$$
$$\Rightarrow \quad \tan\theta_k = \frac{2t}{1 - t^2} = \tan 2A \qquad [\text{let } \tan A = t]$$
$$\Rightarrow \quad \theta_k = 2A$$
$$\Rightarrow \quad A = \frac{\theta_k}{2} = \frac{k\pi}{4n}$$
$$FP_k = \sqrt{(t^2 - 1)^2 + (2t)^2}$$
$$t^2 + 1 = 1 + \tan^2 A = \sec^2 A = \sec^2\frac{k\pi}{4n}$$
$$\because \frac{k}{n} = x$$

Now, $$\int_0^1 FP_k = \int_0^1 \sec^2\frac{\pi x}{4}\,dx = \left[\frac{4}{\pi}\tan\frac{\pi x}{4}\right]_0^1 = \frac{4}{\pi}$$

54. *(d)* The given function, $f(x)=[x]^2-[x^2]$

$\therefore [x]$ is continuous for all non-integers.

Let $x=n\in Z$

$f(x)=x^2-x^2=0$

Now, if $h>0$ is small

$$f(x+h)=[x+h]^2-[(x+h)^2]$$
$$=x^2-[x^2+2xh+h^2] \quad [\because h\to 0]$$
$$=x^2-x^2=0 \quad \text{...(i)}$$
$$f(x-h)=[x-h]^2-[(x-h)^2]$$
$$=(x-1)^2-[x^2-2xh+h^2]$$
$$=(x-1)^2-(x^2-1)=2(1-x) \quad \text{...(ii)}$$

From Eqs. (i) and (ii), we get

$$1-x=0 \Rightarrow x=1$$

$\Rightarrow f(x)$ is continuous at $x=1$.

55. *(a)* Given, $\text{adj}(P)=\begin{bmatrix}1&4&4\\2&1&7\\1&1&3\end{bmatrix}$

$$|\text{adj}(P)|=\begin{vmatrix}1&4&4\\2&1&7\\1&1&3\end{vmatrix}$$
$$=1(3-7)-4(6-7)+4(2-1)$$
$$=-4+4+4=4$$

$\because \quad |\text{adj}\, P|=|P|^{3-1} \quad [\because \text{adj}\, A=|A|^{n-1}]$

$\Rightarrow \quad 4=|P|^2$

$\Rightarrow \quad |P|=\pm 2$

56. *(d)* Given, expression is $(x+\sqrt{x^2-1})^8+(x-\sqrt{x^2-1})^8$

$$=2[{}^8C_0x^8+{}^8C_2x^6(x^2-1)+{}^8C_4x^4(x^2-1)^2+{}^8C_6x^2(x^2-1)^3+{}^8C_8(x^2-1)^4]$$

The coefficients of x^6 in above expansion

$$\alpha=2[{}^8C_2(-1)+2\cdot{}^8C_4{}^2C_1(-1)+{}^8C_6{}^3C_1(-1)+{}^8C_8{}^4C_1(-1)]$$
$$=2[-28-280-84-4]=-792$$

and the coefficient of x^4 in the above expansion

$$\beta=2[{}^8C_4+{}^8C_6(3)+{}^8C_8(6)]$$
$$=2[70+84+6]=320$$

$\therefore \alpha+\beta=-472$

57. *(c)* $\because [2\sin x]=\begin{cases}1, & \text{if } \frac{\pi}{2}<x\le\frac{5\pi}{6}\\ 0, & \text{if } \frac{5\pi}{6}<x\le\pi\\ -1, & \text{if } \pi<x\le\frac{7\pi}{6}\\ -2, & \text{if } \frac{7\pi}{6}<x\le\frac{3\pi}{2}\end{cases}$

$$\therefore \int_{\pi/2}^{3\pi/2}[2\sin x]dx=\int_{\pi/2}^{5\pi/6}1\,dx+\int_{5\pi/6}^{\pi}0\,dx+\int_{\pi}^{7\pi/6}(-1)dx+\int_{\frac{7\pi}{6}}^{3\pi/2}(-2)dx$$
$$=\left(\frac{5\pi}{6}-\frac{\pi}{2}\right)+0-1\left(\frac{7\pi}{6}-\pi\right)-2\left[\frac{3\pi}{2}-\frac{7\pi}{6}\right]$$
$$=\frac{2\pi}{6}-\frac{\pi}{6}-\frac{4\pi}{6}=\frac{-3\pi}{6}=\frac{-\pi}{2}$$

Study Tactics

First, define the function $[2\sin x]$ in the interval $\frac{\pi}{2}$ to $\frac{3\pi}{2}$.

Then, integrate in the interval

$\frac{\pi}{2}$ to $\frac{5\pi}{6}$, $\frac{5\pi}{6}$ to π, π to $\frac{7\pi}{6}$ and $\frac{7\pi}{6}$ to $\frac{3\pi}{2}$.

58. *(c)* Given, curves $y^2=x$ ...(i)

$$2y+3=x \quad \text{...(ii)}$$
$$y=0 \quad \text{...(iii)}$$

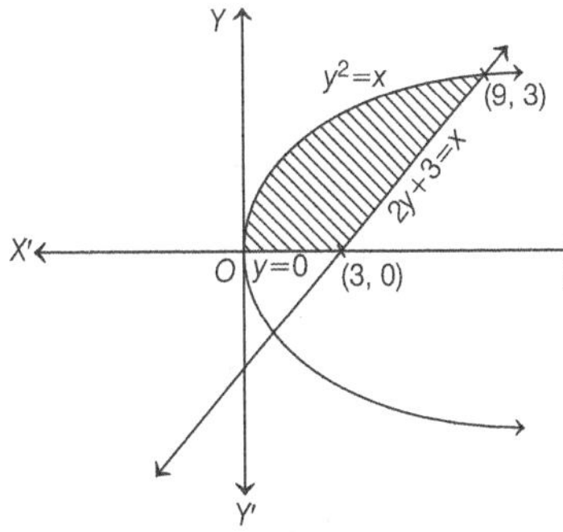

On solving Eqs. (i) and (ii), we get

$$2y+3=y^2$$

$\Rightarrow \quad y^2-2y-3=0$

$\Rightarrow \quad (y-3)(y+1)=0$

$\Rightarrow \quad y=3, y=-1$

$\Rightarrow \quad y=3\ (y\ne \text{negative})$

At $y=3, x=9$

Eq. (ii) meets X-axis at $(3,0)$.

$\therefore$ Shaded region is the required area.

$$\therefore \text{Area}=\int_0^9\sqrt{x}dx-\int_3^9\frac{x-3}{2}dx$$
$$=\left[\frac{x^{3/2}}{3/2}\right]_0^9-\frac{1}{2}\left(\frac{x^2}{2}-3x\right)_3^9$$
$$=\frac{2\times 27}{3}-\frac{1}{2}\left(\frac{81}{2}-27-\frac{9}{2}+9\right)$$
$$=\frac{54}{3}-\frac{1}{2}(18)=18-9$$
$$=9 \text{ sq units}$$

59. *(a)* We observed $a_1 = 1$,

$$a_2 = 1+2 = 3,$$
$$a_3 = 3+3 = 6,$$
$$a_4 = 6+4 = 10$$
$$\vdots$$
$$a_n = a_{n-1} + n$$

Sum $S = 1+3+6+10+15+\ldots+a_n$ ...(i)

$S = \quad 1+3+6+10+\ldots+a_{n-1}+a_n$ ...(ii)

[on substracting Eq. (ii) from Eq. (i)]

$$\Rightarrow \quad a_n = 1+2+3+\ldots+n$$
$$a_n = \frac{n(n+1)}{2}$$
$$\therefore \quad a_{100} = \frac{100\times 101}{2} = 5050$$

60. *(b)* $\because \tan\frac{\theta}{2} = \frac{1-\cos\theta}{\sin\theta}$

$$\tan\left(\frac{\pi}{24}\right) = \frac{1-\cos\frac{\pi}{12}}{\sin\frac{\pi}{12}} = \left(\frac{1-\left(\frac{\sqrt{3}+1}{2\sqrt{2}}\right)}{\frac{\sqrt{3}-1}{2\sqrt{2}}}\right)$$
$$= \left(\frac{2\sqrt{2}-\sqrt{3}-1}{\sqrt{3}-1}\right)\left(\frac{\sqrt{3}+1}{\sqrt{3}+1}\right)$$
$$= \frac{2\sqrt{6}-3-\sqrt{3}+2\sqrt{2}-\sqrt{3}-1}{2}$$
$$= \frac{2\sqrt{6}-4-2\sqrt{3}+2\sqrt{2}}{2}$$
$$= \sqrt{6}-2-\sqrt{3}+\sqrt{2}$$
$$= (\sqrt{3}-\sqrt{2})(\sqrt{2}-1)$$

On comparing, we get

$$p = 3, q = 2, r = 2, s = 1$$

Hence, $p+q+r+s = 3+2+2+1 = 8$

61. *(b)* Let R be $(2k, 3k+2, 4k+3)$.

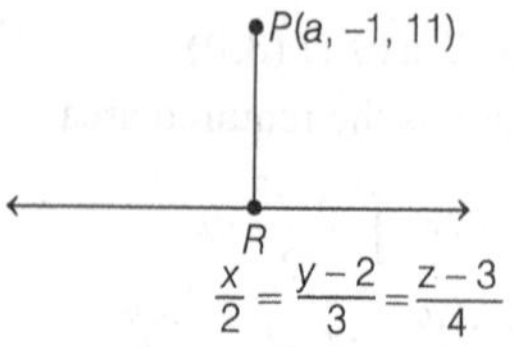

Direction ratios of PR are

$< 2k - a, 3k + 2 + 1, 4k + 3 - 11>$

as PR perpendicular to the given line

$$2(2k-a) + 3(3k+3) + 4(4k-8) = 0$$
$$4k - 2a + 9k + 9 + 16k - 32 = 0$$
$$29k - 23 - 2a = 0$$
$$29k - 2a = 23$$
$$a = \frac{23-29k}{-2}$$

$$PR = \sqrt{(2k-a)^2 + (3k+3)^2 + (4k-8)^2} = \sqrt{53}$$
$$\Rightarrow 4k^2 + a^2 - 4ka + 9k^2 + 9 + 18k + 16k^2 + 64 - 64k = 53$$

Let $k = 1$

$$a^2 - 4a + 56 = 53$$
$$\Rightarrow \quad a^2 - 4a + 3 = 0$$
$$\Rightarrow \quad a = 1, a = 3$$
$$\Rightarrow \quad a = 3 \quad [a = 1 \text{ will discard as } a > 1]$$
$$P \equiv (3, -1, 11), R \equiv (2, 5, 7)$$

Here, $Q(\alpha_1, \alpha_2, \alpha_3)$ be reflection of point P

$$\alpha_1 + 3 = 4, \alpha_2 - 1 = 10, \alpha_3 + 11 = 14$$
$$\alpha_1 = 1, \alpha_2 = 11, \alpha_3 = 3$$
$$\Rightarrow Q(1, 11, 3)$$
$$\therefore a + \sum_{i=1}^{3} \alpha_i = 3 + 1 + 11 + 3 = 18$$

62. *(b)* Let $I = \int_{\frac{-\pi}{4}}^{3\pi/4} \frac{\sin x + \cos x}{e^{x-\frac{\pi}{4}} + 1} dx$ ...(i)

$$= \int_{-\frac{\pi}{4}}^{3\pi/4} \frac{\sin\left(\frac{\pi}{2}-x\right) + \cos\left(\frac{\pi}{2}-x\right)}{e^{\frac{\pi}{4}-x} + 1}$$

$$\left[\because \int_a^b f(x)dx = \int_a^b f(a+b-x)dx\right]$$

$$= \int_{-\frac{\pi}{4}}^{\frac{3\pi}{4}} \frac{\cos x + \sin x}{e^{\frac{\pi}{4}-x}\left(1+e^{x-\frac{\pi}{4}}\right)} dx$$

$$I = \int_{\frac{-\pi}{4}}^{\frac{3\pi}{4}} \frac{(\cos x + \sin x)}{1+e^{x-\frac{\pi}{4}}} e^{x-\frac{\pi}{4}} dx \quad \text{...(ii)}$$

On adding Eqs. (i) and (ii), we get

$$2I = \int_{\frac{-\pi}{4}}^{\frac{3\pi}{4}} \frac{(\cos x + \sin x)\left(1+e^{x-\frac{\pi}{4}}\right)}{\left(1+e^{x-\frac{\pi}{4}}\right)} dx$$
$$= \int_{\frac{-\pi}{4}}^{\frac{3\pi}{4}} \cos x + \sin x \, dx$$
$$= (\sin x - \cos x)_{-\pi/4}^{3\pi/4} = 2\sqrt{2}$$
$$2I = 2\sqrt{2}$$
$$I = \sqrt{2}$$

63. *(a)* First 10 positive integer

$$1,2,3,4,5,6,7,8,9,10$$

On multiplying each number by -1, then adding 1, we get

$$0, -1, -2, -3, -4, -5, -6, -7, -8, -9$$

$\therefore \quad \Sigma x = 0-1-2-3-4-5-6-7-8-9$

$= -45$

and $\Sigma x^2 = 0^2 + (-1)^2 + (-2)^2 + \ldots + (-9)^2$

$= 0+1+4+9+16+25+36+49+64+81$

$= 285$

$$\text{Variance} = \frac{\Sigma x^2}{n} - \left(\frac{\Sigma x}{n}\right)^2$$

$$= \frac{285}{10} - \left(\frac{-45}{10}\right)^2$$

$$= 28.5 - (-4.5)^2$$

$$= 28.5 - 20.25 = 8.25$$

Time Saver Tip

Variance of first 10 natural numbers $= \dfrac{10^2 - 1}{12} = 8.25$

Now, when we multiply each number by a, then the variance becomes $a^2 \times$ old variance.

Hence, new variance $= a^2 \times$ old variance

$= (-1)^2 \times 8.25 = 8.25$

64. *(d)* We have, $\dfrac{d^2x}{dy^2} = \dfrac{d}{dy}\left(\dfrac{dx}{dy}\right) = \left[\dfrac{d}{dx}\left(\dfrac{dx}{dy}\right)\right]\dfrac{dx}{dy}$

$$= \left[\frac{d}{dx}\left\{\frac{1}{\left(\frac{dy}{dx}\right)}\right\}\right] \times \frac{1}{\left(\frac{dy}{dx}\right)}$$

$$= \frac{-1}{\left(\frac{dy}{dx}\right)^2} \times \frac{d^2y}{dx^2} \times \frac{1}{\left(\frac{dy}{dx}\right)}$$

$$= -\left(\frac{dy}{dx}\right)^{-3} \frac{d^2y}{dx^2}$$

65. *(a)* **Statement I** $f(x) = \tan 3x + \cos \dfrac{5x}{2}$

Period of $\tan 3x = \dfrac{\pi}{3}$ and period of $\cos \dfrac{5x}{2} = \dfrac{2\pi}{5/2} = \dfrac{4\pi}{5}$

$\therefore$ Period of $f(x) = \dfrac{\text{LCM of } \pi \text{ and } 4\pi}{\text{HCF of 3 and 5}} = 4\pi$

Statement II $g(x) = \cos x + \{x\} = \cos x + x - [x]$

$\cos x$ is periodic with period 2π and $x - [x]$ is periodic with period 1. But LCM of 2π and 1 is 2π.

66. *(b)* Length of latusrectum $= 4$

$\Rightarrow \quad \dfrac{2b^2}{a} = 4$

$\Rightarrow \quad b^2 = 2a \quad \ldots(i)$

Also, $b^2 = a^2(1-e^2) \quad \ldots(ii)$

Distance between a focus and its nearest vertex on the major axis $= \dfrac{3}{2}$

$\Rightarrow a(1-e) = \dfrac{3}{2} \quad \ldots(iii)$

From Eqs. (i) and (ii), we get

$$2a = a^2(1-e)(1+e)$$

$\Rightarrow \quad 2 = \dfrac{3}{2}(1+e) \quad$ [from Eq. (iii)]

$\Rightarrow \quad 4 = 3 + 3e$

$\Rightarrow \quad e = \dfrac{1}{3}$

67. *(c)* The given equation can be written as

$$\log_{(2x+3)}(2x+3)(3x+7) + \log_{(3x+7)}(2x+3)^2 = 4$$

$\Rightarrow \log_{(2x+3)}(2x+3) + \log_{(2x+3)}(3x+7) + 2\log_{(3x+7)}(2x+3) = 4$

$\Rightarrow \quad 1 + \log_{(2x+3)}(3x+7) + \dfrac{2}{\log_{(2x+3)}(3x+7)} = 4$

Let $\quad \log_{(2x+3)}(3x+7) = t$

$\therefore \quad t + \dfrac{2}{t} = 3$

$\Rightarrow \quad t^2 - 3t + 2 = 0$

$\therefore \quad t = 1, 2$

Case I If $t = 1$, then

$$\log_{(2x+3)}(3x+7) = 1$$

$\Rightarrow \quad 3x + 7 = 2x + 3$

$\therefore x = -4$ but $2x + 3 > 0$ and $3x + 7 > 0$

i.e. $x > -\dfrac{3}{2}$ and $x > -\dfrac{7}{3}$

Hence, no solution for $t = 1$

Case II If $t=2$, then

$$\log_{(2x+3)}(3x+7)=2$$

$$\Rightarrow \quad 3x+7=(2x+3)^2$$

$$\Rightarrow \quad 4x^2+9x+2=0$$

$$\Rightarrow \quad x=-2 \text{ and } x=-\frac{1}{4}$$

$$\left[\because x>-\frac{3}{2} \text{ and } x>-\frac{7}{3}\right]$$

$$\therefore \quad x=-\frac{1}{4}$$

Combining all cases, the given equation has only one solution $x=-\frac{1}{4}$

68. *(d)* Given, $|z_1 z_2|=\left|2\cos\frac{\pi}{4}-2i\sin\frac{\pi}{4}\right|$

$$=2\left|\cos\frac{\pi}{4}-i\sin\frac{\pi}{4}\right|=2$$

$$\therefore \quad |z_1^2\,\bar{z}_2^2|=|z_1|^2\,|z_2|^2$$

$$=|z_1 z_2|^2=4$$

$$\because \quad \arg\frac{z_1}{z_2}=\frac{\pi}{3}$$

$$\Rightarrow \arg z_1-\arg z_2=\frac{\pi}{3}$$

$$\therefore \arg[z_1^2(\bar{z}_2)^2]=2\arg(z_1\bar{z}_2)$$

$$=2(\arg z_1-\arg z_2)=\frac{2\pi}{3}$$

$$\therefore \quad z_1^2\bar{z}_2^2=4\left(\cos\frac{2\pi}{3}+i\sin\frac{2\pi}{3}\right)$$

$$=4\left(-\frac{1}{2}+i\frac{\sqrt{3}}{2}\right)$$

$$=2(-1+i\sqrt{3})$$

69. *(b)* We have,

$T_1=2$ and for $n\geq 2$

$$T_n=\left(1+\frac{1}{n}\right)^n$$

$$={}^nC_0+{}^nC_1\left(\frac{1}{n}\right)+{}^nC_2\left(\frac{1}{n}\right)^2+\ldots+{}^nC_r\left(\frac{1}{n}\right)^r+\ldots+{}^nC_n\left(\frac{1}{n}\right)^n$$

$$=1+1+\frac{n(n-1)}{2!}\cdot\frac{1}{n^2}+\ldots+\frac{n(n-1)\ldots(n-r+1)}{r!}\cdot\frac{1}{n^r}+\ldots+\frac{n(n-1)\ldots21}{n!}\cdot\frac{1}{n^n}$$

$$=2+\frac{1}{2!}\left(1-\frac{1}{n}\right)+\frac{1}{3!}\left(1-\frac{1}{n}\right)\left(1-\frac{2}{n}\right)+\ldots$$
$$+\frac{1}{r!}\left(1-\frac{1}{n}\right)\left(1-\frac{2}{n}\right)\ldots\left(1-\frac{r-1}{n}\right)+\ldots$$
$$+\frac{1}{n!}\left(1-\frac{1}{n}\right)\left(1-\frac{2}{n}\right)\ldots\left(1-\frac{n-1}{n}\right)$$

$$T_n\leq 1+1+\frac{1}{2!}+\frac{1}{3!}+\ldots+\frac{1}{r!}+\ldots+\frac{1}{n!}$$

$$[\text{for } 2\leq r\leq n\Rightarrow r!\geq 2^{r-1}]$$

Then, $T_n\leq 1+1+\frac{1}{2}+\frac{1}{2^2}+\ldots+\frac{1}{2^{r-1}}+\ldots+\frac{1}{2^{n-1}}$

$$\Rightarrow \quad T_n\leq 1+\frac{1-\left(\frac{1}{2}\right)^n}{1-\frac{1}{2}}$$

$$\Rightarrow \quad T_n\leq 3-\frac{1}{2^{n-1}}$$

$$\Rightarrow \quad T_n\leq 3$$

$$\therefore \quad 2\leq T_n<3$$

70. *(b)* Given, $pqr=5$

$$A\,A^T=4I$$

and $$A=\begin{bmatrix}3p & q & r\\ q & 3r & p\\ r & p & 3q\end{bmatrix}=A^T$$

$$\therefore \quad A\,A^T=4I$$

$$\Rightarrow \quad A\cdot A=4I$$

$$\Rightarrow \quad A^2=4I$$

$$\Rightarrow \quad |A|^2=|4I|=4^3|I|$$

$$\Rightarrow \quad |A|=\pm 8$$

But $|A|$ is positive and real number.

$$\Rightarrow \quad |A|=8$$

Now, $$\begin{vmatrix}3p & q & r\\ q & 3r & p\\ r & p & 3q\end{vmatrix}=8$$

$$\Rightarrow \quad 29pqr-3(p^3+q^3+r^3)=8$$

$$\Rightarrow \quad 3(p^3+q^3+r^3)=29\times 5-8=137$$

71. *(9)* Consider the function

$$f(x)=\frac{x^2}{x^3+100} \qquad [x\in[1,\infty)]$$

$$\Rightarrow \quad f'(x)=\frac{2x(x^3+100)-x^2(3x^2)}{(x^3+100)^2}=\frac{x(200-x^3)}{(x^3+100)^2}$$

For maximum or minimum of $f(x)$,

$$f'(x)=0 \Rightarrow x=(200)^{1/3} \quad [\because x \neq 0]$$

$\Rightarrow \quad f'(x)>0$, if $1<x<(200)^{1/3}$

and $f'(x)<0$, if $x>(200)^{1/3}$

The function $f(x)$ increases in $(1,(200)^{1/3})$

and decreases in $((200)^{1/3},\infty)$ or $5<(200)^{1/3}<6$.

It follows the largest term in the given sequence can be either a_5 or a_6. Since, $a_6>a_5$. Hence, the largest term in the given sequence is

$$a_6=\frac{36}{316}=\lambda$$

So, $79\lambda=79\times\frac{36}{316}=9$

Study Tactics

Differentiate the function $f(x)$ for maximum or minimum value of $f(x)$. Put $f'(x)=0$, we get the value of x for which function is maximum, then observe the function in terms of decreasing and increasing and the largest term is nearer to the maximum value of the function.

72. *(60)* The word PATALIPUTRA has eleven letters in which 2P, 3A, 2T, 1L, 1U, 1R and 1I.

Vowels are AAIUA

Number of ways vowels can be arranged $=\frac{5!}{3!}=20$ ways

The consonants are PTLPTR these consonants can be arranged themselves in $\frac{6!}{2!\times 2!}=180$ ways

$\therefore$Required number of words $=20\times 180=3600$

$$=(60)^2=k^2$$

$\therefore\ k=60$

73. *(0)* Given, $|\mathbf{r}\times\mathbf{b}|=|\mathbf{r}||\mathbf{b}|$

$\Rightarrow$ Angle between $\mathbf{r}$ and $\mathbf{b}=\frac{\pi}{2}$

i.e. $\quad \mathbf{r}\perp\mathbf{b}$

$\mathbf{r}\cdot\mathbf{a}=0$

$\Rightarrow \quad \mathbf{r}\perp\mathbf{a}$

and $\quad |\mathbf{r}\times\mathbf{c}|=|\mathbf{r}||\mathbf{c}|$

$\Rightarrow$ Angle between $\mathbf{r}$ and $\mathbf{c}$ is $\frac{\pi}{2}$, i.e. $\mathbf{r}\perp\mathbf{c}$

Thus, $\mathbf{r}$ is perpendicular to $\mathbf{a}$, $\mathbf{b}$ and $\mathbf{c}$.

Hence, $\mathbf{a}\times\mathbf{b}=|\mathbf{a}||\mathbf{b}|\sin 90^\circ=|\mathbf{a}||\mathbf{b}|$

$\Rightarrow (\mathbf{a}\times\mathbf{b})\cdot\mathbf{c}=|\mathbf{a}||\mathbf{b}||\mathbf{c}|\cos 90^\circ=0$

74. *(24)* Given family of parabola is,

$y^2=8ax,\ (a\in R^+)$

and given line is $y=2x+3$

On solving above equations, we get

$$(2x+3)^2-8ax=0$$

$\Rightarrow 4x^2+9+12x-8ax=0$

$\Rightarrow 4x^2+x(12-8a)+9=0$

According to the question,

$D\leq 0$ [$\because$ the line does not intersect any member of given family of parabolas at two distinct points]

$\therefore \quad (12-8a)^2-4\times 4\times 9\leq 0$

$\Rightarrow 144+64a^2-192a-144\leq 0$

$\Rightarrow \quad 64a^2-192a\leq 0$

$\therefore \quad 64a\leq 192 \Rightarrow 8a\leq 24$

Hence, the maximum length of latus rectum is 24.

75. *(6)* Let $P(H)=P=\frac{\sqrt{A}+B}{C}$ (given)

$$P(T)=1-P$$

A wins, if A throws a tail before B tosses a head

$P(A)=P$ (T or HTT or $HTHTT$ or....)

Since, game is fair, so chance of winning both is equal

$\Rightarrow \quad \frac{1-P}{1-P(1-P)}=\frac{1}{2}$

$\Rightarrow \quad 1-P+P^2=2-2P$

$\Rightarrow \quad P^2+P-1=0$

$\Rightarrow \quad P=\frac{\sqrt{5}-1}{2}$

$$=\frac{\sqrt{A}+B}{C}$$

So, $A=5, B=-1$ and $C=2$

$\therefore\ A+B+C=6$

SECTION II

Mastery Builders

PRACTICE SET 7

Total Marks : 300	Total Time : 3 hrs	No. of Questions : 75

Instructions

1. There are **three** parts in the questions paper consisting of **Physics**, **Chemistry** and **Mathematics** having 25 questions in each part of equal weightage. Each part (subject) has two sections.
2. **Section A** This section contains 20 multiple choice questions which have only one correct answer. Each question carries **4 marks** for correct answer and **– 1 mark** for wrong answer.
3. **Section B** This section contains 5 questions. The answer to each of the questions is a numerical value. Each question carries **4 marks** for correct answer and **– 1 mark** for wrong answer. For Section B, the answer should be rounded off to the nearest integer.

PHYSICS

Section A : Objective Type Questions

1. In an electromagnetic wave, the equation of electric field vector is

$\mathbf{E} = 9\cos[2\pi \times 10^5 \times t - \pi \times 10^{-3} \times x]\hat{\mathbf{i}}$,

then the rms value of magnetic field is

(a) 2.12×10^{-8} T (b) 21.21×10^{-8} T

(c) 212.2×10^{-8} T (d) 0.2121×10^{-8} T

2. A capillary tube whose inside radius is 0.5 mm is dipped in water having surface tension 7.0×10^{-2} Nm^{-1}. To what height is the water raised above the normal water level? Angle of contact of water is 0°. (Take, density of water is 10^3 kg m^{-3} and $g = 9.8$ ms^{-2})

(a) 3.25 cm (b) 1.25 cm (c) 2.86 cm (d) 13.76 cm

3. Which of the following statements is correct ?

(a) By inserting a dielectric slab between the plates of an isolated capacitor, the force between the plates will increase.

(b) The capacitance of a capacitor does not depend on the shape of the plates.

(c) By inserting a dielectric slab between the plates of a capacitor, the energy density will increase.

(d) By inserting a dielectric slab between the plates of a capacitor, the capacitance will decrease.

4. In the experiment of photoelectric effect, consider the following statements.

I. Photocurrent linearly increases with frequency of incident light.

II. The minimum wavelength of incident radiation below which no photoelectric emission takes place is called threshold wavelength.

III. The minimum negative potential for which the photocurrent stops is called stopping potential.

Choose the correct statement(s) from the given options.

(a) Both II and III (b) Both I and III

(c) Only III (d) All of these

5. Assertion The correctness of an equation is verified by principle of homogeneity of dimensions.

Reason The relation among distance(s), speed (u) time (t), acceleration (a) is given by $s = ut + \frac{1}{2}at^2$.

Choose the correct option.

(a) Both Assertion and Reason are correct and Reason is the correct explanation of Assertion.

(b) Both Assertion and Reason are correct but Reason is not the correct explanation of Assertion.

(c) Assertion is correct but Reason is incorrect.

(d) Assertion is incorrect but Reason is correct.

6. Isothermal expansion of two different ideal gases are shown in same p-V diagram. If number of moles of gases are same and maintained at temperatures T_1 and T_2, then the correct relation between T_1 and T_2 is

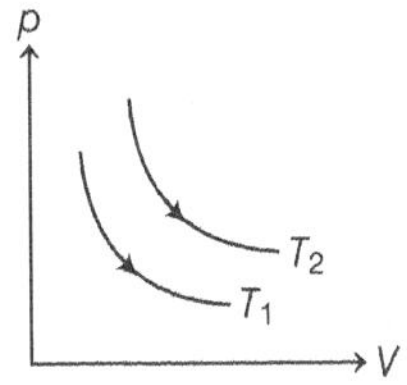

(a) $T_1 > T_2$ (b) $T_1 < T_2$
(c) $T_1 = T_2$ (d) Cannot be predicted

7. For the given case, choose the correct statement.

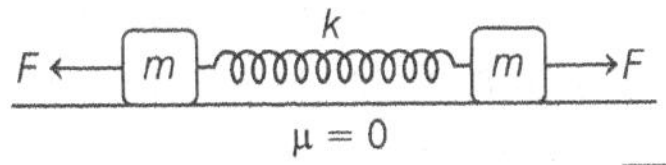

(a) The time period of SHM is $T = 2\pi\sqrt{\frac{m}{k}}$.
(b) The magnitude of force on both the blocks is not same.
(c) Both (a) and (b) are correct.
(d) Both (a) and (b) are incorrect.

8. A square of side 15 m in the plane of the paper is placed in a uniform electric field of $30\left(\frac{\text{V}}{\text{m}}\right)$ acting along the same plane at an angle of θ with horizontal side of the square as shown in figure. The electric flux linked to the surface in unit of V-m is

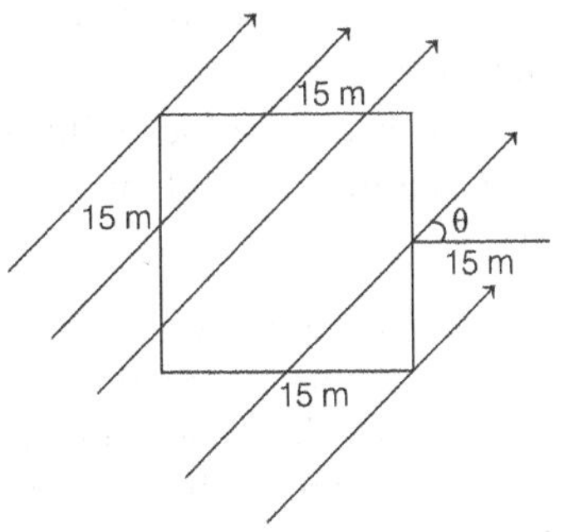

(a) zero (b) 225 V-m
(c) 220 V-m (d) 200 V-m

9. A star initially has 10^{40} deuterons. It produces energy *via* the processes ${}_1^2\text{H} + {}_1^2\text{H} \longrightarrow {}_1^3\text{H} + p$ and ${}_1^2\text{H} + {}_1^3\text{H} \longrightarrow {}_2^4\text{He} + n$, where the masses of the nuclei are $m({}_1^2\text{H}) = 2.014$ amu, $m(p) = 1.007$ amu, $m(n) = 1.008$ amu and $m({}_2^4\text{He}) = 4.001$ amu. If the average power radiated by the star is 10^{16} W, then deuteron supply of the star is exhausted in a time of the order of

(a) 10^{18} s (b) 10^{28} s (c) 10^{12} s (d) 10^{16} s

10. Two balloons are filled, one with pure helium gas and the other by air, respectively. If the pressure and temperature of these balloons are same, then the number of molecules per unit volume is

(a) more in the helium filled balloon
(b) same in both balloons
(c) more in air filled balloon
(d) None of the above

11. If n identical drops of mercury are combined to form a bigger drop, then the capacity of bigger drop, if the capacity of each smaller drop of mercury C, is

(a) $n^{1/3}C$ (b) $n^{2/3}C$
(c) $n^{1/4}C$ (d) nC

12. A particle is moving in a force field given by $\mathbf{F} = y^2\hat{\mathbf{i}} - x^2\hat{\mathbf{j}}$. Starting from A, the particle has to reach C either along ABC or ADC. Let, the work done along the two paths be W_1 and W_2 respectively, then, (W_1, W_2) are

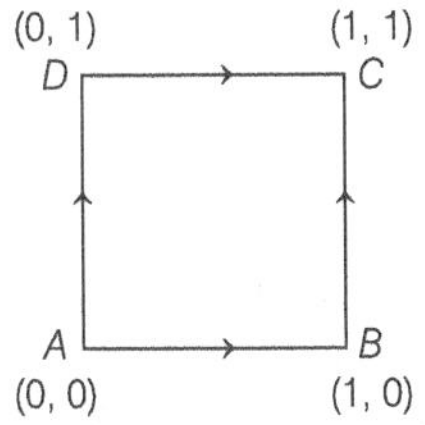

(a) $(-1, 1)$ (b) $(-1, 0)$
(c) $(1, 1)$ (d) $(-1, -1)$

13. Two long and parallel straight wires A and B carrying currents of 8.0 A and 5.0 A in the same direction are separated by a distance of 4.0 cm. Estimate the force on a 10 cm section of wire A.

(a) 1.5×10^{-5} N (b) 2×10^{-5} N
(c) 4×10^{-5} N (d) 3.2×10^{-5} N

14. The gravitational potential difference between the surface of a planet and a point 20 m above it is 14 Jkg^{-1}. The work done in moving a 2.0 kg mass by 8.0 m on a slope of 60° form the horizontal, is equal to

(a) 7 J (b) 9.6 J
(c) 16 J (d) 32 J

15. A plane wave of wavelength 6250 Å is incident normally on a slit of width 2×10^{-2} cm. The width of the central maxima on a screen distant 50 cm will be

(a) 312.5×10^{-3} cm (b) 312.5×10^{-4} cm
(c) 312 cm (d) 312.5×10^{-5} cm

16. Which of the following logic gate is represented by the combination of logic gates ?

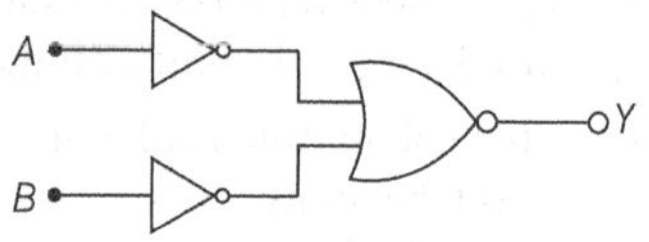

(a) NAND gate (b) NOR gate
(c) AND gate (d) OR gate

17. In the given arrangement, the resistance measured is given by $R_{\text{measured}} = \frac{V}{I}$, V = voltage reading of voltmeter, I = current reading of ammeter. The ammeter and the voltmeter used are not ideal, but having resistances R_A and R_V, respectively. For arrangement, the measured resistance is

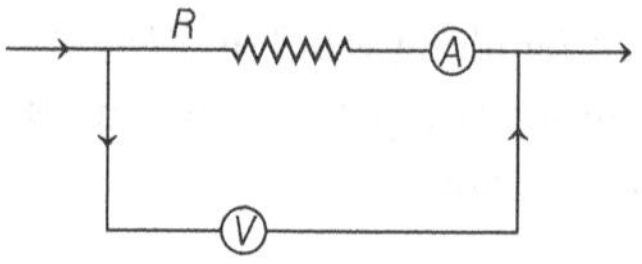

(a) $R + R_V$ (b) $R + R_A$
(c) $\frac{RR_V}{R + R_V}$ (d) $\frac{RR_V}{R + R_V} + R_A$

18. **Statement I** Alternating current cannot be measured by DC ammeter because alternating current changes direction continuously.

Statement II If wattless current flows in the AC circuit, then the circuit is *L-C-R* series circuit.

In the light of the above statements, choose the correct answer from the options given below

(a) Both Statement I and Statement II are true.
(b) Both Statement I and Statement II are false.
(c) Statement I is true but Statement II is false.
(d) Statement I is false but Statement II is true.

19. Match Column I with Column II and select the correct answer from the codes given below.

	Column I		Column II
A.	Maxwell's Ist equation	1.	$\oint \mathbf{B} \cdot d\mathbf{S} = 0$
B.	Maxwell's IInd equation	2.	$\oint \mathbf{E} \cdot d\mathbf{l} = \frac{-d\phi_B}{dt}$
C.	Maxwell's IIIrd equation	3.	$\oint \mathbf{E} \cdot d\mathbf{S} = q / \varepsilon_0$
D.	Maxwell's IVth equation	4.	$\oint \mathbf{B} \cdot d\mathbf{l} = \mu_0 (I_C + I_D)$

Codes

	A	B	C	D
(a)	3	1	2	4
(b)	2	1	3	4
(c)	4	3	2	1
(d)	1	2	3	4

20. A boy is going at a uniform speed of 4.5 km/h under a mango tree on which a monkey is sitting at a height of 29.4 m. At any particular instant, the monkey drops a mango. Boy will receive the mango whose distance from the tree at time of drop is (Take, $g = 9.8 \text{ m/s}^2$)

(a) 3 m (b) 4 m
(c) 10 m (d) 29.4 m

Section B : Numerical Value Type Questions

21. A massive platform of mass M is moving with speed $v = 6 \text{ ms}^{-1}$. At $t = 0$, a body of mass $m (m << M)$ is gently placed on the platform. If coefficient of friction between the body and platform is $\mu = 0.3$ and $g = 10 \text{ m/s}^2$, then the distance covered by body is ...... m.

22. An open pipe is suddenly closed at one end with the result that the frequency of third harmonic of the closed pipe is found to be higher by 100 Hz, than the fundamental frequency of the open pipe. The fundamental frequency of the open pipe is .... Hz.

23. A ball is moving towards a wall (which is not fixed and free to slide on the smooth floor). If after the collision, wall starts sliding with the speed of 2 m/s, then the recoil speed of the ball ($e = 1$) will be ...... m/s.

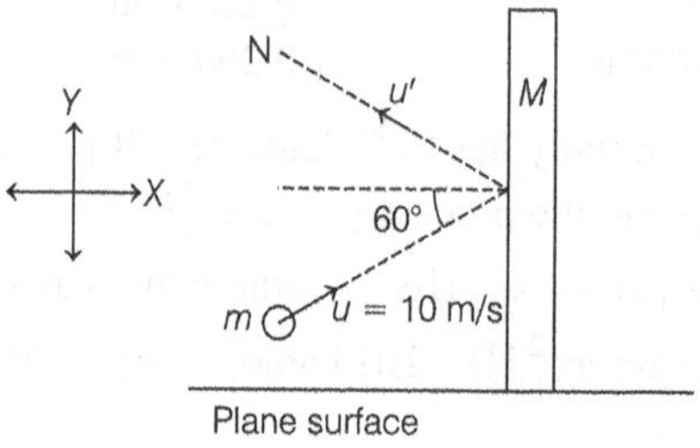

24. A bob of mass 2 kg is tied by a massless string whose other end is wound on a flywheel (disc) of radius 12 m and mass 2 kg. When the bob is released from the rest, it starts falling vertically. When it has covered a distance of 2.7 m, then the angular speed of the wheel is $\frac{n}{10}$ rad/s. What is the value of n ? (There is no slipping between string and flywheel).

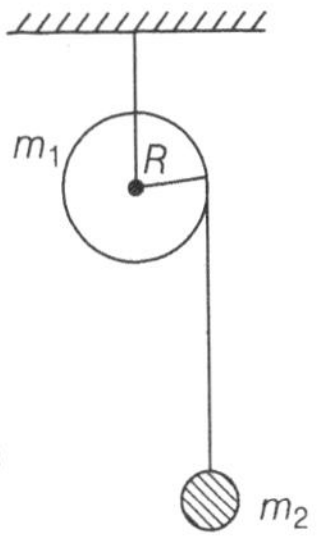

25. A wheel with 30 metallic spokes each 1 m long is rotated with a speed of 150 rpm in a plane perpendicular to a magnetic field of 0.5G. The induced emf between the axle and rim of the wheel will be .... $\times 10^{-4}$ V. $(1\,G = 10^{-4} T)$

CHEMISTRY

SectionA : Objective Type Questions

26. Consider the following reactions,

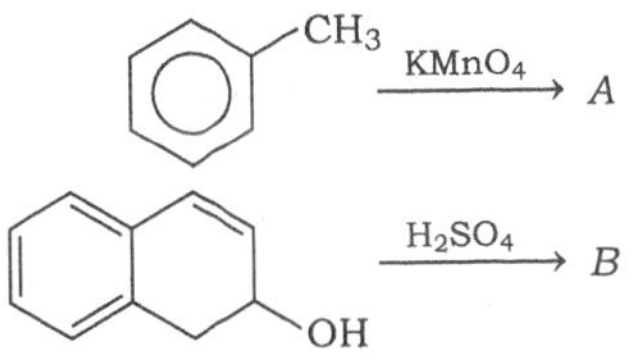

The product (A and B) obtained can be purified by which of the following method?

(a) Sublimation (b) Steam distillation
(c) Fractional distillation (d) Chromatography

27. Match List-I with List-II.

List-I (Name of oxo acid)	List-II (Oxidation state of 'P')
A. Hypophosphorous	I. +5
B. Orthophosphoric acid	II. +4
C. Hypophosphoric acid	III. +3
D. Orthophosphorous acid	IV. +1

Choose the correct answer from the options given below.

(a) A-II, B-I, C-III, D-IV (b) A-II, B-III, C-I, D-IV
(c) A-III, B-IV, C-I, D-II (d) A-IV, B-I, C-II, D-III

28. Select the correct statement among the following.

(a) Co (III) is stabilised in presence of weak field ligands, while Co (II) is stabilised in presence of strong field ligand.

(b) Four coordinated complexes of Pd (II) and Pt (II) are paramagnetic and square planar.

(c) $[Ni(CN)_4]^{2-}$ ion and $[Ni(CO)_4]$ are diamagnetic, tetrahedral and square planar respectively.

(d) Ni^{2+} ion does not form inner orbital octahedral complexes.

29. An amino acid contain a carboxyl group and amino group whose pK_a values are 3.30 and 9.0, respectively. The value of pH at this isoelectric point is ...... and it is a ...... amino acid. Choose the correct option that fill in the given blanks.

(a) 6.15, neutral (b) 2.34, acidic
(c) 9.60, basic (d) 6.97, basic

30. The incorrect statement is

(a) The gemstone, ruby has Cr^{3+} ions occupying the tetrahedral sites of beryl.

(b) The spin-only magnetic moment of $[Ni(NH_3)_4(H_2O)_2]^{2+}$ is 2.83 B.M.

(c) The colour of $[CoCl(NH_3)_5]^{2+}$ is violet as it absorbs the yellow light.

(d) The spin-only magnetic moments of $[Fe(H_2O)_6]^{2+}$ and $[Cr(H_2O)_6]^{2+}$ are nearly similar.

31. The major product C obtained in the following reaction scheme is

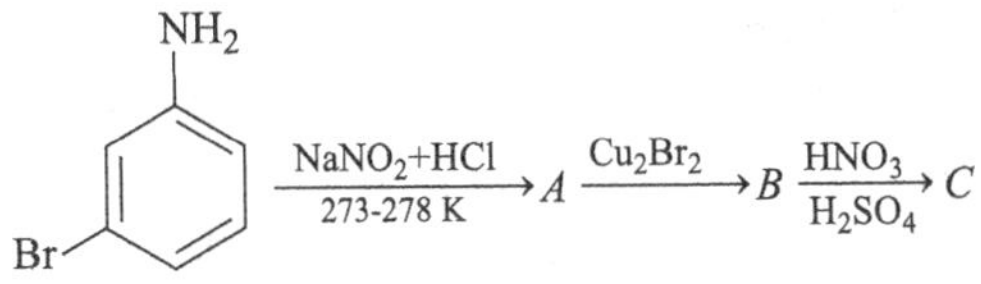

(a)
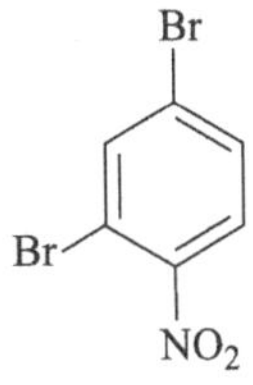

(b)
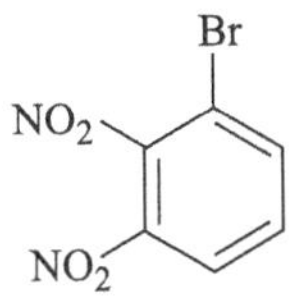

(c)
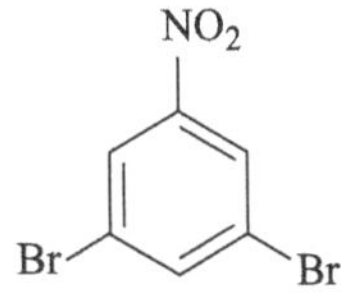

(d)
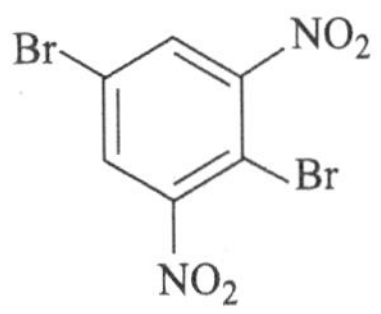

32. Consider the following reaction.

$$\text{Black coloured solution } (A) \xrightarrow[\Delta]{KNO_3 + KOH}$$

$$\text{Green colour solution } (B) \xrightarrow{CO_2} \underset{\text{Purple}}{(C)} + (A)$$

Pink compound (*C*) is decolourised by Fe^{2+}. The compounds *A*, *B* and *C* are

(a) MnO_2, K_2MnO_4, $KMnO_4$
(b) MnO_2, $KMnO_4$, K_2MnO_4
(c) $KMnO_4$, MnO_2, K_2MnO_4
(d) K_2MnO_4, MnO_4, $KMnO_4$

33. Which of the following reaction is elimination reaction?

(a) 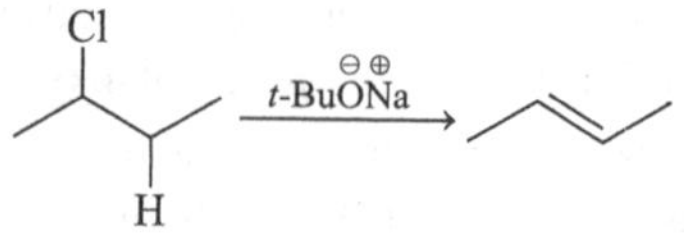

(b) 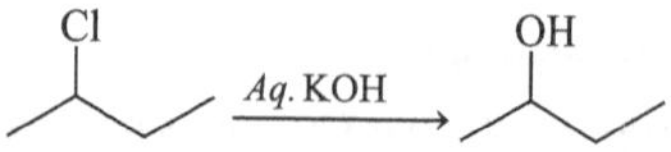

(c) 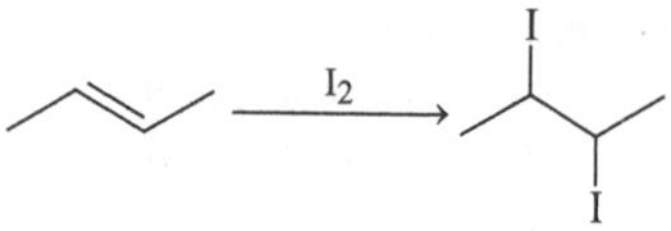

(d) 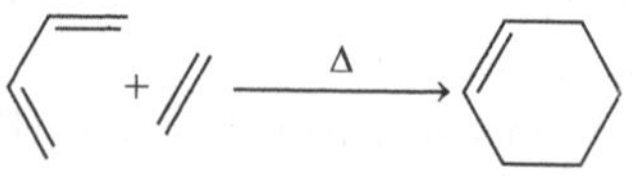

34. What will be the product when an optically active compound having IUPAC name pent-3-ene-2-ol is treated with Jones reagent?

(a) Pentanone
(b) Pent-3-ene-2-one
(c) Pentan-2-ol
(d) Propanoic acid

35. The magnetic moment of $K_3[Fe(CN)_6]$ is found to be 1.73 B.M. How many unpaired electron(s) is/are present per molecule?

(a) 1
(b) 2
(c) 3
(d) 4

36. The number of *meta*-directing functional group from the following is $—OCH_3$, $—NO_2$, —CN, $—CH_3$, $—NHCOCH_3$, —COR, $—SO_3H$

(a) 3
(b) 4
(c) 6
(d) 2

37. Consider the following statements.

A. Alcohols are weaker acids than water.

B. Acid strength of alcohols decreases in the following.

$$RCH_2OH > R_2CHOH > R_3COH$$

C. Carbon-oxygen bond length in methanol, CH_3OH is shorter than that of C—O bond length in phenol.

D. The bond angle C–O–H in methanol is 108.9°.

Choose the correct answer from the given options below.

(a) A, B and D
(b) A and B
(c) Only C
(d) B and D

38. Which of the following is least soluble in ethanol?

(a) HgF_2
(b) $HgCl_2$
(c) $HgBr_2$
(d) HgI_2

39. A solution of $Ni(NO_3)_2$ is electrolysed between platinum electrodes using 0.1 faraday electricity. How many moles of Ni will be deposited at cathode?

(a) 0.05
(b) 0.10
(c) 0.20
(d) 0.40

40. The given plots represents the variation of the concentration of a reactant *R* with time for two different reactions (i) and (ii). The respective orders of the reactions are

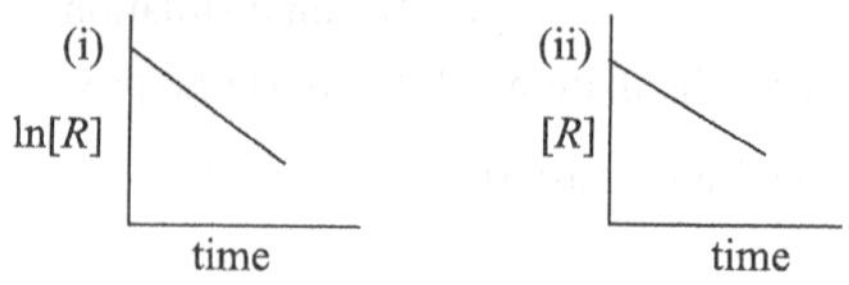

(a) 1, 0
(b) 2, 0
(c) 1, 2
(d) 2, 2

41. Given below are two statements.

Statement I Iron (III) catalyst, acidified $K_2Cr_2O_7$ and neutral $KMnO_4$ have the ability to oxidise I^- to I_2 independently.

Statement II Manganate ion is paramagnetic in nature and involves $p\pi$- $p\pi$ bonding.

In the light of the above statements, choose the most appropriates answer from the options given below.

(a) Statement I is false but Statement II is true.
(b) Both Statement I and Statement II are false.
(c) Both Statement I and Statement II are true.
(d) Statement I is true but StatementII is false.

42. Four samples of acids and bases are taken for an experiment.

A. 100 mL of 1 M NaOH and 100 mL of 1 M HCl.

B. 100 mL of 2 M KOH and 100 mL of 1 M H_2SO_4.

C. 100 mL of 1 M $CH_3COOH$and 100 mL of 1 M NaOH.

D. 100 mL of 0.5 M KOH and 100 mL of 0.5 M HNO_3.

Now, for each sample enthalpy of neutralisation is calculated. Which of the following represents the correct result?

(a) Enthalpy of neutralisation calculated in each case is found same.
(b) In cases (A) and (D), the value of enthalpy of neutralisation is same.
(c) In cases (A), (B) and (D) the value of enthalpy calculated is same.
(d) The value of enthalpy calculated is different for each sample.

43. Which of the following compounds will behave as a reducing sugar in an aqueous KOH solution?

(a)
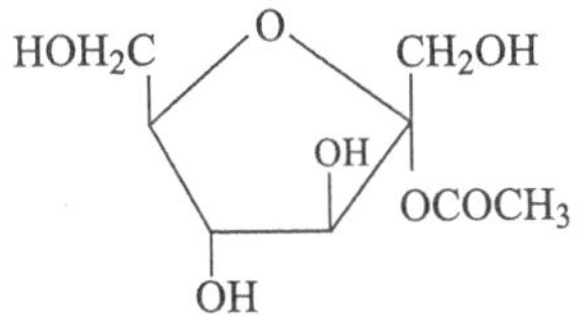

(b)
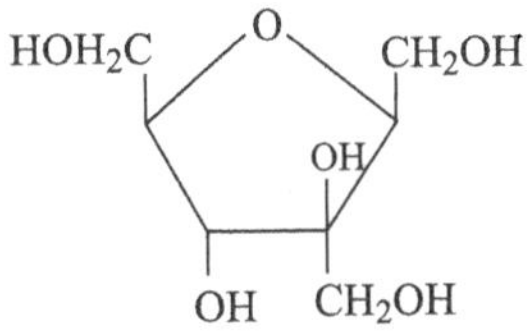

(c)
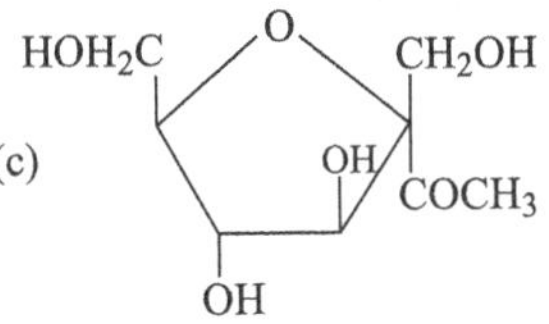

(d)
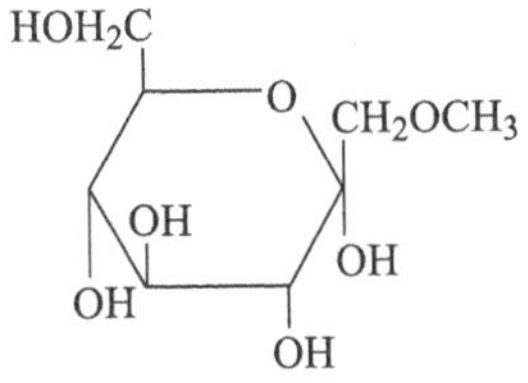

44. Match List-I with List-II.

	List-I		List-II
A.	XeO_3	I.	Square pyramidal
B.	$XeOF_4$	II.	Bent T-shaped
C.	XeF_2	III.	Linear
D.	ClF_3	IV.	Pyramidal

Choose the most appropriate answer from the codes given below:

(a) A-(I), B-(III), C-(II), D-(IV)
(b) A-(III), B-(II), C-(IV), D-(I)
(c) A-(IV), B-(I), C-(III), D-(II)
(d) A-(I), B-(IV), C-(II), D-(III)

45. The decreasing order of the reactivity of the following halides for the S_N1 reaction is

$CH_3(CH_2)_2Cl$ (I) $\quad$ $CH_3—CH_2—\underset{|}{C}H(Cl)—CH_3$ (II)

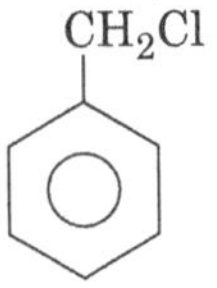

(III)

(a) I > II > III $\qquad$ (b) II > I > III
(c) III > II > I $\qquad$ (d) I > III > II

Section B : Numerical Value Type Questions

46. If the equilibrium constant for the reaction, $H^+(aq) + OH^-(aq) \rightleftharpoons H_2O(l)$ is 10^{12} at certain temperature, then the $E°$ for the reaction, is (−) ......$\times 10^{-2}$ V.

$2H_2O(l) + 2e^- \rightleftharpoons H_2(g) + 2\,OH^-(aq)$.

[Given, $\frac{2.303\,RT}{F} = 0.06$]

47. 1.0 g of monobasic acid HA in 100 g water lowers the freezing point by 0.145 K. If 0.45 g of same acid requires 27 mL of $\frac{N}{9}$ NaOH solution for complete neutralisation. Then % degree of dissociation of acid is ......% (K_f (H_2O) = 1.86 K kg mol^{-1}).

48. For a first order reaction, $A \longrightarrow$ products, the concentration of A changes from 0.1 M to 0.025 M in 40 min. The rate of reaction when the concentration of A is 0.01 M is ...... $\times 10^{-4}$ M min^{-1}.

[Nearest integer]

49. If dissociation constant of a weak acid is 1×10^{-4}, equilibrium constant of its reaction with strong base is .... $\times 10^{10}$.

50. In the cobalt-carbonyl complex: $[Co_2(CO)_8]$, number of Co—Co bonds is 'A' and terminal CO ligands is B. $A + B =$ ......

MATHEMATICS

Section A : Objective Type Questions

51. Let complex number z_0 and $\frac{1}{\bar{z}_0}$ lies on circles $(x-\alpha)^2+(y-\beta)^2=r^2$ and $(x-\alpha)^2+(y-\beta)^2=4r^2$, respectively. If $z_1=\alpha+i\beta$ satisfies the equation $3|z_1|^2=r^2+3$, then $|z_0|^2$ is equal to

(a) $\frac{1}{11}$ (b) $\frac{1}{2}$ (c) $\frac{2}{11}$ (d) $\frac{1}{6}$

52. The area of the region bounded by the curve $a^4y^2=(2a-x)x^5$ is to that of the circle, whose radius is a is given by the ratio $k:\lambda$, then the value of $k+\lambda$ is

(a) 13 (b) 14 (c) 15 (d) 16

53. If the equation $6x^2+2kxy+12y^2+22x+31y+20=0$ represents two straight lines, if the sum of values of k is V, then the value of $20V$ is equal to

(a) 341 (b) 171 (c) 1 (d) 17

54. Consider the function $f(x)=x^2+\int_0^1 (x+t)f(t)\,dt$, then $f'(1)$ is equal to

(a) 1 (b) 3 (c) -3 (d) -1

55. The eccentricity of the hyperbola whose latusrectum is $6\sqrt{2}$ and conjugate axis is equal to half the distance between the foci is

(a) 1 (b) $\frac{1}{\sqrt{3}}$
(c) $\frac{2}{\sqrt{3}}$ (d) None of these

56. Suppose a population A has 100 observations 51, 52, ... 150 and another population B has 100 population 151, 152, ... 250. If σ_A and σ_B represent the variances of the two populations, respectively, then $\sigma_A:\sigma_B$ is equal to

(a) 1 : 2 (b) 2 : 1
(c) 1 : 1 (d) None of these

57. If the variance of the first n natural number is 10 and the variance of first m even numbers is 16, then $m+n$ is equal to

(a) 16 (b) 20 (c) 18 (d) 22

58. If a, b and c are prime numbers and x, y are the positive integers such that LCM $(x, y)=a^2b^3c^4$, then the number of ordered pairs (x, y) is

(a) 315 (b) 325
(c) 300 (d) 310

59. An ellipse is drawn with the major axis and latus rectum of length 10 and $\frac{32}{5}$, respectively. Using one focus as centre, a circle is drawn that is touch to the ellipse, with no part of the circle being outside the ellipse, then the radius of circle is

(a) 2 (b) 4
(c) 6 (d) 12

60. **Statement I** Let g be a differentiable function satisfying $g(x+y)=g(x)+g(y)+2xy-1, \forall x, y\in R$ and $g'(0)=\alpha$, where $0<\alpha<1$, then $g(x)>0\ \forall x$.

Statement II $g(x)$ defined in Statement I is of the form $x^2+\alpha x+1$.

(a) Statement I is true and Statement II is false.
(b) Statement I is false and Statement II is true.
(c) Statement I and Statement II both are true.
(d) Statement I and Statement II both are false.

61. If $A=\{(x, y): x^2+y^2=16\}$ and $B=\{(x, y): 36x^2+9y^2=144\}$, then $A\cap B$ contains

(a) one point (b) three points
(c) two points (d) four points

62. If $I=\int \log(\sqrt{x-a}+\sqrt{x-b})dx$

$$=\frac{1}{2}[\alpha-\beta]\log(\sqrt{x-a}+\sqrt{x-b})-\frac{1}{2}\sqrt{(x-a)(x-b)}+C$$

Then, (α, β) is equal to

(a) $(x, a-b)$ (b) $(2x, a+b)$
(c) $(a-b, x)$ (d) $(a+b, 2x)$

63. If $\tan\theta+\tan\left(\theta+\frac{\pi}{3}\right)+\tan\left(\theta+\frac{2\pi}{3}\right)=3$, then

(a) $\tan 3\theta=1$ (b) $\tan 3\theta=0$
(c) $\tan\theta=1$ (d) $\tan 2\theta=1$

64. Let $f: R\to R$ be a function such that $f(x)=x^3+ax^2+5x+\sin 2x$. If $f(x)$ is invertible, then the range of a is

(a) $(-3, 3)$ (b) $[-3, 3)$
(c) $[-3, 3]$ (d) $(-2, 3)$

65. If the letters of the word 'REGULATIONS' are arranged at random, then the probability that there will be exactly four letters between R and E is

(a) $\frac{{}^9C_4\ 6!}{11!}$ (b) $\frac{{}^9C_5\ 6!2!}{12!}$

(c) $\frac{{}^9P_4\ 6!2!}{11!}$ (d) $\frac{{}^9C_4\ 6!2!}{11!}$

66. If $3|\mathbf{a}| = \sqrt{\lambda}$, where $\mathbf{a}$ is a vector such that $2\mathbf{a} + \mathbf{a} \times (\hat{\mathbf{i}} + 2\hat{\mathbf{j}}) = 2\hat{\mathbf{i}} + \hat{\mathbf{j}}$, then λ is equal to

(a) 3 (b) 9 (c) 12 (d) 24

67. Let P and Q be any symmetric and skew-symmetric matrices respectively of order 3. Then, which of the following is not true?

(a) $P^4 - Q^4$ is a symmetric matrix.
(b) $PQ - QP$ is a symmetric matrix.
(c) $Q^5 - P^5$ is a skew-symmetric matrix.
(d) $PQ + QP$ is a skew-symmetric matrix.

68. Let $y(x)$ be a continuous solution of the initial value problem $y' + 2y = f(x)$, $y(0) = 0$, where $f(x) = \begin{cases} 1, & 0 \le x \le 1 \\ 0, & x > 1 \end{cases}$. Then,

(a) $y\left(\frac{3}{2}\right) = \frac{\sin h(1)}{e^3}$ (b) $y\left(\frac{3}{2}\right) = \frac{\cos h(1)}{e^3}$

(c) $y\left(\frac{3}{2}\right) = \frac{\sin h(1)}{e^2}$ (d) $y\left(\frac{3}{2}\right) = \frac{\cos h(1)}{e^2}$

69. If a, b and c are three positive real numbers, then the minimum value of $\frac{a^2+1}{b+c} + \frac{b^2+1}{c+a} + \frac{c^2+1}{a+b}$ is

(a) 1 (b) 0
(c) 2 (d) 3

70. For $n \in R - \{0, 1\}$, if $f_0(n) = \frac{1}{1-n}$ and $f_{k+1}(n) = f_0[f_k(n)]$, where $k = 0, 1, 2, \ldots$, then the value of $f_{100}(3) + f_1\left(\frac{2}{3}\right) + f_2\left(\frac{3}{2}\right)$ is equal to

(a) $\frac{4}{3}$ (b) $\frac{1}{3}$ (c) $\frac{5}{3}$ (d) $\frac{8}{3}$

Section B : Numerical Value Type Questions

71. Suppose that $\lim_{x \to 0} \frac{F(x)}{x^3}$ exist and is equal to L, where

$$F(x) = \begin{vmatrix} a - \sin\frac{x}{2} & b\cos x & 0 \\ b\cos x & 0 & a - \sin\frac{x}{2} \\ 0 & a - \sin\frac{x}{2} & b\cos x \end{vmatrix}.$$ Then, $144L$ is equal to ....

72. $[(\sqrt{2}+1)^6]$ is equal to ......, where $[\cdot]$ represents greatest integer function.

73. Let $\mathbf{a} = 6\hat{\mathbf{i}} + 2\hat{\mathbf{j}} - 2\hat{\mathbf{k}}$ and $\mathbf{c} = 4\hat{\mathbf{i}} - 6\hat{\mathbf{j}} + 6\hat{\mathbf{k}}$. If $\mathbf{b}$ is a vector such that $\mathbf{a} = \mathbf{b} \times \mathbf{c}$ and $|\mathbf{b}|^2 = 50$, then $|138 - |\mathbf{b} + \mathbf{c}|^2|$ is equal to ....

74. Let 3 lines l_1, l_2 and l_3 forms a triangle in which, l_1 is passing through the origin with direction cosines $(1, 1, 1)$; $l_2 : x = \frac{y}{2} = \frac{z}{3}$ and the third line passing through $(1, 1, 1)$. The area of the triangle is $\sqrt{6}$. Then, the sum of square of coordinates of intersection point between third and second line is ...

75. The value of $\sum_{k=1}^{\infty} \sum_{n=1}^{\infty} \frac{k}{2^{n+k}}$ is ....

DETAILED SOLUTIONS

Physics

1. ***(a)*** The electric field vector for electromagnetic wave is

$$\mathbf{E} = 9\cos[2\pi \times 10^5 \times t - \pi \times 10^{-3} \times x]\hat{\mathbf{i}}$$

$\therefore \quad E_0 = 9$

$\therefore \quad B_0 = \frac{E_0}{c} = \frac{9}{3 \times 10^8} = 3 \times 10^{-8}\,\text{T}$

$\therefore$ rms value of magnetic field is

$$B_{\text{rms}} = \frac{B_0}{\sqrt{2}} = \frac{3 \times 10^{-8}}{\sqrt{2}} = 2.12 \times 10^{-8}\,\text{T}$$

2. ***(c)*** Height raised is given by

$$h = \frac{2T\cos\theta}{r\rho g}$$

Given, $T = 7.0 \times 10^{-2}\,\text{Nm}^{-1}$

$\theta = 0°,\ r = 0.5\,\text{mm} = 0.5 \times 10^{-3}\,\text{m}$

and $\rho = 10^3\,\text{kg m}^{-3}$

$\therefore \quad h = \frac{2 \times 7 \times 10^{-2} \times \cos 0°}{0.5 \times 10^{-3} \times 10^3 \times 9.8}$

$= 2.86 \times 10^{-2}\,\text{m} = 2.86\,\text{cm}$

3. ***(c)*** For an isolated capacitor (battery is disconnected)

Charge on capacitor, Q = constant

$\therefore$ Force between the plates, $F = \frac{Q^2}{2\varepsilon_0 A}$ = constant

$\therefore$ Force between plates will remain unchanged.

By inserting dielectric slab, the capacitance and energy density increases. As, $C = \frac{\varepsilon_0 A}{d} \Rightarrow$ Capacitance depends on size of the plates.

4. ***(c)*** Photocurrent linearly increases with intensity of incident light.

Threshold wavelength is the maximum wavelength above which no photoelectric emission takes place.

Stopping potential is minimum negative potential for which photocurrent stops.

5. ***(b)*** The correctness of equation, $s = ut + \frac{1}{2}at^2$ is verified by principle of homogeneity of dimensions.

$\therefore$ Both Assertion and Reason are individually correct, but the Reason does not explain Assertion.

6. ***(b)*** Because p-V curve at T_2 is above than at T_1 and $pV = nRT$, so for same number of moles $T_2 > T_1$.

7. ***(d)*** The net external force on the system (two blocks + spring),

$$F_{\text{ext}} = F - F = 0$$

$\therefore$ Centre of mass remains at rest.

$\therefore$ The spring will equally elongate by x on both sides of centre of mass

Let total extension of spring $= x_0$

$\therefore \quad x + x = x_0 \Rightarrow x_0 = 2x$

$\therefore$ Spring force, $F_S = kx_0 = 2kx$ ...(i)

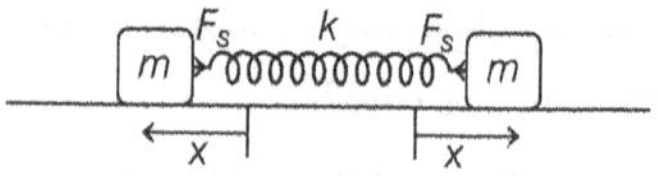

$F_{\text{SHM}} = m\omega^2 x$ ...(ii)

[as the displacement of the block is x]

From Eqs. (i) and (ii), we get

$$m\omega^2 = 2k$$

$\Rightarrow \quad \omega = \sqrt{\frac{2k}{m}} \Rightarrow T = 2\pi\sqrt{\frac{m}{2k}}$

The magnitude of force on two blocks is same and equal to tension in the spring $F_S = 2kx$.

Study Tactics

Students should not be confused between the force expression for SHM and the restoring force (spring force). Here, spring force is providing the force that is necessary for SHM.

8. ***(a)*** Flux of electric field E through any area A is defined as $\phi = E \cdot A\cos\theta$

or $\quad EA\cos 90° = 0 \quad [\because \theta = 90°]$

9. ***(c)*** Net process is

$$3\,{}_1\text{H}^2 \longrightarrow {}_2\text{He}^4 + p + n$$

Mass defect, $\Delta m = (3 \times 2.014) - 1.007 - 1.008 - 4.001$

$= 0.026\,\text{amu}$

Energy released by the fusion of three deuterons

$= \Delta m \times 931.48 = 0.026 \times 931.48\,\text{MeV}$

$= 0.026 \times 931.48 \times 1.6 \times 10^{-13}\,\text{J}$

$= 3.87 \times 10^{-12}\,\text{J}$

$\therefore$ Total energy released $= \frac{10^{40}}{3} \times 3.87 \times 10^{-12}\,\text{J}$

$= 1.29 \times 10^{28}\,\text{J}$

Total time $= \frac{1.29 \times 10^{28}}{10^{16}} = 1.29 \times 10^{12}\,\text{s} \quad \left(\because \text{power} = \frac{\text{energy}}{\text{time}}\right)$

10. *(b)* Ideal gas equation can be written as

$$pV = nRT \quad \ldots\text{(i)}$$

In this equation,

n = number of moles of the gas,
p = pressure of the gas,
V = volume of the gas,
R = universal gas constant
and T = temperature of the gas

From Eq. (i), we have

$$\frac{n}{V} = \frac{p}{RT} = \text{constant}$$

Hence, from above expression, if the pressure and temperature of ballons are same, then the number of molecules per unit volume is same in both the balloons.

11. *(a)* Let, radius of small drop $= r$

Volume of n drops $= n\frac{4}{3}\pi r^3$

Let, R be the radius of bigger drop

$$\Rightarrow \frac{4}{3}\pi R^3 = \frac{4}{3}\pi n r^3 \text{ or } R = n^{1/3} r$$

$$\therefore C = 4\pi\varepsilon_0 r \text{ and } C' = 4\pi\varepsilon_0 R$$

Hence, $C' = Cn^{1/3}$ $\quad (\because R = n^{1/3} r)$

Study Tactics

The volume of the bigger drop will be equal to the summation of the volume of all the smaller drops. This point will give the relation between the radii of bigger and smaller drops.

12. *(a)* $W_{ABC} = W_{AB} + W_{BC} = \int_{x=0}^{x=1} (y^2\hat{i} - x^2\hat{j})\cdot dx\,\hat{i} + \int_{y=0}^{y=1} (y^2\hat{i} - x^2\hat{j})\cdot dy\,\hat{j}$

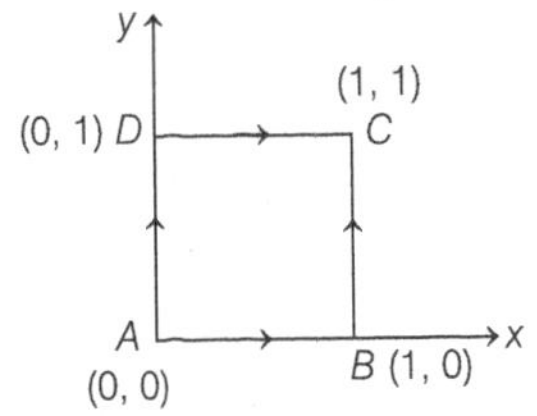

For AB, $y = 0$ and BC, $x = 1$

$$W_{ABC} = \int_{x=0}^{x=1} (0\hat{i} - x^2\hat{j})\cdot dx\,\hat{i} + \int_{y=0}^{y=1} (y^2\hat{i} - \hat{j})\cdot dy\,\hat{j}$$

$$= 0 + (-1)\int_0^1 dy = -1$$

Now, $W_{ADC} = W_{AD} + W_{DC}$

$$= \int_{y=0}^{y=1} (y^2\hat{i} - x^2\hat{j})\cdot dy\,\hat{j}\,\text{d} + \int_{x=0}^{x=1} (y^2\,\hat{i} - x^2\hat{j})\cdot dx\,\hat{i}$$

For AD, $x = 0$ and for DC, $y = 1$

$$\Rightarrow W_{ADC} = \int_0^1 (y^2\hat{i} - 0)\cdot dy\,\hat{j} + \int_0^1 (1\hat{i} - x^2\hat{j})\cdot dx\,\hat{i}$$

$$= 0 + \int_0^1 dx = 1$$

Hence, $(W_1, W_2) = (-1, 1)$

13. *(b)* Given, $i_1 = 8\,\text{A}, i_2 = 5\,\text{A}$ and $r = 4\text{ cm} = 0.04\text{ m}$

Force per unit length on two parallel wire carrying current, $F = \frac{\mu_0}{4\pi}\cdot\frac{2i_1 \cdot i_2}{r} = \frac{10^{-7}\times 2\times 8\times 5}{0.04} = 2\times 10^{-4}\,\text{N}$

The force on wire A of length 10 cm, $F' = F \times 0.1$ $\quad (\because 1\text{ m} = 100\text{ cm})$

$$\Rightarrow F' = 2\times 10^{-4}\times 0.1 = 2\times 10^{-5}\,\text{N}$$

14. *(b)* Gravitational field intensity,

$$I = \frac{dV}{dx} = \frac{14}{20} = 0.7\,\text{Nkg}^{-1}$$

Acceleration due to gravity, $g = I = 0.7\,\text{Nkg}^{-1}$

Work done under this field in displacing a body on a slope of 60° through a distance s.

$$= m(g\sin 60^\circ)s = 2\times(0.7\times\sqrt{3}/2)\times 8 = 9.6\,\text{J}$$

15. *(a)* Here, $\lambda = 6250\,\text{Å} = 6250\times 10^{-10}$ m

$$a = 2\times 10^{-2}\,\text{cm} = 2\times 10^{-4}\,\text{m}$$

$$D = 50\,\text{cm} = 0.5\,\text{m}$$

$\therefore$ Width of central maxima $= \frac{2\lambda D}{a}$

$$= \frac{2\times 6250\times 10^{-10}\times 0.5}{2\times 10^{-4}} = 312.5\times 10^{-3}\text{ cm}$$

16. *(c)* The truth table of given combination of logic gates is represented as

A	B	Y
0	0	0
0	1	0
1	0	0
1	1	1

This truth table is of AND gate. Therefore, correct option is (c).

Study Tactics

In this problem, we can use this expression to solve it.

$Y = \overline{\overline{A} + \overline{B}} = \overline{\overline{A}}\cdot\overline{\overline{B}} = AB$, which is AND gate.

17. *(b)*

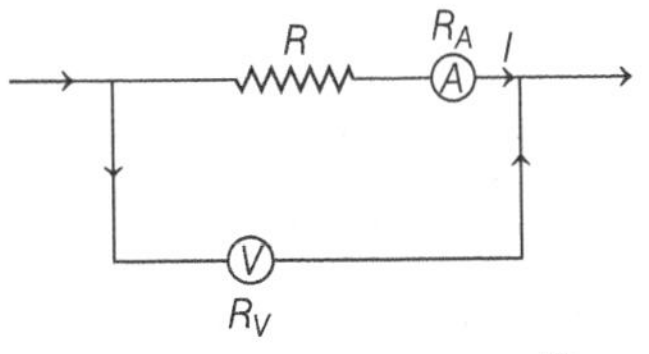

$$(R + R_A)I = V \Rightarrow \frac{V}{I} = R + R_A$$

Measured resistance, $R_{measured} = \frac{V}{I} = R + R_A$

18. *(c)* In AC current, average value of current in complete cycle is zero and DC ammeter measure only the average value of current. Therefore, statement I is correct. Wattless current flows in a circuit only when circuit is purely capacitive or inductive. Therefore, statement II is incorrect.

19. *(a)* Maxwell's Ist equation,

$$\oint \mathbf{E} \cdot d\mathbf{S} = q / \varepsilon_0$$

Maxwell's IInd equation,

$$\oint \mathbf{B} \cdot d\mathbf{S} = 0$$

Maxwell's IIIrd equation,

$$\oint \mathbf{E} \cdot d\mathbf{l} = -\frac{d}{dt}(\phi_B)$$

Maxwell's IVth equation,

$$\oint \mathbf{B} \cdot d\mathbf{l} = \mu_0 (I_C + I_D)$$

Therefore, correct option is (a).

20. *(a)* As we know that, $h = ut + \frac{1}{2}gt^2$

$$\Rightarrow \quad t = \sqrt{\frac{2h}{g}} = \sqrt{\frac{2 \times 29.4}{9.8}} = 2.45\,\text{s}$$

Distance $= \frac{4.5 \times 5}{18} \times 2.45 = 3.0\,\text{m}$

21. (6) Since, $M >> m$, the velocity of M remains unchanged after m is placed on to it. Acceleration of m is

$$a = \frac{\mu mg}{m} = \mu g$$

Initial relative acceleration

$$a_{mM} = a_{mg} - a_{Mg} = a - 0 = a$$

and initial relative velocity,

$$v_{mM} = v_{mg} - v_{Mg} = 0 - v = -v$$

Hence, $$s_{mM} = \frac{v^2}{2\mu g} = \frac{(6)^2}{2 \times 0.3 \times 10} = 6\,\text{m}$$

Study Tactics

When one body slides on the other body, then this type of problem is solved by calculating the relative terms (displacement, velocity, acceleration) between two sliding bodies.

22. (200) For fundamental mode in open pipe,

$$L = \frac{\lambda}{2} \Rightarrow \lambda = 2L \quad \text{and} \quad f_F = \frac{v}{\lambda} = \frac{v}{2L} \quad \ldots\text{(i)}$$

For third harmonic in closed pipe,

$$L = \frac{3\lambda}{4} \Rightarrow \lambda = \frac{4L}{3}$$

$$f_H = \frac{v}{\lambda} = \frac{v}{\frac{4}{3}L} = \frac{3v}{4L} \quad \ldots\text{(ii)}$$

From Eqs. (i) and (ii), we get

$$\frac{f_H}{f_F} = \frac{3}{2} \Rightarrow f_H = f_F \times \frac{3}{2}$$

$$f_H - f_F = 100 \Rightarrow \frac{3}{2}f_F - f_F = 100$$

$$f_F = 200\,\text{Hz}$$

23. (9) $e = \frac{|v_{wall} + (u_x')|}{|(u_x) - 0|}$

$$\Rightarrow \quad 1 = \frac{|2 + (u_x')|}{|u\cos 60° - 0|} \Rightarrow 1 = \frac{2 + u_x'}{5} \text{ (given, } u = 10\,\text{m/s)}$$

$\Rightarrow u_x' = 3\,\text{m/s}$

$\Rightarrow$ Recoil velocity, $u' = \sqrt{(u_x')^2 + (u_y')^2}$

$$= \sqrt{3^2 + \left(\frac{10\sqrt{3}}{2}\right)^2}$$

$$u' = 9.2\,\text{m/s} \approx 9\,\text{m/s}$$

24. (5) Given, $m_1 = m_2 = 2\,\text{kg}$, $R = 12\,\text{m}$, $h = 2.7\,\text{m}$

Applying conservation of mechanical energy, $E_i = E_f$

$\Rightarrow (\text{KE})_i + (\text{PE})_i = (\text{KE})_f + (\text{PE})_f$

$$\Rightarrow \quad 0 + 0 = -m_2gh + \frac{1}{2}m_2v^2 + \frac{1}{2}I\omega^2$$

$$\Rightarrow \quad m_2gh = \frac{1}{2}m_2v^2 + \frac{1}{2}\left(\frac{1}{2}m_1R^2\right)\cdot\left(\frac{v}{R}\right)^2$$

[$\because$ For no slipping, $v = \omega R$]

$$\Rightarrow \quad gh = \frac{1}{2}v^2 + \frac{1}{4}v^2 \quad [\because m_1 = m_2 = 2\,\text{kg}]$$

$$\therefore \quad v = \sqrt{\frac{4gh}{3}} = \sqrt{\frac{4 \times 10 \times 2.7}{3}} = 6\,\text{m/s}$$

$\therefore$ Angular speed of the flywheel, $\omega = \frac{v}{R} = \frac{6}{12} = 0.5\,\text{rad/s}$

According to question, $0.5 = \frac{n}{10}$

$$\therefore \quad n = 5$$

25. (4) Magnetic field, $B = 0.5\,\text{G} = 0.5 \times 10^{-4}\,\text{T}$

$$l = 1\,\text{m}$$

Frequency, $f = 150\,\text{rpm} = \frac{150}{60} = 2.5\,\text{Hz}$

Induced emf between the axle and rim of the wheel is given as

$$e = \frac{1}{2}B\omega l^2 = \frac{1}{2}B(2\pi f)l^2 = \pi Bfl^2$$

$$= 3.14 \times 0.5 \times 10^{-4} \times 2.5 \times (1)^2$$

$$= 3.14 \times 0.5 \times 10^{-4} \times 2.5$$

$$= 3.925 \times 10^{-4}\,\text{V} \approx 4 \times 10^{-4}\,\text{V}$$

Chemistry

26. (*a*) Chemical reaction given in the question produce the following product. It is as follows

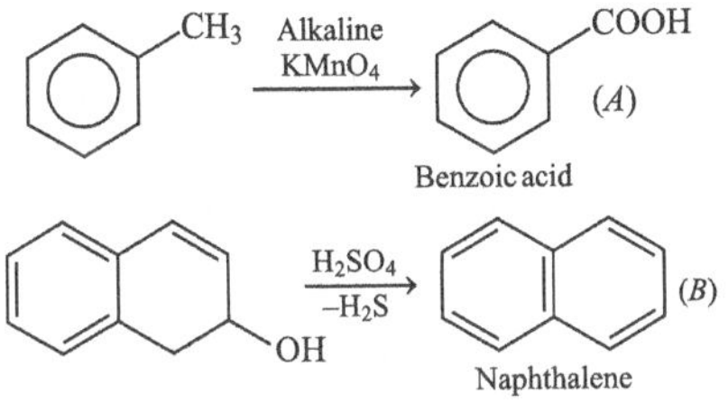

Both the product can be easily purified by sublimation because naphthalene sublimes easily.

27. (*d*) The correct match is A-IV, B-I, C-II, D-III

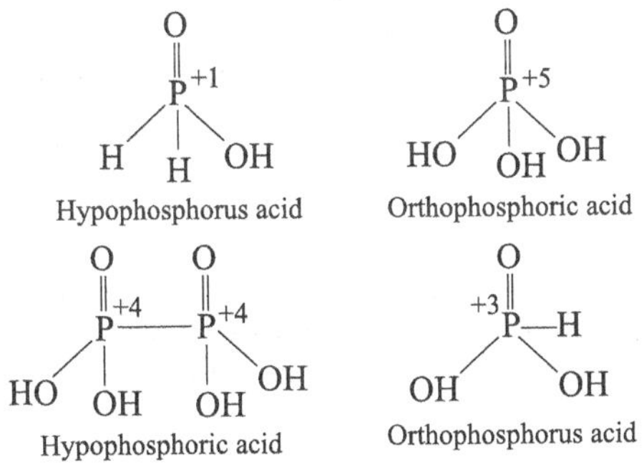

28. (*d*) (a) Both Co^{2+} and Co^{3+} are stabilised in presence of strong field ligands due to high CFSE values.

(b) Due to higher Z_{eff} values on the valence shell, Pd^{2+} and Pt^{2+} cations always form inner orbital complex but are diamagnetic in nature.

(c) $[Ni(CO)_4]$; sp^3, tetrahedral, $\mu = 0$ BM

$[Ni(CN)_4]^{2-}$; dsp^2, square planar, $\mu = 0$ BM

(d) Ni^{2+} ion does not form inner orbital octahedral complexes as their d_{z^2} and $d_{x^2-y^2}$ are occupied and can not participate in bonding which is required for octahedral.

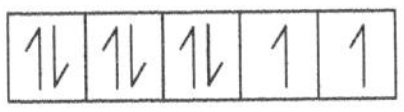

29. (*a*) Amino acid has two functional groups carboxylic (COOH) and amino (NH_2). It means that, it has an acidic and a basic group. The acidic group loses hydrogen and basic group gains H^+. The nature of pH at which there is no net charge on the amino acid is known as isoelectric point.

Isoelectric point may be calculated as

$$\text{Isoelectric point} = \frac{pK_{a_1} + pK_{a_2}}{2} = \frac{9.00 + 3.30}{2} = 6.15$$

∴ This is a neutral amino acid. The range of pH of neutral amino acid is 5.5 to 6.3.

30. (*a*) The gemstone ruby has Cr^{3+} ions occupying the octahedral sites of aluminium oxide (Al_2O_3) while beryl is beryllium aluminium cyclosilicate ($Be_3Al_2Si_6O_{18}$) which is a variety of emerald and aquamarine.

31. (*a*) The given reaction is represented as

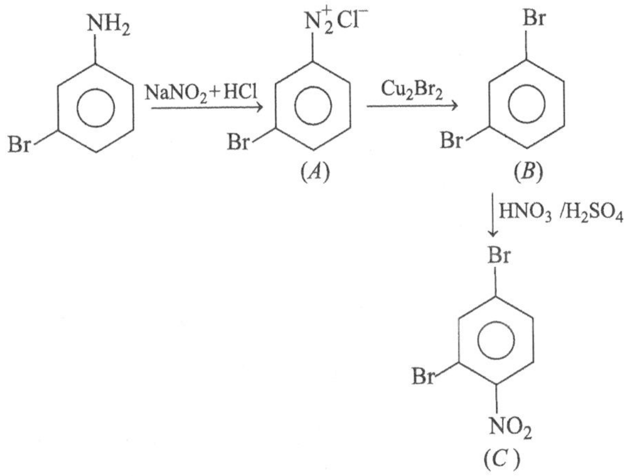

32. (*a*) Black colour compound is MnO_2 (*A*).

When it is fused with KOH and KNO_3, potassium manganate (K_2MnO_4) is formed. It is indicated by its green colour.

In acidic medium, K_2MnO_4 is unstable and disproportionates.

$$\underset{(A)}{MnO_2} + KOH + KNO_3 \xrightarrow{\Delta} \underset{\text{Green colour }(B)}{K_2MnO_4}$$

$$CO_2 + H_2O \rightleftharpoons H_2CO_3 \rightleftharpoons H^+ + HCO_3^-$$

$$3MnO_4^{2-} + 4H^+ \longrightarrow \underset{(A)}{MnO_2} + \underset{(C)\text{Purple}}{2MnO_4^-} + 2H_2O$$

33. (*a*) (i) **Elimination reaction** The reaction in which weak σ bond break and π-bond forms are known as elimination reaction.

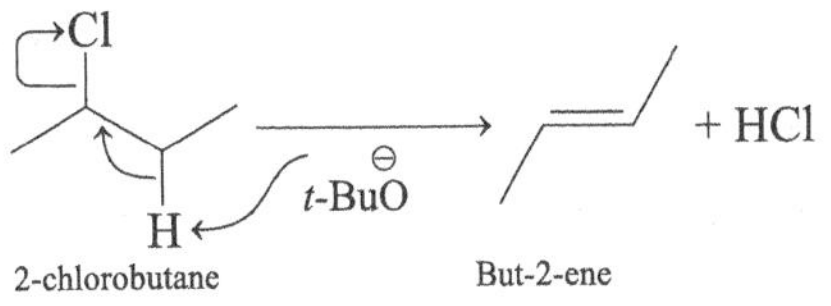

(ii) **Substitution reaction** The reaction in which one nucleophilic group is replaced by another nucleophile is known as nucleophilic substitution reaction.

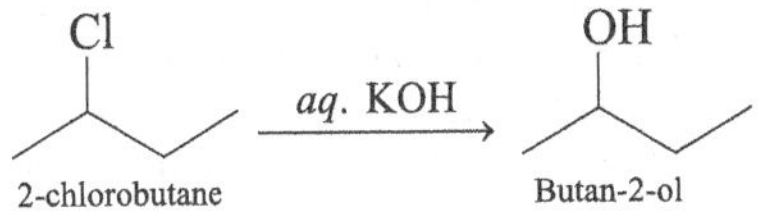

Here, OH^- substitutes Cl^-.

(iii) **Addition reaction** The reaction in which reactant undergoes addition with reagent to give a single product.

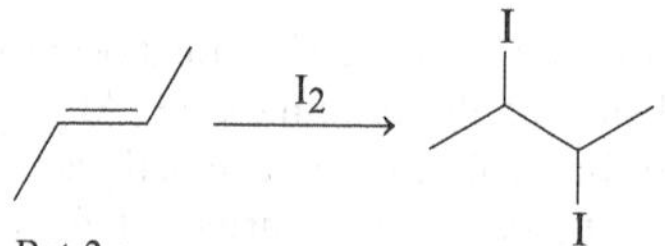

But-2-ene 2,3-di-iodobutane

(iv) **Addition reaction**

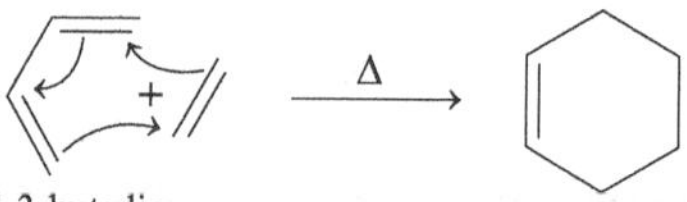

1,3-butadiene Cyclohexene

Hence, correct option is (a) in which elimination reaction takes place.

34. (*b*) Structure of pent-3-ene-2-ol is

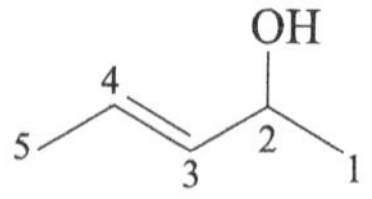

Jones reagent is CrO_3 dissolved in aqueous sulphuric acid.
On treatment with allylic alcohol it produces alkylic ketone. Jones reagent does not react with double bond, it only oxidises the alcoholic group.

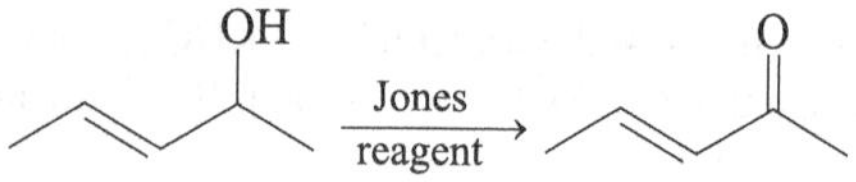

Study Tactics

This problem includes conceptual mixing of nomenclature' and 'oxidation by Jones reagent.
This problem can be solved by using following steps.
Write the structural formula of compound.
Now look at the effective site where the reagent will attack then complete the reaction.

35. (*a*) Magnetic moment of the given complex can be calculated by using the formula

$$\mu = \sqrt{n(n+2)}, \text{ where } n = \text{number of unpaired } e^-$$

$$1.73 = \sqrt{n(n+2)}$$

On squaring both sides,

$$(1.73)^2 = n(n+2)$$

$$n = 1$$

So, the total number of unpaired electron is 1.

36. (*b*) The group that directs an incoming group to the *meta* position of a benzene during an electrophilic reaction. Among the given species, only—NO_2, —CN, —SO_3H and —CO*R* are *meta* directing functional group.

37. (*a*) Among the given statements A, B and D are correct while C is incorrect.

The corrected form is given as :
Carbon-oxygen bond length in methanol CH_3OH is longer than that of C—O bond length in phenol because of the resonance occurs in phenol.

38. (*d*) HgF_2 is ionic in nature, hence dissolves in polar solvents like water and ethanol. HgI_2, is more covalent in nature hence dissolves in aprotic-polar solvents.

Hence, HgI_2 is least soluble in ethanol.

39. (*a*) $Ni^{2+} + 2e^- \longrightarrow Ni$

2×96500 C or 2F deposits 1 mole of Ni

$\therefore$ 0.1F will deposit $= \frac{1}{2} \times 0.1$ mole of N $= \frac{0.1}{2} = 0.05$ mole

40. (*a*) In graph (i), ln [*R*] *vs* time is linear with positive intercept and negative slope. Hence, it is I[st] order. In graph (ii), [Reactant] *vs* time is linear with positive intercept and negative slope. Hence, it is zero order.

41. (*b*) Both Statement I and Statement II are false.
Neutral $KMnO_4$ have the ability to oxidises I^- to IO_3^-.
Manganate ion has $d\pi$-$p\pi$ bonding.

42. (*c*) As in cases (A), (B) and (D), the acids and bases taken are strong and the enthalpy of neutralisation of all strong acids with strong bases is same, i.e. – 57.3 kJ.

43. (*a*)

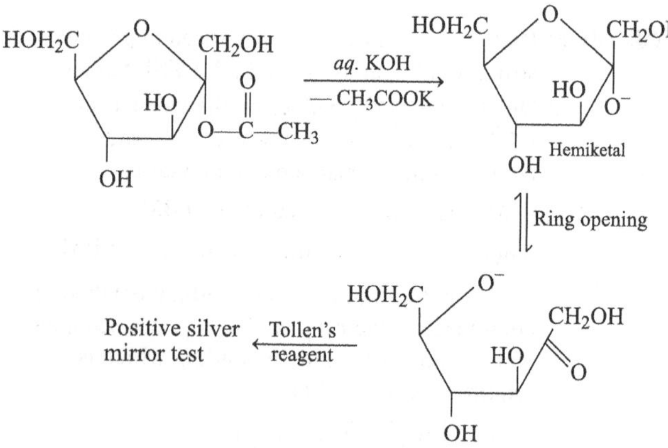

$\therefore$ This compound only behave as reducing sugar.

44. (*c*) **Shape of molecule and VSEPR theory** According to this theory, the exact shape of any molecule is determined on the basis of repulsion between lone pairs, and bond pairs. The shape having lesser value of energy is the acquired shape of compound/ion. The repulsion between lone pair and bond pair are as follows

$$bp - bp < bp - lp < lp - lp$$

where, bp = bond pair
lp = lone pair

According to this theory, exact shape of molecule can be determined from the hybridisation as follows,

$XeO_3 \longrightarrow \frac{8+0}{2} = 4, sp^3$; $XeOF_4 \longrightarrow \frac{8+4}{2} = 6, sp^3d^2$

$XeF_2 \longrightarrow \frac{8+2}{2} = 5, sp^3d$; $ClF_3 \longrightarrow \frac{7+3}{2} = 5, sp^3d$

Molecule		Shape
A. XeO_3	I.	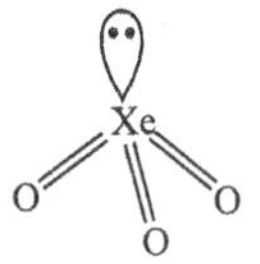 Pyramidal
B. $XeOF_4$	II.	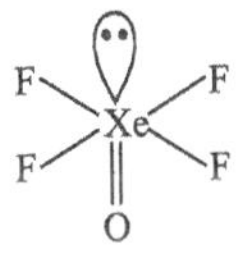 Square pyramidal
C. XeF_2	III.	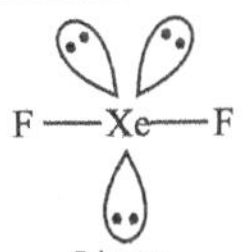 Linear
D. ClF_3	IV.	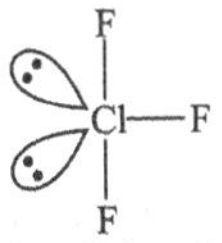 Bent T-shaped

Hence, the correct match is A-(IV), B-(I), C-(III), D-(II).

Study Tactics

Calculate value of $\left[H = \frac{V+M-C+A}{2}\right]$

where, V = valence electrons, M = Mono- valent atoms, $C = (+)$ charges, $A = (-)$ charges and then determine the state of hybridisation.

Now draw the structure using concept of bent rule. Then, interpret their shape.

45. *(c)* (I) is primary alkyl halide and (II) is secondary alkyl halide. Primary alkyl halide is less reactive towards S_N1 reaction than secondary alkyl halide.

However, in (III) compound, the carbocation generated is stabilised due to resonance. Hence, (III) is most reactive towards S_N1 reaction. So, the decreasing order of the reactivity of given halides will be III > II > I.

46. *(72)* $2H_2O\,(l) \longrightarrow 2H^+\,(aq) + 2OH^-\,(aq);\ \Delta G_1^\circ$

$2H^+\,(aq) + 2e^- \longrightarrow H_2\,(g);\ \Delta G_2^\circ$

$2H_2O(l) + 2e^- \rightleftharpoons H_2(g) + 2OH^-(aq);\ \Delta G_3^\circ$

$$\Delta G_3^\circ = \Delta G_1^\circ + \Delta G_2^\circ$$

$$-2 \times F \times E^\circ = -RT \ln (10^{-24}) - 2 \times F \times 0$$

$$E^\circ = \frac{0.06}{2} \log 10^{-24}$$

$$E^\circ = -0.03 \times 24 = -0.72\,V = -72 \times 10^{-2}\,V$$

47. *(17)* Depression in freezing point,

$$\Delta T_f = K_f \times \frac{W_{acid}}{M_{acid} \times W_{H_2O}} \times 1000$$

$$0.145 = 1.86 \times \frac{1.0}{M_{acid} \times 100} \times 1000$$

$$0.145 = 1.86 \times \frac{10}{M_{acid}}$$

$$M_{acid} = \frac{18.6}{0.145} = 128.28$$

M_{acid} (observed) = 128.28

m eq. of acid = m eq. of alkali

$$M_{acid}\ (\text{normal}) \begin{cases} \frac{0.45}{M_{Acid}} \times 1000 = 27 \times \frac{1}{9} \\ M_{Acid}\ (\text{normal}) = 150\,g\ mol^{-1} \end{cases}$$

$$i = \frac{150}{128.28} = 1.169 \approx 1.17$$

$$i = 1 + \alpha$$

$$1 + \alpha = 1.17$$

$$\alpha = 0.17,\ \%\,\alpha = 17$$

48. *(3)* By first order kinetics,

$$\text{rate constant, } k = \frac{2.303}{t} \log\left(\frac{a}{a-x}\right)$$

$$a = 0.1\ M$$

$$(a-x) = 0.025\ M,\ t = 40\ \text{min}$$

$$\therefore \quad k = \frac{2.303}{40} \log \frac{0.1}{0.025} = 0.0346\ \text{min}^{-1}$$

$$\text{Rate} = \left(\frac{dx}{dt}\right) = k[A]^1$$

$$= 0.0346 \times 0.01$$

$$= 3.46 \times 10^{-4}\ M\ min^{-1} \simeq 3 \times 10^{-4}\ M\ min^{-1}.$$

49. *(1)* $K_H = \frac{K_a}{K_w} = \frac{10^{-4}}{10^{-14}} = \frac{1}{10^{-10}} = 1 \times 10^{10}$

50. *(7)* Stucture of $Co_2(CO)_8$

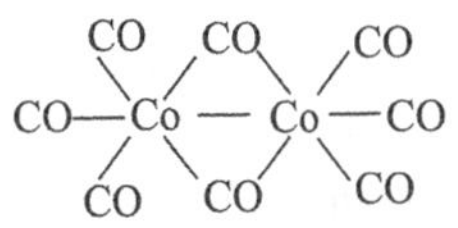

Number of Co—Co bonds = 1 = A

Number of terminal CO ligands = 6 = B

$\therefore A + B = 1 + 6 = 7$

Mathematics

51. *(c)* Given circle are

$$(x-\alpha)^2+(y-\beta)^2=r^2$$

$$\Rightarrow |z-z_1|^2=r^2$$

$$\Rightarrow (z-z_1)(\bar{z}-\bar{z}_1)=r^2$$

$$\Rightarrow z\bar{z}-z\bar{z}_1-z_1\bar{z}+z_1\bar{z}_1=r^2$$

$$\Rightarrow |z|^2+|z_1|^2-z\bar{z}_1-\bar{z}z_1=r^2 \quad ...(i)$$

and $$(x-\alpha)^2+(y-\beta)^2=4r^2$$

$$\Rightarrow |z-z_1|^2=4r^2$$

$$\Rightarrow |z|^2+|z_1|^2-z\bar{z}_1-\bar{z}z_1=4r^2 \quad ...(ii)$$

z_0 lies on circle (i),

$$\therefore |z_0|^2+|z_1|^2-z_0\bar{z}_1-\bar{z}_0z_1=r^2 \quad ...(iii)$$

and $\frac{1}{\bar{z}_0}$ lies on circle (ii),

$$\Rightarrow \left|\frac{1}{\bar{z}_0}\right|^2+|z_1|^2-\frac{1}{\bar{z}_0}\cdot\bar{z}_1-\overline{\left(\frac{1}{\bar{z}_0}\right)}z_1=4r^2$$

$$\Rightarrow \frac{1}{|z_0|^2}+|z_1|^2-\frac{z_0\bar{z}_1}{|z_0|^2}-\frac{\bar{z}_0z_1}{|z_0|^2}=4r^2$$

$$\Rightarrow 1+|z_0|^2|z_1|^2-z_0\bar{z}_1-\bar{z}_0z_1=4r^2|z_0|^2 \quad ...(iv)$$

On subtracting Eq. (iii) from Eq. (iv), we get

$$1+|z_0|^2|z_1|^2-|z_0|^2-|z_1|^2=r^2(4|z_0|^2-1)$$

$$\Rightarrow 1-|z_0|^2+|z_1|^2(|z_0|^2-1)=r^2(4|z_0|^2-1)$$

$$\Rightarrow (|z_0|^2-1)(|z_1|^2-1)=r^2(4|z_0|^2-1)$$

Using $3|z_1|^2=r^2+3$

$$\Rightarrow |z_1|^2=\frac{r^2+3}{3}, \text{ we get}$$

$$(|z_0|^2-1)\left(\frac{r^2}{3}\right)=r^2(4|z_0|^2-1)$$

$$\Rightarrow |z_0|^2-1=12|z_0|^2-3$$

$$\Rightarrow 11|z_0|^2=2 \Rightarrow |z_0|^2=\frac{2}{11}$$

52. *(a)* Given curve, $a^4y^2=(2a-x)x^5$

Cut off X-axis when

$$y=0 \Rightarrow 0=(2a-x)x^5$$

$$\therefore x=0, 2a$$

Hence, the area bounded by the curve

$$a^4y^2=(2a-x)x^5 \text{ is } A_1=\int_0^{2a}\frac{\sqrt{(2a-x)}\,x^{\frac{5}{2}}}{a^2}dx$$

On putting $x=2a\sin^2\theta$

$$\Rightarrow dx=4a\sin\theta\cos\theta\, d\theta$$

$$\therefore A_1=\int_0^{\frac{\pi}{2}}\frac{\sqrt{2a}\cos\theta\,(2a)^{\frac{5}{2}}\sin^5\theta\, 4a\sin\theta\cos\theta}{a^2}d\theta$$

$$=32a^2\int_0^{\frac{\pi}{2}}\sin^6\theta\cos^2\theta\, d\theta$$

$$=32a^2\cdot\frac{(5\cdot3\cdot1)(1)}{8\cdot6\cdot4\cdot2}\cdot\frac{\pi}{2} \quad \text{[by Walli's formula]}$$

$$=\frac{5\pi a^2}{8}$$

Area of circle,

$$A_2=\pi a^2$$

$$\therefore \frac{A_1}{A_2}=\frac{5}{8}=\frac{k}{\lambda}$$

$$\therefore k+\lambda=5+8=13$$

53. *(a)* Given equation is

$$6x^2+2kxy+12y^2+22x+31y+20=0 \quad ...(i)$$

On comparing Eq. (i) with equation

$ax^2+2hxy+by^2+2gx+2fy+c=0$, we have

$$a=6, h=k, b=12, g=11, f=\frac{31}{2} \text{ and } c=20$$

Eq. (i) represents two straight lines

$$abc+2fgh-af^2-bg^2-ch^2=0$$

or $$6\cdot12\cdot20+2\cdot\frac{31}{2}\cdot11\cdot k-6\cdot\frac{31^2}{4}-12\cdot11^2-20k^2=0$$

or $$1440+341k-\frac{2883}{2}-1452-20k^2=0$$

$$\Rightarrow 40k^2-682k+2907=0$$

$$\therefore k=\frac{682\pm\sqrt{465124-465120}}{80}=\frac{682\pm2}{80}=\frac{171}{20},\frac{17}{2}$$

$$\therefore \text{Sum of values of } k=\frac{171+170}{20}$$

$$\Rightarrow V=\frac{341}{20}$$

Hence, $20V=341$

Study Tactics

Compare the equation with the standard equation of pair of straight line and condition for the pair of straight line is

$$abc+2fgh-af^2-bg^2-ch^2=0$$

Then, use sum of roots of quadratic equation.

54. *(c)* Given, $f(x)=x^2+\int_0^1 (x+t)\,f(t)\,dt$

$$f(x)=x^2+x\int_0^1 f(t)\,dt+\int_0^1 tf(t)\,dt$$

Let $f(x)=x^2+ax+b$

where, $a=\int_0^1 f(t)\,dt$ and $b=\int_0^1 tf(t)\,dt$

$\Rightarrow \quad a=\int_0^1 (t^2+at+b)\,dt$ and $b=\int_0^1 t\,(t^2+at+b)\,dt$

$\Rightarrow \quad a=\frac{1}{3}+\frac{a}{2}+b$ and $b=\frac{1}{4}+\frac{a}{3}+\frac{b}{2}$

$\Rightarrow \quad \frac{a}{2}=b+\frac{1}{3}$ and $\frac{b}{2}=\frac{a}{3}+\frac{1}{4}$

On solving, we get $a=-5$

$$b=\frac{-17}{6}$$

$\therefore \quad f(x)=x^2-5x-\frac{17}{6}$

$\therefore f'(x)=2x-5$

$f'(1)=2-5=-3$

55. *(c)* We have,

Latus rectum of hyperbola, $\frac{2b^2}{a}=6\sqrt{2}$

and $\quad 2b=\frac{1}{2}(2ae)$

$\therefore \quad \frac{2}{a}\left(\frac{ae}{2}\right)^2=6\sqrt{2}$

$\Rightarrow \quad ae^2=12\sqrt{2} \quad \ldots\text{(i)}$

Also, $2\times\frac{b^2}{a}=6\sqrt{2}\Rightarrow b^2=3a\sqrt{2}$

$\therefore \quad 3a\sqrt{2}=a^2(e^2-1)$

$\Rightarrow \quad 3\sqrt{2}=ae^2-a \quad \ldots\text{(ii)}$

From Eqs. (i) and (ii), we get

$$3\sqrt{2}=12\sqrt{2}-\frac{12\sqrt{2}}{e^2}$$

$\Rightarrow \quad \frac{12\sqrt{2}}{e^2}=9\sqrt{2}$

$\Rightarrow \quad e^2=\frac{12\sqrt{2}}{9\sqrt{2}}$

$\Rightarrow \quad e^2=\frac{4}{3}\Rightarrow e=\frac{2}{\sqrt{3}}$

Hence, the eccentricity of hyperbola is $\frac{2}{\sqrt{3}}$.

56. *(c)* For population A,

$$\bar{x}=\frac{51+52+\ldots+150}{100}=\frac{100\,(51+150)}{2\times 100}$$

$$=\frac{201}{2}=100.5$$

Now,

$$\sigma_A=\frac{(51-100.5)^2+(52-100.5)^2+\ldots+(150-100.5)^2}{100}$$

$$=\frac{(49.5)^2+(48.5)^2+\ldots+(0.5)^2+(0.5)^2+\ldots+(49.5)^2}{100}$$

$$=\frac{2[(0.5)^2+(1.5)^2+(2.5)^2+\ldots+(49.5)^2]}{100}$$

For population B,

$$\bar{x}=\frac{151+152+\ldots+250}{100}=200.5$$

$$\sigma_B=\frac{(151-200.5)^2+(152-200.5)^2+\ldots+(250-200.5)^2}{100}$$

$$=\frac{(49.5)^2+(48.5)^2+\ldots+(0.5)^2+(0.5)^2+\ldots+(49.5)^2}{100}$$

$$=\frac{2[(0.5)^2+(1.5)^2+(2.5)^2+\ldots+(49.5)^2]}{100}$$

$\therefore \sigma_A:\sigma_B=1:1$

57. *(c)* Observations 1, 2, 3, ..., n

Variance of first n natural numbers $=\frac{n^2-1}{12}=10$

$\Rightarrow \quad n^2=121\Rightarrow n=11$

Observations $=2, 4, 6, .., 2m$

Variance of $2, 4, 6, ..., 2m=\text{Var}\,(2, 4, 6, \ldots 2m)$

$=2^2\ \text{var}\,(1, 2, 3, .., m)$

$=4\cdot\frac{m^2-1}{12}=16$

$\Rightarrow \quad m^2-1=48$

$\Rightarrow \quad m=7$

$\therefore \quad m+n=11+7=18$

58. *(a)* Since a, b and c are prime numbers.

Selection of x and y are as

To select *a*

x	y	Number of ways
a^0	a^2	1
a^1	a^2	1
a^2	a^0, a^1, a^2	3

Total number of ways = 5

To select b

x	y	Number of ways
b^0	b^3	1
b^1	b^3	1
b^2	b^3	1
b^3	b^0, b^1, b^2, b^3	4

Total number of ways = 7

To select c

x	y	Number of ways
c^0	c^4	1
c^1	c^4	1
c^2	c^4	1
c^3	c^4	1
c^4	c^0, c^1, c^2, c^3, c^4	5

Total number of ways = 9

$\therefore$ The number of ordered pair $(x, y) = 5 \times 7 \times 9 = 315$

59. *(a)* Given, length of major axis = 10

$\Rightarrow \quad 2a = 10$

$\Rightarrow \quad a = 5$

and length of latus rectum $= \dfrac{2b^2}{a} = \dfrac{32}{5}$

$\Rightarrow \quad 2b^2 = 32 \quad [\because a = 5]$

$\Rightarrow \quad b = 4$

Therefore, the equation of ellipse is

$$\frac{x^2}{25} + \frac{y^2}{16} = 1 \quad \ldots(i)$$

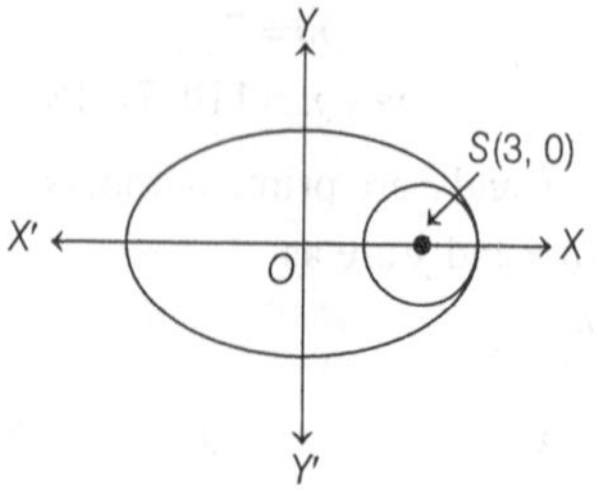

$\therefore \quad e^2 = 1 - \dfrac{16}{25} \Rightarrow e = \dfrac{3}{5}$

Then, foci are $(\pm ae, 0) \equiv (\pm 3, 0)$

Now, the circle having centre (3, 0) is

$$(x-3)^2 + y^2 = r^2 \quad \ldots(ii)$$

On eliminating y^2 from Eqs. (i) and (ii), we get

$$\frac{x^2}{25} + \frac{r^2 - (x-3)^2}{16} = 1$$

$\Rightarrow 16x^2 + 25r^2 - 25(x^2 - 6x + 9) = 400$

$\Rightarrow \quad -9x^2 + 150x + 25r^2 - 625 = 0$

Since, the circle touch the ellipse, the above equation has equal roots.

Hence, $D = 0$

i.e. $22500 - 4(-9)(25r^2 - 625) = 0 \Rightarrow r = 0$,

which is not possible.

Then, the circle will touch the ellipse at the end of the major axis (5, 0).

Hence, the radius is 2.

60. *(c)* $\because g(x+y) = g(x) + g(y) + 2xy - 1, \forall x, y \in R$

Let $y = 0$

Then, $g(x) = g(x) + g(0) - 1 \Rightarrow g(0) = 1$

Now, $g(x)$ is differentiable function such that $g'(0) = \alpha$

Then, $g'(x) = \lim\limits_{h \to 0} \dfrac{g(x+h) - g(x)}{h}$

$= \lim\limits_{h \to 0} \dfrac{g(x) + g(h) + 2hx - 1 - g(x)}{h}$

$= \lim\limits_{h \to 0} \left[\dfrac{2hx}{h} + \dfrac{g(h) - 1}{h}\right]$

$= 2x + \lim\limits_{h \to 0} \dfrac{g(h) - g(0)}{h - 0}$

$\Rightarrow \quad g'(x) = 2x + g'(0)$

$\Rightarrow \quad g'(x) = 2x + \alpha$

On integrating both sides, we get

$$g(x) = x^2 + \alpha x + b$$

$\therefore \quad g(0) = 1 \Rightarrow b = 1$

$\therefore \quad g(x) = x^2 + \alpha x + 1$

So, Statement II is true.

Now, $g(x) = x^2 + \alpha x + 1 = \left(x + \dfrac{\alpha}{2}\right)^2 + \left(1 - \dfrac{\alpha^2}{4}\right)$

$\because 0 < \alpha < 1 \Rightarrow g(x) > 0, \forall x$

Statement I is true.

61. *(c)* Given, $x^2 + y^2 = 16 \quad \ldots(i)$

and $\quad 36x^2 + 9y^2 = 144 \quad \ldots(ii)$

From Eq. (i), $y^2 = 16 - x^2$ on putting in Eq. (ii), we get

$$36x^2 + 9(16 - x^2) = 144$$

$\Rightarrow \quad 36x^2 + 144 - 9x^2 = 144$

$\Rightarrow \quad 27x^2 = 0$

$\Rightarrow \quad x = 0$

On putting $x=0$ in Eq. (ii), we get

$$36\times 0+9y^2=144$$
$$y^2=16$$
$$y=\pm 4$$

Hence, $A\cap B=\{(0,4),(0,-4)\}=$ Set of two points.

62. *(b)* $I=\int 1\cdot\log(\sqrt{x-a}+\sqrt{x-b})dx$

$$=x\log(\sqrt{x-a}+\sqrt{x-b})-\int\frac{x}{2}\left(\frac{1}{\sqrt{x-a}+\sqrt{x-b}}\right)\left(\frac{1}{\sqrt{x-a}}+\frac{1}{\sqrt{x-b}}\right)dx$$

$$=x\log(\sqrt{x-a}+\sqrt{x-b})-\frac{1}{2}I_1$$

where, $I_1=\int\frac{x}{\sqrt{(x-a)(x-b)}}dx$

Let $u=x-\frac{a+b}{2}$ and $v=\frac{a-b}{2}$

Then, $I_1=\int\frac{u+\frac{a+b}{2}}{\sqrt{(u-v)(u+v)}}du$

$$=\int\frac{u+\frac{a+b}{2}}{\sqrt{u^2-v^2}}du$$

$$=\int\frac{u}{\sqrt{u^2-v^2}}du+\frac{a+b}{2}\int\frac{du}{\sqrt{u^2-v^2}}$$

$$=\sqrt{u^2-v^2}+\frac{a+b}{2}\log|u+\sqrt{u^2-v^2}|$$

Thus, $I=x\log(\sqrt{x-a}+\sqrt{x-b})-\frac{1}{2}\sqrt{(x-a)(x-b)}$

$$-\frac{1}{4}(a+b)\log\left|x-\frac{a+b}{2}+\sqrt{(x-a)(x-b)}\right|+C_1$$

$$=x\log(\sqrt{x-a}+\sqrt{x-b})-\frac{1}{2}\sqrt{(x-a)(x-b)}-\frac{1}{4}(a+b)\log(\sqrt{x-a}+\sqrt{x-b})^2+C$$

$$=\frac{1}{2}[2x-(a+b)]\log(\sqrt{x-a}+\sqrt{x-b})-\frac{1}{2}\sqrt{(x-a)(x+b)}+C$$

On comparing, we get

$$\alpha=2x$$

and $\beta=a+b$

63. *(a)* The given equation is

$$\tan\theta+\tan\left(\theta+\frac{\pi}{3}\right)+\tan\left(\theta+\frac{2\pi}{3}\right)=3$$

$$\Rightarrow\quad \tan\theta+\frac{\tan\theta+\tan\frac{\pi}{3}}{1-\tan\theta\cdot\tan\frac{\pi}{3}}+\frac{\tan\theta+\tan\frac{2\pi}{3}}{1-\tan\theta\cdot\tan\frac{2\pi}{3}}=3$$

$$\Rightarrow\quad \tan\theta+\frac{\tan\theta+\sqrt{3}}{1-\sqrt{3}\tan\theta}+\frac{\tan\theta-\sqrt{3}}{1+\sqrt{3}\tan\theta}=3$$

$$\Rightarrow\quad \tan\theta+\frac{(\tan\theta+\sqrt{3})(1+\sqrt{3}\tan\theta)+(\tan\theta-\sqrt{3})(1-\sqrt{3}\tan\theta)}{1-3\tan^2\theta}=3$$

$$\Rightarrow\quad \tan\theta+\frac{8\tan\theta}{1-3\tan^2\theta}=3$$

$$\Rightarrow\quad \frac{\tan\theta-3\tan^3\theta+8\tan\theta}{1-3\tan^2\theta}=3$$

$$\Rightarrow\quad \frac{3(3\tan\theta-\tan^3\theta)}{1-3\tan^2\theta}=3$$

$$\Rightarrow\quad \tan 3\theta=1$$

64. *(a)* We have, $f(x)=x^3+ax^2+5x+\sin 2x$

$\therefore\quad f'(x)=3x^2+2ax+5+2\cos 2x$

Given $f(x)$ is invertible, then

$$f'(x)>0\ \forall\ x \text{ or } f'(x)<0\ \forall\ x.$$

$\Rightarrow 3x^2+2ax+5+2\cos 2x>0\ \forall\ x$

or $3x^2+2ax+5+2\cos 2x<0\ \forall\ x$

$\Rightarrow\quad 3x^2+2ax+5-2>0\ \forall\ x$ [$\because$ min $(\cos 2x)=-1$]

or $\quad 3x^2+2ax+5+2<0\ \forall\ x$ [$\because$ max $(\cos 2x)=1$]

$\Rightarrow\quad 3x^2+2ax+3>0\ \forall\ x$

or $\quad 3x^2+2ax+7<0\ \forall\ x$

$$4a^2-4\cdot 3\cdot 3<0 \text{ or } \left(x+\frac{a}{3}\right)^2-\frac{a^2}{9}+\frac{7}{3}<0$$

$$a^2<9 \text{ or } a^2>21$$
$$a^2<9$$

$\Rightarrow\quad a\in(-3,3)$

$\therefore\quad a\in(-3,3)$

65. *(c)* Total number of ways of arranging 11 letters $=11!$ ...(i)

Selection of 4 letters to be placed between R and E from the remaining $9={}^9C_4$.

These 4 letters can be permuted in 4! ways.

Now, permutation of

$$\underbrace{(R\_\_\_\_E)}_{4\text{ Letters}}\underbrace{\_\_\_\_\_}_{5\text{ Letters}} = 6!$$

R and E can be interchanged in 2! ways.

$\therefore$ Number of favourable ways $= {}^9C_4\ 6!2!4!$

Hence, required probability $= \dfrac{{}^9C_4\ 6!4!2!}{11!} = \dfrac{{}^9P_4\ 6!2!}{11!}$

66. *(b)* $\because \quad 2\mathbf{a} + \mathbf{a} \times (\hat{\mathbf{i}} + 2\hat{\mathbf{j}}) = 2\hat{\mathbf{i}} + \hat{\mathbf{j}}$ ...(i)

$\Rightarrow \quad 2\mathbf{a} \cdot (\hat{\mathbf{i}} + 2\hat{\mathbf{j}}) + [\mathbf{a} \times (\hat{\mathbf{i}} + 2\hat{\mathbf{j}})] \cdot (\hat{\mathbf{i}} + 2\hat{\mathbf{j}})$

$= (2\hat{\mathbf{i}} + \hat{\mathbf{j}}) \cdot (\hat{\mathbf{i}} + 2\hat{\mathbf{j}})$

$\Rightarrow \quad 2\mathbf{a} \cdot (\hat{\mathbf{i}} + 2\hat{\mathbf{j}}) + 0 = 4$

$\Rightarrow \quad \mathbf{a} \cdot (\hat{\mathbf{i}} + 2\hat{\mathbf{j}}) = 2$

$\Rightarrow \quad \left|\mathbf{a} \cdot (\hat{\mathbf{i}} + 2\hat{\mathbf{j}})\right|^2 = 4$

$\Rightarrow |\mathbf{a}|^2 \left|\hat{\mathbf{i}} + 2\hat{\mathbf{j}}\right|^2 \cos^2\theta = 4$

$\Rightarrow \quad |\mathbf{a}|^2\, 5(1 - \sin^2\theta) = 4$

$\Rightarrow \quad |\mathbf{a}|^2\, 5\sin^2\theta = 5|\mathbf{a}|^2 - 4$ ...(ii)

Again, from Eq. (i),

$$\left|2\mathbf{a} + \mathbf{a} \times (\hat{\mathbf{i}} + 2\hat{\mathbf{j}})\right|^2 = \left|2\hat{\mathbf{i}} + \hat{\mathbf{j}}\right|^2$$

$\Rightarrow \quad 4|\mathbf{a}|^2 + \left|\mathbf{a} \times (\hat{\mathbf{i}} + 2\hat{\mathbf{j}})\right|^2 = 5$

$\Rightarrow 4|\mathbf{a}|^2 + |\mathbf{a}|^2 \left|\hat{\mathbf{i}} + 2\hat{\mathbf{j}}\right|^2 \sin^2\theta = 5$

$\Rightarrow \quad 4|\mathbf{a}|^2 + |\mathbf{a}|^2\, 5\sin^2\theta = 5$

$\Rightarrow \quad |\mathbf{a}|^2\, 5\sin^2\theta = 5 - 4|\mathbf{a}|^2$ ...(iii)

From Eqs. (ii) and (iii), we get

$$5|\mathbf{a}|^2 - 4 = 5 - 4|\mathbf{a}|^2$$

$\Rightarrow \quad 9|\mathbf{a}|^2 = 9$

$\Rightarrow \quad |\mathbf{a}|^2 = 1$

$\Rightarrow \quad 3|\mathbf{a}| = 1$

$\Rightarrow \quad 3|\mathbf{a}| = 3 = \sqrt{\lambda}$

$\Rightarrow \quad \sqrt{\lambda} = 3$

$\Rightarrow \quad \lambda = 9$

67. *(c)* $\because P$ is symmetric matrix $\Rightarrow P^T = P$

and Q is skew-symmetric matrix $\Rightarrow Q^T = -Q$

(a) Let $A = P^4 - Q^4$

$\therefore A^T = (P^4 - Q^4)^T$

$= (P^4)^T - (Q^4)^T$

$= (P^T)^4 - (Q^T)^4$

$= P^4 - Q^4$

$= A$

(b) Let $A = PQ - QP$

$A^T = (PQ - QP)^T$

$= (PQ)^T - (QP)^T$

$= Q^T P^T - P^T Q^T$

$= -QP + PQ$

$= PQ - QP$

$= A$

(c) Let $A = Q^5 - P^5$

$A^T = (Q^5 - P^5)^T$

$= (Q^T)^5 - (P^T)^5$

$= (-Q)^5 - (P)^5$

$= -Q^5 - P^5$

$= -(Q^5 + P^5) \neq -A$

(d) Let $A = PQ + QP$

$A^T = (PQ + QP)^T$

$= (PQ)^T + (QP)^T$

$= Q^T P^T + P^T Q^T$

$= -QP - PQ$

$= -(PQ + QP)$

$= -A$

68. *(c)* Given, $y' + 2y = f(x)$, $y(0) = 0$,

where $f(x) = \begin{cases} 1, & 0 \le x \le 1 \\ 0, & x > 1 \end{cases}$

Then, $y' + 2y = 1,\ 0 \le x \le 1$ ...(i)

$y' + 2y = 0,\ x > 1$ ...(ii)

On solving Eq. (i), we get

$$\text{IF} = e^{2\int dx} = e^{2x}$$

$\Rightarrow \quad y \cdot e^{2x} = \int e^{2x}\, dx + C_1$

$\Rightarrow \quad y \cdot e^{2x} = \dfrac{e^{2x}}{2} + C_1$

$\Rightarrow \quad y = \dfrac{1}{2} + C_1 e^{-2x}$

$\Rightarrow \quad y(0) = 0 \Rightarrow C_1 = -\dfrac{1}{2}$

$\Rightarrow \quad y(x) = -\dfrac{e^{-2x}}{2} + \dfrac{1}{2}$ in $0 \le x \le 1$

On solving Eq. (ii), we get

$$y' + 2y = 0$$

$$\Rightarrow \quad \frac{y'}{y} = -2$$

$$\Rightarrow \quad y(x) = C_2 e^{-2x} \text{ in } x > 1$$

Since, solution $y(x)$ are continuous.

$\therefore y(x)$ is continuous at $x = 1$.

$$\Rightarrow \quad \lim_{x \to 1^+} y(x) = \lim_{x \to 1^-} y(x)$$

$$\Rightarrow \quad \lim_{x \to 1^+} C_2 e^{-2x} = \lim_{x \to 1^-} \left(\frac{-e^{-2x}}{2} + \frac{1}{2} \right)$$

$$\Rightarrow \quad C_2 = \frac{1}{2}(e^2 - 1)$$

$$y(x) = \frac{1}{2}(e^2 - 1)e^{-2x} \text{ for } x > 1$$

$$y\left(\frac{3}{2}\right) = \frac{(e^2 - 1)}{2} e^{-3}$$

$$= \frac{e - e^{-1}}{2e^2} = \frac{1}{e^2} \sin h(1)$$

Study Tactics

We define the given differential equation into two intervals as $0 \le x \le 1$ and $x > 0$. Then, get the solutions of differential equation by integrating factors in linear differential equation. After that, find the limit of functions at left and right hand side at 1 for the constant of integration.

69. *(d)* Using AM $\ge$ GM, we get

$$a^2 + 1 \ge 2a$$

$$\Rightarrow \quad \frac{a^2 + 1}{b + c} \ge \frac{2a}{b + c} \quad \text{...(i)}$$

Similarly, $$\frac{b^2 + 1}{a + c} \ge \frac{2b}{a + c} \quad \text{...(ii)}$$

and $$\frac{c^2 + 1}{a + b} \ge \frac{2c}{a + b} \quad \text{...(iii)}$$

On adding Eqs. (i), (ii) and (iii), we get

$$\frac{a^2 + 1}{b + c} + \frac{b^2 + 1}{a + c} + \frac{c^2 + 1}{a + b} \ge \frac{2a}{b + c} + \frac{2b}{a + c} + \frac{2c}{a + b}$$

Now, $$\frac{2a}{b + c} + \frac{2b}{a + c} + \frac{2c}{a + b}$$

$$= \left(\frac{2a}{b + c} + 2 \right) + \left(\frac{2b}{a + c} + 2 \right) + \left(\frac{2c}{a + b} + 2 \right) - 6$$

$$= 2\left(\frac{a + b + c}{b + c} + \frac{a + b + c}{a + c} + \frac{a + b + c}{a + b} \right) - 6$$

Now, $$\frac{\frac{a + b + c}{b + c} + \frac{a + b + c}{a + c} + \frac{a + b + c}{a + b}}{3}$$

$$\ge \frac{3}{\frac{b + c}{a + b + c} + \frac{a + c}{a + b + c} + \frac{a + b}{a + b + c}}$$

$$\Rightarrow \frac{a + b + c}{b + c} + \frac{a + b + c}{a + c} + \frac{a + b + c}{a + b} \ge \frac{9}{2}$$

$$\Rightarrow \quad \frac{2a}{b + c} + \frac{2b}{a + c} + \frac{2c}{a + b} \ge 9 - 6 = 3$$

$$\Rightarrow \quad \frac{a^2 + 1}{b + c} + \frac{b^2 + 1}{a + c} + \frac{c^2 + 1}{a + b} \ge 3$$

Hence, the minimum value of $\frac{a^2 + 1}{b + c} + \frac{b^2 + 1}{c + a} + \frac{c^2 + 1}{a + b}$ is 3.

70. *(c)* $f_0(n) = \dfrac{1}{1 - n}$

$$f_1(n) = f_0[f_0(n)] = f_0\left(\frac{1}{1 - n}\right), \quad [n \ne 1]$$

$$= \frac{1}{1 - \frac{1}{1 - n}} = \frac{1 - n}{-n}, \quad [n \ne 0]$$

$$f_2(n) = f_0[f_1(n)]$$

$$= \frac{1}{1 - f_1(n)} = \frac{1}{1 - \left(\frac{1 - n}{-n}\right)} = n$$

Similarly, $f_3(n) = f_0(n)$

$$f_4(n) = f_1(n)$$

$$f_5(n) = f_2(n)$$

$$\vdots \qquad \vdots$$

Now, $f_{100}(3) + f_1\left(\frac{2}{3}\right) + f_2\left(\frac{3}{2}\right)$

$$= f_1(3) + f_1\left(\frac{2}{3}\right) + f_2\left(\frac{3}{2}\right)$$

$$= 1 - \frac{1}{3} + 1 - \frac{3}{2} + \frac{3}{2}$$

$$= \frac{5}{3}$$

71. *(18)* Given, $F(x)=\begin{vmatrix} a-\sin\frac{x}{2} & b\cos x & 0 \\ b\cos x & 0 & a-\sin\frac{x}{2} \\ 0 & a-\sin\frac{x}{2} & b\cos x \end{vmatrix}$

Expand w.r.t to R_1

$$F(x)=\left(a-\sin\frac{x}{2}\right)\left(-\left(a-\sin\frac{x}{2}\right)^2\right)$$

$$-b\cos x\,((b\cos x)^2)+0$$

$$F(x)=-\left(a-\sin\frac{x}{2}\right)^3-b^3\cos^3 x$$

$$\because \lim_{x\to 0}\frac{F(x)}{x^3}=-\lim_{x\to 0}\frac{\left(a-\sin\frac{x}{2}\right)^3+b^3\cos^3 x}{x^3}$$

$$=-\lim_{x\to 0}\frac{\left(a-\frac{\sin\frac{x}{2}}{\frac{x}{2}}\cdot\frac{x}{2}\right)^3+b^3}{x^3}$$

$$=-\lim_{x\to 0}\frac{\left(a-\frac{x}{2}\right)^3+b^3}{x^3}$$

Applying L' Hospitals rule three times, we get

$$=-\left(-\frac{1}{2}\right)\left(-\frac{1}{2}\right)\left(-\frac{1}{2}\right)$$

$$\lim_{x\to 0}\frac{F(x)}{x^3}=\frac{1}{8}$$

$\because \quad 144L=\frac{1}{8}$

$$L=\frac{144}{8}$$

$\Rightarrow \quad L=18$

72. *(197)* Let $(\sqrt{2}+1)^6=I+F$,

where $I=$ Integer, $F=$ Fractional part

$\therefore \quad 0\le F<1$

Let $\quad f=(\sqrt{2}-1)^6$

$\Rightarrow \quad \sqrt{2}-1=\frac{1}{\sqrt{2}+1}$

$\Rightarrow \quad 0<\sqrt{2}-1<1 \Rightarrow 0<f<1$

Also, $I+F+f=(\sqrt{2}+1)^6+(\sqrt{2}-1)^6$

$$=2[{}^6C_0\cdot 2^3+{}^6C_2\cdot 2^2+{}^6C_4\cdot 2+{}^6C_6]$$

$$=2(8+60+30+1)$$

$$=198$$

Hence, $F+f=198-I$ is an integer.

On putting $0<F+f<2$

$\Rightarrow \quad F+f=1$

$\therefore \quad I=197$

$[(\sqrt{2}+1)^6]=197$

73. *(132)* Given, $\mathbf{a}=6\hat{\mathbf{i}}+2\hat{\mathbf{j}}-2\hat{\mathbf{k}}$, $\mathbf{c}=4\hat{\mathbf{i}}-6\hat{\mathbf{j}}+6\hat{\mathbf{k}}$

and $\mathbf{a}=\mathbf{b}\times\mathbf{c}$

$$|\mathbf{a}|=\sqrt{44},\ |\mathbf{c}|=\sqrt{88}$$

$$|\mathbf{a}|=|\mathbf{b}\times\mathbf{c}|=|\mathbf{b}||\mathbf{c}|\sin\theta$$

$$\sqrt{44}=\sqrt{50}\cdot\sqrt{88}\sin\theta$$

$\Rightarrow \quad \sin\theta=\sqrt{\frac{44}{50\times 88}}$

$$|\mathbf{b}+\mathbf{c}|^2=|\mathbf{b}|^2+|\mathbf{c}|^2+2\mathbf{b}\cdot\mathbf{c}$$

$$=|\mathbf{b}|^2+|\mathbf{c}|^2+2|\mathbf{b}|\cdot|\mathbf{c}|\cos\theta$$

$$=50+88+2\sqrt{50}\sqrt{88}\times\frac{\sqrt{99}}{10}$$

$$=138+\frac{660}{5}$$

$$=138+132=270$$

Now, $|138-|\mathbf{b}+\mathbf{c}|^2|=132$

74. *(56)* We have, $x=y=z$ ...(i)

$$x=\frac{y}{2}=\frac{z}{3}$$ ...(ii)

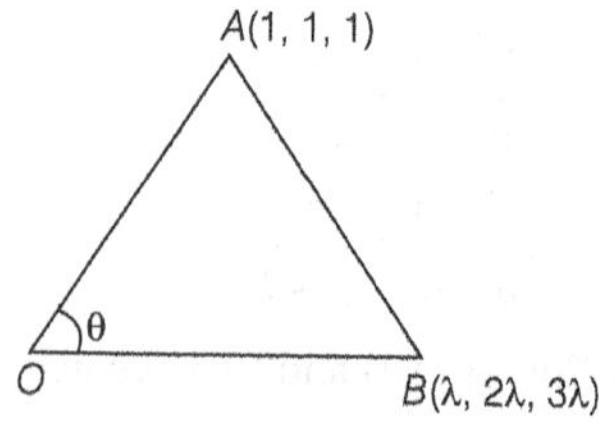

$\therefore$ Point of intersection of Eqs. (i) and (ii) are (0, 0, 0).

Direction ratios of Eq. (i) are (1, 1, 1).

Direction ratios of Eq. (ii) are (1, 2, 3).

Let θ be the angle between Eqs. (i) and (ii),

$$\cos\theta=\frac{6}{\sqrt{42}},\ \sin\theta=\frac{\sqrt{6}}{\sqrt{42}}$$

Let any point on second line be $(\lambda, 2\lambda, 3\lambda)$.

Third line passing through (1, 1, 1).

(1, 1, 1) lies on Eq. (i).

$$\text{Area of } \Delta OAB = \frac{1}{2} \times OA \times OB \sin\theta$$

$$= \frac{1}{2}\sqrt{3}\lambda\sqrt{14} \times \frac{\sqrt{6}}{\sqrt{42}} = \sqrt{6}$$

$\Rightarrow \quad \lambda = 2$

So, B is (2, 4, 6)

$\Rightarrow (2)^2 + (4)^2 + (6)^2 = 56$

75. *(2)* Given, $S = \sum_{k=1}^{\infty}\sum_{n=1}^{\infty} \frac{k}{2^{n+k}}$

$$= \sum_{k=1}^{\infty}\sum_{n=1}^{\infty} \frac{k}{2^k}\cdot\frac{1}{2^n}$$

$$= \sum_{k=1}^{\infty}\left(\frac{k}{2^k}\sum_{n=1}^{\infty}\frac{1}{2^n}\right)$$

$$= \sum_{k=1}^{\infty}\frac{k}{2^k}\left(\frac{1}{2} + \frac{1}{2^2} + \frac{1}{2^3} + \ldots \infty\right)$$

$$= \sum_{k=1}^{\infty}\frac{k}{2^k}\frac{\frac{1}{2}}{1-\frac{1}{2}} = \sum_{k=1}^{\infty}\frac{k}{2^k}$$

So, $S = \frac{1}{2} + \frac{2}{2^2} + \frac{3}{2^3} + \ldots$ ...(i)

$\Rightarrow \frac{S}{2} = \frac{1}{2^2} + \frac{2}{2^3} + \frac{3}{2^4} + \ldots$ ...(ii)

On subtracting Eq. (ii) from Eq. (i), we get

$$\frac{S}{2} = \frac{1}{2} + \frac{1}{2^2} + \frac{1}{2^3} + \ldots = 1$$

$\Rightarrow S = 2$

SECTION II

Mastery Builders

PRACTICE SET 8

Total Marks : 300	Total Time : 3 hrs	No. of Questions : 75

Instructions

1. There are **three** parts in the questions paper consisting of **Physics**, **Chemistry** and **Mathematics** having 25 questions in each part of equal weightage. Each part (subject) has two sections.
2. **Section A** This section contains 20 multiple choice questions which have only one correct answer. Each question carries **4 marks** for correct answer and **–1 mark** for wrong answer.
3. **Section B** This section contains 5 questions. The answer to each of the questions is a numerical value. Each question carries **4 marks** for correct answer and **–1 mark** for wrong answer. For Section B, the answer should be rounded off to the nearest integer.

PHYSICS

Section A : Objective Type Questions

1. Given below are two statements.

Statement I The current which consume no power for its maintenance in the circuit is called idle current.

Statement II For a series *L-C-R* circuit, resonant frequency is inversely proportional to resistance.

In the light of the above statements, choose the correct answer from the options given below

(a) Statement I is true but Statement II is false.
(b) Statement I is false but Statement II is true.
(c) Both Statement I and Statement II are false.
(d) Both Statement I and Statement II are true.

2. A particle of mass m is moving in a horizontal circle of radius r under a centripetal force $-\frac{K}{r^3}$, where K is constant. Total energy of the particle is

(a) $\frac{K}{2r}$ (b) $-\frac{K}{2r^2}$

(c) zero (d) $-\frac{K}{3r^2}$

3. Which one is the correct option for the two different thermodynamic processes?

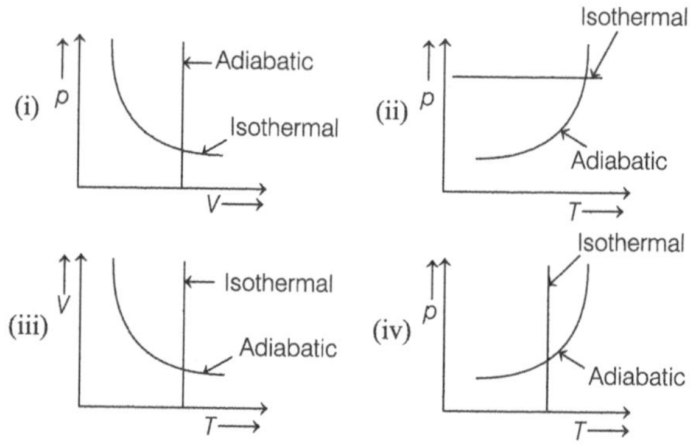

(a) (iii) and (i) (b) (iii) and (iv)
(c) (i) only (d) (ii) and (iii)

4. Oxygen is filled in a closed metal jar of volume 10^{-3} m^3 at a pressure of 1.5×10^5 Pa and temperature 400 K. The jar has a small leak in it. The atmospheric pressure is 10^5 Pa and the atmospheric temperature is 300 K. The mass of the gas that leaks out by the time, the pressure and the temperature inside the jar equalise with the surrounding is

(a) 0.16 g (b) 16 g
(c) 0.18 g (d) 0.14 g

5. A wire of resistance 9Ω is broken into two parts. The length ratio being $1:2$. The two wires are then connected in parallel. The net resistance will be

(a) 1Ω (b) 105Ω (c) 2Ω (d) 4Ω

6. If 24.6 eV of energy is required to remove one electron from neutral He-atom, then the energy required to create an α-particle from neutral helium atom shall be

(a) 58.2 eV (b) 49.2 eV (c) 51.8 eV (d) 79 eV

7. A force F acting on a particle of mass m is indicated by the force-time graph as shown. The change in momentum of the particle over the time interval from zero to 10 s is

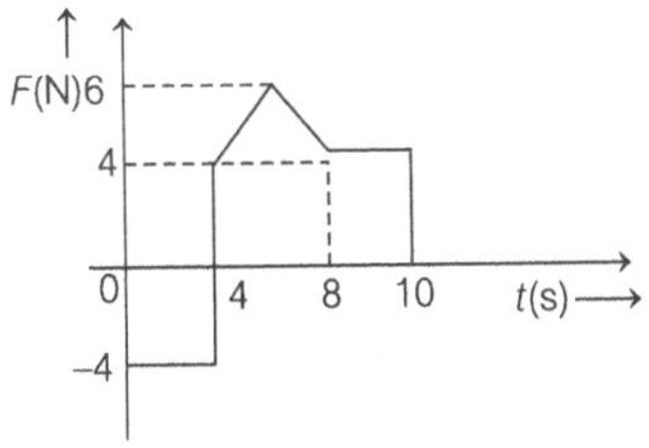

(a) 12 N-s (b) 14 N-s (c) 16 N-s (d) 10 N-s

8. Match the Column I with Column II and select the correct answer from the codes given below.

Column I	Column II
A. Initial velocity of charge particle is parallel to magnetic field.	1. Parabolic path
B. Initial velocity of charge particle is perpendicular to magnetic field.	2. Helical path
C. Initial velocity of charge particle is perpendicular to electric field.	3. Circular path
D. Initial velocity of charge particle inclined to B.	4. Straight path

Codes

	A	B	C	D		A	B	C	D
(a)	4	3	2	1	(b)	4	3	1	2
(c)	1	3	4	2	(d)	2	3	4	1

9. **Assertion** If a circular loop is entering into a magnetic field perpendicular to the plane of loop, then the flux through the loop will increase.

Reason During the motion of the loop, its area increases.

Choose the correct option.

(a) Both Assertion and Reason are correct and Reason is the correct explanation of Assertion.
(b) Both Assertion and Reason are correct but Reason is not the correct explanation of Assertion.
(c) Assertion is correct but Reason is incorrect.
(d) Assertion is incorrect but Reason is correct.

10. An electromagnetic wave is propagating along X-axis such that $\mathbf{E} = E_0 \sin(kx - \omega t)\hat{\mathbf{j}}$. If a charge q is projected with speed u along Y-axis, then the maximum magnetic force experienced by the charge particle is

(a) $\frac{quc}{E_0}$ (b) $\frac{qE_0}{uc}$ (c) $\frac{uc}{qE_0}$ (d) $\frac{quE_0}{c}$

11. The diameter of a cylinder is measured by using vernier callipers with no zero error. It is found that zero of the vernier scale lies between 5.15 cm and 5.20 cm of the main scale. The vernier scale has 30 divisions which is equivalent to 1.44 cm. The 16th division of vernier scale exactly coincides with one of the main scale division. The diameter of the cylinder is

(a) 5.172 cm (b) 5.187 cm (c) 5.182 cm (d) 5.287 cm

12. The gravitational field g inside the Earth is proportional to (where, r is distance from the centre)

(a) $\frac{1}{r}$ (b) $\frac{1}{r^2}$ (c) r^2 (d) r

13. Poise is the CGS unit of coefficient of viscosity. Suppose, we employ a system of units in which unit of mass is α kg, the unit of length is β metre and unit of time is γ second. In this new system, 1 poise is equal to

(a) $1000\,\alpha\beta^{-1}\gamma^{-1}$ (b) $10\,\alpha\beta^{-1}\gamma^{-1}$
(c) $0.1\,\alpha\beta^{-1}\gamma^{-1}$ (d) $0.01\,\alpha^{-1}\beta\gamma$

14. A ball is moving in uniform circular motion in a horizontal plane. Some parallel rays of light are coming from behind. The motion of the shadow of the ball on the screen will be

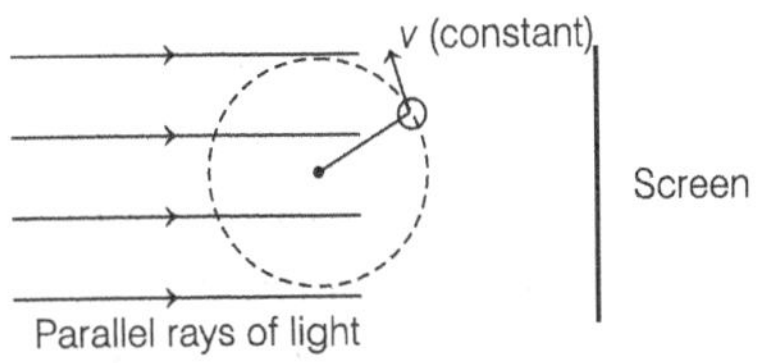

(a) rotatory motion
(b) oscillatory motion (SHM)
(c) non-uniform circular motion
(d) uniform circular motion

15. A 100 W light bulb is placed at the centre of a spherical chamber of radius 0.10 m. Assume that 66% of the energy supplied to the bulb is converted into light and that the surface of chamber is perfectly absorbing. The pressure exerted by the light on the surface of the chamber is

(a) 0.87×10^{-6} Pa
(b) 1.75×10^{-6} Pa
(c) 3.50×10^{-6} Pa
(d) None of the above

16. Consider the arrangement of three metal plates *A, B* and *C* of equal surface area and separation *d* as shown in figure. The energy stored in the arrangement, when the plates are fully charged, is

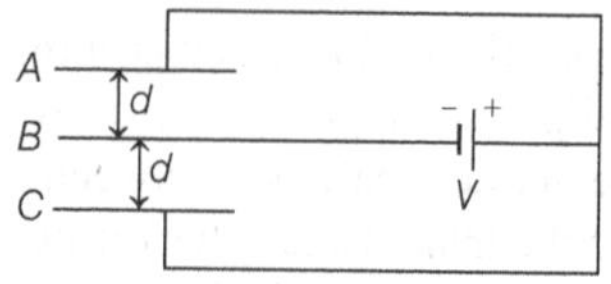

(a) $\frac{\varepsilon_0 AV^2}{2d}$ (b) $\frac{\varepsilon_0 AV^2}{d}$
(c) $\frac{2\varepsilon_0 AV^2}{d}$ (d) $\frac{3\varepsilon_0 AV^2}{2d}$

17. Two soap bubbles of radii r_1 and r_2 equal to 4 cm and 5 cm respectively are touching each other over a common surface *AB* (shown in figure). Its radius will be

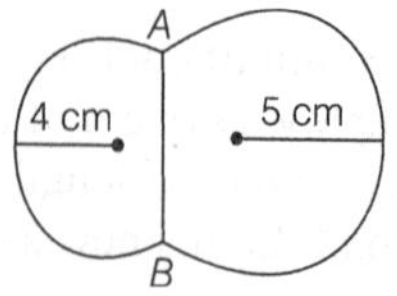

(a) 4 cm (b) 4.5 cm (c) 5 cm (d) 20 cm

18. Magnetic dipole moment of a current carrying square loop is *m*. If this loop is changed into circular loops and having same current, then its magnitude of magnetic dipole moment will be

(a) $\frac{2m}{\pi}$ (b) $\frac{2\pi m}{3}$
(c) $\frac{4m}{\pi}$ (d) $\frac{3m}{2\pi}$

19. 22g of carbon dioxide at 27°C is mixed in a closed container with 16 g of oxygen at 37°C. If both gases are considered as ideal gases, then the temperature of the mixture is

(a) 24.2°C (b) 28.5°C
(c) 31.5°C (d) 33.5°C

20. A bullet emerges from a barrel of length 1.2 m with a speed of 640 m/s. Assuming constant acceleration, the approximate time that it spends in the barrel after gun is fired, is

(a) 4 ms (b) 40 ms
(c) 400 ms (d) 1 s

Section B : Numerical Value Type Questions

21. In the circuit shown, the cell is ideal with emf = 15V, each resistance is of 3Ω. The potential difference across the capacitor is ...... V.

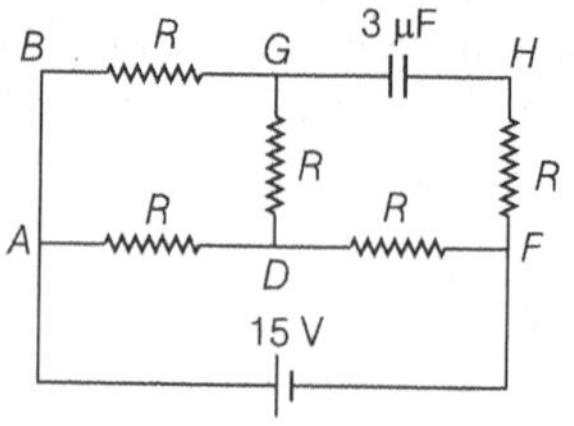

22. There is an electric circuit shown in the figure. The current through 15Ω resistor is $x\times10^{-1}$ A. The value of *x* is ....A.

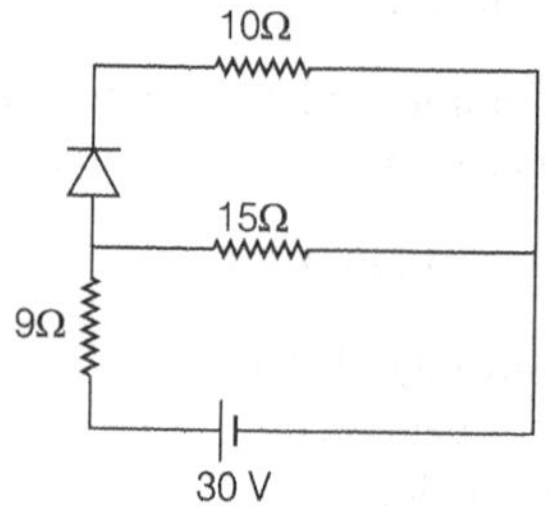

23. A fish looking up through the water sees the outside world contained in a circular horizon. If the refractive index of water is 4/3 and the fish is 12 cm below the surface, the radius of this circle is $\frac{36}{\sqrt{x}}$ cm. The value of *x* is ....

24. A mixture of gas consists of 3 moles of O_2 and 5 moles of Ar at temperature *T*. Considering only rotational and translational modes, the total internal energy of the system is found to be *xRT*. The value of *x* is ....

25. A rubber cube of each side 7 cm has one side fixed, while a tangential force equal to the weight of 300 kgf is applied to the opposite face. If the modulus of rigidity for rubber is 2×10^7 dyne cm^{-2} and $g=10$ ms^{-2}, then the distance (in cm) through which the strain side moves will be approximately .... .

CHEMISTRY

Section A : Objective Type Questions

26. Which of the following statement is incorrect about partition chromatography?

(a) Mobile phase can be a gas.
(b) Separation depends upon equilibration of solute between a mobile and a stationary phase.
(c) Stationary phase is a finely divided solid adsorbent.
(d) Paper chromatography is an example of partition chromatography.

27. In XeF_2, XeF_4 and XeF_6 the number of lone pair of electrons on Xe respectively is

(a) 1, 2, 3 (b) 2, 3, 1 (c) 3, 2, 1 (d) 4, 1, 1

28. The major product of the following reaction is

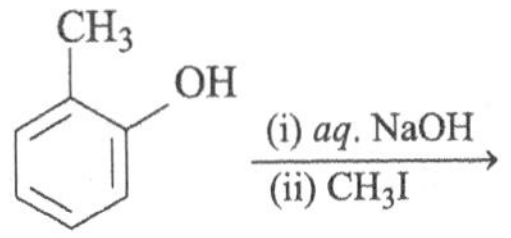

(a) 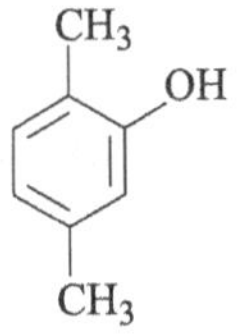 (b) 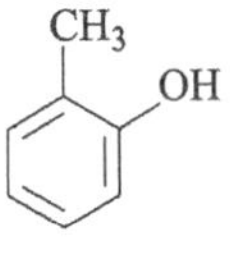

(c) 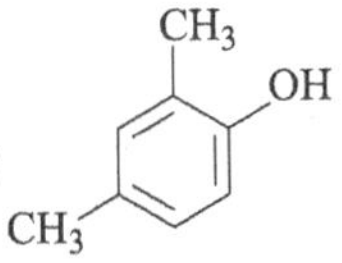 (d) 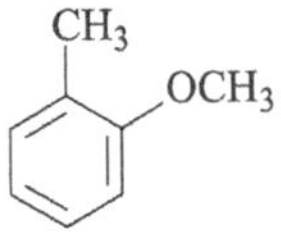

29. Among the following, the correct isostructural pairs are NF_3, NO_3^-, BF_3, H_3O^+, HN_3.

(a) $[NF_3, NO_3^-]$ and $[BF_3, H_3O^+]$
(b) $[NF_3, HN_3]$ and $[NO_3^-, BF_3]$
(c) $[NF_3, H_3O^+]$ and $[NO_3^-, BF_3]$
(d) $[NF_3, H_3O^+]$ and $[HN_3, BF_3]$

30. Glucose when heated with CH_3OH in presence of dry HCl gas gives α and β-methyl glucosides because it contains

(a) an aldehyde group (b) a —CH_2OH group
(c) a ring structure (d) five hydroxyl groups

31. Equivalent conductivity of $BaCl_2$, H_2SO_4 and HCl are y_1, y_2 and y_3 $S\,cm^{-1}\,eq^{-1}$, respectively at infinite dilution. If conductivity of saturated $BaSO_4$ solution is y $S\,cm^{-1}$, then K_{sp} of $BaSO_4$ will be

(a) $\dfrac{y^2}{(y_1 + y_2 - 2y_3)^2}$ (b) $\dfrac{2.5\,y^{-2}}{(y_1 + y_2 - 2y_3)^2}$

(c) $\dfrac{500}{y_1 + y_2 - 2y_3}$ (d) $\dfrac{2.5 \times 10^5\,y^2}{(y_1 + y_2 - 2y_3)^2}$

32. Consider the following compounds.

A. H_2SO_3 B. $H_2S_2O_5$ C. $H_2S_2O_6$

Choose the correct sequence of increasing number of π-bond in structures of the following molecules.

(a) $C < B < A$ (b) $A < B < C$
(c) $B = C < A$ (d) $B < A < C$

33. Match List-I with List-II.

	List-I (Bases)		List-II (Structure)
A.	Uracil	I.	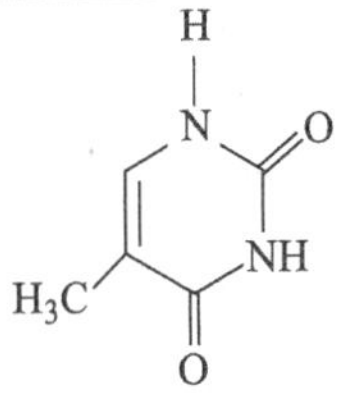
B.	Adenine	II.	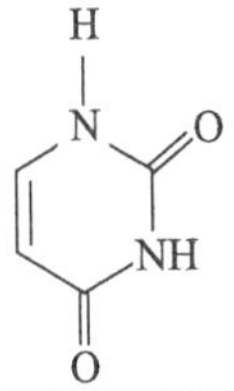
C.	Thymine	III.	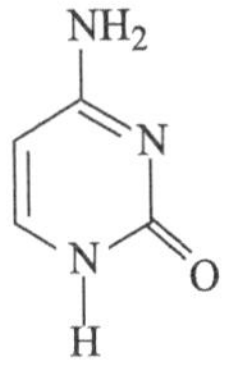
D.	Cytosine	IV.	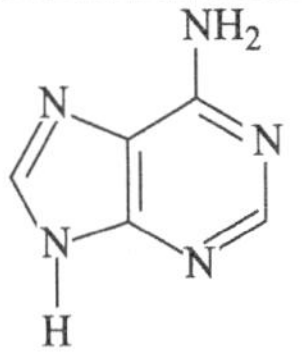

Choose the correct answer from the options given below.

(a) A-IV, B-I, C-III, D-II
(b) A-II, B-IV, C-I, D-III
(c) A-II, B-III, D-I, C-IV
(d) A-III, B-IV, C-II, D-I

34. The shape of SF_4 is

(a) *T*-shaped
(b) linear
(c) *see-saw*
(d) octahedral

35. Order of paramagnetism.

A. $[Fe(CN)_6]^{3-} > [Co(NH_3)_6]^{3+}$

B. $[Fe(H_2O)_6]^{2+} > [Zn(H_2O)_6]^{2+}$

C. $K_3[Cu(CN)_4] > K_3[Cr(oxalate)_3]$

D. $[Fe(CN)_6]^{3-} < [Cr(H_2O)_6]^{3+}$

E. $K_3[Cu(CN)_4] > [CuF_6]^{3-}$

Choose the correct answer from the options given below.

(a) A, B and D (b) D and E
(c) B, A and C (d) B, C and D

36. Which of the following is correct regarding the major product (B) of reaction?

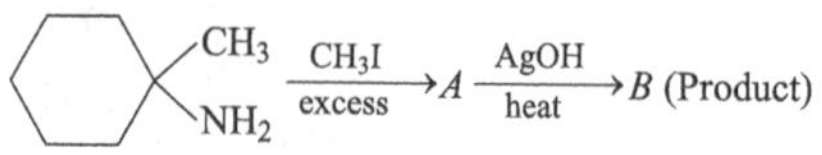

(a)

which shows optical isomerism

(b)

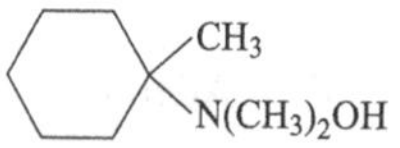

which shows optical isomerism

(c)

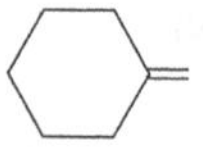

which does not show geometrical isomerism

(d)

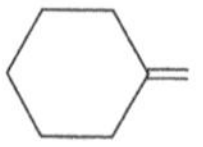

which shows geometrical isomerism

37. In which of the following transformations, the bond order has increased and the magnetic behaviour has changed?

(a) $C_2 \rightarrow C_2^+$ (b) $NO \rightarrow NO^+$
(c) $O_2 \rightarrow O_2^+$ (d) $N_2 \rightarrow N_2^+$

38. Identify the (A) and (B) in the following reaction.

$$\underset{\substack{|\\ CH_3}}{Ph—CH}—CH{=}CH—CH_3 \xrightarrow[\Delta]{NBS} (A) \xrightarrow{Alc.\ KOH} (B)$$

(a) $A = Ph—C(Br)(CH_3)CH = CH—CH_3$,
$B = Ph—HC(CH_3)—CH{=}CH—CH_3$

(b) $A = Ph—C(Br)(CH_3)—CH{=}CH—CH_3$,
$B = Ph—\underset{\substack{\|\\ CH_2}}{C}—CH{=}CH—CH_3$

(c) $A = Ph—C(Br)(CH_3)CH—CH_2—CH_3$,
$B = Ph—C(CH_3)(Br)CH{=}CH—CH_3$

(d) $A = Ph—C(Br)(CH_3)CH—CH_2—CH_3$,
$B = Ph—C(OH)(CH_3)CH{=}CH—CH_3$

39. Given below are two statements.

Statement I $[FeCl_4]^-$ has higher spin-only magnetic moment than $[Co(en)(NH_3)_2Cl_2]^+$.

Statement II The cobalt ion in $[Co(en)(NH_3)_2Cl_2]^+$ has sp^3d^2-hybridisation.

In the light of the above statements choose the correct answer from the options given below.

(a) Both Statement I and Statement II are correct.
(b) Both Statement I and Statement II are incorrect.
(c) Statement I is correct but Statement II is incorrect.
(d) Statement I is incorrect but Statement II is correct.

40. Which of the following represents correctly the changes in thermodynamic properties during the formation of 1 mol of an ideal binary solution?

(a)

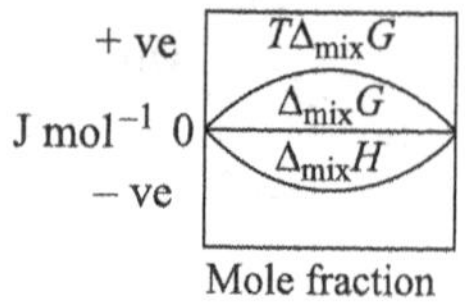

(b)

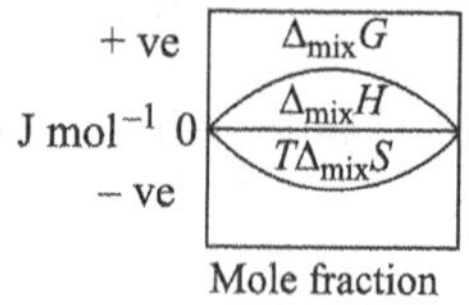

(c)

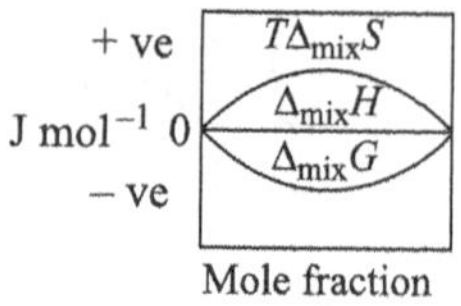

(d)

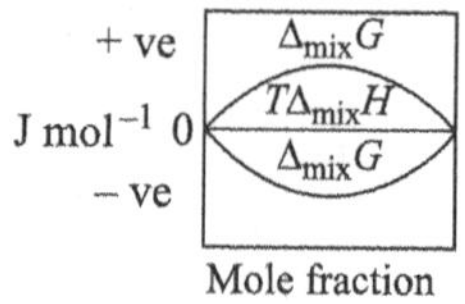

41. The amount of H_2O_2 present in 1 L of 1.5 N H_2O_2 solution is

(a) 2.5 g (b) 25.5 g (c) 3.0 g (d) 8.0 g

42. Given below are two statements one is labelled as Assertion (A) and other is labelled as Reason (R).

Assertion (A) The electron gain enthalpy of S or Cl is more negative than the preceding element in a group.

Reason (R) The added electron goes to the smaller $n = 2$ and suffer significant repulsion from other electron present in this level.

In the light of the above statements choose the correct answer from the options given below.

(a) Both (A) and (R) are correct but (R) is the correct explanation of (A).

(b) Both (A) and (R) are correct but (R) is not the correct explanation of (A).
(c) (A) is correct but (R) is incorrect.
(d) (A) is incorrect but (R) is correct.

43. Identify the product formed (*D*) in the following reaction.

$$CH_3COOH \xrightarrow{SOCl_2} [A] \xrightarrow[\text{Anhyd. } AlCl_3]{\text{Benzene}} [B] \xrightarrow{HCN} [C] \xrightarrow{H_3O^+} [D]$$

(a) 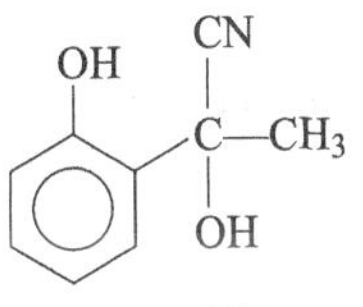
(b) 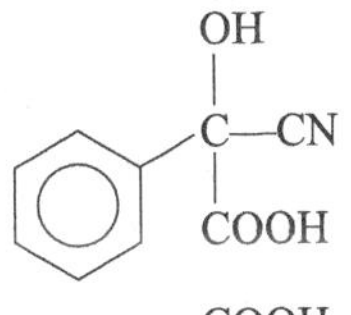
(c) 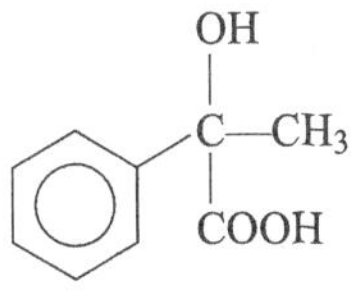
(d) 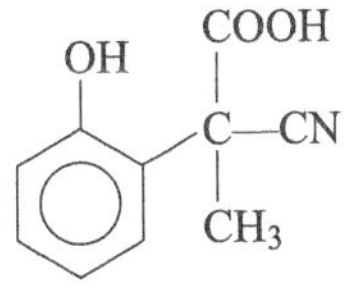

44. Identify the correct statement(s).

A. The time taken for the completion of 75% reaction is thrice the $t_{1/2}$ of the first order reaction.

B. For the first order reaction, the degree of dissociation is equal to $(1 - e^{-kt})$.

C. The pre-exponential factor in the Arrhenius equation has the dimension of time, T^{-1}.

D. A plot of $\log [X]$ *versus* time is linear for a first order reaction, $X \rightarrow P$.

Choose the correct answer from the options given below.

(a) B, A and D (b) B, C and D
(c) Only C (d) A and D

45. Match List-I with List-II.

	List-I		List-II
A.	Nessler's reagent	I.	Halogen specifically
B.	Cu(II) oxide	II.	Ni^{2+}
C.	Silver nitrate	III.	NH_3 or NH_4^+
D.	Dimethyl glyoxime	IV.	Carbon

Choose the most appropriate answer from the options given below.

(a) A-II, B-IV, C-III, D-I
(b) A-I, B-III, C-IV, D-II
(c) A-III, B-IV, C-I, D-II
(d) A-III, B-II, C-IV, D-I

Section B : Numerical Value Type Questions

46. The mass of 112 cm^3 of CH_4 gas at STP is .... $\times 10^{-2}$ g.

47. In a measurement of quantum, efficiency of photosynthesis in green plants, it was found that 10 quanta of red light of wavelength 6850 Å were needed to release one molecule of O_2. The average energy stored in this process is 112 kcal/mole O_2 evolved. The energy conversion efficiency in this experiment is ......%.
[Given, 1 cal = 4.18 J; $N_A = 6 \times 10^{23}$, $h = 6.62 \times 10^{-34}$ Js^{-1}] [Nearest integer]

48. The cell potential for the following cell
$Pt|H_2(g)|H^+(aq)||Cu^{2+}(0.01M)|Cu(s)$
is 0.576 V at 298 K. The pH of the solution is .... (Nearest integer)

49. A solution of glucose in water is labelled as 10% *w*/*W*. If the density of solution is 1.2 g mL^{-1}, then the molality of the solution should be ......$\times 10^{-3}$m.

50. Consider the decomposition of HNO_3 as follows

$$4HNO_3(g) \rightleftharpoons 4NO_2(g) + 2H_2O(g) + O_2(g)$$

The equation approaches to equilibrium at 800 K temperature and 15 atm pressure. At equilibrium, partial pressure of HNO_3 is 1 atm. The value of K_C at 800 K is ......$\times 10^{-1}$.
(Given, R = 0.08 L atm $mol^{-1}K^{-1}$).

MATHEMATICS

Section A : Objective Type Questions

51. At each point (x, y) of the curve intercept of the straight line $Y - y = \frac{dy}{dx}(X - x)$ on Y-axis is equal to $2xy^2$, then equation of curve satisfying the condition $y(2) = 1$, is

(a) $x^2y - x = 2y$ (b) $x^2 - x = 2y$
(c) $x - x^2y = 2y$ (d) $x - x^2 = 2y$

52. If $\int \frac{5\tan x}{\tan x - 2} dx = x + A \ln|\sin x + B\cos x| + C$, then (A, B) is equal to

(a) $(-2, -2)$ (b) $(2, 2)$
(c) $(2, -2)$ (d) $(-2, 2)$

53. The value of the integral $\int_{1/e}^{e^2} |f(x)|\, dx$ is, where $f(x) = \frac{\log_e x}{x}$

(a) $\frac{3}{2}$ (b) $\frac{5}{2}$
(c) $\frac{2}{5}$ (d) $\frac{2}{3}$

54. Let

$$f(x) = \begin{cases} \frac{1+\cos x}{(\pi - x)^2} \cdot \frac{\sin^2 x}{\log(1+\pi^2 - 2\pi x + x^2)}, & x \neq \pi \\ k, & x = \pi \end{cases}.$$

If $f(x)$ is continuous at $x = \pi$, then k is equal to

(a) 0.1 (b) 0.3
(c) 0.4 (d) 0.5

55. The square bounded by coordinate axes and the lines $x = 4$ and $y = 4$ divided by two parabolas $x^2 = 4y$ and $y^2 = 4x$. If S_1, S_2 and S_3 are respectively, the areas of these parts numbered from top to bottom, then $S_1 : S_2 : S_3$ is

(a) 1 : 2 : 1 (b) 1 : 2 : 3
(c) 2 : 1 : 2 (d) 1 : 1 : 1

56. Let $A = \left\{x \in R; f(x) = \sqrt{\frac{x - |x|}{x - [x]}}\right\}$ and B be the range of real function $g(x) = \frac{2x}{4 + x^2}$, then $A \cap B$ is equal to

(a) $\left[\frac{-1}{2}, \frac{1}{2}\right]$ (b) $\left(0, \frac{1}{2}\right]$
(c) $(-\infty, \infty)$ (d) None of these

57. One vertex of rectangular parallelopiped is at the origin O and length of its edges along X, Y and Z-axis 13, 12 and 5, respectively. Let P be the vertex (13, 12, 5).

Then, the shortest distance between the diagonal OP and an edge parallel to Z-axis, not passing through O or P, is

(a) $\frac{612}{\sqrt{41}}$ units (b) $\frac{156}{\sqrt{313}}$ units
(c) $\frac{819}{\sqrt{416}}$ units (d) $\frac{915}{\sqrt{515}}$ units

58. One set containing 5 numbers has mean 8 and variance 18 and second set containing 3 numbers has mean 8 and variance 24. Then, the variance of combined set of number is

(a) 21.25 (b) 19.25
(c) 22.25 (d) 20.25

59. Consider $A = \begin{bmatrix} a & 2 & 1 \\ 0 & b & 0 \\ 0 & -3 & c \end{bmatrix}$, where a, b and c are the roots of the equation $x^3 - 3x^2 + 2x - 1 = 0$, if matrix B is such that $AB = BA$, $A + B - 2I \neq 0$ and $A^2 - B^2 = 4I - 4B$, then the value of $\det(B)$ is equal to

(a) 25 (b) 26
(c) 27 (d) 28

60. If $\sin(2\sin x) = \cos(2\cos x)$ for $x \in \left(0, \frac{\pi}{2}\right)$ and $\tan x + \cot x = \frac{a}{\pi^b - c}$, where a, b and c are positive integers, then $\frac{a+b+c}{c}$ is equal to

(a) $\frac{25}{8}$ (b) $\frac{50}{7}$
(c) $\frac{100}{3}$ (d) $\frac{25}{2}$

61. Let a point $P(\lambda,\lambda,\lambda)$, where PQ and PR are drawn perpendicular respectively on the lines $\frac{x-1}{1}=\frac{y-2}{1}=\frac{z-4}{0}$ and $\frac{x-2}{1}=\frac{y-3}{-1}=\frac{z-4}{0}$

If P is such that $\angle QPR$ is a right angle, then possible value of λ is

(a) ± 6 (b) ± 8
(c) ± 4 (d) ± 2

62. If a circle cuts a rectangular hyperbola $xy=1$ at four points A, B, C and D. The parameters of these four points be t_1, t_2, t_3 and t_4, respectively and $-20t_1t_2t_3t_4=\lambda$, then the value of λ equals

(a) -4 (b) -10
(c) -20 (d) -40

63. Consider a ΔABC, whose vertices are $A(0, 6, 6)$, $B(6, 0, 6)$ and $C(6, 6, 0)$. Let D be a point moving on the line $x+z-3=0=y$ and G be the centroid of ΔABC. If the minimum length of GD is α, then the value of $\sqrt{\frac{2}{57}}\alpha$ is equal to

(a) 4 (b) 3
(c) 2 (d) 1

64. Consider the following statements with respect to functions $f(x)=\begin{cases} 2-|x|, & -1\le x\le 1 \\ |x-2|-x, & 1<x\le 3 \end{cases}$ and

$$g(x)=\begin{cases} \sin x-1, & 0\le x<\frac{\pi}{2} \\ [x]-\cos(x-2), & \frac{\pi}{2}\le x\le \pi \end{cases}$$

where, $[\cdot]$ denotes greatest integer function.

Statement I $\lim_{x\to 1^+} g[f(x)]=-1$

Statement II $\lim_{x\to 0^+} \frac{g[f(x)]}{[f(x)-2]^2}=\frac{1}{2}$

(a) Statement I is true and Statement II is false
(b) Statement I is false and Statement II is true
(c) Statement I and Statement II both are true
(d) Statement I and Statement II both are false

65. $f(x)=\begin{vmatrix} \cos x & 1 & 0 \\ 1 & 2\cos x & 1 \\ 0 & 1 & 2\cos x \end{vmatrix}$, then $\int_{-\frac{\pi}{2}}^{\frac{\pi}{2}} f(x)\,dx$ is equal to

(a) $-\frac{2}{3}$ (b) $\frac{2}{3}$
(c) 0 (d) 1

66. If the probability of choosing an integer λ out of $2n$ integers $1, 2, 3, \ldots, 2n$ is inversely proportional to $\lambda^4 (1\le\lambda\le n)$. If P_1 is the probability that chosen number is odd and P_2 is the probability that chosen number is even, then

(a) $P_1>\frac{1}{2}$ (b) $P_1<\frac{1}{2}$
(c) $P_1<\frac{2}{3}$ (d) $P_1>\frac{2}{3}$

67. There are x seats in the first row of a cinema hall of which y are to be occupied. Then, the number of ways of arranging y persons, if no two persons sit side by side, is

(a) $\frac{(x-y+1)!}{(x-3y+1)!}$ (b) $\frac{(x-y+1)!}{(x-2y)!}$
(c) $\frac{(x-y+1)!}{(x-2y+1)!}$ (d) $\frac{(x-y+2)!}{(x-2y-1)!}$

68. The coefficient of the middle term in the binomial expansion in powers of $(1+\alpha x)^4$ and of $(1-\alpha x)^6$ is the same, then $|[\alpha]|$ equals (where $[\cdot]$ is greatest integer function)

(a) 0 (b) 1
(c) 2 (d) 3

69. Suppose $f(x)=\frac{2x+7}{9}$ and $g(x)=\frac{9x+5}{7}$, then we define $h(x)=f^{-1}(x)+g^{-1}(x)-5x$. If $h^{-1}(x)=\frac{px+q}{r}$, where r is prime, then $p+q+r$ is equal to

(a) 48 (b) 96
(c) 86 (d) 106

70. If $\lambda \in R$ is such that the sum of the cubes of the roots of the equation $x^2 + (2-\lambda)x + (10-\lambda) = 0$ is minimum, then the modulus of the difference of the roots of this equation is

(a) $4\sqrt{2}$ (b) 20 (c) $2\sqrt{7}$ (d) $2\sqrt{5}$

Section B : Numerical Value Type Questions

71. A spherical balloon is being inflated at the rate of 70 cc/min. The rate of increase in the surface area (in cm^2/min) of the balloon when its diameter is 14 cm is ....

72. Let $y^2 = 32x$ be a given parabola and L be an extremity of its latus rectum in the first quadrant. A chord is drawn through L with slope -1 if the length of this chord is $K\sqrt{2}$, then K is equal to ....

73. If $\sum_{\theta=1}^{89} \frac{1}{1+\cot\theta}$, then the sum of the series is S_n, then $2S_n$ is equal to ....

74. Let $x = p + q + r$, $y = p + q\omega + r\omega^2$ and $z = p + q\omega^2 + r\omega$, where p, q, r, x, y and z be non-zero complex numbers and $\omega = e^{i2\pi/3}$, then the value of $\frac{|x|^2 + |y|^2 + |z|^2}{|p|^2 + |q|^2 + |r|^2}$ is equal to ....

75. The number of words, with or without meaning, that can be formed by taking 4 letters at a time from the letters of the word SYLLABUS such that two letters are distinct and two letters are like, is ....

DETAILED SOLUTIONS

Physics

1. *(a)* Statement I is true but Statement II is false.

In a circuit, if power consumption is zero, then the current flow in this circuit is called idle current.

The resonance frequency is given by

$$f_0 = \frac{1}{2\pi\sqrt{LC}}$$

Hence, it is independent of resistance.

2. *(c)* $\frac{mv^2}{r} = -\frac{K}{r^3}$

$\Rightarrow \frac{1}{2}mv^2 = \frac{1}{2}\frac{K}{r^2} = \text{KE}$

Potential energy, $\text{PE} = -\int_{\infty}^{r} F \cdot dr$

$$= -\int_{\infty}^{r}\left(-\frac{K}{r^3}\right)dr = -\frac{1}{2}\frac{K}{r^2}$$

Total energy, $E = \text{KE} + \text{PE} = 0$

3. *(b)* Graph (i) is incorrect, since in adiabatic process $V \neq$ constant.

Graph (ii) is incorrect, since in isothermal process $T =$ constant.

Graphs (iii) and (iv) match isotherms and adiabatic formula.

$TV^{\gamma - 1} = \text{constant}$ and $\frac{T^{\gamma}}{p^{\gamma - 1}} = \text{constant}$

4. *(a)* Given, volume, $V = 10^{-3}\ \text{m}^3$,

Pressure, $p_1 = 1.5 \times 10^5$ Pa, $p_2 = 10^5$ Pa

Temperature, $T_1 = 400\,\text{K}$, $T_2 = 300\,\text{K}$,

From ideal gas equation, $p_1 V = n_1 R T_1$

$$\Rightarrow \quad n_1 = \frac{p_1 V}{R T_1}$$

Similarly, $p_2 V = n_2 R T_2$

$$\Rightarrow \quad n_2 = \frac{p_2 V}{R T_2}$$

$\therefore$ Number of mole leaked,

$$\Delta n = n_1 - n_2 = \left(\frac{p_1}{T_1} - \frac{p_2}{T_2}\right)\frac{V}{R}$$

$$= \left(\frac{1.5}{400} - \frac{1}{300}\right) \times \frac{10^5 \times 10^{-3}}{8.3} = 5.02 \times 10^{-3}$$

Mass of gas leaked $= \Delta n M_0 = 5.02 \times 10^{-3} \times 32 = 0.16\,\text{g}$

5. *(c)* Resistance, $R = \frac{\rho l}{A}$

$$\therefore \quad R \propto l$$

$$\therefore \quad \frac{R_1}{R_2} = \frac{l_1}{l_2} = \frac{1}{2}$$

As, $R_1 + R_2 = 9\,\Omega$ (initial resistance)

$$3R_1 = 9\,\Omega \qquad (\because 2R_1 = R_2)$$

$$\therefore \quad R_1 = \frac{9}{3} = 3\,\Omega \text{ and } R_2 = 6\,\Omega$$

Now, R_1 and R_2 are connected in parallel.

$$\therefore \quad R_{\text{net}} = \frac{R_1 R_2}{R_1 + R_2}$$

$$= \frac{3 \times 6}{3 + 6} = 2\,\Omega$$

6. *(d)* α-particle is the nucleus of ${}_2\text{He}^4$ atom, i.e. doubly ionised He-atom.

To remove one electron, energy required is 24.6 eV.

After removing electron, the energy of remaining electron in He^+ is given by

$$E_n = \frac{-13.6}{n^2} \cdot Z^2$$

[for hydrogen like atoms/ions]

To remove second electron, put $n = 1$ and $Z = 2$

$$\therefore \quad E_{\text{ionisation}} = -[E_1]$$

$$= \frac{13.6}{(1)^2}(2)^2$$

$$= 54.4\ \text{eV}$$

$\therefore$ Total energy required $= 24.6 + 54.4 = 79\,\text{eV}$

7. *(a)* Change in momentum = Area under F-t graph in the given time interval.

$$= (-4 \times 4) + \left(4 \times 4 + \frac{1}{2} \times 4 \times 2\right) + (2 \times 4)$$

$$= 12\,\text{N-s}$$

8. *(b)* A. When initial velocity of the charged particle is parallel to magnetic field,

$$\theta = 0^\circ$$

$$F = qvB \sin 0^\circ = 0$$

The parallel magnetic field does not exert any magnetic force on moving charged particle. The charge particle will continue to move along the line of force.

B. When the initial velocity of the charged particle is perpendicular to the magnetic field, then $\theta = 90°$, so $F = qvB \sin 90° = qvB$, a maximum force. This force continuously deflects the particle sideways with changing its speed and particle will move along a circle.

C. When the initial velocity of the charged particle is perpendicular to electric field, then path of the charged particle is parabolic.

D. When the initial velocity makes an arbitrary angle with the field direction, then its path is helical.

9. *(a)*

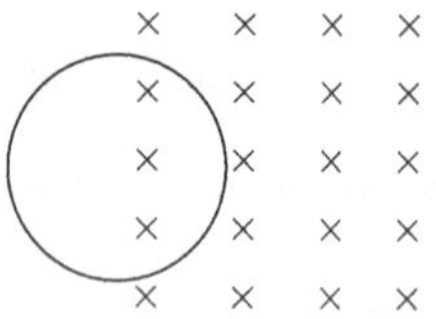

Magnetic flux, $\phi = BA\cos\theta = \mathbf{B}\cdot\mathbf{A}$ $\quad [\because \theta = 0°]$

$\therefore \quad \phi \propto A$

Hence, when the circular loop is entering into uniform magnetic field, then its area increases, so flux through the circular loop increases.

10. *(d)* Electric field vector, $\mathbf{E} = E_0 \sin(kx - \omega t)\hat{\mathbf{j}}$

$$\therefore \quad B_0 = \frac{E_0}{c} \quad \ldots(\text{i})$$

The maximum magnetic force on charge q moving with velocity u along Y-axis is

$$F_{\max} = quB_0 = \frac{qu \cdot E_0}{c} \quad [\text{using Eq. (i)}]$$

11. *(c)* 1 MSD = (5.20 − 5.15) cm = 0.05 cm

$$1 \text{ VSD} = \frac{1.44}{30}\text{ cm} = 0.048 \text{ cm}$$

$\therefore$ Least count (LC) = 1 MSD − 1 VSD

$$= 0.05 - 0.048 = 0.002 \text{ cm}$$

$\therefore$Diameter = MSR + (LC × VSD)

$$= 5.15 + 0.002 \times 16$$
$$= 5.182 \text{ cm}$$

12. *(d)* Here, Earth is considered as a solid sphere of mass M_e. The intensity of gravitational field is equal to $\frac{GM_e}{R_e^3} r$

$$\therefore \quad g_{\text{inside}} = \frac{GM_e r}{R_e^3} \Rightarrow g_{\text{inside}} \propto r$$

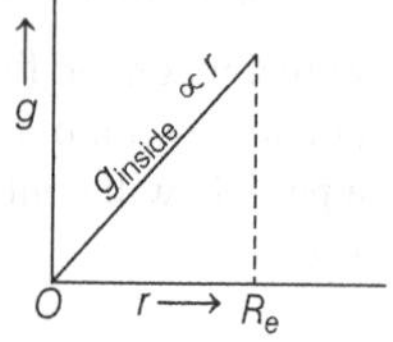

13. *(c)* 1 Poise = 1 g/cm-s

$$= 10^{-3} \text{ kg}/10^{-2} \text{ m-s}$$
$$= 10^{-1} \text{ kg/m-s}$$
$$= 0.1\alpha\beta^{-1}\gamma^{-1}$$

14. *(b)* The ball will appear to execute to and fro motion along a horizontal line with the point of rotation as mid point. Thus, the shadow will execute SHM on the screen.

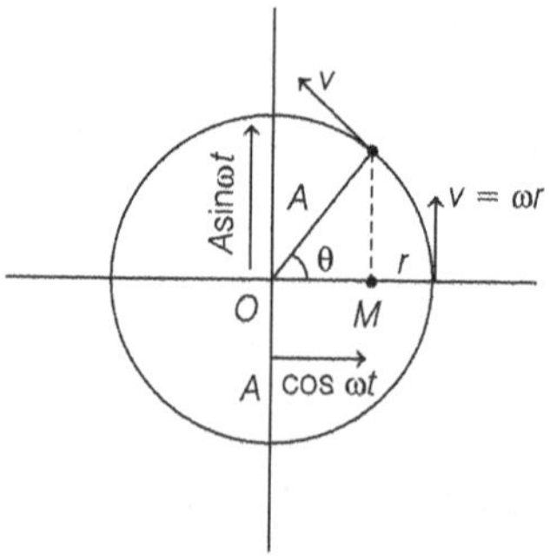

15. *(b)* Energy of light falling per second on the surface of sphere ,

$$E = \frac{66}{100} \times 100 = 66 \text{ W}$$

Momentum of the light falling per second on the surface of sphere $= \frac{E}{c}$

Momentum of the reflected light = 0, as the light is completely absorbed.

Force exerted by light, $F = \frac{E}{c} - 0 = \frac{E}{c}$

Pressure on surface, $p = \frac{F}{4\pi r^2} = \frac{E/c}{4\pi r^2}$

$$= \frac{66/(3 \times 10^8)}{4 \times (22/7) \times (0.10)^2}$$
$$= 1.75 \times 10^{-6} \text{ Pa}$$

16. *(b)* The arrangement behaves as a parallel combination of two capacitors each of capacitance, $C = \frac{\varepsilon_0 A}{d}$

Thus, equivalent capacity = $2C$

$\therefore$Total energy stored, $U = \frac{1}{2} \times (2C)V^2 = CV^2 = \frac{\varepsilon_0 A}{d} V^2$

17. *(d)* Excess pressure, $\frac{4S}{r_1} - \frac{4S}{r_2} = \frac{4S}{r}$

or $\quad \frac{1}{r} = \frac{1}{r_1} - \frac{1}{r_2} = \frac{1}{4} - \frac{1}{5} = \frac{1}{20}$ or $r = 20$ cm

18. *(c)* Magnetic dipole moment of square loop having current I, $m = Ia^2$, where a is length of its side.

If r be the radius of circular loop, then

$$4a = 2\pi r \Rightarrow r = \frac{2a}{\pi}$$

$$\therefore \quad m' = I \cdot \pi r^2 = I \cdot \pi \left(\frac{2a}{\pi}\right)^2$$

$$= I \cdot \frac{4a^2}{\pi} = \frac{4}{\pi}(Ia^2) = \frac{4}{\pi} m$$

$$\therefore \quad m' = \frac{4m}{\pi}$$

19. *(c)* For CO_2 gas, $n_1 = \frac{22}{44} = \frac{1}{2}$

Molar specific heat, $(C_V)_1 = 3R$

For O_2 gas, $n_2 = \frac{16}{32} = \frac{1}{2}$

Molar specific heat $(C_V)_2 = \frac{5R}{2}$

Assume that, t is the temperature of mixture.

Now, Heat lost by O_2 = Heat gain by CO_2

$$n_2 (C_V)_2 \Delta T_2 = n_1 (C_V)_1 \Delta T_1$$

$$\frac{1}{2} \times \frac{5}{2} R(310 - T) = \frac{1}{2} \times 3R(T - 300)$$

$$T = 304.54 \text{ K}$$

$$T = 31.5^\circ\text{C}$$

20. *(a)* Here, $u = 0$, $v = 640 \text{ ms}^{-1}$, $s = 1.2 \text{ m}$

As, we know that

$$v^2 = u^2 + 2as$$

$$\Rightarrow \quad a = \frac{v^2 - u^2}{2s} = \frac{(640)^2}{2 \times 1.2} = \frac{(640)^2}{2.4} \text{ m/s}^2 \quad \ldots\text{(i)}$$

Now, $t = \frac{v - u}{a} = \frac{640 - 0}{\frac{(640)^2}{2.4}} = \frac{2.4}{640}$ [using Eq. (i)]

$$t = 3.75 \times 10^{-3} \text{ s} \approx 4 \text{ ms}$$

21. (12) A fully charged capacitor draws no current. Therefore, no current flows in arm GHF. So, the R of arm HF is ineffective. The total resistance of other resistors in circuit is

$$R' = \frac{(R + R) \times R}{(R + R) + R} + R$$

$$= \frac{6 \times 3}{6 + 3} + 3 = 5\Omega$$

Total current, $I = \frac{E}{R'} = \frac{15}{5} = 3 \text{ A}$

Since in parallel circuit, the current divides in the inverse ratio of resistance, so current in arm $ABGD = 1$ A and current in arm $AD = 2$ A.

So, potential difference between G and D

$$V_{GD} = V_G - V_D = 1 \times 3 = 3 \text{ V}$$

Potential difference between D and F

$$V_{DF} = V_D - V_F = 3 \times 3 = 9 \text{ V}$$

$$\therefore \quad V_G - V_F = (V_G - V_D) + (V_D - V_F)$$

$$= 3 + 9 = 12 \text{ V}$$

Potential difference across capacitor

$$= V_G - V_F = 12 \text{ V}$$

22. (8) As the diode is forward biased, so it will behave as simple wire. Hence, the modified circuit becomes.

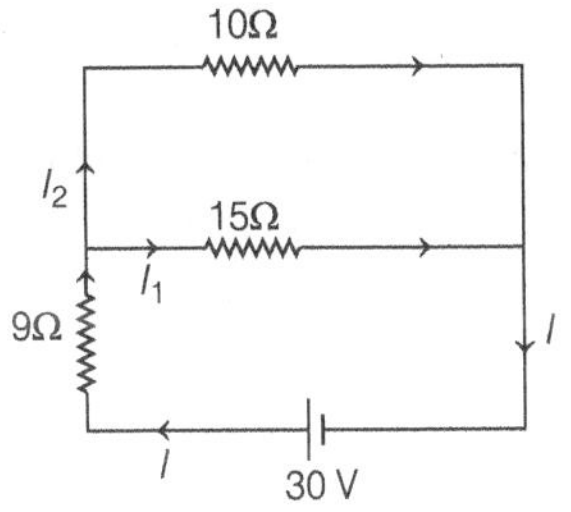

$$\therefore \quad R_{\text{net}} = 9 + \frac{10 \times 15}{10 + 15} = 15\Omega$$

$$\therefore \quad I = \frac{V}{R_{\text{net}}} = \frac{30}{15} \text{ A} = 2\text{A}$$

$$\therefore \quad I_1 = \left(\frac{10}{10 + 15}\right) \times I$$

$$= \frac{10}{25} \times 2 = 0.8 = 8 \times 10^{-1} \text{ A} = x \times 10^{-1} \text{ A}$$

So, $\quad x = 8$

23. (7)

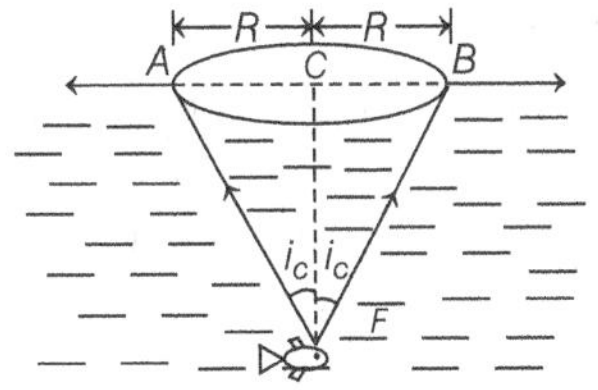

$$FC = 12 \text{ cm}, \ \mu_w = \frac{4}{3},$$

$$\sin i_c = \frac{1}{\mu_w} = \frac{3}{4}$$

$$\therefore \quad \cos i_c = \frac{\sqrt{7}}{4}$$

$$\therefore \quad \tan i_c = \frac{3}{\sqrt{7}} \quad \ldots\text{(i)}$$

From figure,

$$\tan i_c = \frac{CB}{CF} = \frac{R}{12} \quad \ldots\text{(ii)}$$

From Eqs. (i) and (ii),

$$\frac{R}{12} = \frac{3}{\sqrt{7}} = \frac{36}{\sqrt{7}} = \frac{36}{\sqrt{x}}$$

So, $\quad x = 7$

24. (15) We know that, internal energy of gas with f degree of freedom,

$$U = \frac{nfRT}{2}$$

where, n is number of moles.

Internal energy of O_2 gas,

$$U_1 = \frac{n_1 f_1 RT}{2}$$

$$= 3 \times \frac{5}{2} RT \qquad [\because n_1 = 3 \text{ and } f_1 = 5]$$

Internal energy due to Ar gas, which is monoatomic gas,

$$U_2 = \frac{n_2 f_2 RT}{2}$$

$$= 5 \times \frac{3}{2} RT \qquad [\because n_2 = 5, f_2 = 3]$$

$\therefore$ Total internal energy,

$$U = U_1 + U_2$$

$$= 3 \times \frac{5}{2} RT + 5 \times \frac{3}{2} RT = 15RT$$

$$= xRT \qquad \text{(given)}$$

$$\therefore \quad x = 15$$

25. (2) Given, $l = 7\,\text{cm} = 7 \times 10^{-2}$ m

$$F = 300\,\text{kgf} = 300 \times 10\,\text{N}$$

$$\eta = 2 \times 10^7 \text{dyne cm}^{-2}$$

$$= 2 \times 10^6 \text{ Nm}^{-2}$$

As, $$\eta = \frac{F/A}{\theta}$$

or $$\theta = \frac{F}{A\eta} = \frac{F}{l^2\eta}$$

$$= \frac{300 \times 10}{(7 \times 10^{-2})^2 \times 2 \times 10^6} = 0.3 \text{ rad}$$

$$\theta = \frac{\Delta l}{l}$$

or $$\Delta l = l\theta = 7 \times 0.3 = 2.1\text{cm} \simeq 2\text{ cm}$$

Chemistry

26. (*c*) Partition chromatography is the process of separation whereby the component of the mixture get distributed into two phases that may be liquid-liquid, liquid-gas but not solid-gas. Thus, it is not possible that the stationary phase is a finely divided solid adsorbent in partition chromatography.

27. (*c*) The structures of XeF_2, XeF_4 and XeF_6 is given as

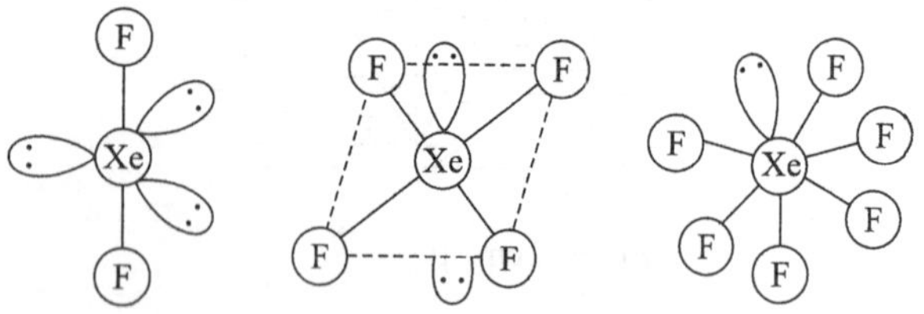

In $XeF_2 \rightarrow 2bp + 3lp$,

In $XeF_4 \rightarrow 4bp + 2lp$

In $XeF_6 \rightarrow 6bp + 1lp$

28. (*d*)

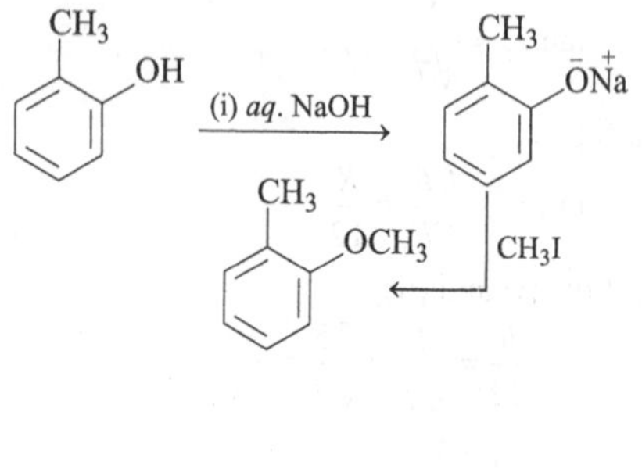

29. (*c*) **Structure of NF_3**

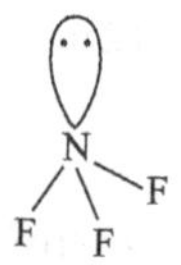

$$H = \frac{5+3}{2} = 4$$

Hybridisation $= sp^3$, according to VSEPR, shape of NF_3 is trigonal pyramidal.

Structure of H_3O^+

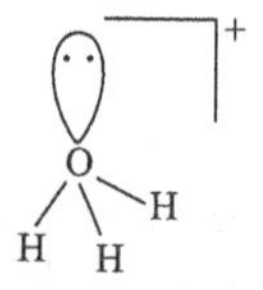

$$H = \frac{6+3-1}{2} = 4$$

Hybridisation $= sp^3$, shape = trigonal pyramidal

Structure of NO_3^-

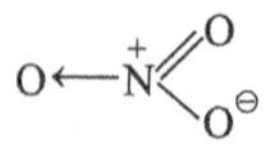

$$H = \frac{5+1}{2} = 3$$

Hybridisation $= sp^2$, shape = trigonal planar

Structure of BF_3

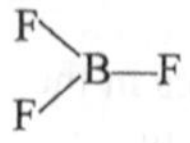

$$H = \frac{3+3}{2} = 3$$

Hybridisation $= sp^2$ shape = trigonal planar.

Structure of N_3H

The hybridisation of central N-atom is *sp*, but due to irregular distribution of electron cloud its structure in slightly bent.

Hence, option (c) is correct regarding isostructural pairs.

Note In case of NO_3^-, π-bond is not considered in structural determination of molecule/ion.

30. (*c*) Glucose reacts with methanol, in presence of dry HCl to form *alpha* and *beta* methyl glucosides.

$$C_6H_{12}O_6 + CH_3OH \xrightarrow{HCl} (\alpha)C_6H_{11}O_6CH_3 + (\beta)C_6H_{11}O_6CH_3$$

The formation of *alpha* and *beta* methyl glucosides is due to the ring structure of glucose, as the methyl alcohol can attack the anomeric carbon in two ways, leading to form these two configuration.

31. (*d*) $\Lambda_m^\infty(BaSO_4) = 2\Lambda_{eq}^\infty(BaSO_4)$

$$\Lambda_{eq}^\infty(BaSO_4) = \Lambda_{eq}^\infty(Ba^{2+}) + \Lambda_{eq}^\infty(SO_4^{2-})$$

$$= \Lambda_{eq}^\infty(BaCl_2) + \Lambda_{eq}^\infty(H_2SO_4) - 2\Lambda_{eq}^\infty(HCl)$$

$$\Lambda_{eq}^\infty(BaSO_4) = y_1 + y_2 - 2y_3$$

$$\therefore \quad \Lambda_m^\infty = 2(y_1 + y_2 - 2y_3)$$

For sparingly soluble salt, $\Lambda_m^\infty = \dfrac{\kappa}{M} \times 1000$

$$\text{or } M = \frac{y}{2(y_1 + y_2 - 2y_3)} \times 1000 = \frac{500\,y}{y_1 + y_2 - 2y_3}$$

$$K_{sp} = M^2 = \frac{2.5 \times 10^5\, y^2}{(y_1 + y_2 - 2y_3)^2}$$

Study Tactics

To solve this problem, firstly write the dissociation equation for $BaSO_4$. After that, calculate the equivalent conductivity for each ion at infinite dilution. Since $BaSO_4$ is sparingly soluble, we can assume that the concentration of Ba^{2+} and SO_4^{2-} is approximately equal.

Now you have to calculate expression for K_{sp} in terms of given equivalent conductivities (y_1, y_2, y_3).

32. (*b*) A. 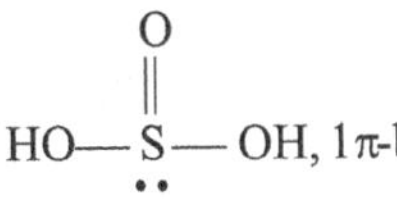 1π-bond

B. 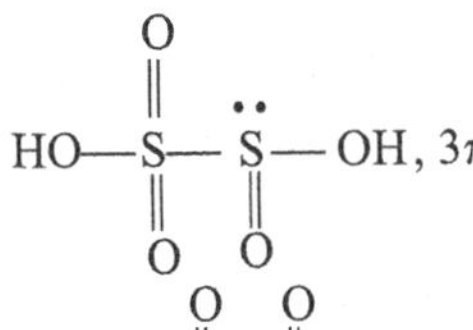 3π-bond

C. 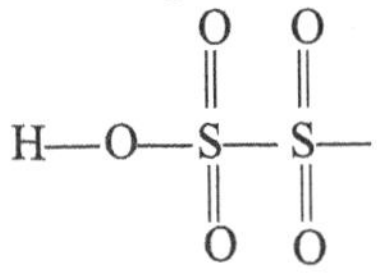 4π-bond

Hence, the correct order is A < B < C

33. (*b*) The correct order is A-II, B-IV, C-I, D-III.

The structure of uracil, thymine cytosine and adenine are

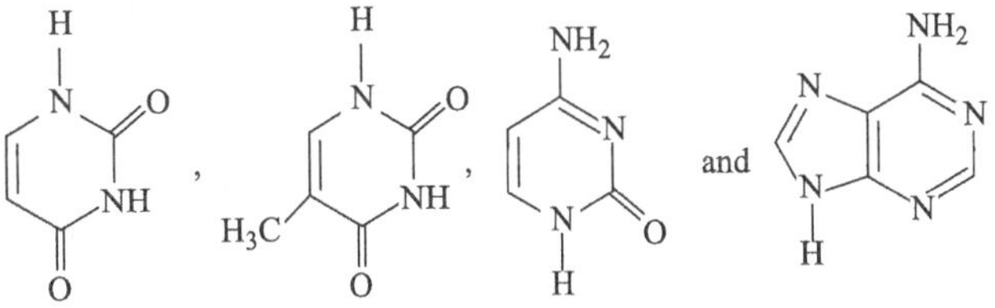

34. (*c*) The hybridisation of iodine in SF_4 is sp^3d and its structure is

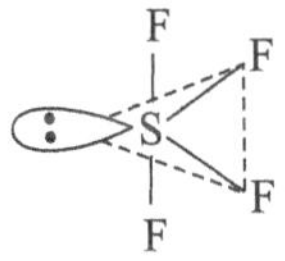

The shape of SF_4 is *see-saw*.

35. (*a*) Paramagnetism is due to the presence of unpaired electrons of metal.

A. $[Fe(CN)_6]^{3-}$, $[Co(NH_3)_6]^{3+}$
$Fe^{3+} = 3d^5$ $\quad Co^{3+} = 3d^6$

B. $[Fe(H_2O)_6]^{2+}$, $[Zn(H_2O)_6]^{2+}$
$Fe^{2+} = 3d^6$ $\quad Zn^{2+} = 3d^{10}$

C. $K_3[Cu(CN)_4]$, $K_3[Cr(oxalate)_3]$
$Cu^+ = 3d^{10}$ $\quad Cr^{3+} = 3d^3$

D. $[Fe(CN)_6]^{3-}$, $[Cr(H_2O)_6]^{3+}$
$Fe^{3+} = 3d^5$ $\quad Cr^{3+} = 3d^3$

E. $K_3[Cu(CN)_4]$, $[CuF_6]^{3-}$
$Cu^+ = 3d^{10}$ $\quad Cu^{3+} = 3d^8$

Hence, the correct option is (a).

36. (*c*) **Hofmann elimination** The reaction of amine with CH_3I produces quaternary ammonium salt which on further treatment with AgOH undergo Hofmann elimination to produce less substituted product in major amount.

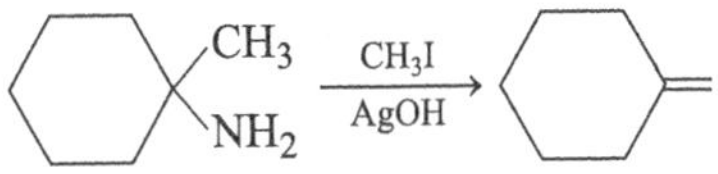

37. (*b*)

Process	Change in magnetic nature	Bond order change
$C_2 \rightarrow C_2^+$	Diamagnetic-Paramagnetic	2-1·5
$O_2 \rightarrow O_2^+$	Paramagnetic-Paramagnetic	2-2·5
$NO \rightarrow NO^+$	Paramagnetic-Diamagnetic	2·5-3
$N_2 \rightarrow N_2^+$	Diamagnetic-Paramagnetic	3-2·5

Hence, in $NO \rightarrow NO^+$ the bond order increased and the magnetic behaviour also changed.

38. *(b)*

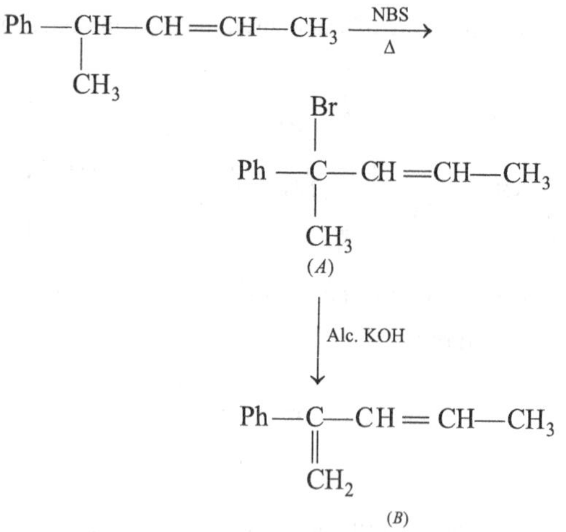

In this reaction, compound (A) is optically active and compound (B) shows geometrical isomerism.

However, both (A) and (B) are stereoisomers.

39. *(c)* In $[FeCl_4]^-$: Cl^- is a weak ligand. So, it will not cause pairing.

$$Fe^{3+} = [Ar]3d^5$$

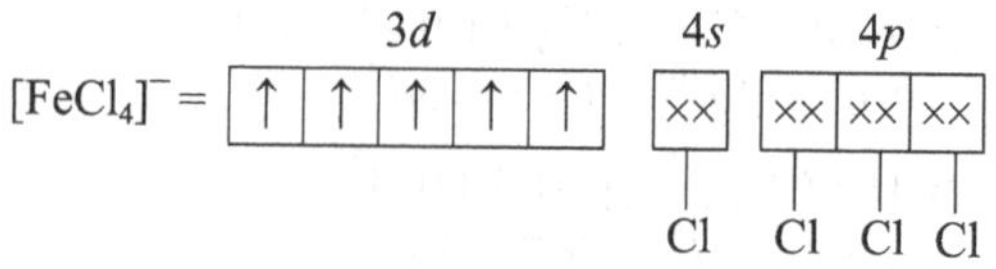

Hybridisation : sp^3

No. of unpaired electrons $(n) = 5$

Spin only magnetic moment $= \sqrt{n(n+2)}$

$$= \sqrt{5(5+2)} = \sqrt{35} \text{ BM}$$

$[Co(en)(NH_3)_2Cl_2]^+$

$$Co^{3+} = [Ar]3d^6$$

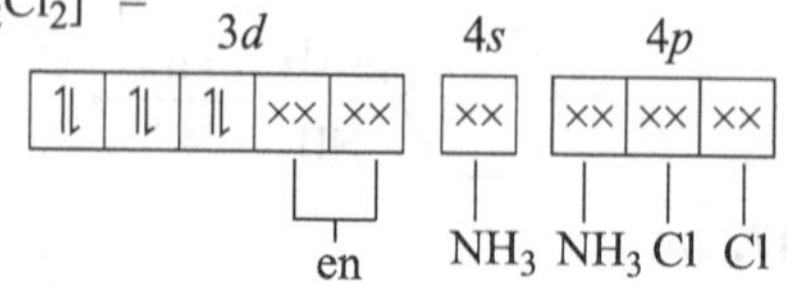

Hybridisation : d^2sp^3

Number of unpaired electron $(n) = 0$

Spin-only magnetic moment $(\mu) = \sqrt{n(n+2)} = 0$

Hence, $[FeCl_4]^-$ has higher spin only magnetic moment than $[Co(en)(NH_3)_2Cl_2]^+$ and Co^{3+} in $[Co(en)(NH_3)_2Cl_2]^+$ has d^2sp^3-hybridisation.

40. *(c)* With increase in the mole fraction,

- The free energy change of mixing first decreases, reaches a minimum value and then increases.
- The value of term $T\Delta S_{mix}$ first increases, reaches a maximum value and then decreases.

Hence, the option (c) correctly represents the changes in the thermodynamic properties during the formation of 1 mole of an ideal binary solution.

41. *(b)* Using the formula

$$\text{Normality} = \text{Molarity} \times n\text{-factor}$$

$$1.5 = \frac{\text{Mass} \times n\text{-Factor}}{\text{Molar mass of } H_2O_2 \times \text{volume (L)}}$$

$$1.5 = \frac{w}{34} \times \frac{2}{1}$$

$$w = 17 \times 1.5$$

$$= 25.5 \text{ g}$$

42. *(c)* (A) is correct but (R) is incorrect. The electron gain enthalpy of S or Cl is more negative than that of preceding element because the added electron goes to $n = 3$ and occupies a larger region of space and electron-electron repulsion is much less.

43. *(c)* The given reaction is represented as

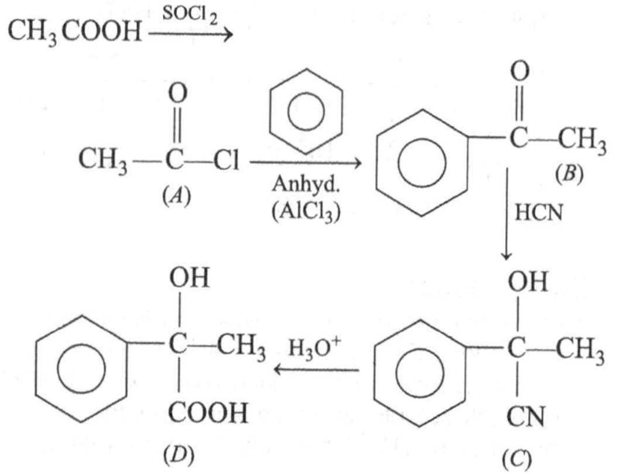

44. *(b)* Among the given statements, B, C and D are correct while the statement A is incorrect.

The correct form is : The time taken for the completion of 75% reaction is twice of the $t_{1/2}$ of the first order reaction.

$$t_{75\%} = 2 \times t_{1/2}$$

45. *(c)* The correct order is A-III, B-IV, C-I, D-II.

- Nessler's reagent is used to detect NH_3 or NH_4^+.
- Cu(II) oxide is used to detect carbon and hydrogen.
- Silver nitrate is used to detect specifically halogens.
- Dimethyl glyoxime is used to detect nickel ion (Ni^{2+}).

46. (*8*) $n = \frac{w}{M} = \frac{V}{22400}$ (at STP)

$$\frac{w}{16} = \frac{112}{22400}$$

$$w = 0.08\,g$$

$$= 8 \times 10^{-2}\ g$$

47. (*27*) $E = \frac{hc}{\lambda} = 2.9 \times 10^{-19}$ J

Total energy of 10 quanta $= 29 \times 10^{-19}$ J

$$\text{Energy stored} = 112\,\text{k cal} = \frac{112 \times 4.18 \times 10^3}{6 \times 10^{23}}$$

$$= 7.80 \times 10^{-19}\,\text{J}$$

$$\%\ \text{efficiency} = \frac{7.8 \times 10^{-19}}{29 \times 10^{-19}} \times 100$$

$$= 26.9$$

$$\approx 27\%$$

48. (*5*) Anode $H_2 \longrightarrow 2H^+ + 2e^-$

Cathode $Cu^{2+} + 2e^- \longrightarrow Cu$

$$Cu^{2+} + H_2 \longrightarrow 2H^+ + Cu$$

$$E_{cell} = E^\circ_{cell} - \frac{0.06}{2} \log \frac{[H^+]^2}{[Cu^{2+}]}$$

$$0.576 = 0.34 - \frac{0.06}{2} \log \frac{[H^+]^2}{[0.01]}$$

$$7.87 = -\log [H^+]^2 + \log 10^{-2}$$

$$\Rightarrow pH = 4.93 \approx 5$$

49. (*617*) 10% *w/W* means

w_B = Mass of glucose = 10 g

Mass of solution = 100 g

W_A = Mass of water (solvent)

= 100 − 10 = 90 g = 0.09 kg

M_B = Molar mass of glucose ($C_6H_{12}O_6$)

$= (12 \times 6) + (1 \times 12) + (16 \times 6)$

$= 72 + 12 + 96$

$= 180\,g\ mol^{-1}$

$$\text{Molality of solution } (m) = \frac{w_B / M_B}{W_A\ (\text{in kg})}$$

$$= \frac{(10\,g)/(180\,g\ mol^{-1})}{(0.09\,kg)}$$

$$= \frac{(10\,g)}{(180\,g\ mol^{-1}) \times (0.09\,kg)}$$

$$= 0.617\ mol\ kg^{-1} \text{ or } 0.617\ m$$

$$= 617 \times 10^{-3}\ m$$

50. (*5*) Using

$$K_p = K_C (RT)^{\Delta n}$$

$$p_{total} = p_{HNO_3} + p_{NO_2} + p_{H_2O} + p_{O_2}$$

$$\frac{p_{NO_2}}{p_{O_2}} = 4$$

$$\Rightarrow p_{NO_2} = 4 p_{O_2}$$

and $\frac{p_{H_2O}}{p_{O_2}} = 2$

$$\Rightarrow p_{H_2O} = 2 p_{O_2}$$

$$p_{total} = p_{HNO_3} + 7 p_{O_2}$$

$$15 = 1 + 7 p_{O_2}$$

$$p_{O_2} = \frac{14}{7} = 2$$

$$K_p = \frac{[p_{NO_2}]^4 \cdot [p_{H_2O}]^2 \cdot p_{O_2}}{p^4_{HNO_3}}$$

$$= \frac{(4 \times 2)^4 \cdot (2 \times 2)^2 \times 2}{1^4} = 2^{17}$$

Now, $\Delta n_g = 7 - 4 = 3$

$$K_p = K_C (RT)^{\Delta n}$$

$$= K_C (0.08 \times 800)^3$$

$$K_C = \frac{2^{17}}{2^{18}} = 2^{-1} = \frac{1}{2} = 0.5$$

$$\approx 5 \times 10^{-1}.$$

Mathematics

51. *(a)* Equation of straight line at (x, y) is

$$Y - y = \frac{dy}{dx}(X - x)$$

Intercept of the straight line on Y-axis is $y - x\frac{dy}{dx}$.

According to the question,

$$y - x\frac{dy}{dx} = 2xy^2 \Rightarrow -\frac{1}{y^2}\frac{dy}{dx} + \frac{1}{xy} = 2$$

On putting $\frac{1}{y} = t \Rightarrow \frac{-1}{y^2}\frac{dy}{dx} = \frac{dt}{dx}$

$$\therefore \quad \frac{dt}{dx} + \frac{1}{x}t = 2$$

which is linear differential equation in t and x.

Now, $\quad \text{IF} = e^{\int \frac{1}{x}dx} = e^{\log x} = x$

$$tx = \int 2x\,dx + C$$

$$\Rightarrow \quad tx = x^2 + C$$

$$\Rightarrow \quad \frac{x}{y} = x^2 + C \qquad \left[\because t = \frac{1}{y}\right]$$

Given at $x = 2$, $y = 1$

$$\Rightarrow \quad \frac{2}{1} = 4 + C \Rightarrow C = -2$$

$\therefore$ Equation of curve is

$$x = x^2y - 2y$$

or $\quad x^2y - x = 2y$

52. *(c)* $\displaystyle\int \frac{5\tan x}{\tan x - 2}dx = \int \frac{5\frac{\sin x}{\cos x}}{\frac{\sin x}{\cos x} - 2}dx$

$$= \int \frac{5\sin x}{\cos x} \times \frac{\cos x}{\sin x - 2\cos x}dx$$

$$= \int \frac{5\sin x}{\sin x - 2\cos x}dx$$

$$= \int \frac{4\sin x + \sin x + 2\cos x - 2\cos x}{\sin x - 2\cos x}dx$$

$$= \int \frac{(\sin x - 2\cos x) + (4\sin x + 2\cos x)}{\sin x - 2\cos x}dx$$

$$= \int dx + 2\int \frac{\cos x + 2\sin x}{\sin x - 2\cos x}dx$$

$$= x + 2\ln|\sin x - 2\cos x| + C$$

$$\left[\because \int \frac{f'(x)}{f(x)}dx = \ln|f(x)| + C\right]$$

But $\displaystyle\int \frac{5\tan x}{\tan x - 2}dx = x + A\ln|\sin x + B\cos x| + C$

$$\therefore \quad A = 2 \text{ and } B = -2$$

$$(A, B) = (2, -2)$$

Time Saver Tip

RHS $= x + A\ln|\sin x + B\cos x| + C$

On differentiating, we get

$$1 + \frac{A}{(\sin x + B\cos x)} \cdot (\cos x - B\sin x)$$

Dividing both numerator and denominator by $\cos x$, we get

$$\frac{(1 - AB)\tan x + (A + B)}{\tan x + B}$$

Now, on comparing coefficients with LHS, we get

$$B = -2$$

$$A = 2$$

Hence, $(A, B) \equiv (2, -2)$

53. *(b)* Given, $f(x) = \frac{\log_e x}{x}$

We know that $\frac{1}{e} < x < 1$,

$$\therefore \quad \log_e x < 0 \Rightarrow \frac{\log_e x}{x} < 0$$

and $\quad 1 < x < e^2$, $\log_e x > 0$

$$\Rightarrow \quad \frac{\log_e x}{x} > 0$$

$$\therefore \quad \int_{1/e}^{e^2} |f(x)|dx = \int_{1/e}^{e^2}\left|\frac{\log_e x}{x}\right|dx$$

$$= \int_{1/e}^{1}\left(\frac{-\log_e x}{x}\right)dx + \int_1^{e^2}\frac{\log_e x}{x}dx$$

$$= -\frac{1}{2}[(\log_e x)^2]_{1/e}^1 + \frac{1}{2}[(\log_e x)^2]_1^{e^2}$$

$$= \frac{1}{2} + 2$$

$$= \frac{5}{2}$$

54. *(d)* Since, $f(x)$ is continuous at $x = \pi$.

$$\therefore \quad \lim_{x \to \pi} f(x) = f(\pi)$$

$$\Rightarrow \quad \lim_{h \to 0} f(\pi - h) = k$$

$$\Rightarrow \quad k = \lim_{h \to 0} \frac{1 + \cos(\pi - h)}{(\pi - \pi + h)^2} \cdot \frac{\sin^2(\pi - h)}{\log[1 + \pi^2 - 2\pi(\pi - h) + (\pi - h)^2]}$$

$$= \lim_{h \to 0} \frac{1-\cos h}{h^2} \times \frac{\sin^2 h}{\log(1+h^2)}$$

$$= \lim_{h \to 0} \frac{1}{2}\left(\frac{\sin \frac{h}{2}}{\frac{h}{2}}\right)^2 \cdot \frac{h^2}{\log(1+h^2)} \cdot \left(\frac{\sin h}{h}\right)^2$$

$$= \frac{1}{2} \times 1 \times 1 = \frac{1}{2}$$

$$\Rightarrow \quad k = \frac{1}{2} = 0.5$$

55. *(d)* Given, parabolas $x^2 = 4y$ and $y^2 = 4x$

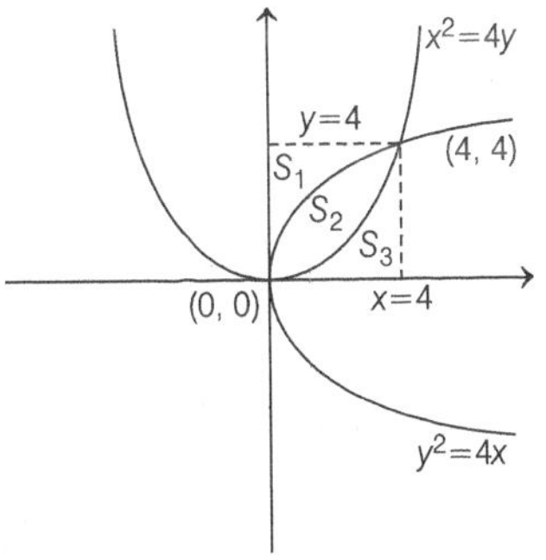

Point of intersection are (0, 0) and (4, 4).
By symmetry,

$$S_1 = S_3 = \int_0^4 y\,dx$$

$$= \int_0^4 \frac{x^2}{4} dx$$

$$= \frac{1}{4} \cdot \frac{x^3}{3}\Bigg|_0^4 = \frac{16}{3}$$

and $$S_2 = \int_0^4 \left(2\sqrt{x} - \frac{x^2}{4}\right) dx$$

$$= \left(2 \cdot \frac{x^{3/2}}{3/2} - \frac{x^3}{12}\right)_0^4$$

$$= \frac{4}{3} \times 8 - \frac{16}{3} = \frac{16}{3}$$

$$\therefore \; S_1 : S_2 : S_3 = 1:1:1$$

56. *(b)* Given, $f(x) = \sqrt{\dfrac{x - |x|}{x - [x]}}$

$\because \quad x - [x] = \{x\} \in [0, 1)$

$x - [x] = 0$, when $x \in Z$

$\Rightarrow \quad x \notin Z$

Also, $x - |x| = \begin{cases} 0, & \text{if } x \geq 0 \\ 2x, & \text{if } x < 0 \end{cases}$

$\Rightarrow \quad x > 0$

$\therefore A = R^+ - N$

Let $y = g(x) = \dfrac{2x}{4+x^2} \Rightarrow yx^2 - 2x + 4y = 0$

$\because \quad x \in R$

$\Rightarrow$ Discriminant ≥ 0

$\Rightarrow \quad 4 - 16y^2 \geq 0$

$\Rightarrow \quad y^2 - \dfrac{1}{4} \leq 0$

$\Rightarrow \quad \dfrac{-1}{2} \leq y \leq \dfrac{1}{2}$

$\Rightarrow \quad B = \left[\dfrac{-1}{2}, \dfrac{1}{2}\right]$

$\therefore \quad A \cap B = \left(0, \dfrac{1}{2}\right]$

Study Tactics

Use property of modulus function and greatest integer function for the value of A. Then, use discriminant > 0 (as $x \in R$) for the value of B. Then, final answer will be $A \cap B$.

57. *(b)*

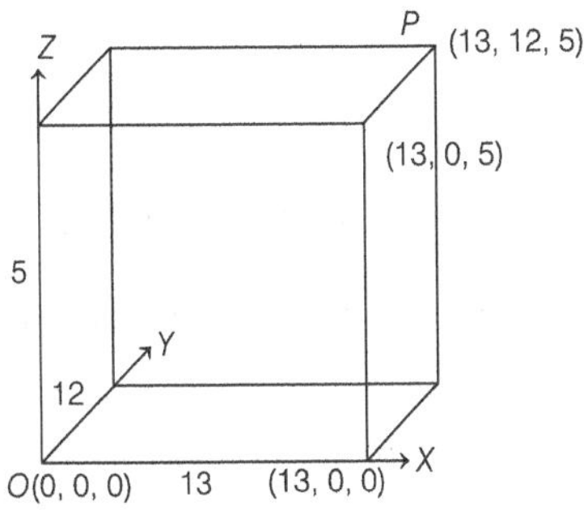

Equation of OP is

$$\frac{x-0}{13-0} = \frac{y-0}{12-0} = \frac{z-0}{5-0}$$

$$\Rightarrow \quad \frac{x}{13} = \frac{y}{12} = \frac{z}{5} \quad \ldots\text{(i)}$$

Equation of edge parallel to Z-axis is

$$\frac{x-13}{0} = \frac{y-0}{0} = \frac{z-5}{1} \quad \ldots\text{(ii)}$$

$\therefore$ Shortest distance

$$= \frac{\begin{vmatrix} x_2 - x_1 & y_2 - y_1 & z_2 - z_1 \\ a_1 & b_1 & c_1 \\ a_2 & b_2 & c_2 \end{vmatrix}}{\sqrt{(a_1b_2 - a_2b_1)^2 + (b_1c_2 - b_2c_1)^2 + (c_1a_2 - c_2a_1)^2}}$$

Here $x_1 = 0$, $y_1 = 0$ and $z_1 = 0$

$$x_2 = 13, y_2 = 0 \text{ and } z_2 = 5$$
$$a_1 = 13, b_1 = 12, c_1 = 5$$
$$a_2 = 0, b_2 = 0, c_2 = 1$$

$$\therefore \begin{vmatrix} x_2 - x_1 & y_2 - y_1 & z_2 - z_1 \\ a_1 & b_1 & c_1 \\ a_2 & b_2 & c_2 \end{vmatrix} = \begin{vmatrix} 13 & 0 & 5 \\ 13 & 12 & 5 \\ 0 & 0 & 1 \end{vmatrix} = 13(12) = 156$$

and
$$\sqrt{(a_1b_2 - a_2b_1)^2 + (b_1c_2 - b_2c_1)^2 + (c_1a_2 - c_2a_1)^2}$$
$$= \sqrt{0 + (12)^2 + (0-13)^2} = \sqrt{313}$$

$\therefore$ Shortest distance $= \dfrac{156}{\sqrt{313}}$ unit

58. *(d)* Given, $\bar{x}_1 = 8, \sigma_1^2 = 18, n_1 = 5$

and $\bar{x}_2 = 8, \sigma_2^2 = 24, n_2 = 3$

Combined mean, $\bar{x} = \dfrac{n_1\bar{x}_1 + n_2\bar{x}_2}{n_1 + n_2} = \dfrac{5\times 8 + 3\times 8}{5+3} = 8$

Deviation, $d_1 = \bar{x}_1 - \bar{x} = 8 - 8 = 0$

and $d_2 = \bar{x}_2 - \bar{x} = 8 - 8 = 0$

Combined variance,
$$\sigma^2 = \frac{1}{n_1 + n_2}[n_1(\sigma_1^2 + d_1^2) + n_2(\sigma_2^2 + d_2^2)]$$
$$= \frac{1}{8}[5\times 18 + 3\times 24] = 20.25$$

59. *(a)* Given that a, b and c are roots of equation $x^3 - 3x^2 + 2x - 1 = 0.$

Let $x^3 - 3x^2 + 2x - 1 = (x-a)(x-b)(x-c)$ ...(i)

Now, $A^2 - B^2 = 4I - 4B$

$\therefore$ $A^2 = B^2 - 4B + 4I$

$\Rightarrow$ $A^2 = (B - 2I)^2 \Rightarrow A^2 - (B-2I)^2 = 0$

$(A + B - 2I)(A - B + 2I) = 0$ [as $AB = BA$]

Now, $A + B - 2I \neq 0$

$\therefore$ $A - B + 2I = 0$

[as otherwise $(A - B + 2I)^{-1}$ exists and $A + B - 2I = 0$]

$$\therefore \quad B = A + 2I = \begin{bmatrix} a+2 & 2 & 1 \\ 0 & b+2 & 0 \\ 0 & -3 & c+2 \end{bmatrix}$$

$\therefore |B| = (a+2)(b+2)(c+2)$ ...(ii)

On putting $x = -2$ in Eq. (i), we get
$$-8 - 12 - 4 - 1 = (-2-a)(-2-b)(-2-c)$$
$\therefore$ $25 = (a+2)(b+2)(c+2)$

$\therefore$ $|B| = 25$

60. *(a)* $\because$ $\sin(2\sin x) = \cos(2\cos x)$

$\Rightarrow$ $\sin(2\sin x) = \sin\left(\dfrac{\pi}{2} - 2\cos x\right)$

$\Rightarrow$ $2\sin x = \dfrac{\pi}{2} - 2\cos x$

$\Rightarrow$ $\sin x + \cos x = \dfrac{\pi}{4}$

On squaring both sides, we get
$$1 + \sin 2x = \frac{\pi^2}{16}$$

$\Rightarrow$ $\sin 2x = \dfrac{\pi^2 - 16}{16}$

$\therefore$ $\tan x + \cot x = \dfrac{\sin x}{\cos x} + \dfrac{\cos x}{\sin x}$
$$= \frac{1}{\sin x \cos x} = \frac{2}{\sin 2x}$$

$\Rightarrow$ $\dfrac{32}{\pi^2 - 16} = \dfrac{a}{\pi^b - c}$ [given]

$\therefore$ $a = 32, b = 2$ and $c = 16$
$$\frac{a+b+c}{c} = \frac{32+2+16}{16} = \frac{50}{16} = \frac{25}{8}$$

61. *(c)* We have,
$$L_1 : \frac{x-1}{1} = \frac{y-2}{1} = \frac{z-4}{0} = r, Q(r+1, r+2, 4)$$
$$L_2 : \frac{x-2}{1} = \frac{y-3}{-1} = \frac{z+4}{0} = k, R(k+2, -k+3, -4)$$

$\therefore$ **PQ** $= (\lambda - r - 1)\hat{\mathbf{i}} + (\lambda - r - 2)\hat{\mathbf{j}} + (\lambda - 4)\hat{\mathbf{k}}$

$\therefore$ **PQ** is perpendicular to line L_1.
$$\lambda - r - 1 + \lambda - r - 2 = 0$$
$$2\lambda - 2r - 3 = 0$$
$$2\lambda = 2r + 3$$
$$\Rightarrow \quad r = \frac{2\lambda - 3}{2}$$

PR $= (\lambda - k - 2)\hat{\mathbf{i}} + (\lambda + k - 3)\hat{\mathbf{j}} + (\lambda + 4)\hat{\mathbf{k}}$

PR is perpendicular to the line L_2.
$$(\lambda - k - 2) - (\lambda + k - 3) = 0$$
$$\lambda - k - 2 - \lambda - k + 3 = 0$$
$$-2k + 1 = 0$$
$$2k = 1$$
$$k = \frac{1}{2}$$

Now, **PQ** $\perp$ **PR**

$\Rightarrow (\lambda - r - 1)(\lambda - k - 2) + (\lambda - r - 2)(\lambda + k - 3) + (\lambda - 4)(\lambda + 4) = 0$

$\Rightarrow (\lambda - r - 1)\dfrac{(2\lambda - 5)}{2} + (\lambda - r - 2)\dfrac{(2\lambda - 5)}{2} + \lambda^2 - 16 = 0$

$$\Rightarrow \left(\lambda - \frac{(2\lambda-3)}{2} - 1\right)\frac{(2\lambda-5)}{2} + \left(\lambda - \frac{(2\lambda-3)}{2} - 2\right)\frac{(2\lambda-5)}{2} + \lambda^2 - 16 = 0$$

$$\Rightarrow \frac{(2\lambda-2\lambda+3-2)}{2} \times \frac{(2\lambda-5)}{2} + \frac{(2\lambda-2\lambda+3-4)}{2} \times \frac{(2\lambda-5)}{2} + \lambda^2 - 16 = 0$$

$$\Rightarrow \frac{1}{2} \times \frac{2\lambda-5}{2} + \frac{(-1)}{2} \times \frac{(2\lambda-5)}{2} + \lambda^2 - 16 = 0$$

$$\Rightarrow \frac{2\lambda-5-2\lambda+5+2\lambda^2-32}{4} = 0$$

$$\Rightarrow 2\lambda^2 = 32$$

$$\Rightarrow \lambda^2 = 16$$

Hence, $\lambda = \pm 4$

62. *(c)* Let the equation of circle be

$$x^2 + y^2 + 2gx + 2fy + c = 0 \quad \ldots\text{(i)}$$

and the equation of hyperbola is

$$xy = 1 \quad \ldots\text{(ii)}$$

On putting $x = t$ and $y = \frac{1}{t}$ in Eq. (i), we get

$$t^2 + \frac{1}{t^2} + 2gt + \frac{2f}{t} + c = 0$$

$$t^4 + 2gt^3 + ct^2 + 2ft + 1 = 0,$$

which is fourth degree equation in t.

Let its roots be t_1, t_2, t_3 and t_4, then

$$t_1 t_2 t_3 t_4 = \frac{\text{Constant term}}{\text{Coefficient of } t^4} = \frac{1}{1}$$

$$\Rightarrow -20 t_1 t_2 t_3 t_4 = -20$$

$$\therefore \lambda = -20$$

63. *(d)* Centroid of $\Delta ABC = G\left(\frac{12}{3}, \frac{12}{3}, \frac{12}{3}\right) = G(4, 4, 4)$

Equation of line, $\frac{x}{1} = \frac{z-3}{-1} = \frac{y}{0} = K$

$\therefore D(K, 0, -K+3)$ be any point on given line

$\therefore GD = \sqrt{(K-4)^2 + 4^2 + (-K+3-4)^2}$

Let $P = GD^2 = (K-4)^2 + 4^2 + (-K-1)^2$

$$\frac{dP}{dK} = 2(K-4) + 2(-K-1) \times -1 = 2K - 8 + 2K + 2$$

$$\Rightarrow 4K - 6 = 0$$

$$K = \frac{3}{2}$$

For minimum value of GD,

$$GD = \sqrt{\left(\frac{3}{2} - 4\right)^2 + 4^2 + \left(-\frac{3}{2} - 1\right)^2} = \sqrt{\frac{57}{2}}$$

Hence, $\sqrt{\frac{2}{57}}\,\alpha = \sqrt{\frac{2}{57}} \times \sqrt{\frac{57}{2}} = 1$

64. *(b)* $\lim_{x \to 1^+} g[f(x)] = \lim_{x \to 0^-} g(x)$

$=$ does not exist

$\therefore$ Statement I is false.

Also, $\lim_{x \to 0^+} \frac{g[f(x)]}{[f(x)-2]^2} = \lim_{x \to 0^+} \frac{g(2-x)}{(2-x-2)^2}$

$$= \lim_{x \to 0^+} \frac{[2-x] - \cos(2-x-2)}{x^2}$$

$$= \lim_{x \to 0^+} \frac{1-\cos x}{x^2} = \frac{1}{2}$$

$\Rightarrow$ Statement II is true.

65. *(a)* Given, $f(x) = \begin{vmatrix} \cos x & 1 & 0 \\ 1 & 2\cos x & 1 \\ 0 & 1 & 2\cos x \end{vmatrix}$

$$= \cos x\,(4\cos^2 x - 1) - (2\cos x)$$

$$= 4\cos^3 x - \cos x - 2\cos x$$

$$f(x) = 4\cos^3 x - 3\cos x$$

$$f(-x) = 4\cos^3 x - 3\cos x = f(x)$$

[$\because$ $\cos x$ is even function]

Then, $2\int_0^{\frac{\pi}{2}} (4\cos^3 x - 3\cos x)\,dx$

$$= 2\int_0^{\frac{\pi}{2}} (\cos 3x)\,dx \quad [\because \cos 3x = 4\cos^3 x - 3\cos x]$$

$$= 2\left[\frac{\sin 3x}{3}\right]_0^{\frac{\pi}{2}}$$

$$= \frac{2}{3}\left[\sin\frac{3\pi}{2} - \sin 0\right] = \frac{2}{3}[(-1)] = -\frac{2}{3}$$

66. *(a)* Given, $P(\lambda) \propto \frac{1}{\lambda^4}$

$$\Rightarrow P(\lambda) = \frac{k}{\lambda^4} \quad [k \text{ is proportionality constant}]$$

Then, $\sum_{\lambda=1}^{2n} \frac{k}{\lambda^4} = 1$

Let P_1 be the probability of choosing odd number, then

$$P_1 = \sum_{\lambda=1}^{n} P(2\lambda-1) = k\sum_{\lambda=1}^{n} \frac{1}{(2\lambda-1)^4}$$

Also, $1 - P_1 = \sum_{\lambda=1}^{n} P(2\lambda) = k\sum_{\lambda=1}^{n} \frac{1}{(2\lambda)^4} < k\sum_{\lambda=1}^{n} \frac{1}{(2\lambda-1)^4}$

$$\Rightarrow 1 - P_1 < P_1 \Rightarrow P_1 > \frac{1}{2}$$

67. *(c)* Let m_0 denote the number of empty seats to the left of the first person, $m_i (1 \le i \le y-1)$ the number of empty seats between the ith and $(i+1)$th person and m_y is the number of empty seats to the right of the yth person.

Then, $m_0, m_y \ge 0$ and $m_1 \ge 1$ for $1 \le i \le y-1$

$m_0 + m_1 + \ldots + m_y = x - y$

On putting $m_i = n_i + 1$, where $n_i \ge 0$

$$m_0 + n_1 + n_2 + \ldots\ldots. n_{y-1} + m_y + (1+1+1+\ldots y \text{ times}) = x - y$$

$$\Rightarrow m_0 + n_1 + \ldots + n_{y-1} + m_y = x - y - (y-1)$$

$$\Rightarrow m_0 + n_1 + \ldots + n_{y-1} + m_y = x - 2y + 1$$

Permutation of y persons is $y!$ ways

The required number of ways is

$$^{x-y+1}C_y \cdot y! = \frac{(x-y+1)!}{y!(x-2y+1)!} \cdot y!$$

$$= \frac{(x-y+1)!}{(x-2y+1)!}$$

68. *(b)* The middle term in the expansion of $(1+\alpha x)^4$ is $T_3 = {}^4C_2(\alpha x)^2 = 6\alpha^2x^2$ and middle term in the expansion of $(1-\alpha x)^6$ is $T_4 = {}^6C_3(-\alpha x)^3 = -20\alpha^3x^3$

According to the question,

$$6\alpha^2 = -20\alpha^3$$

$$\Rightarrow \quad \alpha = -\frac{3}{10}, \text{ then } |[\alpha]| = \left|\left[\frac{-3}{10}\right]\right| = |-1| = 1$$

69. *(b)* Let $f(x) = \dfrac{ax+b}{c}$

$$f^{-1}(x) = \frac{cx-b}{a}$$

$$f^{-1}(x) = \frac{9x-7}{2}, g^{-1}(x) = \frac{7x-5}{9}$$

Given, $h^{-1}(x) = \dfrac{px+q}{r}$

$$\therefore \quad h(x) = \frac{rx-q}{p}$$

$$h(x) = f^{-1}(x) + g^{-1}(x) - 5x$$

$$\frac{rx-q}{p} = \frac{9x-7}{2} + \frac{7x-5}{9} - 5x$$

On comparing coefficient of x and constant term, then

$$\frac{r}{p} = \frac{9}{2} + \frac{7}{9} - 5 = \frac{5}{18} \text{ and } \frac{q}{p} = \frac{73}{18}$$

$$r = \frac{5p}{18}, q = \frac{73}{18}p$$

r is prime.

$$p = 18, q = 73 \text{ and } r = 5$$

$$\therefore \quad p+q+r = 18+73+5 = 96$$

70. *(d)* Let α and β be the roots of

$$x^2 + (2-\lambda)x + (10-\lambda) = 0 \quad \ldots\text{(i)}$$

$$\therefore \quad \alpha + \beta = \lambda - 2$$

$$\text{and} \quad \alpha\beta = 10 - \lambda \quad \ldots\text{(ii)}$$

Let $A = \alpha^3 + \beta^3$

$$= (\alpha+\beta)^3 - 3\alpha\beta(\alpha+\beta)$$

$$= (\lambda-2)^3 - 3(10-\lambda)(\lambda-2)$$

$$= \lambda^3 - 8 - 6\lambda^2 + 12\lambda + 3(\lambda^2 - 12\lambda + 20)$$

$$= \lambda^3 - 3\lambda^2 - 24\lambda + 52$$

Now, on differentiating w.r.t. λ, we get

$$\frac{dA}{d\lambda} = 3\lambda^2 - 6\lambda - 24$$

On putting $\dfrac{dA}{d\lambda} = 0 \Rightarrow \lambda^2 - 2\lambda - 8 = 0$

$$\Rightarrow \quad (\lambda+2)(\lambda-4) = 0$$

$$\Rightarrow \quad \lambda = -2, 4$$

Now, $\dfrac{d^2A}{d\lambda^2} = 6\lambda - 6$

For $\lambda = -2$, $6\lambda - 6 = 6(-2) - 6 = -18 < 0$ (maximum)

For $\lambda = 4$, $6\lambda - 6 = 6(4) - 6 = 18 > 0$ (minimum)

$\therefore$ Eq. (i) becomes, $x^2 - 2x + 6 = 0$

Thus, $(\alpha-\beta)^2 = (\alpha+\beta)^2 - 4\alpha\beta$

$$= (2)^2 - 4(6) = 4 - 24 = -20$$

$$\Rightarrow \quad \alpha - \beta = \pm 2\sqrt{5}i \Rightarrow |\alpha - \beta| = 2\sqrt{5}$$

Study Tactics

From given, we find $\alpha^3 + \beta^3$ which forms the cubic equation in variable λ. Then, on differentiating w.r.t. λ, we get $\dfrac{dA}{d\lambda}$. Now, put $\dfrac{dA}{d\lambda} = 0$ for maximum and minimum value of λ.

71. *(20)* $\dfrac{dV}{dt} = 70$ cc/min

$$\Rightarrow 4\pi r^2 \frac{dr}{dt} = 70 \Rightarrow \frac{dr}{dt} = \frac{70}{4\pi r^2}$$

$$\because \quad S = 4\pi r^2$$

$$\Rightarrow \quad \frac{dS}{dt} = 8\pi r \frac{dr}{dt} = \frac{8\pi r \times 70}{4\pi r^2}$$

$$\Rightarrow \quad \frac{dS}{dt} = \frac{2 \times 70}{r}$$

At $r = 7$,

$$\Rightarrow \quad \frac{dS}{dt} = 20 \text{ cm}^2/\text{min}$$

72. *(64)* Coordinates of L are $(a, 2a) = (8, 16)$

Equation of the chord through L with slope -1 is

$$x + y = 24 \quad \ldots(\text{i})$$

Any point on parabola is $(at^2, 2at)$

$= (8t^2, 16t)$ which lies on Eq. (i) if

$$8t^2 + 16t - 24 = 0 \Rightarrow t^2 + 2t - 3 = 0 \Rightarrow t = 1, -3$$

If $L(8, 16)$ be one end of chord at $t = 1$ and $M(72, -48)$ be the other end of the chord, then the length of the chord is

$$\sqrt{(72-8)^2 + (-48-16)^2} = \sqrt{4096 + 4096} = 64\sqrt{2}$$

$\Rightarrow \quad K\sqrt{2} = 64\sqrt{2}$

$\therefore \quad K = 64$

73. *(89)* Given, $S_n = \sum_{\theta=1}^{89} \frac{1}{1+\cot\theta} = \frac{1}{1+\cot 1^\circ} + \frac{1}{1+\cot 2^\circ} + \ldots + \frac{1}{1+\cot 89^\circ}$

$$= \left(\frac{1}{1+\cot 1^\circ} + \frac{1}{1+\cot 89^\circ}\right) + \left(\frac{1}{1+\cot 2^\circ} + \frac{1}{1+\cot 88^\circ}\right) + \ldots$$

$$= \left(\frac{1}{1+\cot 1^\circ} + \frac{1}{1+\tan 1^\circ}\right) + \left(\frac{1}{1+\cot 2^\circ} + \frac{1}{1+\tan 2^\circ}\right) + \ldots$$

$$= \left(\frac{1}{1+\cot 1^\circ} + \frac{\cot 1^\circ}{1+\cot 1^\circ}\right) + \left(\frac{1}{1+\cot 2^\circ} + \frac{\cot 2^\circ}{1+\cot 2^\circ}\right) + \ldots$$

$$= (1 + 1 + \ldots + 1)\,44 \text{ times} + \frac{1}{1+\cot 45^\circ}$$

$$= 44 + \frac{1}{2} = \frac{89}{2} \Rightarrow 2S_n = 89$$

74. *(3)* $\because \omega = e^{i2\pi/3}$

$\Rightarrow \omega^2 = e^{i4\pi/3}$

$\therefore 1 + \omega + \omega^2 = 0$

Now,

$$\frac{|x|^2 + |y|^2 + |z|^2}{|p|^2 + |q|^2 + |r|^2} = \frac{x\bar{x} + y\bar{y} + z\bar{z}}{|p|^2 + |q|^2 + |r|^2}$$

$$= \frac{(p+q+r)(\bar{p}+\bar{q}+\bar{r}) + (p + q\omega + r\omega^2)(\bar{p} + \bar{q}\omega^2 + \bar{r}\omega) + (p + q\omega^2 + r\omega)(\bar{p} + \bar{q}\omega + \bar{r}\omega^2)}{|p|^2 + |q|^2 + |r|^2}$$

$$= \frac{3|p|^2 + 3|q|^2 + |r|^2 + (p\bar{q} + \bar{p}q + q\bar{r} + \bar{q}r + r\bar{p} + \bar{r}p)(1 + \omega + \omega^2)}{|p|^2 + |q|^2 + |r|^2}$$

$= 3 \quad [\because 1 + \omega + \omega^2 = 0]$

75. *(240)* The given word is SYLLABUS having letters SS, LL, ABUY.

Now, number of ways to select two like letters are ${}^2C_1$ and number of ways to select two distinct letters are ${}^5C_2$ and number of ways to permutate the 2 like letters and 2 distinct letters is $\frac{4!}{2!}$.

So, number of required words $= {}^2C_1 \times {}^5C_2 \times \frac{4!}{2!}$

$= 2 \times 10 \times 12 = 240$

SECTION II
Mastery Builders

PRACTICE SET 9

Total Marks : 300	Total Time : 3 hrs	No. of Questions : 75

Instructions

1. There are **three** parts in the questions paper consisting of **Physics**, **Chemistry** and **Mathematics** having 25 questions in each part of equal weightage. Each part (subject) has two sections.
2. **Section A** This section contains 20 multiple choice questions which have only one correct answer. Each question carries **4 marks** for correct answer and **–1 mark** for wrong answer.
3. **Section B** This section contains 5 questions. The answer to each of the questions is a numerical value. Each question carries **4 marks** for correct answer and **–1 mark** for wrong answer. For Section B, the answer should be rounded off to the nearest integer.

PHYSICS

Section A : Objective Type Questions

1. In an experiment to determine the period of a simple pendulum of length 100 cm, it is attached to different spherical bobs of radii r_1 and r_2. The two spherical bobs have uniform mass distribution. If the relative difference in the periods is found to be 5×10^{-4} s, then the difference in radii $|r_1 - r_2|$ is best given by

(a) 10^{-2} m (b) 10^{-3} m
(c) 10^{-1} m (d) 10^{-4} m

2. In Rutherford's experiments, a silver foil is used. Let kinetic energy of α-particles is 4.8 MeV, mass of α-particle is 6.64×10^{-27} kg and charge on silver nucleus is $47e$. For a head-on-collision, impact parameter for an α-particle, which is scattered at an angle of 45° is

(a) 2.8×10^{-14} m (b) 3.36×10^{-14} m
(c) 1.41×10^{-14} m (d) 5.60×10^{-14} m

3. A pith ball A of mass 9×10^{-5} kg carries a charge of 5 μC. What must be the magnitude and sign of the charge on a pith ball B held 2 cm directly above the pitch ball A, such that the pith ball A, remains stationary ?

(a) 5×10^{-6}C (b) -5×10^{-12} C
(c) 8×10^{-6}C (d) -7.84×10^{-12}C

4. The ratio of magnitude of magnetic field at a distance $2d$ from short bar magnet in longitudinal direction to the field at distance d in transverse direction is

(a) 1 : 1 (b) 1 : 2 (c) 2 : 1 (d) 1 : 4

5. The bob of a simple pendulum is of mass 10 g. It is suspended with a thread of 1 m. If we hold the bob, so as to stretch the string horizontally and release it, tension at lowest point is

(a) zero (b) 0.1 N
(c) 0.3 N (d) 1.0 N

6. A cord is used to lower vertically a block of mass M by a distance d with constant downward acceleration $g/4$. Work done by the cord on the block is

(a) $Mg\frac{d}{4}$ (b) $3Mg\frac{d}{4}$
(c) $-3Mg\frac{d}{4}$ (d) Mgd

7. A cube is floating in liquid as shown in figure.

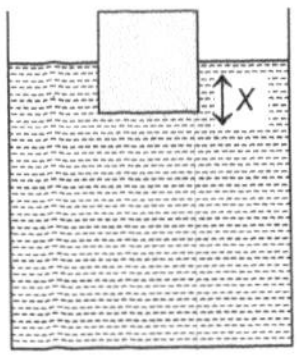

Match the following columns.

	Column I		Column II
A.	If density of liquid decreases, x will	1.	Increase
B.	If height of the cube is decreased keeping base area and density same, x will	2.	Decrease
C.	If the whole system is accelerated upwards, x will	3.	Remain same
D.	If the cube is replaced by another cube of same size but leasser density and liquid is also replaced by another liquid of leasser density, x will	4.	May increase or decrease

Codes

	A	B	C	D		A	B	C	D
(a)	1	2	3	4	(b)	3	2	1	4
(c)	2	1	4	3	(d)	2	3	4	1

8. If ε_0, μ_0 and E represents the permittivity, permeability and electric field respectively, then the dimension of $\frac{1}{2}\varepsilon_0^{3/2}\mu_0^{1/2}E^2$ is

(a) $[ML^2T^{-3}]$ (b) $[M^{-1}L^2T^{-1}]$

(c) $[ML^{-2}T^{-1}]$ (d) $[ML^3T^{-2}]$

9. Two blocks of equal mass m are tied with a light string which passes over a massless pulley as shown in the figure. The magnitude of acceleration of centre of mass of both the blocks is (neglect friction everywhere)

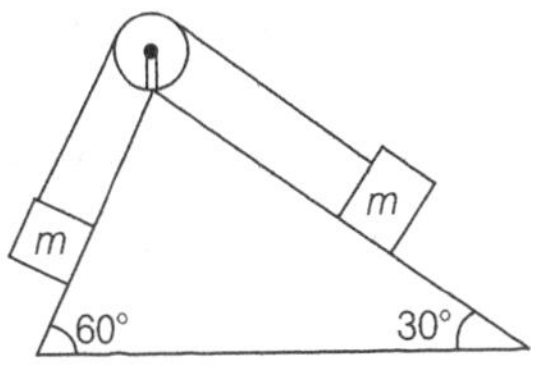

(a) $\left(\frac{\sqrt{3}-1}{4\sqrt{2}}\right)g$ (b) $(\sqrt{3}-1)g$

(c) $\frac{g}{2}$ (d) $\left(\frac{\sqrt{3}-1}{\sqrt{2}}\right)g$

10. A parallel plate capacitor has circular plates of radius r and d is the separation between the plates. E is a uniform electric field in the space between the plates. The energy stored in capacitor is

(a) $\frac{1}{2}\varepsilon_0 E^2$ (b) $\frac{E^2\pi r^2 d}{\varepsilon_0}$

(c) $\frac{\pi}{2}\varepsilon_0 E^2 r^2 d$ (d) $\varepsilon_0 \pi r^2 d$

11. The magnetic field in electromagnetic waves is given by $B = 5 \times 10^{-7}$ T $\sin(\omega t - kx)$, then the energy contained in a cubical box of dimension 1 cm is

(a) 5×10^{-12} J (b) 10^{-12} J

(c) 10^{-13} J (d) 5×10^{-13} J

12. A particle starts from the origin at time $t = 0$ and moves along positive X-axis. The velocity-time graph is shown in figure below. The average velocity of the particle during first 40 s is

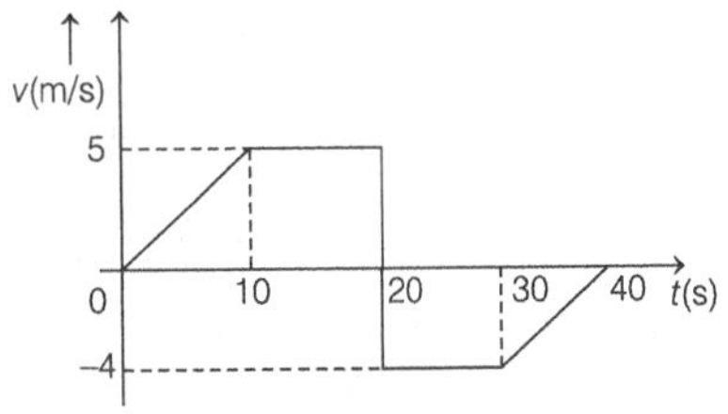

(a) 0.386 m/s (b) 0.375 m/s

(c) 0.576 m/s (d) 0.477 m/s

13. Which of the following option is correct ?

(a) Boyle's law is applicable for adiabatic process.

(b) The value of $(C_p - C_V)$ as well as C_p / C_V are same for any type of ideal gas.

(c) The molecules of monoatomic gas have only translational degree of freedom.

(d) Any real gas behaves as an ideal gas at low temperature and high pressure.

14. Assertion In a cycle, initial volume is equal to final volume, then work done by the gas in that cycle is zero.

Reason Work done in isochoric process is zero.

Choose the correct option.

(a) Both Assertion and Reason are correct and Reason is the correct explanation of Assertion.

(b) Both Assertion and Reason are correct but Reason is not the correct explanation of Assertion

(c) Assertion is correct but Reason is incorrect

(d) Assertion is incorrect but Reason is correct

15. A uniform disc of radius r and mass m is rolling without slipping as shown in figure.

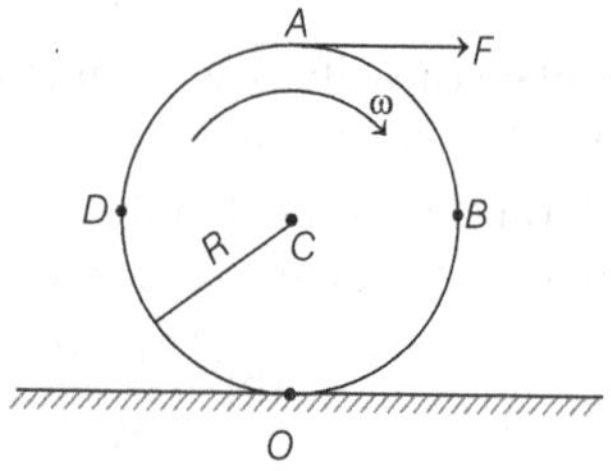

I. Velocity of point of contact O is zero.

II. The frictional force at point O acts in the direction of F.

III. $v_B = v_D > v_A$; (where, $v =$ velocity)

Which of the following statement is correct ?

(a) Only I (b) Both II and III
(c) Both I and II (d) Only III

16. The work done in bringing three particles each having mass m from a large distance to the vertices of an equilateral triangle of side a is

(a) $\frac{3Gm^2}{r}$ (b) $-\frac{3Gm^2}{r}$
(c) $\frac{3Gm}{r}$ (d) $-\frac{3Gm}{r}$

17. An L-C-R series circuit with $100\,\Omega$ resistance is connected to an AC source 400 V and angular frequency 300 rad/s. When only the capacitance is removed, the current lags behind the voltage by 60°, and if when only inductance is removed, the current leads the voltage by 60°. The current and power dissipated in the L-C-R circuit is

(a) 400 W, 2 A (b) 800 W, 4 A
(c) 800 W, 2 A (d) 1600 W, 4 A

18. In the circuit given below, the charge on the left plate of capacitor at steady state is

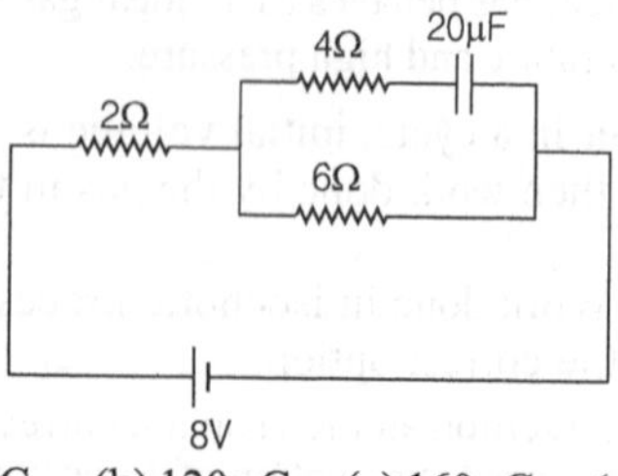

(a) 100μC (b) 120μC (c) 160μC (d) 200μC

19. Binding energy per nucleon *versus* mass number curve for nuclei is shown in the figure below. A, B, C and D are four nuclei indicated on the curve. The process that would release energy is

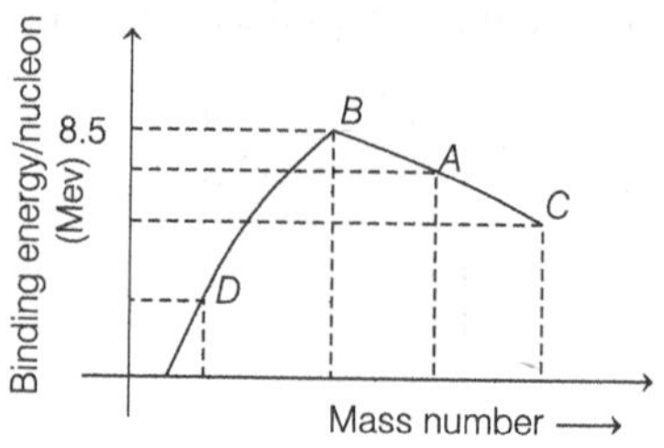

(a) $B \to 2D$ (b) $C \to A + D$
(c) $C \to 2B$ (d) $A \to B + D$

20. **Statement I** The barrier potential of germanium based p-n junction diode is 0.3 V.

Statement II The net charge on n-type semiconductor is negative.

Choose the correct option.

(a) Statement I is true but Statement II is false
(b) Statement I is false but Statement II is true
(c) Both Statement I and Statement II are true
(d) Both Statement I and Statement II are false

Section B : Numerical Value Type Questions

21. The frictional force on 30 kg block will be ...... N.

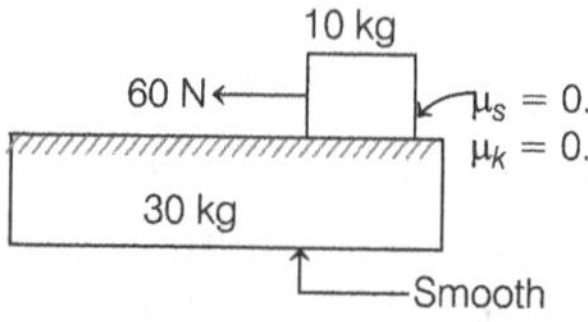

22. Light of wavelength 488 nm is produced by an argon laser, which is used in the photoelectric effect. When light from this spectral line is incident on the emitter, the stopping (cut-off) potential of photoelectrons is 0.38 V. Then, the work-function of the material is ......meV.

23. An object is placed at a distance 30 cm away from centre of curvature of a concave mirror, its image is formed at a distance 20 cm from the centre of curvature towards the mirror. The radius of curvature of the mirror is ....cm.

24. In the circuit shown in the figure, the current through 1Ω resistor is ....mA.

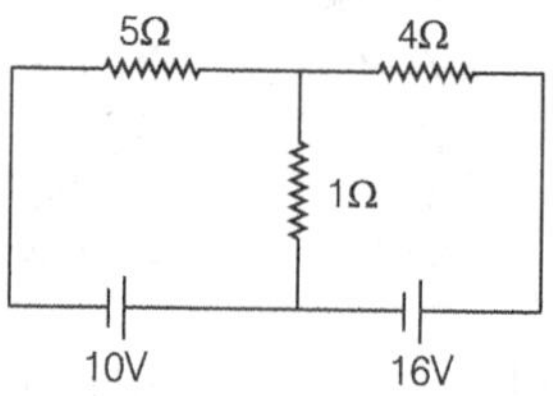

25. A circular coil radius 5 cm is placed in a uniform magnetic field of 4.0×10^{-5}T with its plane perpendicular to the field initially. It is rotated at constant angular speed about an axis along the diameter of coil and perpendicular to magnetic field, so that it undergoes half of rotation in 0.1 s. The maximum value of emf induced (in μV) in the coil will be close to the integer ....

CHEMISTRY

Section A : Objective Type Questions

26. Choose the correct order of bond angles as marked in the following species.

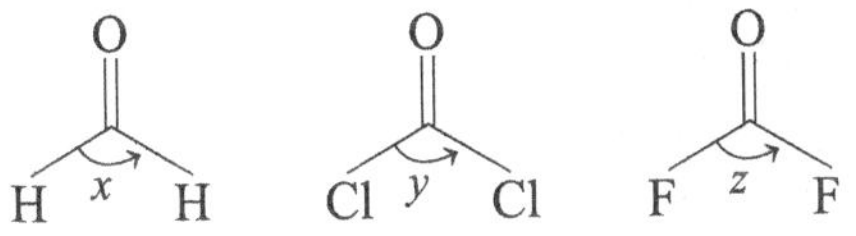

(a) $x > z > y$ (b) $x > y > z$
(c) $z > y > x$ (d) $y > x > z$

27. Consider the following statements regarding Arrhenius equation.

A. A high activation energy usually implies a fast reaction.

B. Rate constant increases with increase in temperature.

C. Higher the magnitude of activation energy, stronger is the temperature dependence of the rate constant.

D. The pre-exponential factor is a measure of the rate at which collisions occur, irrespective of their energy.

Choose the correct statements and mark the option accordingly.

(a) A, B and D (b) A, C and D
(c) B, C and D (d) A and C

28. Two consecutive irreversible first order reaction can be represented by $A \xrightarrow{k_1} B \xrightarrow{k_2} C.$

The rate equation for A is integrated to obtain

$[A]_t = [A]_0 e^{-k_1 t}$ and $[B]_t = \dfrac{k_1 [A_0]}{k_2 - k_1}[e^{-k_1 t} - e^{-k_2 t}].$

At what time will B be present in the greatest concentration?

(a) $t_{\max} = \dfrac{1}{k_1 + k_2} \ln \dfrac{k_2}{k_1}$ (b) $t_{\max} = \dfrac{1}{k_1 - k_2} \ln \dfrac{k_1}{k_2}$

(c) $t_{\max} = \dfrac{1}{k_2 - k_1} \ln \dfrac{k_1}{k_2}$ (d) $t_{\max} = \dfrac{k_1}{k_2 - k_1} \ln \dfrac{k_2}{k_1}$

29. Which of the following reaction establishes difference between glucose and fructose?

(a) Tollen's reagent reaction
(b) Phenyl hydrazine
(c) P/HI
(d) Conc. HNO_3

30. Match List-I with List-II.

	List-I (Complex)	List-II (Hybridization)
A.	$Ni(CO)_4$	I. sp^3
B.	$[PtCl_4]^{2-}$	II. sp^3d^2
C.	$[Co(CN)_6]^{3-}$	III. d^2sp^3
D.	$[CoF_6]^{3-}$	IV dsp^2

Choose the correct answer from the options given below.

(a) A-IV, B-I, C-III, D-II (b) A-I, B-IV, C-III, D-II
(c) A-I, B-IV, C-II, D-III (d) A-IV, B-I, C-II, D-III

31. Identify the incorrect statement for PCl_5 from the following.

(a) The three equatorial bonds of PCl_5 lie in a plane.
(b) PCl_5 has two axial bonds stronger than three equatorial bonds.
(c) The geometry of PCl_5 is not square pyramidal.
(d) The orbitals of phosphorous are assumed to undergo sp^3d hybridisation.

32. Consider the following reaction,

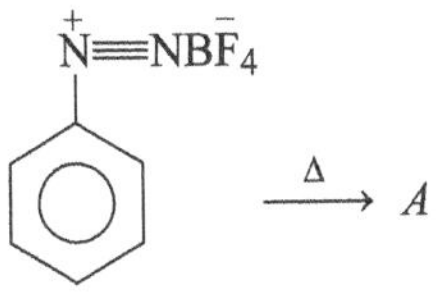

In the above process product A is

(a) fluorobenzene
(b) benzene
(c) 1, 4-difluorobenzene
(d) 1, 3-difluorobenzene

33. When bromocyclohexane and sodium propynide are heated together, the product will be

A. 2-cyclohexylpropyne
B. 1-cyclohexylpropyne
C. propyne
D. cyclohexene

Choose the correct answer from the options given below.

(a) A and B
(b) B, C and D
(c) A, B and C
(d) All four products are formed

34. The pyrimidine bases present in DNA are

(a) cytosine and adenine
(b) cytosine and guanine
(c) cytosine and thymine
(d) cytosine and uracil

35. Apatite is a mineral that is found in tooth enamel. When fluoride tooth pastes are used, this mineral is converted to fluoroapatite, which is more resistant to tooth decay. The formula of apatite is $Ca_5(PO_4)_xF$. What is the value of x, if the compound contains 18.45% by weight of phosphorus? (Atomic weight of P = 31, Ca = 40, F = 19 and O = 16)

(a) 2 (b) 4 (c) 5 (d) 3

36. Among the following boron halides, which has the greater tendency to behave as Lewis acid?

(a) BF_3 (b) BCl_3
(c) BBr_3 (d) BI_3

37. Which combination is best explained by the co-ordinate covalent bond?

(a) $H^+ + H_2O$ (b) $Cl + Cl$
(c) $Mg + \frac{1}{2}O_2$ (d) $H_2 + I_2$

38. The decreasing order of basicity for the following intermediates is

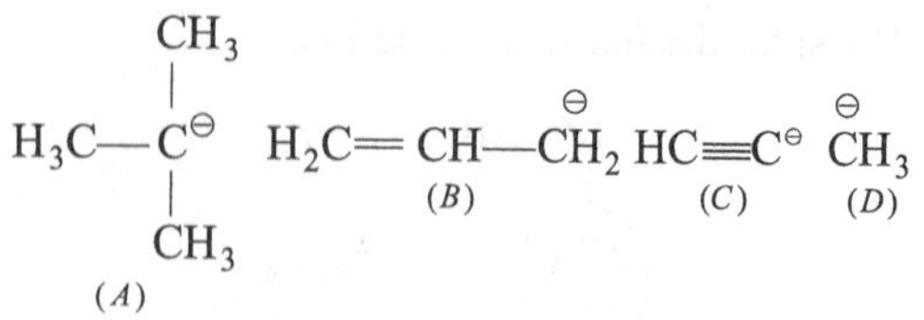

(a) $A > D > B > C$ (b) $A > B > C > D$
(c) $B > A > D > C$ (d) $C > B > A > D$

39. The number of white coloured salts among the salts $SrSO_4$, $Mg(NH_4)PO_4$, $Mn(OH)_2$, $PbSO_4$, PbI, CaC_2O_4 and $[Fe(OH)_2(CH_3COO)]$

(a) 6 (b) 5
(c) 7 (d) 4

40. Given below are two statements one is labelled as Assertion (A) and other is labelled as Reason (R).

Assertion (A) $(I.E.)_n$ of an atom is always greater than $(I.E.)_{n-1}$, where n = integer number.

Reason (R) ne/Z ratio decreases on the successive elimination of electrons.

In the light of the above statements. Choose the correct answer from the options given below.

(a) Both (A) and (R) are true and (R) is the correct explanation of (A).
(b) Both (A) and (R) are true but (R) is not the correct explanation of (A).
(c) (A) is true but (R) is false.
(d) (A) is false but (R) is true.

41. Match the List-I with List-II.

List-I	List-II
A. Sandmeyer reaction	I. Dry acetone
B. Hunsdiecker reaction	II. CuCN/KCN
C. Finkelstein reaction	III. CCl_4/Reflux
D. Dow's process	IV NaOH, 623 K, 300 atm

Choose the appropriate answer from the following options.

(a) A-II, B-I, C-IV, D-III
(b) A-IV, B-II, C-I, D-III
(c) A-II, B-III, C-I, D-IV
(d) A-I, B-IV, C-III, D-II

42. Given below are two statements.

Statement I Marshall's acid is more stronger oxidising agent than $H_2S_2O_6$.

Statement II Oxidation state of sulphur atom(s) in peroxydisulphuric acid is +5.

In the light of the above statements. Choose the correct answer from the options given below.

(a) Both Statement I and Statement II are correct.
(b) Both Statement I and Statement II are incorrect.
(c) Statement I is correct but Statement II is incorrect.
(d) Statement I is incorrect but Statement II is correct.

43. Which of the following will undergo oxidation reaction with MnO_2?

(a) Hex-3-ene-1-ol (b) Cyclohex-3-ene-1-ol
(c) Hex-2-ene-1-ol (d) Propanol

44. Consider the following reaction,

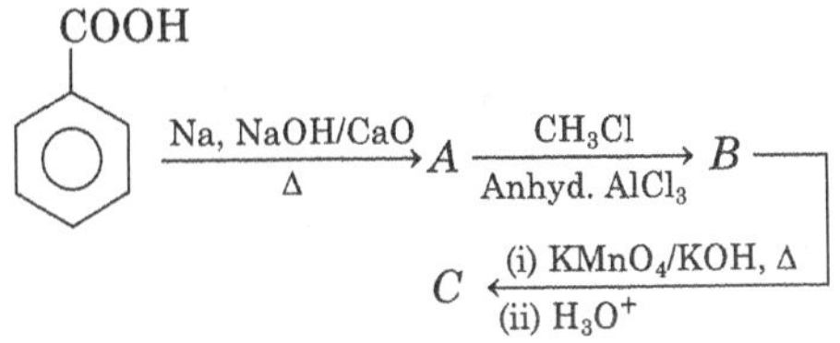

Identify the final product (*C*) formed.

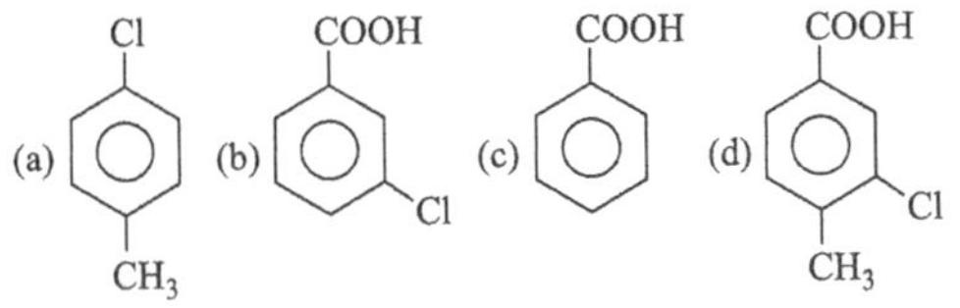

45. Consider the following sequence of reactions:

$$\text{Compound } [A] \xrightarrow{\text{Reduction}} [B] \xrightarrow{HNO_2} CH_3CH_2OH$$

The compound $[A]$ is

(a) CH_3CH_2CN
(b) CH_3NO_2
(c) CH_3NC
(d) CH_3CN

Section B : Numerical Value Type Questions

46. For the cell,

$Cd(s)\ |Cd(OH)_2(s)\ ||\ NaOH\ (aq, 0.1\ M)|$
$H_2(g)(1\ atm), Pt(s)$ if $E_{cell} = 0.0$ V and
$E^{\circ}_{Cd^{2+}/Cd} = -0.39$ V, then calculate

K_{sp} of $Cd(OH)_2$ will be 10^{-x}. The value of x

is ....

47. Following figure shows spectrum of an ideal black body at four different temperatures. The number of correct statements from the following is

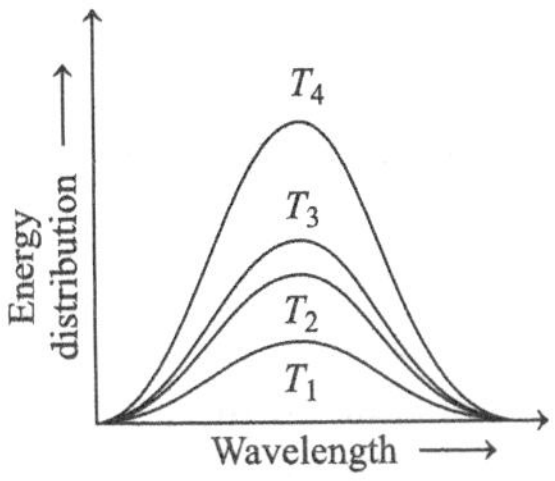

A. $T_1 > T_2 > T_3 > T_4$
B. The black body consists of particles performing simple harmonic motion.
C. The peak of the spectrum shifts to shorter wavelength as temperature increases.
D. $\frac{T_4}{V_4} = \frac{T_3}{V_3} = \frac{T_2}{V_2} \neq$ constant.
E. The given spectrum could be explained using quantisation of energy.

48. A solution containing 28 g of phosphorus in 315 g of CS_2 (boiling point 46.3°C) boils at 47.98°C. The molal elevation constant for CS_2 is 2.34 K kg mol^{-1}. The molecular mass of phosphorus in CS_2 is ...... grams/mole.

49. A solution is 0.10 M $Ba(NO_3)_2$ and 0.10 M $Sr(NO_3)_2$. If solid, Na_2CrO_4 is added to the solution, $SrCrO_4$ begins to precipitate at $x \times 10^{-6}$ M $[Ba^{2+}]$.

The value of x is ....

[Given : $K_{sp}(BaCrO_4) = 6 \times 10^{-9}$,
$K_{sp}(SrCrO_4) = 7 \times 10^{-6}$]

50. 4 g of ethyl alcohol is burnt in excess of O_2 is bomb calorimeter and product analyzed are $H_2O(l)$ and $CO_2(g)$ and unreacted O_2. If the heat capacity of the system is 3600 JK^{-1} and the rise in temperature is 15°. The heat evolved in this reaction is − .... kJ mol^{-1}.

MATHEMATICS

Section A : Objective Type Questions

51. If $\int \sin^{-1}\left(\sqrt{\frac{x}{1+x}}\right)dx = (x+1)P(x) - \sqrt{Q(x)} + R$,

where R is a constant of integration, then the ordered pair $(P(x), Q(x))$ can be

(a) $(\tan\sqrt{x}, x)$ (b) $(x, \tan\sqrt{x})$

(c) $(\tan^{-1}\sqrt{x}, x)$ (d) $(x, \tan^{-1}\sqrt{x})$

52. For $a \in (0, 1)$, let $g(a) = \lim_{h \to 0^+} \int_h^{1-h} x^{-a}(1-x)^{a-1}dx$,

where $g(a)$ is differentiable on (0, 1). Then, the value of $g'\left(\frac{1}{2}\right)$ is equal to

(a) 0 (b) 1 (c) −1 (d) 2

53. Let **c** be resultant of two vectors **a** and **b**. **c** is doubled, when **b** is doubled or direction of **b** is reversed. Then, $a^2 : b^2 : c^2$ is equal to

(a) 3:1:1 (b) 3:2:3 (c) 2:3:2 (d) 2:2:3

54. If $f(x) = \sum_{n=0}^{\infty} \frac{x^n}{n!}(\log a)^n$, at $x = 0$ then $f(x)$

(a) has no limit. (b) is discontinuous.

(c) is continuous but not differentiable .

(d) is differentiable.

55. The equation of curve which satisfy the equation $\frac{dy}{dx} = \frac{y(x+y^3)}{x(y^3-x)}$ and passing through the point (1, 1) is

(a) $y^3 + 2x + 3x^2y = 0$ (b) $y^3 + 2x - 3x^2y = 0$

(c) $y^3 + x + x^2y = 0$ (d) $y^3 + 3x + 2x^2y = 0$

56. If matrix $A = \begin{bmatrix} a & b & c \\ b & c & a \\ c & a & b \end{bmatrix}$, where a, b and c are real positive number, if $abc = 1$ and $A^T A = I$, then which of the following is true?

(a) $a^3 + b^3 + c^3 = 4$ (b) $a^3 + b^3 + c^3 = 3$

(c) $a^3 + b^3 + c^3 = 6$ (d) $a^3 + b^3 + c^3 = 0$

57. Let $\mathbf{a} = 6\hat{i} + 4\hat{j} + 4\hat{k}$ and $\mathbf{b} = 2\hat{i} + 4\hat{j} - 4\hat{k}$ be two vectors. If a vector perpendicular to both the vectors **a** + **b** and **a** − **b** has the magnitude 12, then one such vector is

(a) $6(2\hat{i} + 2\hat{j} + \hat{k})$ (b) $4(2\hat{i} - 2\hat{j} - \hat{k})$

(c) $4(2\hat{i} + 2\hat{j} - \hat{k})$ (d) $4(-2\hat{i} - 2\hat{j} - \hat{k})$

58. If $f(x+1) + f(x-1) = 2f(x)$, $f(0) = 0$, then $\frac{f(1)}{f(10)}$ is equal to

(a) 10 (b) $\frac{1}{11}$ (c) $\frac{1}{10}$ (d) $\frac{1}{9}$

59. If $L = \lim_{x \to 0} \frac{x\sqrt[3]{z^2 - (z-x)^2}}{(\sqrt[3]{8xz - 4x^2} + \sqrt[3]{8xz})^4}$, then $\frac{1}{(Lz)^3}$ is equal to

(a) 2 (b) 2^9 (c) 2^{21} (d) 2^{23}

60. If the last term in the binomial expansion of $\left(\sqrt[3]{2} - \frac{1}{\sqrt{2}}\right)^n$ is $\left(\frac{1}{3 \cdot \sqrt[3]{9}}\right)^{\log_3 8}$, then the fifth term is

(a) 210 (b) 420 (c) 105 (d) 220

61. Let the observation x_i $(1 \le i \le 10)$ satisfy the equations, $\sum_{i=1}^{10}(x_i - 6) = 20$ and $\sum_{i=1}^{10}(x_i - 6)^2 = 50$. If μ and λ are the mean and variance of the observations, $x_1 - 2, x_2 - 2, x_3 - 2, \ldots, x_{10} - 2$, then the ordered pair (μ, λ) is

(a) (5, 1) (b) (6, 1) (c) (3, 2) (d) (8, 9)

62. If $a\cos\theta + b\sin\theta = 3$ and $a\sin\theta - b\cos\theta = 4$, then the value of $a^2 + b^2$ is

(a) 25 (b) 28 (c) 16 (d) 4

63. Let α, β and γ be positive real numbers and

$x = \tan^{-1}\sqrt{\frac{\alpha(\alpha+\beta+\gamma)}{\beta\gamma}} + \tan^{-1}\sqrt{\frac{\beta(\alpha+\beta+\gamma)}{\gamma\alpha}} + \tan^{-1}\sqrt{\frac{\gamma(\alpha+\beta+\gamma)}{\alpha\beta}}$, then $\tan x$ is equal to

(a) 1 (b) −1 (c) ∞ (d) 0

64. Consider an ellipse $\frac{x^2}{9} + \frac{y^2}{4} = 1$ with O as the origin, n points denoted as $A_1, A_2, A_3, \ldots$ are taken on the ellipse such that the angle $A_iOA_{i+1} = \frac{\pi}{n}$, where $i \in (1, n-1)$. A_1 is at positive end of major axis. Then, the value of $\sum_{i=1}^{18}\frac{1}{OA_i^2}$ is equal to

(a) $\frac{13}{4}$ (b) $\frac{9}{4}$ (c) $\frac{1}{2}$ (d) $\frac{1}{3}$

65. After striking the floor a certain ball rebounds to $\frac{2}{3}$ of the height from which it has fallen. Then, the total distance it travels before coming to rest if it is dropped from a height of 600 m, is

(a) 3000 m (b) 6000 m
(c) 1500 m (d) 750 m

66. If α_1, β_1 are the roots of the equation $ax^2 + bx + c = 0$ and α_2, β_2 are the roots of the equation $px^2 + qx + r = 0$ if a, b and c are in GP as well as $\alpha_1, \beta_1, \alpha_2, \beta_2$ are in GP, then p, q and r are in

(a) arithmetic progression
(b) Both (a) and (c)
(c) geometric progression
(d) None of the above

67. Four fair dice D_1, D_2, D_3 and D_4 each having six faces numbered 1, 2, 3, 4, 5 and 6 are rolled, simultaneously. The probability that D_4 shows a number appearing on one of D_1, D_2 and D_3 is $\frac{k_1}{k_2}$, then $k_2 - k_1$ is

(a) 125 (b) -120
(c) 135 (d) 307

68. If $f(x) = \left(4\cos^4 x - 2\cos 2x - \frac{1}{2}\cos 4x - x^7\right)^{\frac{1}{7}}$, then the value of $f[f(100)]$ is equal to

(a) 0 (b) 1
(c) -1 (d) 100

69. If the shortest distance between the lines $L_1 : \frac{x+\lambda}{-6} = \frac{y-6}{7} = \frac{z-0}{1}$, $L_2 : \frac{x-\lambda}{-2} = \frac{y-2}{1} = \frac{z-6}{1}$ is $2\sqrt{29}$, then the value of λ^2 $(\lambda > 0)$ is

(a) 82 (b) 64 (c) 49 (d) 52

70. An ellipse intersects the hyperbola $2x^2 - 2y^2 = 1$ orthogonally. The eccentricity of the ellipse is reciprocal of that of the hyperbola. If the axes of the ellipse are along the coordinate axes, then the equation of ellipse is

(a) $\frac{x^2}{4} + \frac{y^2}{2} = 1$ (b) $\frac{x^2}{2} + \frac{y^2}{1} = 2$
(c) $\frac{x^2}{2} + \frac{y^2}{1} = 1$ (d) None of these

Section B : Numerical Value Type Questions

71. If $\alpha(p)$ and $\beta(p)$ be the roots of the equation $(\sqrt[3]{1+p} - 1)x^2 + (\sqrt{1+p} - 1)x + (\sqrt[6]{1+p} - 1) = 0$ where $p > -1$ and $\lim_{p \to 0^+} \alpha(p) = \lambda$, $\lim_{p \to 0^+} \beta(p) = \mu$, then $|[\lambda + \mu]|$ is equal to (where $[\cdot]$ is GIF) ....

72. If $A = \sum_{r=1}^{50} \frac{{}^{50+r}C_r (2r-1)}{{}^{50}C_r (50+r)}$ and $B = {}^{100}C_{50}$, then $|A - B|$ is equal to ....

73. The locus of the mid-points of those chords of the parabola $7y^2 = 8x$ which are parallel to the line $2x + 3y - 5 = 0$ is the straight line $ax + by + c = 0$ (where $a, b, c \in Q$ and GCD $(a, b, c) = 1$). Then, $|bc - b - c - a^2|$ is equal to ....

74. If $\Delta(x) = \begin{vmatrix} x-2 & (x-1)^2 & x^3 \\ x-1 & x^2 & (x+1)^3 \\ x & (x+1)^2 & (x+2)^3 \end{vmatrix}$, and coefficient of x is α, then $|\alpha|$ is equal to ....

75. The least value of $f(x) = \frac{\tan\left(x + \frac{\pi}{6}\right)}{\tan x}$, $\forall x \in \left(0, \frac{\pi}{3}\right)$, is equal to .....

DETAILED SOLUTIONS

Physics

1. *(b)* Given, $\Delta T = 5\times 10^{-4}$ s

We know that,

$$T = 2\pi\sqrt{\frac{l}{g}} \Rightarrow T \propto \sqrt{l}$$

$$\therefore \quad \Delta T = \frac{1}{2}\Delta l$$

$$\Delta l = |r_1 - r_2|$$

$$\Rightarrow \quad 5\times 10^{-4} = \frac{|r_1 - r_2|}{2}$$

$$\Rightarrow \quad |r_1 - r_2| = 2\times 5\times 10^{-4} = 10^{-3} \text{ m}$$

2. *(b)* Impact parameter,

$$b = \frac{Z_1 Z_2 e^2}{4\pi\varepsilon_0}\times\frac{1}{2K}\times\left(\cot\frac{\theta}{2}\right) \quad (K = \text{kinetic energy})$$

For a head-on-collision, distance of closest approach

$$d_0 = \frac{Z_1 Z_2 e^2}{4\pi\varepsilon_0}\times\frac{1}{K}$$

So, from above expressions, we have

$$b = d_0 \times \frac{1}{2\tan\left(\frac{\theta}{2}\right)}$$

$$\text{Now, } d_0 = \frac{1}{4\pi\varepsilon_0}\times Z_1 Z_2 \times e^2 \times \frac{1}{K}$$

$$= \frac{9\times 10^9 \times 2\times 47\times (1.6\times 10^{-19})^2}{4.8\times 10^6 \times 1.6\times 10^{-19}}$$

$$= 2.8\times 10^{-14} \text{ m}$$

$$\therefore \quad b = \frac{2.8\times 10^{-14}}{2\times(\sqrt{2}-1)} = 3.36\times 10^{-14} \text{ m}$$

$$[\because \tan 22.5^\circ = (\sqrt{2}-1)]$$

3. *(d)* Here, charge on pitch ball A, $q_1 = 5\ \mu\text{C} = 5\times 10^{-6}$ C
Mass of pith ball A, $m_1 = 9\times 10^{-5}$ Kg

The weight $m_1 g$ of the pitch ball A acts vertically downwards.

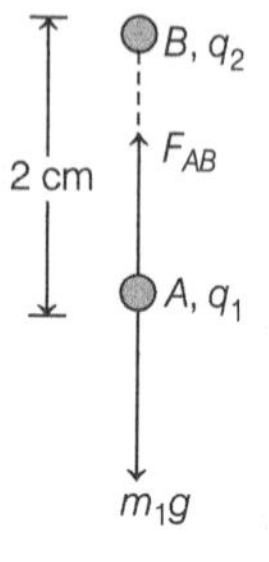

Let q_2 be charge on the pith ball B held 2 cm above the pitch ball A, so that the pith ball A remains stationary. It can be possible only, if the charges on two pith balls are of opposite signs,

i.e. If charge on pith ball A is positive, charge on B must be negative. Then, the force on pith ball A due to B, i.e., F_{AB} will act vertically upwards (figure).

For charge q_1 to remain stationary,

$$F_{AB} = m_1 g \text{ or } \frac{1}{4\pi\varepsilon_0}\cdot\frac{q_1 q_2}{(AB)^2} = m_1 g$$

Here, $\quad AB = 2\text{ cm} = 0.02\text{ m}$

$$\therefore \quad 9\times 10^9 \times \frac{5\times 10^{-6}\times q_2}{(0.02)^2} = 9\times 10^{-5}\times 9.8$$

or $\quad q_2 = 7.84\times 10^{-12}$ C

4. *(d)* For longitudinal direction,

$$B = \frac{\mu_0}{4\pi}\frac{2M}{r^3}$$

$$\text{So, for given case, } B_l = \frac{\mu_0}{4\pi}\times\frac{2M}{(2d)^3} = \frac{\mu_0 M}{4\pi\times(4d^3)}$$

$$\text{For transverse direction, } B_t = \frac{\mu_0}{4\pi}\times\frac{M}{r^3}$$

$$\Rightarrow \quad = \frac{\mu_0}{4\pi}\times\frac{M}{d^3}$$

$$\text{Hence,} \quad \frac{B_l}{B_t} = \frac{1}{4} = 1:4$$

5. *(c)* When, the bob falls through a vertical height h, the velocity acquired at lowest point, $v = \sqrt{2gh} = \sqrt{20}$ m/s

$$\text{Centrifugal force} = \frac{mv^2}{r} = \frac{0.01\times 20}{1} = 0.20$$

So, net tension at lowest point,

$$= \text{Weight} + \text{Centrifugal force}$$

$$= mg + 0.20 = 0.01\times 10 + 0.20 = 0.30\text{ N}$$

Study Tactics

The tension in the thread at an angle θ from lowest point is

$$T = mg\cos\theta + \frac{mv^2}{r}$$

For lowest point, $\theta = 0^\circ \Rightarrow T = mg + \dfrac{mv^2}{r}$

6. *(c)* When, the block moves vertically downward with acceleration $\frac{g}{4}$, then tension in the cord,

$$T = M(g - \frac{g}{4}) = \frac{3}{4}Mg$$

Work done by the cord,

$$W = \mathbf{F} \cdot \mathbf{s} = Fs\cos\theta = Td\cos 180°$$
$$= \left(\frac{-3}{4}Mg\right)d = -3Mg\frac{d}{4}$$

7. *(a)* A-(1), B-(2), C-(3), D-(4)

$$F_B = W$$
$$\rho_l A \times x \times g = \rho_s A \times h \times g$$
$$x = \frac{\rho_s}{\rho_l} \times h \quad \ldots\text{(i)}$$

As density of liquid (ρ_l) decreases, x increases. From equation (i), as height of cube is decreased keeping base area and density same, x will decrease.

If the whole system is accelerated upward, x will remain same because no acceleration term in equation (i). If cube is replaced by another cube of same size but lesser density and liquid is replaced with lesser density liquid, then x may increase or decrease.

8. *(c)* Dimensions of $\frac{1}{2}\varepsilon_0^{3/2}\mu_0^{1/2}E^2$ is

$$\left[\frac{1}{2}\varepsilon_0^{3/2}\mu_0^{1/2}E^2\right] = \frac{\left[\frac{1}{2}\varepsilon_0 E^2\right]}{\left[\frac{1}{\sqrt{\mu_0\varepsilon_0}}\right]} = \frac{[\text{Energy density}]}{[\text{Speed}]}$$
$$= \frac{[\text{ML}^{-1}\text{T}^{-2}]}{[\text{LT}^{-1}]} = [\text{ML}^{-2}\text{T}^{-1}]$$

9. *(a)* The FBD of the system is shown below

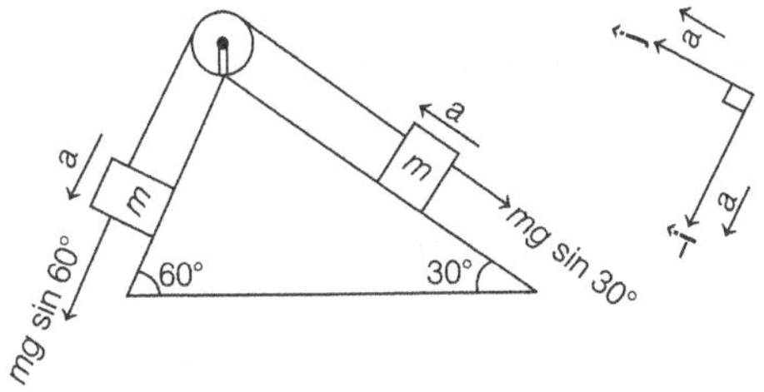

As, $mg\sin 60° > mg\sin 30°$

$\therefore$ Both the blocks move with common acceleration a towards left.

$$\therefore \quad a = \frac{mg\sin 60° - mg\sin 30°}{m+m} = \left(\frac{\sqrt{3}-1}{4}\right)g$$

Both acceleration can be represented by

$$\therefore \quad \mathbf{a}_1 = -a\hat{\mathbf{i}}, \quad \mathbf{a}_2 = a\hat{\mathbf{j}}$$
$$\therefore \quad \mathbf{a}_{CM} = \frac{m_1\mathbf{a}_1 + m_2\mathbf{a}_2}{m_1 + m_2}$$
$$= \frac{-ma\hat{\mathbf{i}} + ma\hat{\mathbf{j}}}{2m} = \frac{a}{2}(-\hat{\mathbf{i}} + \hat{\mathbf{j}})$$
$$a_{CM} = \frac{a}{2} \times \sqrt{2} = \frac{a}{\sqrt{2}} = \left(\frac{\sqrt{3}-1}{4\sqrt{2}}\right)g$$

10. *(c)* The energy stored in a capacitor due to electric field E between the plates,

$$U = \frac{1}{2}\varepsilon_0 E^2 \times \text{Volume} = \frac{1}{2}\varepsilon_0 E^2 \times \pi r^2 d = \frac{\pi}{2}\varepsilon_0 E^2 r^2 d$$

11. *(c)* $B_0 = 5 \times 10^{-7}$ T

$\therefore$ Energy in the magnetic field,

$$E = \frac{B_0^2}{2\mu_0} \times \text{Volume} = \frac{(5 \times 10^{-7})^2}{2 \times 4\pi \times 10^{-7}} \times 10^{-6}$$
$$\approx 10^{-13} \text{ J}$$

12. *(b)* The distance travelled by the particle during time 40 s,

d = Area of v-t graph from time 0 to 40 s

$$= \frac{1}{2} \times (10+20) \times 5 - \frac{1}{2} \times (20+10) \times 4 = 15\,\text{m}$$

$\therefore$ Average velocity during time 40s $= \frac{d}{t} = \frac{15}{40} = 0.375$ m/s

13. *(c)* Boyle's law, pV = constant

$\Rightarrow \quad nRT = \text{constant} \Rightarrow T = \text{constant}$

$\therefore$ It is applicable for isothermal process.

$C_p - C_V = R$, is universal gas constant. So, it is same for any type of ideal gas.

$\frac{C_p}{C_V} = \gamma$, is different for different ideal gas. Any real gas behaves more closely as an ideal gas at low pressure and high temperature.

14. *(d)*

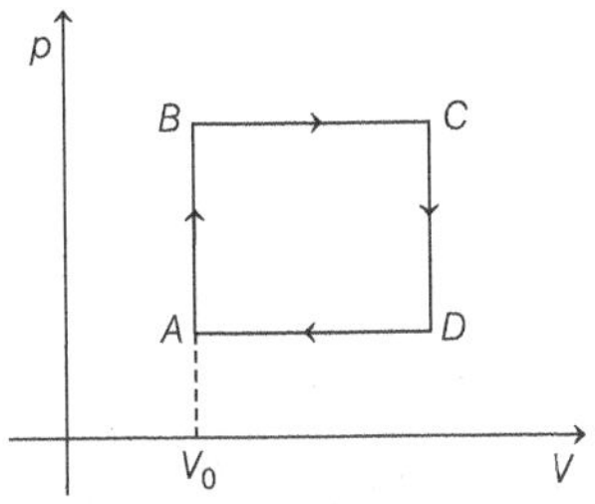

$$V_i = V_f = V_0$$

But, work done = Area of p-V diagram = positive

$\therefore$ Assertion is incorrect.

For isochoric process, V = constant

Work done $= p\Delta V = p \times 0 = 0$

$\therefore$ Reason is correct.

15. *(c)*

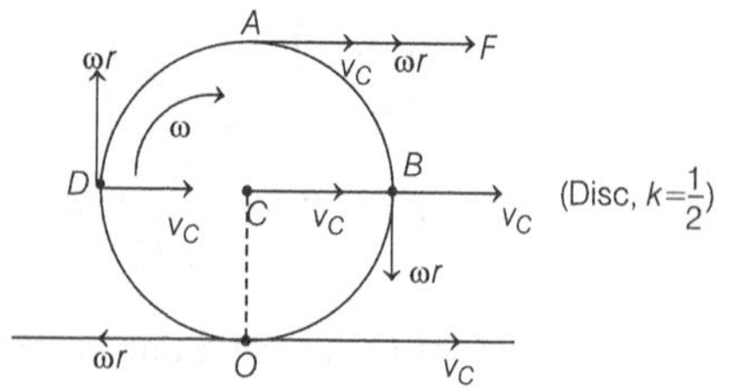

For pure rolling or rolling without slipping, $V_c = \omega r$

$\therefore$ Velocity of point O, $v_0 = V_c - \omega r = 0$

Velocity of points B and D,

$$v_B = v_D = \sqrt{V_C^2 + (\omega r)^2} = \sqrt{2}\omega r$$

Velocity of point A,

$$v_A = V_C + \omega r = 2\omega r$$

$$\therefore \quad v_A > v_B = v_D$$

The frictional force at the point of contact O is given by

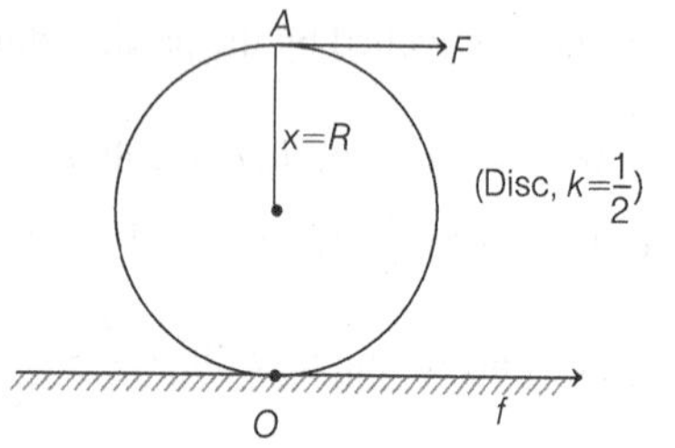

$$f = \left(\frac{x - kR}{R + kR}\right) \cdot F = \left(\frac{R - \frac{1}{2}R}{R + \frac{1}{2}R}\right) \cdot F = \frac{F}{3}$$

As, $f = \frac{F}{3}$ = positive, so the direction of frictional force is in the direction of F.

16. *(b)* Work done by external force = Change in PE

$\therefore$ Work done $= U_f - U_i$

$$= 3\left[-G\frac{m^2}{r}\right] - 0 \qquad [\because U_i = U(\infty) = 0]$$

$$= -\frac{3Gm^2}{r}$$

17. *(d)* Given, $R = 100\Omega$, $E_0 = 400\,V$, $\omega = 300$ rad/s

When capacitance is removed, then circuit is *R-L* circuit.

$$\therefore \quad \tan\phi = \tan 60° = \frac{X_L}{R} \qquad \ldots(i)$$

When inductance is removed, then circuit is *R-C* circuit.

$$\therefore \quad \tan\phi = \tan 60° = \frac{X_C}{R} \qquad \ldots(ii)$$

From Eqs. (i) and (ii), we get

$$\frac{X_L}{R} = \frac{X_C}{R} \Rightarrow X_L = X_C$$

$\therefore$ Impedance, $Z = \sqrt{R^2 + (X_L - X_C)^2}$

$$= R = 100\Omega$$

$\therefore$ Current in *L-C-R* circuit,

$$I_0 = \frac{E_0}{Z} = \frac{400}{100} = 4\,\text{A}$$

$\therefore$ Power dissipated in *L-C-R* circuit, when $X_L = X_C$

$$P = I^2 Z = (4)^2 \times 100$$

$$= 1600\,\text{W}$$

18. *(b)* At steady state, the capacitor behaves as an open circuit,

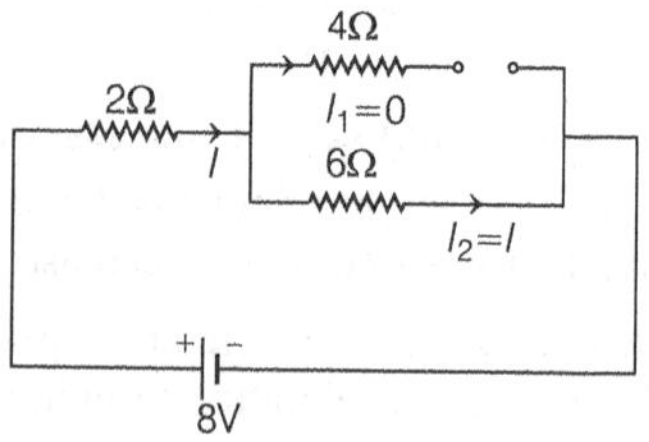

$$I = \frac{E}{R_{net}} = \frac{8}{8} = 1\text{A}$$

$\therefore$ Potential difference across 6Ω resistor is equal to potential difference across capacitor.

$\therefore$ Potential difference across capacitor,

$$V = I \times 6 = 1 \times 6 = 6\text{V}$$

$\therefore$ Charge on the left plate of capacitor,

$$Q = CV = 20 \times 10^{-6} \times 6 = 120\mu\text{C}$$

19. *(c)* The energy is released in a process in which binding energy per nucleon increases and it becomes stable.

For the process, $C \rightarrow 2B$, binding energy per nucleon increases. So, energy is released in this process.

20. *(a)* The knee voltage (barrier potential) of germanium based *p-n* junction diode is 0.3 V.

The net charge on *n*-type or *p*-type semiconductor is zero.

21. (30) As given, $m = 10$ kg

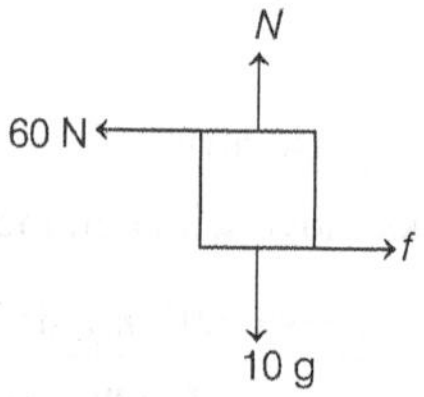

As maximum static friction

$$= \mu_s N = 0.5 \times 10 \times 10 = 50\,\text{N} \qquad [\because g = 10\,\text{m/s}^2]$$

As 60 N is greater than force of static friction (maximum). So, block will slip over the lower one and hence force of kinetic friction f_k has to be considered in that case.

and $\quad f_k = \mu_k N = 0.3 \times 10 \times 10 = 30\,\text{N}$

22. (2170) $\lambda = 488\,\text{nm} = 4880\,\text{Å}, V_0 = 0.38\,\text{V}$

By Einstein's photoelectric equation,

$$\phi = \frac{hc}{\lambda} - eV_0$$

$$= \left(\frac{12420}{\lambda(\text{in Å})} - V_0\right)\text{eV}$$

$$= \left(\frac{12420}{4880} - 0.38\right)\text{eV} = 2.17\,\text{eV}$$

$$= 2170\,\text{meV}$$

23. (120) The position of object and image formed by a concave mirror is shown below.

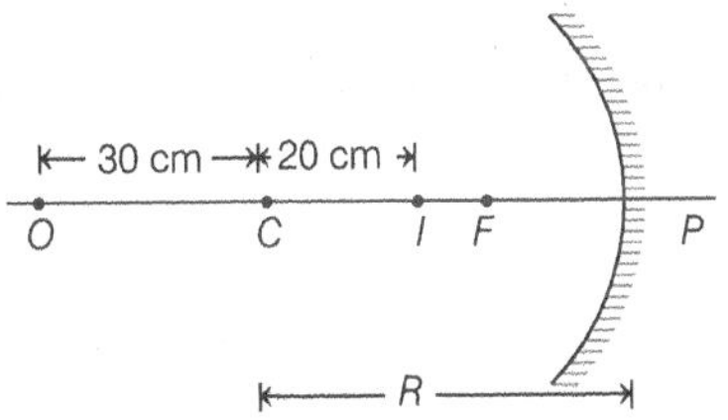

$u = -(R+30)$ cm, $v = -(R-20)$ cm,

From mirror formula, $\frac{1}{f} = \frac{1}{v} + \frac{1}{u}$

$$\Rightarrow \quad \frac{2}{-R} = \frac{1}{-(R-20)} + \frac{1}{-(R+30)}$$

$$\Rightarrow \quad R = 120\,\text{m}$$

24. (1380) The given circuit can be redrawn as follows

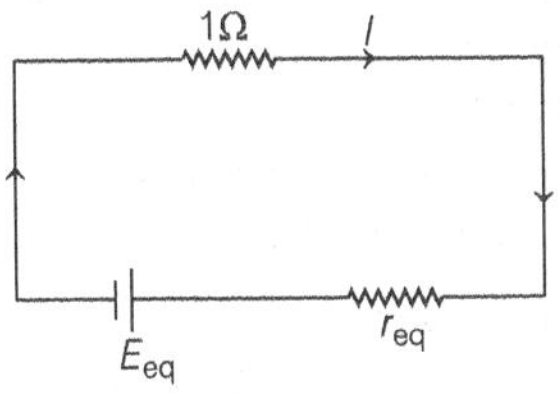

$$E_{eq} = \frac{\frac{E_1}{r_1} + \frac{E_2}{r_2}}{\frac{1}{r_1} + \frac{1}{r_2}}$$

$$= \frac{-\frac{10}{5} + \frac{16}{4}}{\frac{1}{5} + \frac{1}{4}} = \frac{40}{9}\,\text{V}$$

$$r_{eq} = \frac{r_1 r_2}{r_1 + r_2} = \frac{5 \times 4}{5+4} = \frac{20}{9}$$

$$\therefore \quad I = \frac{E_{eq}}{1 + r_{eq}} = \frac{40/9}{1 + \frac{20}{9}} = 1.38\,\text{A} = 1380\,\text{mA}$$

25. (10) Here, $B = 4.0 \times 10^{-5}\,\text{T}$,

$$r = 5\,\text{cm} = 0.05\,\text{m},$$

$$T = 2 \times 0.1 = 0.2\,\text{s}$$

Flux linked with coil,

$$\phi = \mathbf{B} \cdot \mathbf{A} = BA \cos \omega t$$

Magnitude of emf induced in coil,

$$\frac{d\phi}{dt} = BA\omega \sin \omega t$$

$$\left|\frac{d\phi}{dt}\right|_{\text{max}} = BA\omega = \frac{B\pi r^2\, 2\pi}{T}$$

$$= \frac{4.0 \times 10^{-5} \times \pi \times (0.05)^2 \times 2 \times \pi}{0.2}$$

$$= 9.8 \times 10^{-6} \approx 10\,\mu\text{V}$$

Chemistry

26. (b) According to bent rule, as the electronegativity of terminal atom increases, bond angle decrease. This is due to presence of more electron density towards electronegative atom.

$$\text{Bond angle} \propto \frac{1}{\text{Electronegativity of terminal atom}}$$

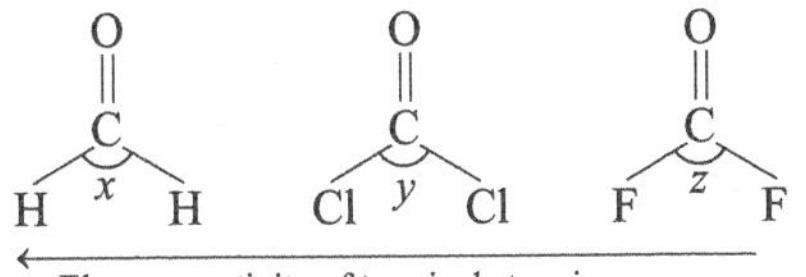

Electronegativity of terminal atom increases.
Bond angle decreases.

Hence, the correct order is $x > y > z$.

27. (c) The correct statements are B, C and D.

28. (b) For maximum concentration of B, $\frac{d[B]}{dt} = 0$

$$\frac{d}{dt}\left[\frac{k_1 [A]_0}{k_2 - k_1}(e^{-k_1 t} - e^{-k_2 t})\right] = 0$$

$$\frac{k_1 [A]_0}{k_2 - k_1}\left[\frac{d}{dt}(e^{-k_1 t} - e^{-k_2 t})\right] = 0$$

Solving (differentiating), we get $t_{\text{max}} = \frac{1}{k_1 - k_2} \ln \frac{k_1}{k_2}$

29. (d) Glucose/Fructose both give different products with conc. HNO_3.

$$\text{Glucose} \xrightarrow{\text{Conc. } HNO_3} \underset{\text{Saccharic acid}}{COOH\,(CHOH)_4{-}COOH}$$

$$\text{Fructose} \xrightarrow{\text{Conc. } HNO_3} \underset{\text{Tartaric acid}}{COOH(CHOH)_2{-}COOH} + \underset{\text{Glycolic acid}}{HOCH_2COOH}$$

30. (*b*) The correct match is A-I, B-IV, C-III, D-II.
$Ni(CO)_4 - sp^3, [PtCl_4]^{2-} - dsp^2, [Co(CN)_6]^{3-} - d^2sp^3$, $[CoF_6]^{3-} - sp^3d^2$.

31. (*b*) The sp^3d-hybridisation in PCl_5 molecule is a combination of $(sp^2 + pd)$ mixing. sp^2-hybrid orbitals form equatorial bonds and pd-hybrid orbitals forms axial bonds. Hence, axial bonds are longer and weaker than equatorial bonds.

32. (*a*)

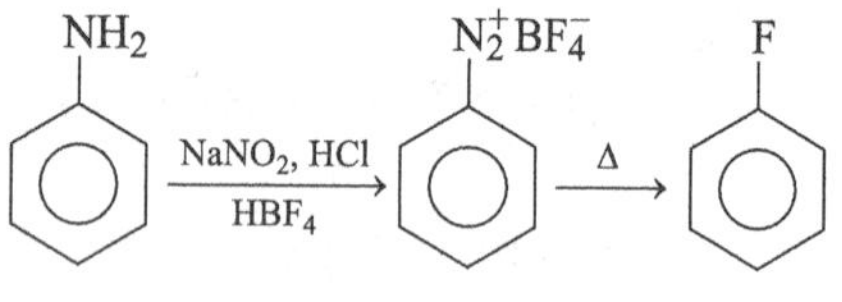

33. (*b*) Among the given compounds B, C, D are correct while *A* is incorrect.

2-cyclohexylpropyne is wrong name. Compound with such name does not exist.

The other three compounds are formed as

In elimination,

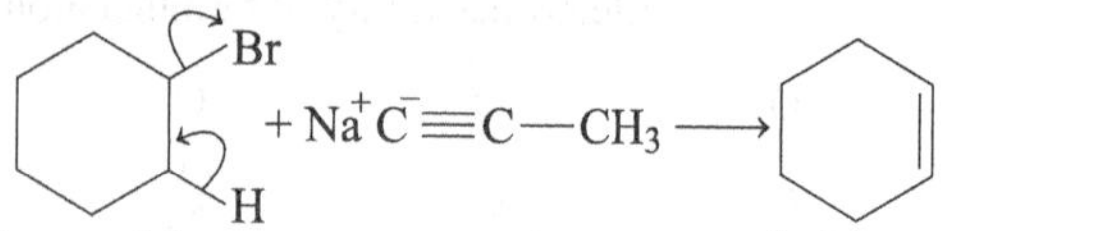

Bromocyclohexane Cyclohexene

$+ H—C\equiv C—CH_3$
Propyne

In substitution,

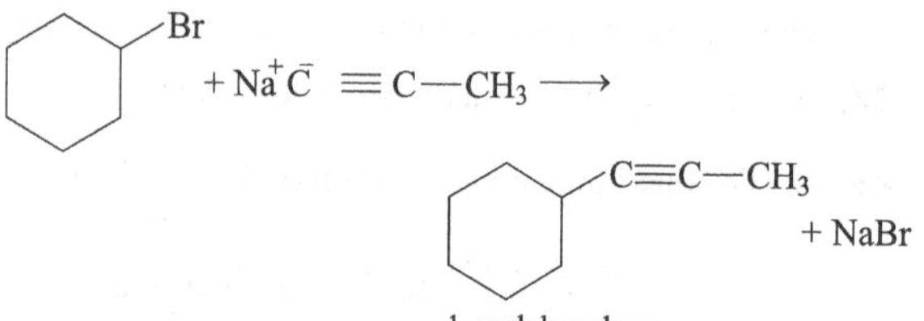

1-cyclohexylpropyne

The yield of substituting product will be greater than that of elimination products.

34. (*c*) The two pyrimidine bases present in DNA are cytosine and thymine.

35. (*d*) $\% P = \dfrac{x \times 31}{5 \times 40 + 31x + 64x + 19} \times 100$

$\Rightarrow \quad 18.45 = \dfrac{3100x}{219 + 95x}$

$\Rightarrow \quad (18.45 \times 219) + (18.45 \times 95x) = 3100x$

$\Rightarrow \quad 1347.25x = 4040.5$

$\Rightarrow \quad x = 2.999$

$\Rightarrow \quad x = 3$

36. (*d*) Boron has an empty p-orbital, which can accommodate a lone pair of electrons. In boron trihalides, depending on which period the halogen belongs to, there is a varied scope for back-bonding from the lone pairs of the halogen atoms.

The order of extent of back-bonding from halogens decrease as we go down the group. The order of Lewis acidity is

$$BI_3 > BBr_3 > BCl_2 > BF_3$$

37. (*a*) $H—\ddot{O}(H): + H^+ \longrightarrow H—\ddot{O}(H)\longrightarrow H$

38. (*a*) The more stable is the carbanion, lesser is its basicity. Basicity order of the intermediates is

$$H_3C—\overset{CH_3}{\underset{CH_3}{C^{\ominus}}} > \overset{\ominus}{C}H_3 > \underbrace{H_2C{=}CH—\overset{\ominus}{C}H_2}_{\text{Resonance stabilised}} > HC{\equiv}\overset{\ominus}{C}$$

(Least stable due to $+I$ effect of CH_3 group)

39. (*b*) The white coloured salt among the given salts are $SrSO_4$, $Mg(NH_4)(PO_4)$, $Mn(OH)_2$, $PbSO_4$ and CaC_2O_4.

40. (*a*) Successive ionisation energy of an atom increases than previous one because as the number of electrons remove from an atom, ratio of ne/Z decrease and Z_{eff} increases, hence, more energy will be required to remove an electrons.

Therefore, both (A) and (R) are true and (R) is the correct explanation of (A).

41. (*c*) The correct match is
$A \rightarrow (II)$, $B \rightarrow (III)$, $C \rightarrow (I)$, $D \rightarrow (IV)$.

In Sandmeyer's reaction, CuCN/KCN are used to convert benzene diazonium chloride into cyanobenzene.

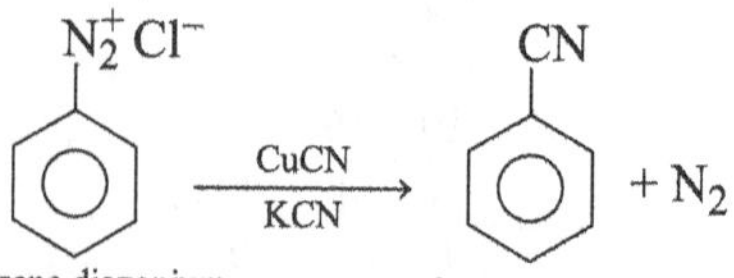

Benzene diazonium chloride Cyanobenzene

In Hunsdiecker reaction, silver salts of carboxylic acids reacts with halogen in the presence of CCl_4 and on reflux, alkyl halide is obtained.

$$R—COOAg + Br_2 \xrightarrow[\text{Reflux}]{CCl_4} R—Br + AgBr + CO_2$$

In Finkelstein reaction, alkyl halide is treated with sodium iodide in the presence of dry acetone to produce alkyl iodide.

$$R\text{—Cl} + \text{NaI} \xrightarrow{\text{Dry acetone}} R\text{—I} + \text{NaCl}$$

In Dow's process, phenol is obtained from chlorobenzene and sodium hydroxide at high temperature (623 K) and pressure (300 atm).

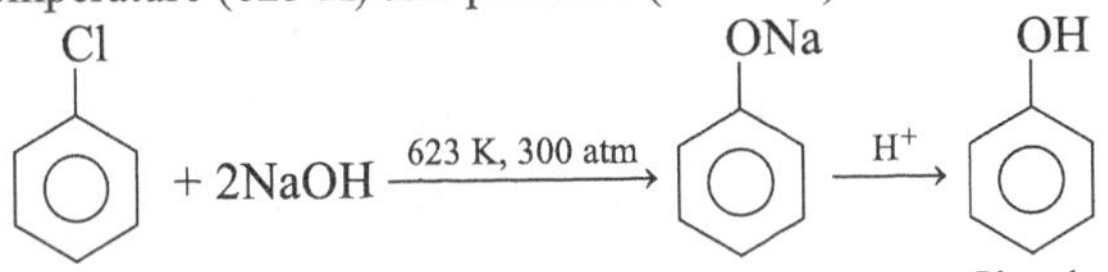

Chlorobenzene Phenol

42. *(c)* Peroxydisulphuric acid is also known as Marshall's acid, and it has molecular formula as $H_2S_2O_8$.

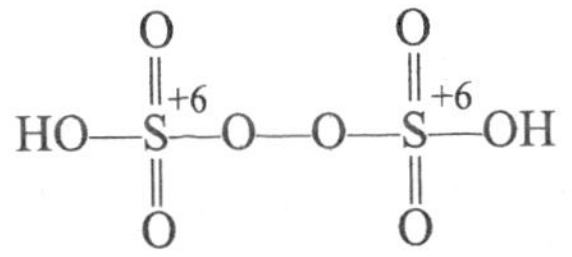

The oxidation state of 2-sulphur atoms is + 6 each.

As Marshall's acid have sulphur atoms with + 6 oxidation state, it is the more stronger oxidising agent than $H_2S_2O_6$ which has sulphur with + 5 oxidation state. Hence, Statement I is correct but Statement II is incorrect.

43. (*c*) MnO_2 causes allylic oxidation, i.e. oxidation of allylic alcohol.

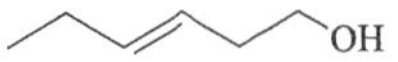

(a) Hex 3-ene-1-ol

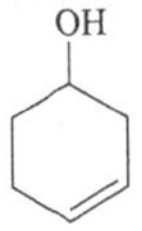

(b) cyclohex 3-ene-1-ol

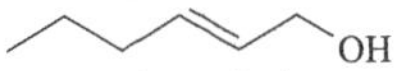

(c) Hex 2-ene-1-ol

↑

This is allylic alcohol.
Hence, it will undergo oxidation.

44. *(c)* The reaction is represented as

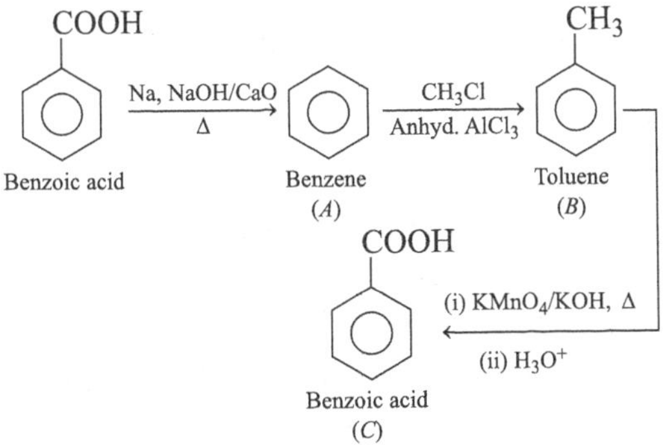

45. (*d*) $\underset{(A)}{CH_3CN} \xrightarrow{\text{Reduction}} \underset{(B)}{CH_3—CH_2—NH_2} \xrightarrow{\text{HONO}} CH_3CH_2OH$

46. (*15*)

$$E^\circ_{\text{cell}} = E^\circ_{H^+/H_2} - E^\circ_{OH^-/Cd(OH)_2|Cd}$$

$$= E^\circ_{H^+/H_2} - E^\circ_{Cd^{2+}/Cd}$$

$$= 0 - (-0.39) = +0.39\text{ V}$$

$$E_{\text{cell}} = E^\circ_{\text{cell}} - \frac{0.059}{2}\log\frac{[Cd^{2+}]}{[H^+]^2}$$

$$E^\circ_{\text{cell}} = 0.03\log\frac{[Cd^{2+}][OH^-]^2}{[H^+]^2[OH^-]^2}$$

$$[\because E_{\text{cell}} = 0]$$

$$= 0.03\log\frac{[Cd^{2+}][OH^-]^2}{(K_w)^2}$$

$$\log\frac{K_{sp}}{K_w^2} = \frac{0.39}{0.03} = 13$$

$$K_{sp} = 10^{13} \times (10^{-14})^2 \qquad [\because K_w = 10^{-14}]$$

$$K_{sp} = 10^{-15}$$

Hence, $x = 15$

47. (*3*) The number of correct statements are A, C and E.

48. (*124*) $w_B = 28\text{ g}$

$W_A = 315\text{ g}$

$T_b = 46.3\,°\text{C}$

$T_S = 47.98\,°\text{C}$

K_b of $CS_2 = 2.34\text{ K kg mol}^{-1}$

Let P_x be the molecular formula of the molecule.

Molar mass of P_x,

$$M_B = (x \times 31)\text{ g/mol}$$

Now, $\Delta T_b = T_s - T_b = \dfrac{1000 \times K_b \times w_B}{M_B \times W_A}$

$$M_B = \frac{1000 \times K_b \times w_B}{W_A \times (T_S - T_b)}$$

$$= \frac{1000 \times 2.34 \times 28}{315 \times (47.98 - 46.3)}$$

$$= 123.8\text{ g/mol}$$

$$\approx 124\text{ g/mol}$$

49. *(85)* $K_{sp}(SrCrO_4) = [Sr^{2+}][CrO_4^{2-}]$

$$[CrO_4^{2-}] = \frac{7 \times 10^{-6}}{0.1} = 7 \times 10^{-5}$$

$$K_{sp}(BaCrO_4) = [Ba^{2+}][CrO_4^{2-}]$$

$[CrO_4^{2-}]_{Total} \approx [CrO_4^{2-}]$ from $SrCrO_4$

$$[Ba^{2+}] = \frac{K_{sp}[BaCrO_4]}{[CrO_4^{2-}]}$$

$$[Ba^{2+}] = \frac{6 \times 10^{-9}}{7 \times 10^{-5}} = 0.85 \times 10^{-4} = 85 \times 10^{-6} \text{ M}$$

50. (623) The chemical equation is

$$C_2H_5OH(l) + 3O_2(g) \longrightarrow 2CO_2(g) + 3H_2O(l)$$

So, $\Delta n_g = -1$

Heat evolved here is at constant volume.

So, $\Delta E = Q_V = C \cdot \Delta T = (3600 \text{ JK}^{-1})(15 \text{ K}) = 54000 \text{ J}$

This heat is given by 4 g of C_2H_5OH,

so, 1 mol will give,

$$\frac{54000 \text{ J} \times 46}{4} = 621000 \text{ J mol}^{-1}$$

i.e. $\Delta E = -621000 \text{ J mol}^{-1} = -621 \text{ kJ mol}^{-1}$

Using equation,

$$\Delta H = \Delta E + RT\,\Delta n_g$$
$$= -621 + (-1)(8.314 \times 10^{-3})(298)$$
$$= -623.47 \text{ kJ mol}^{-1}$$
$$\simeq -623 \text{ kJ mol}^{-1}$$

Mathematics

51. *(c)* Let $I = \int \sin^{-1} \sqrt{\frac{x}{1+x}}\, dx$

On putting $x = \tan^2 t$, we get

$\Rightarrow \quad t = \tan^{-1} \sqrt{x}$

Then, $\sin^{-1} \sqrt{\frac{x}{1+x}} = \sin^{-1} \sqrt{\frac{\tan^2 t}{1 + \tan^2 t}}$

$$= \sin^{-1} \frac{\tan t}{\sec t}$$
$$= \sin^{-1} \sin t = t$$
$$= \tan^{-1} \sqrt{x}$$

$\therefore \quad I = \int \tan^{-1} \sqrt{x}\, dx$

Using integration by parts, we get

$$I = \tan^{-1} \sqrt{x} \int dx - \int \left(\frac{1}{1+x} \times \frac{1}{2\sqrt{x}} \int dx \right) dx$$
$$= x \tan^{-1} \sqrt{x} - \int \frac{x}{2\sqrt{x}\,(1+x)}\, dx + R$$
$$= x \tan^{-1} \sqrt{x} - \frac{1}{2} \int \frac{\sqrt{x}}{1+x}\, dx + R$$

On putting $x = u^2 \Rightarrow dx = 2u\, du$, we get

$$I = x \tan^{-1} \sqrt{x} - \int \frac{u^2}{1+u^2}\, du + R$$
$$= x \tan^{-1} \sqrt{x} - \int \frac{1 + u^2 - 1}{1+u^2}\, du + R$$
$$= x \tan^{-1} \sqrt{x} - \int du + \int \frac{du}{1+u^2} + R$$
$$= x \tan^{-1} \sqrt{x} - u + \tan^{-1} u + R$$
$$= x \tan^{-1} \sqrt{x} - \sqrt{x} + \tan^{-1} \sqrt{x} + R$$
$$= (x+1) \tan^{-1} \sqrt{x} - \sqrt{x} + R$$

$\Rightarrow P(x) = \tan^{-1} \sqrt{x},\ Q(x) = x$

Time Saver Tip

Put $x = \tan^2 \theta \Rightarrow dx = 2 \tan \theta \cdot \sec^2 \theta\, d\theta$

$$\int \underset{\text{I}}{\theta} \cdot \underset{\text{II}}{(2 \tan \theta \cdot \sec^2 \theta)}\, d\theta = \theta \cdot \tan^2 \theta - \int \tan^2 \theta \cdot d\theta$$
$$= \theta \cdot \tan^2 \theta - \int (\sec^2 \theta - 1)\, d\theta$$
$$= \theta(1 + \tan^2 \theta) - \tan \theta + R$$
$$= \tan^{-1}(\sqrt{x})(1 + x) - \sqrt{x} + R$$

$P(x) = \tan^{-1}(\sqrt{x}),\ Q(x) = x$

52. *(a)* Given, $g(a) = \lim_{h \to 0^+} \int_h^{1-h} x^{-a} (1-x)^{a-1}\, dx$

Using $\int_a^b f(x)\, dx = \int_a^b f(a + b - x)\, dx$

$$g(a) = \lim_{h \to 0^+} \int_h^{1-h} (1-x)^{-a} x^{a-1} dx$$

Also, $g(1-a) = \lim_{h \to 0^+} \int_h^{1-h} x^{a-1} (1-x)^{-a}\, dx$

Thus, $\quad g(a) = g(1-a)$

$\Rightarrow \quad g'(a) = -g'(1-a)$

$\Rightarrow \quad g'(a) + g'(1-a) = 0$

On putting $a = \frac{1}{2}$, we get

$$g'\left(\frac{1}{2}\right) + g'\left(\frac{1}{2}\right) = 0 \Rightarrow g'\left(\frac{1}{2}\right) = 0$$

53. *(c)* Given, $\mathbf{c} = \mathbf{a} + \mathbf{b}$

$\Rightarrow c^2 = a^2 + b^2 + 2\mathbf{a} \cdot \mathbf{b}$ ...(i)

Also, $\mathbf{c}$ is doubled when $\mathbf{b}$ is doubled.

$\Rightarrow \quad |\mathbf{a} + 2\mathbf{b}| = 2|\mathbf{c}|$

$\Rightarrow a^2 + 4b^2 + 4\mathbf{a} \cdot \mathbf{b} = 4c^2$ ...(ii)

and $\mathbf{c}$ is doubled when direction of $\mathbf{b}$ is reversed.

$\Rightarrow \quad |\mathbf{a} - \mathbf{b}| = 2|\mathbf{c}|$

$\Rightarrow a^2 + b^2 - 2\mathbf{a} \cdot \mathbf{b} = 4c^2$ ...(iii)

On adding Eqs. (i) and (iii), we get

$$2a^2+2b^2-5c^2=0 \quad \text{...(iv)}$$

On multiplying Eq. (iii) by 2 and then adding with Eq. (ii), we get

$$a^2+2b^2-4c^2=0 \quad \text{...(v)}$$

On solving Eqs. (iv) and (v), we get

$$\therefore \quad \frac{a^2}{2}=\frac{b^2}{3}=\frac{c^2}{2}$$

$$\therefore \quad a^2:b^2:c^2=2:3:2$$

54. *(d)* $f(x)=\sum_{n=0}^{\infty}\frac{x^n}{n!}(\log a)^n$

$$=\sum_{n=0}^{\infty}\frac{(x\log a)^n}{n!}$$

$$=1+(x\log a)+\frac{(x\log a)^2}{2!}+\frac{(x\log a)^3}{3!}+\dots$$

$$=e^{x\log a} \quad \left[\because e^x=1+x+\frac{x^2}{2!}+\frac{x^3}{3!}+\dots\right]$$

$f(x)=a^x$,

which is exponential functions and exponential function is continuous everywhere.

Now, $Lf'(0)=\lim_{h\to 0^-}\frac{f(0-h)-f(0)}{-h}$

$$=\lim_{h\to 0^-}\frac{a^{-h}-1}{-h}=\log_e a$$

$$Rf'(0)=\lim_{h\to 0^+}\frac{f(0+h)-f(0)}{h}$$

$$=\lim_{h\to 0^+}\frac{a^h-1}{h}=\log_e a$$

Since, $Lf'(0)=Rf'(0)$

$\Rightarrow f(x)$ is differentiable at $x=0$.

55. *(b)* Given, $\frac{dy}{dx}=\frac{y(x+y^3)}{x(y^3-x)}$

$$\Rightarrow x(y^3-x)dy=y(x+y^3)dx$$

$$\Rightarrow (xy^3-x^2)dy=(xy+y^4)dx$$

$$\Rightarrow xydx+y^4dx-xy^3dy+x^2dy=0$$

$$\Rightarrow x(ydx+xdy)+y^3(ydx-xdy)=0$$

$$\Rightarrow xd(xy)+\frac{y^3(ydx-xdy)}{x^2}\times x^2=0$$

$$\Rightarrow xd(xy)-x^2y^3d\left(\frac{y}{x}\right)=0$$

$$\Rightarrow \frac{d(xy)}{x^2y^2}=\frac{y}{x}d\left(\frac{y}{x}\right)$$

On integrating both sides, we get

$$\Rightarrow \int\frac{d(xy)}{(xy)^2}=\int\frac{y}{x}d\left(\frac{y}{x}\right)$$

$$\Rightarrow -\frac{1}{xy}=\left(\frac{y}{x}\right)^2\times\frac{1}{2}+C$$

$$\Rightarrow y^3+2x+2Cx^2y=0$$

At (1, 1),

$$1+2+2C=0\Rightarrow C=-\frac{3}{2}$$

Then, $y^3+2x-3x^2y=0$

56. *(a)* Given, $A=\begin{bmatrix}a&b&c\\b&c&a\\c&a&b\end{bmatrix}$, $abc=1$ and $A^TA=I$ ...(i)

$\because A^TA=I$

$$\Rightarrow \begin{bmatrix}a&b&c\\b&c&a\\c&a&b\end{bmatrix}\begin{bmatrix}a&b&c\\b&c&a\\c&a&b\end{bmatrix}=\begin{bmatrix}1&0&0\\0&1&0\\0&0&1\end{bmatrix}$$

$$\Rightarrow \begin{bmatrix}a^2+b^2+c^2&ab+bc+ca&ab+bc+ca\\ab+bc+ca&a^2+b^2+c^2&ab+bc+ca\\ab+bc+ca&ab+bc+ca&a^2+b^2+c^2\end{bmatrix}=\begin{bmatrix}1&0&0\\0&1&0\\0&0&1\end{bmatrix}$$

$$\Rightarrow a^2+b^2+c^2=1,\ ab+bc+ca=0 \quad \text{...(ii)}$$

We know, $a^3+b^3+c^3-3abc=(a+b+c)(a^2+b^2+c^2-ab-bc-ca)$

$$\Rightarrow a^3+b^3+c^3=(a+b+c)(1-0)+3$$

[from Eqs. (i) and (ii)]

$$\Rightarrow a^3+b^3+c^3=a+b+c+3 \quad \text{...(iii)}$$

Now, $(a+b+c)^2=a^2+b^2+c^2+2(ab+bc+ca)$

$$=1+2(0)=1$$

$$\Rightarrow (a+b+c)^2=1 \quad \text{...(iv)}$$

From Eq. (iii), $a^3+b^3+c^3=1+3=4$

57. *(b)* Given, $\mathbf{a}=6\hat{\mathbf{i}}+4\hat{\mathbf{j}}+4\hat{\mathbf{k}}$, $\mathbf{b}=2\hat{\mathbf{i}}+4\hat{\mathbf{j}}-4\hat{\mathbf{k}}$

Now, $\mathbf{a}+\mathbf{b}=8\hat{\mathbf{i}}+8\hat{\mathbf{j}}$, $\mathbf{a}-\mathbf{b}=4\hat{\mathbf{i}}+8\hat{\mathbf{k}}$

and a vector which is perpendicular to both the vectors $\mathbf{a}+\mathbf{b}$ and $\mathbf{a}-\mathbf{b}$ is given by

$$(\mathbf{a}+\mathbf{b})\times(\mathbf{a}-\mathbf{b})=\begin{vmatrix}\hat{\mathbf{i}}&\hat{\mathbf{j}}&\hat{\mathbf{k}}\\8&8&0\\4&0&8\end{vmatrix}$$

$$=64\hat{\mathbf{i}}-64\hat{\mathbf{j}}-32\hat{\mathbf{k}}=32(2\hat{\mathbf{i}}-2\hat{\mathbf{j}}-\hat{\mathbf{k}})$$

Now, vectors along $(\mathbf{a}+\mathbf{b})\times(\mathbf{a}-\mathbf{b})$ having magnitude 12, are

$$\pm\frac{12\times 32(2\hat{\mathbf{i}}-2\hat{\mathbf{j}}-\hat{\mathbf{k}})}{32\sqrt{4+4+1}}=\pm 4(2\hat{\mathbf{i}}-2\hat{\mathbf{j}}-\hat{\mathbf{k}})$$

Thus, one required vector is $4(2\hat{\mathbf{i}}-2\hat{\mathbf{j}}-\hat{\mathbf{k}})$.

58. *(c)* Given, $f(x+1)+f(x-1)=2f(x)$ ...(i)

and $f(0)=0$

On putting $x=1$ in Eq. (i), we get

$$f(2)+f(0)=2f(1)$$

$$\Rightarrow \quad f(2)=2f(1)$$

On putting $x=2$ in Eq. (i), we get

$$f(3)+f(1)=2f(2)$$

$$\Rightarrow \quad f(3)=2f(2)-f(1)=4f(1)-f(1)$$

$$\Rightarrow \quad f(3)=3f(1)$$

Continuing in this way, we get

$$f(10)=10f(1)$$

$$\Rightarrow \quad \frac{f(1)}{f(10)}=\frac{1}{10}$$

59. *(d)* $L=\lim\limits_{x\to 0}\dfrac{x\sqrt[3]{z^2-(z-x)^2}}{(\sqrt[3]{8xz-4x^2}+\sqrt[3]{8xz})^4}$

$$=\lim_{x\to 0}\frac{x\sqrt[3]{2zx-x^2}}{(\sqrt[3]{8xz-4x^2}+\sqrt[3]{8xz})^4}$$

$$=\lim_{x\to 0}\frac{x\sqrt[3]{2zx}}{(\sqrt[3]{8xz}+\sqrt[3]{8xz})^4}$$

$$=\lim_{x\to 0}\frac{x^{\frac{4}{3}}(2z)^{\frac{1}{3}}}{x^{\frac{4}{3}}\times z^{\frac{4}{3}}\times 4^4}$$

$$\Rightarrow \quad L=\frac{1}{z\times 2^{\frac{23}{3}}}\Rightarrow \frac{1}{Lz}=2^{\frac{23}{3}}\Rightarrow \frac{1}{(Lz)^3}=2^{23}$$

60. *(a)* Last term of $\left(\sqrt[3]{2}-\dfrac{1}{\sqrt{2}}\right)^n$ is $\left(\dfrac{1}{3\sqrt[3]{9}}\right)^{\log_3 8}$

$$\Rightarrow \quad {}^nC_n\left(2^{\frac{1}{3}}\right)^{n-n}\left(\frac{-1}{2^{\frac{1}{2}}}\right)^n=\left(\frac{1}{3\sqrt[3]{9}}\right)^{\log_3 8}$$

$$\Rightarrow \quad \frac{(-1)^n}{2^{\frac{n}{2}}}=\left(\frac{1}{3^{\frac{5}{3}}}\right)^{\log_3 8}=\frac{1}{3^{\frac{5}{3}\log_3 8}}$$

$$=3^{-\frac{5}{3}\log_3 2^3}$$

$$=3^{-5\log_3 2}$$

$$\frac{(-1)^n}{2^{\frac{n}{2}}}=2^{-5}=\frac{1}{2^5}$$

$$\Rightarrow \quad \frac{n}{2}=5$$

$$\Rightarrow \quad n=10$$

Now, $T_5=T_{4+1}={}^{10}C_4\left(2^{\frac{1}{3}}\right)^{10-4}\left(-\dfrac{1}{\sqrt{2}}\right)^4$

$$=\frac{10!}{4!6!}\left(2^{\frac{1}{3}}\right)^6\frac{(-1)^4}{(\sqrt{2})^4}=210$$

61. *(b)* For the observations, $x_i\ (1\le i\le 10)$, we have

$$\sum_{i=1}^{10}(x_i-6)=20$$

$$\Rightarrow \quad \sum_{i=1}^{10}x_i-60=20$$

$$\Rightarrow \quad \sum_{i=1}^{10}x_i=80 \quad \text{...(i)}$$

and $\quad \sum_{i=1}^{10}(x_i-6)^2=50$

$$\Rightarrow \sum_{i=1}^{10}x_i^2-12\sum_{i=1}^{10}x_i+10(6)^2=50$$

$$\Rightarrow \quad \sum_{i=1}^{10}x_i^2-12(80)+360=50$$

$$\Rightarrow \quad \sum_{i=1}^{10}x_i^2=650 \quad \text{...(ii)}$$

Now, mean $(\mu)=\dfrac{\sum_{i=1}^{10}(x_i-2)}{10}=\dfrac{\sum_{i=1}^{10}x_i-20}{10}$

$$=\frac{80-20}{10}=6$$

and variance $(\lambda)=\dfrac{\sum_{i=1}^{10}(x_i-2)^2}{10}-(\mu)^2$

$$=\frac{\sum_{i=1}^{10}x_i^2-4\sum_{i=1}^{10}x_i+10(2^2)}{10}-(6)^2$$

$$=\frac{650-4(80)+40}{10}-36$$

$$=\frac{650-320+40}{10}-36$$

$$=37-36=1$$

So, ordered pair $(\mu,\lambda)=(6,1)$.

62. *(a)* We have,

$$a\cos\theta + b\sin\theta = 3 \quad ...(i)$$

$$a\sin\theta - b\cos\theta = 4 \quad ...(ii)$$

On squaring Eqs. (i) and (ii) and add them, we get

$$(a\cos\theta + b\sin\theta)^2 + (a\sin\theta - b\cos\theta)^2 = 3^2 + 4^2$$

$$a^2\cos^2\theta + b^2\sin^2\theta + 2ab\cos\theta\sin\theta + a^2\sin^2\theta + b^2\cos^2\theta - 2ab\cos\theta\sin\theta = 9 + 16$$

$$a^2(\cos^2\theta + \sin^2\theta) + b^2(\sin^2\theta + \cos^2\theta) = 25$$

$$a^2 + b^2 = 25$$

$$[\because \sin^2\theta + \cos^2\theta = 1]$$

63. *(d)* Let $\alpha + \beta + \gamma = a$, then

$$x = \tan^{-1}\sqrt{\frac{\alpha a}{\beta\gamma}} + \tan^{-1}\sqrt{\frac{\beta a}{\gamma\alpha}} + \tan^{-1}\sqrt{\frac{\gamma a}{\alpha\beta}}$$

We know that

$$\tan^{-1}x + \tan^{-1}y = \pi + \tan^{-1}\left(\frac{x+y}{1-xy}\right), \text{ when } xy > 1$$

$$\sqrt{\frac{\alpha a}{\beta\gamma}} \times \sqrt{\frac{\beta a}{\gamma\alpha}} = \frac{a}{\gamma} = \frac{\alpha+\beta+\gamma}{\gamma} > 1,$$

where α, β and γ are positive real numbers.

$$\therefore\ x = \pi + \tan^{-1}\left[\frac{\sqrt{\frac{a\alpha}{\beta\gamma}} + \sqrt{\frac{\beta a}{\gamma\alpha}}}{1 - \sqrt{\frac{a\alpha}{\beta\gamma}}\sqrt{\frac{\beta a}{\gamma\alpha}}}\right] + \tan^{-1}\sqrt{\frac{\gamma a}{\alpha\beta}}$$

$$= \pi + \tan^{-1}\left[\frac{\frac{(\alpha+\beta)}{\sqrt{\alpha\beta\gamma}}\sqrt{a}}{1 - \frac{a}{\gamma}}\right] + \tan^{-1}\sqrt{\frac{\gamma a}{\alpha\beta}}$$

$$= \pi + \tan^{-1}\left[\frac{(a-\gamma)\sqrt{a}}{\sqrt{\alpha\beta\gamma}} \times \frac{\gamma}{-(a-\gamma)}\right] + \tan^{-1}\sqrt{\frac{\gamma a}{\alpha\beta}}$$

$$= \pi - \tan^{-1}\sqrt{\frac{\gamma a}{\alpha\beta}} + \tan^{-1}\sqrt{\frac{\gamma a}{\alpha\beta}}$$

$$x = \pi$$

$$\therefore\ \tan x = \tan\pi = 0$$

64. *(a)* Let $A_i(3\cos\theta_i, 2\sin\theta_i)$

Then, $3\cos\theta_i = r_i\cos\left[\frac{\pi}{n}(i-1)\right]$

and $\quad 2\sin\theta_i = r_i\sin\left[\frac{\pi(i-1)}{n}\right]$

where, r_i = Radius vector $= OA_i$

A_i lies on the ellipse $\frac{x^2}{9} + \frac{y^2}{4} = 1$

$$\Rightarrow \frac{\left[r_i\cos\left\{\frac{\pi}{n}(i-1)\right\}\right]^2}{9} + \frac{\left[r_i\sin\left\{\frac{\pi}{n}(i-1)\right\}\right]^2}{4} = 1$$

$$\Rightarrow \frac{\cos^2\left[\frac{\pi}{n}(i-1)\right]}{9} + \frac{\sin^2\left[\frac{\pi}{n}(i-1)\right]}{4} = \frac{1}{r_i^2}$$

Using $\cos^2\theta = \frac{1+\cos 2\theta}{2}$, $\sin^2\theta = \frac{1-\cos 2\theta}{2}$

and simplyfing, we get

$$\frac{13}{72} - \frac{5}{72}\cos\left[\frac{2\pi(i-1)}{n}\right] = \frac{1}{r_i^2}$$

$$\sum_{i=1}^{n}\frac{1}{r_i^2} = \frac{13n}{72} - \frac{5}{72}\sum_{i=1}^{n}\cos\left[\frac{2\pi}{n}(i-1)\right]$$

$$= \frac{13n}{72} - \frac{5}{72} \times \frac{\sin\left(n \times \frac{2\pi}{2n}\right)}{\sin\frac{2\pi}{2n}}\cos\left(\frac{(n-1)\pi}{n}\right) = \frac{13n}{72}$$

For $n = 18$, $\sum_{i=1}^{n}\frac{1}{(OA_i)^2} = \frac{13\times 18}{72} = \frac{13}{4}$

65. *(a)* Let the height from which the ball is dropped be h and let it rebound to the height eh.

Total distance covered before ball comes to rest

$$= h + (eh + eh) + (e^2h + e^2h) + \ldots\infty$$

$$= h + 2eh + 2e^2h + \ldots\infty$$

$$= h + 2eh(1 + e + e^2 + \ldots\infty)$$

$$= h + 2eh \times \frac{1}{1-e} \quad [\because e < 1]$$

$$= h\left(1 + \frac{2e}{1-e}\right) = \left(\frac{1+e}{1-e}\right)h$$

For given problem, $h = 600\,\text{m}$, $e = \frac{2}{3}$

Thus, the total distance $= \left(\frac{1+\frac{2}{3}}{1-\frac{2}{3}}\right) \times 600$

$$= 5 \times 600 = 3000\,\text{m}$$

66. *(c)* Since, α_1 and β_1 are roots of $ax^2 + bx + c = 0$.

$$\Rightarrow \alpha_1 + \beta_1 = -\frac{b}{a} \text{ and } \alpha_1\beta_1 = \frac{c}{a}$$

and α_2, β_2 are roots of $px^2 + qx + r = 0$.

$$\Rightarrow \alpha_2 + \beta_2 = -\frac{q}{p} \text{ and } \alpha_2\beta_2 = \frac{r}{p}$$

Since, $\alpha_1, \beta_1, \alpha_2, \beta_2$ are in GP.

$\Rightarrow \quad \frac{\beta_1}{\alpha_1} = \frac{\beta_2}{\alpha_2}$

$\Rightarrow \quad \frac{\beta_1 + \alpha_1}{\alpha_1} = \frac{\beta_2 + \alpha_2}{\alpha_2}$

$\Rightarrow \frac{(\alpha_1 + \beta_1)^2}{\alpha_1^2} = \frac{(\alpha_2 + \beta_2)^2}{\alpha_2^2}$

$\Rightarrow \frac{(\alpha_1 + \beta_1)^2}{(\alpha_2 + \beta_2)^2} = \frac{\alpha_1\alpha_1}{\alpha_2\alpha_2} = \frac{\alpha_1\beta_1}{\alpha_2\beta_2}$

On putting the values,

$$\frac{\left(-\frac{b}{a}\right)^2}{\left(\frac{-q}{p}\right)^2} = \frac{\frac{c}{a}}{\frac{r}{p}} \Rightarrow b^2 rp = acq^2$$

Since, a, b, c are in GP.

$\Rightarrow \quad b^2 = ac \Rightarrow q^2 = pr$

$\Rightarrow p, q, r$ are in GP.

Study Tactics

Given equations are

$ax^2 + bx + c = 0$ and $px^2 + qx + r = 0$

Find sum of roots and product of roots of above equations.

$\therefore \alpha_1, \beta_1, \alpha_2, \beta_2$ are in geometric progression, then

$$\frac{\beta_1}{\alpha_1} = \frac{\beta_2}{\alpha_2} \quad \ldots(i)$$

Arrange Eq. (i) and use sum of roots and product of roots in this equation and solve.

67. *(a)* Sample space $= 6 \times 6 \times 6 \times 6 = 6^4$

Favourable events = Case I or Case II or Case III

Case I First, we should select one number for D_4 which appears on all i.e. ${}^6C_1 \times 1$.

Case II For D_4 there are ${}^6C_1$ ways. Now, it appears on any one of D_1, D_2, D_3 i.e. ${}^3C_1 \times 1$.

For other two there are 5×5 ways

$\Rightarrow {}^6C_1 \times {}^3C_1 \times 1 \times 5 \times 5$

Case III For D_4 there are ${}^6C_1$ ways. Now, it appears on any two of D_1, D_2, D_3

$\Rightarrow {}^3C_2 \times 1^2$

For other one there are 5 ways.

$\Rightarrow {}^6C_1 \times {}^3C_2 \times 1^2 \times 5$

$\therefore$ Required probability

$$= \frac{{}^6C_1 + {}^6C_1 \times {}^3C_1 \times 5^2 + {}^6C_1 \times {}^3C_2 \times 5}{6^4}$$

$$= \frac{6(1 + 75 + 15)}{6^4} = \frac{91}{216}$$

$\Rightarrow k_1 = 91$ and $k_2 = 216$

$\therefore k_2 - k_1 = 216 - 91 = 125$

68. *(d)* We have, $4\cos^4 x - 2\cos 2x - \frac{1}{2}\cos 4x - x^7$

$= 4\cos^4 x - 2(2\cos^2 x - 1) - \frac{1}{2}(2\cos^2 2x - 1) - x^7$

$= 4\cos^4 x - 4\cos^2 x + 2 - (2\cos^2 x - 1)^2 + \frac{1}{2} - x^7$

$= \left(\frac{3}{2} - x^7\right)$

$\Rightarrow f(x) = \left(\frac{3}{2} - x^7\right)^{\frac{1}{7}}$

$f(f(x)) = \left[\frac{3}{2} - [f(x)]^7\right]^{\frac{1}{7}} = \left[\frac{3}{2} - \left(\frac{3}{2} - x^7\right)\right]^{\frac{1}{7}} = x$

Hence, $f(f(100)) = 100$

69. *(c)* We know that the shortest distance formula

$$d = \left|\frac{(\mathbf{b}_1 \times \mathbf{b}_2)\cdot(\mathbf{a}_2 - \mathbf{a}_1)}{|\mathbf{b}_1 \times \mathbf{b}_2|}\right|$$

$$\therefore \quad \mathbf{b}_1 \times \mathbf{b}_2 = \begin{vmatrix} \hat{\mathbf{i}} & \hat{\mathbf{j}} & \hat{\mathbf{k}} \\ -6 & 7 & 1 \\ -2 & 1 & 1 \end{vmatrix}$$

$= (7 - 1)\hat{\mathbf{i}} - (-6 + 2)\hat{\mathbf{j}} + (-6 + 14)\hat{\mathbf{k}}$

$= 6\hat{\mathbf{i}} + 4\hat{\mathbf{j}} + 8\hat{\mathbf{k}}$

$\Rightarrow \quad |\mathbf{b}_1 \times \mathbf{b}_2| = \sqrt{36 + 16 + 64} = \sqrt{116}$

$\Rightarrow \quad \mathbf{a}_2 - \mathbf{a}_1 = -2\lambda\hat{\mathbf{i}} + 4\hat{\mathbf{j}} - 6\hat{\mathbf{k}}$

$$\Rightarrow \quad d = \left|\frac{(6\hat{\mathbf{i}} + 4\hat{\mathbf{j}} + 8\hat{\mathbf{k}})\cdot(-2\lambda\hat{\mathbf{i}} + 4\hat{\mathbf{j}} - 6\hat{\mathbf{k}})}{\sqrt{116}}\right|$$

$\Rightarrow 2\sqrt{29} \times \sqrt{116} = |-12\lambda + 16 - 48|$

$\Rightarrow \quad 116 = |-12\lambda - 32|$

$\Rightarrow \quad \pm 116 = 12\lambda + 32$

$\Rightarrow \quad \frac{-148}{12}, \frac{84}{12} = \lambda$

$\Rightarrow \quad \lambda = 7 \quad$ [as $\lambda > 0$]

Then, $\quad \lambda^2 = 49$

70. *(c)* The ellipse and hyperbola will be confocal.

Foci of given hyperbola are $(\pm 1, 0)$.

Let the equation of ellipse be $\frac{x^2}{a^2}+\frac{y^2}{b^2}=1$.

$$\therefore \left(\pm a \times \frac{1}{\sqrt{2}}, 0\right) \equiv (\pm 1, 0)$$

$$\Rightarrow \quad a=\sqrt{2} \text{ and } e=\frac{1}{\sqrt{2}}$$

[$\because$ eccentricity of hyperbola $=\sqrt{2}$, so eccentricity of ellipse $=\frac{1}{\sqrt{2}}$]

$$\because \quad b^2=a^2(1-e^2)$$

$$\Rightarrow \quad b^2=2\left(1-\frac{1}{2}\right)$$

$$\Rightarrow \quad b^2=1$$

Therefore, the equation of ellipse is $\frac{x^2}{2}+\frac{y^2}{1}=1$.

71. *(2)* Given equation is

$$(\sqrt[3]{1+p}-1)x^2+(\sqrt{1+p}-1)x+(\sqrt[6]{1+p}-1)=0$$

Let $ax^2+bx+c=0$, $\forall\, x \in R$

where $a=\sqrt[3]{1+p}-1$, $b=\sqrt{1+p}-1$

and $c=\sqrt[6]{1+p}-1$

When $p \to 0$

$a=b=c \to 0$

First, we make above equation independent from coefficients as 0.

Let $p+1=t^6$

Then, when $p \to 0$, $t \to 1$

$$\therefore \quad (t^2-1)x^2+(t^3-1)x+(t-1)=0$$

$$\Rightarrow (t-1)\{(t+1)x^2+(t^2+t+1)x+1\}=0$$

As $t \to 1$

$$\Rightarrow \quad 2x^2+3x+1=0$$

$$\Rightarrow \quad (2x+1)(x+1)=0$$

Thus, $x=-1, -\frac{1}{2}$

or $\lambda=\lim_{p \to 0^+} \alpha(p)=-\frac{1}{2}$

and $\mu=\lim_{p \to 0^+} \beta(p)=-1$

$$\therefore |[\lambda+\mu]|=\left|\left[-\frac{1}{2}+(-1)\right]\right|$$

$$=\left|\left[\frac{-3}{2}\right]\right|$$

$$=|-2|=2$$

72. *(1)* Let $T_r=\frac{{}^{50+r}C_r(2r-1)}{{}^{50}C_r\,(50+r)}$

$$=\frac{{}^{50+r}C_r}{{}^{50}C_r}-\frac{{}^{50+r}C_r}{{}^{50}C_r}\left(\frac{50-r+1}{50+r}\right)$$

$$=\frac{{}^{50+r}C_r}{{}^{50}C_r}-\frac{{}^{50+r-1}C_{r-1}}{{}^{50}C_{r-1}}$$

$$=u(r)-u(r-1),$$

where $u(r)=\frac{{}^{50+r}C_r}{{}^{50}C_r}$

Now, $A=\sum_{r=1}^{50} T_r$

$$=\sum_{r=1}^{50}[u(r)-u(r-1)]$$

$$=u(1)-u(0)+u(2)-u(1)+\ldots+u(50)-u(49)$$

$$=u(50)-u(0)$$

$$=\frac{{}^{100}C_{50}}{{}^{50}C_{50}}-1={}^{100}C_{50}-1$$

$\therefore \quad B={}^{100}C_{50}$ [given]

$\Rightarrow |A-B|=|-1|=1$

73. *(29)* If $M(h,k)$ is the mid-point of a chord of the parabola $7y^2=8x$, then the equation of the chord is given by

$$7ky-8\left(\frac{x+h}{2}\right)=7k^2-8h$$

$$\Rightarrow 7ky-4x-4h=7k^2-8h$$

$$\Rightarrow \quad 7ky-4x=7k^2-4h$$

Its slope is $\frac{4}{7k}$.

Now, $\frac{4}{7k}=-\frac{2}{3}$ $\left[\because \text{slope of } 2x+3y-5=0 \text{ is } -\frac{2}{3}\right]$

$$\Rightarrow \quad 7k=-6$$

$\therefore$Locus of the mid-points is $7y+6=0$.

On comparing with $ax+by+c=0$, we get

$$a=0,\ b=7 \text{ and } c=6$$

$$\therefore \left|bc-b-c-a^2\right|=|7\times 6-7-6-0^2|$$

$$=29$$

74. *(2)* $\Delta(x)$ is a polynomial of degree at most 6 in x.

If $\Delta(x)=a_0+a_1x+a_2x^2+\ldots+a_6x^6$, then

$\Delta'(x)=a_1+2a_2x+\ldots+6a_6x^5$

$\Rightarrow a_1=\Delta'(0)$

$$\text{Now, } \Delta'(x) = \begin{vmatrix} 1 & (x-1)^2 & x^3 \\ 1 & x^2 & (x+1)^3 \\ 1 & (x+1)^2 & (x+2)^3 \end{vmatrix} + \begin{vmatrix} x-2 & 2(x-1) & x^3 \\ x-1 & 2x & (x+1)^3 \\ x & 2(x+1) & (x+2)^3 \end{vmatrix} + \begin{vmatrix} x-2 & (x-1)^2 & 3x^2 \\ x-1 & x^2 & 3(x+1)^2 \\ x & (x+1)^2 & 3(x+2)^2 \end{vmatrix}$$

$$\Rightarrow \Delta'(0) = \begin{vmatrix} 1 & 1 & 0 \\ 1 & 0 & 1 \\ 1 & 1 & 8 \end{vmatrix} + \begin{vmatrix} -2 & -2 & 0 \\ -1 & 0 & 1 \\ 0 & 2 & 8 \end{vmatrix} + \begin{vmatrix} -2 & 1 & 0 \\ -1 & 0 & 3 \\ 0 & 1 & 12 \end{vmatrix}$$

$$= -8 - 12 + 18 = -2$$

$$\therefore \quad |\alpha| = |-2| = 2$$

75. *(3)* $f(x) = \dfrac{\tan\left(x + \dfrac{\pi}{6}\right)}{\tan x}$

$$= \cot x \tan\left(x + \frac{\pi}{6}\right), \forall x \in \left(0, \frac{\pi}{3}\right)$$

$$f'(x) = \cot x \sec^2\left(x + \frac{\pi}{6}\right) - \operatorname{cosec}^2 x \tan\left(x + \frac{\pi}{6}\right)$$

$$\therefore f''(x) = 2\cot x \sec^2\left(x + \frac{\pi}{6}\right)\tan\left(x + \frac{\pi}{6}\right)$$

$$-\operatorname{cosec}^2 x \sec^2\left(x + \frac{\pi}{6}\right) - \operatorname{cosec}^2 x \sec^2\left(x + \frac{\pi}{6}\right)$$

$$+ 2\operatorname{cosec}^2 x \cot x \tan\left(x + \frac{\pi}{6}\right)$$

Now, $f'(x) = 0$

$$\Rightarrow \frac{1}{2}\sin 2x = \frac{1}{2}\sin\left(2x + \frac{\pi}{3}\right)$$

$$\Rightarrow \quad 2x = \pi - 2x - \frac{\pi}{3}$$

$$\Rightarrow \quad x = \frac{\pi}{6}$$

Then, $f''\left(\dfrac{\pi}{6}\right) > 0$

$\therefore$ At $x = \dfrac{\pi}{6}$, $f(x)$ is minimum and there is no other minimum in $\left(0, \dfrac{\pi}{3}\right)$.

$\therefore$ The least value of $f(x) = f\left(\dfrac{\pi}{6}\right)$

$$= \frac{\tan\left(\frac{\pi}{6} + \frac{\pi}{6}\right)}{\tan\frac{\pi}{6}}$$

$$= \frac{\sqrt{3}}{\frac{1}{\sqrt{3}}} = 3$$

SECTION II

Mastery Builders

PRACTICE SET 10

Total Marks : 300 | Total Time : 3 hrs | No. of Questions : 75

Instructions

1. There are **three** parts in the questions paper consisting of **Physics**, **Chemistry** and **Mathematics** having 25 questions in each part of equal weightage. Each part (subject) has two sections.
2. **Section A** This section contains 20 multiple choice questions which have only one correct answer. Each question carries **4 marks** for correct answer and **– 1 mark** for wrong answer.
3. **Section B** This section contains 5 questions. The answer to each of the questions is a numerical value. Each question carries **4 marks** for correct answer and **– 1 mark** for wrong answer. For Section B, the answer should be rounded off to the nearest integer.

PHYSICS

Section A : Objective Type Questions

1. The wavelength of the radiation emitted when the electron in a hydrogen atom jumps from $n = \infty$ to $n = 3$.

(a) 365 nm (b) 828 nm
(c) 627 nm (d) 487 nm

2. A small coin is resting on the bottom of a tank filled with a liquid. A ray of light from the coin travels upto the surface of the liquid and moves along the surface (figure). How fast is light travelling in the liquid ?

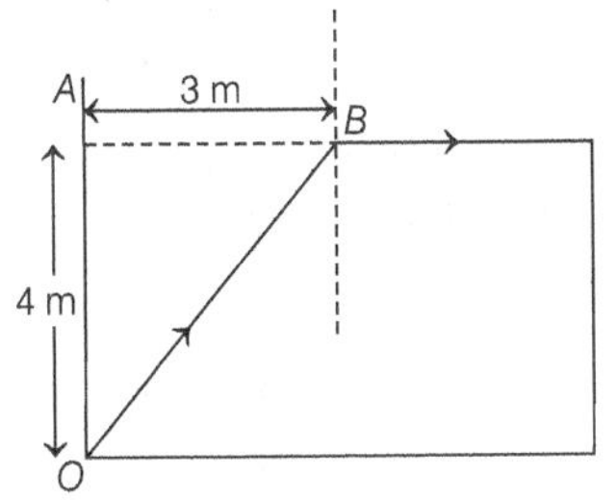

(a) 2.4×10^8 m/s (b) 1.8×10^8 m/s
(c) 3.0×10^8 m/s (d) 5.0×10^8 m/s

3. Two electric dipoles of moment p and $8p$ are placed in opposite direction on a line at a distance of 25 cm. The electric field will be zero at point between the dipoles whose distance from dipole of moment p is

(a) 10 cm (b) 5 cm
(c) 8 cm (d) (25/3) cm

4. An equilateral triangular loop having a resistance R and length of each side l is placed in a magnetic field which is varying at $\frac{dB}{dt} = 1$ T/s. The induced current in the loop will be

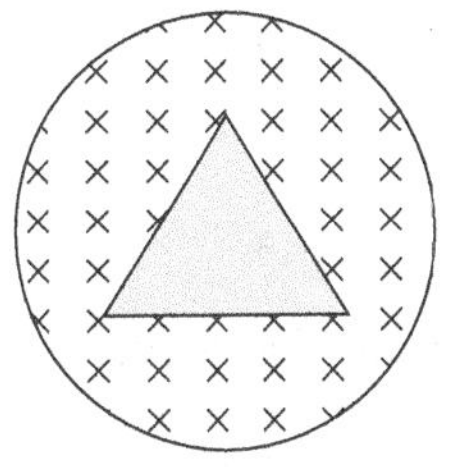

(a) $\frac{\sqrt{3}}{4} \cdot \frac{l^2}{R}$ (b) $\frac{4}{\sqrt{3}} \cdot \frac{l^2}{R}$ (c) $\frac{\sqrt{3}}{4} \cdot \frac{R}{l^2}$ (d) $\frac{4}{\sqrt{3}} \cdot \frac{R}{l^2}$

5. A submarine experience a pressure of 7.07×10^6 Pa at a depth of d_1 in a sea. When it goes further to a depth of d_2, it experiences a pressure of 9.09×10^6 Pa, then $d_2 - d_1$ is approximately (density of water $= 10^3$ kg/m^3 and acceleration due to gravity $= 10 \text{ms}^{-2}$)

(a) 100 m (b) 300 m
(c) 400m (d) 200 m

6. A wooden box of mass 8 kg slides down an inclined plane of inclination of 45° to the horizontal with a constant acceleration of 0.5 m/s^2. What is the force of friction between the box and inclined plane ? (Take, $g = 10 \text{ m/s}^2$)

(a) 37 N (b) 38 N
(c) 53 N (d) 52 N

7. The acceleration of a particle increasing linearly with time t is bt. The particle starts from the origin with an initial velocity v_0. The distance travelled by the particle in time t will be

(a) $v_0 t + \frac{1}{6} bt^3$ (b) $v_0 t + \frac{1}{6} bt^2$
(c) $v_0 t + \frac{1}{3} bt^3$ (d) $v_0 t + \frac{1}{3} bt^2$

8. Match Column I with Column II and select the correct answer from the codes given below.

Column I (Name of Wave)		Column II (Detection)	
A.	Radio waves	1.	Photocell
B.	Microwaves	2.	Thermopiles bolometer
C.	Infrared waves	3.	Point contact diode
D.	Ultraviolet rays	4.	Receiver's aerials

Codes

	A	B	C	D
(a)	1	2	3	4
(b)	3	2	1	4
(c)	4	3	2	1
(d)	2	1	4	3

9. Two blocks are placed over a rough pulley of moment of inertia I. When the block of mass M comes down by a distance h, the angular speed of the pulley is

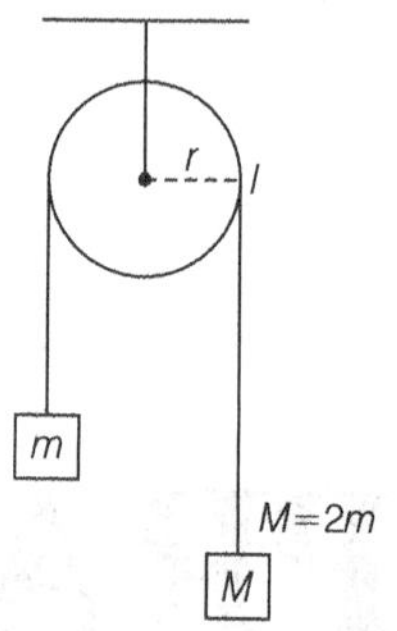

(a) $\sqrt{\frac{2gh}{(M+m)r^2 + I}}$ (b) $\sqrt{\frac{2(M+m)gh}{(M-m)r^2 + I}}$

(c) $\sqrt{\frac{2Mgh}{(M-m)r^2 + I}}$ (d) $\sqrt{\frac{2(M-m)gh}{(M+m)r^2 + I}}$

10. An ideal ammeter and an ideal voltmeter are connected as shown. The ammeter and voltmeter reading for the circuit having $R_1 = 5\ \Omega$, $R_2 = 15\ \Omega$, $R_3 = 1.25\ \Omega$ and $E = 20$ V are given as

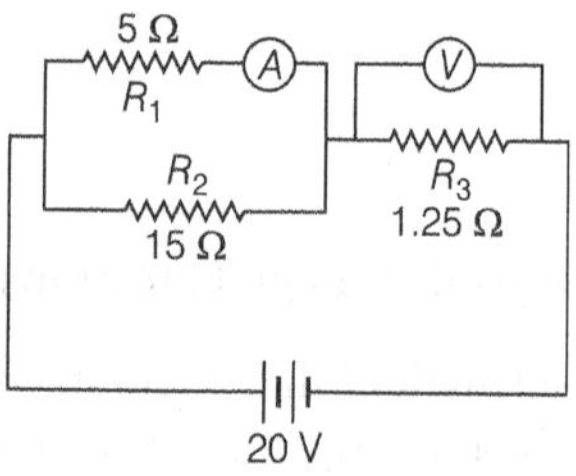

(a) 6.25 A, 3.75 V
(b) 3.00 A, 5 V
(c) 3.75 A, 3.75 V
(d) 3.75 A, 6.25 V

11. An AC source rated 200V, 50 Hz is connected to a resistor. The time taken by the current to change from its maximum to the rms value is

(a) 3.5 ms (b) 2 ms
(c) 2.5 ms (d) 0 ms

12. For a system of two springs and block system, choose the correct option.

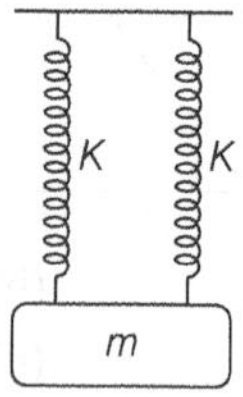

(a) For the equilibrium position (about which oscillation occur), the stretch is $x = \frac{mg}{k}$

(b) The time period of oscillation is $2\pi\sqrt{\frac{m}{2k}}$

(c) Both (a) and (b) are correct

(d) Both (a) and (b) are incorrect

13. Figure shows the variation of the internal energy U with density ρ of one mole of an ideal monoatomic gas for thermodynamic cycle $ABCA$. Here, process AB is a part of rectangular hyperbola.

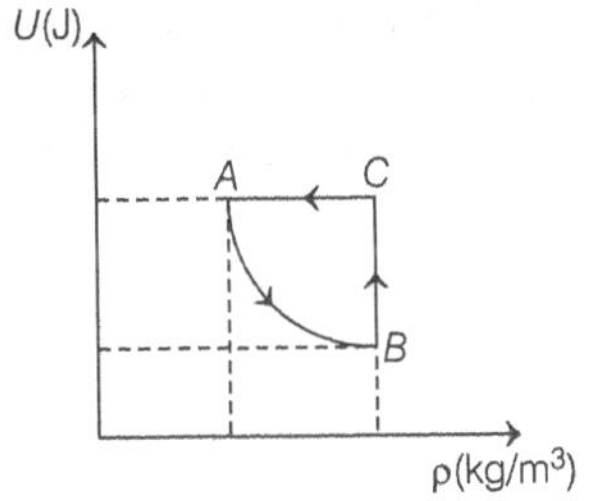

(a) process AB is isothermal and net work in cycle is done by gas

(b) process AB is isobaric and net work in cycle is done by gas

(c) process AB is isobaric and net work in cycle is done on the gas

(d) process AB is adiabatic and net work in cycle is done by gas

14. The de-Broglie wavelength of a particle moving with a velocity 2.0×10^8 m/s is equal to the wavelength of photon. The ratio of kinetic energy of the particle to the energy of the photon is (velocity of light is 3×10^8 m/s)

(a) $\frac{1}{8}$ (b) $\frac{1}{3}$ (c) $\frac{3}{8}$ (d) $\frac{8}{3}$

15. The velocity (v) of water waves may depend on their wavelength (λ), density of water (ρ) and acceleration due to gravity (g). The method of dimensions gives the relation between these quantities as

(a) $v \propto \lambda^{-1}\delta^{-1} g$ (b) $v^2 \propto \lambda g \delta^{-1}$

(c) $v \propto \lambda g^{-3}$ (d) $v^2 \propto \lambda g$

16. A boy's catapult is made of rubber cord which is 84 cm long, with 12 mm diameter of cross-section and of negligible mass. The boy keeps a stone weighing 0.04 kg on it and stretches the cord by 40 cm by applying a constant force. When released, the stone flies off with a velocity of $20\, \text{ms}^{-1}$. Neglect the change in the area of cross-section of the cord while stretched. The Young's modulus of rubber is closest to

(a) $10^6\, \text{Nm}^{-2}$ (b) $10^4\, \text{Nm}^{-2}$

(c) $10^8\, \text{Nm}^{-2}$ (d) $10^4\, \text{Nm}^{-3}$

17. I. Forward bias of p-n junction diode increases the width of depletion region.

II. The barrier potential of p-n junction depends on diode design.

III. Forward bias of p-n junction diode decreases the width of potential barrier

Which of the following statement(s) is/are correct ?

(a) Both I and II

(b) Both II and III

(c) Both I and III

(d) Only III

18. **Statement I** The zero error in a vernier caliper can be either positive or negative and must be consider when making measurements.

Statement II A zero error in a vernier caliper occurs when the zero mark on the vernier scale have left side of zero mark on the main scale.

Choose the correct answer from the options given below.

(a) Both Statement I and Statement II are true.

(b) Statement I is true but Statement II is false.

(c) Both Statement I and Statement II are false.

(d) Statement I is false but Statement II is true.

19. The ρ-p diagram of an ideal gas is shown below.

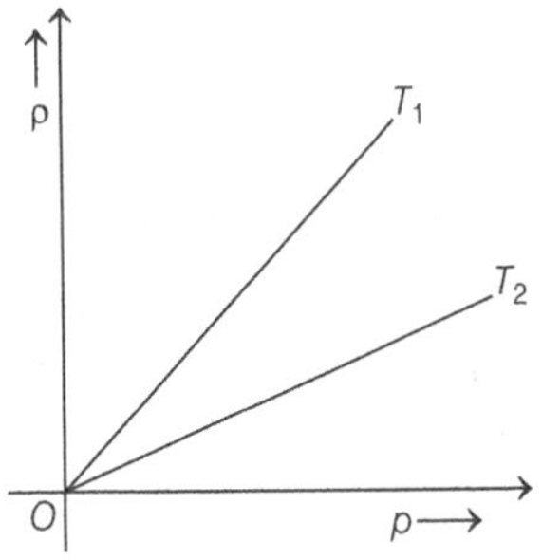

Which of the following option is correct ?

(a) $T_1 > T_2$

(b) $T_1 = T_2$

(c) $T_1 < T_2$

(d) None of the above

20. **Assertion** For an isolated capacitor, if dielectric material is inserted between its plate, then the potential difference across the capacitor increases.

Reason The potential difference across the capacitor is directly proportional to the separation between the plates.

Choose the correct option.

(a) Both Assertion and Reason are correct and Reason is the correct explanation of Assertion
(b) Both Assertion and Reason are correct but Reason is not the correct explanation of Assertion
(c) Assertion is correct but Reason is incorrect
(d) Assertion is incorrect but Reason is correct

Section B : Numerical Value Type Questions

21. Carbon monoxide is carried around a closed cycle *abca* in which *bc* is an isothermal process as shown in the figure. The gas absorbs 7000 J of heat as its temperature increases from 300 K to 1000 K in going from *a* to *b*.

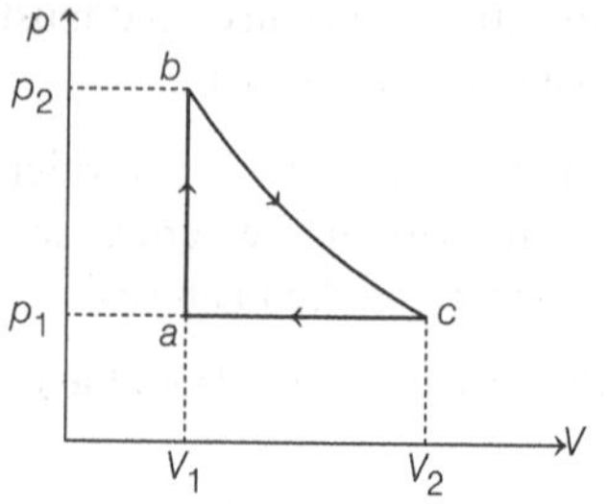

The quantity of heat rejected by the gas during the process *ca* is .... J. (rounded off to nearest integer)

22. Two balls *A* and *B* are projected from same height as shown in the figure. If the horizontal range of both the balls remains same, then the speed *u* of second ball is ...... m/s (rounded off to nearest integer)

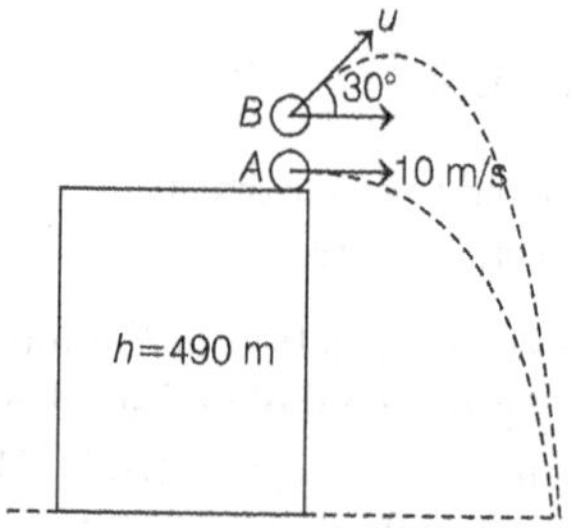

23. The potential difference between points *A* and *B* for the given figure is *x* volt, then what will be the value of *x* (rounded off to nearest integer)?

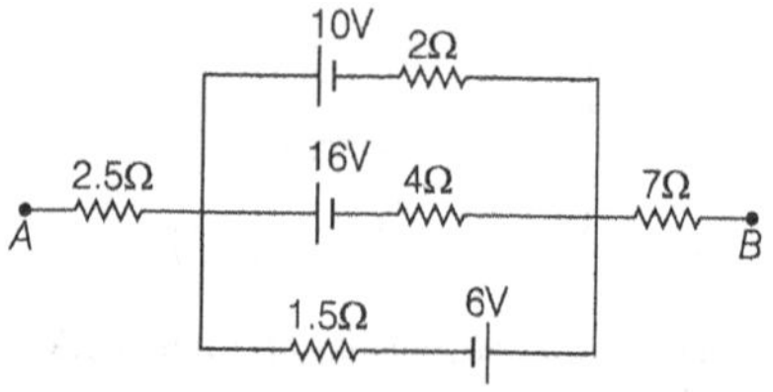

24. In YDSE, the 10th maxima is 3 cm from the central maxima. The YDSE apparatus is immersed in a liquid of refractive index 1.5, then the 5th maxima from the central maxima will be *x* cm. What is the value of *x* (rounded of to nearest integer)?

25. If the weight of an object at equator is 192 N, then the weight of the object at the pole is *x* N. The value of *x* is .... (rounded off to nearest integer)

CHEMISTRY

Section A : Objective Type Questions

26. How many isomeric forms are possible for a tripeptides equally composed of L-valine, L-tyrosine and L-alanine?

(a) 3 (b) 4 (c) 6 (d) 8

27. Match List-I with List-II.

List-I		List-II	
A.	Glucose + HI	I.	Glucose pentaacetate
B.	Glucose + Br_2 water	II.	Hexane
C.	Glucose + acetic anhydride	III.	Saccharic acid
D.	Glucose + HNO_3	IV.	Gluconic acid

Choose the correct answer from the options given below.

(a) A-II, B-IV, C-I, D-III (b) A-I, B-II, C-III, D-IV
(c) A-III, B-IV, C-II, D-I (d) A-IV, B-III, C-II, D-I

28. Which of the following compound does not form during Lassaigne's test of unknown organic compound?

(a) Ferriferrocyanide (b) Ferricthiocyanate
(c) Silver chloride (d) Ferric cyanide

29. What is the systematic IUPAC name of the following compound?

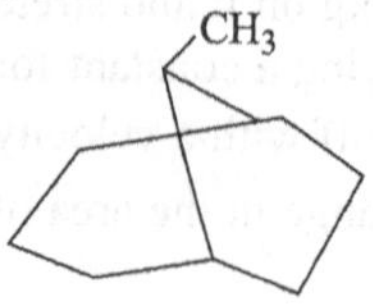

(a) 1-methyl-1-bicyclo [3, 2, 1] octane
(b) 6-methyl-1-bicyclo [3, 2, 1] octane
(c) 8-methyl-1-bicyclo [3, 2, 1] octane
(d) 8-methyl-1-tricyclo [3, 2, 1] octane

30. Which of the following have X—O—X linkage? (where, X is central atom)

I. Pyrosilicate II. $Cr_2O_7^{2-}$
III. $S_2O_3^{2-}$ IV. Hyponitrous acid

Choose the most appropriate answer from the options given below.
(a) II and III
(b) I and II
(c) I, III and IV
(d) I, II, III, IV

31. A_2B_3 is sparingly soluble salt of molecular weight M (gmol^{-1}) and solubility x g/L. The ratio of the molar concentration of $[B^{2-}]$ to the solubility product of the solution will be

(a) $\frac{7M^4}{54x^4}$ (b) $\frac{3}{28}\frac{M^4}{x^4}$
(c) $\frac{M^4}{36x^4}$ (d) $\frac{36M^4}{x^4}$

32. Among the following complexes (K-P)
$K_3[Fe(CN)_6](K)$, $[Ni(CN)_4]^{2-}(L)$, $[Co(NH_3)_6]Cl_3(M)$, $Na_3[Co(oxalate)_3](N)$, $[Ni(H_2O)_6]Cl_2(O)$ and $[Zn(H_2O)_6](NO_3)_2(P)$
The number of diamagnetic complexes are
(a) 4 (b) 3
(c) 2 (d) 5

33. Consider the following reaction,
$$[Ni(NH_3)_6]Cl_2 + Ag^+ \longrightarrow X + Y$$
Where, X is complex ion and Y is precipitate.
The geometry of 'X' will be
(a) square planar
(b) octahedral
(c) trigonal bipyramidal
(d) pentagonal bipyramidal

34. In the following reaction,
$$3Br_2 + 6CO_3^{2-} + 3H_2O \longrightarrow 5Br^- + BrO_3^- + 6HCO_3$$
Choose the correct option.
(a) Bromine is oxidised and carbonate is reduced.
(b) Bromine is reduced and water is oxidised.
(c) Bromine is neither oxidised nor reduced.
(d) Bromine is oxidised and reduced simultaneously.

35. A white precipitate was formed when $BaCl_2$ was added to water extract of an inorganic salt. Further, a gas 'X' with characteristic odour was released, when the formed white precipitate was dissolved in dilute HCl. The anion present in the inorganic salt is
(a) SO_3^{2-}
(b) S^{2-}
(c) SO_4^{2-}
(d) NO_2^-

36. Identify the correct product of following reaction.

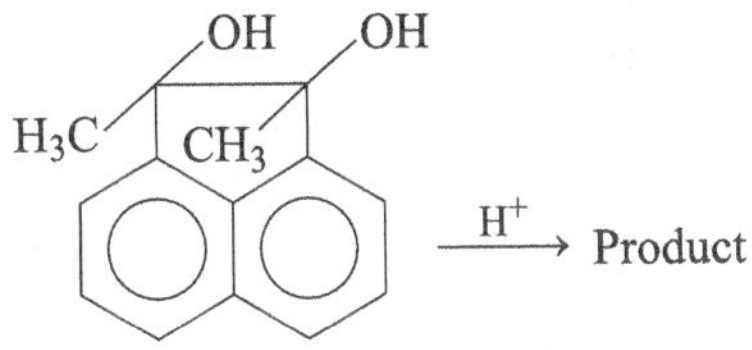

(a) 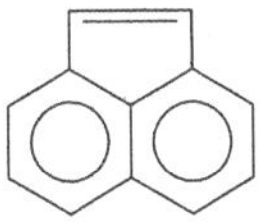

(b) 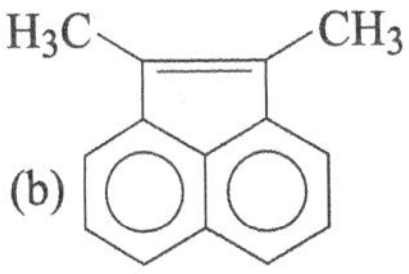

(c) 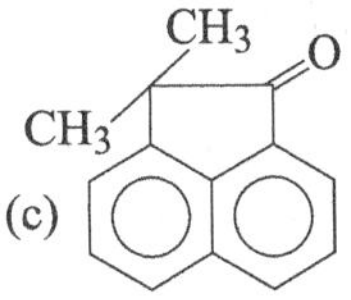

(b) 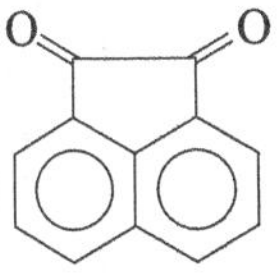

37. Some half-cell reactions and their respective standard potentials are given as below.
(i) $MnO_4^-(aq) + 8H^+(aq) + 5e^- \longrightarrow Mn^{2+}(aq) + 4H_2O(l); E° = 1.51$ V
(ii) $Cr_2O_7^{2-}(aq) + 14H^+(aq) + 6e^- \longrightarrow 2Cr^{3+}(aq) + 7H_2O(l); E° = 1.38$ V
(iii) $Fe^{3+}(aq) + e^- \longrightarrow Fe^{2+}(aq); E° = 0.77$ V
(iv) $Cl_2(g) + 2e^- \longrightarrow 2Cl^-(aq); E° = 1.40$ V
Identify the incorrect statement.
(a) MnO_4^- is weaker oxidising agent than Cl_2.
(b) $Cr_2O_7^{2-}$ is weaker oxidising agent than Cl_2.
(c) MnO_4^- is stronger oxidising agent than Fe^{3+}.
(d) $Cr_2O_7^{2-}$ is weaker oxidising agent than MnO_4^-.

38. What will be the degree of unsaturation in the product when a compound having molecular formula C_4H_6BrCl is treated with metallic sodium in ether? (Given that halogens are present on opposite corner in a cyclic compound).
(a) 2 (b) 1
(c) 0 (d) 3

39.

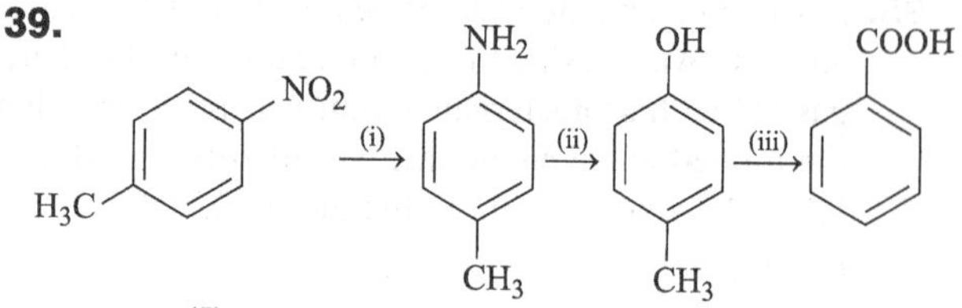

(P) (Q) (R) (S)

The correct sequence of reagents for the preparation of *P* to *S* is

(a) (i) Pd/H_2, ethanol (ii) $NaBH_4$ (iii) CH_3CO_2H, H_2O
(b) (i) CuX_2 (ii) $NaNO_2$/HCl, H_2O
(iii) (a) H_3O^+ (b) H_3PO_2
(c) (i) Pd/H_2, ethanol (ii) $NaNO_2$/HCl, H_2O
(iii) (a) H_3PO_2 or ethanol (b) $KMnO_4$- KOH/Δ
(d) (i) $KMnO_4$, H_2 (ii) CH_3CO_2H, H_3O^+
(iii) $H_2SO_4 + SO_3$, O_3

40. An ester used as medicine is

(a) methyl acetate (b) ethyl acetate
(c) methyl salicylate (d) ethyl benzoate

41. Given below are two statements.

Statement I Basic strength of $PH_3 > AsH_3$ but reverse is true for their bond angles.

Statement II Acidic strength of HBr > HCl but reverse is true for their reducing property.

In the light of the above statements choose the correct answer from the options given below.

(a) Both statement I and Statement II are true.
(b) Both statement I and Statement II are false.
(c) Statement I is true but Statement II is false.
(d) Statement I is false but Statement II is true.

42. Match List-I with List-II.

	Molecule		Bond order
A.	O_2^+	I.	1.5
B.	O_2	II.	2
C.	O_2^-	III.	2.5
D.	O_2^{2-}	IV.	1

Choose the correct answer from the options given below.

(a) A-(I), B-(II), C-(III), D-(IV)
(b) A-(III), B-(II), C-(I), D-(IV)
(c) A-(IV), B-(III), C-(II), D-(I)
(d) A-(III), B-(IV), C-(II), D-(I)

43. Given below are two statements, one is labelled as Assertion (A) and other is labelled as Reason (R).

Assertion (A) Conjugate acid of 3° amine are less stable as compared to 2° amine.

Reason (R) Basic strength depends on the electron donating tendency.

In the light of the above statements, choose the correct answer from the options given below.

(a) Both (A) and (R) are true and (R) is the correct explanation of (A).
(b) Both (A) and (R) are true but (R) is not the correct explanation of (A).
(c) (A) is true but (R) is false.
(d) (A) is false but (R) is true.

44. Arrange the following in decreasing order of their reactivity towards S_N2 reaction.

I. Me_3CO^- II. MeO^- III. $MeCH_2O^-$

IV. Me_2CHO^- V.

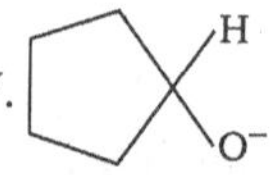

(a) II > III > V > IV > I (b) V > III > II > I > IV
(c) I > V > II > III > IV (d) III > V > I > II > III

45. Which of the following are isoelectronic and isostructural?

(a) NO_3^-, CO_3^{2-} (b) SO_3, NO_3^-
(c) ClO_3^-, CO_3^{2-} (d) CO_3^{2-}, SO_3

Section B : Numerical Value Type Questions

46. Consider the following reactions and their respective enthalpies ($\Delta H°$) values,

$$C\text{ (graphite)} + \frac{1}{2}O_2 \longrightarrow CO(g), \Delta H_1^\circ = -110.5 \text{ kJ}$$

$$CO(g) + \frac{1}{2}O_2(g) \longrightarrow CO_2(g), \Delta H_2^\circ = -282.9 \text{ kJ}$$

$$H_2(g) + \frac{1}{2}O_2(g) \longrightarrow H_2O(l), \Delta H_3^\circ = -285.8 \text{ kJ}$$

$$C\text{ (graphite)} + 2H_2(g) \longrightarrow CH_4(g), \Delta H_4^\circ = -74.8 \text{ kJ}$$

The value of $\Delta U°$ at 400 K for the reaction.

$$4CO(g) + 8H_2(g) \longrightarrow 3CH_4(g) + CO_2(g) + 2H_2O(l)$$

will be (–) ...... kJ. [Nearest integer]

47. An organic compound undergoes first order decomposition. The time taken for its decomposition to 1/8th and 1/10th of its initial concentration are $t_{1/8}$ and $t_{1/10}$ respectively.

The value of $\frac{[t_{1/8}]}{[t_{1/10}]} \times 10$ will be ...... ($\log_{10}2 = 0.3$)

48. x g of non-electrolyte compound of molecular mass 200 g mol^{-1}, when dissolved in 1L of 0.05 M NaCl solution, the osmotic pressure of the resulting solution is found to be 4.92 atm at 27°C. Consider the complete dissociation of NaCl and ideal behaviour of solution, the value of x would be ....

49. The crystalline salt $Na_2SO_4 \cdot xH_2O$ on heating loses 56% of its mass and becomes anhydrous. The value of x will be .... [Nearest integer]

50. The energy absorbed by each molecule (A_2) of substance is 4.4×10^{-19} J and bond energy per molecule is 4×10^{-19} J. The kinetic energy of the molecules per atom will be ....$\times 10^{-20}$ J.

MATHEMATICS

Section A : Objective Type Questions

51. The solution of $x^3 \dfrac{dy}{dx} + 4x^2 \tan y = e^x \sec y$ satisfying $y(1) = 0$ is $\sin y = e^{\alpha x}(\beta x + \gamma)x^{-k}$, then the value of $(\alpha + \beta + \gamma)k$ is equal to

(a) 0 (b) 2 (c) 4 (d) 1

52. The least positive integral value of c for which equation $e^x = cx^2$ has three distinct real roots, is

(a) 1 (b) 3 (c) 2 (d) 6

53. Let $A = \begin{bmatrix} 3 & a & 0 \\ a & a^2+1 & a \\ 0 & a & 3 \end{bmatrix}$, where $a > 0$. Then, the minimum value of $\dfrac{\det(A)}{a}$ is

(a) $3\sqrt{3} + \dfrac{9}{\sqrt{3}}$ (b) $3\sqrt{3} - \dfrac{9}{\sqrt{3}}$

(c) 0 (d) 1

54. The value of $\int_{\pi}^{2\pi} [2\cos x]\, dx$, where $[\cdot]$ represents the greatest integer function, is

(a) $-\pi$ (b) $-\dfrac{\pi}{2}$ (c) $\dfrac{\pi}{2}$ (d) $-\dfrac{5\pi}{6}$

55. Which of the following have not same bounded area?

(a) $f(x) = \sin x, g(x) = \sin^2 x, 0 \to x \to 10\pi$

(b) $f(x) = \sin x, g(x) = |\sin x|, 0 \to x \to 20\pi$

(c) $f(x) = |\sin x|, g(x) = \sin^3 x, 0 \to x \to 10\pi$

(d) $f(x) = \sin x, g(x) = \sin^4 x, 0 \to x \to 10\pi$

56. Let $C(0,1)$ is the centre of the circle and radius unity. P is the parabola $y = 4ax^2$. The set of values of a for which they meet at a point other than the origin is

(a) $a > 0$ (b) $a \in \left(0, \dfrac{1}{4}\right)$

(c) $a \in \left(\dfrac{1}{4}, \dfrac{1}{2}\right)$ (d) $a \in \left(\dfrac{1}{8}, \infty\right)$

57. If $f(x) = \begin{cases} \dfrac{p|x^2 - x - 2|}{-(x^2 - x - 2)}, & x < 2 \\ q, & x = 2 \\ \dfrac{x - [x]}{x - 2}, & x > 2 \end{cases}$

is continuous at $x = 2$, then $p + q$ is equal to

(a) 6 (b) 4 (c) -4 (d) 2

58. If ΔABC is an equilateral triangle inscribed in the auxiliary circle of the ellipse $\dfrac{x^2}{25} + \dfrac{y^2}{16} = 1$ and ΔPQR is the corresponding triangle inscribed within the ellipse, then the centroid of ΔPQR lies at

(a) focus of ellipse

(b) centre of ellipse

(c) between focus and centre an major axis

(d) None of the above

59. If the angle between the lines $\dfrac{x}{2} = \dfrac{y}{2} = \dfrac{z}{1}$ and $\dfrac{5-x}{-2} = \dfrac{7y-14}{p} = \dfrac{z-3}{4}$ is $\cos^{-1}\left(\dfrac{2}{3}\right)$, then p is equal to

(a) $-\dfrac{7}{4}$ (b) -2 (c) 2 (d) $\dfrac{7}{2}$

60. Consider $\hat{\mathbf{a}}$ and $\hat{\mathbf{b}}$ be two unit vectors such that $\hat{\mathbf{a}} + \hat{\mathbf{b}}$ is also a unit vector. If the angle between $\hat{\mathbf{a}}$ and $\hat{\mathbf{b}}$ is $K\pi$, then K is equal to

(a) $\dfrac{1}{3}$ (b) $\dfrac{2}{3}$ (c) $\dfrac{4}{3}$ (d) $\dfrac{3}{2}$

61. Let $f(x) = \sin x + \cos x + \tan x + \text{arc}\sin x + \text{arc}\cos x + \text{arc}\tan x$. If M and m are maximum and minimum values of $f(x)$, then their arithmetic mean is equal to

(a) $\dfrac{\pi}{2} + \cos 1$ (b) $\dfrac{\pi}{2} + \sin 1$

(c) $\dfrac{\pi}{4} + \tan 1 + \cos 1$ (d) $\dfrac{\pi}{4} + \tan 1 + \sin 1$

62. The last four digits of the natural number 3^{100} are *pqrs*, then $p+q+r+s$ is equal to

(a) 5 (b) 3 (c) 7 (d) 2

63. The number of complex number Z such that $|z-1|=|z-i|=|z+1|$ equals

(a) 0 (b) 1
(c) 2 (d) infinite

64. If $\begin{bmatrix} \cos^2\theta & \cos\theta\sin\theta \\ \cos\theta\sin\theta & \sin^2\theta \end{bmatrix}\begin{bmatrix} \cos^2\phi & \cos\phi\sin\phi \\ \cos\phi\sin\phi & \sin^2\phi \end{bmatrix}$ is zero matrix, then θ and ϕ differ by

(a) even multiple of $\frac{\pi}{2}$ (b) odd multiple of $\frac{\pi}{2}$
(c) even multiple of π (d) odd multiple of π

65. Let the sum of the first n terms of a non-constant AP $a_1, a_2, a_3 \dots$ be $50n + \frac{n(n-7)}{2}A$, where A is a constant. If d is the common difference of this AP, then the value of $(a_{40} - 36d)$ is

(a) 50 (b) 18 (c) 60 (d) 40

66. A box contains 2 white balls, 3 black balls and 4 red balls. What is the number of ways of drawing 3 balls from the box with at least one black ball?

(a) 84 (b) 72 (c) 64 (d) 48

67. The probability that England wins a match against India in a hockey match is $\frac{1}{3}$. India and England play 3 matches.

Statement I The probability that England wins atleast one match is $\frac{4}{27}$.

Statement II The probability that England looses all the three matches is $\frac{8}{27}$.

(a) Statement I is true and Statement II is false
(b) Statement I is false and Statement II is true
(c) Statement I and Statement II both are true
(d) Statement I and Statement II both are false

68. If $f(x) + 2f\left(\frac{1}{x}\right) = 3x, x \neq 0$

$S = \{x \in R : f(x) = f(-x)\}$, then S

(a) contains more than two elements
(b) is an empty set
(c) contains exactly one element
(d) contains exactly two elements

69. If both the mean and standard deviation of 50 observations $x_1, x_2 \dots x_{50}$ are equal to 16, then the mean of $(x_1-4)^2, (x_2-4)^2, \dots, (x_{50}-4)^2$ is K^2, then $\frac{K^2}{20}$ equals

(a) 14 (b) 20 (c) 22 (d) 28

70. Let α and β be two roots of the equation $x^2+2x+2=0$, then $\alpha^{15}+\beta^{15}+515$ is equal to

(a) 259 (b) -256 (c) -259 (d) 512

Section B : Numerical Value Type Questions

71. Let $\alpha_r = r \cdot {}^7C_r, \beta_r = (7-r){}^7C_r$ and $A_r = \begin{pmatrix} \alpha_r & 0 \\ 0 & \beta_r \end{pmatrix}$. If $A = \sum_{r=0}^{7} A_r = \begin{pmatrix} a & 0 \\ 0 & b \end{pmatrix}$, then the value of $a+b$ is equal to ....

72. If $a>0, b>0, c>0$ and the minimum value of $a(b^2+c^2)+b(c^2+a^2)+c(a^2+b^2)$ is $\frac{K}{2}abc$, then K is equal to ....

73. A pack of playing cards was found to contain only 51 cards. If the first 13 cards, which are examined are all red, the probability that the missing card is black is $\frac{\alpha}{\beta}$, then $\alpha\beta$ is equal to ....

74. If $\lim_{x\to 0}\left(\frac{p^x+q^x+r^x}{3}\right)^{\lambda/x}$ $(p, q, r$ and $\lambda > 0)$ is equal to pqr, then λ is ....

75. Let the shortest distance between lines
$L_1 : \mathbf{r} = 6\hat{\mathbf{i}} + \alpha\hat{\mathbf{j}} + \alpha\hat{\mathbf{k}} + \lambda(\hat{\mathbf{i}} - 2\hat{\mathbf{j}} + 2\hat{\mathbf{k}}), \alpha > 0$ and
$L_2 : \mathbf{r} = -4\hat{\mathbf{i}} - \hat{\mathbf{k}} + \mu(3\hat{\mathbf{i}} - 2\hat{\mathbf{j}} - 2\hat{\mathbf{k}})$ be 9 units.
If $(\alpha_1, \alpha_2, \alpha_3)$ lies on L_1, then the value of $(\alpha_2 + \alpha_3)$ is

DETAILED SOLUTIONS

Physics

1. *(b)* The energy of $n = 3$ state is

$$E_3 = \frac{-13.6\text{ eV}}{9} = -1.5\text{ eV}$$

The energy of $n = \infty$ state is zero.

The wavelength emitted in the given transition is

$$\lambda = \frac{hc}{\Delta E}$$

$$= \frac{1242\text{ eV nm}}{1.5\text{ eV}}$$

$$= 828\text{ nm}$$

2. *(b)*

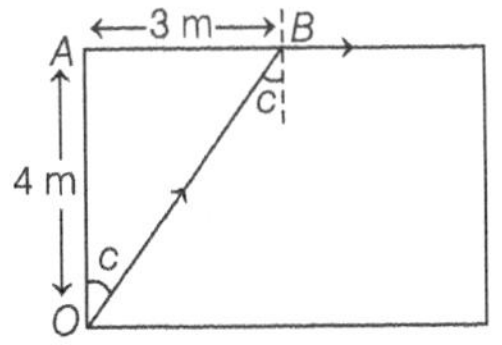

From figure, $OA = 4$ m, $AB = 3$ m

$\therefore \quad OB = \sqrt{4^2 + 3^2} = 5\text{ m}$

Now, $\quad \mu = \frac{1}{\sin C} = \frac{1}{AB/OB} = \frac{OB}{AB} = \frac{5}{3}$

From $\quad \mu = \frac{c}{v} \Rightarrow v = \frac{c}{\mu} = \frac{3 \times 10^8}{5/3}$

$$= 1.8 \times 10^8 \text{ m/s}$$

3. *(d)* Let us consider a neutral point O lies at a distance x from the dipole of moment p or at a distance $(25 - x)$ from dipole of $8p$.

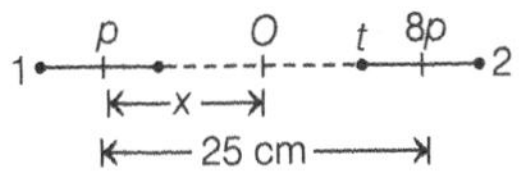

At O, electric field due to dipole (1) = electric field due to dipole (2)

$$\Rightarrow \frac{1}{4\pi\varepsilon_0}\left(\frac{2p}{x^3}\right) = \frac{1}{4\pi\varepsilon_0}\left(\frac{2(8p)}{(25-x)^3}\right)$$

$$\Rightarrow \frac{1}{x^3} = \frac{8}{(25-x)^3}$$

$$\Rightarrow x = (25/3)\text{ cm}$$

4. *(a)* $\frac{dB}{dt} = 1\text{ T/s}, A = \frac{\sqrt{3}}{4}l^2, \phi = BA$

By Faraday's law of EMI,

Induced emf, $\quad \varepsilon = \left|\frac{d\phi}{dt}\right|$

Induced current, $i = \frac{\varepsilon}{R} = \frac{1}{R} \cdot \left|\frac{d\phi}{dt}\right|$

$$= \frac{1}{R} \cdot \frac{d}{dt}(B.A)$$

$$= \frac{A}{R} \cdot \frac{dB}{dt} = \frac{1}{R} \cdot \frac{\sqrt{3}}{4} l^2 \times 1 = \frac{\sqrt{3}}{4} \cdot \frac{l^2}{R}$$

5. *(d)* Pressure inside a fluid volume open to atmosphere,

$$p = p_0 + h\rho g$$

where, p_0 = pressure at the fluid's surface,

h = depth,

ρ = density of fluid

and $\quad g$ = acceleration due to gravity.

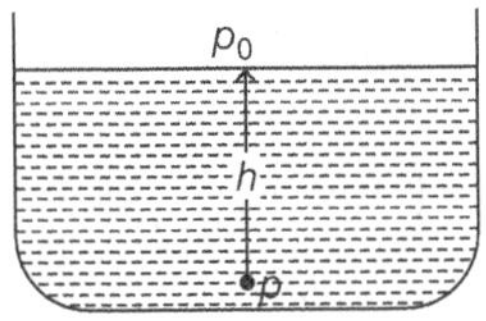

In the given problem,

When $h = d_1$, pressure, $p_1 = 7.07 \times 10^6$ Pa and when, $h = d_2$, pressure, $p_2 = 9.09 \times 10^6$ Pa

So, we have

$$p_1 = p_0 + d_1\rho g = 7.07 \times 10^6$$

$$p_2 = p_0 + d_2\rho g = 9.09 \times 10^6$$

$$\Rightarrow p_2 - p_1 = (d_2 - d_1)\rho g = 2.02 \times 10^6$$

$$d_2 - d_1 = \frac{2.02 \times 10^6}{\rho g}$$

$$\rho = 10^3 \text{ kgm}^{-3}$$

$$g = 10 \text{ ms}^{-2}$$

$$d_2 - d_1 = \frac{2.02 \times 10^6}{10^3 \times 10} = 2.02 \times 10^2 \text{ m}$$

$$= 202 \text{ m} \approx 200 \text{ m}$$

6. *(d)*

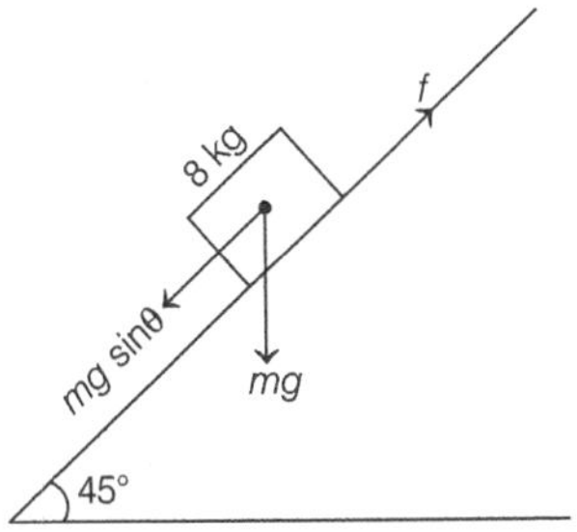

From the above diagram,

$$ma = mg\sin\theta - f$$
$$f = mg\sin\theta - ma$$
$$= m(g\sin\theta - a)$$
$$= 8\left(10\times\frac{1}{\sqrt{2}} - 0.5\right)$$
$$= 8(7.0 - 0.5) = 8\times 6.5 = 52\text{ N}$$

7. *(a)* Given, $\frac{dv}{dt} = bt \Rightarrow dv = bt\,dt$

$$\int_{v_0}^{v} dv = \int_0^t bt\,dt$$
$$[v]_{v_0}^{v} = b\left[\frac{t^2}{2}\right]_0^t \Rightarrow [v - v_0] = \frac{bt^2}{2}$$
$$v = v_0 + \frac{bt^2}{2}$$
$$\frac{dx}{dt} = v_0 + \frac{bt^2}{2}$$
$$dx = v_0\,dt + \frac{bt^2}{2}dt$$
$$\int_0^x dx = \int_0^t v_0\,dt + \int_0^t \frac{bt^2}{2}dt$$
$$x = v_0 t + \frac{bt^3}{6}$$

Hence, correct option is (a).

8. *(c)* Radio waves are detected by receiver's aerials, microwaves are detected by point contact diode, infrared waves are detected by thermopiles bolometer and ultraviolet rays are detected by photocell.

So, correct option is (c).

9. *(d)* In this question, as mass moves and pulley rotates, then by energy conservation,

Loss in potential energy = Gain in linear kinetic energy + Gain in rotational kinetic energy

Loss in PE $= (Mgh - mgh) = (M - m)gh$

Gain in linear KE $= \frac{1}{2}(m + M)v^2$

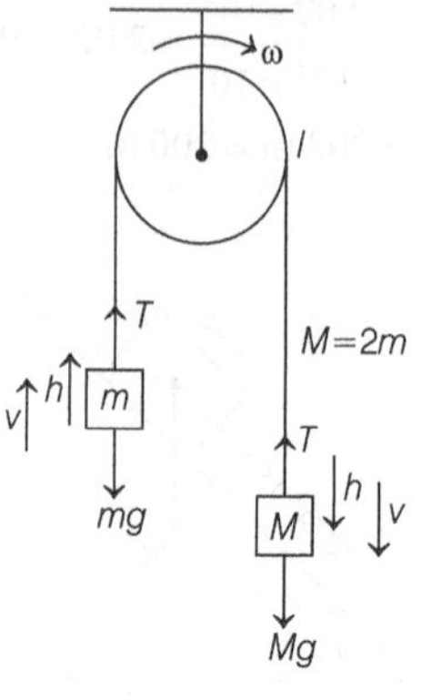

Gain in rotational KE $= \frac{1}{2}I\omega^2$

$$(M - m)gh = \frac{1}{2}(m + M)v^2 + \frac{1}{2}I\omega^2$$

We know that, $v = r\omega$

$$(M - m)gh = \frac{1}{2}(M + m)r^2\omega^2 + \frac{1}{2}I\omega^2$$
$$\Rightarrow \omega^2 = \frac{2(M - m)gh}{[(M + m)r^2 + I]}$$
$$\omega = \sqrt{\frac{2(M - m)gh}{(M + m)r^2 + I}}$$

10. *(b)* R_{eq} of the circuit $= \frac{R_1 \times R_2}{R_1 + R_2} + R_3$

$$= \frac{5\times 15}{5 + 15} + \frac{125}{100}$$
$$= \frac{75}{20} + \frac{5}{4} = 5\,\Omega$$
$$I = \frac{E}{R_{eq}} = \frac{20}{5} = 4\text{ A}$$

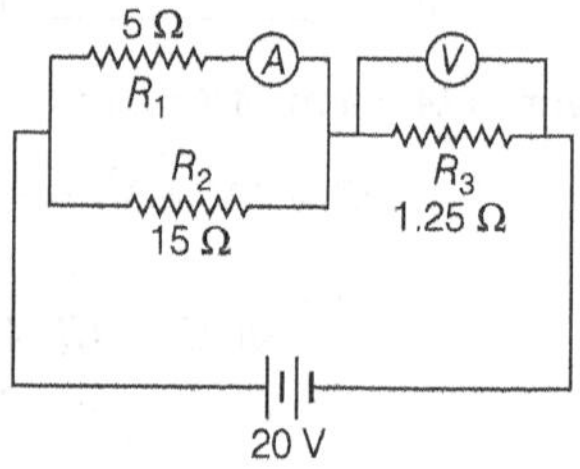

Current in $R_1 = \frac{IR_2}{R_1 + R_2} = 3\text{ A}$

Potential difference across $R_3 = IR_3$

$$= 4\times 1.25 = 5\text{ V}$$

11. *(c)* $i = i_0\cos(\omega t)$

$$i = i_0 \text{ at } t = 0$$
$$i = \frac{i_0}{\sqrt{2}} \text{ at } \omega t = \pi/4$$
$$t = \frac{\pi}{4\omega} = \frac{\pi}{4(2\pi f)} = \frac{1}{8f}$$
$$t = \frac{1}{8\times 50} = \frac{1}{400} = 2.5\text{ ms}$$

12. *(b)* When spring of spring constants $k_1, k_2, k_3, \ldots$ are connected in parallel, then their equivalent spring constant is given by

$$k_{eq} = k_1 + k_2 + k_3 \ldots$$

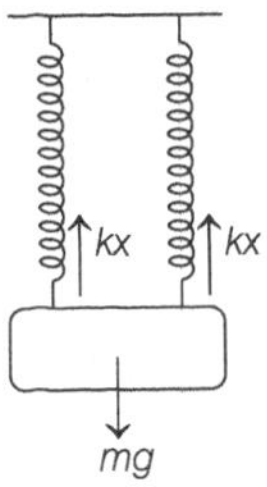

For two springs system,

$\therefore \quad k_{eq} = k + k = 2k$

So, $T = 2\pi\sqrt{\dfrac{m}{k_{eq}}}, T = 2\pi\sqrt{\dfrac{m}{2k}}$ and $2kx = mg \Rightarrow x = \dfrac{mg}{2k}$

13. *(b)* For process AB,

$$U \propto \frac{1}{\rho} \Rightarrow U\rho = \text{constant} \quad \ldots\text{(i)}$$

Also, $\rho = \dfrac{pM}{RT}$ and $U = C_V T \quad \ldots\text{(ii)}$

From Eqs. (i) and (ii), we get

$$C_V \cdot T \cdot \frac{pM}{RT} = \text{constant} = k$$

$$\Rightarrow \quad p = k \cdot \left(\frac{R}{MC_V}\right) = k \cdot \frac{R}{M.\frac{3R}{2}}$$

$$\left[\because (C_V)_{\text{monoatomic}} = \frac{3R}{2}\right]$$

$$= \frac{2k}{3M} = \text{constant}$$

$\therefore \quad p = \text{constant} \Rightarrow$ Process is isobaric.

p-V graph for the cycle is

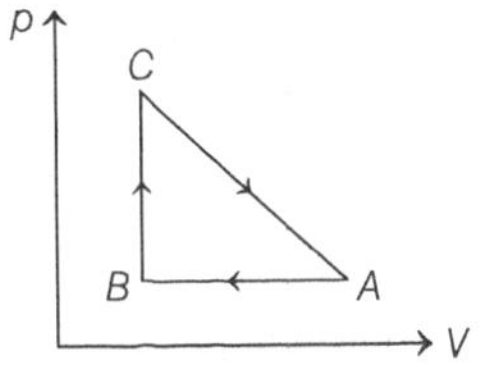

Work done in cycle is positive, hence net work in cycle is done by the gas.

14. *(b)* de-Broglie wavelength of the particle,

$$\lambda = \frac{h}{mv} \quad \ldots\text{(i)}$$

Wavelength of the photon,

$$\lambda = \frac{hc}{E} \quad \ldots\text{(ii)}$$

From Eqs. (i) and (ii), we get

$$\frac{h}{mv} = \frac{hc}{E} \Rightarrow E = mvc$$

Now, kinetic energy of the particle,

$$E_K = \frac{1}{2}mv^2$$

$$\therefore \quad \frac{E_K}{E} = \frac{\frac{1}{2}mv^2}{mvc} = \frac{v}{2c} = \frac{2.0 \times 10^8}{2 \times 3 \times 10^8} = \frac{1}{3}$$

15. *(d)* According to question,

$$v \propto \lambda^a \rho^b g^c \quad \ldots\text{(i)}$$

$$[v] = [k\lambda^a \rho^b g^c]$$

$$[v] = [\text{LT}^{-1}]$$

$$[k] = [\text{M}^0\text{L}^0\text{T}^0], [\rho] = [\text{ML}^{-3}]$$

$$[g] = [\text{LT}^{-2}], [\lambda] = [\text{L}]$$

$$\therefore \quad [\text{LT}^{-1}] = [\text{M}^0\text{L}^0\text{T}^0][\text{L}]^a[\text{ML}^{-3}]^b[\text{LT}^{-2}]^c$$

$$\Rightarrow \quad [\text{M}^0\text{LT}^{-1}] = [\text{M}^b\text{L}^{a-3b+c}\text{T}^{-2c}]$$

Comparing the exponents of M, L and T on both sides,

$$b = 0, \quad \left| \begin{array}{l} a - 3b + c = 1 \\ \Rightarrow a + \frac{1}{2} = 1 \\ \Rightarrow a = \frac{1}{2} \end{array} \right|, -2c = -1 \Rightarrow c = \frac{1}{2}$$

$\therefore$ From Eq. (i), we get

$$v \propto \lambda^{1/2}\rho^0 g^{1/2} \Rightarrow v^2 \propto \lambda g$$

16. *(a)* When rubber cord is stretched, it stores potential energy and when released, this potential energy is given to the stone as kinetic energy.

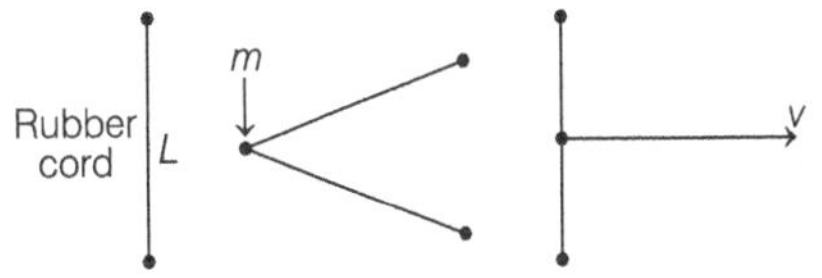

So, potential energy of stretched cord = Kinetic energy of stone

$$\Rightarrow \frac{1}{2}Y\left(\frac{\Delta L}{L}\right)^2 A \cdot L = \frac{1}{2}mv^2$$

Here, $\Delta L = 40\,\text{cm} = 0.4\text{ m}, L = 84\text{ cm} = 0.84\text{ m}$

$v = 20\,\text{ms}^{-1}, m = 0.04\text{ kg}$

$d = 12\,\text{mm} = 12 \times 10^{-3}\text{ m}$

$$\therefore \quad A = \pi r^2 = \pi\left(\frac{d}{2}\right)^2 = \pi\left(\frac{12 \times 10^{-3}}{2}\right)^2 = 36\pi \times 10^{-6}\,\text{m}^2$$

On substituting values, we get

$$Y = \frac{mv^2 L}{A(\Delta L)^2} = \frac{0.04 \times (20)^2 \times 0.84}{36\pi \times 10^{-6} \times (0.4)^2} = 0.74 \times 10^6\,\text{Nm}^{-2}$$

So, the closest value of Young's modulus is $10^6\,\text{Nm}^{-2}$.

17. *(d)* Reverse biasing of p-n junction diode increases the width of depletion region.

The barrier potential of p-n junction depends on doping density, temperature and type of semiconductor material and is independent of diode design.

Forward biasing of p-n junction decreases the width of potential barrier.
Hence, only statement III is correct.

18. *(b)* Statement I is correct but Statement II is not correct.
Statement II is incorrect as, a zero error occurs when the zero mark on vernier scale does not align perfectly with the zero mark of the main scale. It can be either positive or negative.

19. *(c)* The density of an ideal gas is

$$\rho = \frac{pM}{RT} \Rightarrow \rho = CP \text{ where, } C = \text{slope} = \frac{M}{RT}$$

i.e., $\quad \text{Slope} \propto \frac{1}{T}$

As, $\quad \theta_1 > \theta_2 \Rightarrow C_1 > C_2 \Rightarrow T_1 < T_2$

20. *(d)* The potential difference across the capacitor after inserting dielectric material, $V' = \frac{Q}{KC} = \frac{V}{K}$

($\because$potential difference decreases)

Also, potential difference across the capacitor,

$$V' = Ed \Rightarrow V' \propto d$$

21. (9792) For path ab, $(\Delta U)_{ab} = 7000\,\text{J}$

By using $\Delta U = nC_V \Delta T$

$$7000 = n \times \frac{5}{2} R \times 700 \Rightarrow n = 0.48$$

For path ca,

$$(\Delta Q)_{ca} = (\Delta U)_{ca} + (\Delta W)_{ca} \quad \text{... (i)}$$

$\because (\Delta U)_{ab} + (\Delta U)_{bc} + (\Delta U)_{ca} = 0$

$\therefore 7000 + 0 + (\Delta U)_{ca} = 0$

$\Rightarrow \quad (\Delta U)_{ca} = -7000\,\text{J} \quad \text{... (ii)}$

Also, $(\Delta W)_{ca} = p_1(V_1 - V_2) = nR(T_1 - T_2)$

$= 0.48 \times 8.31 \times (300 - 1000)] \quad \text{... (iii)}$

By solving Eqs. (i), (ii) and (iii), we get

$$(\Delta Q)_{ca} = -7000 - 2792.16 = -9792.16\,\text{J} \simeq -9792$$

22. (11) For ball A,

$$u_x = 10\,\text{m/s},\ u_y = 0,\ h = 490\,\text{m}$$

$$\therefore \quad h = u_y t_A + \frac{1}{2} g t_A^2 = \frac{1}{2} g t_A^2$$

$$\Rightarrow t_A = \sqrt{\frac{2h}{g}} = \sqrt{\frac{2 \times 490}{10}} = 7\sqrt{2}\,\text{s}$$

$$R_A = u_x \cdot t_A = 10 \times 7\sqrt{2} = 70\sqrt{2} \quad \text{...(i)}$$

For ball B,

$$u_x = u\cos 30° = \frac{\sqrt{3}u}{2},\ u_y = u\sin 30° = \frac{u}{2},\ h = 490\,\text{m}$$

$$\therefore \quad R_B = u_x t_B = \frac{\sqrt{3}u}{2} \cdot t_B \quad \text{...(ii)}$$

$$h = -u_y t_B + \frac{1}{2} g t_B^2 = -\frac{u}{2} t_B + \frac{1}{2} g t_B^2 \quad \text{...(iii)}$$

From Eqs.(i) and (ii), we get

$$R_A = R_B$$

$$\Rightarrow \quad 70\sqrt{2} = \frac{\sqrt{3}u}{2} \cdot t_B$$

$$\Rightarrow \quad u t_B = \frac{70 \times 2\sqrt{2}}{\sqrt{3}} \quad \text{...(iv)}$$

$\therefore$From Eq. (iii), we get

$$490 = -\frac{1}{2}\left(\frac{70 \times 2 \times \sqrt{2}}{\sqrt{3}}\right) + \frac{1}{2} \times 10 \times t_B^2$$

$$\Rightarrow \quad t_B = 10.46$$

$\therefore$ From Eq. (iv), $u = 10.9\,\text{m/s} \simeq 11\,\text{m/s}$

23. (3) As current in $2.5\,\Omega$ and $7\,\Omega$ resistors are zero.

$$\therefore \quad V_{AB} = \frac{\frac{E_1}{r_1} + \frac{E_2}{r_2} + \frac{E_3}{r_3}}{\frac{1}{r_1} + \frac{1}{r_2} + \frac{1}{r_3}} = \frac{\frac{10}{2} + \frac{16}{4} - \frac{6}{1.5}}{\frac{1}{2} + \frac{1}{4} + \frac{1}{1.5}} = 3.52\,\text{V} \simeq 3\,\text{V}$$

24. (1) In YDSE,
The distance of nth maxima from the central maxima,

$$x_n = \frac{n\lambda D}{d}$$

$$\Rightarrow \quad 3 = \frac{10\lambda D}{d} \Rightarrow \frac{\lambda D}{d} = \frac{3}{10} \quad \text{...(i)}$$

When the apparatus is immersed in a liquid of refractive index $\mu = 1.5$,

$$x_n' = \frac{n\lambda D}{\mu d} = \frac{n}{\mu}\left(\frac{\lambda D}{d}\right)$$

$$\Rightarrow \quad x_5' = \frac{5}{1.5} \times \frac{3}{10} \quad \text{[from Eq. (i)]}$$

$$= 1\,\text{cm}$$

$$x = 1\,\text{cm}$$

25. (193) Weight of object at pole,

$$w_p = mg \quad \text{...(i)}$$

Weight of object at equator,

$$w_e = mg - m\omega^2 R$$

$$= w_p - m\omega^2 R$$

$$\therefore \quad w_p = w_e + m\omega^2 R \quad \left[\because m = \frac{w_e}{g}\right]$$

$$\omega = \frac{2\pi}{T}$$

$$w_p = 192 + \frac{192}{10} \times \left(\frac{2\pi}{24 \times 3600}\right)^2 \times 6400 \times 10^3$$

$$= 192.64\,\text{N}$$

According to the question,

$$x\text{N} = 192.64\,\text{N}$$

$$\therefore \quad x = 192.64$$

$$\simeq 193$$

Chemistry

26. (*c*)

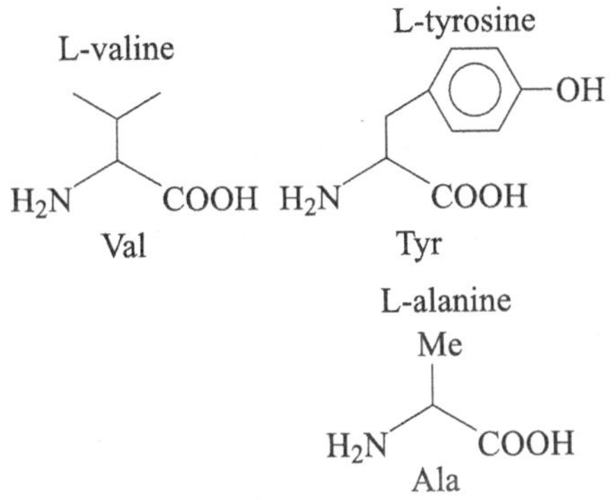

six various possible isomers are

Val–Tyr–Ala Tyr–Ala–Val
Val–Ala–Tyr Ala–Tyr–Val
Tyr–Val–Ala Ala–Val–Tyr

27. (***a***) The correct match is A-II, B-IV, C-I, D-III.

(A) $CHO(CHOH)_4CH_2OH \xrightarrow{HI,\ \Delta} CH_3—CH_2—CH_2—CH_2—CH_3$ (*n*-hexane)

(B) $CHO(CHOH)_4CH_2OH \xrightarrow{Br_2\ water} COOH(CHOH)_4CH_2OH$ (Gluconic acid)

(C) $CHO(CHOH)_4CH_2OH \xrightarrow{Acetic\ anhydride} CHO(CH—O—\overset{O}{\overset{\|}{C}}—CH_3)_4CH_2—O—\overset{O}{\overset{\|}{C}}—CH_3$ (Glucose pentaacetate)

(D) $CHO(CHOH)_4CH_2OH \xrightarrow{HNO_3} COOH(CHOH)_4COOH$ (Saccharic acid)

28. (***d***) Organic compound $\xrightarrow[Fused]{Na}$ sodium extract

For nitrogen detection Sodium extract is taken and freshly prepared $FeSO_4$ is added to it followed by few drops of $FeCl_3$. The resulting solution is acidified with conc.HCl. Prussian blue colour appears due to formation of ferriferrocyanide.

$$8NaCN + 2FeSO_4 \xrightarrow[-2Na_2SO_4]{} Fe(CN)_2 + Na_4[Fe(CN)_6]$$

$$\xrightarrow{xH_2O \downarrow\ 4FeCl_3} \underset{\text{Ferriferrocyanide (Prussian blue colour)}}{Fe_4[Fe(CN)_6]_3 xH_2O}$$

Detection of sulphur and nitrogen When sodium extract is taken and $FeCl_3$ is added to it, this lead to the formation of ferric thiocyanate (blood red colouration).

$$\underset{\text{Sodium thiocyanate}}{NaSCN} \xrightarrow{FeCl_3} \underset{\text{Ferricthiocyanate (blood red)}}{[Fe(SCN)]Cl_2}$$

Detection of chloride Sodium extract is taken, and a solution of HNO_3 is added to it followed by heating and addition of $AgNO_3$ then we get a white precipitate of AgCl.

$$NaCl + AgNO_3 \xrightarrow{\Delta} \underset{\text{White ppt. (soluble in } NH_4OH)}{AgCl\downarrow} + NaNO_3$$

But $Fe(CN)_3$ is never formed during Lassaigne's test.

29. (*c*) In bicyclo compounds, numbering starts from bridgehead carbon. Cycle is completed from larger ring to smaller ring.

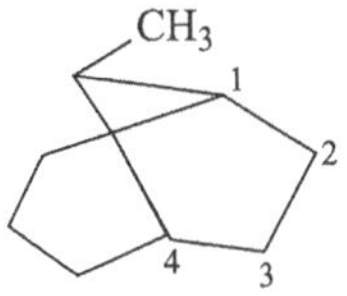

8-methyl-1-bicyclo [3, 2, 1] octane

30. (***b***)

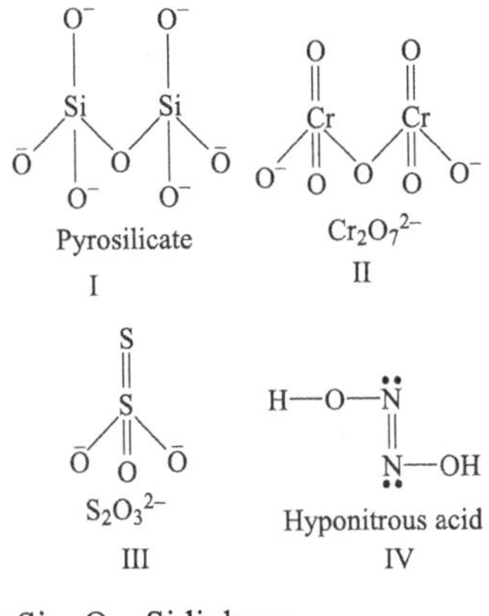

(I) have Si—O—Si linkage
(II) have Cr—O—Cr linkage.

31. (*c*) We know that, $S = \dfrac{x}{M}$

Ionisation reaction,

$$A_2B_3 \rightleftharpoons \underset{2S}{2A^{3+}} + \underset{3S}{3B^{2-}}$$

$$K_{sp} = (2S)^2(3S)^3 = 4S^2 \times 27S^3 = 108S^5$$

$$\therefore \frac{[B^{2-}]}{K_{sp}} = \frac{3S}{108S^5} = \frac{1}{36S^4}$$

$$= \frac{1}{36 \times \left(\frac{x}{M}\right)^4} = \frac{M^4}{36x^4}$$

32. (*a*)

Complex	No. of electrons in outer *d*-orbital	No. of unpaired electron(s)
$[Fe(CN)_6]^{3-}$ (*K*)	$3d^5$	1(CN^- causes pairing of electrons)
$[Ni(CN)_4]^{2-}$ (*L*)	$3d^8$	0(CN^- causes pairing of electrons)
$[Co(NH_3)_6]^{3+}$ (*M*)	$3d^6$	0(NH_3 causes pairing of electrons)
$[Co(ox)_3]^{3-}$ (*N*)	$3d^6$	0 (oxalate causes pairing of electrons)
$[Ni(H_2O)_6]^{2+}$ (*O*)	$3d^8$	2
$[Zn(H_2O)_6]^{2+}$ (*P*)	$3d^{10}$	0

Thus, *L, M, N* and *P* are diamagnetic.

33. (*b*) The complete reaction can be written as

$$[Ni(NH_3)_6]Cl_2 + 2Ag^+ \longrightarrow \underset{(X)}{[Ni(NH_3)_6]^{2+}} + \underset{(Y)}{2AgCl}$$

The oxidation state of Ni in $[Ni(NH_3)_6]^{2+}$ is + 2.

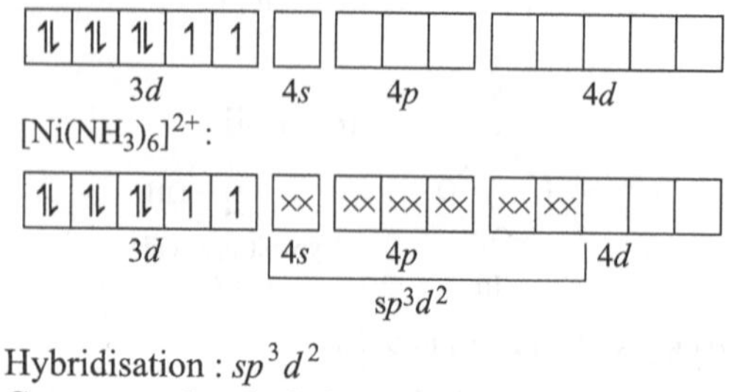

Hybridisation : sp^3d^2

Geometry : Octahedral

34. (*d*)

$$3\overset{0}{Br_2} + 6\overset{2-}{CO_3^{2-}} + 3H_2O \longrightarrow 5Br^- + \overset{+5}{BrO_3^-} + 6\overset{+4}{HCO_3^{-1}}$$

(Reduction: $Br_2 \to Br^-$; Oxidation: $Br_2 \to BrO_3^-$)

In this reaction Br is oxidised as well as reduced.

35. (*a*) Salt is dissolved in water to form a solution. Barium chloride test is performed for the confirmation of sulphite ions. If the precipitate formed after addition of $BaCl_2$ are dissolved in dil. HCl then sulphite ions are present and if the precipitate remain undissolved then the ion is sulphate.

36. (*c*) **Pinacol-pinacolone rearrangement** The conversion of diol to α-hydroxy ketone on reaction with protic acid is known as pinacol-pinacolone rearrangement and this reaction proceeds as

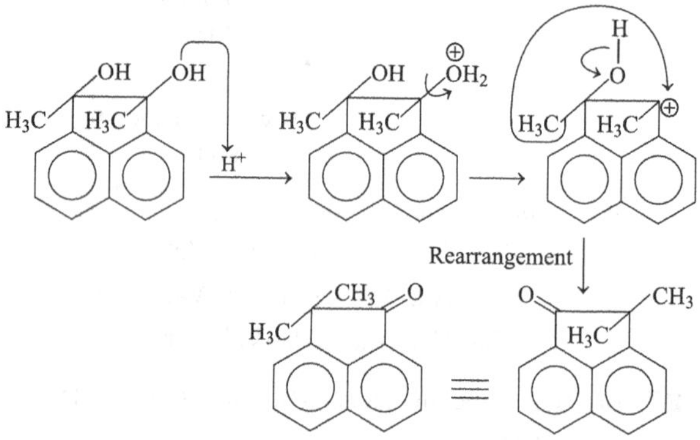

37. (*a*) Greater the value of standard reduction potential, higher will be the oxidising character. Standard reduction potential of MnO_4^- is greater than that of other three. So it is a stronger oxidising agent than the others.

38. (*a*) **Determination of molecular structure of** C_4H_6ClBr

Degree of unsaturation (*u*)

$$u = (C + 1) - \frac{H}{2} - \frac{X}{2} + \frac{N}{2}$$

$$= (4+1) - \frac{6}{2} - \frac{2}{2} = (4+1) - \frac{8}{2} = 1$$

According to question, the compound is cyclic which means the compound is cyclobutane derivative having Cl and Br at diagonal position. Hence, the structure of compound is

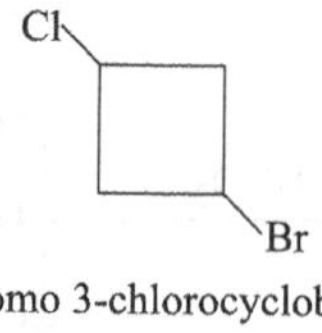

1-bromo 3-chlorocyclobutane

This compound on reaction with metallic sodium produces bicyclic product as follows

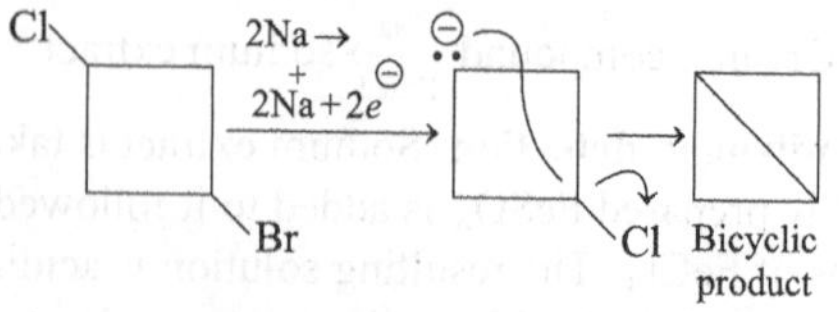

Saturated monocyclic compound (reactant) has $u = 1$

Saturated bicyclic compound (product) has $u = 2$

39. (*c*) The mechanism of the given reaction sequence is as follow

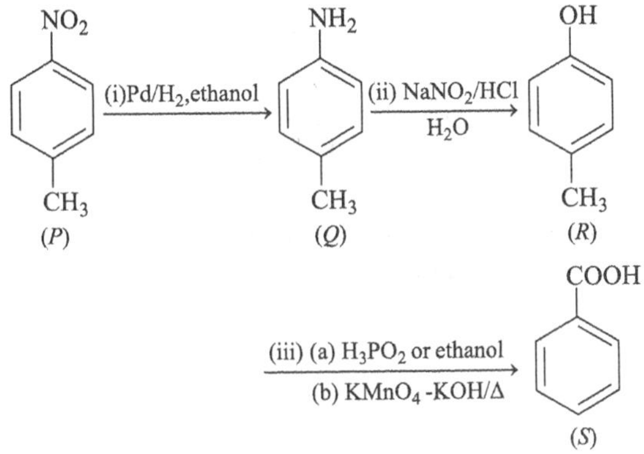

40. (*c*) Oil of winter green or methyl salicylate is used as medicine.

41. (*b*) The acidic strength of HBr > HCl because down the group, electronegativity decreases and size increase. Thus, it becomes easy to lose H^+ ion. Similarly, their reducing property is also in same order.

The basic strength of $PH_3 > AsH_3$ because down the group, the size of central atom increases, thereby their orbital become more diffused, hence the nucleophilicity of the lone pair is less dense over the bigger atoms, so basicity decreases. Therefore, PH_3 is more basic than AsH_3.

Similarly, their bond angles are in the order of

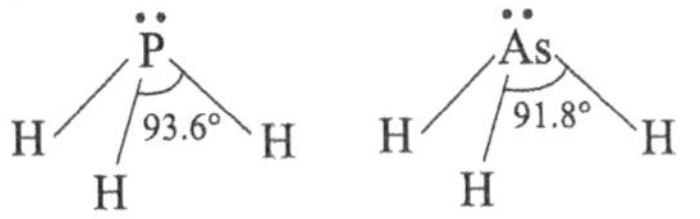

42. (*b*) $O_2 \to 2, O_2^+ \to 2.5, O_2^- \to 1.5, O_2^{2-} \to 1.$

Hence, the correct match is A-(III), B-(II), C-(I), D-(IV).

43. (*b*) Conjugate acid of 3° amines are less stable than 2° amines due to low hydration.

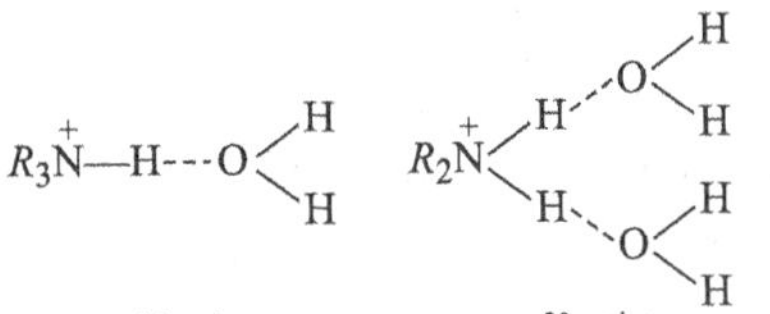

3° amine (Protonated 3° amine can form H-bonding with water molecule only at one) [Less stable]

2° amine (Protonated 2° amine can form H-bonding with water molecule at two) [More stable]

44. (*a*) The reagent having greater value of nucleophilic character undergo faster nucleophilic substitution reaction.

Steric hindrance play an important role according to which greater the steric hindrance lower the reactivity towards S_N2.

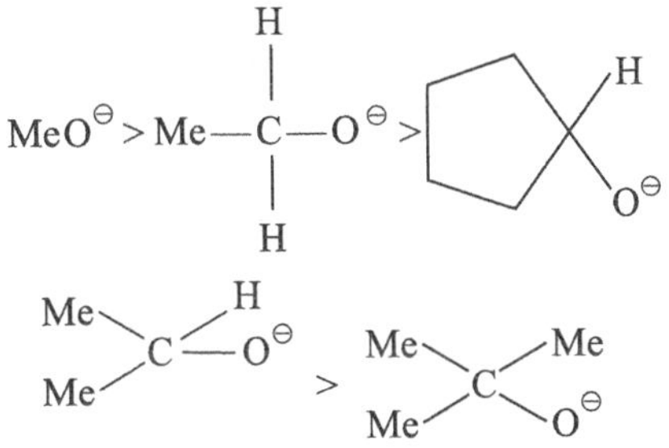

Number of electron donating group increases, electron density on oxygen increases. Thus, bulkiness of nucleophile increases.Extent of S_N2 reaction decreases.

45. (*a*) Isoelectronic means same number of electrons and isostructural means same structure.

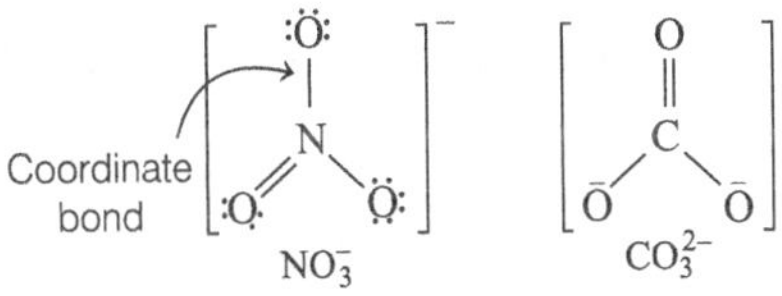

46. (*721*) For calculating $\Delta H°$ for the reaction

$$4CO(g)+8H_2(g) \longrightarrow 3CH_4(g)+CO_2(g)+2H_2O(l)$$

$$C+\frac{1}{2}O_2 \longrightarrow CO(g), \Delta H_1^\circ = -110.5\,kJ$$

$$3C+\frac{3}{2}O_2 \longrightarrow 3CO, \Delta H_I^\circ = 3\times(-110.5)\,kJ \quad \ldots (i)$$

$$CO+\frac{1}{2}O_2 \longrightarrow CO_2, \Delta H_2^\circ = -282.9\,kJ \quad \ldots (ii)$$

$$2H_2 + O_2 \longrightarrow 2H_2O, \Delta H_{III}^\circ = 2\times(-285.8)\,kJ \ldots (iii)$$

$$3C + 6H_2 \longrightarrow 3CH_4, \Delta H_{IV}^\circ = 3\times(-74.8)\,kJ \ldots (iv)$$

$$\Delta H° = -3\Delta H_1^\circ + \Delta H_2^\circ + 2\Delta H_3^\circ + 3\Delta H_4^\circ$$

$$= -3\times(-110.5)+(-282.9)+2(-285.8)+3(-74.8)$$

$$= -747.4\,kJ$$

$$\Delta H° = \Delta U° + \Delta n_g RT$$

$$\because \Delta n_g = -8$$

$$-747.4 = \Delta U° - \frac{8\times 8.314\times 400}{1000}$$

$$= 747.4 + 26.60 = \Delta U°$$

$$\Delta U° = -720.79\,kJ \simeq -721\,kJ$$

47. (*9*) For a first order reaction,

$$kt = \ln\frac{[A]_0}{[A]}$$

where, $[A]_0$ = initial concentration.

$[A]$ = concentration of reactant remaining at time *t*.

$$kt_{1/8} = \ln\frac{[A]_0}{[A]_0/8} = \ln 8 \quad \ldots (i)$$

$$kt_{1/10} = \ln \frac{[A]_0}{[A]_0/10} = \ln 10 \quad \ldots\text{(ii)}$$

Therefore,

$$\frac{t_{1/8}}{t_{1/10}} = \frac{\ln 8}{\ln 10} = \log 8 = 3 \log 2$$

$$\frac{t_{1/8}}{t_{1/10}} = 3 \times 0.3 = 0.9$$

$$\frac{t_{1/8}}{t_{1/10}} \times 10 = 0.9 \times 10 = 9.0$$

48. (*20*) Molar mass of non electrolyte $M = 200\text{ g mol}^{-1}$

No. of mole $= \frac{x}{200}$

Total no. of mole of Na^+ and $Cl^- = 2 \times 1\text{L} \times 0.05\text{M}$

$= 0.1$ mole

$$C = \left(\frac{x}{200} + 0.1\right) \Big/ 1\text{litre}$$

$$= \left(\frac{x}{200} + 0.1\right)$$

Osmotic pressure,

$$\pi = iCRT$$

$$4.92 = \left\{\frac{x}{200} + 0.1\right\} \times 0.0821 \times 300$$

$$\Rightarrow \quad \frac{x}{200} + 0.1 = \frac{4.92}{0.0821 \times 300} = 0.199 \approx 0.2$$

$$\therefore \quad \frac{x}{200} + 0.1 = 0.2$$

$$\Rightarrow \quad x = 0.1 \times 200 = 20\text{ g}$$

49. (*10*) Molecular mass of given salt

$$= 23 \times 2 + 32 + 64 + 18x$$
$$= 142 + 18x$$

Mass of water of hydration $= 18x$

% of water in hydrate $= 56\%$

$$\therefore \frac{18x}{142 + 18x} \times 100 = 56$$

$$1800x = 56(142 + 18x)$$
$$1800x = 7952 + 1008x$$
$$792x = 7952$$
$$x = \frac{7952}{792}$$
$$x = 10.04$$
$$x \approx 10$$

So, the formula of crystalline salt $= Na_2SO_4 \cdot 10H_2O$.

50. (*2*) As we know,

$$\text{K.E. per atom} = \frac{\text{energy absorbed} - \text{bond energy}}{\text{no. of atoms in the molecules}}$$

and given molecules is diatomic (A_2)

$\therefore$ Energy absorbed $= 4.4 \times 10^{-19}$ J

Bond energy $= 4 \times 10^{-19}$ J

$$\therefore \text{KE per atom} = \frac{(4.4 \times 10^{-19}) - (4 \times 10^{-19})}{2} = 2 \times 10^{-20}\text{ J}$$

Mathematics

51. (c) Given, differential equation

$$x^3 \frac{dy}{dx} + 4x^2 \tan y = e^x \sec y$$

$$\Rightarrow \quad \frac{dy}{dx} + \frac{4 \tan y}{x} = \frac{e^x \sec y}{x^3}$$

$$\Rightarrow \quad \cos y \frac{dy}{dx} + \frac{4 \sin y}{x} = \frac{e^x}{x^3} \quad \ldots\text{(i)}$$

Let $t = \sin y$

$$\Rightarrow \quad \frac{dt}{dx} = \cos y \frac{dy}{dx}$$

Eq.(i) becomes, $\frac{dt}{dx} + \frac{4t}{x} = \frac{e^x}{x^3}$,

which is linear differential equation in t and x.

Then, integrating factor

$$\text{IF} = e^{\int \frac{4}{x} dx} = e^{4 \log x} = x^4$$

Now, solution is given by

$$t \cdot x^4 = \int x^4 \cdot \frac{e^x}{x^3} dx + C$$

$$x^4 \cdot \sin y = e^x (x - 1) + C$$

At $x = 1$, $y = 0 \Rightarrow C = 0$

$\therefore$ Solution is $\sin y = e^x (x - 1) x^{-4}$

On comparing it with $\sin y = e^{\alpha x} (\beta x + \gamma) x^{-k}$, we get

$\alpha = 1, \beta = 1, \gamma = -1$ and $k = 4$

$\therefore (\alpha + \beta + \gamma)k = (1 + 1 - 1)4 = 4$

52. (c) We have, equation $f(x) = e^x - cx^2 = 0$

If equation has three distinct real roots, then $f'(x) = 0$ must have two distinct roots as between two roots of the equation $f(x) = 0$, there lies at least one root of $f'(x) = 0$.

$$f'(x) = e^x - 2cx$$

Roots of $f'(x)=0$ occur for values of x, where graphs of $y=e^x$ and $y=2cx$ intersect.

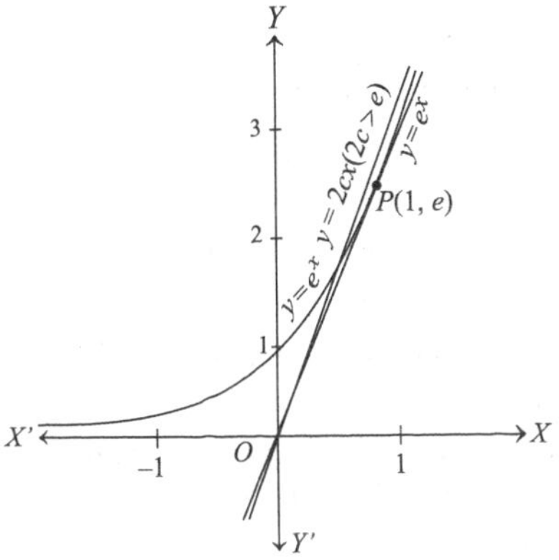

For the curve, $\frac{dy}{dx}=e^x$

Also, slope of the line is $2c$.

Let the line touch the curve at $P(x_1, y_1)$.

Then, $2c=e^{x_1}$, $y_1=e^{x_1}$ and $y_1=2cx_1$.

$\therefore x_1=1$ and $y_1=e$

So, point P lies at $(1, e)$.

The line intersects the curve at two points, if slope of line is greater than e.

$\Rightarrow \quad 2c>e$

$\therefore \quad c>e/2$

Therefore, least positive integral value of c is 2.

53. *(a)* Given, $A=\begin{bmatrix}3 & a & 0\\ a & a^2+1 & a\\ 0 & a & 3\end{bmatrix}$

$$|A|=\begin{vmatrix}3 & a & 0\\ a & a^2+1 & a\\ 0 & a & 3\end{vmatrix}$$

$$\Rightarrow \quad |A|=3(3a^2+3-a^2)-a(3a)+(0)$$

$$\Rightarrow \quad |A|=9a^2+9-3a^2-3a^2$$

$$\Rightarrow \quad |A|=3a^2+9$$

We have to calculate, $\min\left(\frac{\det(A)}{a}\right)$

$$\Rightarrow \quad f(a)=\frac{3a^2+9}{a}=3a+\frac{9}{a}$$

$$\Rightarrow f'(a)=3-\frac{9}{a^2}$$

$$\Rightarrow f''(a)=0+\frac{18}{a^3}$$

For maxima and minima,

$$f'(a)=0$$

$$\therefore \quad 3-\frac{9}{a^2}=0$$

$$\Rightarrow \quad \frac{9}{a^2}=3 \Rightarrow a=\sqrt{3} \qquad [\because a>0]$$

and $f''(\sqrt{3})>0$

Therefore, $f(a)$ is minimum at $a=\sqrt{3}$.

Therefore, minimum value of $f(a)=\frac{\det(A)}{a}$ is

$$f(\sqrt{3})=3\sqrt{3}+\frac{9}{\sqrt{3}}$$

54. *(b)* $\because I=\int_{\pi}^{2\pi}[2\cos x]dx$

$$\Rightarrow I=\int_{\pi}^{\frac{4\pi}{3}}[2\cos x]dx+\int_{\frac{4\pi}{3}}^{\frac{3\pi}{2}}[2\cos x]dx+\int_{\frac{3\pi}{2}}^{\frac{5\pi}{3}}[2\cos x]\,dx+\int_{\frac{5\pi}{3}}^{2\pi}[2\cos x]\,dx$$

$$\Rightarrow I=\int_{\pi}^{\frac{4\pi}{3}}-2dx+\int_{\frac{4\pi}{3}}^{\frac{3\pi}{2}}-1dx+\int_{\frac{3\pi}{2}}^{\frac{5\pi}{3}}0dx+\int_{\frac{5\pi}{3}}^{2\pi}1dx$$

$$\Rightarrow I=-2\left(\frac{4\pi}{3}-\pi\right)-\left(\frac{3\pi}{2}-\frac{4\pi}{3}\right)+0+\left(2\pi-\frac{5\pi}{3}\right)$$

Hence, $I=\frac{-2\pi}{3}-\frac{\pi}{6}+\frac{\pi}{3}=-\frac{\pi}{2}$

55. *(b)*

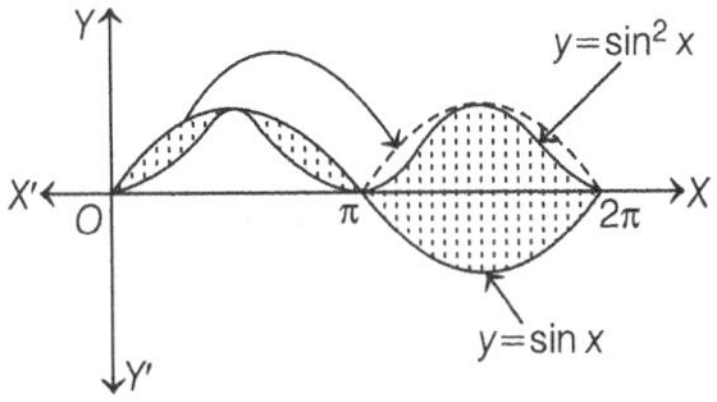

We know that the area bounded by $y=\sin x$ and X-axis for $x\in[0,\pi]$ is 2 sq units.

Then, the area bounded by $y=\sin x$ and $y=\sin^2 x$ is 4 sq units for $x\in[0,2\pi]$.

Then, for $x\in[0,10\pi]$, the area bounded is 20 sq units.

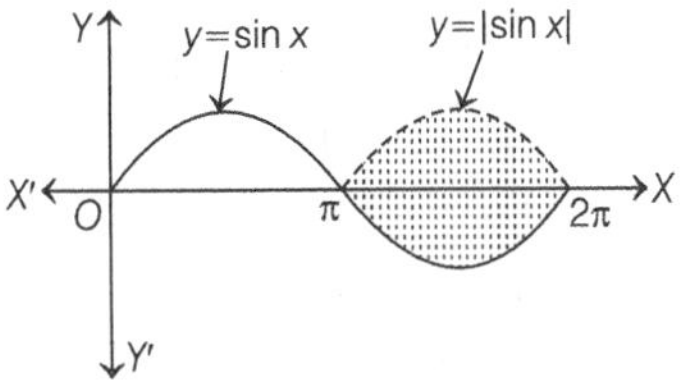

The area bounded by $y=\sin x$ and $y=|\sin x|$ for $x\in[0,2\pi]$ is 4 sq units.

Then, for $x\in[0,20\pi]$, the bounded area is 40 sq units.

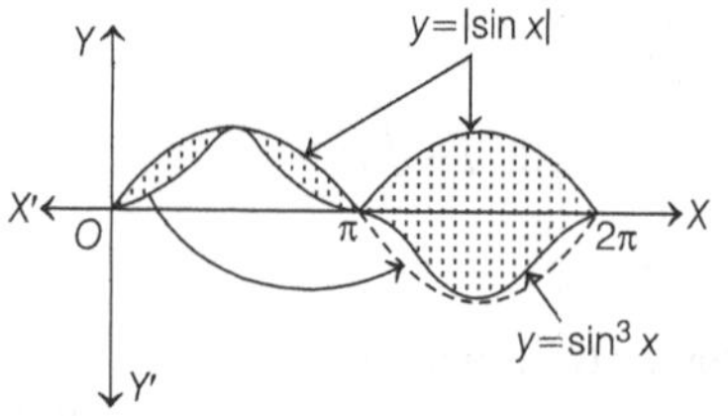

The area bounded by $y=|\sin x|$ and $y=\sin^3 x$ for $x\in[0,2\pi]$ is 4 sq units.

Then, for $x\in[0,10\pi]$ the bounded area is 20 sq units.

56. *(d)* Given parabola is $y=4ax^2$

$$\Rightarrow \quad x^2=\frac{y}{4a}$$

On putting $x^2=\frac{y}{4a}$ in the circle $x^2+(y-1)^2=1$,

we get

$$\Rightarrow \frac{y}{4a}+y^2-2y=0$$

(Note that for $4a<0$ they cannot intersect other than the origin)

Hence, we get

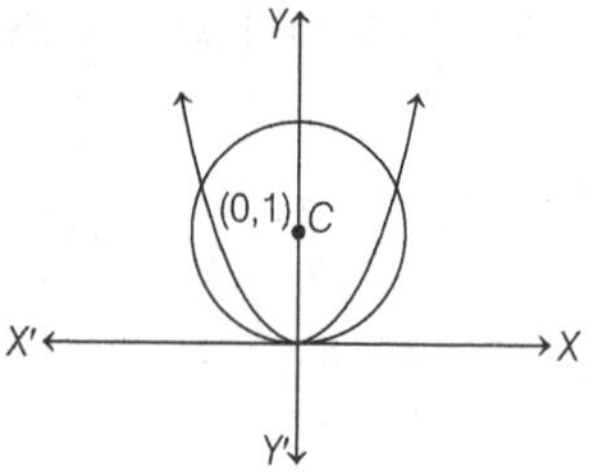

$$y=0 \text{ or } y=2-\frac{1}{4a}$$

On substituting $y=2-\frac{1}{4a}$ in $y=4ax^2$, we get

$$2-\frac{1}{4a}=4ax^2$$

$$\Rightarrow \quad x^2=\frac{8a-1}{16a^2}>0$$

$$\Rightarrow \quad 8a-1>0$$

$$\Rightarrow \quad a>\frac{1}{8}$$

57. *(d)* Since, $x^2-x-2=(x+1)(x-2)$

For $x<2$, $(x+1)(x-2)<0$

$\therefore |x^2-x-2|=-(x^2-x-2)$

and for $x\in(2,3)$, $x-[x]=x-2$

$$\text{Therefore, } f(x)=\begin{cases} p, & x<2 \\ q, & x=2 \\ 1, & x>2 \end{cases}$$

Since, $f(x)$ is continuous at $x=2$

$\Rightarrow \quad p=q=1$

$\therefore \quad p+q=2$

58. *(b)* We have, the ellipse

$$\frac{x^2}{25}+\frac{y^2}{16}=1$$

Let the auxiliary circle has equation

$$x^2+y^2=a^2$$

Parametric coordinates on auxiliary circle are $(a\cos\theta, a\sin\theta)$

$\because$ Parametric coordinates of the same point when it is projected on the ellipse becomes $(a\cos\theta, b\sin\theta)$.

So, consider

$$A(\theta), B\left(\theta+\frac{2\pi}{3}\right) \text{ and } C\left(\theta+\frac{4\pi}{3}\right)$$

Then, $P\equiv(5\cos\theta, 4\sin\theta)$

$$Q\equiv\left(5\cos\left(\theta+\frac{2\pi}{3}\right), 4\sin\left(\theta+\frac{2\pi}{3}\right)\right)$$

$$R\equiv\left(5\cos\left(\theta+\frac{4\pi}{3}\right), 4\sin\left(\theta+\frac{4\pi}{3}\right)\right)$$

Let the centroid of ΔPQR be (x, y).

$$x=5\left[\frac{\cos\theta+\cos\left(\theta+\frac{2\pi}{3}\right)+\cos\left(\theta+\frac{4\pi}{3}\right)}{3}\right]$$

$$=\frac{5}{3}\left[\cos\theta+2\cos(\theta+\pi)\cos\frac{\pi}{3}\right]=0$$

$$y=\frac{4}{3}\left[\sin\theta+\sin\left(\theta+\frac{2\pi}{3}\right)+\sin\left(\theta+\frac{4\pi}{3}\right)\right]$$

$$=\frac{4}{3}\left[\sin\theta+2\sin(\theta+\pi)\cos\frac{\pi}{3}\right]=0$$

$$=0$$

Hence, centroid of ΔPOQ lies at centre of ellipse.

59. *(d)* Given lines are

$$\frac{x}{2}=\frac{y}{2}=\frac{z}{1} \quad \text{...(i)}$$

$$\text{and} \quad \frac{x-5}{2}=\frac{y-2}{\frac{p}{7}}=\frac{z-3}{4} \quad \text{...(ii)}$$

Direction ratios of line (i) are 2,2 and 1.

Direction ratios of line (ii) are $2, \frac{p}{7}$ and 4.

Angle between line (i) and line (ii) is $\cos^{-1}\left(\frac{2}{3}\right)$.

$$\therefore \cos^{-1}\frac{2}{3}=\cos^{-1}\frac{4+\frac{2p}{7}+4}{\sqrt{2^2+2^2+1^2}\sqrt{2^2+\left(\frac{p}{7}\right)^2+4^2}}$$

$$\Rightarrow \quad \frac{2}{3}=\frac{56+2p}{3\sqrt{p^2+980}}$$

$$\Rightarrow \sqrt{p^2+980}=p+28$$

$$\Rightarrow \quad p^2+980=p^2+56p+784$$

$$\Rightarrow \quad 56p=196 \Rightarrow p=\frac{7}{2}$$

60. *(b)* Given that

$|\mathbf{a}|=1, |\mathbf{b}|=1$ and $|\mathbf{a}+\mathbf{b}|=1$

Then, $\quad |\mathbf{a}+\mathbf{b}|^2=1$

$\Rightarrow \quad (\mathbf{a}+\mathbf{b})\cdot(\mathbf{a}+\mathbf{b})=1$

$\Rightarrow \mathbf{a}\cdot\mathbf{a}+\mathbf{a}\cdot\mathbf{b}+\mathbf{b}\cdot\mathbf{a}+\mathbf{b}\cdot\mathbf{b}=1$

$\Rightarrow \quad 1+1+2\mathbf{a}\cdot\mathbf{b}=1$

$\Rightarrow \quad 2\mathbf{a}\cdot\mathbf{b}=-1$

$\Rightarrow \quad \mathbf{a}\cdot\mathbf{b}=-\frac{1}{2}$

$\Rightarrow \quad |\mathbf{a}|\cdot|\mathbf{b}|\cos\theta=-\frac{1}{2}$

$\Rightarrow \quad \cos\theta=-\frac{1}{2}$

$\Rightarrow \quad \cos\theta=\cos 120^\circ$

$\Rightarrow \quad \theta=\frac{2\pi}{3}=K\pi \quad$ [given]

Hence, $\quad K=\frac{2}{3}$

61. *(a)* Domain of f is $[-1,1]$;

$$f(x)=\sin x+\cos x+\tan x+\sin^{-1}x+\cos^{-1}x+\tan^{-1}x$$

$$f'(x)=\cos x-\sin x+\underbrace{\sec^2 x}_{>1}+0+\underbrace{\frac{1}{1+x^2}}_{[1/2,1]}$$

Hence, $f'(x)>0 \Rightarrow f$ is increasing.

$\Rightarrow$ Range is $[f(-1), f(1)]$

$$\therefore \quad f(x)|_{\min}=f(-1)=-\sin 1+\cos 1-\tan 1-\frac{\pi}{2}+\pi-\frac{\pi}{4}$$

$$=\frac{\pi}{4}+\cos 1-\sin 1-\tan 1$$

and $f(x)|_{\max}=f(1)=\sin 1+\cos 1+\tan 1+\frac{\pi}{2}+\frac{\pi}{4}$

$$=\frac{3\pi}{4}+\cos 1+\sin 1+\tan 1$$

Now, $\frac{M+m}{2}=\frac{\pi}{2}+\cos 1$

62. *(b)* $\because 3^{100}=9^{50}=(1-10)^{50}$

$=1-{}^{50}C_1\cdot 10+{}^{50}C_2\cdot 10^2-{}^{50}C_3\cdot 10^3+$ Multiple of 10^4

$=1-50\times 10+\frac{50\times 49}{2}\times 100-\frac{50\times 49\times 48}{1\times 2\times 3}\times 1000$ + Multiple of 10^4

$=1-500+122500+$ Multiple of 10^4

$=122001+$ Multiple of 10^4

$\Rightarrow$ Last four digits $=2001$

$p=2, q=0, r=0$ and $s=1$

Therefore, $p+q+r+s=2+0+0+1=3$

63. *(b)* Since,

$|z-1|=|z-i|=|z+1|$

$\therefore \quad |z-1|=|z+1|$

$|z-1|^2=|z+1|^2$

$\Rightarrow \quad (z-1)\overline{(z-1)}=(z+1)\overline{(z+1)}$

$\Rightarrow \quad (z-1)(\bar{z}-1)=(z+1)(\bar{z}+1)$

$\Rightarrow \quad z\bar{z}-z-\bar{z}+1=z\bar{z}+z+\bar{z}+1$

$\Rightarrow \quad z+\bar{z}=0$

$\Rightarrow \quad x+iy+x-iy=0 \Rightarrow x=0$

Again, $\quad |z-1|=|z-i|$

$\Rightarrow \quad |z-1|^2=|z-i|^2$

$\Rightarrow \quad (x-1)^2+y^2=x^2+(y-1)^2$

$\because \quad x=0 \Rightarrow 1+y^2=(y-1)^2$

$\Rightarrow \quad 1+y^2=y^2+1-2y \Rightarrow y=0$

$\therefore z=x+iy=0+0i$ is the only complex number that satisfies the given condition.

64. *(b)*
$$\because \begin{bmatrix}\cos^2\theta & \cos\theta\sin\theta\\ \cos\theta\sin\theta & \sin^2\theta\end{bmatrix}\begin{bmatrix}\cos^2\phi & \cos\phi\sin\phi\\ \cos\phi\sin\phi & \sin^2\phi\end{bmatrix}$$

$$=\begin{bmatrix}\cos\theta\cos\phi\cos(\theta-\phi) & \cos\theta\sin\phi\cos(\theta-\phi)\\ \cos\phi\sin\theta\cos(\theta-\phi) & \sin\theta\sin\phi\cos(\theta-\phi)\end{bmatrix}$$

$$=\begin{bmatrix}0 & 0\\ 0 & 0\end{bmatrix}=0 \quad \left[\text{if } \theta-\phi=(2n+1)\frac{\pi}{2}, n\in I\right]$$

65. *(a)* Given, $S_n=50n+\frac{n(n-7)}{2}A$

Then, $T_n=S_n-S_{n-1}$

$$=50n+\frac{n(n-7)}{2}A-50(n-1)-\frac{(n-1)(n-8)}{2}A$$

$$=50+\frac{A}{2}[n(n-7)-(n-1)(n-8)]$$

$$=50+\frac{A}{2}(n^2-7n-n^2+9n-8)$$

$$T_n = 50 + A(n-4)$$

Also, $d = T_n - T_{n-1}$

$$= 50 + A(n-4) - 50 - A(n-5) = A$$

$$a_{40} = T_{40} = 50 + A(40-4)$$

$$= 50 + 36A$$

$$\therefore a_{40} - 36d = 50 + 36A - 36A = 50$$

66. *(c)* Total balls = 9

The number of ways of selecting 3 balls out of 9 balls

$$= {}^9C_3 = \frac{9 \times 8 \times 7}{1 \times 2 \times 3} = 84$$

The number of ways of selecting 3 balls out of non-black 6 balls

$$= {}^6C_3 = \frac{6 \times 5 \times 4}{1 \times 2 \times 3} = 20$$

$\therefore$The number of ways of drawing 3 balls with at least one black ball $= 84 - 20 = 64$

67. *(b)* Let $A =$ England wins

$B =$ India wins

$$P(A) = \frac{1}{3} \text{ and } P(B) = \frac{2}{3}$$

I. P (England wins at least one match) $= 1 - \left(\frac{2}{3}\right)^3$

$$= 1 - \frac{8}{27} = \frac{19}{27}$$

Statement I is false.

II. P (England losses all three matches) $= \left(\frac{2}{3}\right)^3 = \frac{8}{27}$

Statement II is true.

68. *(d)* Given, $f(x) + 2f\left(\frac{1}{x}\right) = 3x$ ...(i)

On replacing x by $\frac{1}{x}$, we get

$$f\left(\frac{1}{x}\right) + 2f(x) = \frac{3}{x} \quad \text{...(ii)}$$

On multiplying Eq. (ii) by 2 and subtracting Eq. (i) from it, we get

$$3f(x) = \frac{6}{x} - 3x \Rightarrow \quad f(x) = \frac{2}{x} - x$$

Then, $\quad f(-x) = -\frac{2}{x} + x$

$\because \quad f(x) = f(-x)$

$\Rightarrow \quad \frac{2}{x} - x = \frac{-2}{x} + x$

$\Rightarrow \quad x^2 = 2$

$\Rightarrow \quad x = \pm\sqrt{2}$

$\therefore \quad S = \{\sqrt{2}, -\sqrt{2}\}$

69. *(b)* $\dfrac{\sum_{i=1}^{50} x_i}{50} = 16 \Rightarrow \sum_{i=1}^{50} x_i = 800$

Standard deviation = 16

$\Rightarrow$ Variance = 256

$$\therefore \frac{\sum_{i=1}^{50} x_i^2}{50} - (16)^2 = 256 \Rightarrow \sum_{i=1}^{50} x_i^2 = 25600$$

$$\frac{\sum_{i=1}^{50} (x_i - 4)^2}{50} = \frac{\sum_{i=1}^{50} x_i^2 - 8\sum_{i=1}^{50} x_i + \sum_{i=1}^{50} 16}{50}$$

$$= \frac{25600 - 8 \times 800 + 16 \times 50}{50} = 400$$

$$K^2 = 400$$

$$\therefore \quad \frac{K^2}{20} = 20$$

70. *(a)* We have, $x^2 + 2x + 2 = 0$

$\Rightarrow \quad (x+1)^2 - (i)^2 = 0$

$\Rightarrow (x + 1 + i)(x + 1 - i) = 0$

$\therefore \quad x = -(1+i), -(1-i)$

Let $\alpha = -(1+i)$ and $\beta = -(1-i)$

$\Rightarrow \alpha^2 = 2i$ and $\beta^2 = -2i$

Now, $\alpha^{15} + \beta^{15} + 515 = (\alpha^2)^7 \cdot \alpha + (\beta^2)^7 \cdot \beta + 515$

$$= (2i)^7 \alpha + (-2i)^7 \beta + 515$$

$$= (2i)^7 [\alpha - \beta] + 515$$

$$= -128i[-1 - i + 1 - i] + 515$$

$$= -256 + 515$$

$$= 259$$

71. *(896)* $A = \sum_{r=0}^{7} A_r = \sum_{r=0}^{7} \begin{pmatrix} \alpha_r & 0 \\ 0 & \beta_r \end{pmatrix} = \begin{pmatrix} a & 0 \\ 0 & b \end{pmatrix}$

$$\therefore \quad a = \sum_{r=0}^{7} \alpha_r = \sum_{r=0}^{7} r \cdot {}^7C_r$$

$$b = \sum_{r=0}^{7} \beta_r = \sum_{r=0}^{7} (7-r) \cdot {}^7C_r$$

$$a + b = \sum_{r=0}^{7} r \cdot {}^7C_r + \sum_{r=0}^{7} (7-r) \cdot {}^7C_r = \sum_{r=0}^{7} 7({}^7C_r)$$

$$= 7({}^7C_0 + {}^7C_1 + \ldots + {}^7C_7)$$

$$= 7 \times (1 + 7 + 21 + 35 + 35 + 21 + 7 + 1)$$

$$= 7 \times 128 = 896$$

72. *(12)* Since, AM $\geq$ GM

$$\therefore \frac{\frac{b}{c} + \frac{c}{b} + \frac{c}{a} + \frac{a}{c} + \frac{a}{b} + \frac{b}{a}}{6} \geq \left(\frac{b}{c} \times \frac{c}{b} \times \frac{c}{a} \times \frac{a}{c} \times \frac{a}{b} \times \frac{b}{a}\right)^{1/6}$$

$$\Rightarrow \frac{b^2+c^2}{bc}+\frac{c^2+a^2}{ac}+\frac{a^2+b^2}{ab}\geq 6$$

$$\Rightarrow a(b^2+c^2)+b(c^2+a^2)+c(a^2+b^2)\geq 6abc$$

$\therefore$ Minimum value of

$a(b^2+c^2)+b(c^2+a^2)+c(a^2+b^2)$ is $6abc$.

$$\Rightarrow \frac{K}{2}abc=6abc \Rightarrow K=12$$

73. *(6)* The probability of getting 13 red cards is

$P(R)=$ Lost card is black, then red card is selected + Lost card is red, then red card is selected

$$P(R)=P(LB)\,P\left(\frac{R}{LB}\right)+P(LR)\cdot P\left(\frac{R}{LR}\right)$$

$$=\frac{26}{52}\cdot\frac{^{26}C_{13}}{^{51}C_{13}}+\frac{26}{52}\cdot\frac{^{25}C_{13}}{^{51}C_{13}}$$

$$=\frac{1}{2}\cdot\frac{^{26}C_{13}}{^{51}C_{13}}+\frac{1}{2}\cdot\frac{^{25}C_{13}}{^{51}C_{13}}$$

$$\therefore P\left(\frac{LB}{R}\right)=\frac{P(LB)\cdot P\left(\frac{R}{LB}\right)}{P(R)}$$

$$=\frac{\frac{1}{2}\cdot\frac{^{26}C_{13}}{^{51}C_{13}}}{\frac{1}{2}\cdot\frac{^{26}C_{13}}{^{51}C_{13}}+\frac{1}{2}\cdot\frac{^{25}C_{13}}{^{51}C_{13}}}=\frac{2}{3}=\frac{\alpha}{\beta}$$

$$\therefore \quad \alpha\beta=2\times 3=6$$

74. *(3)* $\lim\limits_{x\to 0}\left(\frac{p^x+q^x+r^x}{3}\right)^{\frac{\lambda}{x}}$

$$\lim_{x\to 0}\left[\frac{3+(p^x-1)+(q^x-1)+(r^x-1)}{3}\right]^{\lambda/x}$$

$$=\lim_{x\to 0}\left[1+\frac{k}{3}\right]^{\frac{3}{k}\cdot\frac{k}{3}\cdot\frac{\lambda}{x}}$$

where, $k=(p^x-1)+(q^x-1)+(r^x-1)$

$$=e^{\lim\limits_{x\to 0}\frac{\lambda}{x}\left(\frac{k}{3}\right)}$$

$$=e^{\lim\limits_{x\to 0}\frac{\lambda}{3}\left\{\frac{(p^x-1)+(q^x-1)+(r^x-1)}{x}\right\}}$$

$$=e^{\frac{\lambda}{3}[\ln p+\ln q+\ln r]} \qquad [\because \text{as } x\to 0 \text{ and } k\to 0]$$

$$=e^{\frac{\lambda}{3}\ln pqr}$$

$$=(pqr)^{\frac{\lambda}{3}}$$

$$=pqr, \text{ if } \lambda=3$$

Time Saver Tip

Let $y=\lim\limits_{x\to 0}\left(\frac{p^x+q^x+r^x}{3}\right)^{\frac{\lambda}{x}}$

Taking log both sides, we get

$$\log y=\lim_{x\to 0}\frac{\lambda\cdot\log(p^x+q^x+r^x)-\log 3}{x} \qquad \left[\frac{0}{0}\text{ form}\right]$$

Thus, using L-Hospital Rule,

$$\log y=\lambda\frac{(p^x\log p+q^x\log q+r^x\log r)}{(p^x+q^x+r^x)\cdot 1}$$

$$=\frac{\lambda}{3}\log(pqr)$$

$\therefore \log y=\log(pqr)^{\frac{\lambda}{3}}$ or $y=(pqr)^{\frac{\lambda}{3}}$

So, $\lambda=3$

75. *(4)* $\mathbf{a}_1=6\hat{\mathbf{i}}+\alpha\hat{\mathbf{j}}+\alpha\hat{\mathbf{k}}$, $\mathbf{b}_1=\hat{\mathbf{i}}-2\hat{\mathbf{j}}+2\hat{\mathbf{k}}$

$\mathbf{a}_2=-4\hat{\mathbf{i}}-\hat{\mathbf{k}}$, $\mathbf{b}_2=3\hat{\mathbf{i}}-2\hat{\mathbf{j}}-2\hat{\mathbf{k}}$

Now, shortest distance $=\left|\frac{(\mathbf{a}_2-\mathbf{a}_1)\cdot(\mathbf{b}_1\times\mathbf{b}_2)}{|\mathbf{b}_1\times\mathbf{b}_2|}\right|$

$$\mathbf{a}_2-\mathbf{a}_1=(-4\hat{\mathbf{i}}-\hat{\mathbf{k}})-(6\hat{\mathbf{i}}+\alpha\hat{\mathbf{j}}+\alpha\hat{\mathbf{k}})$$

$$=-10\hat{\mathbf{i}}-\alpha\hat{\mathbf{j}}+(-1-\alpha)\hat{\mathbf{k}}$$

$$\mathbf{b}_1\times\mathbf{b}_2=\begin{vmatrix}\hat{\mathbf{i}} & \hat{\mathbf{j}} & \hat{\mathbf{k}}\\ 1 & -2 & 2\\ 3 & -2 & -2\end{vmatrix}$$

$$=\hat{\mathbf{i}}(4+4)-\hat{\mathbf{j}}(-2-6)+\hat{\mathbf{k}}(-2+6)$$

$$=8\hat{\mathbf{i}}+8\hat{\mathbf{j}}+4\hat{\mathbf{k}}$$

$$|\mathbf{b}_1\times\mathbf{b}_2|=\sqrt{64+64+16}=12$$

$\therefore$ Shortest distance,

$$9=\left|\frac{(-10\hat{\mathbf{i}}-\alpha\hat{\mathbf{j}}+(-1-\alpha)\hat{\mathbf{k}})\cdot(8\hat{\mathbf{i}}+8\hat{\mathbf{j}}+4\hat{\mathbf{k}})}{12}\right|$$

$$\Rightarrow \quad 9=\left|\frac{-80-8\alpha+4(-1-\alpha)}{12}\right|$$

$$\Rightarrow \quad 9\times 12=|-80-8\alpha-4-4\alpha|$$

$$\Rightarrow \quad 108=|-84-12\alpha|$$

$$\Rightarrow \quad \pm 108=12\alpha+84$$

$$\Rightarrow \quad \alpha=2,-16$$

$$\Rightarrow \quad \alpha=2 \qquad [\because \alpha>0]$$

$(\alpha_1,\alpha_2,\alpha_3)$ lies on L_1,

where $L:\frac{x-6}{1}=\frac{y-2}{-2}=\frac{z-2}{2}=k$ (let)

$$(\alpha_1,\alpha_2,\alpha_3)=(k+6,-2k+2,2k+2)$$

$$\therefore \alpha_3+\alpha_2=2k+2-2k+2=4$$

SECTION II
Mastery Builders

PRACTICE SET 11

Total Marks : 300	Total Time : 3 hrs	No. of Questions : 75

Instructions

1. There are **three** parts in the questions paper consisting of **Physics**, **Chemistry** and **Mathematics** having 25 questions in each part of equal weightage. Each part (subject) has two sections.
2. **Section A** This section contains 20 multiple choice questions which have only one correct answer. Each question carries **4 marks** for correct answer and – **1 mark** for wrong answer.
3. **Section B** This section contains 5 questions. The answer to each of the questions is a numerical value. Each question carries **4 marks** for correct answer and – **1 mark** for wrong answer. For Section B, the answer should be rounded off to the nearest integer.

PHYSICS

Section A : Objective Type Questions

1. A charged particle is moving in a circle of radius R with constant speed v under the influence of a uniform magnetic field. The time period of the motion

(a) depends on v and not on R
(b) depends on both R and v
(c) is independent of both R and v
(d) depends on R and not on v

2. A copper block of mass 3.35 kg is heated at 300°C and then placed on a large block of ice. What is the maximum mass of ice that can melt ?
(Specific heat capacity of copper $= 390\ \text{J kg}^{-1}\text{k}^{-1}$ and latent heat of fusion of water $= 335\ \text{J kg}^{-1}$).

(a) 2.5×10^3 kg
(b) 1.17×10^3 kg
(c) 3.9×10^3 kg
(d) 3×10^3 kg

3. The p-V diagram of 2 g of helium gas for a certain process $A \to B$ is shown in the figure. What is the heat given to the gas during process $A \to B$?

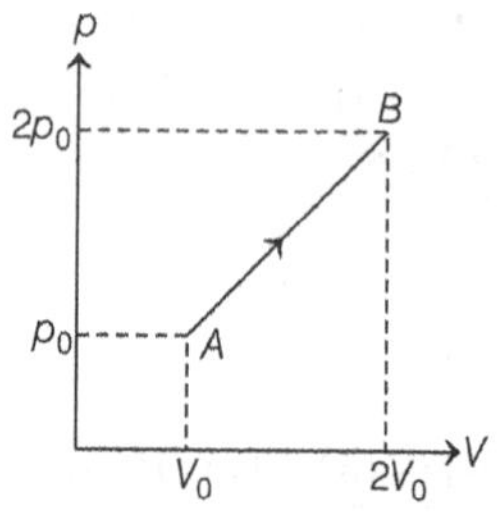

(a) $4p_0V_0$ (b) $6p_0V_0$ (c) $4.5\,p_0V_0$ (d) $2p_0V_0$

4. A nucleus with mass numbers 240 breaks into two fragments of mass numbers 110 and 130 having binding energy per nucleon 7.8 MeV and 8.0 MeV, respectively. If binding energy per nucleon of unfragmented nuclei is 7.5 MeV, then the energy released in the process is

(a) 216 MeV (b) 153 MeV
(c) 124 MeV (d) 98 MeV

5. In a region of space, the electric field is in the x-direction and proportional to x i.e., $\mathbf{E} = 56x_0\hat{\mathbf{i}}$. Consider an imaginary cubical volume of edge 0.6 m with its edges parallel to axes of coordinates. The charge (in pC) inside this volume will be

(a) 107pC (b) 214 pC (c) 56pC (d) 12pC

6. Two spherical rain drops of equal size are falling vertically through air with a terminal velocity of 1 ms^{-1}. What would be the terminal speed, if these two drops were to coalesce to form a large spherical drop?

(a) 1.125 m/s (b) 2.225 m/s
(c) 1.587 m/s (d) 3.125 m/s

7. On the basis of photoelectric effect, which of the following statements is correct?

I. The current in a photocell increases with increasing frequency of light.
II. The photocurrent increases with increase in applied voltage upto saturation stage.
III. The photocurrent increases with increasing intensity of light.
IV. The stopping potential increases with increasing intensity of light.

Choose the correct option given below.

(a) Both II and III (b) Both II and IV
(c) I, II and III (d) All of these

8. A rocket is fired from the Earth towards the Sun. At what point on its path is the gravitational force on the rocket is zero?
(Take, mass of Sun $= 2 \times 10^{30}$ kg,
mass of Earth $= 6 \times 10^{24}$ kg,
orbital radius of Earth $= 1.5 \times 10^{11}$ m
neglect the effect of other planets)

(a) 4.2×10^{7} m (b) 2.59×10^{8} m
(c) 7.21×10^{6} m (d) 3.2×10^{6} m

9. A cricket ball is projected from the ground at a speed of 20 ms^{-1} making an angle of 30° with the horizontal. Another ball is simultaneously released from a point on the vertical line along the maximum height of the projectile. First ball collides with the second ball at its maximum height. The initial height of the second ball is

(a) 2.5 m (b) 5 m (c) 10 m (d) 12.5 m

10. The value of power dissipated across the zener diode of breakdown voltage 10 V connected in the circuit as shown in figure is

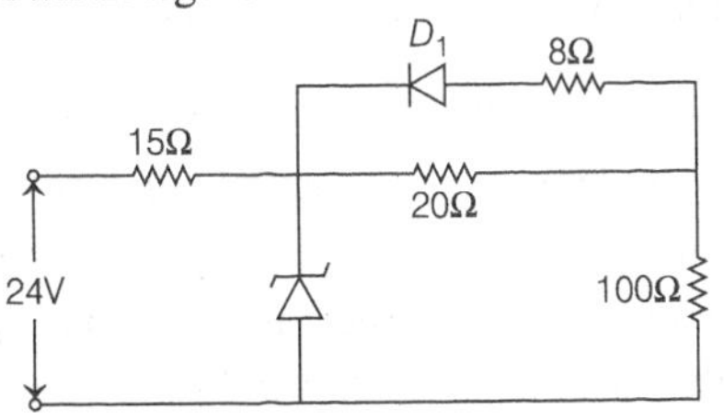

(a) 8.5 W (b) 8.0 W (c) 6.5 W (d) 9.5 W

11. A charge $-q$ is placed at the axis of a charged ring of radius r at a distance of $2\sqrt{2}\,r$ as shown in figure. If ring is fixed and carrying a charge Q, the kinetic energy of charge $-q$ when it is released and reaches the centre of ring will be

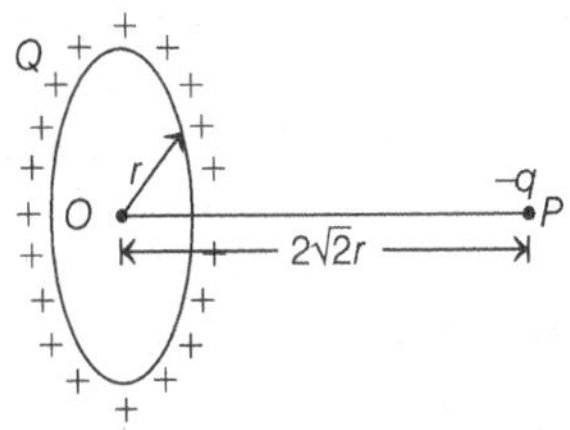

(a) $\frac{qQ}{4\pi\varepsilon_0 r}$ (b) $\frac{qQ}{12\pi\varepsilon_0 r}$
(c) $\frac{qQ}{6\pi\varepsilon_0 r}$ (d) $\frac{qQ}{2\pi\varepsilon_0 r}$

12. When two resistances are connected in parallel, then the equivalent resistance is $\frac{6}{5}\,\Omega$. When one of the resistance is removed, then the effective resistance is 2 Ω. The resistance of the wire removed will be

(a) 3 Ω (b) 2 Ω
(c) $\frac{3}{5}\,\Omega$ (d) $\frac{6}{5}\,\Omega$

13. A thermodynamic cycle $ABCA$ is shown on p-T diagram. The corresponding p - V diagram is

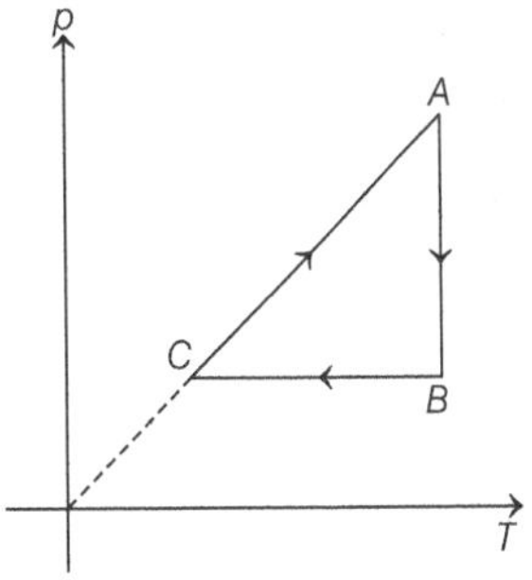

(a)
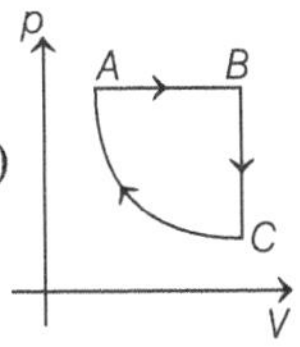

(b)
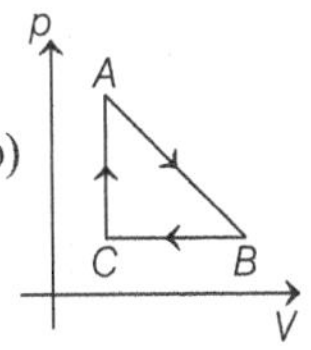

(c)
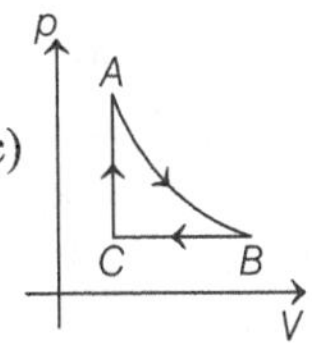

(d)
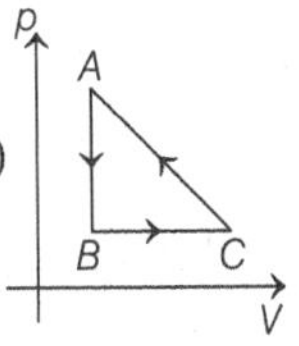

14. **Assertion** The direction of propagation of EM wave is perpendicular to both electric field (**E**) and magnetic field (**B**).

Reason Energy of EM waves is divided equally between electric and magnetic fields.

(a) Both Assertion and Reason are correct and Reason is the correct explanation of Assertion.
(b) Both Assertion and Reason are correct but Reason is not the correct explanation of Assertion.
(c) Assertion is correct but Reason is incorrect.
(d) Assertion is incorrect but Reason is correct.

15. **Statement I** If c_s is the velocity of sound in air ($\gamma_{air} = 1.4$) and c is the rms velocity, then $c > c_s$.

Statement II The value of molar specific heat at constant pressure for one mole of triatomic gas is $3R$.

(a) Statement I is true but Statement II is false.
(b) Statement I is false but Statement II is true.
(c) Both Statement I and Statement II are true.
(d) Both Statement I and Statement II are false.

16. If Planck's constant (h), speed of light (c), gravitational constant (G) and current (I) are taken as fundamental units to express the stopping potential (V_0) in photoelectric effect as

$$V_0 = \frac{h^p c^q G^r}{I^s}$$

then match Column I (exponents) with Column II (values) and select the correct answer from the codes given below.

	Column I		Column II
A.	Value of p	1.	5
B.	Value of q	2.	−1
C.	Value of r	3.	1
D.	Value of s	4.	0

Codes

	A	B	C	D
(a)	1	3	4	2
(b)	1	4	3	2
(c)	4	1	2	3
(d)	4	3	1	2

17. An ideal spring of spring constant $k = 100$ N/m is attached to the bottom of wedge which is placed at angle 30° with horizontal. A mass $m = 10$ kg is released from height h on the wedge which compressed the spring by 2 m. Then, what is the value of h, if the wedge is frictionless?

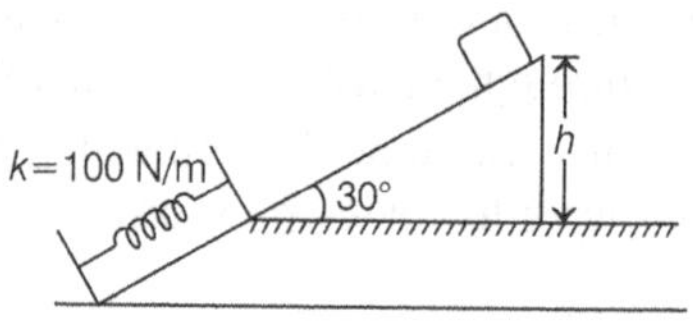

(a) 2m (b) 1m (c) 3m (d) 4m

18. A simple pendulum has time period T_1. The point of suspension is now moved upward according to the relation $y = kt^2$, ($k = 1\text{ m/s}^2$) where, y is the vertical displacement. The time period now becomes T_2. The ratio of $\frac{T_1^2}{T_2^2}$ is ($g = 10\text{ m/s}^2$)

(a) 6/5 (b) 5/6 (c) 1 (d) 4/5

19. The image of an object, formed by a planoconvex lens at a distance of 8 m behind the lens, is real and is one-third the size of the object. The wavelength of light inside the lens is $\frac{2}{3}$ times the wavelength in free space. The radius of the curved surface of the lens is

(a) 1 m (b) 2 m (c) 3 m (d) 6 m

20. A car is moving in a circular horizontal path with increasing speed, then choose the most probable answer.

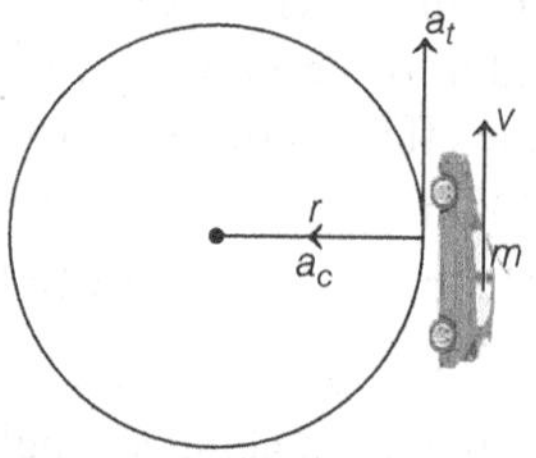

(a) $f_s = m \times a_c$ (b) $f_s = m \times a_t$
(c) $f_s = m \times \sqrt{a_t^2 + a_c^2}$ (d) $(f_s)_{max} = m \times \sqrt{a_t^2 + a_c^2}$

Section B : Numerical Value Type Questions

21. The image of an illuminated square is obtained on a screen with the help of a converging lens. The distance of the square from the lens is 40 cm. The area of the image is 9 times that of the square. The focal length of the lens is ...... cm.

22. A particle is at the origin in XY-plane at $t = 0$. The particle moves with initial velocity $8\hat{j}$ m/s with a constant acceleration of $(5\hat{i} + 2\hat{j})$ m/s². After time t the position of the particle in XY-plane is given by coordinate (40 m, y_0 m), then the value of y_0 is ....

23. The number density of free electrons in a conductor is $7.5 \times 10^{30}\ m^{-3}$ and the mean free time 30 fs, the resistivity of the conductor is $n \times 10^{-12}\ \Omega$-m, then the value of n is .... (Take, $m_e = 9.1 \times 10^{-31}$ kg)

24. A homogeneous disc of radius 2 m and mass 10 kg is pulled on a cricket pitch by a horizontal force 60 N as shown in figure. Assuming rolling without slipping, angular acceleration of the cylinder is .... rad/s^2.

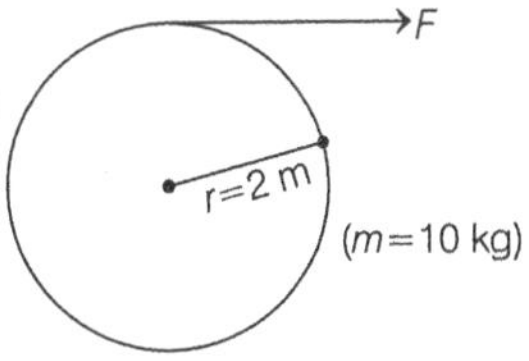

25. If a body of mass 0.5 kg rests on an inclined plane, inclined at angle of 30° with the horizontal. The force required to move the body up the inclined plane is .... N.

(Answer should be rounded off to the nearest integer. Given, coefficient of friction $\mu = 0.02$)

CHEMISTRY

Section A : Objective Type Questions

26. Consider the following compounds.

(I) $H-C(CH_3)=CH-CH_2-C(C_2H_5)=CH_2$

(II) $C_2H_5-CH=C(CH_3)-CH_3$

(III) $C_2H_5-CH_2-C(CH_3)=CH_2$

(IV) $C_2H_5-CH_2-C(CH_3)_2-OCH_3$

(V) $CH_3-CH(CH_3)-CH=CH_2$

The total number of compound will be formed when 2-chloro-2-methyl pentane reacts with sodium methoxide in methanol.

(a) 2 (b) 1 (c) 3 (d) 4

27. Which of the following statements is correct regarding vitamin-B_{12}?

I. It is also known as cyanocobalamin.

II. Metal ion present in vitamin-B_{12} is cobalt.

III. It is a CO_2 carrier in carboxylation reaction.

IV. It catalyses isomerism reaction.

Choose the most appropriate answer from the options given below.

(a) I, II and III (b) I, II and IV
(c) I and II (d) I, II, III and IV

28. The following mechanism has been proposed for the exothermic catalysed complex reaction,

$$A + B \underset{\text{Fast}}{\overset{\text{Slow}}{\rightleftharpoons}} IAB \xrightarrow{k_1} AB + I \xrightarrow{k_2} P + A$$

If k_1 is much smaller than k_2, the most suitable qualitative plot of potential energy (PE) *versus* reaction coordinate (R) for the above reaction is

(a)
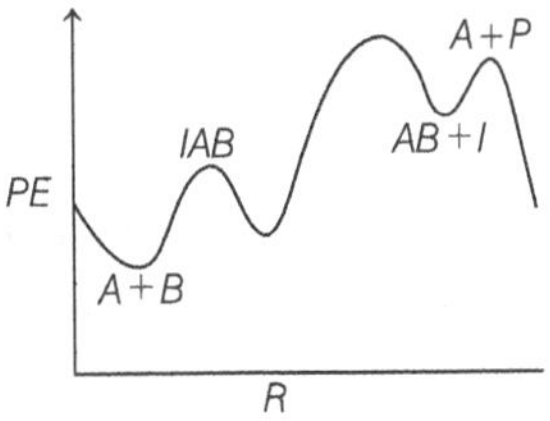

(b)
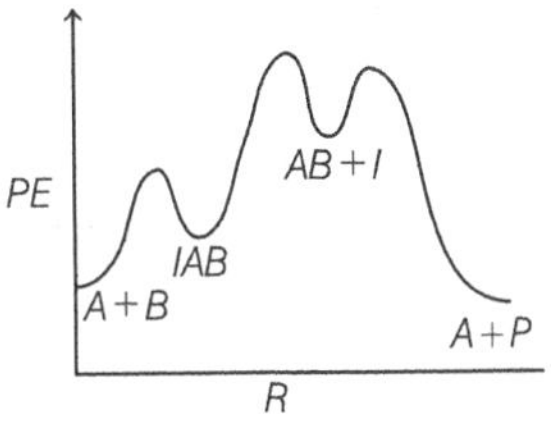

(c)
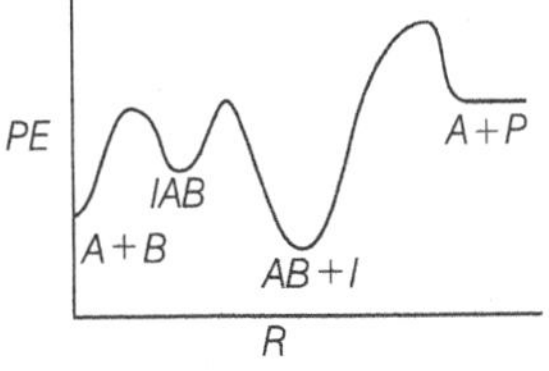

(d)
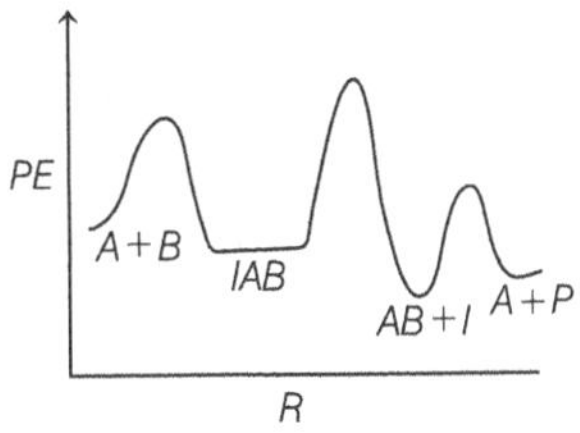

29. Energy of an electron is calculated as,

$$E_n = \frac{-2.18\times10^{-18} Z^2}{n^2}$$

Which of the following correctly states the relationship between the $n=2$ level of He^+ atom ($Z=2$) and $n=2$ level of Li^{2+} in ($Z=3$)?

(a) $E_{He^+} = \frac{9}{4} E_{Li^{2+}}$

(b) $E_{He^+} = \frac{4}{9} E_{Li^{2+}}$

(c) $E_{He^+} = \frac{9}{2} E_{Li^{2+}}$

(d) $E_{He^+} = \frac{2}{9} Li^{2+}$

30. Consider the following reactions,

$$Br_2 + OH^- \xrightarrow[\text{Hot}]{} (X) + (Y)$$

$$X + Y + H^+ \longrightarrow Br_2$$

If (X) gives yellow precipitate with $AgNO_3$, then (X) and (Y) are

(a) Br^-, BrO_3^-

(b) BrO^-, BrO_4^-

(c) BrO_2, Br^-

(d) BrO_4^-, BrO^-

31. The value of x in the partial redox equation is

$$MnO_4^- + 8H^+ + xe^- \rightleftharpoons Mn^{2+} + 4H_2O$$

(a) 5 (b) 3
(c) 1 (d) 0

32. In a bimolecular reaction, the steric factor P was experimentally determined to be 5.0.

A. The value of frequency factor predicted by Arrhenius equation is higher than that determined experimentally.

B. Since, $P = 5.0$, the reaction will not proceed unless an effective catalyst is used.

C. The activation energy of the reaction is unaffected by the value of the steric factor.

D. Experimentally determined value of frequency factor is higher than that predicted by Arrhenius equation.

Choose the correct statements and mark the option accordingly.

(a) C and D (b) A, C and D
(c) B, C and D (d) A and C

33. What will be the product(s) when *cis* 2,3-dihydroxybutan-1-al will be treated with $NaIO_4$?

(a) $2HCOOH + CH_3CHO$

(b) $HCOOH + 2CH_3CHO$

(c) $HCOOH + HCHO + CH_3CHO$

(d) $3HCOOH$

34. The standard Gibbs energy change at 300 K for the reaction $2A \longrightarrow B + C$ is 249.2 J. At a given time, the composition of the reaction mixture is $[A] = \frac{1}{2}$, $[B] = 2$ and $[C] = \frac{1}{2}$. The reaction proceeds in the $[R = 8.314$ J/K mol, $e = 2.78]$

(a) Forward direction because $Q > K_C$

(b) Backward direction because $Q > K_C$

(c) Forward direction because $Q < K_C$

(d) Backward direction because $Q < K_C$

35. Which of the following correctly represents the variation of melting point for the first transition series?

(a)

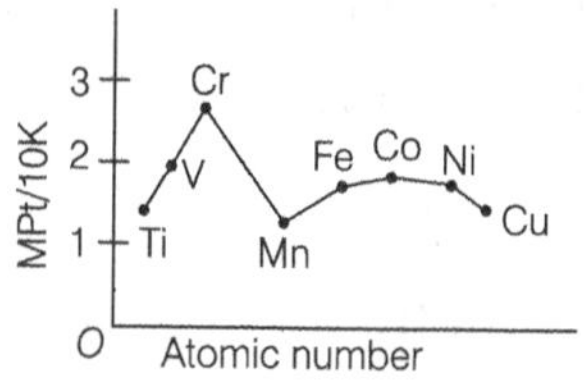

(b)

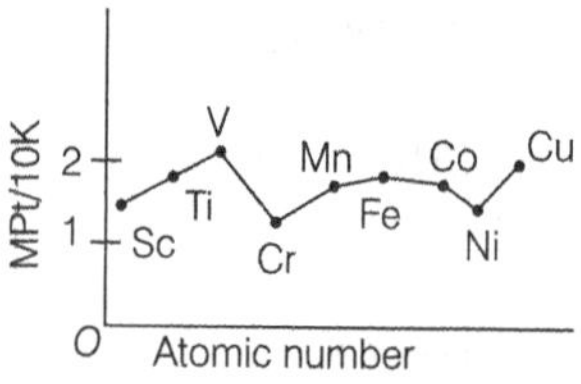

(c)

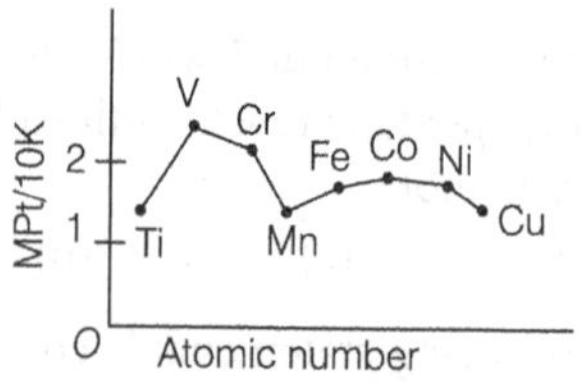

(d)

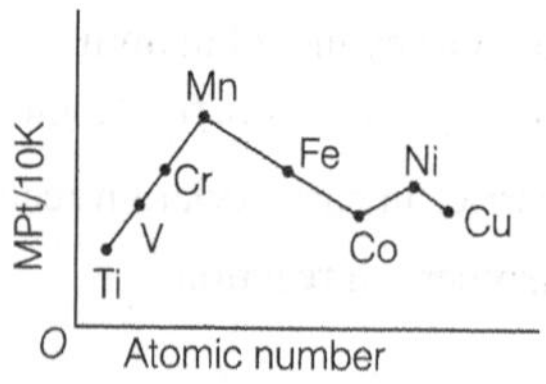

36. Consider the following compounds,

A. 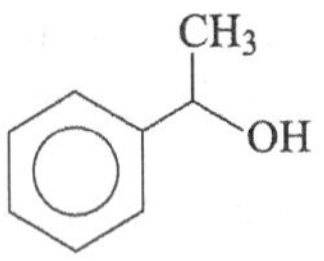

B. 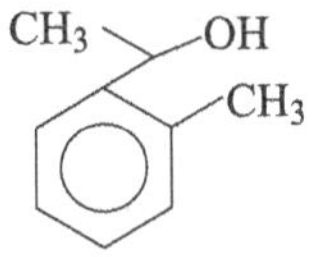

C. 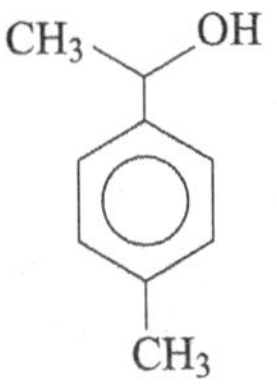

D. 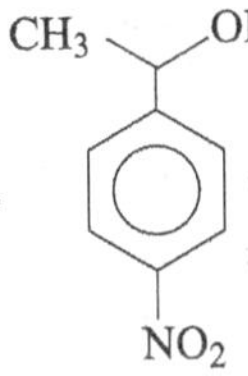

The correct decreasing order of dehydration is

(a) A > B > C > D (b) B > C > A > D
(c) D > A > C > B (d) D > A > B > C

37. In which of the following pairs the two species are not isostructural?

(a) CO_3^{2-} and NO_3^- (b) PCl_4^+ and $SiCl_4$
(c) PF_5 and BrF_5 (d) AlF_6^{3-} and SF_6

38. Among the following, the maximum covalent character is shown by compound.

(a) $FeCl_2$ (b) $SnCl_2$ (c) $AlCl_3$ (d) $FeCl_3$

39. Which of the following sets of species have planar structures?

(a) I_3^+, ICl_4^-, Al_2Cl_6, $TeCl_4$
(b) SCl_2, N_2O_5, SF_4, $XeOF_4$
(c) I_2Cl_6, XeF_2, BrF_4^-, XeF_5^-
(d) I_3^-, $\bar{C}H_3$, ClO_3^-, SiF_6^{2-}

40. A compound CH_3CH_2Br is when treated with KCN produces a compound *A* which on hydrolysis produces *B*. *A* on reduction with DIBAL-H gives *C* which on hydrolysis produces *D*. *D* on oxidation produces *B*. Identify the *D*.

(a) CH_3CH_2COOH (b) CH_3CH_2CHO
(c) $CH_3CH_2CH_2COOH$ (d) CH_3CH_2CN

41. Given below are two statements.

Statement I The atomic radii of fluorine and neon in angstrom units are 1.60 and 0.72 respectively.

Statement II The pairs Rb, Cs and Ar, Kr have nearly same electron gain enthalpy.

In the light of the above statements, choose the correct answer from the options given below.

(a) Both Statement I and Statement II are true.
(b) Both Statement I and Statement II are false.
(c) Statement I is true but Statement II is false.
(d) Statement I is false but Statement II is true.

42. The Fischer presentation of D-glucose is given below.

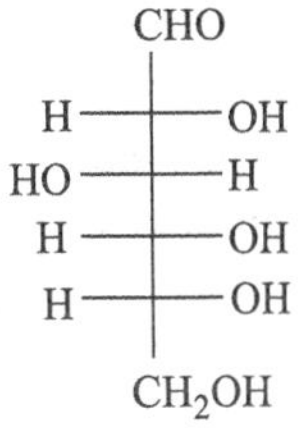

D- glucose

The correct structure(s) of β-L-glucopyranose is (are)

(a) 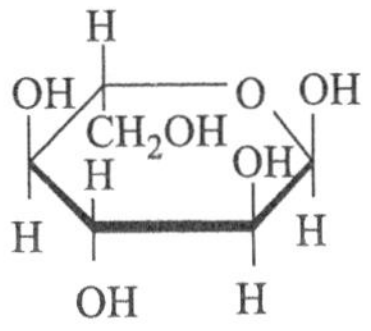

(b) 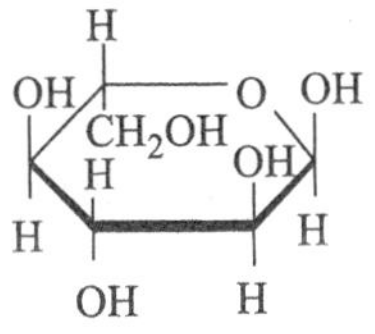

(c) 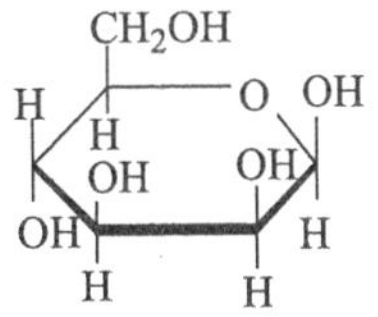

(d) 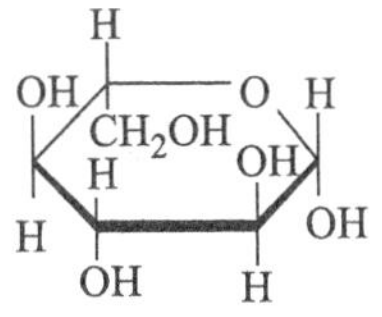

43. Given below are two statements, one is labelled as Assertion A and the other is labelled as Reason R.

Assertion (A) Among the carbon allotropes, diamond is an insulator whereas, graphite is a good conductor of electricity.

Reason (R) Hybridisation of carbon in diamond and graphite are sp^3 and sp^2, respectively.

In the light of above statements, choose the correct answer from the options given below.

(a) Both (A) and (R) are true and (R) is the correct explanation of (A).
(b) Both (A) and (R) are true but (R) is not the correct explanation of (A).
(c) (A) is true but (R) is false.
(d) (A) is false but (R) is true.

44. Match List-I with List-II.

List-I		List-II	
A.	$[Ni(CN)_4]^{2-}$	I.	sp^3d
B.	BrF_5	II.	d^2sp^3
C.	PCl_5	III.	dsp^2
D.	$[Co(NH_3)_6]^{3+}$	IV.	sp^3d^2

Choose the most appropriate answer from the options given below.

(a) A-II, B-IV, C-I, D-III (b) A-III, B-IV, C-I, D-II
(c) A-III, B-I, C-IV, D-II (d) A-II, B-I, C-IV, D-III

45. Match List-I with List-II.

	List-I (Organic compounds containing nitrogen)		List-II (pK_b value)
A.	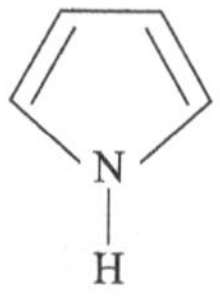	I.	8.8
B.	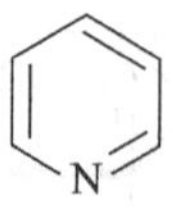	II.	5.64
C.	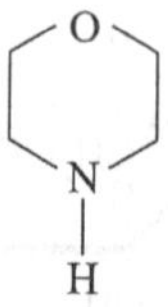	III.	13.36
D.	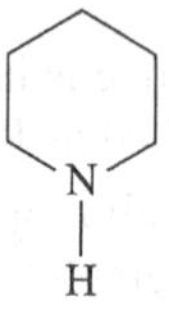	IV.	2.78

Choose the most appropriate option given below.

(a) A-II, B-IV, C-III, D-I (b) A-II, B-I, C-IV, D-III
(c) A-I, B-III, C-II, D-IV (d) A-III, B-I, C-II, D-IV

Section B : Numerical Value Type Questions

46. When a certain conductivity cell was filled with 0.01 M solution of KCl, it had a resistance of 160 ohm at 25° C and when filled with 0.005 M NaOH, it had a resistance of 190 ohm. If specific resistance of KCl solution is 700 ohm-cm, specific conductance of NaOH solution is ......10^{-4} ($ohm^{-1}cm^{-1}$).

47. H-atom is exposed to electromagnetic radiation of wavelength 1051.2 Å and excited atom gives out induced radiations. The minimum wavelength of these induced radiations is ......nm. [Nearest integer]

48. The standard enthalpies of formation of $CO_2(g)$, $H_2O(l)$ and glucose (s) at room temperature are – 400 kJ/mol, – 300 kJ/mol and – 1200 kJ/mol respectively. The value of standard enthalpy of combustion per gram of glucose at 273 K is (–) ....kJ/g. [Nearest integer]

49. The number of given molecules (in pure form) which are unstable at the room temperature is ....

(I)

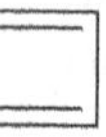
(II)

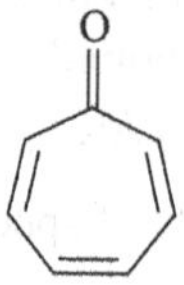
(III)

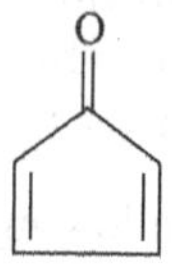
(IV)

50. *n*-butane is obtained by the monobromination of ethane followed by Wurtz reaction. The volume of ethane at NTP required to produce 56 g *n*-butane is '*x*' L. If the bromination takes place with 90% yield and Wurtz reaction with 85% yield. The value of *x* is ....

MATHEMATICS

Section A : Objective Type Questions

51. Let the population of honey-bees surviving at a time is given by the differential equation $\frac{dx}{dt}=\frac{1}{2}x+50$, if $x(0)=200$, then $x(t)$ is equal to $ae^{t/2}-b$. The value of $a-b$ is

(a) 200 (b) 100 (c) 150 (d) 250

52. If $f(x)=\begin{cases}5[x]+3\frac{|x|}{x}, & x\neq 0\\ 3, & x=0\end{cases}$, then $\int_{-1/2}^{2} f(x)dx$ is equal to (where $[\cdot]$ denotes the greatest integer function)

(a) 4 (b) 6 (c) 7 (d) 9

53. Let A be an interval in which $f(x)=\int_0^x \frac{dt}{1+t^5}$ is an invertible function. Suppose, g is the inverse of f, then which of the following is true?

(a) $g'(x)=\frac{1}{f[f'(x)]}$ (b) $g'(x)=\frac{1}{g[g'(x)]}$

(c) $g'(x)=\frac{1}{g[f'(x)]}$ (d) $g'(x)=\frac{1}{f'[g(x)]}$

54. If z and ω are two complex numbers such that $|z\omega|=1$ and $\arg(z)-\arg(\omega)=\frac{3\pi}{2}$, then $\arg\left(\frac{1-2\bar{z}\omega}{1+3\bar{z}\omega}\right)$ is equal to

(a) $\frac{\pi}{4}$ (b) $\frac{3\pi}{4}$

(c) $\frac{-3\pi}{4}$ (d) $\frac{-\pi}{4}$

55. If the coefficient of x^7 in $\left(bx^2+\frac{1}{ax}\right)^{11}$ and x^{-7} in $\left(bx-\frac{1}{ax^2}\right)^{11}$ are equal, then $ab=k$. The value of k is

(a) -1 (b) 0

(c) 1 (d) 2

56. Suppose $a_1, a_2, ..., a_n, ...$ be an arithmetic progression of natural numbers. If the ratio of sum of first six terms to the sum of first ten terms of progression is $\frac{33}{95}$ and $110<a_{15}<120$, then the sum of first eleven terms of the progression is equal to

(a) 380 (b) 824

(c) 462 (d) 921

57. Let $P=\begin{bmatrix}4 & -1 & -3\\ 3 & 0 & \alpha\\ 4 & -5 & 0\end{bmatrix}$, where $\alpha\in R$.

Suppose, $Q=[a_{ij}]$ is a matrix, satisfying $PQ=KI_3$ for some non-zero $K\in R$. If $q_{32}=-\frac{K}{7}$ and $|Q|=\frac{K^2}{2}$, then α^2+K^2 is equal to

(a) $\frac{225}{2}$ (b) 225

(c) $\frac{63}{2}$ (d) 0

58. If $\alpha=\frac{\pi}{6}$ and $A=\begin{bmatrix}\cos\alpha & \sin\alpha\\ -\sin\alpha & \cos\alpha\end{bmatrix}$, then $B=A+A^2+A^3+A^4+A^5$ is

(a) singular (b) non-singular

(c) symmetric (d) $|B|=1$

59. If $f(x)=\frac{e^{2x}-(1+4x)^{1/2}}{\log(1+x^2)}$ for $x\neq 0$, then f has

(a) an irremovable discontinuity at $x=0$.

(b) a removable discontinuity at $x=0$ and $f(0)=\frac{1}{4}$.

(c) a removable discontinuity at $x=0$ and $f(0)=-\frac{1}{4}$.

(d) a removable discontinuity at $x=0$ and $f(0)=4$.

60. The area bounded by the curve $y=\sin^2 x-2\sin x$ and the X-axis, where $x\in[0,2\pi]$ is

(a) 16 (b) 4 (c) 8 (d) 0

61. If the probability distribution of a random variable is $P(x)=\begin{cases}\frac{x}{10}, & x=1,2,3,4\\ 0, & \text{otherwise}\end{cases}$, then $P\left\{\frac{1}{2}<x<\frac{7}{2} \text{ and } x>1\right\}$ is equal to

(a) $\frac{4}{10}$ (b) $\frac{5}{9}$ (c) $\frac{3}{2}$ (d) $\frac{1}{10}$

62. Let $f:R\to R$ be a function such that $f(x+y)=f(x)+f(y)\ \forall x, y\in R$. Then, consider two statements

$S_1: f(x)=xf(1), \forall x\in Z$ (Z = set of integers)

$S_2 : f\left(\frac{p}{q}\right) = f(p) - f(q)$, where p and q are positive integers.

(a) Both S_1 and S_2 are true
(b) S_1 is false and S_2 is true
(c) Both S_1 and S_2 are false
(d) S_1 is true and S_2 is false

63. Let $a, b \in R$. If the mirror image of the point $P(a, 6, 3)$ with respect to the line $\frac{x}{1} = \frac{y-1}{2} = \frac{z-2}{3}$ is $Q(1, b, a+6)$, then $|a+b|$ is equal to

(a) 2 (b) 3 (c) 1 (d) 0

64. The lines $x + y = |c|$ and $cx - y = 1$ intersect each other in the first quadrant, then the set of all possible values of c is in the interval

(a) $(0, \infty)$ (b) $(1, \infty)$
(c) $(-1, \infty)$ (d) $[-1, 1]$

65. Let PQ be a double ordinates of the parabola $y^2 = -4x$, where P lies in the second quadrant. If R divides PQ in the ratio 5 : 2, then the locus of R is

(a) $7y^2 + 6x = 0$
(b) $49y^2 + 36x = 0$
(c) $49y^2 - 36x = 0$
(d) None of the above

66. A line of the shortest distance between the lines $\frac{x}{2} = \frac{y}{-3} = \frac{z}{1}$ and $\frac{x-2}{3} = \frac{y-1}{-5} = \frac{z+2}{2}$ is perpendicular to the line $\frac{x-3}{\alpha} = \frac{y-2}{2} = \frac{z-4}{3}$, then $|3+\alpha|^2$ is equal to

(a) $\frac{22}{100}$ (b) $\frac{82}{100}$
(c) 58 (d) 2

67. The mean of the numbers $\frac{^{50}C_0}{1}, \frac{^{50}C_2}{3}, \frac{^{50}C_4}{5}, \ldots, \frac{^{50}C_{50}}{51}$, is

(a) $\frac{2^{49}}{(39)(17)}$ (b) $\frac{2^{50}}{51}$
(c) $\frac{2^{49}}{51}$ (d) $\frac{2^{49}}{54}$

68. The truth set of the open statement $p(x) = x^2 - x(3+\sqrt{2}) + 3\sqrt{2} = 0, \forall x \in Q$, is

(a) $\{\sqrt{2}\}$ (b) $\{3\}$
(c) $\{3, \sqrt{2}\}$ (d) $\{\}$

69. If a function $f : R \to R$ be such that $f[x - f(y)] = f[f(y)] + xf(y) + f(x) - 1$ $\forall x, y \in R$, then $f(2)$ is

(a) 1 (b) -1 (c) 0 (d) 2

70. If the distance between the foci of an ellipse of 8 and the distance between its directrices is 16, then the length of its latus rectum is

(a) $4\sqrt{2}$ (b) $2\sqrt{3}$ (c) $3\sqrt{2}$ (d) $\frac{5}{\sqrt{2}}$

Section B : Numerical Value Type Questions

71. The number of positive integral solutions of the equation $abcde = 1050$ is $375\,k$, when $k \in N$. Then, k is equal to ………

72. The minimum integral value of λ, for which the equation $e^x = \lambda x^2$ has exactly three real distinct solution is ……

73. If the shortest distance between the lines $\mathbf{r}_1 = \alpha\hat{\mathbf{i}} + 2\hat{\mathbf{j}} + 2\hat{\mathbf{k}} + \lambda(\hat{\mathbf{i}} - 2\hat{\mathbf{j}} + 2\hat{\mathbf{k}}), \lambda \in R, \alpha > 0$ and $\mathbf{r}_2 = -4\hat{\mathbf{i}} - \hat{\mathbf{k}} + \mu(3\hat{\mathbf{i}} - 2\hat{\mathbf{j}} - 2\hat{\mathbf{k}}), \mu \in R$ is 7, then α is equal to

74. Let $\mathbf{a}$ and $\mathbf{b}$ be two vector such that $|\mathbf{a} + \mathbf{b}|^2 = |\mathbf{a}|^2 + 2|\mathbf{b}|^2$, $\mathbf{a} \cdot \mathbf{b} = 6$ and $|\mathbf{a} \times \mathbf{b}|^2 = 65$. Then, $12|\mathbf{a}|^2$ is equal to ………..

75. The value of $\sum_{k=1}^{12} \frac{1}{\cos\left(\frac{\pi}{4} + \frac{(k-1)\pi}{3}\right)\cos\left(\frac{\pi}{4} + \frac{k\pi}{3}\right)}$ is equal to …….

DETAILED SOLUTIONS

Physics

1. *(c)* To move on circular path in a magnetic field, a centripetal force is provided by the magnetic force. When magnetic field is perpendicular to motion of charged particle, then

Centripetal force = Magnetic force

$$\Rightarrow \quad \frac{mv^2}{R} = Bqv$$

$$\Rightarrow \quad R = \frac{mv}{Bq}$$

Time period, $T = \frac{2\pi R}{v}$

$$= 2\pi \frac{\left(\frac{mv}{Bq}\right)}{v}$$

$$\Rightarrow \quad T = \frac{2\pi m}{Bq}$$

So, T is independent of R and v

2. *(b)* At 300°C, the heat energy is $Q = ms\Delta T$

$$= (3.35 \times 390 \times 300)\,\text{J}$$

Given, $L = 335\,\text{J kg}^{-1}$

The maximum mass of ice that can melt is

$$m = \frac{Q}{L} = \frac{3.35 \times 390 \times 300}{335}$$

$$= 1.17 \times 10^3 \text{ kg}$$

3. *(b)* Change in internal energy from $A \to B$ is

$$\Delta U = \frac{f}{2} nR\Delta T = \frac{f}{2}(p_f V_f - p_i V_i)$$

[$\because n$ = number of moles]

$$= \frac{3}{2}(2p_0 \times 2V_0 - p_0 V_0) = \frac{9}{2} p_0 V_0$$

Work done in process $A \to B$ is equal to the area covered by the graph with volume axis, i.e.,

$$W_{A \to B} = \frac{1}{2}(p_0 + 2p_0) \times (2V_0 - V_0) = \frac{3}{2} p_0 V_0$$

Hence, $\Delta Q = \Delta U + \Delta W = \frac{9}{2} p_0 V_0 + \frac{3}{2} p_0 V_0 = 6 p_0 V_0$

4. *(d)* $X^{240} \to Y^{110} + Z^{130}$

$$Q = (\text{BE})_{\text{product}} - (\text{BE})_{\text{reactant}}$$

$$= [(7.8 \times 110) + (8 \times 130)] - [240 \times 7.5]$$

$$= (858 + 1040) - 1800$$

$$= 1898 - 1800 = 98\,\text{MeV}$$

5. *(a)* Field at face $ABCD = 56x_0\hat{\mathbf{i}}$ and

area vector, $\mathbf{A} = -a^2\hat{\mathbf{i}}$

Given, side, $a = 0.6\,\text{m}$

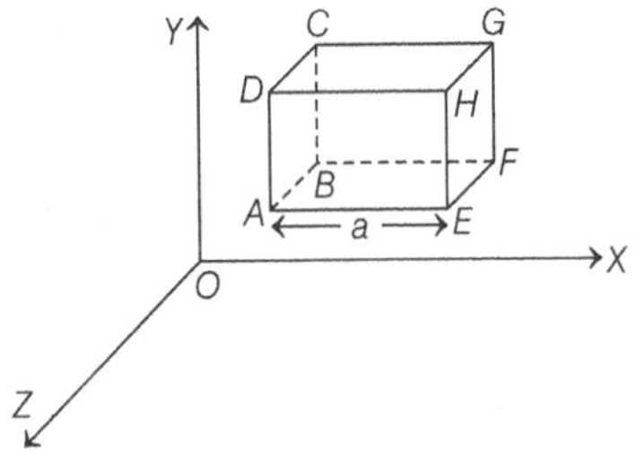

Flux over the face $ABCD$, ϕ_1

$$= -56x_0(a^2) = -20.16x_0$$

Here, negative sign arises as the area vector is in negative x-direction.

Field at face $EFGH$, $\phi_2 = 56(x_0 + a)\hat{\mathbf{i}}$

$$= 56(x_0 + 0.6)\hat{\mathbf{i}}$$

Flux over the face $EFGH$

$$= 56(x_0 + 0.6) \times (0.6)^2 = 20.16(x_0 + 0.6)$$

Flux over the other four faces is zero as the field is not parallel to the surfaces.

Total flux over the cube = $\phi_1 + \phi_2$

$$= 20.16(x_0 + 0.6) - 20.16x_0$$

$$= 20.16 \times 0.6 = 12.096$$

As, $\phi = \frac{q}{\varepsilon_0} \Rightarrow q = \phi\varepsilon_0$

Hence, $q = 12\varepsilon_0$

$$= 12 \times 8.85 \times 10^{-12} = 107\,\text{pC}$$

Study Tactics

The area vector always points outside from the body. Hence, area vector of *ABCD* will point in negative *x*-direction and area vector of *EFGH* will point in positive *x*-direction.

After calculating the total flux, do not forget to use $\phi = \frac{q}{\varepsilon_0}$ to calculate the charge enclosed by the box.

6. *(c)* As, $v_t \propto r^2$ ...(i)

Let r be the radius of small rain drops and R is the radius of large drop.

Equating the volumes, we have

$$\frac{4}{3}\pi R^3 = 2\left(\frac{4}{3}\pi r^3\right)$$

$\therefore \quad R = (2)^{1/3} \cdot r$

or $\quad \dfrac{R}{r} = (2)^{1/3}$

$\therefore \quad \dfrac{v_t'}{v_t} = \left(\dfrac{R}{r}\right)^2 = (2)^{2/3}$

$\therefore \quad v_t' = (2)^{2/3}\, v_t = (2)^{2/3}(1.0) = 1.587 \text{ ms}^{-1}$

7. *(a)* Photoelectric current $\propto$ Intensity of light.

Current in a photocell remains constant as frequency of light increases.

Photocurrent increases with an increase in applied voltage until it becomes maximum upto saturation stage and then does not increase on increasing the applied voltage.

The stopping potential does not depend on the intensity of light.

8. *(b)* Let x be the distance of a point from the Earth where gravitational force on the rocket due to Sun and Earth becomes equal and opposite. The distance of rocket from the Sun $= (r - x)$. If m is the mass of rocket, then

$$\frac{GM_s m}{(r-x)^2} = \frac{GM_e m}{x^2}$$

$\Rightarrow \quad \dfrac{(r-x)^2}{x^2} = \dfrac{M_s}{M_e}$

or $\quad \dfrac{r-x}{x} = \sqrt{\dfrac{M_s}{M_e}}$

$$= \sqrt{\frac{2\times 10^{30}}{6\times 10^{24}}} = \frac{10^3}{\sqrt{3}}$$

or $\quad \dfrac{r}{x} - 1 = \dfrac{10^3}{\sqrt{3}}$

On solving, we get

$$x = \frac{r\sqrt{3}}{\sqrt{3}+10^3}$$

Or $\quad x = \dfrac{1.5\times 10^{11}\times\sqrt{3}}{\sqrt{3}+10^3} = 2.59\times 10^8 \text{ m}$

9. *(c)* Maximum height of projectile

$$H = \frac{u^2 \sin^2\theta}{2g} = \frac{(20)^2 \sin^2 30^\circ}{2\times 10} = 5\text{ m}$$

Time to reach maximum height,

$$t = \frac{u\sin\theta}{g} = \frac{20\times \sin 30^\circ}{10} = 1\text{ s}$$

So, distance of vertical fall in 1 s

$$h = \frac{1}{2}gt^2 = \frac{1}{2}\times 10\times (1)^2 = 5\text{ m}$$

$\therefore$ Initial height of second ball = 5 m + 5 m = 10 m

10. *(a)* The diode D_1 is reversed biased. So, the circuit can be re-drawn as

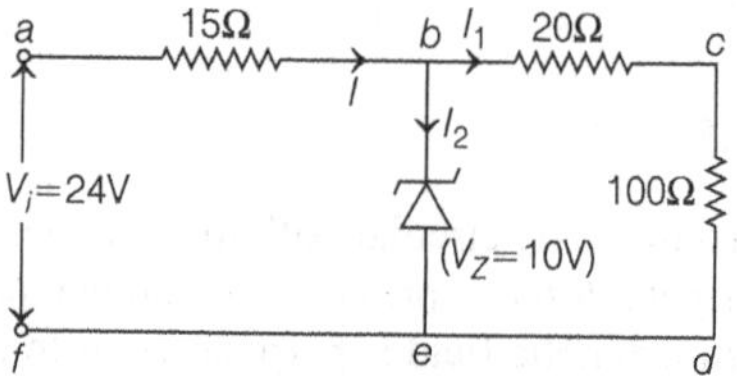

Voltage across $15\,\Omega$ resistor is $V_{ab} = 24 - 10 = 14$ V

Current through $15\,\Omega$ resistor is

$$I = \frac{V_{ab}}{15} = \frac{14}{15}\text{ A}$$

Current through $20\,\Omega$ and $100\,\Omega$ resistors is

$$I_1 = \frac{V_z}{120} = \frac{10}{120} = \frac{1}{12}\text{ A}$$

$\therefore$ Current through zener diode is

$$I_2 = I - I_1 = \frac{14}{15} - \frac{1}{12} = \frac{51}{60}\text{ A}$$

$\therefore$ Power dissipated through zener diode is

$$P = V\cdot I_2 = 10\times\frac{51}{60} = 8.5\text{ W}$$

11. *(c)* Potential at point P due to a charged ring Q,

$$V_P = \frac{1}{4\pi\varepsilon_0}\frac{Q}{\sqrt{r^2+(2\sqrt{2}r)^2}} = \frac{1}{4\pi\varepsilon_0}\cdot\frac{Q}{3r}$$

Applying conservation of energy at point P and O.

$$U_P + K_P = U_0 + K_0$$

$\Rightarrow \quad \dfrac{1}{4\pi\varepsilon_0}\cdot\dfrac{Q(-q)}{3r} + 0 = \dfrac{1}{4\pi\varepsilon_0}\cdot\dfrac{Q(-q)}{r} + K_0$

$\therefore \quad K_0 =$ Kinetic energy at point O

$$K_0 = \frac{qQ}{6\pi\varepsilon_0 r}$$

12. *(a)* $\dfrac{R_1 R_2}{R_1 + R_2} = \dfrac{6}{5}\,\Omega$

Now, one resistance (R_2) is removed, then effective resistance is R_1.

$\therefore \quad R_1 = 2\,\Omega$

$\therefore \quad \dfrac{2R_2}{2+R_2} = \dfrac{6}{5}$

$\Rightarrow \quad 10R_2 = 12 + 6R_2$

$\Rightarrow \quad 4R_2 = 12$

$\Rightarrow \quad R_2 = 3\,\Omega$

Hence, the resistance of the wire removed will be $3\,\Omega$.

13. *(c)* For p-T diagram, for process $A \rightarrow B$,

$$T = \text{constant} \Rightarrow p \propto \frac{1}{V}$$

$\therefore \quad pV = RT = \text{constant} \Rightarrow p \propto \dfrac{1}{V}$

$\therefore$ Isothermal process

For process $B \rightarrow C$

$$p = \text{constant}$$

For process $C \rightarrow A$

$$p \propto T \Rightarrow p = kT = \frac{RT}{V}$$

$$\therefore \quad k = \frac{R}{V}$$

$$\Rightarrow \quad V = \text{constant}$$

$\therefore$ Option (c) is correct.

14. ***(b)*** An accelerated charge radiates EM wave. This charge produces electric field perpendicular to propagation of EM wave. The produced electric field further produces magnetic field perpendicular to both electric field and direction of EM wave.

The average energy density of electric field,

$$u_E = \frac{1}{2}\varepsilon_0 E_{rms}^2$$

$$= \frac{1}{2}\varepsilon_0 \left(\frac{E_0}{\sqrt{2}}\right)^2 = \frac{1}{4}\varepsilon_0 E_0^2$$

The average energy density of magnetic field,

$$u_B = \frac{B_{rms}^2}{2\mu_0} = \frac{(B_0/\sqrt{2})^2}{2\mu_0} = \frac{B_0^2}{4\mu_0}$$

Now,
$$\frac{u_E}{u_B} = \frac{\frac{1}{4}\varepsilon_0 E_0^2}{\frac{B_0^2}{4\mu_0}} = \varepsilon_0 . E_0^2 . \frac{\mu_0}{B_0^2}$$

$$= \frac{(E_0/B_0)^2}{\left(\frac{1}{\sqrt{\mu_0 \varepsilon_0}}\right)^2} = \frac{c^2}{c^2} = 1$$

$$\therefore \quad u_E = u_B \qquad \left[\because c = \frac{1}{\sqrt{\mu_0\varepsilon_0}} = \frac{E_0}{B_0}\right]$$

Hence, Reason is correct but it does not explain Assertion.

15. ***(a)*** As, velocity of sound in air, $c_s = \sqrt{\frac{\gamma P}{\rho}}$

RMS velocity of air, $c = \sqrt{\frac{3RT}{M}} = \sqrt{\frac{3P}{\rho}}$

$$\therefore \quad \frac{c_s}{c} = \sqrt{\frac{\gamma}{3}} = \sqrt{\frac{1.4}{3}} < 1$$

$$\Rightarrow \quad c_s < c$$

Now, for triatomic gas, $f = 6$, $n = 1$ mole

$$C_P = C_V + R$$

$$= n\frac{f}{2}R + R = 3R + R = 4R$$

16. ***(c)*** Given,

$$V_0 = \frac{h^p c^q G^r}{I^s} = h^p c^q G^r I^{-s}$$

$$\therefore \quad [V_0] = [h^p c^q G^r I^{-s}] \qquad \ldots\text{(i)}$$

$$\therefore \quad [V_0] = \left[\frac{\text{Potential energy}}{\text{Charge}}\right]$$

$$= \frac{[\text{ML}^2\text{T}^{-2}]}{[\text{AT}]} = [\text{ML}^2\text{T}^{-3}\text{A}^{-1}]$$

$$[h] = \left[\frac{\text{Energy}}{\text{Frequency}}\right] = \frac{[\text{ML}^2\text{T}^{-2}]}{[\text{T}^{-1}]} = [\text{ML}^2\text{T}^{-1}]$$

$$[c] = [\text{Speed}] = [\text{LT}^{-1}], [I] = [\text{A}]$$

$$[G] = \left[\frac{\text{Force} \times (\text{Distance})^2}{(\text{Mass})^2}\right]$$

$$= \frac{[\text{MLT}^{-2}][\text{L}^2]}{[\text{M}^2]}$$

$$= [\text{M}^{-1}\text{L}^3\text{T}^{-2}]$$

$\therefore$ From Eq. (i), we get

$$[\text{ML}^2\text{T}^{-3}\text{A}^{-1}] = [\text{ML}^2\text{T}^{-1}]^p . [\text{LT}^{-1}]^q . [\text{M}^{-1}\text{L}^3\text{T}^{-2}]^r [\text{A}]^{-s}$$

$$\therefore \quad p - r = 1 \qquad \ldots\text{(ii)}$$

$$\therefore \quad 2p + q + 3r = 2 \qquad \ldots\text{(iii)}$$

$$\therefore \quad -p - q - 2r = -3 \qquad \ldots\text{(iv)}$$

$$-s = -1$$

$$\Rightarrow \quad s = 1 \qquad \ldots\text{(v)}$$

On solving Eqs. (ii), (iii), (iv) and (v), we get

$$p = 0, q = 5, r = -1, s = 1$$

17. ***(b)***

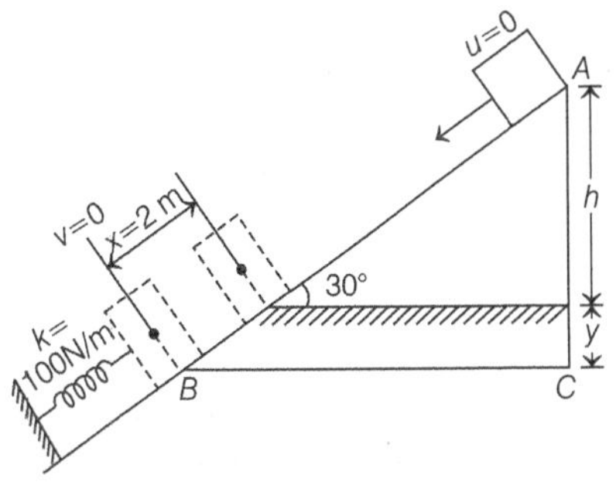

Given, $m = 10\,\text{kg}$, $k = 100\,\text{N/m}$,

$$y = x \sin 30° = 2 \sin 30° = 1\,\text{m}$$

Applying conservation of energy at points A and B,

$$mg\,(h + y) = \frac{1}{2}kx^2$$

$$\Rightarrow \quad 10 \times 10 \times (h + 1) = \frac{1}{2} \times 100 \times (2)^2$$

$$\therefore \quad h = 1\,\text{m}$$

18. *(a)* Given, $y = kt^2, \frac{dy}{dt} = 2kt$

and $\frac{d^2y}{dt^2} = 2k = 2\times 1 = 2\,m/s^2$

So, point of suspension of pendulum is moving upwards with acceleration $a = 2\,m/s^2$.

Then, effective acceleration due to gravity on pendulum,

$$g = (g + a) = 10 + 2 = 12\,m/s^2$$

$$T_1 = 2\pi\sqrt{\frac{l}{g}},\ T_2 = 2\pi\sqrt{\frac{l}{g'}}$$

$$\Rightarrow \quad \frac{T_1^2}{T_2^2} = \frac{g'}{g} = \frac{12}{10} = \frac{6}{5}$$

19. *(c)* $\mu = \frac{\lambda}{\lambda_m} = \frac{\lambda}{\frac{2}{3}\lambda} = \frac{3}{2}$

The focal length of a plano-convex lens is

$$\frac{1}{f} = \frac{\mu - 1}{R} = \frac{\frac{3}{2} - 1}{R} = \frac{1}{2R} \quad \text{...(i)}$$

From lens formula,

$$\frac{1}{f} = \frac{1}{v} - \frac{1}{u} \quad \text{...(ii)}$$

$$\Rightarrow \quad \frac{1}{v} - \frac{1}{u} = \frac{1}{2R}$$

$$\Rightarrow \quad m = -\frac{v}{u} = \frac{h}{h_0}$$

$$\Rightarrow \quad -\frac{v}{u} = \frac{\frac{1}{3}h_0}{h_0} = \frac{1}{3}$$

$$u = -3v = -3\times 8 = -24$$

$$\Rightarrow \quad \frac{1}{2R} = \frac{1}{8} - \frac{1}{-24}$$

$$\Rightarrow \quad \frac{1}{2R} = \frac{3+1}{24}$$

$$\Rightarrow \quad R = 3\,m$$

20. *(c)* Static friction will provide the net acceleration as it is the only force in the horizontal plane. But it is not given in the question that whether maximum value of friction is acting or not, hence we should not use $(f_s)_{max}$. That is why we have used f_s.

Here, friction is the only real force that is acting on the horizontal plane.

$$\Rightarrow \quad f_s = m\times a_{net} = m\times\sqrt{a_c^2 + a_t^2}$$

Thus, friction will provide both $\mathbf{a}_c$ and $\mathbf{a}_t$.

Here, we can write

$$f_s \le (f_s)_{max}$$

21. (30) Given, $u = -40\,cm, \frac{A_I}{A_0} = 9$

∴ Linear magnification

$$= \sqrt{\frac{A_I}{A_0}} = \sqrt{9} = 3$$

$$m = -3 = \frac{v}{u} \quad (\because \text{real image})$$

$$\therefore \quad v = -3u = -3\times(-40) = 120\,cm$$

From lens formula,

$$\frac{1}{f} = \frac{1}{v} - \frac{1}{u}$$

$$= \frac{1}{120} - \frac{1}{-40} = \frac{1+3}{120}$$

$$\therefore \quad f = 30\,cm$$

22. (48)

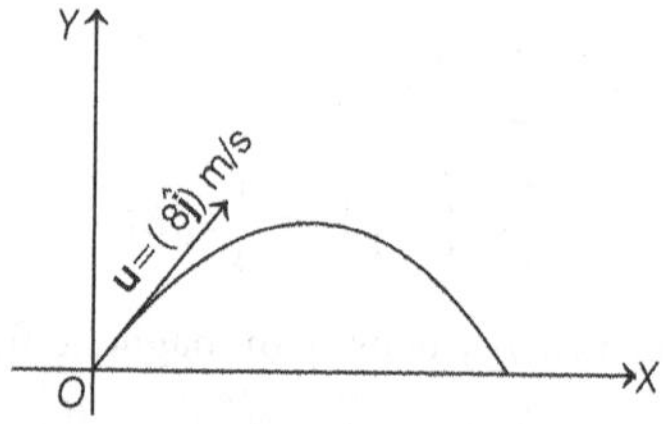

Given, acceleration of the particle, $\mathbf{a} = (5\hat{\mathbf{i}} + 2\hat{\mathbf{j}})\,m/s^2$

Considering motion of the particle in x-direction.

$u_x = 0\,m/s,\ a_x = 5\,m/s^2$

$$\therefore \quad x_0 = u_x t + \frac{1}{2}a_x t^2$$

$$40 = 0 + \frac{1}{2}\times 5\times t^2$$

$$\therefore \quad t = 4\,s$$

Considering motion of the particle in y-direction.

$$u_y = 8\,m/s,\ a_y = 2\,m/s^2$$

$$\therefore \quad y_0 = u_y t + \frac{1}{2}a_y t^2$$

$$= 8\times 4 + \frac{1}{2}\times 2\times(4)^2 = 48\,m$$

23. (157) Given, $n = 7.5\times 10^{30}\,m^{-3}, \tau = 30\,fs = 30\times 10^{-15}\,s$, $m_e = 9.1\times 10^{-31}\,kg$

The resistivity of the conductor is given by

$$\rho = \frac{m_e}{ne^2\tau}$$

$$= \frac{9.1\times 10^{-31}}{7.5\times 10^{30}\times(1.6\times 10^{-19})^2\times 30\times 10^{-15}}$$

$$= 1.57\times 10^{-10}\,\Omega\text{-m}$$

$$\therefore \quad \rho = 157\times 10^{-12}$$

$$\Rightarrow \quad n = 157$$

24. (4) For pure rolling, $a_0 = r\alpha$

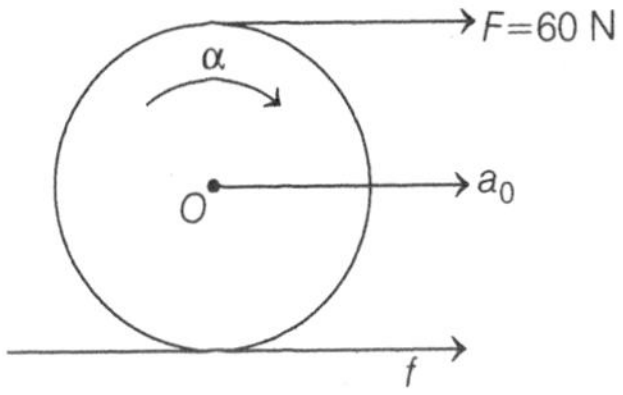

For translatory motion,

$$F + f = ma_0 = mr\alpha \quad \ldots(i)$$

For rotational motion,

$$\sum \tau_{net} = I\alpha$$

$$\Rightarrow \quad (F - f)r = \left(\frac{1}{2}mr^2\right).\alpha$$

$$\therefore \quad F - f = \frac{1}{2}mr\alpha \quad \ldots(ii)$$

From Eqs. (i) and (ii), we get

$$\therefore \quad \alpha = \frac{4F}{3mr}$$

$$= \frac{4 \times 60}{3 \times 10 \times 2}$$

$$= 4 \text{ rad/s}^2$$

25. (3)

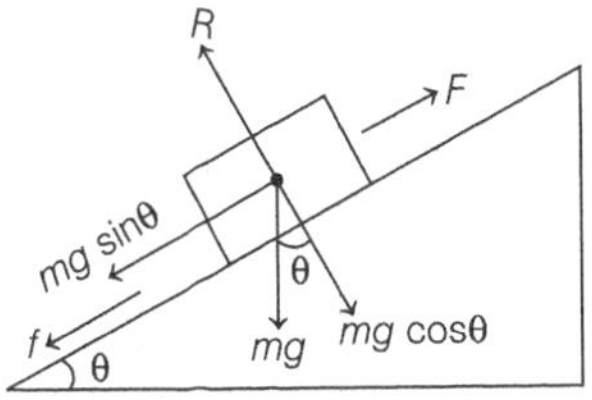

When the body is made to move up the plane, it moves with retardation.

From the FBD diagram,

$$F = -(mg\sin\theta + f)$$

$$F = -(mg\sin\theta + \mu R)$$

$$ma = -m(g\sin\theta + \mu g\cos\theta)$$

$$(\because R = mg\cos\theta)$$

$$a = -g(\sin\theta + \mu\cos\theta)$$

$$= -9.8(\sin 30° + 0.2\cos 30°)$$

$$= -6.597 \text{ m/s}^2$$

Retardation $= 6.597 \text{ m/s}^2$

Retardation force = Mass × Retardation

$$= 0.5 \times 6.597$$

$$= 3.299 \text{ N}$$

$$= 3.3 \text{ N}$$

$$\approx 3 \text{ N (rounded off answer)}$$

Chemistry

26. (c) Among the given compounds only three i.e. II, III and IV will be obtained as a product upon reaction with sodium methoxide in methanol.

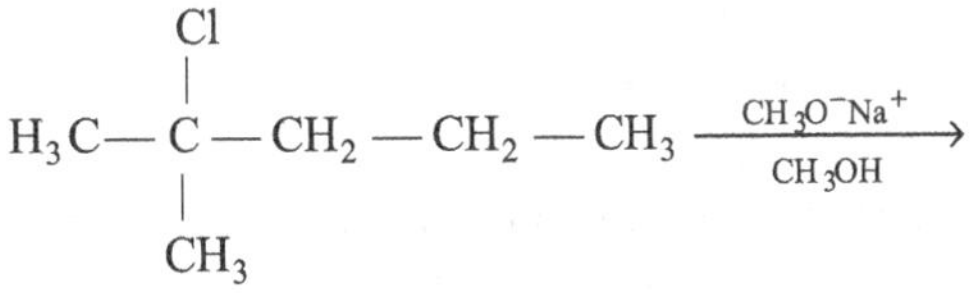

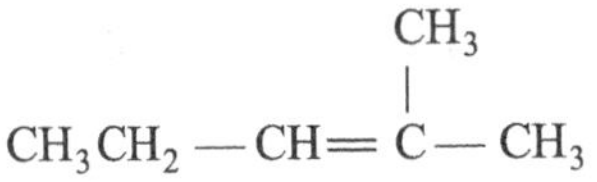

(II)

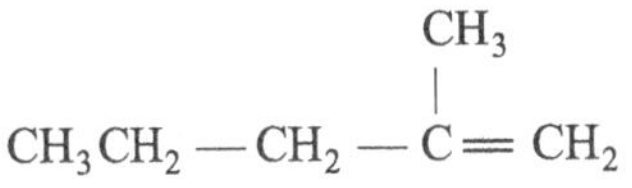

(III)

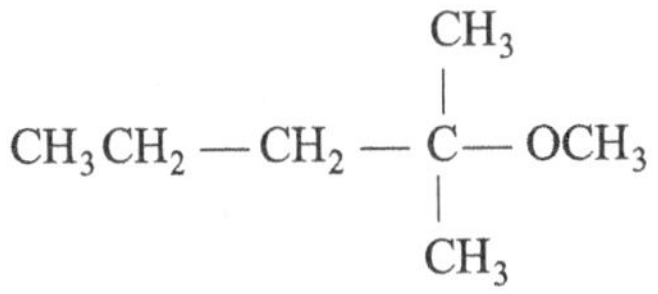

(IV)

27. (b) Characteristic of Vitamin-B_{12}

(i) Vitamin-B_{12} is also known as cyanocobalamin.

(ii) Metal present in cyanocobalamin is cobalt.

(iii) It catalyses isomerism reaction.

28. (b) $A + B \underset{\text{Fast}}{\overset{\text{Slow}}{\rightleftharpoons}} IAB$ so $E_{a(f)}$ is high and $E_{a(b)}$ is low.

Now, $(k_1 << k_2)$; E_a for the step with k_1 is very high and step (k_2) is low and overall reaction is exothermic.

29. (b) $E_{He^+} = \dfrac{-2.18 \times 10^{-18} \times 2^2}{2^2} = \dfrac{-2.18 \times 10^{-18}}{4} \times 4$

$$E_{Li^{2+}} = \frac{-2.18 \times 10^{-18} \times 3^2}{2^2} = \frac{-2.18 \times 10^{-18}}{4} \times 9$$

$$\frac{E_{He^+}}{E_{Li^{2+}}} = \frac{4}{9} \text{ or } E_{He^+} = \frac{4}{9}E_{Li^{2+}}$$

30. (a) $3Br_2 + 6OH^- \longrightarrow \underset{(X)}{5Br^-} + \underset{(Y)}{BrO_3^-} + 3H_2O$

$$5Br^- + BrO_3^- + 6H^+ \longrightarrow 3Br_2 + 3H_2O$$

Since, X is $Br^{\ominus}$, it gives yellow ppt. with $AgNO_3$ solution.

31. (a) $MnO_4^- + 8H^+ + 5e^- \rightleftharpoons Mn^{2+} + 4H_2O$

32. *(a)* Among the given statements C and D are correct while B and A are incorrect.

$$\frac{K_{\text{exp}}}{K_{\text{Theoretical}}} = P$$

Since, $P > 1$

$$K_{\text{exp}} > K_{\text{Theoretical}}$$
$$A_{\text{exp}} > A_{\text{Theoretical}}$$

So, there is no need of an effect catalyst to reduce activation energy as $K_{\text{exp}} > K_{\text{Theoretical}}$.

33. *(a)* Structure of *cis*-2, 3-dihydroxybutan-1-al is

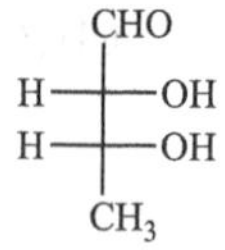

This reactant on reaction with $NaIO_4$ undergo malaprade oxidation to produce 2 mole of formic acid and one mole of acetaldehyde.

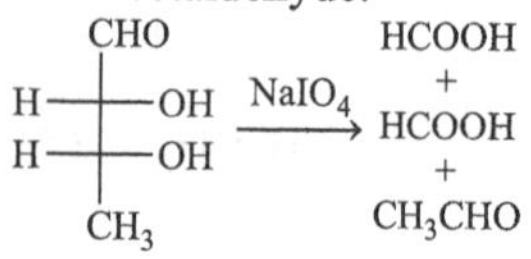

Study Tactics

This reaction includes conceptual mixing of oxidation of α-hydroxy carbonyl compound and nomenclature.

Follow the tips while solving the question.

I. Draw the structure of given compound.

II. Perform malaprade oxidation by breaking bond between vicinal OH containing carbons.

34. *(b)* $\Delta G = \Delta G° + RT \ln Q$

$$\Delta G = 249.2 + 8.314 \times 300 \times \ln 4$$
$$\Delta G = +\text{ve}$$

So, $Q > K_C$

Therefore, reaction shift in reverse direction.

35. *(a)* The melting point increases from Sc to Cr as the number of unpaired electrons increases. Then, Mn shows a low melting point which break the series.

This is because the metallic bond is weak in comparison to the bond of other transition metals. Then, fall regularly as the atomic number increases.

36. *(b)* **Ease of Dehydration** Greater the stability of carbocation, more will be reactivity of alcohol towards dehydration, the stability of carbocation is explained by presence of substituent around carbon containing —OH group.

The carbocation generated are as follows

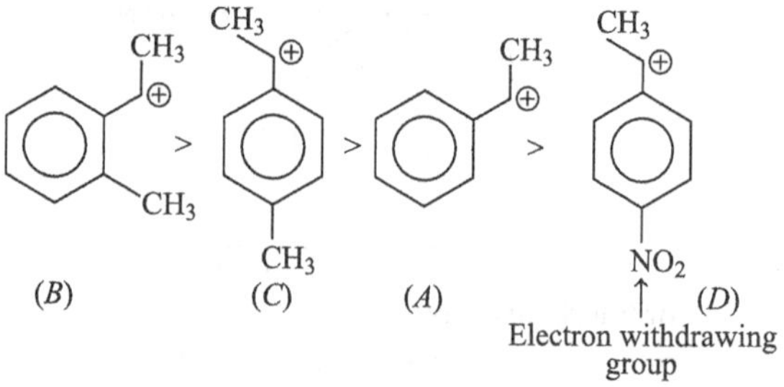

Electron releasing group present on ring increases electron density on carbon atom of carbocation. Nearer the electron releasing group from carbocation, more will be the stability of carbocation.

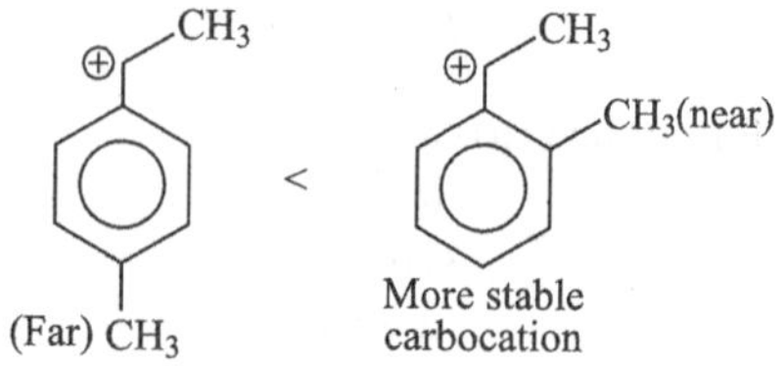

Electron withdrawing group decreases electron density to the ring hence causes decrease in stability of carbocation.

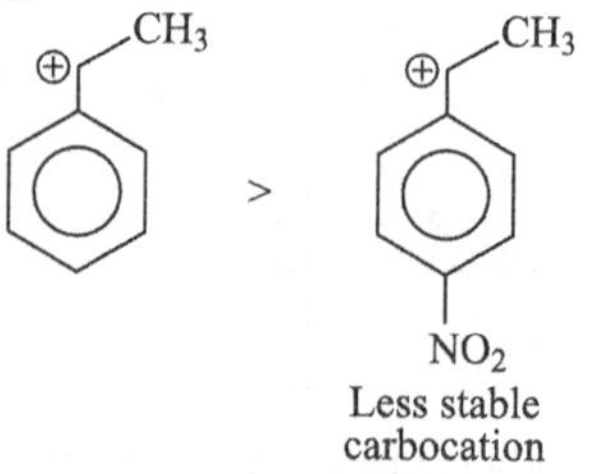

Study Tactics

This problem includes conceptual mixing of ease of dehydration and stability of carbocation. To solve the problem write the structure of carbocation obtained after dehydration and then analyse the stability of carbocation using inductive effect, resonance and hyperconjugation.

37. *(c)* (a) Both are sp^2-hybridised and trigonal planar.

(b) Both are sp^3-hybridised and tetrahedral.

(c) PF_5 is in sp^3d hybridisation state so geometry will be trigonal bipyramidal and in BrF_5, Br is in sp^3d^2 hybridised state so geometry will be **octahedral.**

(d) Both are sp^3d^2-hybridised and octahedral.

38. *(c)* According to fajan's rule; smaller the size of the cation, the larger the size of the anion and greater the covalent character of the ionic bond.

39. *(c)* To solve this problem follow the following steps

- Determine the structure of molecule using VSEPR theory
- Use bent's rule to draw the correct structure of molecule

I_2Cl_6 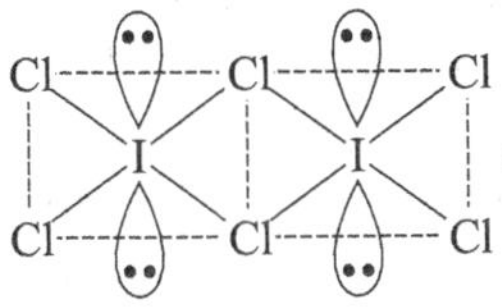 sp^3d^2-planar

XeF_2 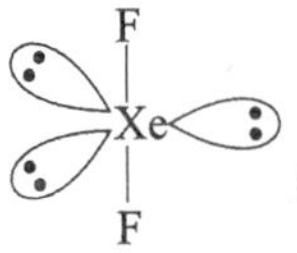 sp^3d (Planar) It is planar as three points are in same plane.

BrF_4^- 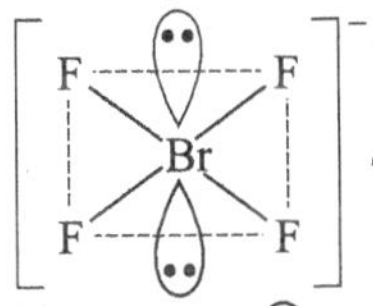 sp^3d^2-square planar

XeF_5^- 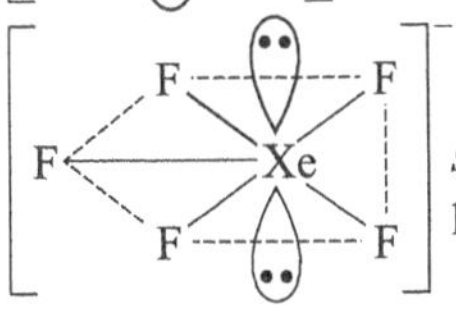 sp^3d^3-pentagonal planar

40. *(b)* $CH_3CH_2Br \xrightarrow{KCN} \underset{(A)}{CH_3CH_2CN}$

$\underset{(A)}{CH_3CH_2CN} \xrightarrow{H_3O^+} \underset{(B)}{CH_3CH_2COOH}$

$CH_3CH_2CN \xrightarrow[\text{Reduction}]{\text{DIBAL-H}} \underset{\text{Imine }(C)}{CH_3—CH_2—\overset{\overset{NH}{\|}}{C}—H}$

$\xrightarrow[H_3O^{\oplus}]{\text{Hydrolysis}} \underset{(D)}{CH_3—CH_2—\overset{\overset{O}{\|}}{C}—H}$

$\xrightarrow{\text{Oxidation}} \underset{(B)}{CH_3CH_2COOH}$

41. *(d)* The atomic radii of F and Ne is 0.72 Å and 1.6 Å. Electron gain enthalpy of Rb, Cs pair is – 46 and Ar, Kr is + 96 kJ/mol which is same, so statement II is true but Statement I is false.

42. *(d)*

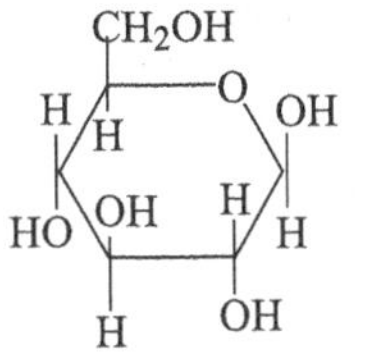

β-(D)-glucopyranose

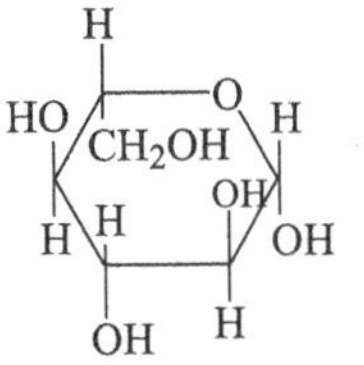

β-(L)-glucopyranose

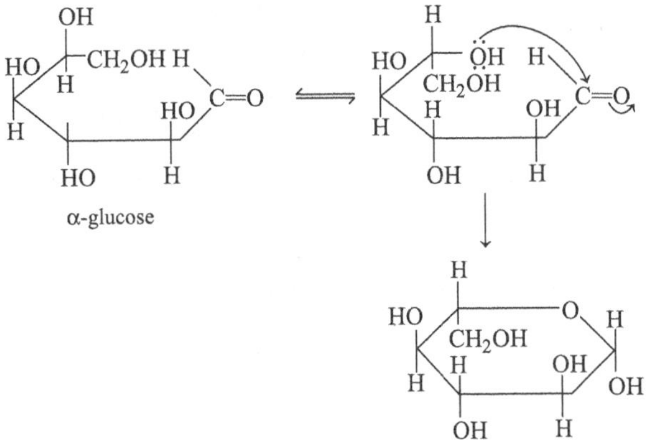

α-glucose

β-(L) Glucopyranose

43. *(b)* Both (A) and (R) are true but (R) is not the correct explanation of (A).

As in diamond the hybridisation of carbon is sp^3-hybridised due to which there is non-availability of free electrons. This makes diamond a bad conductor of electricity. Whereas, in graphite, each carbon is sp^2-hybridised i.e. one electron of each carbon atom is free to move. Hence, graphite is said to be a good conductor of electricity.

44. *(b)* The correct match is A-III, B-IV, C-I, D-II.

$[Ni(CN)_4]^{2-} - dsp^2$; $BrF_5 - sp^3d^2$; $PCl_5 - sp^3d$; $[Co(NH_3)_6]^{3+} - d^2sp^3$.

45. *(d)* The correct match is A-(III), B-(I), C-(II), D-(IV).

More the pK_b value, lower will be the basicity.

Among the given compounds, pyrrole (A) is a very weak base. The lone pair present on N is involved in resonance. Protonation results in the loss of aromaticity and this makes it an unfavourable process.

In pyridine (B), the lone pair present on nitrogen atom is sp^2-hybridised (more electronegative atom). Higher is the electronegativity of the atom, lesser is the donation tendency of the N-atom and hence, lesser is the basicity.

In morpholine (C), the lone pair present on N is in sp^3-hybridised state and there is no resonance. However, the electron withdrawing effect of O-atom from ring makes it less available for donation.

Lastly, the piperidine (D) which is the most basic among the given compound have lowest pK_b value. The lone pair present on N is in sp^3-hybridised and there is no resonance.

So, the order of basicity is D > C > B > A

and the decreasing order of pK_b value is

A > B > C > D.

46. *(12)* Cell constant $= \dfrac{\text{Resistance}}{\text{Specific resistance}}$

$$= \frac{160 \text{ ohm}}{700 \text{ ohm-cm}} = 0.228 \text{ cm}^{-1}$$

$\because$ Cell constant for every cell is fixed. Thus, specific conductance of NaOH

$$= \frac{1}{\text{resistance of NaOH}} \times \text{cell constant}$$

$$\kappa = \frac{1}{190} \times 0.228 = 0.00120 \text{ ohm}^{-1}\text{cm}^{-1}$$

$$\approx 12 \times 10^{-4} \text{ ohm}^{-1}\text{cm}^{-1}$$

47. *(122)* $\Delta E = \dfrac{hc}{\lambda}$, $hc = 12400 \text{ eV-Å}$

$$= \frac{12400}{1051.2} = 11.796 \text{ eV} \approx 11.8 \text{ eV}$$

Using, $E = -13.6 \dfrac{Z^2}{n^2}$

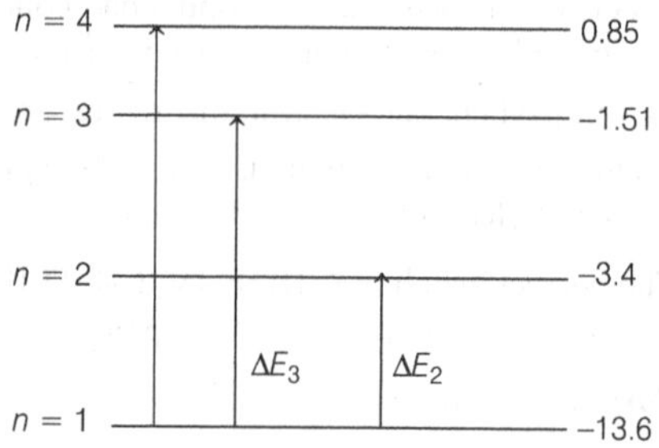

$$\Delta E_3 = 12.09 \text{ eV}$$
$$\Delta E_2 = 10.2 \text{ eV}$$

Since, energy of radiation is less than 12.09. So, only one transition is possible corresponding to ΔE_2.

$$\lambda(\text{nm}) = \frac{1240 \text{ eV-nm}}{10.2 \text{ eV}} = \frac{1240}{10.2} = 121.6$$

So, λ of induced radiation ≈ 122 nm

48. *(17)* The reaction involved in the combustion of glucose is

$$C_6H_{12}O_6(s) + 6O_2(g) \longrightarrow 6CO_2(g) + 6H_2O(l)$$

$$\Delta H° = \Sigma[\Delta H_f^\circ \text{(product)}] - \Sigma[\Delta H_f^\circ \text{(reactant)}]$$

$$\therefore \; \Delta H° = [6 \times (-400) + 6 \times (-300)] - [(-1200 + 0)]$$
$$= -4200 + 1200 = -3000 \text{ kJ/mol}$$

Now, $\Delta H°$ (kJ/g) $= \dfrac{\Delta H° \text{ (kJ/mol)}}{\text{Molecular mass (in g per mol)}}$

Molecular mass of glucose = 180 g/mol

$$\therefore \quad \Delta H° \text{ (kJ/g)} = \frac{-3000}{180} = -16.67 \text{ kJ/g} \sim -17 \text{ kJ/g}$$

Hence, the standard enthalpy of combustion per gram of glucose at 273 K is – 16.67 kJ/g or approximately – 17 kJ/g.

49. *(2)* According to Huckel rule, the compounds that have $4n$ ($n = 0, 1, 2, \ldots$) delocalised π-electrons in a close-loop are anti-aromatic and characteristically unstable. So, in this case, compound II and IV satisfy this criterion of anti-aromaticity and therefore, unstable.

50. *(57)* The reactions involved in given process are

$$C_2H_6 + Br_2 \longrightarrow C_2H_5Br + HBr$$

$$2C_2H_5Br \xrightarrow[\text{ether}]{\text{Na}} \underset{n\text{-butane}}{C_4H_{10}} + NaBr \quad \text{[Wurtz reaction]}$$

Suppose, moles of $C_2H_6 = x$

$\therefore$ Moles of $C_2H_5Br = y$ when 100% yield

Given, 90% yield of C_2H_5Br occurred.

$\therefore$ Moles of $C_2H_5Br = y \times \dfrac{90}{100} = 0.9y$

Now, from 2 moles of $C_2H_5Br \longrightarrow$ 1 mole of C_4H_{10} obtained.

$\therefore$ From 0.9 y moles of C_2H_5Br gives $\dfrac{1}{2} \times 0.9y$ moles of C_4H_{10} at 100% yield.

But, at 85% yield,

Moles of $C_4H_{10} = \dfrac{1}{2} \times 0.9y \times \dfrac{85}{100}$

$$= \frac{(0.9 \times 0.85)y}{2}$$

So, according to given statement,

$$\frac{(0.9 \times 0.85)y}{2} = \frac{56}{58}$$

$$\therefore \quad y = \frac{56}{29 \times 0.9 \times 0.85} = 2.52$$

$\therefore$ Volume of $C_2H_6 = 2.52 \times 22.4$ litres $= 56.448$ litres

$\therefore$ The value of $x \simeq 57$ L.

Mathematics

51. *(a)* Given, $\dfrac{dx}{dt} = \dfrac{1}{2}x + 50 \Rightarrow \dfrac{dx}{x+100} = \dfrac{1}{2}dt$

On integrating both sides, we get

$$\int \frac{dx}{x+100} = \frac{1}{2}\int dt$$

$$\Rightarrow \quad \log|x+100| = \frac{1}{2}t + \log C$$

$$\Rightarrow \quad \log\left|\frac{x+100}{C}\right| = \frac{1}{2}t \Rightarrow \frac{x+100}{C} = e^{\frac{1}{2}t}$$

$$\Rightarrow \quad x(t) = Ce^{\frac{1}{2}t} - 100$$

When $\quad x(0) = 200$

$$200 = C - 100 \Rightarrow C = 300$$

$\therefore$ Solution $\quad x(t) = 300\, e^{t/2} - 100$

Hence, $\quad a - b = 300 - 100 = 200$

52. *(c)* Here, $f(x)=\begin{cases}5[x]+3 & , \text{ if } x>0\\ 5[x]-3 & , \text{ if } x<0\end{cases}$

$$\int_{-1/2}^{2} f(x)\,dx=\int_{-1/2}^{0} f(x)\,dx+\int_{0}^{1} f(x)\,dx+\int_{1}^{2} f(x)\,dx$$
$$=\int_{-1/2}^{0}[5(-1)-3]dx+\int_{0}^{1}[5(0)+3]\,dx+\int_{1}^{2}[5(1)+3]\,dx$$
$$=(-8x)_{-1/2}^{0}+(3x)_{0}^{1}+(8x)_{1}^{2}$$
$$=(-8)\left(\frac{1}{2}\right)+3(1)+8(1)=-4+3+8=7$$

53. *(d)* $\because f'(x)=\dfrac{1}{1+x^5}$ ...(i)

Also, $g=f^{-1}$

$\therefore\quad f[g(x)]=x=\displaystyle\int_0^{g(x)}\frac{dt}{1+t^5}$

$\Rightarrow\quad 1=\dfrac{1}{1+[g(x)]^5}\times g'(x)\Rightarrow g'(x)=1+[g(x)]^5$

$g'(x)=\dfrac{1}{f'[g(x)]}$ [from Eq. (i)]

54. *(c)* Let $z=ae^{i\theta}$ and $\omega=be^{i\phi}$

$\Rightarrow\quad \bar{z}=ae^{-i\theta}$

$\because\quad |z\omega|=1\Rightarrow |abe^{i\theta}e^{i\phi}|=1$

$\Rightarrow\ |abe^{i(\theta+\phi)}|=1\Rightarrow ab=1$ ...(i)

Given, $\arg(z)-\arg(\omega)=\dfrac{3\pi}{2}$

$\Rightarrow\quad \theta-\phi=\dfrac{3\pi}{2}$ ...(ii)

$$\bar{z}\omega=abe^{-i(\theta-\phi)}=e^{-i\frac{3\pi}{2}}$$
$$=\cos\frac{3\pi}{2}-i\sin\frac{3\pi}{2}=i$$
$$\therefore\ \arg\left(\frac{1-2\bar{z}\omega}{1+3\bar{z}\omega}\right)=\arg\left(\frac{1-2i}{1+3i}\right)=\arg\left(-\frac{1}{2}(1+i)\right)$$
$$=-\pi+\frac{\pi}{4}=-\frac{3\pi}{4}$$

55. *(c)* T_{r+1} in expansion of $\left(bx^2+\dfrac{1}{ax}\right)^{11}$

$$={}^{11}C_r\,(bx^2)^{11-r}\left(\frac{1}{ax}\right)^r$$
$$={}^{11}C_r b^{11-r}\,a^{-r}\,x^{22-2r}\cdot x^{-r}$$

For coefficient of x^7,

$22-3r=7$

$\Rightarrow\quad 3r=22-7=15$

$\Rightarrow\quad r=5$

$\therefore$ Coefficient of $x^7={}^{11}C_r\,b^{11-r}\,a^{-r}={}^{11}C_5 b^6 a^{-5}$

Also, T_{r+1} in expansion of $\left(bx-\dfrac{1}{ax^2}\right)^{11}$

$$={}^{11}C_r\,(bx)^{11-r}\left(-\frac{1}{ax^2}\right)^r$$
$$={}^{11}C_r\,b^{11-r}(-a)^{-r}x^{11-r}\cdot x^{-2r}$$
$$={}^{11}C_r\,b^{11-r}(-a)^{-r}x^{11-3r}$$

For coefficient of x^{-7},

$11-3r=-7$

$\Rightarrow\quad 3r=18\Rightarrow r=6$

$\therefore\quad$ The coefficient of $x^{-7}={}^{11}C_6\,b^{11-6}\,(-a)^{-6}$
$={}^{11}C_6\,b^5 a^{-6}$

It is given that coefficient are equal.

$\therefore\quad {}^{11}C_5 b^6 a^{-5}={}^{11}C_6 b^5 a^{-6}$

$\Rightarrow\quad b=\dfrac{1}{a}$

$\Rightarrow\quad ab=1$

Hence, $\quad k=1$

56. *(c)* Let first term $=a$, common difference $=d$

$$\therefore\quad \frac{S_6}{S_{10}}=\frac{\frac{6}{2}(2a+5d)}{\frac{10}{2}(2a+9d)}=\frac{33}{95}$$

$\Rightarrow (6a+15d)95=33(10a+45d)$

$570a+1425d=330a+1485d$

$240a=60d$

$4a=d$

$a_{15}=a+14d=a+56a=57a,\ a\in N$

$\Rightarrow\quad a_{15}=114$

[as $110<a_{15}<120,\ 110<a\times 57<120$]

$\therefore\quad a=\dfrac{114}{57}=2$

$\therefore\quad d=8$

$\therefore\quad S_{11}=\dfrac{11}{2}[4+10\times 8]=11\times 42=462$

57. *(a)* Given, $P=\begin{bmatrix}4 & -1 & -3\\ 3 & 0 & \alpha\\ 4 & -5 & 0\end{bmatrix}$

$$|P|=\begin{vmatrix}4 & -1 & -3\\ 3 & 0 & \alpha\\ 4 & -5 & 0\end{vmatrix}$$
$$=4[5\alpha]+1[-4\alpha]-3[-15]=20\alpha-4\alpha+45$$
$$|P|=16\alpha+45$$

According to the question,

$PQ=KI_3\Rightarrow Q=KP^{-1}I_3$

Now, $Q = \frac{K}{|P|}(\text{adj } P) = I_3$

$$= \frac{K}{|P|}\begin{bmatrix} 5\alpha & 4\alpha & -15 \\ 15 & 12 & 16 \\ -\alpha & -(4\alpha+9) & 3 \end{bmatrix}\begin{bmatrix} 1 & 0 & 0 \\ 0 & 1 & 0 \\ 0 & 0 & 1 \end{bmatrix}$$

$$Q = \frac{K}{16\alpha+45}\begin{bmatrix} 5\alpha & 4\alpha & -15 \\ 15 & 12 & 16 \\ -\alpha & -4\alpha-9 & 3 \end{bmatrix}\begin{bmatrix} 1 & 0 & 0 \\ 0 & 1 & 0 \\ 0 & 0 & 1 \end{bmatrix}$$

$\because$ Given, $q_{32} = -\frac{K}{7}$

$\therefore \frac{K}{16\alpha+45}(-4\alpha-9) = -\frac{K}{7}$

$\Rightarrow (4\alpha+9)\cdot 7 = 16\alpha+45$

$\Rightarrow 28\alpha+63 = 16\alpha+45$

$12\alpha = -18 \Rightarrow \alpha = \frac{-18}{12} = -\frac{3}{2}$

Also, $|Q| = \frac{K^3|I|}{|P|}$

$\frac{K^2}{2} = \frac{K^3|I|}{16\alpha+45}$

$16\alpha+45 = 2K$

$2K = -\frac{3}{2}\times 16+45$

$= -24+45$

$2K = +21 \Rightarrow K = \frac{21}{2}$

$\therefore \alpha^2+K^2 = \frac{3^2}{2^2}+\frac{21^2}{2^2} = \frac{9}{4}+\frac{441}{4} = \frac{225}{2}$

58. *(b)* Given, $A = \begin{bmatrix} \cos\alpha & \sin\alpha \\ -\sin\alpha & \cos\alpha \end{bmatrix}$

We know that

$$A^n = \begin{bmatrix} \cos\alpha & \sin\alpha \\ -\sin\alpha & \cos\alpha \end{bmatrix}^n = \begin{bmatrix} \cos n\alpha & \sin n\alpha \\ -\sin n\alpha & \cos n\alpha \end{bmatrix}$$

$$\therefore \quad A^2 = \begin{bmatrix} \cos 2\alpha & \sin 2\alpha \\ -\sin 2\alpha & \cos 2\alpha \end{bmatrix}; A^3 = \begin{bmatrix} \cos 3\alpha & \sin 3\alpha \\ -\sin 3\alpha & \cos 3\alpha \end{bmatrix}$$

$$A^4 = \begin{bmatrix} \cos 4\alpha & \sin 4\alpha \\ -\sin 4\alpha & \cos 4\alpha \end{bmatrix}; A^5 = \begin{bmatrix} \cos 5\alpha & \sin 5\alpha \\ -\sin 5\alpha & \cos 5\alpha \end{bmatrix}$$

$\therefore \cos\alpha+\cos 2\alpha+\cos 3\alpha+\cos 4\alpha+\cos 5\alpha$

$= \cos\frac{\pi}{6}+\cos\frac{2\pi}{6}+\cos\frac{3\pi}{6}+\cos\frac{4\pi}{6}+\cos\frac{5\pi}{6}$

$= \cos\frac{\pi}{6}+\cos\frac{2\pi}{6}+\cos\frac{\pi}{2}$

$+\cos\left(\pi-\frac{\pi}{3}\right)+\cos\left(\pi-\frac{\pi}{6}\right)$

$= 0$

$\sin\alpha+\sin 2\alpha+\sin 3\alpha+\sin 4\alpha+\sin 5\alpha$

$= \sin\frac{\pi}{6}+\sin\frac{2\pi}{6}+\sin\frac{3\pi}{6}+\sin\frac{4\pi}{6}+\sin\frac{5\pi}{6}$

$= \sin\frac{\pi}{6}+\sin\frac{\pi}{3}+\sin\frac{\pi}{2}+\sin\left(\pi-\frac{\pi}{3}\right)+\sin\left(\pi-\frac{\pi}{6}\right)$

$= 2\sin\frac{\pi}{6}+2\sin\frac{\pi}{3}+\sin\frac{\pi}{2}$

$= 2\times\frac{1}{2}+2\times\frac{\sqrt{3}}{2}+1 = 2+\sqrt{3}$

$\therefore B = \begin{bmatrix} 0 & 2+\sqrt{3} \\ -(2+\sqrt{3}) & 0 \end{bmatrix}$

$|B| = (2+\sqrt{3})(2+\sqrt{3}) \neq 0$

$\therefore B$ is non-singular.

59. *(d)* $\lim_{x\to 0}\frac{e^{2x}-(1+4x)^{1/2}}{\log(1+x^2)}$

$= \lim_{x\to 0}\frac{e^{2x}-(1+4x)^{1/2}}{\frac{\log(1+x^2)}{x^2}\times x^2}$

$= \lim_{x\to 0}\frac{e^{2x}-(1+4x)^{1/2}}{x^2}$ $\left[\frac{0}{0}\text{ form}\right]$

On using L-Hospital's rule, we get

$= \lim_{x\to 0}\frac{2e^{2x}-\frac{1}{2}(1+4x)^{-1/2}\cdot 4}{2x}$

$= \lim_{x\to 0}\frac{e^{2x}-(1+4x)^{-1/2}}{x}$ $\left[\frac{0}{0}\text{ form}\right]$

Again, on using L-Hospital's rule, we get

$= \lim_{x\to 0}\frac{2e^{2x}+\frac{1}{2}(1+4x)^{-3/2}\cdot 4}{1} = 2(1)+2(1) = 4$

60. *(c)* $y = \sin^2 x - 2\sin x$

$= \sin x(\sin x - 2)$

$y = 0$

$\therefore \sin x = 0$

$\therefore x = 0, \pi, 2\pi$

Also, $\sin x - 2 < 0$

$\therefore y > 0$, if $\sin x < 0$ and $y < 0$, if $\sin x > 0$

$\therefore$ Graph of the function $y = \sin^2 x - 2\sin x$ is shown in the figure below

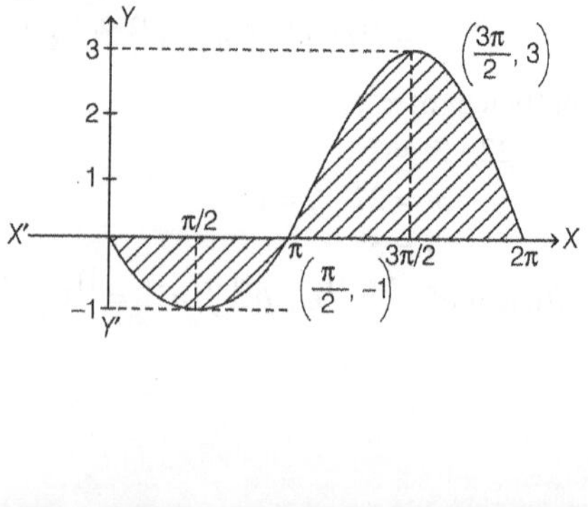

The required area = The shaded area

$$= \left| \int_0^{\pi} (\sin^2 x - 2\sin x)\,dx \right| + \int_{\pi}^{2\pi} (\sin^2 x - 2\sin x)\,dx$$

$$= \left| \frac{1}{2}\int_0^{\pi} (1-\cos 2x)\,dx - 2\int_0^{\pi} \sin x\,dx \right|$$

$$+ \int_{\pi}^{2\pi} \left(\frac{1-\cos 2x}{2} - 2\sin x \right) dx$$

$$= \left| \frac{1}{2}\left(x - \frac{\sin 2x}{2} \right) + 2\cos x \right|_0^{\pi} + \left[\frac{1}{2}\left(x - \frac{\sin 2x}{2} \right) + 2\cos x \right]_{\pi}^{2\pi}$$

$$= \left| \frac{1}{2}\pi - 4 \right| + \frac{2\pi - \pi}{2} + 4 = \left(4 - \frac{\pi}{2} \right) + 4 + \frac{\pi}{2} = 8 \text{ sq units}$$

61. *(b)* $P(x) = \begin{cases} \dfrac{x}{10}, & x = 1, 2, 3, 4 \\ 0, & \text{otherwise} \end{cases}$

Now, $P\left(\dfrac{1}{2} < x < \dfrac{7}{2} \text{ and } x > 1 \right) = \dfrac{P\left\{ \left(\dfrac{1}{2} < x < \dfrac{7}{2} \right) \cap (x > 1) \right\}}{P(x > 1)}$

$$= \frac{P\{(x = 1, 2, 3) \cap (x = 2, 3, 4)\}}{1 - P(x = 1)} = \frac{\{P(x = 2, 3)\}}{1 - P(x = 1)}$$

$$= \frac{P(x = 2) + P(x = 3)}{1 - P(x = 1)} = \frac{\frac{2}{10} + \frac{3}{10}}{1 - \frac{1}{10}} = \frac{\frac{5}{10}}{\frac{9}{10}} = \frac{5}{9}$$

62. *(d)* Given, $f(x + y) = f(x) + f(y)$

Clearly, $f(0 + 0) = f(0) + f(0)$

$\Rightarrow \quad f(0) = f(0) + f(0)$

$\Rightarrow \quad f(0) = 0$

Also, $\quad f(x - x) = f(0)$

$\Rightarrow \quad f(x + (-x)) = 0 \Rightarrow f(x) + f(-x) = 0$

$\Rightarrow \quad f(-x) = -f(x) \quad$ ...(i)

Condition I Let $x > 0$ and x be an integer.

$f(2) = f(1 + 1) = f(1) + f(1) = 2f(1)$

$f(3) = f(2 + 1) = f(2) + f(1) = 3f(1)$

By inspection, $f(x) = xf(1) \quad$ ...(ii)

If x is a negative integer, then let

$x = -p$

$f(+x) = f(-p)$

$= -f(p) \quad$ [using Eq. (i)]

$= -pf(1) = x\,f(1)$

$\therefore$ S_1 is true for all integers.

Condition II

Let $x = \dfrac{p}{q}$, $p, q \in Z^+$

$f(p) = pf(1)$

$\Rightarrow \quad pf(1) = f(qx) \quad$ $[p = xq]$

$\Rightarrow \quad pf(1) = f(x + x + \ldots + q \text{ times})$

$\Rightarrow \quad pf(1) = f(x) + f(x) + \ldots + f(x)$ upto q times

$\Rightarrow \quad pf(1) = q\,f(x)$

$\Rightarrow \quad f(x) = \dfrac{p}{q} f(1) \Rightarrow f\left(\dfrac{p}{q} \right) = \dfrac{p}{q} f(1)$

$\therefore$ S_2 is false.

63. *(c)* Given, $P(a, 6, 3)$

Equation of line: $\dfrac{x}{1} = \dfrac{y - 1}{2} = \dfrac{z - 2}{3}$

Image of point P w.r.t to line is $Q(1, b, a + 6)$ mid-point of P and Q.

$$Q = \left(\frac{a + 1}{2}, \frac{6 + b}{2}, \frac{a + 9}{2} \right)$$

This point lies on the line.

$$\therefore \quad \frac{\frac{a + 1}{2}}{1} \text{ i.e. } \frac{a+1}{2} = \frac{\frac{6 + b}{2} - 1}{2} = \frac{\left(\frac{a + 9}{2} \right) - 2}{3}$$

$\Rightarrow \quad \dfrac{a + 1}{2} = \dfrac{6 + b - 2}{4} = \dfrac{a + 9 - 4}{6}$

$\Rightarrow \quad \dfrac{a + 1}{2} = \dfrac{a + 9 - 2 \times 2}{6}$ and $\dfrac{6 + b - 2}{4} = \dfrac{a + 9 - 2 \times 2}{6}$

On solving these equations, we get

$a = 1, b = 0$

$\therefore \quad |a + b| = 1$

64. *(b)* Given lines $\quad x + y = |c| \quad$ ...(i)

$cx - y = 1 \quad$ ...(ii)

On solving Eqs. (i) and (ii), we get

$$x = \frac{1 + |c|}{1 + c}, \; y = \frac{c|c| - 1}{c + 1}$$

Since, they intersect in first quadrant.

$\therefore \quad x > 0, y > 0$

$\Rightarrow \quad 1 + c > 0$ and $c|c| > 1$

As, $\quad |c| > 0$

$\Rightarrow \quad c > 0$

Thus, $c^2 > 1 \Rightarrow c > 1$

$\Rightarrow \quad c \in (1, \infty)$

65. *(b)* Let the coordinates of P, Q and R be $P(-t_1^2, 2t_1)$, $Q(-t_1^2, -2t_1)$ and $R(h, k)$.

As R divides PQ in the ratio $5:2$, then

$$h = \frac{-2t_1^2 - 5t_1^2}{7} = -t_1^2 \quad \text{...(i)}$$

$$k = \frac{2 \times 2t_1 - 5 \times 2t_1}{7}$$

$$k = \frac{4t_1 - 10t_1}{7} = \frac{-6}{7} t_1$$

$$\Rightarrow \quad k^2 = \frac{36}{49} t_1^2 \quad \text{...(ii)}$$

From Eqs. (i) and (ii), we get

$$k^2 = \frac{-36}{49}h \Rightarrow 49k^2 + 36h = 0$$

$$\therefore \quad 49y^2 + 36x = 0$$

66. *(d)* Let $P(2r_1, -3r_1, r_1)$ and $Q(3r_2 + 2, -5r_2 + 1, 2r_2 - 2)$ be the point on the given lines so that PQ is the line of the shortest distance between the given lines.

Now, direction ratios of QP are

$$2r_1 - 3r_2 - 2, -3r_1 + 5r_2 - 1, r_1 - 2r_2 + 2$$

Since, it is perpendicular to the given lines

$$2(2r_1 - 3r_2 - 2) - 3(-3r_1 + 5r_2 - 1) + (r_1 - 2r_2 + 2) = 0$$

and $3(2r_1 - 3r_2 - 2) - 5(-3r_1 + 5r_2 - 1) + 2(r_1 - 2r_2 + 2) = 0$

$$\Rightarrow \quad 14r_1 - 23r_2 + 1 = 0, 23r_1 - 38r_2 + 3 = 0$$

$$\Rightarrow \quad r_1 = \frac{31}{3}, r_2 = \frac{19}{3}$$

$\Rightarrow P$ is $\left(\frac{62}{3}, -31, \frac{31}{3}\right)$ and Q is $\left(21, -\frac{92}{3}, \frac{32}{3}\right)$ and

direction ratios of PQ are $\frac{1}{3}, \frac{1}{3}, \frac{1}{3}$.

Hence, equation of PQ as it passes through Q is

$$\frac{x-21}{\frac{1}{3}} = \frac{y+\frac{92}{3}}{\frac{1}{3}} = \frac{z-\frac{32}{3}}{\frac{1}{3}} \quad \text{...(i)}$$

Also, this line is

perpendicular to $\frac{x-3}{\alpha} = \frac{6(y-2)}{2} = \frac{9(z-4)}{3}$

$$\frac{1}{3}(\alpha) + \frac{1}{3} \times 2 + \frac{1}{3} \times 3 = 0$$

$$\frac{\alpha}{3} + \frac{2}{3} + 1 = 0 \Rightarrow \alpha + 5 = 0$$

$$\alpha = -5 \Rightarrow |3-5|^2 = 4$$

67. *(a)* Consider

$$(1+x)^{50} = {}^{50}C_0 + {}^{50}C_1 x + \ldots + {}^{50}C_{50} x^{50} \quad \ldots\text{(i)}$$

$$(1-x)^{50} = {}^{50}C_0 - {}^{50}C_1 x + \ldots + {}^{50}C_{50} x^{50} \quad \ldots\text{(ii)}$$

On adding Eqs. (i) and (ii), we get

$${}^{50}C_0 + {}^{50}C_2 x^2 + \ldots + {}^{50}C_{50} x^{50} = \frac{1}{2}[(1+x)^{50} + (1-x)^{50}]$$

Now, on integrating the above equation with limits from zero to one.

$$\int_0^1 ({}^{50}C_0 + {}^{50}C_2 x^2 + \ldots + {}^{50}C_{50} x^{50})\,dx = \int_0^1 \frac{1}{2}[(1+x)^{50} + (1-x)^{50}]\,dx$$

$$\Rightarrow \quad {}^{50}C_0 + \frac{{}^{50}C_2}{3} + \frac{{}^{50}C_4}{5} + \ldots + \frac{{}^{50}C_{50}}{51}$$

$$= \frac{1}{2} \cdot \frac{2^{51}}{51} = \frac{2^{50}}{51}$$

Number of terms from ${}^{50}C_0$ to ${}^{50}C_{50} = 26$

$$\therefore \text{Required mean} = \frac{\frac{2^{50}}{51}}{26} = \frac{2^{49}}{39 \times 17}$$

68. *(b)* $x^2 - x(3+\sqrt{2}) + 3\sqrt{2} = 0$

$$\Rightarrow \quad x(x-3) - \sqrt{2}(x-3) = 0 \Rightarrow (x-\sqrt{2})(x-3) = 0$$

$$\Rightarrow \quad x = 3 \text{ or } x = \sqrt{2}$$

But $x \in Q \Rightarrow x \neq \sqrt{2}$

So, the truth set is $\{3\}$.

69. *(b)* We have, $f(x - f(y)) = f(f(y)) + x f(y) + f(x) - 1$

On putting $x = f(y) = 0$, we get

Then, $f(0) = f(0) + 0 + f(0) - 1$

$$\therefore \quad f(0) = 1$$

On putting $f(y) = x$, we get

$$\therefore \quad f(0) = f(x) + x^2 + f(x) - 1$$

Hence, $\quad f(x) = 1 - \frac{x^2}{2}$

$$\therefore \quad f(2) = 1 - \frac{4}{2} = -1$$

70. *(a)* Let the equation of ellipse is $\frac{x^2}{a^2} + \frac{y^2}{b^2} = 1, (a > b)$

According to the given equation,

distance between foci $= 2ae = 8$ and distance between directrices $= \frac{2a}{e} = 16$

So, $2ae \times \frac{2a}{e} = 8 \times 16$

$$\Rightarrow \quad 4a^2 = 128 \Rightarrow a = 4\sqrt{2} \text{ and } \frac{2ae}{\frac{2a}{e}} = \frac{8}{16} \Rightarrow e^2 = \frac{1}{2}$$

$$\because \quad e^2 = 1 - \frac{b^2}{a^2} \Rightarrow \frac{1}{2} = 1 - \frac{b^2}{a^2} \Rightarrow \frac{b^2}{a^2} = \frac{1}{2}$$

$$\Rightarrow \quad b^2 = \frac{a^2}{2} = \frac{128}{8} = 16 \quad [\because a = 4\sqrt{2}]$$

Now, as length of latusrectum

$$\frac{2b^2}{a} = \frac{2 \times 16}{4\sqrt{2}} = 4\sqrt{2}$$

Hence, option (a) is correct.

71. *(5)* Given, $abcde = 1050 = 2 \times 3 \times 5^2 \times 7$

We can assign 2, 3 or 7 to any of 5 variables a, b, c, d or e.

We can assign 5^2 to just one variable in 5 ways or can assign to two variables in ${}^5C_2$ ways, then 5 can be assigned in ${}^5C_1 + {}^5C_2 = 5 + 10 = 15$ ways

Hence, the required solutions $= 5\times5\times5\times15 = 1875$

Now, according to the question, we have

$$375k = 1875 \Rightarrow k = \frac{1875}{375} = 5$$

72. *(2)* Given, equation is $e^x = \lambda x^2$

$$\lambda = \frac{e^x}{x^2}$$

Now, let $f(x) = \dfrac{e^x}{x^2}$

$$\therefore \quad f'(x) = \frac{(x-2)e^x}{x^3} \Rightarrow f'(x) = 0$$

$\Rightarrow$ $x = 2$, which is point of minima.

Also, $f(2) = \dfrac{e^2}{4}$

$$\lim_{x\to\infty} \frac{e^x}{x^2} \to \infty \qquad \text{[using L-Hospital's rule twice]}$$

$$\lim_{x\to-\infty} \frac{e^x}{x^2} = 0 \Rightarrow \lim_{x\to0} \frac{e^x}{x^2} \to \infty$$

Further, $f(x) > 0 \,\forall\, x \in R$

From this information, graph of $f(x)$ is as shown below

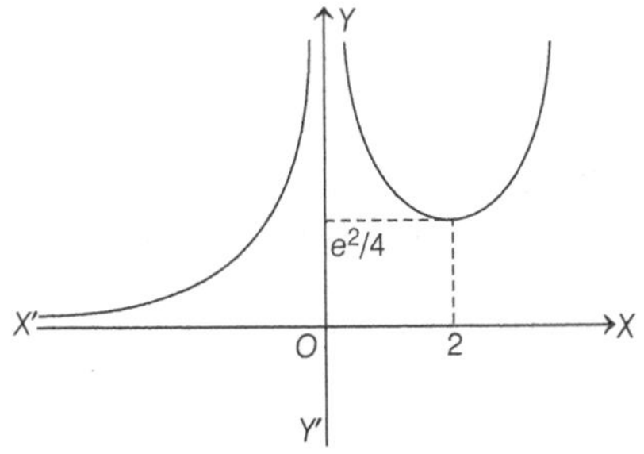

Thus, for three real distinct solution $\lambda > \dfrac{e^2}{4}, \lambda \in I$

So, $\lambda_{\min} = 2$

73. *(3)* If $\mathbf{r}_1 = \mathbf{a} + \lambda\mathbf{b}$ and $\mathbf{r}_2 = \mathbf{c} + \mu\mathbf{d}$

Then, shortest distance

$$d = \frac{|(\mathbf{b}\times\mathbf{d})\cdot(\mathbf{a}-\mathbf{c})|}{|\mathbf{b}\times\mathbf{d}|}$$

$$\therefore \quad \mathbf{a} - \mathbf{c} = (\alpha+4)\hat{\mathbf{i}} + 2\hat{\mathbf{j}} + 3\hat{\mathbf{k}}$$

$$\mathbf{b}\times\mathbf{d} = \begin{vmatrix} \hat{\mathbf{i}} & \hat{\mathbf{j}} & \hat{\mathbf{k}} \\ 1 & -2 & 2 \\ 3 & -2 & -2 \end{vmatrix} = 8\hat{\mathbf{i}} + 8\hat{\mathbf{j}} + 4\hat{\mathbf{k}}$$

$$|\mathbf{b}\times\mathbf{d}| = \sqrt{8^2+8^2+4^2} = 12$$

Shortest distance

$$= \frac{|[(\alpha+4)\hat{\mathbf{i}} + 2\hat{\mathbf{j}} + 3\hat{\mathbf{k}}]\cdot(8\hat{\mathbf{i}} + 8\hat{\mathbf{j}} + 4\hat{\mathbf{k}})|}{12} = 7$$

$$\Rightarrow \quad \frac{2(\alpha+4)+4+3}{3} = 7 \Rightarrow 2\alpha + 15 = 21 \Rightarrow \alpha = 3$$

74. *(101)* $|\mathbf{a}+\mathbf{b}|^2 = |\mathbf{a}|^2 + 2|\mathbf{b}|^2$

$\Rightarrow$ $\mathbf{a}\cdot\mathbf{b} = 6$ and $|\mathbf{a}\times\mathbf{b}|^2 = 65$

$\therefore$ $|\mathbf{a}|^2 + |\mathbf{b}|^2 + 2\mathbf{a}\cdot\mathbf{b} = |\mathbf{a}|^2 + 2|\mathbf{b}|^2$

$\Rightarrow$ $2\mathbf{a}\cdot\mathbf{b} = |\mathbf{b}|^2 \Rightarrow 2\times6 = |\mathbf{b}|^2 \Rightarrow |\mathbf{b}|^2 = 12$

Also, $(\mathbf{a}\cdot\mathbf{b})^2 + |\mathbf{a}\times\mathbf{b}|^2 = |\mathbf{a}|^2\,|\mathbf{b}|^2$

$$\Rightarrow \quad (6)^2 + 65 = |\mathbf{a}|^2 \times 12 \Rightarrow 36 + 65 = 12|\mathbf{a}|^2$$

$$\Rightarrow \quad |\mathbf{a}|^2 = \frac{101}{12} \Rightarrow 12|\mathbf{a}|^2 = 101$$

75. *(0)* Given, $\displaystyle\sum_{k=1}^{12} \frac{1}{\cos\left(\frac{\pi}{4} + \frac{(k-1)\pi}{3}\right)\cos\left(\frac{\pi}{4} + \frac{k\pi}{3}\right)}$

$$= \sum_{k=1}^{12} \frac{1}{\cos\frac{\pi}{3}}\left[\frac{\cos\left(\frac{\pi}{4} + \frac{k\pi}{3} - \left(\frac{\pi}{4} + \frac{(k-1)\pi}{3}\right)\right)}{\cos\left(\frac{\pi}{4} + \frac{(k-1)\pi}{3}\right)\cos\left(\frac{\pi}{4} + \frac{k\pi}{3}\right)}\right]$$

$$= 2\sum_{k=1}^{12} \frac{\cos\left(\frac{\pi}{4} + \frac{k\pi}{3}\right)\cos\left(\frac{\pi}{4} + \frac{(k-1)\pi}{3}\right) + \sin\left(\frac{\pi}{4} + \frac{k\pi}{3}\right)\sin\left(\frac{\pi}{4} + \frac{(k-1)\pi}{3}\right)}{\cos\left(\frac{\pi}{4} + \frac{(k-1)\pi}{3}\right)\cos\left(\frac{\pi}{4} + \frac{k\pi}{3}\right)}$$

$$= 2\sum_{k=1}^{12}\left[1 + \tan\left(\frac{\pi}{4} + \frac{k\pi}{3}\right)\tan\left(\frac{\pi}{4} + \frac{(k-1)\pi}{3}\right)\right]$$

Apply $\tan(A-B) = \dfrac{\tan A - \tan B}{1 + \tan A \tan B}$

$$\therefore \quad 2\sum_{k=1}^{12}\left[\frac{\tan\left(\frac{\pi}{4} + \frac{k\pi}{3}\right) - \tan\left(\frac{\pi}{4} + \frac{(k-1)\pi}{3}\right)}{\tan\left(\frac{\pi}{4} + \frac{k\pi}{3} - \frac{\pi}{4} - \frac{(k-1)\pi}{3}\right)}\right]$$

$$= \frac{2}{\sqrt{3}}\sum_{k=1}^{12}\left[\tan\left(\frac{\pi}{4} + \frac{k\pi}{3}\right) - \tan\left(\frac{\pi}{4} + \frac{(k-1)\pi}{3}\right)\right]$$

$$= \frac{2}{\sqrt{3}}\left[\tan\left(\frac{\pi}{4} + \frac{\pi}{3}\right) - \tan\left(\frac{\pi}{4}\right) + \tan\left(\frac{\pi}{4} + \frac{2\pi}{3}\right) - \tan\left(\frac{\pi}{4} + \frac{\pi}{3}\right) + \tan\left(\frac{\pi}{4} + \frac{3\pi}{3}\right) - \tan\left(\frac{\pi}{4} + \frac{2\pi}{3}\right) + \ldots + \tan\left(\frac{\pi}{4} + \frac{12\pi}{3}\right) - \tan\left(\frac{\pi}{4} + \frac{11\pi}{3}\right)\right]$$

$$= \frac{2}{\sqrt{3}}\left[-\tan\left(\frac{\pi}{4}\right) + \tan\left(\frac{\pi}{4} + \frac{12\pi}{3}\right)\right]$$

$$= \frac{2}{\sqrt{3}}\left[\tan\left(\frac{\pi}{4} + 4\pi\right) - 1\right] = \frac{2}{\sqrt{3}}[1-1] = 0$$

SECTION II

Mastery Builders

PRACTICE SET 12

- Total Marks : 300
- Total Time : 3 hrs
- No. of Questions : 75

Instructions

1. There are **three** parts in the questions paper consisting of **Physics, Chemistry** and **Mathematics** having 25 questions in each part of equal weightage. Each part (subject) has two sections.
2. **Section A** This section contains 20 multiple choice questions which have only one correct answer. Each question carries **4 marks** for correct answer and – **1 mark** for wrong answer.
3. **Section B** This section contains 5 questions. The answer to each of the questions is a numerical value. Each question carries **4 marks** for correct answer and – **1 mark** for wrong answer. For Section B, the answer should be rounded off to the nearest integer.

PHYSICS

Section A : Objective Type Questions

1. A cylinder of mass 1 kg is given heat of 20000 J at atmospheric pressure. If initially, temperature of the cylinder is 20°C, then work done by the cylinder will be (Given that specific heat of cylinder $= 400\ \text{J kg}^{-1}\,°\text{C}$, coefficient of volume expansion $= 9\times10^{-5}\,°\text{C}^{-1}$, atmospheric pressure $= 10^5\ \text{N/m}^2$ and density of cylinder $= 9000\ \text{kg / m}^3$)

(a) 0.02 J (b) 0.05 J (c) 0.08 J (d) 0.1 J

2. In a double slit experiment, a glass plate of refractive index 1.5 of thickness t is introduced in the path of one of the interfering beams of wavelength λ. The position of central maxima has got shifted by $\frac{2\lambda D}{d}$, then what would be the thickness of plate ?

(a) 4λ (b) 2λ (c) λ (d) 3λ

3. The wavelength of first line of Balmer series of hydrogen atom is λÅ. The wavelength of this line of a double ionised lithium atom ($Z = 3$) is

(a) $\frac{\lambda}{3}$ (b) $\frac{\lambda}{9}$ (c) $\frac{\lambda}{8}$ (d) $\frac{\lambda}{27}$

4. 10 resistors, each of resistance R are connected in series to a battery of emf E and negligible internal resistance, then those are connected in parallel to the same battery. If I_p and I_S be the current drawn by the battery in parallel and series combination, then

(a) I_p will become 10 times of I_s
(b) I_p will become 100 time of I_s
(c) I_s will become 10 times of I_p
(d) $I_p = I_s$

5. A block of wood weighs 12 kg and has a relative density 0.6. It is to be in water with 0.9 of its volume immersed. What weighs of a metal is needed, if it is attached below the wood? (Relative density of metal is 14)

(a) 7 kg (b) 6.5 kg (c) 8 kg (d) 16 kg

6. If the velocity v of a particle moving along a straight line decreases linearly with its displacement s, from 20 ms^{-1} to a value approaching to zero at $s = 30$ m, then acceleration of the particle at $s = 15$ m, is

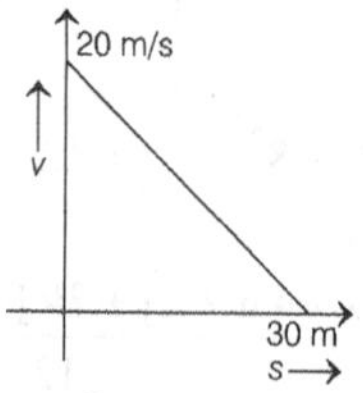

(a) $\frac{2}{3}\text{ms}^{-2}$ (b) $-\frac{2}{3}\text{ms}^{-2}$ (c) $\frac{20}{3}\text{ms}^{-2}$ (d) $-\frac{20}{3}\text{ms}^{-2}$

7. If the least count of a vernier scale is $\frac{1}{2}$ times of main scale division, then the number of division on vernier scale that coincide with 4 divisions on main scale is

(a) 4 (b) 10 (c) 8 (d) 11

8. A body of mass 5 kg is placed on earth's surface. If this body is taken from earth's surface to a height h, where $h = 4R$, the change in gravitational potential energy will be

(a) 2.5×10^8 J (b) 1.5×10^6 J
(c) 1.5×10^8 J (d) 2.5×10^6 J

9. The force acting on a body in SI unit is 65 N. In a certain unit in which the unit of length is 20 cm, unit of mass is 30 g and the unit of time is 15 s, the numerical value of force acting on the body is

(a) 126500 (b) 165200
(c) 125600 (d) 2437500

10. For a medium with permittivity ε and permeability μ, the refractive index of medium would be (μ_0 and ε_0 refer to permeability and permittivity of vacuum)

(a) $\mu\varepsilon / \mu_0\varepsilon_0$ (b) $\mu_0\varepsilon_0 / \mu\varepsilon$
(c) $\sqrt{\frac{\mu\varepsilon}{\mu_0\varepsilon_0}}$ (d) $\sqrt{\frac{\mu_0\varepsilon_0}{\mu\varepsilon}}$

11. A batsman hits a ball as shown in the figure and starts running to catch it. By what speed, he should run so that he could catch the ball?

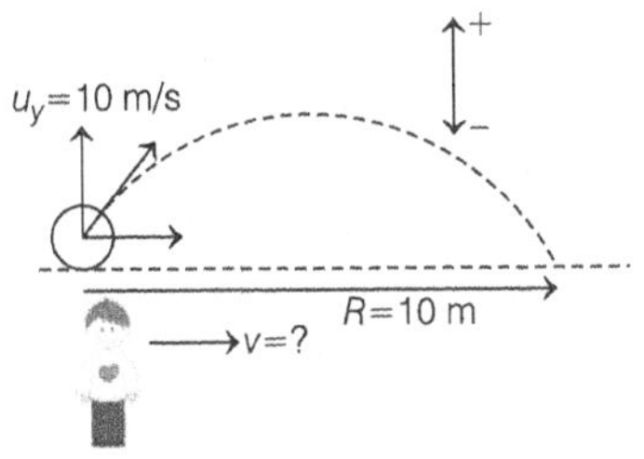

(a) 4 m/s (b) 5 m/s
(c) 10 m/s (d) 15 m/s

12. Which of the following statement is correct ?

(a) Periodic motion is necessarily an oscillatory motion
(b) Every oscillatory motion is SHM
(c) Oscillatory motion may be periodic (without resistive forces)
(d) None of the above

13. In a photoelectric effect experiment, the graph of stopping potential V *versus* frequency of light is shown in the figure. As the intensity of light decreased,

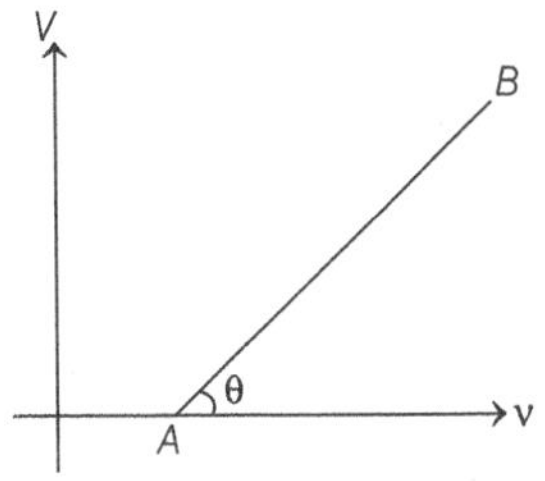

(a) the angle θ remains constant
(b) the point A shifts towards left with same angle θ.
(c) the point A shifts towards right with angle greater than θ
(d) the straight line AB becomes a curve.

14. Assertion The amount of work done in the isothermal expansion is greater than work done in the adiabatic system for the same final volume.

Reason In the adiabatic expansion of a gas, temperature and pressure both decrease due to decrease in internal energy of the system.

In the light of above statements, choose the most appropriate answer from the options given below.

(a) Both Assertion and Reason are correct and Reason is the correct explanation of Assertion.
(b) Both Assertion and Reason and correct but Reason is not the correct explanation of Assertion.
(c) Assertion is correct but Reason is incorrect.
(d) Assertion is incorrect but Reason is correct.

15. Statement I For a massless string, the tension has the same magnitude at all points throughout the string.

Statement II The direction of tension on a body is away from the point of contact and on a pulley, is towards the point of contact.

In the light of above statements, choose the most appropriate answer from the options given below.

(a) Statement I is true but Statement II is false.
(b) Statement I is false but Statement II is true.
(c) Both Statement I and Statement II are true.
(d) Both Statement I and Statement II are false.

16. Match the Column I (Molecule type) with Column II (C_p / C_V) for ideal gases and select the correct answer from the codes given below.

	Column I		Column II
A.	Helium (He)	1.	7/5
B.	Chlorine (Cl_2) rigid	2.	9/7
C.	Oxygen (O_2) non-rigid	3.	4/3
D.	Water (H_2O) rigid	4.	5/3

Codes

	A	B	C	D
(a)	4	1	2	3
(b)	3	4	2	1
(c)	2	3	1	4
(d)	4	2	1	3

17. In the following digital circuit, what will be the output Y, when the input (A, B) are (0, 1), (0, 0), (1, 1), (1, 0) ?

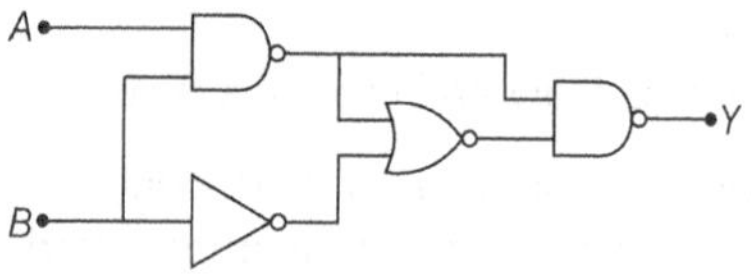

(a) 0, 0, 1, 0
(b) 1, 0, 1, 1
(c) 1, 1, 1, 1
(d) 0, 0, 0, 0

18. A uniform chain has a mass m and length l. It is held on a frictionless table with $\left(\frac{1}{6}\right)$th of its length hanging over the edge. The work done in just pulling the hanging part back on the table is

(a) $\frac{mgl}{72}$ (b) $\frac{mgl}{36}$
(c) $\frac{mgl}{12}$ (d) $\frac{mgl}{6}$

19. An electron and a proton both have same kinetic energy and enter into a magnetic field. If r_e and r_p is the radius of curved trajectory of electron and proton, then

(a) $r_e > r_p$
(b) $r_e = r_p$
(c) $r_e < r_p$
(d) None of the above

20. I. If four equal charges q are placed on the vertices of a square, then the electric field at the centre of square is zero.

II. Electric field inside a uniformly charged spherical shell is non-zero constant.

III. The electric potential outside a uniformly charged non-conducting solid sphere is inversely proportional to the distance.

Which of the following statement(s) is/are correct ?

(a) Both I and II
(b) Both II and III
(c) Both I and III
(d) Only III

Section B : Numerial Value Type Questions

21. A thin convex lens of focal length f is put on a plane mirror as shown in figure. When an object is kept at a distance a from the lens mirror combination, its image is formed at $\frac{a}{3}$ in front of the combination. The value of a is nf, then what is the value of n?

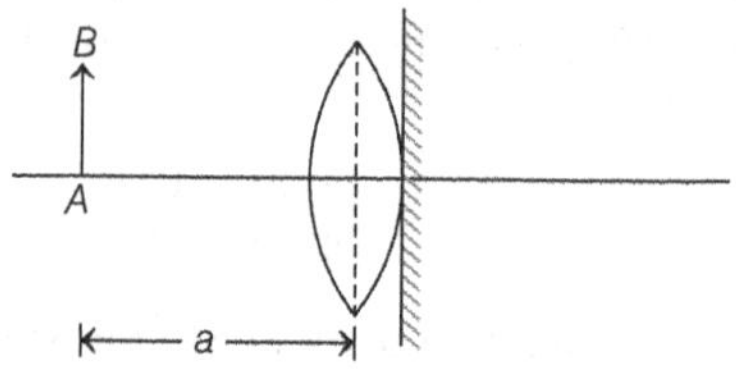

22. Two blocks of mass $M_1 = 20$ kg and $M_2 = 12$ kg are connected by a metal rod of mass 8 kg. The system is pulled vertically up by applying a force of 480 N as shown. The tension at the mid-point of the rod is ...N.

23. An L-C-R circuit as shown in the figure is connected to a voltage source V_{AC} whose frequency can be varied. The frequency, at which the voltage across the resistor is maximum, is ......Hz.

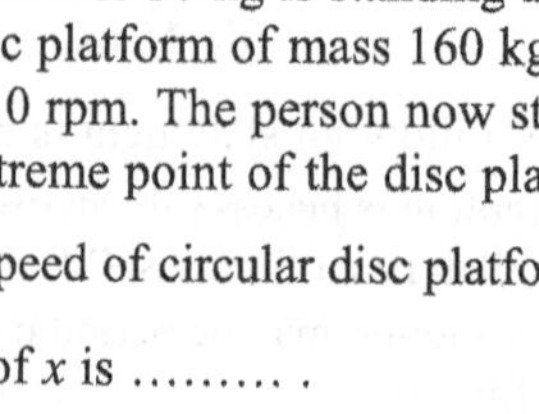

24. A person of mass 50 kg is standing at the centre of a circular disc platform of mass 160 kg rotating about its axis at 10 rpm. The person now starts moving towards extreme point of the disc platform, then the rotational speed of circular disc platform is $\frac{x}{100}$ rad/s. The value of x is ....

25. The equation of travelling wave on a stretched string of linear mass density 4 g/m is $y = 0.05 \sin(512t - 8x)$, where the distance and time is given in SI units. The tension in the string is approximately .... N.

CHEMISTRY

Section A : Objective Type Questions

26. Choose the correct option(s) for the following reaction sequence

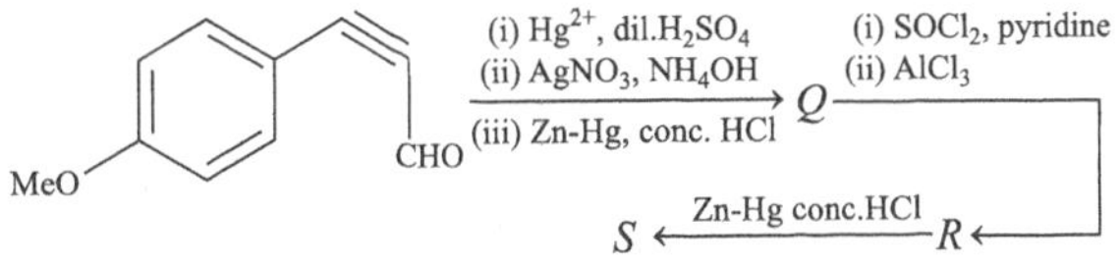

Identify the products formed Q and S.

(a) $Q =$ 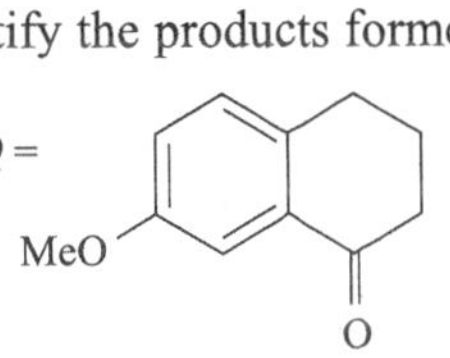 , $S =$ 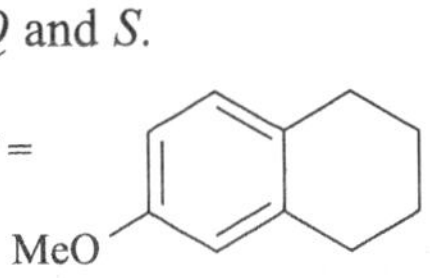

(b) $Q =$ 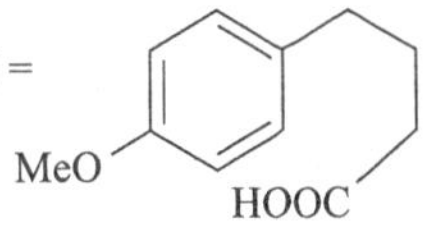 , $S =$ 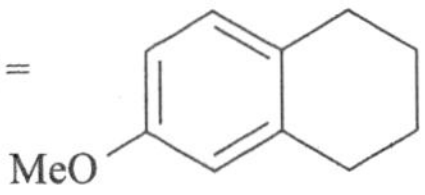

(c) $Q =$ 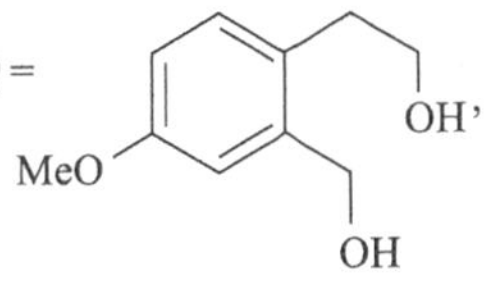 , $S =$ 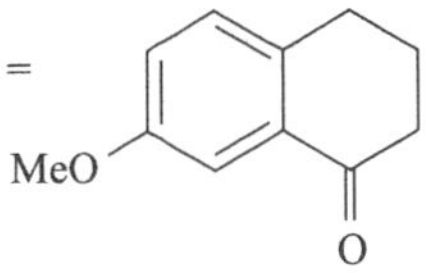

(d) $Q =$ 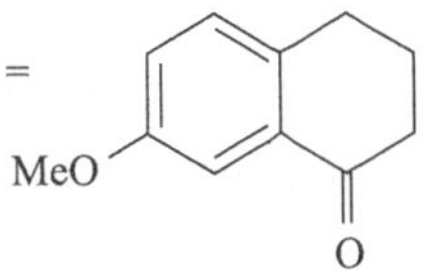 , $S =$ 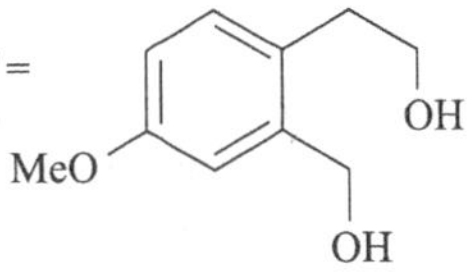

27. What are the products formed when xenon hexafluoride (XeF_6) reacts with oxide of silicon?

(a) $XeO_3 + SiF_4$ (b) $XeOF_4 + SiF_4$
(c) $XeF_4 + SiO_2$ (d) $XeSiO_4 + SiF_4$

28. Consider the following statements one is labelled as Assertion (A) and the other is labelled as Reason (R):

Assertion (A) Glucose gives a reddish brown precipitate with Fehling's solution.

Reason (R) Reaction of glucose with Fehling's solution gives Cu_2O and gluconic acid.

Choose the most appropriate answer from the options given below.

(a) Both A and R are true and R is the correct explanation of A.
(b) A is true but R is false.
(c) A is false but R is true.
(d) Both A and R are true but R is not the correct explanation of A.

29. The bond order and magnetic property of acetylide ion are same as that of

(a) O_2^+ (b) O_2^- (c) N_2^+ (d) NO^+

30. The transformation occurring in Duma's method is given below:

$$C_2H_7N + \left(2x + \frac{y}{2}\right)CuO \longrightarrow xCO_2 + \frac{y}{2}H_2O + \frac{Z}{2}N_2 + \left(2x + \frac{y}{2}\right)Cu$$

The value of y is

(a) 5
(b) 6
(c) 7
(d) 8

31. Which of the following will produce two different carbonyl compounds on ozonolysis of X, out of which one is acetone and another produces diastereomer on reduction with $LiAlH_4$?

(a) 2,3, 4-trimethylhex-2-ene
(b) 2,3-dimethylhex-2-ene
(c) 3,4 -dimethylhex-2-ene
(d) 2-methylhex-3-ene

32. If conc.HCl followed by H_2S is added to aqueous solution containing ions such as Al^{3+}, Ni^{2+}, Fe^{3+}, Ca^{2+} and Cu^{2+}. Then, how many cations will be precipitated during this reaction.

(a) 1 (b) 2
(c) 3 (d) 4

33. One mole of an ideal gas at 800 K, undergoes two reversible process, I followed by II, as shown below. If the work done by the gas in the two processes are same, the value of $\ln V_3 / V_2$ will be,

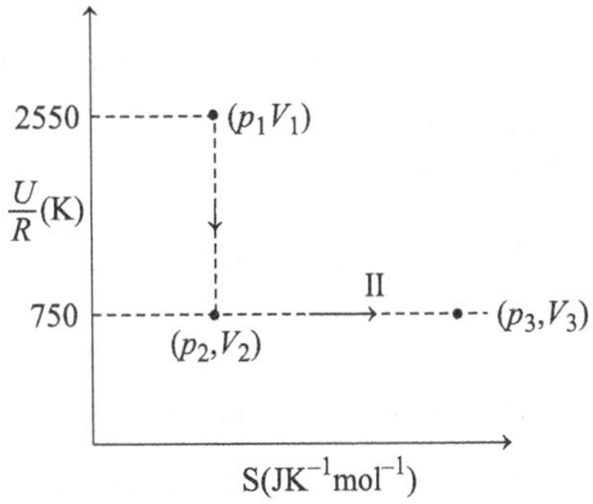

(a) 10 (b) 7
(c) 19·5 (d) 2·30

34. For the reactions,

$A + K_2CO_3 + \text{air} \xrightarrow{\text{Heat}} B$ and $B + Cl_2 \longrightarrow \underset{\text{(pink coloured)}}{C}$

Which of the following is correct?

(a) A is black Bi_2O_3, B is green coloured $KBiO_2$, C is pink K_2MnO_4.

(b) A is green $FeCO_3$, B is yellow K_2CrO_4, C is pink $K_2Cr_2O_7$.

(c) A is black MnO_2, B is yellow K_2CrO_4, C is pink Cr_2O_3.

(d) A is black MnO_2, B is green K_2MnO_4, C is pink $KMnO_4$.

35. Consider the following reaction,

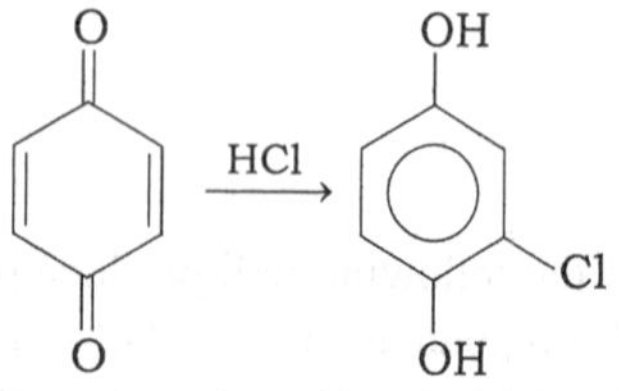

Which of the following statement is correct for the given example?

(a) It is an example of 1, 2-addition of HCl followed by tautomerism.

(b) It is an example of 1, 2-addition followed by reduction.

(c) It is an example of 1, 4-addition followed by tautomerism.

(d) It is an example of 1, 4-addition followed by oxidation.

36. Match List-I with List-II.

List I (Reaction)	List II (Reagents)
A. Cannizzaro reaction	I. Br_2/NaOH
B. Clemmensen reduction	II. $NaNO_2$+HX/CuX
C. Sandmeyer reaction	III. $CHCl_3$, NaOH/H_3O^+
D. Hofmann degradation	IV. Zn-Hg/HCl

Choose the correct answer from the options given below.

(a) A-III, B-IV, C-I, D-II (b) A-III, B-IV, C-II, D-I
(c) A-II, B-III, C-IV, D-I (d) A-II, B-III, C-I, D-IV

37. All of the given substrates undergoes S_N1 reaction. Choose the correct order for decreasing regarding S_N1 reaction.

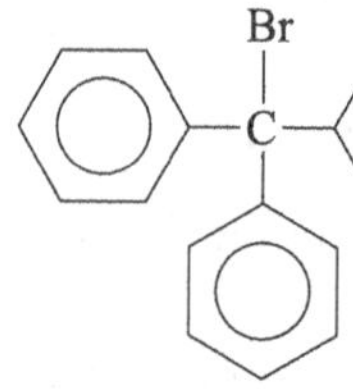

I

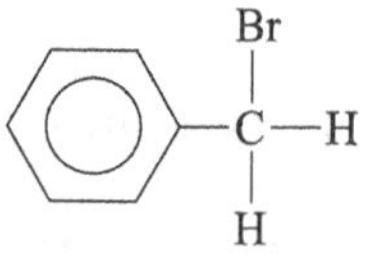

II

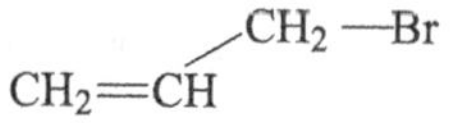

III

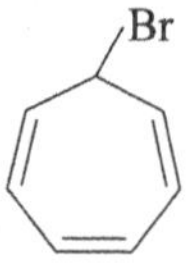

IV

(a) IV > I > II > III (b) I > IV > II > III
(c) I > II > III > IV (d) I > III > IV> II

38. A cyclic compound X having molecular formula C_3H_6O is treated with Tollen's reagent does not produce any colouration. When treated with NH_2OH it produces a compound Y. What will be the product Z when it is treated with H_2SO_4?

(a) $CH_3-\overset{\overset{O}{\|}}{C}-\overset{\ominus}{O}\overset{\oplus}{N}H_4$ (b) $CH_3-\overset{\overset{O}{\|}}{C}-OCH_3$

(c) $CH_3-\overset{\overset{O}{\|}}{C}-\underset{\underset{H}{|}}{N}-CH_3$ (d) $CH_3CH_2-\overset{\overset{O}{\|}}{C}-NH_2$

39. Which of the following is incorrect about tetraamminedithiocyanato-S cobalt (III)*tris*(oxalato)-cobaltate(III)?

(a) It is a chelating complex and show linkage isomerism.
(b) It shows optical isomerism.
(c) It shows geometrical isomerism.
(d) The formula of the complex is, $[Co(SCN)_2\ (NH_3)_4]\ [Co(ox)_3]$.

40. Consider the following compounds.

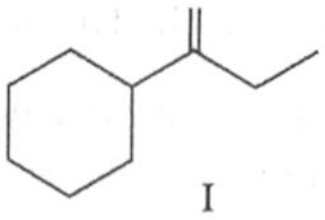

I

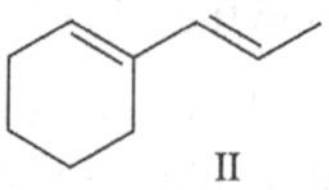

II

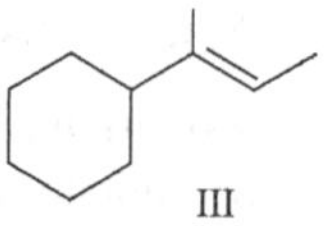

III

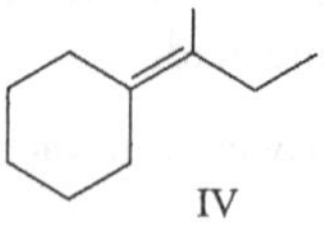

IV

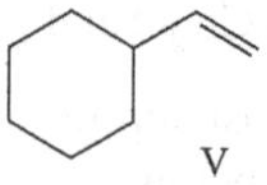

V

Total number of compounds that will formed by dehydration of 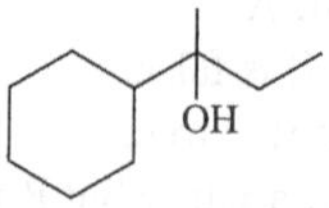

(a) 2 (b) 4 (c) 3 (d) 1

41. Which of the following statements are correct regarding Bohr's model?

A. The concept of electron in Bohr's orbit, does not violate the Heisenberg uncertainty principle.

B. The existence of Bohr's orbits is supported by hydrogen spectrum.

C. Atomic orbital is characterised by the quantum numbers n and l only.

D. It the radius of the 3rd Bohr's orbit of hydrogen atom is r_3 and the radius of 4th Bohr's orbit is r_4 then, $r_4 = \frac{16}{9} r_3$.

Choose the correct answer from the options given below.

(a) A and D
(b) B and D
(c) B and A
(d) C and A

42. Identify the incorrect statement.

(a) An electronic wave function must be symmetrical with respect to the interchange of any two electrons.

(b) ψ^2 must remain unchanged when the spin and space coordinates of spin and space coordinates of paired electrons are interchanged.

(c) For n electron system the wave function ψ will be a function of $3n$ spatial coordinate.

(d) Anti-symmetrical wave function are found to represent the properties of an electron.

43. Consider the following statements.

Statement I The number of P—O—P bridges in the structure of phosphorus pentoxide and phosphorus trioxide are 6.

Statement II Basicity of orthophosphoric acid and phosphorous acid are 2 and 3 respectively.

In light of the above statements, choose the correct answer from the options given below.

(a) Statement I is true but Statement II is false.
(b) Statement I is false but Statement II is true.
(c) Both Statement I and Statement II are true.
(d) Both Statement I and Statement II are false.

44. Consider the separate solution of 0.400 M NaOH (aq), 0.200 M $Mg_3(PO_4)_2(aq)$, 0.300 M KBr (aq) and 0.350 M $Na_3PO_4(aq)$ has osmotic pressure π_1, π_2, π_3 and π_4 respectively at 25°C. Choose the correct option for the osmotic pressure of these solutions, assuming all salts to be strong electrolyte?

(a) $\pi_3 < \pi_1 < \pi_2 < \pi_4$
(b) $\pi_1 < \pi_2 < \pi_4 < \pi_3$
(c) $\pi_2 < \pi_1 < \pi_3 < \pi_4$
(d) $\pi_4 < \pi_3 < \pi_1 < \pi_2$

45. Given below are two statements one is labelled as Assertion (A) and other is labelled as Reason (R).

Assertion (A) Linkage isomerism arises in coordination compounds containing ambidentate ligand.

Reason (R) Ambidentate ligand has two different donor atoms/sites.

In light of the above statements choose the correct answer from the options given below.

(a) Both (A) and (R) are true and (R) is the correct explanation of (A).
(b) Both (A) and (R) are true but (R) is not the correct explanation of (A).
(c) (A) is true but (R) is false.
(d) (A) is false but (R) is true.

Section B : Numerical Value Type Questions

46. The standard electrode potential for the following reaction is + 1.33 V. The potential at pH = 2 will be (+......V). (Nearest integer)

$$Cr_2O_7^{2-}(aq, 1M) + 14H^{\oplus}(aq) + 6e^- \longrightarrow 2Cr^{3+}(aq, 1M) + 7H_2O\ (l)$$

47. The molar heat of formation of $NH_4NO_3(s)$ is −367.57 kJ and those of $N_2O(g)$, $H_2O(l)$ are 81.46 and − 285.8 kJ respectively at 25°C and 1 atm. The change in internal energy for the reaction is (−) ......kJ.

$$NH_4NO_3(s) \longrightarrow N_2O(g) + 2H_2O(l)$$

48. The solubility of AgCl in water at 25°C is 1.79×10^{-3} g/L. The K_{sp} of AgCl at 25°C is ...... 10^{-12}. ($mol^2\ L^{-2}$)

49. A mixture of NH_4NO_3 and $(NH_4)_2HPO_4$ contains 30.40 mass % of nitrogen. The mass ratio of the two components in the mixture is $x : y$. The sum of x and y is ....

50. For the reaction of H_2 with I_2, the rate constant is $2.5 \times 10^{-4}\ dm^3\ mol^{-1}\ s^{-1}$ at 327°C and $1\ dm^3\ mol^{-1}\ s^{-1}$ at 527°C. The activation energy for the reaction is .... kJ mol^{-1}. (Nearest integer) (Given: $R = 8.314\ JK^{-1}\ mol^{-1}$)

MATHEMATICS

Section A : Objective Type Questions

51. Let $P(z_1), Q(z_2)$ and $R(z_3)$ be the vertices of a triangle whose circumcentre is at origin, if altitude from P to QR intersect the circumcircle at $S(z_4)$, then

(a) $z_4 = -\frac{z_1 z_3}{z_2}$ (b) $z_4 = -\frac{z_2 z_3}{z_1}$

(c) $z_4 = -\frac{z_1 z_2}{z_3}$ (d) $z_4 = \frac{z_2 z_3}{z_1}$

52. Consider the following statements

Statement I $\frac{{}^nC_1}{2} - \frac{{}^nC_3}{4} + \frac{{}^nC_5}{6} - \ldots + \frac{{}^nC_n}{42} = \frac{1}{42}$, if $n = 41$

Statement II The maximum value of the expression $\frac{1}{2}\sum_{n=3}^{20} {}^nC_3(\sin^{2n} x + \operatorname{cosec}^{2n} x)$ is ${}^{21}C_4$.

(a) Statement I is true and Statement II is false
(b) Statement I is false and Statement II is true
(c) Statement I and Statement II both are true
(d) Statement I and Statement II both are false

53. If $K(n)^{n-1} = n^{n-1} + 2(n+1)^1 n^{n-2} + 3(n+1)^2 n^{n-3} + \ldots + (n+1)^{n-1} n$, then the value of K if $n = 10$, is

(a) 10 (b) 100
(c) −100 (d) −10

54. The sum of values of α for which

$$\lim_{x \to \infty} \frac{2x^3 - (\tan^{-1} x)^3}{\frac{8}{\pi} x^3 \cot^{-1}|\alpha x| + \alpha^2 x^6 \sin \frac{1}{x^3} - 3\alpha x^3} = \frac{1}{2}, \text{ is}$$

(a) 0 (b) 1
(c) 3 (d) 4

55. In the odd man out game, each of $n \geq 2$ persons tosses a coin to determine who will buy refreshments for the entire group the odd man is the one with a different outcome from the rest, the probability that there is a looser in any game, is

(a) $\frac{n}{2^{n-1}}$ (b) $\frac{n}{2^{n-2}}$ (c) $\frac{1}{2^{n-1}}$ (d) $\frac{3n}{2^{n-1}}$

56. Consider the function

$$f(x) = \begin{cases} [x], & -2 \leq x \leq -1 \\ |x| + 1, & -1 < x \leq 2 \end{cases}$$

$$g(x) = \begin{cases} [x] & , -\pi \leq x < 0 \\ \sin x & , 0 < x \leq \pi \end{cases}$$

(where, $[\cdot]$ denotes the greatest integer function), then the range of $g[f(x)]$ is

(a) $[\sin 3, \sin 1]$ (b) $[\sin 3, 1] \cup \{-2, 0\}$
(c) $[\sin 3, 1] \cup \{-2, -1\}$ (d) $[\sin 1, 1]$

57. Let the shortest distance between the two lines $\frac{x-1}{4} = \frac{y+3}{5} = \frac{z+4}{3}$ and $\frac{x-4}{3} = \frac{y-3}{4} = \frac{z-5}{2}$ is α, where $\alpha > 0$. Then, $\sqrt{6}\alpha$ is equal to

(a) 2 (b) 14 (c) 9 (d) 12

58. An ellipse passes through the foci of the hyperbola $\frac{x^2}{9} - \frac{y^2}{16} = 1$ and it major and minor axes lie along the transverse and conjugate axes of the hyperbola respectively. If the sum of eccentricities of the two conic is $\frac{5\sqrt{17}+3}{3\sqrt{17}}$, then which of the following point does not lie on the ellipse?

(a) (5, 0) (b) $\left(0, \frac{20}{\sqrt{17}}\right)$

(c) $\frac{5}{\sqrt{2}}, \frac{20}{\sqrt{34}}$ (d) $\frac{9}{5\sqrt{2}}, \frac{5}{\sqrt{34}}$

59. If r is inradius of ΔPQR and R is its circumradius, then $\frac{R}{r}$ is equal to, if ΔPQR is equilateral, where $r = \frac{\Delta}{s}$ and $R = \frac{abc}{4\Delta}$

(a) 1 (b) 2 (c) 3 (d) 4

60. For two data sets, each of size 5, the variances are given to be 4 and 5 and the corresponding mean are given to be 2 and 4, respectively. The variance of the combined data set is

(a) $\frac{5}{2}$ (b) $\frac{11}{2}$ (c) 6 (d) $\frac{13}{2}$

61. If a and b are the roots of the quadratic equation $x^2 + x\sin\alpha - 2\sin\alpha = 0, \alpha \in \left(0, \frac{\pi}{2}\right)$, then $\frac{a^{12} + b^{12}}{(a^{-12} + b^{-12})(a-b)^{24}}$ is equal to

(a) $\frac{2^{12}}{(\sin\alpha + 4)^{12}}$ (b) $\frac{2^{12}}{(\sin\alpha + 8)^6}$

(c) $\frac{2^6}{(\sin\alpha + 8)^{12}}$ (d) $\frac{2^{12}}{(\sin\alpha + 8)^{12}}$

62. Let a and b respectively be the semi-transverse and semi-conjugate axes of a hyperbola whose eccentricity satisfies the equation $3e^2 - 6e + \frac{5}{3} = 0$. If $S(4, 0)$ is a focus and $25x = 36$ is corresponding directrix of this hyperbola, then $a^2 + b^2$ is equal to
(a) $25\sqrt{2}$ (b) $12\sqrt{5}$ (c) 16 (d) 24

63. The foot of the perpendicular from the origin to the join of $A(-9, 4, 5)$ and $B(11, 0, -1)$ divides AB in the ratio $k : 1$. Then, which of the following is correct?
(a) $2k + 5 = 8$ (b) $5k + 6 = 11$
(c) $2k + 5 = 10$ (d) $3k + 6 = 12$

64. Let $y_1(x)$ and $y_2(x)$ be the solutions of the differential equation $\frac{dy}{dx} = y + 17$ satisfying the conditions $y_1(0) = 0$ and $y_2(0) = 1$, then
(a) y_1 and y_2 will never intersect.
(b) y_1 and y_2 will interest at $x = 17$.
(c) y_1 and y_2 will intersect at $x = e$.
(d) y_1 and y_2 will intersect at $x = 1$.

65. If $\int \frac{dt}{t^3(1+t^6)^{2/3}} = t f(t)(1+t^6)^{1/3} + K$, where K is integration constant, then the function $f(t)$ is equal to
(a) $\frac{-1}{2t^3}$ (b) $\frac{3}{t^2}$ (c) $-\frac{1}{6t^3}$ (d) $-\frac{1}{2t^2}$

66. Let $P = \begin{bmatrix} 1 & 1.5 \\ 1 & 2 \end{bmatrix}, Q = \begin{bmatrix} 4 & -3 \\ -2 & 2 \end{bmatrix}$ and $R_k = \begin{bmatrix} k3^k & 2^k \\ 0 & (k-1)3^k \end{bmatrix}$

If $\sum_{k=1}^{50} \text{tr}[(PQ)^k R_k] = 3 + a \cdot 3^b$, where tr (A) denotes trace of matrix A, then $(b-a):(b+a)$ is equal to
(a) 1 : 50 (b) 50 : 1
(c) 51 : 49 (d) 49 : 51

67. Let $f : R \to R$ be defined by
$$f(x) = \begin{cases} \alpha_m + \sin \pi x, \text{for } x \in [2m, 2m+1] \\ \beta_m + \cos \pi x, \text{for } x \in (2m-1, 2m) \end{cases}$$
for all integer m, where α_m and β_m are real numbers. If f is continuous $\forall m$, then $\alpha_m - \beta_m$ is equal to
(a) 0 (b) -1
(c) 1 (d) 2

68. Let the image of the point $P(1, 0, 0)$ in the line $L: \frac{x-1}{2} = \frac{y+1}{-3} = \frac{z+10}{8}$ be Q. Let $R(\alpha, \beta, \gamma)$ be a point that divides internally the line segment PQ in the ratio $1:3$. Then, the value of $77(\alpha + \beta + \gamma)$ is equal to
(a) 77 (b) -66
(c) 88 (d) -77

69. Let $H(x) = f(x) + g(x)$, where $f(x) = \int_1^x \frac{\log u}{1+u} du$ and $g(x) = f\left(\frac{1}{x}\right)$, then $H(e)$ is equal to
(a) $\frac{1}{4}$ (b) $\frac{1}{2}$
(c) 0 (d) $\frac{1}{8}$

70. The area bounded by the curves $y = x$, $x = e$, $y = \frac{1}{x}$ and the positive X-axis, is
(a) $\frac{1}{2}$ sq unit (b) $\frac{1}{4}$ sq unit
(c) $\frac{5}{2}$ sq units (d) $\frac{3}{2}$ sq units

Section B : Numerical Value Type Questions

71. In a survey it is to be found that 70% of employees like mangoes and 64% like apples. If x% like both mangoes and apples, then $\lambda \le x \le \mu$. This expression comes as a result. So, the value of $\lambda + \mu$ is equal to ....

72. If $4x + 8\cos x + \tan x - 2\sec x - 4\log\{\cos x(1+\sin x)\} \ge 6, \forall x \in [0, \lambda)$, then the largest value of λ is $\frac{\pi K}{24}$, so the value of K is equal to ....

73. If the equation of the locus of the middle point of a chord of the circle $x^2 + y^2 = 2(x + y)$ such that the pair of lines joining the origin to the point of intersection of the chord and the circle are equally inclined to the X-axis is $px + qy = 1$, then $p + q$ is equal to ....

74. Let $\mathbf{a} = 4\hat{\mathbf{i}} + 3\hat{\mathbf{j}}$ and $\mathbf{b}$ be two vectors perpendicular to each other in XY-plane. All vectors in the same plane having projections 1 and 2 along $\mathbf{a}$ and $\mathbf{b}$ respectively, having magnitude $\sqrt{K}$, then K is equal to ....

75. Two numbers x and y are drawn without replacement from the set of the first 15 natural numbers. The number ways of drawing them such that $x^2 - y^2$ is multiple of 5, is ....

DETAILED SOLUTIONS

Physics

1. *(b)* Here, $\Delta Q = mc\Delta T$

$$\Rightarrow \quad \Delta T = \frac{20000\,\text{J}}{1\,\text{kg} \times (400\,\text{J/kg}^\circ\text{-C})} = 50^\circ\,\text{C}$$

Hence, $\quad W = p_{\text{atm}}\Delta V = p_{\text{atm}} V_0 \gamma \Delta T$

(γ = coefficient of volume expansion)

$$= (10^5\,\text{N/m}^2)\left(\frac{1}{9\times 10^3}\,\frac{\text{m}^3}{\text{kg}}\right)$$

$$(9\times 10^{-5}/^\circ\text{C})(50^\circ\text{C}) = 0.05\,\text{J}$$

2. *(a)* Due to insertion of glass plate of refractive index μ of thickness t, the optical path becomes μt instead of t. So, net increases in path travelled by wave,

$$\mu t - t = (\mu - 1)t$$

This increase would shift the fringe by a distance of $(\mu - 1)t \times \frac{D}{d}$

Now, given shift is $(\mu - 1)t\frac{D}{d} = \frac{2\lambda D}{d}$

$$\Rightarrow \quad t(\mu - 1) = 2\lambda$$

$$\Rightarrow \quad t = \frac{2\lambda}{\mu - 1} = \frac{2\lambda}{1.5 - 1} = 4\lambda$$

3. *(b)* For wavelength, $\frac{1}{\lambda} = RZ^2\left(\frac{1}{n_1^2} - \frac{1}{n_2^2}\right)$

Here, transition is same.

So, $\lambda \propto \frac{1}{Z^2}$

$$\frac{\lambda_{\text{H}}}{\lambda_{\text{Li}}} = \frac{(Z_{\text{Li}})^2}{(Z_{\text{H}})^2} = \frac{(3)^2}{(1)^2} = 9$$

$$\lambda_{\text{Li}} = \frac{\lambda_{\text{H}}}{9} = \frac{\lambda}{9}$$

4. *(b)* When 10 resistors, each of resistance R is connected in series, then equivalent resistance,

$$R_s = R + R + R + \ldots\ldots\ldots 10 \text{ times}$$

$$\Rightarrow \quad R_s = 10R$$

Current drawn from the battery,

$$I_s = \frac{E}{R_s} = \frac{E}{10R}$$

$$\Rightarrow \quad I_s = \frac{E}{10R} \quad \ldots\text{(i)}$$

In second case, when resistors are connected in parallel, then equivalent resistance,

$$\frac{1}{R_p} = \frac{1}{R} + \frac{1}{R} + \frac{1}{R} + \ldots\ldots 10 \text{ times}$$

$$= \frac{10}{R}$$

$$\Rightarrow \quad R_p = \frac{R}{10}$$

$\because$ Current drawn from the battery,

$$I_p = \frac{E}{R_p} = \frac{E}{R/10} = 10 \cdot \frac{E}{R}$$

$$= 100 \cdot \frac{E}{10R} \quad \ldots\text{(ii)}$$

From Eqs. (i) and (ii), we get

$$I_p = 100 I_s$$

Study Tactics

If n resistors are connected in series and then in parallel (resistance of all resistors are same), then the ratio of current in series and parallel can be calculated as

$$\frac{I_P}{I_S} = n^2$$

5. *(b)*

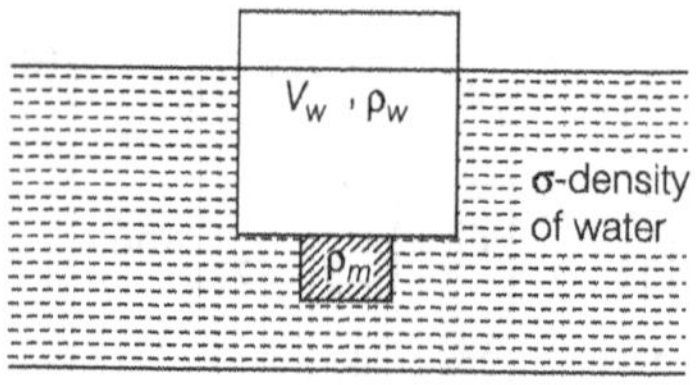

Let mass of wood be M, mass of metal be m, volume of wood be V_w and volume of metal be V_m.

For the body (wood + metal) to be float in water,

Weight = Buoyant force

$$(M + m)g = (0.9V_w + V_m)\sigma g$$

$$\Rightarrow \quad M + m = \left(0.9\frac{M}{\rho_w} + \frac{m}{\rho_m}\right)\sigma \quad \left(\text{As, density} = \frac{\text{mass}}{\text{volume}}\right)$$

$$\Rightarrow \quad m\left(1 - \frac{\sigma}{\rho_m}\right) = M\left(0.9\frac{\sigma}{\rho_w} - 1\right)$$

$$\Rightarrow \quad m\left(1 - \frac{1}{14}\right) = 12\left(\frac{0.9}{0.6} - 1\right)$$

$$\Rightarrow \quad m \approx 6.5\,\text{kg}$$

Study Tactics

In this case, 0.9 part of total volume of wooden box is inside the water and metal box is completely inside the water hence, total volume of displaced water is $0.9V_w + V_m$.

6. *(d)* Slope of line $= -\frac{2}{3}$

Equation of line is

$$(v-20) = -\frac{2}{3}(s-0)$$

$$\Rightarrow \quad v = 20 - \frac{2}{3}s \quad \ldots \text{(i)}$$

Velocity at $s = 15\,\text{m}$,

i.e., $$v = \left.\frac{ds}{dt}\right|_{s=15\,\text{m}} = 20 - \frac{2}{3}(15) = 10\,\text{ms}^{-1}$$

Differentiating Eq. (i) w.r.t time, we get

$$\text{Acceleration} = \frac{dv}{dt} = -\frac{2}{3}\frac{ds}{dt}$$

$$\therefore \quad \left.\frac{dv}{dt}\right|_{s=15\,\text{m}} = -\frac{2}{3}\left.\frac{ds}{dt}\right|_{s=15\,\text{m}}$$

$$= -\frac{20}{3}\,\text{ms}^{-2}$$

7. *(c)* Least count $= \frac{1}{2}$ MSD

Let n division of vernier scale coincides with 4 divisions of main scale.

$$\therefore \quad n\text{VSD} = 4\,\text{MSD}$$

$$1\,\text{VSD} = \frac{4}{n}\,\text{MSD}$$

Now, Least Count,

$$\text{LC} = 1\,\text{MSD} - 1\,\text{VSD}$$

$$\therefore \quad \text{LC} = \left(1 - \frac{4}{n}\right)\text{MSD}$$

$$\frac{1}{2} = 1 - \frac{4}{n}$$

$$\frac{4}{n} = 1 - \frac{1}{2}$$

$$\frac{4}{n} = \frac{1}{2}$$

$$n = 8$$

Hence, 8 divisions of vernier scale coincide with 4 divisions of main scale.

8. *(a)* Given, $m = 5\,\text{kg}$

When body is at earth's surface, the gravitational potential energy,

$$U_e = \frac{-GMm}{r} = -\frac{5GM}{R} \quad \ldots \text{(i)}$$

Now, if this body is taken to height, $h = 4R$, then the gravitational potential energy,

$$U_h = -\frac{GMm}{(R+h)} = \frac{-5GM}{5R} \quad \ldots \text{(ii)}$$

Thus, change in gravitational potential energy,

$$\Delta U = U_h - U_e$$

Using Eqs. (i) and (ii), we get

$$\Delta U = -\frac{5GM}{5R} + \frac{5GM}{R}$$

$$= -\frac{GM}{R} + \frac{5GM}{R}$$

$$= \frac{GM}{R}(-1+5)$$

$$= \frac{4GM}{R} = \frac{4gR^2}{R} \quad (\because GM = gR^2)$$

$$= 4gR = 4 \times 9.8 \times 6.4 \times 10^6$$

$$= 250.8 \times 10^6$$

$$= 2.5 \times 10^8\ \text{J}$$

9. *(d)* To convert a measured value from one system to another system, we have

$$n_1 u_1 = n_2 u_2$$

$$\Rightarrow \quad n_2 = n_1\left[\frac{u_1}{u_2}\right]$$

$$= n_1\left[\frac{\text{M}_1}{\text{M}_2}\right]^a\left[\frac{\text{L}_1}{\text{L}_2}\right]^b\left[\frac{\text{T}_1}{\text{T}_2}\right]^c$$

As, $[F] = [\text{MLT}^{-2}]$

$\therefore \quad a = b = 1,\ c = -2$

$M_1 = 1\,\text{kg},\ L_1 = 1\,\text{m},\ T_1 = 1\,\text{s}$

$M_2 = 30\,\text{g},\ L_2 = 20\,\text{cm},\ T_2 = 15\,\text{s},\ n_1 = 65$

$$\therefore \quad n_2 = 65\left[\frac{1\,\text{kg}}{30\,\text{g}}\right]\left[\frac{1\ \text{m}}{20\ \text{cm}}\right]\left[\frac{1\text{s}}{15\text{s}}\right]^{-2}$$

$$= 65 \times \frac{1000}{30} \times \frac{100}{20} \times 15^2$$

$$= 2437500$$

10. *(c)* Refractive index of a medium is given by

$$\mu = \frac{c}{v} = \frac{\frac{1}{\sqrt{\varepsilon_0\mu_0}}}{\frac{1}{\sqrt{\mu\varepsilon}}} = \sqrt{\frac{\varepsilon\mu}{\varepsilon_0\mu_0}}$$

11. *(b)* Given, $R = 10\,\text{m},\ u_y = 10\,\text{m/s}$

Applying equation of motion in y-direction,

$$v_y = u_y - g\frac{T}{2}$$

At maximum height, $v_y = 0$

$$\therefore \quad 0 = 10 - 10 \times \frac{T}{2}$$

$\therefore T =$ Time of flight $= 2\,\text{s}$

Now, $R = vT \Rightarrow 10 = v \times 2$

$$\therefore \quad v = 5\,\text{m/s}$$

12. *(d)* Periodic motion is not necessarily an oscillatory motion.

For example, motion of the earth around the sun is a periodic motion, but not an oscillatory motion.

An oscillatory motion in which acceleration of the particle is directly proportional to displacement and acts in a direction opposite to the direction of displacement in called SHM. So, all oscillatory motions are not SHM.

An oscillatory motion without resistive force must be a periodic motion.

13. *(a)* From the photoelectric equation,

$$h\nu = \phi_0 + eV$$

$$V = \left(\frac{h}{e}\right)\cdot\nu - \frac{\phi_0}{e}$$

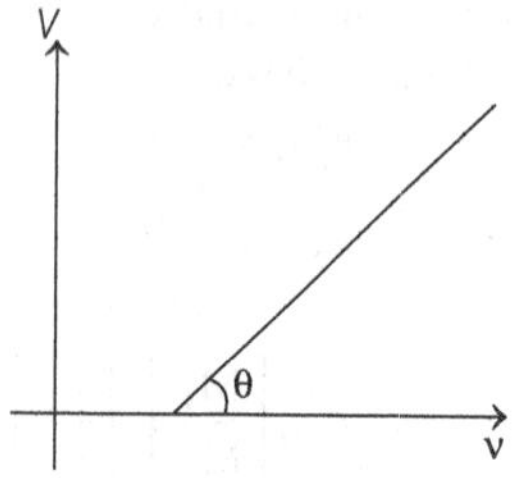

$\text{Slope} = \tan\theta = \frac{h}{e}$,

$Y\text{-intercept} = -\frac{\phi_0}{e}$

On decreasing the intensity of light, the value of h, e, ϕ_0 remain constant.

So, slope = $\tan\theta$ = constant

$\therefore$ θ remains constant.

14. *(a)* In the isothermal expansion, the gas absorbs heat from the surrounding and does work to expand the volume.

In an adiabatic expansion, when an ideal gas expands adiabatically $(Q = 0)$, work done on it is only due to change in internal energy of the system. Hence, its temperature and pressure both decrease.

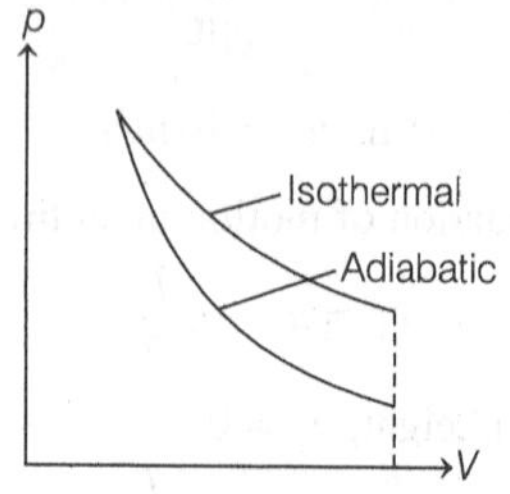

So, work done in reversible adiabatic is less.

Both Assertion and Reason are correct and Reason is the correct explanation of Assertion.

15. *(a)* For a massless string, the tension at all the points on the string is same in magnitude.

The direction of tension is away from the point of contact in a body or a pulley.

16. *(a)* As, $\gamma = \frac{C_p}{C_V} = 1 + \frac{2}{f}$

A. For helium (He) – monoatomic gas, $f = 3$

$\therefore \quad \gamma = 1 + \frac{2}{3} = \frac{5}{3}$

B. For chlorine (Cl_2) rigid-diatomic, rigid gas, $f = 5$

$\therefore \quad \gamma = 1 + \frac{2}{5} = \frac{7}{5}$

C. For oxygen (O_2) non-rigid-diatomic, non-rigid gas, $f = 7$

$\therefore \quad \gamma = 1 + \frac{2}{7} = \frac{9}{7}$

D. For water (H_2O) rigid-triatomic rigid molecule. $f = 6$

$\therefore \quad \gamma = 1 + \frac{2}{6} = \frac{4}{3}$

Hence, $A \to 4, B \to 1, C \to 2, D \to 3$.

17. *(c)*

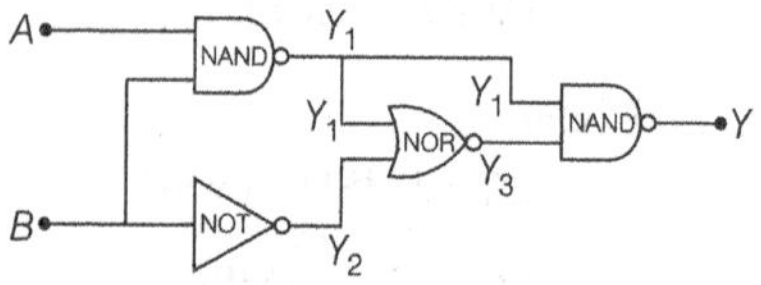

The truth table of the above circuit is as follows

A	B	$Y_1 = \overline{A \cdot B}$	$Y_2 = \overline{B}$	$Y_3 = \overline{Y_1 + Y_2}$	$Y = \overline{Y_1 \cdot Y_3}$
0	1	1	0	0	1
0	0	1	1	0	1
1	1	0	0	1	1
1	0	1	1	0	1

18. *(a)* Weight of overhanging part of the chain $= \frac{mg}{6}$

The weight acts at the centre of gravity of the overhanging part, i.e. $\frac{l}{12}$ below the surface of the table.

Gain in potential energy $= \frac{mg}{6} \times \frac{l}{12} = \frac{mgl}{72}$

$\therefore$ Work done = Gain in potential energy $= \frac{mgl}{72}$

19. *(c)* As, $E = \frac{1}{2}mv^2$

$\therefore \quad v = \sqrt{\frac{2E}{m}}$

Now, radius of circular path in a magnetic field,

$$r = \frac{mv}{Bq} = \frac{m}{Bq}\sqrt{\frac{2E}{m}} = \frac{\sqrt{2mE}}{Bq}$$

$\therefore \quad r \propto \sqrt{m}$

As, $\quad m_e < m_p$

$\Rightarrow \quad r_e < r_p$

20. *(c)* When four equal charges of magnitude q are placed on the vertices of a square,

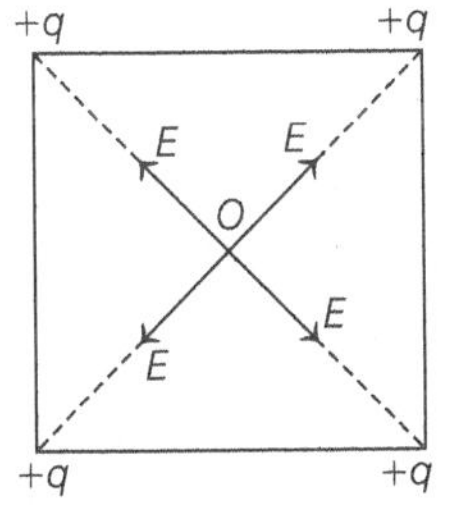

Hence, net electric field at the centre O is zero.

Electric field inside a uniformly charged spherical shell is zero.

The electric potential outside a uniformly charged non-conducting solid sphere is given by

$$V = \frac{1}{4\pi\varepsilon_0}\cdot\frac{Q}{r}$$

$$\Rightarrow \quad V \propto \frac{1}{r}$$

21. (2) $\quad u = -a$

$$\frac{1}{f} = \frac{1}{v} - \frac{1}{u}$$

$$\frac{1}{v} = \frac{1}{f} - \frac{1}{a} \quad \text{...(i)}$$

After reflection from mirror, $u' = v$

$$\frac{1}{v'} - \frac{1}{u'} = \frac{1}{f}$$

$$v' = \frac{a}{3} \quad \text{[given]}$$

$$\frac{3}{a} - \frac{1}{f} + \frac{1}{a} = \frac{1}{f}$$

$$\frac{4}{a} = \frac{2}{f}$$

$$a = 2f$$

So, $nf = 2f$

$\therefore \quad n = 2$

22. (192) FBD of combined system is

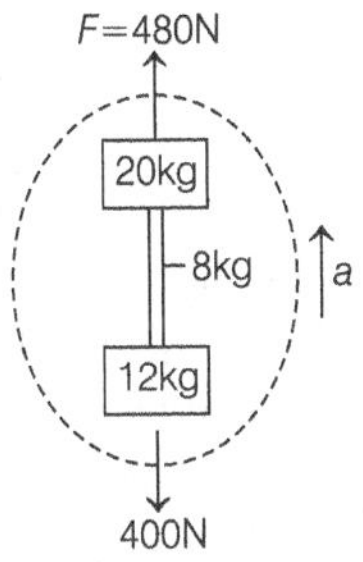

Total mass of the system $= (20 + 8 + 12) = 40\,\text{kg}$

We know, $ma = f - mg$

$\therefore$ Common acceleration of the system,

$$a = \frac{480 - 400}{40} = 2\,\text{m/s}^2$$

The tension at the mid-point of the rod is shown in FBD as shown below.

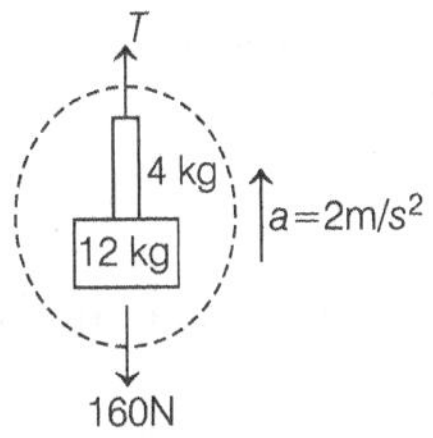

$$\therefore \quad T - 160 = 16 \times 2 \Rightarrow T = 192\,\text{N}$$

23. (23) The impedance of L-C-R circuit,

$$Z = \sqrt{R^2 + (X_L - X_C)^2}$$

$\therefore$ Current through the circuit,

$$I = \frac{V_0}{Z} = \frac{V_0}{\sqrt{R^2 + (X_L - X_C)^2}}$$

$\therefore$ Voltage across resistor,

$$V = IR = \frac{V_0 R}{\sqrt{R^2 + (X_L - X_C)^2}}$$

For V to be maximum,

$$X_L = X_C$$

$$\Rightarrow \quad \omega L = \frac{1}{\omega C}$$

$$\Rightarrow \quad (2\pi f)^2 = \frac{1}{LC}$$

$$\Rightarrow \quad f = \frac{1}{2\pi\sqrt{LC}}$$

$$\Rightarrow \quad f = \frac{1}{2\pi\sqrt{24 \times 2 \times 10^{-6}}}$$

$$= 23\,\text{Hz}$$

24. (64)

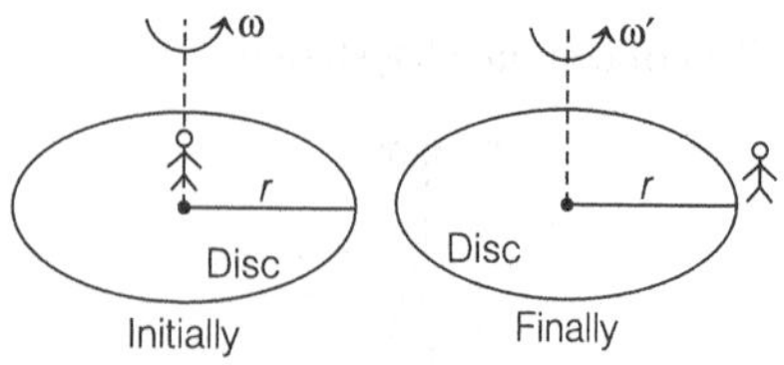

Mass of disc, $M = 160\,\text{kg}$

Mass of man, $m = 50\,\text{kg}$ and $N = 10\,\text{rpm}$

Applying conservation of angular momentum,

$$I_1\omega_1 = I_2\omega_2$$

$$\Rightarrow \left(\frac{1}{2}Mr^2\right)\left(\frac{2\pi N}{60}\right) = \left(\frac{1}{2}Mr^2 + mr^2\right)\cdot\omega'$$

$$\Rightarrow \frac{160\times\pi\times10}{60} = \left(\frac{1}{2}\times160+50\right)\cdot\omega'$$

$$\therefore \omega' = 0.64\ \text{rad/s}$$

$$\because \omega' = \frac{x}{100} \quad \text{[given]}$$

$$\therefore x = 64$$

25. (16) $\mu = 4\ \text{g/m} = 4\times10^{-3}\ \text{kg/m}$

The equation of travelling wave is $y = 0.05\sin(512t - 8x)$

$\therefore\ \omega = 512\,\text{rad/s},\ k = 8\,\text{m}^{-1}$

$\therefore$ Velocity of wave, $v = \frac{\omega}{k} = \frac{512}{8} = 64\ \text{m/s}$

The velocity of wave on a stretched string is $v = \sqrt{\frac{T}{\mu}}$

$\therefore\ T = \mu v^2 = 4\times10^{-3}\times64\times64 = 16.38\,\text{N} \simeq 16\,\text{N}$

Chemistry

26. (*b*) The mechanism of the given reaction is as follow:

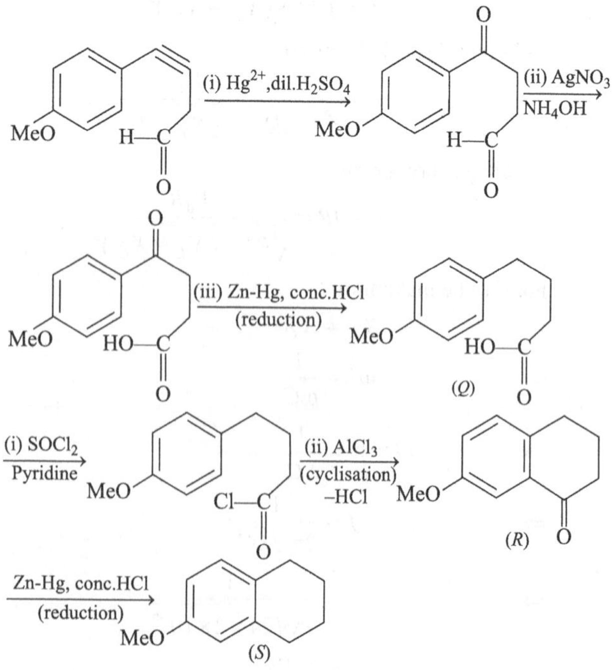

27. (*b*) $2XeF_6 + SiO_2 \longrightarrow 2XeOF_4 + SiF_4$

28. (*a*) Both A and R are true and R is the correct explanation of A.

29. (*d*) C_2^{2-} (acetylide ion) Configuration is

$$\sigma 1s^2\ \sigma^* 1s^2\ \sigma 2s^2\ \sigma^* 2s^2\ \pi 2p_x^2 = \pi 2p_y^2\ \sigma 2p_z^2$$

$$\text{Bond order} = \frac{10-4}{2} = \frac{6}{2} = 3$$

Thus, the correct option with bond order 3 is NO^+.

30. (*c*) On balancing the equation

$$C_2H_7N + \frac{15}{2}CuO \longrightarrow 2CO_2 + \frac{7}{2}H_2O + \frac{1}{2}N_2 + \frac{15}{2}Cu$$

On comparing $y = 7$

31. (*a*) Ozonolysis of alkene is done by treating alkene with O_3 which produces ozonide followed by reduction with Zn / CH_3COOH to give a carbonyl compound. The reaction takes place as follows

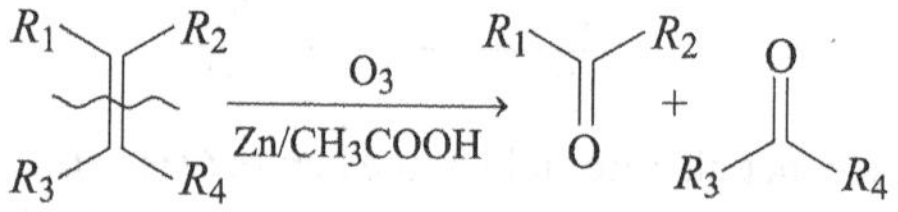

Now, lets see the ozonolysis of given alkenes

I.
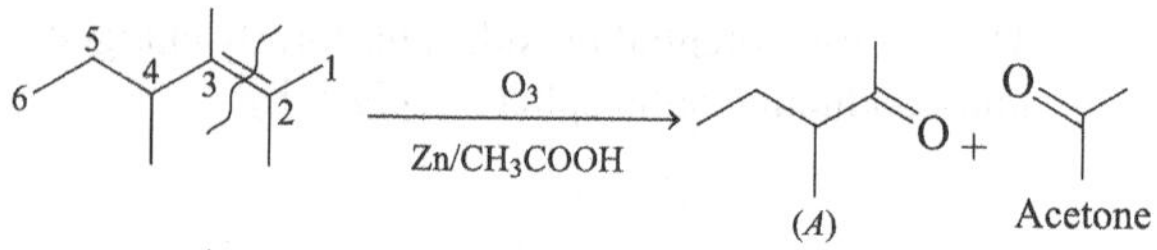

II.
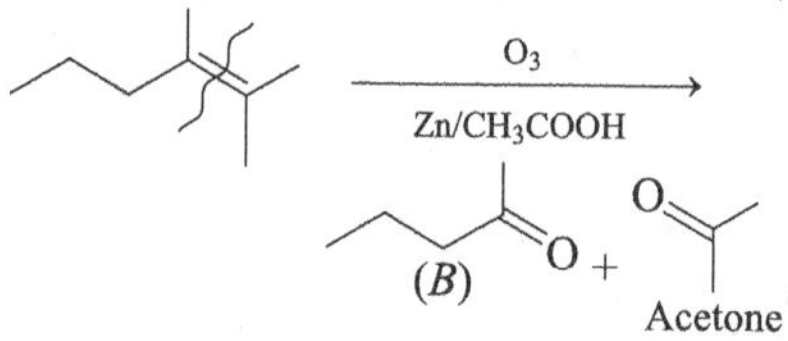

III.
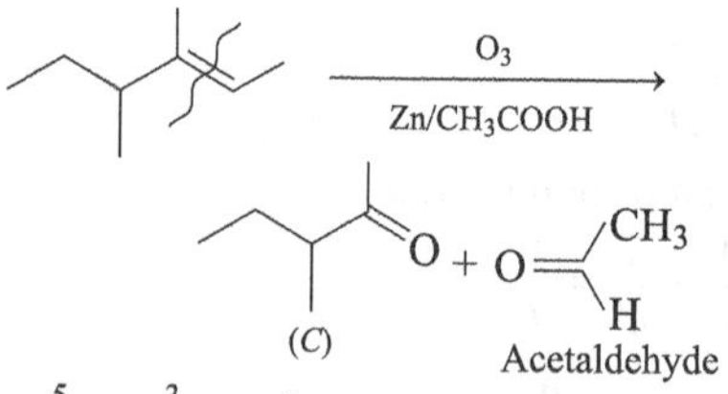

IV.
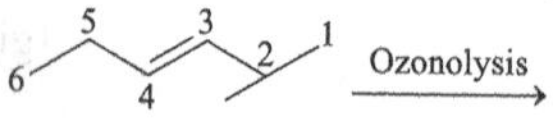

Acetone is not formed

Out of the above three reactions, I and II are producing acetone along with *A* and *B* respectively. But III is not producing acetone as a byproduct. Hence, III could be eliminated as acetone is not one of the products.

On reduction of the carbonyl compound *A* and *B* with $LiAlH_4$ they produce respective products such as

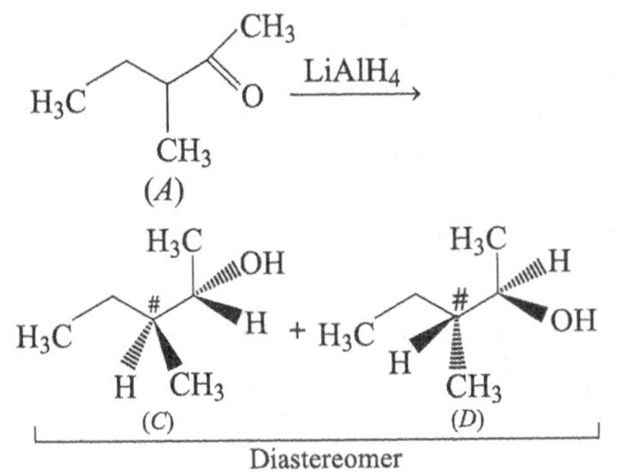

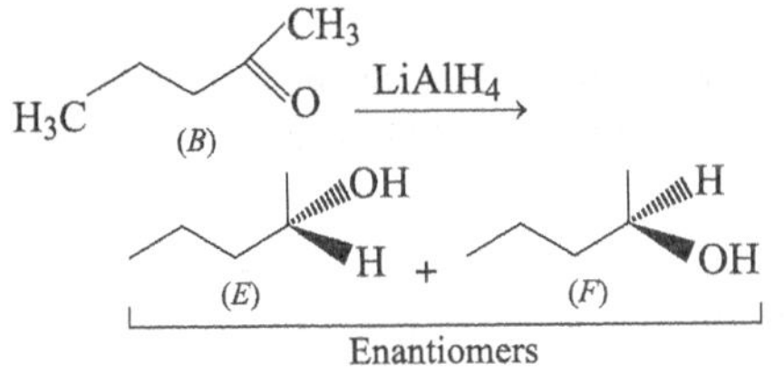

Here, C and D are diastereomer because of the presence of chiral centre at adjacent carbon of carbonyl group, while in B there is no chiral carbon adjacent to carbonyl centre.

32. (*a*) Al^{3+} and Fe^{3+} sulphide hydrolyse in water. Ni^{2+} requires basic medium with H_2S to form precipitate. Ca^{2+} sulphide is soluble. Only CuS ppt. will be formed by Cu^{2+}.

33. (*a*) Process (I) = (Adiabatic reversible)

$$\frac{\Delta U}{R} = 750 - 2550$$

$$\Delta U = -1800R$$

$$W_1 = \Delta U$$

$$= -1800R$$

Process (II)

$\Rightarrow$ (Reversible isothermal process)

$$T_1 = 800\,K$$

Calculation of T_2 after reversible adiabatic process use,

$$\Delta U = nC_v dT$$

$$\Rightarrow -1800R = 1 \times \frac{5}{2} R(T_2 - 800)$$

$$T_2 = 180K$$

$$W_{II} = -nRT_2 \ln \frac{V_3}{V_2} = W_I$$

$$\Rightarrow \quad -1 \times R \times 180 \ln \frac{V_3}{V_2} = -1800R$$

$$\ln \frac{V_3}{V_2} = 10$$

34. (*d*) $\underset{\text{(Black)}\atop (A)}{MnO_2} + K_2CO_3 + \underset{\text{(Air)}}{\frac{1}{2} O_2} \xrightarrow{\Delta} \underset{(B)\atop \text{Green}}{K_2MnO_4} + CO_2$

$$2K_2MnO_4 + Cl_2 \longrightarrow \underset{\text{(Pink)}\atop (C)}{2KMnO_4} + 2KCl$$

35. (*c*) **1, 4-addition reaction** The reagent HCl attacks quinone to give 1, 4-addition product as follows

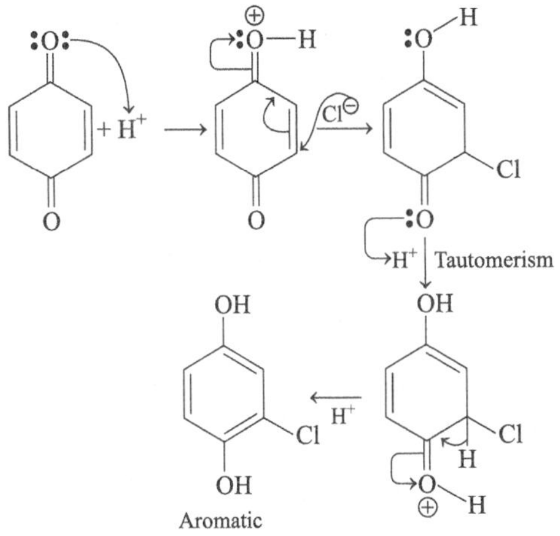

36. (*b*) The correct order is A-III, B-IV, C-II, D-I.

- The reagent $CHCl_3$, $NaOH/H_3O^+$ is used in Cannizzaro reaction.
- The reagent Zn-Hg/HCl is used in Clemmensen reduction.
- The reagent $NaNO_2$+HX/CuX is used in Sandmeyer reaction.
- The reagent Br_2/NaOH is used in Hofmann degradation.

37. (*a*) Use concept of resonance and aromaticity to check stability order. The S_N1 reaction proceeds through carbocation mechanism. Hence, more stable carbocation undergoes S_N1 reaction more easily.

The stability of carbocation are as follows

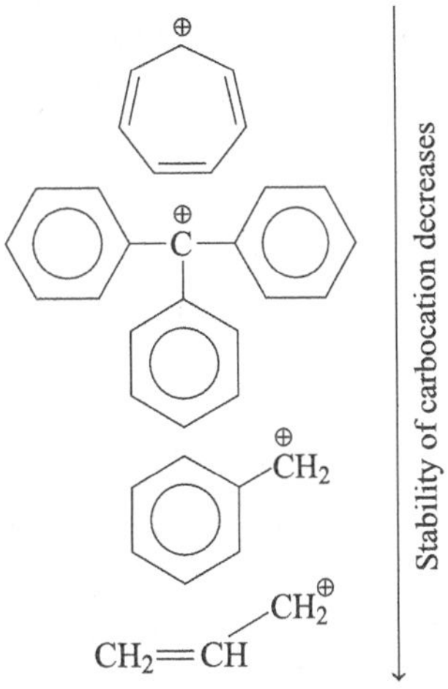

Thus, the correct order of reactants towards S_N1 reaction is as follows

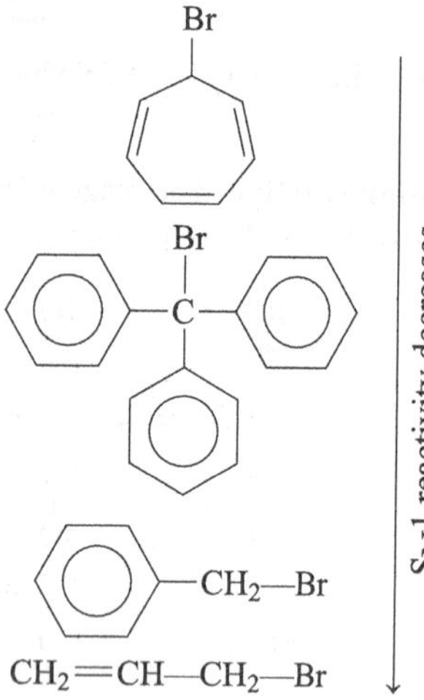

38. (*c*) **Determination of molecular structure**

Molecular formula = C_3H_6O

$$u = (C+1) - \frac{H}{2} + \frac{N}{2} \quad \text{(degree of unsaturation)}$$

$$= (3+1) - \frac{6}{2} + \frac{0}{2} = 4 - 3 = 1$$

It means that it may contain double bond or ring but possibility of ring is ruled out due to acyclic nature of compound.

The compound does not reduce Tollen's reagent, hence it is a ketone.

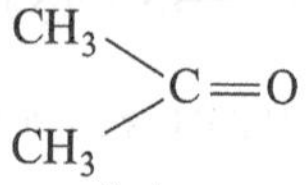

Acetone

Now, acetone on reaction with NH_2OH produces *Y* as,

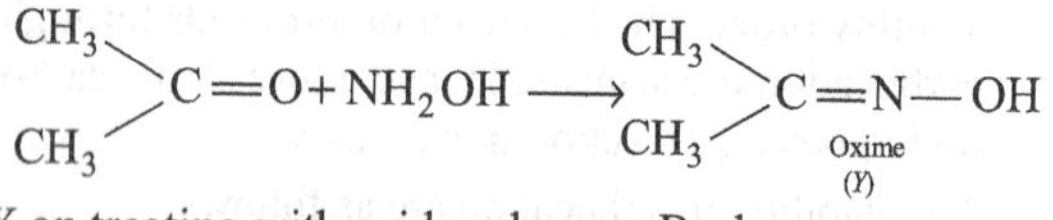

Y on treating with acid undergoes Beckmann rearrangement. This is as follows

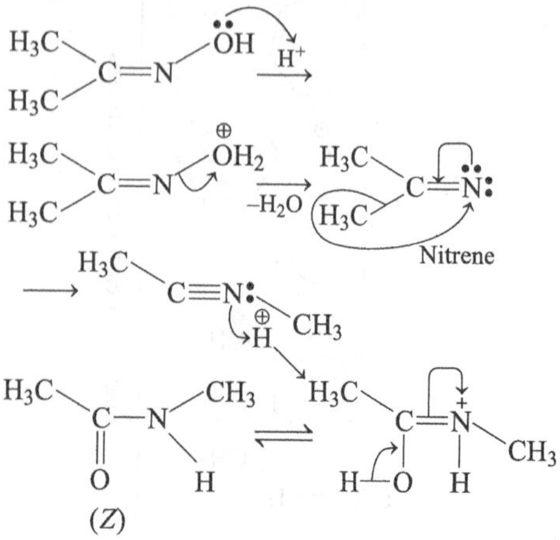

39. (*d*) Formula will be, $[Co(SCN)_2(NH_3)_4]_3[Co(ox)_3]$

It shows linkage isomerism due to presence of SCN^-.

Optical isomerism is exhibited due to presence of $[Co(ox)_3]^{3-}$ complex ion, which has asymmetric structure.

Geometrical isomerism exhibited due to cationic part.

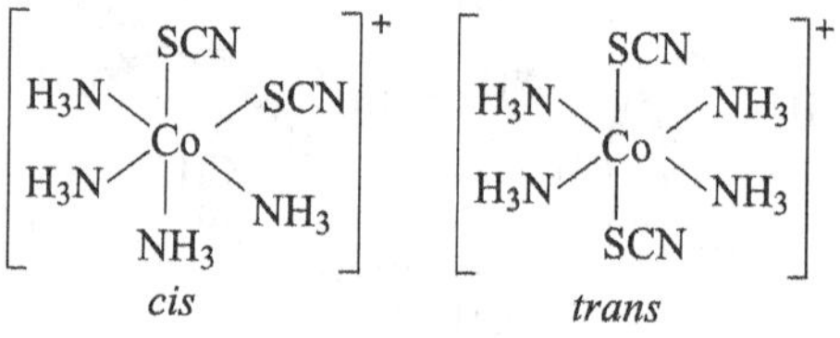

40. (*c*) Among the given compounds only three i.e. I, III and IV will be formed as a product of dehydration of given compound.

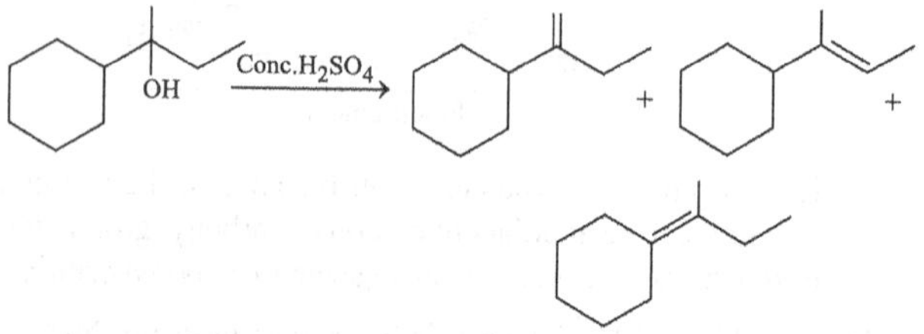

41. (*b*) Statements B and D are correct while statements A and C are incorrect. The corrected form is as,

The concept of electron in Bohr's orbit, violates the Heisenberg's uncertainty principle.

Atomic orbital is characterised by n, l and m.

42. (*a*) ψ^2 remain unchanged w.r.t spin and space coordinate and that is possible only if ψ changes its sign (anti- symmetrical).

43. (*a*) Statement I is true but Statement II is false.

The corrected form is as,

Orthophosphoric acid (H_3PO_4) is a tribasic acid because it has three replaceable hydrogen atoms. Hence, the basicity of H_3PO_4 is 3 while phosphorous acid (H_3PO_3) is a dibasic acid because it has two replaceable hydrogen atom. Hence, the basicity of H_3PO_3 is 2.

44. (*a*) As we know,

Osmotic pressure, $\pi = i \times MRT$

For, 0.400 M NaOH, $i = 2$

So, $\pi_1 = 0.8RT$

For, 0.200 M $Mg_3(PO_4)_2$ (*aq*), $i = 5$

So, $\pi_2 = 0.200 \times 5RT = RT$

For, 0.300 M KBr (*aq*), $i = 2$

So, $\pi_3 = 0.300 \times 2 \times RT = 0.6RT$

For, 0.350 M Na_3PO_4 (*aq*), $i = 4$

So, $\pi_4 = 0.350 \times 4 \times RT = 1.4RT$

Hence, $\pi_3 < \pi_1 < \pi_2 < \pi_4$, is the correct order for the osmotic pressure of these solutions.

45. (a) (A) and (R) both are true and (R) is the correct explanation of (A).

Linkage isomerism arises in coordination compounds containing ambidentate ligands because ambidentate ligand has two different donor atoms. e.g. CN, SCN, NCS.

46. (1) $E_{Cr_2O_7^{2-}/Cr^{3+}}$

$$= E^{\circ}_{Cr_2O_7^{2-}/Cr^{3+}} - \frac{0.0591}{6}\log\frac{[Cr^{3+}]^2}{[Cr_2O_7^{2-}]\,[H^+]^{14}}$$

$$= 1.33 - \frac{0.0591}{6}\log\frac{1}{(0.01)^{14}}$$

$$= 1.05\,V \simeq 1\,V.$$

47. (125) $NH_4NO_3(s) \longrightarrow N_2O(g) + 2H_2O(l)$

$$\Delta H_{Reaction} = \Delta H_{Products} - \Delta H_{Reactants}$$
$$= [\Delta H_{N_2O} + 2\times\Delta H_{H_2O}] - \Delta H_{NH_4NO_3}$$
$$= 81.46 + 2(-285.8) - (-367.57)$$
$$\Delta H = -122.57\text{ kJ}$$
$$\Delta H = \Delta E + \Delta n_g RT$$
$$-122570 = \Delta E + 1\times 8.314\times 298$$
$$\Delta E = -125047.57\text{ J} = -125.047\text{ kJ} \approx -125$$

48. (156) Solubility of AgCl in water

$$= 1.79\times10^{-3}\text{ g/L}$$
$$= \frac{1.79\times10^{-3}}{143.32}\text{ mol/L}$$
$$= 1.249\times10^{-5}\text{ mol/L}$$

For AgCl, $AgCl \longrightarrow Ag^+ + Cl^-$

$$K_{sp} = S\times S = S^2 = (1.249\times10^{-5})^2\text{ mol}^2\text{L}^{-2}$$
$$= 1.56\times10^{-10}\text{ mol}^2\text{ L}^{-2}$$
$$= 156\times10^{-12}\text{ mol}^2\text{ L}^{-2}$$

49. (3) Let the weight of NH_4NO_3 and $(NH_4)_2HPO_4$ be x and y, respectively.

So, the mass % of $N_2 = \dfrac{\frac{x}{80}\times2\times14 + \frac{y}{132}\times2\times14}{x+y}\times100$

$$= 30.4\,\%$$

$\Rightarrow \quad x : y = 2 : 1$

So, the sum of x and $y = 3$

50. (166) $H_2 + I_2 \longrightarrow 2HI$

$$\log\frac{k_2}{k_1} = \frac{E_a}{2.303R}\left(\frac{1}{600} - \frac{1}{800}\right)$$

$$\log\frac{1}{2.5\times10^{-4}} = \frac{E_a}{2.303\times8.314}\left(\frac{200}{600\times800}\right)$$

$\therefore \quad E_a \approx 166\text{ kJ/mol}$

Mathematics

51. *(b)* Let r be the radius of the circle.

$$|z_1| = |z_2| = |z_3| = |z_4| = r$$
$$\Rightarrow z_1\bar{z}_1 = z_2\bar{z}_2 = z_3\bar{z}_3 = z_4\bar{z}_4 = r^2$$

$\because PS \perp QR$

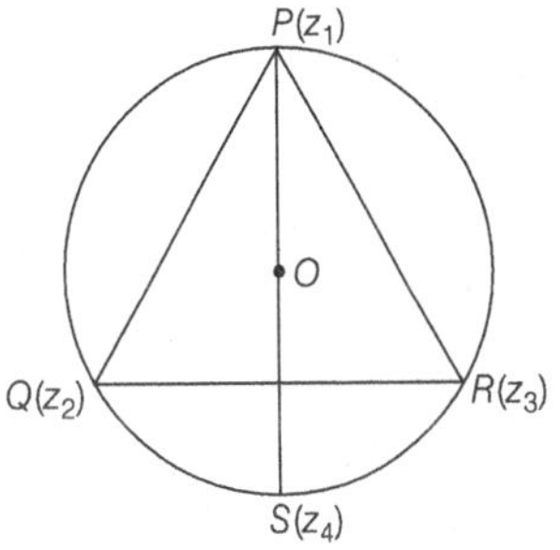

$$\Rightarrow \quad \arg QR + \arg PS = 0$$

$$\Rightarrow \quad \frac{z_1 - z_4}{\bar{z}_1 - \bar{z}_4} + \frac{z_2 - z_3}{\bar{z}_2 - \bar{z}_3} = 0$$

$$\frac{z_1 - z_4}{\bar{z}_1 - \bar{z}_4} = -\left(\frac{z_2 - z_3}{\bar{z}_2 - \bar{z}_3}\right)$$

$$\Rightarrow \quad \frac{z_1 - z_4}{\frac{r^2}{z_1} - \frac{r^2}{z_4}} = -\left(\frac{z_2 - z_3}{\frac{r^2}{z_2} - \frac{r^2}{z_3}}\right)$$

$$\Rightarrow \quad \frac{z_1 - z_4}{\frac{r^2(z_4 - z_1)}{z_1 z_4}} = -\frac{1}{r^2}\left(\frac{z_2 - z_3}{\frac{z_3 - z_2}{z_2 z_3}}\right)$$

$$\Rightarrow \quad -\frac{z_1 z_4}{r^2} = \frac{z_2 z_3}{r^2} \Rightarrow z_4 = -\frac{z_2 z_3}{z_1}$$

52. *(a)* I. $\because (1+x)^n = {}^nC_0 + {}^nC_1x + {}^nC_2x^2 + \ldots + {}^nC_nx^n$

On integrating w.r.t. x from 0 to n to the both sides, we get

$$\frac{(1+x)^{n+1}}{n+1} = {}^nC_0x + {}^nC_1\frac{x^2}{2} + {}^nC_2\frac{x^3}{3} + \ldots + {}^nC_n\frac{x^{n+1}}{n+1}$$

On putting $n = 41$ and $x = i$ and comparing real parts, we get

$$-\frac{{}^{41}C_1}{2} + \frac{{}^{41}C_3}{4} - \frac{{}^{41}C_5}{6} + \ldots - \frac{{}^{41}C_{41}}{42} = -\frac{1}{42}$$

$$\Rightarrow \frac{{}^{41}C_1}{2} - \frac{{}^{41}C_3}{4} + \frac{{}^{41}C_5}{6} - \ldots + \frac{{}^{41}C_{41}}{42} = \frac{1}{42}$$

Statement I is true.

II. $\frac{1}{2}\sum_{n=3}^{20} {}^nC_3(\sin^{2n}x+\text{cosec}^{2n}x)\ge\sum_{n=3}^{20} {}^nC_3$

$= {}^3C_3 + {}^4C_3 + \ldots + {}^{20}C_3$

$= {}^{21}C_4$

$\therefore \frac{1}{2}\sum_{n=3}^{20} {}^nC_3(\sin^{2n}x+\text{cosec}^{2n}x)\ge {}^{21}C_4$

Statement II is false.

53. *(b)* Given,

$K(n)^{n-1} = n^{n-1} + 2(n+1)n^{n-2} + 3(n+1)^2 n^{n-3} + \ldots + (n+1)^{n-1} n$

On putting $n = 10$,

$K(10)^9 = 10^9 + 2(11)(10)^8 + 3(11)^2(10)^7 + \ldots + 10(11)^9$

$$\Rightarrow \quad K = 1 + 2\left(\frac{11}{10}\right) + 3\left(\frac{11}{10}\right)^2 + \ldots + 10\left(\frac{11}{10}\right)^9 \quad \ldots(i)$$

$$\therefore \quad \frac{11K}{10} = \frac{11}{10} + 2\left(\frac{11}{10}\right)^2 + 3\left(\frac{11}{10}\right)^3 + \ldots + 10\left(\frac{11}{10}\right)^{10} \quad \ldots(ii)$$

On subtracting Eq. (ii) from Eq. (i), we get

$$\frac{-K}{10} = 1 + \frac{11}{10} + \left(\frac{11}{10}\right)^2 + \ldots + \left(\frac{11}{10}\right)^9 - 10\left(\frac{11}{10}\right)^{10}$$

$$= \frac{\left(\frac{11}{10}\right)^{10} - 1}{\frac{11}{10} - 1} - 10\left(\frac{11}{10}\right)^{10}$$

$$= 10\left(\frac{11}{10}\right)^{10} - 10 - 10\left(\frac{11}{10}\right)^{10}$$

$$\Rightarrow \quad \frac{-K}{10} = -10$$

$\Rightarrow \quad K = 100$

54. *(c)* Given,

$$L = \frac{1}{2} = \lim_{x\to\infty} \frac{2x^3 - (\tan^{-1}x)^3}{\frac{8}{\pi}x^3\cot^{-1}|\alpha x| + \alpha^2 x^6 \sin\frac{1}{x_3} - 3\alpha x^3}$$

$$= \lim_{x\to\infty} \frac{2 - \left(\frac{\tan^{-1}x}{x}\right)^3}{\frac{8}{\pi}\cot^{-1}|\alpha x| + \frac{\alpha^2\sin\frac{1}{x^3}}{\frac{1}{x^3}} - 3\alpha}$$

Here, two cases arises.

(i) Either $\alpha = 0$ Or (ii) $\alpha \ne 0$

If $\alpha = 0$, then

$$L = \frac{2-0}{\frac{8}{\pi}\times\frac{\pi}{2}} = \frac{1}{2}$$

and if $\alpha \ne 0$, then

$$L = \frac{2}{0 + \alpha^2 - 3\alpha} = \frac{1}{2}$$

$\Rightarrow \quad \alpha^2 - 3\alpha - 4 = 0$

$\Rightarrow \quad \alpha = 4, -1$

$\therefore$ Possible value of $\alpha = 0, -1, 4$

Sum of values of $\alpha = 0 + (-1) + 4 = 3$

55. *(a)* Let S = Sample space

$n(S) = 2^n$

and E_1 = One person gets a head, whereas remaining $(n-1)$ persons gets a tail

E_2 = One person gets a tail, whereas remaining $(n-1)$ persons gets a head

$\therefore$ Required event, $E = E_1 \cup E_2$

$\Rightarrow \quad n(E) = n(E_1) + n(E_2)$

$= {}^nC_1\,{}^{n-1}C_{n-1} + {}^nC_1\,{}^{n-1}C_{n-1}$

$= n + n = 2n$

Required probability,

$$P(E) = \frac{n(E)}{n(S)} = \frac{2n}{2^n} = \frac{n}{2^{n-1}}$$

56. *(c)* Given,

$$f(x) = \begin{cases} [x], & -2 \le x \le -1 \\ |x|+1, & -1 < x \le 2 \end{cases}$$

$$\text{and } g(x) = \begin{cases} [x], & -\pi \le x < 0 \\ \sin x, & 0 < x \le \pi \end{cases}$$

Then,

$$g[f(x)] = \begin{cases} [f(x)], & -\pi \le f(x) < 0 \\ \sin f(x), & 0 \le f(x) \le \pi \end{cases}$$

$$= \begin{cases} [[x]], & -\pi \le [x] < 0, & -2 \le x \le -1 \\ [|x|+1], & -\pi \le |x|+1 < 0, & -1 < x \le 2 \\ \sin[x], & 0 \le [x] \le \pi, & -2 \le x \le -1 \\ \sin(|x|+1), & 0 \le |x|+1 \le \pi, & -1 < x \le 2 \end{cases}$$

$$= \begin{cases} [x], & -2 \le x \le -1 \\ \sin(|x|+1), & -1 < x \le 2 \end{cases}$$

For $-2 \le x \le -1$

$[x] = -2, -1$

For $-1 < x \le 2$

$|x|+1 \in [1, 3]$

$\therefore \sin(|x|+1) \in [\sin 3, 1]$

Hence, the range of $g[f(x)]$ is $\{-2, -1\} \cup [\sin 3, 1]$.

57. *(c)* Given lines,

$$\frac{x-1}{4}=\frac{y+3}{5}=\frac{z+4}{3}=\lambda \quad \text{(let)}$$

$$\frac{x-4}{3}=\frac{y-3}{4}=\frac{z-5}{2}=\mu \quad \text{(let)}$$

Vector form of lines

$$L_1: \mathbf{r}=(\hat{\mathbf{i}}-3\hat{\mathbf{j}}-4\hat{\mathbf{k}})+\lambda(4\hat{\mathbf{i}}+5\hat{\mathbf{j}}+3\hat{\mathbf{k}})$$

$$L_2: \mathbf{r}=(4\hat{\mathbf{i}}+3\hat{\mathbf{j}}+5\hat{\mathbf{k}})+\mu(3\hat{\mathbf{i}}+4\hat{\mathbf{j}}+2\hat{\mathbf{k}})$$

Here, $\mathbf{a}_1=\hat{\mathbf{i}}-3\hat{\mathbf{j}}-4\hat{\mathbf{k}}$, $\mathbf{b}_1=4\hat{\mathbf{i}}+5\hat{\mathbf{j}}+3\hat{\mathbf{k}}$

$\mathbf{a}_2=4\hat{\mathbf{i}}+3\hat{\mathbf{j}}+5\hat{\mathbf{k}}$, $\mathbf{b}_2=3\hat{\mathbf{i}}+4\hat{\mathbf{j}}+2\hat{\mathbf{k}}$

Now, $\mathbf{a}_2-\mathbf{a}_1=3\hat{\mathbf{i}}+6\hat{\mathbf{j}}+9\hat{\mathbf{k}}$

$$\Rightarrow \quad \mathbf{b}_1\times\mathbf{b}_2=\begin{vmatrix}\hat{\mathbf{i}} & \hat{\mathbf{j}} & \hat{\mathbf{k}}\\ 4 & 5 & 3\\ 3 & 4 & 2\end{vmatrix}=-2\hat{\mathbf{i}}+\hat{\mathbf{j}}+\hat{\mathbf{k}}$$

$$\Rightarrow |\mathbf{b}_1\times\mathbf{b}_2|=\sqrt{(-2)^2+1^2+1^2}=\sqrt{6}$$

Now, $(\mathbf{a}_2-\mathbf{a}_1)\cdot(\mathbf{b}_1\times\mathbf{b}_2)=-6+6+9=9$

$$\text{Shortest distance}=\left|\frac{(\mathbf{a}_2-\mathbf{a}_1)\cdot(\mathbf{b}_1\times\mathbf{b}_2)}{|\mathbf{b}_1\times\mathbf{b}_2|}\right|=\frac{9}{\sqrt{6}}$$

Hence, $\sqrt{6}\alpha=\frac{9}{\sqrt{6}}\times\sqrt{6}=9$

58. *(d)* Given equation of hyperbola is

$$\frac{x^2}{9}-\frac{y^2}{16}=1$$

If e_1 is the eccentricity of the hyperbola, then

$$16=9(e_1^2-1)\Rightarrow e_1^2=\frac{16}{9}+1\Rightarrow e_1=\frac{5}{3}$$

$\therefore$ Foci of the hyperbola are $(\pm 3e_1, 0)=(\pm 5, 0)$

If e_2 is the eccentricity of the ellipse, then

$$e_1+e_2=\frac{5\sqrt{17}+3}{3\sqrt{17}}$$

$$\Rightarrow \quad e_2=\frac{5\sqrt{17}+3}{3\sqrt{17}}-\frac{5}{3}$$

$$\Rightarrow \quad e_2=\frac{1}{\sqrt{17}}$$

Let the equation of the ellipse be

$$\frac{x^2}{a^2}+\frac{y^2}{b^2}=1$$

As it passes through $(\pm 5, 0)$

$$\frac{25}{a^2}=1\Rightarrow a^2=25$$

Also, $b^2=a^2(1-e_2^2)$

$$\Rightarrow \quad b^2=25\left(1-\frac{1}{17}\right)=\frac{400}{17}$$

$\therefore$ Equation of ellipse is

$$\frac{x^2}{25}+\frac{17y^2}{400}=1$$

Hence, $\left(\frac{9}{5\sqrt{2}}, \frac{5}{\sqrt{34}}\right)$ does not lie on above ellipse.

59. *(b)* Let $QR=a$, $PR=b$

and $\quad PQ=c$

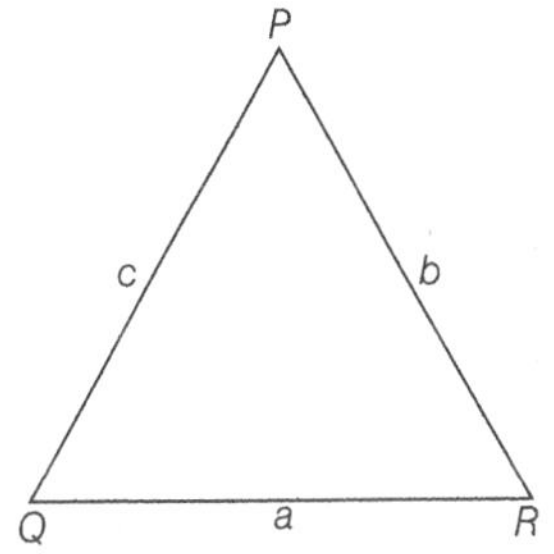

Then, $\quad s=\frac{a+b+c}{2}$

and $\quad \Delta=\sqrt{s(s-a)(s-b)(s-c)}$

We have, $r=\frac{\Delta}{s}$ and $R=\frac{abc}{4\Delta}$

$$\text{Now,} \quad \frac{r}{R}=\frac{4\Delta^2}{s(abc)}=\frac{4s(s-a)(s-b)(s-c)}{s\cdot(abc)} \quad \ldots\text{(i)}$$

$$\text{Consider,} \ \frac{(s-a)+(s-b)}{2}\geq\sqrt{(s-a)(s-b)}$$

[$\because$ AM $\geq$ GM]

$$\Rightarrow \quad \frac{c}{2}\geq\sqrt{(s-a)(s-b)} \quad \ldots\text{(ii)}$$

$$\text{Similarly,} \ \frac{b}{2}\geq\sqrt{(s-a)(s-c)} \quad \ldots\text{(iii)}$$

$$\text{and} \quad \frac{a}{2}\geq\sqrt{(s-b)(s-c)} \quad \ldots\text{(iv)}$$

From Eqs. (ii), (iii) and (iv), we get

$$\Rightarrow (s-a)(s-b)(s-c)\leq\frac{abc}{8}$$

$$\Rightarrow \frac{4(s-a)(s-b)(s-c)}{abc}\leq\frac{1}{2}\Rightarrow\frac{r}{R}\leq\frac{1}{2} \quad \text{[using Eq. (i)]}$$

$$\Rightarrow \quad \frac{R}{r}\geq 2$$

For equilateral triangle, $\frac{R}{r}=2$

60. *(b)* Given,

$n_1=n_2=5$, [number of observation]

$\bar{x}_1=2, \bar{x}_2=4$ [mean]

and $\sigma_1^2=4, \sigma_2^2=5$ [variances]

Combined mean,

$$\bar{x} = \frac{n_1\bar{x}_1 + n_2\bar{x}_2}{n_1 + n_2} = \frac{5\times 2 + 5\times 4}{10} = \frac{10+20}{10} = 3$$

Deviation, $d_1^2 = (\bar{x}_1 - \bar{x})^2 = (2-3)^2 = 1$

and $d_2^2 = (\bar{x}_2 - \bar{x})^2 = (4-3)^2 = 1$

Combined variance,

$$\sigma^2 = \frac{1}{n_1 + n_2}[n_1(\sigma_1^2 + d_1^2) + n_2(\sigma_2^2 + d_2^2)]$$

$$= \frac{1}{10}[5(4+1) + 5(5+1)] = \frac{11}{2}$$

Time Saver Tip

We know that

$$\theta^2 = \frac{n_1(\sigma_1)^2 + n_2(\sigma_2)^2}{n_1 + n_2} + \frac{n_1 . n_2 (\bar{x}_1 - \bar{x}_2)^2}{(n_1 + n_2)^2}$$

$$= \frac{5\times 4 + 5\times 5}{5+5} + \frac{5\times 5(2-4)^2}{(5+5)^2} = \frac{45}{10} + 1 = 5.5$$

61. *(d)* We have, $x^2 + x\sin\alpha - 2\sin\alpha = 0$

Since, a and b are the roots of the given quadratic equation.

$\therefore\ a + b = -\sin\alpha$ and $ab = -2\sin\alpha$ ...(i)

Now, $\dfrac{a^{12} + b^{12}}{\dfrac{1}{a^{12}} + \dfrac{1}{b^{12}}(a-b)^{24}}$

$$= \frac{a^{12}b^{12}(a^{12} + b^{12})}{(a^{12} + b^{12})(a-b)^{24}} = \frac{a^{12}b^{12}}{(a-b)^{24}}$$

$$= \frac{(ab)^{12}}{[(a+b)^2 - 4ab]^{12}} = \left[\frac{ab}{(a+b)^2 - 4ab}\right]^{12}$$

$$= \left[\frac{-2\sin\alpha}{\sin^2\alpha + 8\sin\alpha}\right]^{12} = \frac{2^{12}}{(8+\sin\alpha)^{12}}$$

62. *(c)* Given, $3e^2 - 6e + \frac{5}{3} = 0$

$\Rightarrow\ 9e^2 - 18e + 5 = 0 \Rightarrow 3e(3e-5) - (3e-5) = 0$

$\Rightarrow\ (3e-1)(3e-5) = 0 \Rightarrow e = \frac{1}{3}, \frac{5}{3}$

As $e > 1$, $e = \frac{5}{3}$

We have, $ae = 4 \Rightarrow a\times\frac{5}{3} = 4 \Rightarrow a = \frac{12}{5}$

Also, $a^2 + b^2 = a^2 + a^2(e^2 - 1)$

$$= \frac{144}{25}\left(1 + \frac{25}{9} - 1\right) = \frac{144}{25}\times\frac{25}{9} = \frac{144}{9}$$

$\Rightarrow\ a^2 + b^2 = 16$

63. *(b)* Let D be the foot of the perpendicular from the origin O to the join of A and B and divide AB in the ratio $k:1$, then coordinates of D are

$$\left(\frac{11k-9}{k+1}, \frac{0k+4}{k+1}, \frac{-k+5}{k+1}\right).$$

So that the direction cosines of OD are proportional to $11k-9, 4, 5-k$

and direction cosines of AB are proportional to

$11+9, 0-4, -1-5$

i.e. $20, -4, -6$ or $10, -2, -3$

Since, OD is perpendicular to AB.

$10(11k-9) - 2(4) - 3(5-k) = 0$

$\Rightarrow\ 110k - 90 - 8 - 15 + 3k = 0$

$\Rightarrow\ 113k = 113$

$\Rightarrow\ k = 1$

64. *(a)* Given, $\dfrac{dy}{dx} = y + 17$

On solving by separation of variables, we get

$$\frac{dy}{y+17} = dx$$

$\Rightarrow\ \log(y+17) = x + C_1$

$\Rightarrow\ y + 17 = e^{x+C_1}$

$\Rightarrow\ y + 17 = Ce^x$

$\Rightarrow\ y(x) = Ce^x - 17$

$\therefore\ y_1(0) = 0 \Rightarrow y_1(x) = 17(e^x - 1)$

and $y_2(0) = 1 \Rightarrow y_2(x) = 18e^x - 17$

If y_1 and y_2 intersect at x_0

$\Rightarrow\ y_1(x_0) = y_2(x_0)$

$\Rightarrow 17e^{x_0} - 17 = 18e^{x_0} - 17$

$\Rightarrow e^{x_0} = 0$, which is a contradiction.

$\Rightarrow y_1$ and y_2 does not intersect.

65. *(a)* Given,

$$tf(t)(1+t^6)^{1/3} + K = \int\frac{dt}{t^3(1+t^6)^{2/3}} \quad \text{...(i)}$$

$$= \int\frac{dt}{t^7\left(\frac{1}{t^6} + 1\right)^{2/3}}$$

On putting $1 + \frac{1}{t^6} = x$, we get

$$-\frac{6}{t^7}dt = dx \Rightarrow \frac{dt}{t^7} = \frac{-dx}{6}$$

$$\int\frac{dt}{t^7\left(\frac{1}{t^6}+1\right)^{2/3}} = -\frac{1}{6}\int\frac{dx}{x^{2/3}} = -\frac{1}{6}3\cdot x^{1/3} + K$$

$$= -\frac{1}{2}\left(\frac{1}{t^6}+1\right)^{1/3} + K$$

$$= -\frac{1}{2}\left(\frac{t^6+1}{t^6}\right)^{1/3} + K \qquad \text{...(ii)}$$

On comparing Eqs. (i) and (ii), we get

$$f(t) = -\frac{1}{2t^3}$$

66. *(a)* Given, $P = \begin{bmatrix} 1 & 1.5 \\ 1 & 2 \end{bmatrix}$ and $Q = \begin{bmatrix} 4 & -3 \\ -2 & 2 \end{bmatrix}$

We have, $PQ = \begin{bmatrix} 1 & 1.5 \\ 1 & 2 \end{bmatrix}\begin{bmatrix} 4 & -3 \\ -2 & 2 \end{bmatrix} = \begin{bmatrix} 1 & 0 \\ 0 & 1 \end{bmatrix} = I$

$(PQ)^k R_k = I^k R_k = R_k$ for $k = 1, 2, \ldots 50$

Now, $\sum_{k=1}^{50} \text{tr}((PQ)^k R_k) = \text{tr}(PQ)^1 R_1 + \text{tr}(PQ)^2 R_2$

$+ \ldots + \text{tr}(PQ)^{50} R_{50} = S$ [let]

$S = \text{tr}(R_1) + \text{tr}(R_2) + \ldots + \text{tr}(R_{50})$

$S = 1\cdot 3^1 + 3\cdot 3^2 + 5\cdot 3^3 + \ldots + 99\cdot 3^{50}$

$\Rightarrow 3S = 1\cdot 3^2 + 3\cdot 3^3 + 5\cdot 3^4 + \ldots + 99\cdot 3^{51}$

On subtracting, we get

$-2S = 1\cdot 3 + 2\cdot 3^2 + 2\cdot 3^3 + \ldots + 2\cdot 3^{50} - 99\cdot 3^{51}$

$= -3 + 2\cdot 3 + 2\cdot 3^2 + \ldots + 2\cdot 3^{50} - 99\cdot 3^{51}$

$= -3 + 2\cdot \frac{3\cdot(3^{50}-1)}{3-1} - 99\cdot 3^{51}$

$= -3 + 3^{51} - 3 - 99\cdot 3^{51}$

$-2S = -6 - 98\cdot 3^{51}$

$\Rightarrow S = 3 + 49\cdot 3^{51} = 3 + a3^b$ [given]

$\Rightarrow a = 49$ and $b = 51$

$\therefore \frac{b-a}{b+a} = \frac{51-49}{51+49} = \frac{2}{100}$

$\Rightarrow (b-a):(b+a) = 1:50$

67. *(c)* Given, $f(x) = \begin{cases} \alpha_m + \sin \pi x, & x \in [2m, 2m+1] \\ \beta_m + \cos \pi x, & x \in (2m-1, 2m) \end{cases}$

$\because f$ is continuous for all m.

At $x = 2m$,

LHL = RHL = $f(2m)$

$\Rightarrow \beta_m + \cos 2m\pi = \alpha_m + \sin 2m\pi = \alpha_m + \sin 2m\pi$

$\Rightarrow \beta_m + 1 = \alpha_m$

$\Rightarrow \alpha_m - \beta_m = 1$

68. *(d)* Let $\frac{x-1}{2} = \frac{y+1}{-3}$

$= \frac{z+10}{8} = \lambda$

M be the mid-point of PQ and coordinates of M are $(2\lambda+1, -3\lambda-1, 8\lambda-10)$

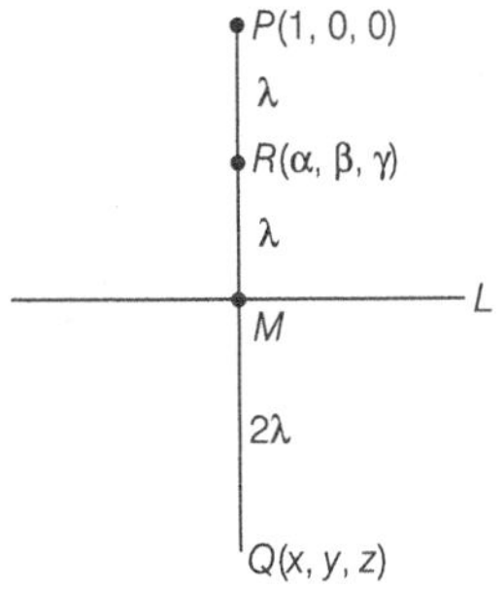

$\therefore PM = (2\lambda+1)\hat{\mathbf{i}} + (-3\lambda-1)\hat{\mathbf{j}} + (8\lambda-10)\hat{\mathbf{k}}$

Here, $PM = (2\lambda)\hat{\mathbf{i}} + (-3\lambda-1)\hat{\mathbf{j}} + (8\lambda-10)\hat{\mathbf{k}}$

Direction ratios of line L are $(2, -3, 8)$.

$\because$ **PM** $\perp (2\hat{\mathbf{i}} - 3\hat{\mathbf{j}} + 8\hat{\mathbf{k}})$

[$\because 2\hat{\mathbf{i}} - 3\hat{\mathbf{j}} + 8\hat{\mathbf{k}}$ is the normal vector of line l]

$\therefore (2\lambda)2 - 3(-3\lambda-1) + 8(8\lambda-10) = 0$

$4\lambda + 9\lambda + 3 + 64\lambda - 80 = 0$

$\Rightarrow \lambda = \frac{77}{77} \Rightarrow \lambda = 1$

$\Rightarrow M(2+1, -3-1, 8-10) \Rightarrow M(3, -4, -2)$

Since, R is the mid-point of PM.

Coordinates of R are $(2, -2, -1)$

$\therefore \alpha = 2, \beta = -2$ and $\gamma = -1$

$\therefore 77(\alpha+\beta+\gamma) = 77(2-2-1) = -77$

69. *(b)* Given, $H(x) = f(x) + g(x)$

$= f(x) + f\left(\frac{1}{x}\right)$

$H(e) = f(e) + f\left(\frac{1}{e}\right) = \int_1^e \frac{\log u}{1+u}\,du + \int_1^{1/e} \frac{\log u}{1+u}\,du$

$H(e) = I_1 + I_2$ ...(i)

where, $I_2 = \int_1^{1/e} \frac{\log u}{1+u}\,du$

Let $\frac{1}{u} = z$

$\Rightarrow du = -\frac{dz}{z^2}$

and where $u = 1$, $z = 1$ and $u = \frac{1}{e}$, $z = e$

$$I_2 = \int_1^e \frac{\log\left(\frac{1}{z}\right)}{1+\frac{1}{z}}\left(-\frac{dz}{z^2}\right)$$

$$= \int_1^e \frac{(-\log z)}{1+z}\cdot z\cdot \frac{(-dz)}{z^2}$$

$$= \int_1^e \frac{\log z}{z(z+1)}\,dz$$

$$= \int_1^e \frac{\log u}{u(u+1)}\,du$$

Eq. (i) becomes, we get

$$H(e)=\int_1^e \frac{\log u}{1+u}du+\int_1^e \frac{\log u}{u(u+1)}du$$

$$=\int_1^e \frac{u\log u+\log u}{u(u+1)}du=\int_1^e \frac{\log u}{u}du$$

Let $\log u = t$

$\Rightarrow \frac{1}{u}du = dt$

When $u=1, t=0$ and $u=e, t=1$

$\therefore \quad H(e)=\int_0^1 t\,dt=\left[\frac{t^2}{2}\right]_0^1=\frac{1}{2}$

70. *(d)* Given curves are

$y=x$ ...(i)

$x=e$ ...(ii)

$y=\frac{1}{x}$ ...(iii)

$\therefore$ Required area = Region $OPQR$

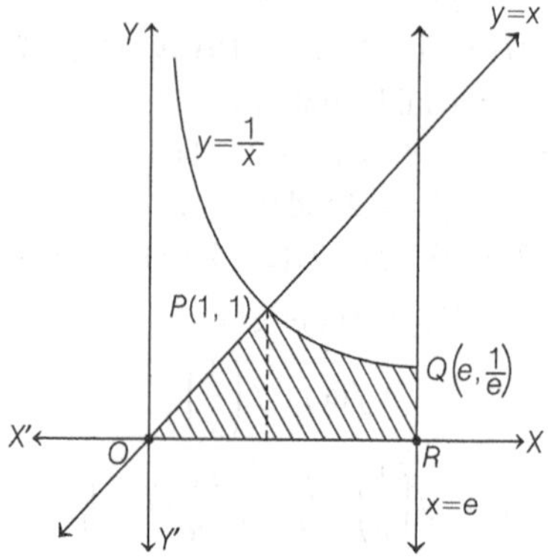

$$=\int_0^1 x\,dx+\int_1^e \frac{1}{x}dx=\frac{1}{2}+1$$

$$=\frac{3}{2}\text{ sq units}$$

71. *(98)* Let A and B denote the set of employees who like mangoes and apples respectively and total number of employees are 100.

$\because n(A)=70, n(B)=64, n(U)=100$ and $n(A\cap B)=x$

Now, $n(A\cup B)\le 100$

$\therefore n(A)+n(B)-n(A\cap B)\le 100$

$70+64-x\le 100$

$x\ge 34$ ...(i)

Again, $A\cap B\le B$

$\therefore n(A\cap B)\le n(B)$

$\Rightarrow \quad x\le 64$ ...(ii)

From Eqs. (i) and (ii), we get

$34\le x\le 64$

Then, $\lambda = 34$ and $\mu = 64$

$\therefore \lambda+\mu = 98$

72. *(4)* Let $f(x)=4x+8\cos x+\tan x-2\sec x-4\log[\cos x(1+\sin x)]-6$

Now, $f(0)=0+8+0-2-4\log 1-6=0$ and

$$f'(x)=4-8\sin x+\sec^2 x-2\sec x\tan x+4\tan x-\frac{4\cos x}{1+\sin x}$$

$$=4(1-2\sin x)+\sec^2 x(1-2\sin x)+\frac{4\sin x}{\cos x}-\frac{4\cos x[1-\sin x]}{\cos^2 x}$$

$$=(4+\sec^2 x)(1-2\sin x)+\frac{4}{\cos x}(\sin x-1+\sin x)$$

$$=(1-2\sin x)(4+\sec^2 x-4\sec x)$$

$$=(1-2\sin x)(\sec x-2)^2$$

Thus, $f'(x)>0$, if $1-2\sin x>0$ i.e. $x\in(0,\pi/6)$

Therefore, f increases on $[0,\pi/6]$.

But $f(0)=0$, so $f(x)\ge f(0), \forall x\in[0,\pi/6]$.

Thus, the largest such value of λ is $\frac{\pi}{6}$.

$\therefore K=4$

73. *(1)* Given circle

$$x^2+y^2-2(x+y)=0$$

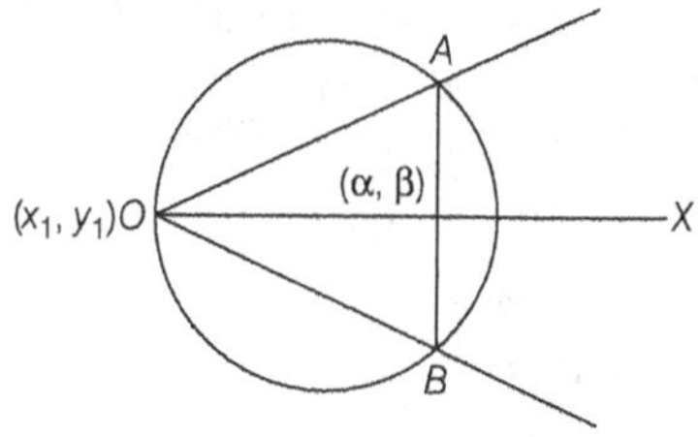

According to the question,

$$S_1=T$$

$\Rightarrow xx_1+yy_1-(x+x_1)-(y+y_1)=0$

$\Rightarrow \quad (x_1-1)x+(y_1-1)y=x_1+y_1$

$$\therefore x^2+y^2-2(x+y)\left[\frac{(x_1-1)x+(y_1-1)y}{x_1+y_1}\right]=0$$

This is equally inclined to X-axis.

$\Rightarrow$ Coefficient of $xy=0$

$\Rightarrow \frac{-2}{x_1+y_1}(y_1-1+x_1-1)=0 \Rightarrow x_1+y_1=2$

Locus of (x_1, y_1)

$$x+y=2$$

On comparing with $px+qy=1$

$\because \quad p=\frac{1}{2}, q=\frac{1}{2}$

$\therefore \quad p+q=\frac{1}{2}+\frac{1}{2}=1$

Study Tactics

The equation of chord of the circle

$S \equiv x^2 + y^2 + 2gx + 2fy + c = 0$

bisected at the point (x_1, y_1) is given by $T = S_1$

i.e. $xx_1 + yy_1 + g(x + x_1) + f(y + y_1) + c$

$= x_1^2 + y_1^2 + 2gx_1 + 2fy_1 + c$

74. *(5)* Let $\mathbf{b} = \alpha\hat{\mathbf{i}} + \beta\hat{\mathbf{j}}$ and $\mathbf{a} \perp \mathbf{b}$

$\Rightarrow \quad \mathbf{a} \cdot \mathbf{b} = 0$

$\Rightarrow (4\hat{\mathbf{i}} + 3\hat{\mathbf{j}}) \cdot (\alpha\hat{\mathbf{i}} + \beta\hat{\mathbf{j}}) = 0$

$\Rightarrow \quad 4\alpha + 3\beta = 0$

$\alpha = \frac{-3\beta}{4}$

$\Rightarrow \quad \frac{\alpha}{3} = \frac{\beta}{-4} = \lambda$ [let]

$\Rightarrow \quad \alpha = 3\lambda$ and $\beta = -4\lambda$ ...(i)

Let $\mathbf{c} = x\hat{\mathbf{i}} + y\hat{\mathbf{j}}$ be the required vectors, then as per question

Projection of $\mathbf{c}$ along $\mathbf{a} = \frac{\mathbf{c} \cdot \hat{\mathbf{a}}}{|\mathbf{a}|}$

$= \frac{4x + 3y}{\sqrt{4^2 + 3^2}} = 1 \Rightarrow 4x + 3y = 5$ ...(ii)

Also, projection of $\mathbf{c}$ along $\mathbf{b} = 2$

$\Rightarrow \quad \frac{\mathbf{c} \cdot \hat{\mathbf{b}}}{|\mathbf{b}|} = 2$

$\Rightarrow \quad \frac{\alpha x + \beta y}{\sqrt{\alpha^2 + \beta^2}} = 2$

$\Rightarrow \quad \frac{3\lambda x - 4\lambda y}{\sqrt{(3\lambda)^2 + (-4\lambda)^2}} = 2$

$\Rightarrow \quad 3\lambda x - 4\lambda y = 10\lambda$

$\Rightarrow \quad 3x - 4y = 10$ ...(iii)

On solving Eqs. (ii) and (iii), we get

$x = 2$ and $y = -1$

$\therefore \quad \mathbf{c} = 2\hat{\mathbf{i}} - \hat{\mathbf{j}}$

$|\mathbf{c}| = \sqrt{5} = \sqrt{K}$ (given)

$\Rightarrow \quad K = 5$

75. *(33)* The numbers are

1	6	11
2	7	12
3	8	13
4	9	14
5	10	15

$x^2 - y^2 = (x - y)(x + y)$ is (divisible by 5) multiple of 5.

$\Rightarrow$ Both x and y are from any of these rows or one from row 1 and the other from row 4 or one from row 2 and other from row 3.

$\therefore$ Desired number $= 5 \times {}^3C_2 + 2 \times {}^3C_1 \times {}^3C_1$

$= 15 + 18 = 33$

SECTION II

Mastery Builders

PRACTICE SET 13

Total Marks : 300	Total Time : 3 hrs	No. of Questions : 75

Instructions

1. There are **three** parts in the questions paper consisting of **Physics**, **Chemistry** and **Mathematics** having 25 questions in each part of equal weightage. Each part (subject) has two sections.
2. **Section A** This section contains 20 multiple choice questions which have only one correct answer. Each question carries **4 marks** for correct answer and – **1 mark** for wrong answer.
3. **Section B** This section contains 5 questions. The answer to each of the questions is a numerical value. Each question carries **4 marks** for correct answer and – **1 mark** for wrong answer. For Section B, the answer should be rounded off to the nearest integer.

PHYSICS

Section A : Objective Type Questions

1. When a H-atom in ground state is excited by monochromatic radiation of photon energy 12.75 eV. Then, the number of spectral lines will be (given its ionisation energy is 13.6 eV)

(a) one (b) two (c) four (d) six

2. Statement I When two charged spheres are connected to each other by thin conducting wire, charge flow from bigger sphere to smaller sphere, if initial charges on them are same.

Statement II Electrostatic energy will be lost in redistribution.

In the light of the above statements, choose the most appropriate answer from the options given below.

(a) Both Statement I and Statement II are incorrect .
(b) Statement I is correct but Statements II is incorrect.
(c) Statement I is incorrect but Statement II is correct.
(d) Both Statement I and Statement II are correct.

3. A gaseous mixture enclosed in a vessel consists of 1 gm mole of a gas A with $\gamma = 5/3$ and some amount of gas B with $\gamma = 7/5$ at a temperature T. The gases A and B do not react with each other and are assumed to be ideal. If γ for the gaseous mixture is 19/13, then the number (in gm moles) of the gas B is

(a) 2 (b) 4 (c) 3 (d) 8

4. Match the Column I with Column II and select the correct answer from the codes given below.

	Column I		Column II
A.	Energy density	1.	$C^2/N\text{-}m^2$
B.	Young's modulus	2.	Jm^{-3}
C.	Permittivity of free space	3.	Nm^{-2}
D.	Force constant of a spring	4.	Nm^{-1}

Codes

	A	B	C	D
(a)	2	4	1	3
(b)	1	2	3	4
(c)	3	2	4	1
(d)	2	3	1	4

5. Two wires carrying currents I_1 and I_2 are placed in a horizontal plane as given below. Then, magnetic field at point P in the same plane will be (given $OP = d$)

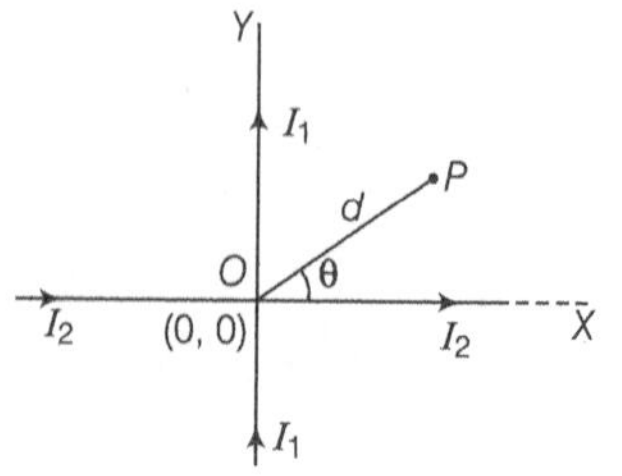

(a) $\frac{\mu_0}{2\pi d}[-I_1 \sin\theta + I_2 \cos\theta]\hat{k}$

(b) $\frac{\mu_0}{2\pi d}\left[\frac{I_1}{\sin\theta}+\frac{I_2}{\cos\theta}\right]\hat{k}$

(c) $\frac{\mu_0}{2\pi d}[-I_1 \sin\theta - I_2 \cos\theta]\hat{k}$

(d) $\frac{\mu_0}{2\pi d}\left[\frac{-I_1}{\cos\theta}+\frac{I_2}{\sin\theta}\right]\hat{k}$

6. Which of the following statement is correct ?
(a) Least count of screw gauge of pitch 0.5 mm and having 50 divisions on circular scale is 0.02 mm.
(b) Least count of vernier callipers with main scale division value of 1 mm and having 25 divisions on sliding scale is 0.04 mm.
(c) Both (a) and (b).
(d) None of the above

7. A gasoline engine takes in 5 moles of air at 20°C and 1 atm and compresses it adiabatically to (1/10) of the original volume. Assume, air to be diatomic. The work done and change in internal energy is
(a) 46 kJ, – 46 kJ (b) 36 kJ, – 36 kJ
(c) – 46 kJ, 46 kJ (d) 36 kJ, – 46 kJ

8. At 0.3V and 0.7V, the diodes Ge and Si become conductor respectively. In given figure, if ends of diode Ge overturned, the change in potential V_0 will be

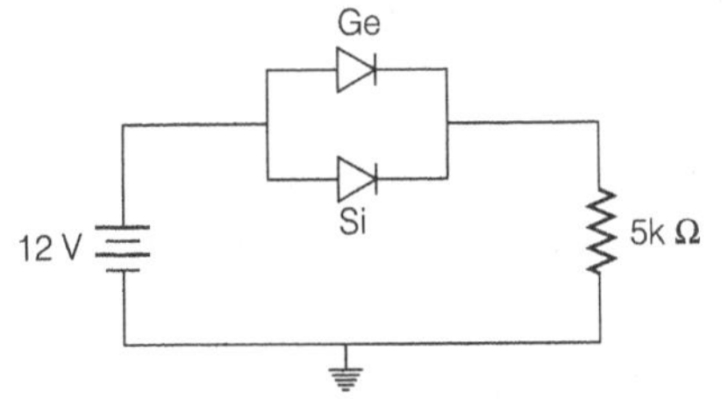

(a) 0.7 V (b) 0.6 V (c) 0.4 V (d) 0.8 V

9. A particle of mass m is executing oscillations about the origin on the X-axis. Its potential energy is $U(x) = k|x^3|$, where k is a positive constant. If the amplitude of oscillation is A, then its time period T is
(a) proportional to $\frac{1}{\sqrt{A}}$ (b) independent of A
(c) proportional to $\sqrt{A}$ (d) proportional to $A^{3/2}$

10. The following questions consist of two statements Assertion (A) and Reason (R).

Assertion A person standing on a rotating platform suddenly stretched his arms, the platform slows down.

Reason A person by stretching his arms increases the moment of inertia of the body and decreases angular velocity.

Answer these questions by selecting the appropriate option given below.
(a) Both A and R are true and R is the correct explanation of A.
(b) Both A and R are true but R is not the correct explanation of A.
(c) A is true but R is false.
(d) A is false but R is true.

11. The electric field in a region is given by

$$E = (6\hat{i} + 8\hat{j})\text{ N/kg}$$

If the potential at the origin is taken to be zero, then the ratio of the potential at the points (8m, 0) and (0, 6m) is
(a) 0 (b) 1 (c) 2 (d) $\frac{1}{2}$

12. A star has a supply of 10^{50} deuterons. It produces energy *via* a fusion reaction

$${}_1^2\text{H} + {}_1^2\text{H} \longrightarrow {}_1^3\text{H} + p$$

$$\text{and } {}_1^2\text{H} + {}_1^3\text{H} \longrightarrow {}_2^4\text{He} + n$$

where, masses of nuclei are
$m(^2\text{H}) = 2.014$ amu, $m(p) = 1.007$ amu,
$m(n) = 1.008$ amu and $m(^4\text{He}) = 4.001$ amu.
If the average power radiated by the star is 10^{16} W, then the deuteron supply of the star is exhausted in a time of the order of (1 amu = 931 MeV)
(a) 10^{22} s (b) 10^{18} s (c) 10^{8} s (d) 10^{12} s

13. In two different experiments, an object of mass 10 kg moving with a speed of 25 ms^{-1} hits two different walls and come to rest within
(i) 4 s and
(ii) 8 s respectively, some statements are given below.
Choose the correct set of statements.
I. Change in momentum is same in both cases.
II. Impulse imparted is same in both cases.
III. F_{avg} in case (ii) is more than in case (i)
(a) Only I is true (b) Only II is true
(c) I and II both are true (d) I, II and III all are true

14. The current flowing through the zener diode is

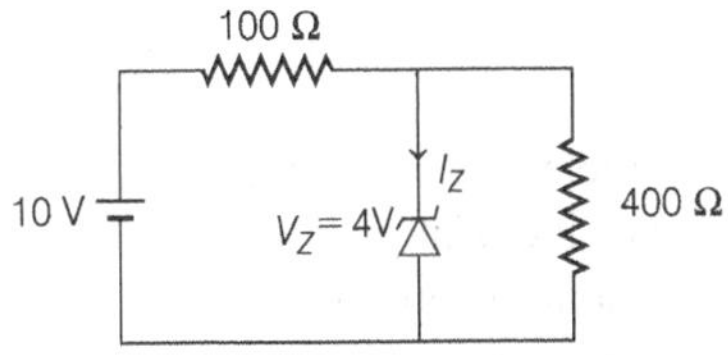

(a) 50 mA (b) 100 mA
(c) 25 mA (d) 40 mA

15. A small body of mass m slides over a hemisphere of radius 3 m. At what height from the ground, the body will detach itself from hemisphere?

(a) $\frac{2}{3}$ m (b) 2 m (c) $\frac{3}{5}$ m (d) $\frac{5}{3}$ m

16. Two light bulbs shown in the circuit are having ratings P (12 V, 12 W) and Q (12 V, 36 W) as shown in the figure. When the switch is closed, then

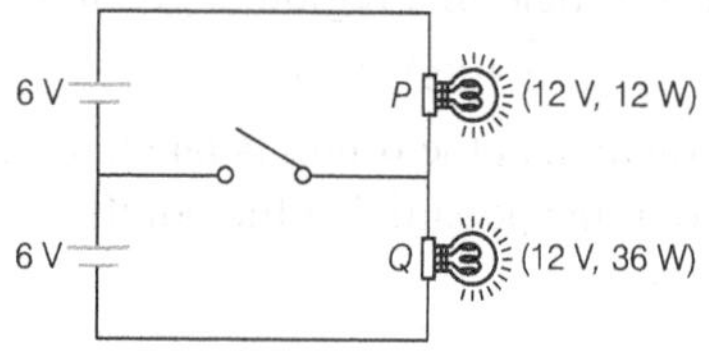

(a) the intensity of light bulb P increases
(b) the intensity of light bulb P decreases
(c) the intensity of light bulb Q remains unchanged
(d) the intensity of light bulb Q decreases

17. A plane electromagnetic wave has frequency 5.0×10^{10} Hz and its energy density is 2.04×10^{-8} J / m^3 in vacuum. The amplitude of magnetic field of the wave is

(Take, $\frac{1}{4\pi\varepsilon_0}=9\times10^9$ N-m^2/C^2, speed of light $=3\times10^8$ m/s)

(a) 225.4 nT (b) 226.4 nT
(c) 375.8 nT (d) 250.6 nT

18. Figure shows an electric line of force which curves along a circular arc. The magnitude of electric field intensity is same at all points on this curve and is equal to E. If the potential at A is V, then the potential at B is

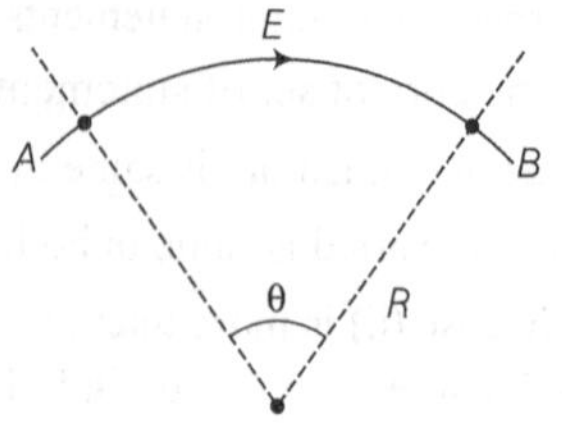

(a) $V - ER\theta$ (b) $V - 2ER\sin\frac{\theta}{2}$
(c) $V + ER\theta$ (d) $V + 2ER\sin\frac{\theta}{2}$

19. If in a photoelectric experiment, intensity and frequency of incident photons are both doubled, then the saturation current

(a) remains unchanged (b) increases slightly
(c) decreases slightly (d) doubles

20. A light ray is incident perpendicular to one face of a 90° prism and is totally internally reflected at the glass-air interface. If the angle of reflection is 45°, we conclude that the refractive index n,

(a) $n>\sqrt{2}$ (b) $n<\sqrt{2}$
(c) $n>\frac{1}{\sqrt{2}}$ (d) $n<\frac{1}{\sqrt{2}}$

Section B : Numerical Value Type Questions

21. A metal cube of side 15 cm is subjected to a stress of 10^5 N/m^2. If the top of the cube is displaced by 0.1 cm with respect to the bottom, then the modulus of rigidity is $p\times10^6$ N/m^2. The value of p is ... .

22. For a optical arrangement shown in the figure, the height of the image (in mm) is ....

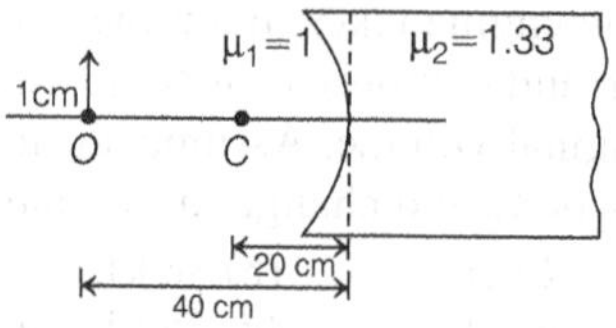

23. If a body is projected from surface of Earth with a velocity $\frac{1}{Z\text{th}}$ of escape velocity, such that maximum height achieved is $\frac{R}{8}$ (where, R is radius of Earth), then the value of Z is ....

24. Two cricket balls P and Q thrown with speeds in the ratio $1:\sqrt{3}$ acquired the same heights. If P is thrown at an angle of 45° with the horizontal, then the angle of projection is found to be $\sin^{-1}\frac{1}{\sqrt{m}}$, then the value of m is ....

25. A metallic rod of length l is tied to a string of length $2l$ and made to rotate with angular speed ω on a horizontal table with one end of the string fixed. If there is a vertical magnetic field B in the region, the emf induced across the ends of the rod is $N\frac{B\omega l^2}{2}$. Then, the value of N is ....

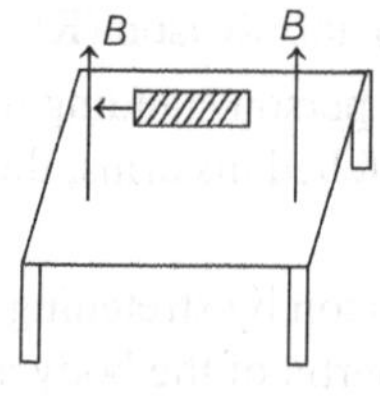

CHEMISTRY

Section A : Objective Type Questions

26. Find out the major products from the following reactions.

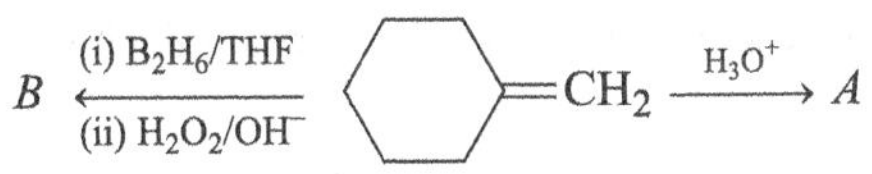

(a) $A=$ 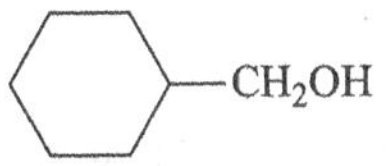 , $B=$ 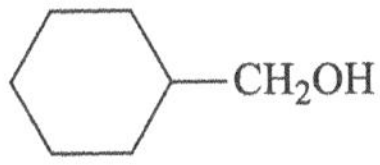

(b) $A=$ 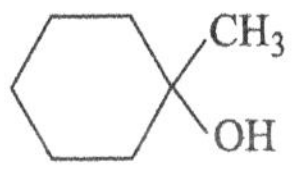 , $B=$ 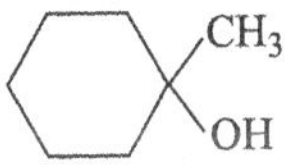

(c) $A=$ 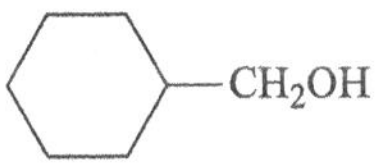 , $B=$ 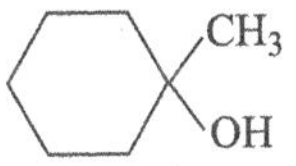

(d) $A=$ 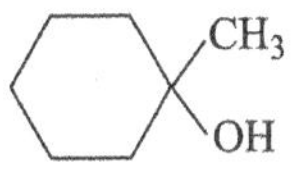 , $B=$ 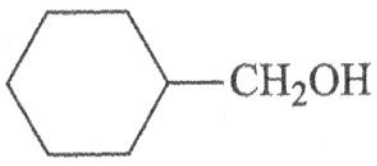

27. Which of the following carbohydrate will form same osazone?

(a) Glucose and fructose
(b) Glucose and gluconic acid
(c) Mannose and sucrose
(d) Fructose and maltose

28. Which one of the following compounds has the electron pair geometry as the trigonal bipyramidal with three equatorial positions occupied by lone pair of electrons?

(a) XeF_2
(b) $AlCl_3$
(c) $[Pt(NH_3)_2Cl_2]$
(d) $CH_3—Mg—Br$

29. Which of the following is incorrect statement?

(a) Ions of d-block elements are coloured due to d - d transition.
(b) Ions of f-block elements are coloured due to f - f transition.
(c) $[Sc(H_2O)_6]^{3+}$, $[Ti(H_2O)_6]^{4+}$ are coloured complexes.
(d) Cu^+ is colourless ion.

30. Consider the following reaction,

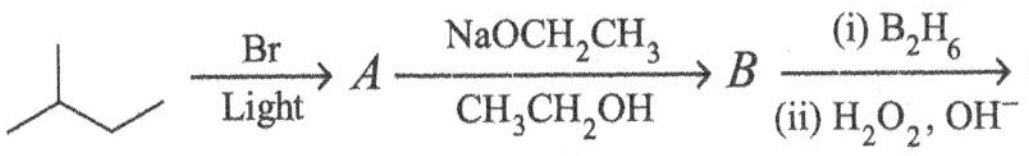

Major product formed by the above sequence of reactions is

(a) 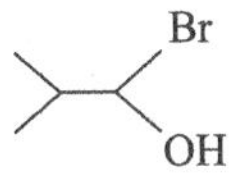 (b) 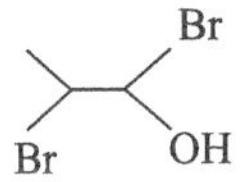

(c) 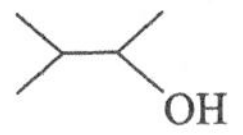 (d) 

31. A student has studied the decomposition of a gas XY_3 at 25°C. She obtained the following data.

P(mm Hg)	25	50	100	200
Relative $t_{1/2}(s)$	8	4	2	1

The order of the reaction is

(a) 1 (b) 0 (c) 2 (d) 3

32. For the following reactions, identify the salt (X) that satisfy these properties.

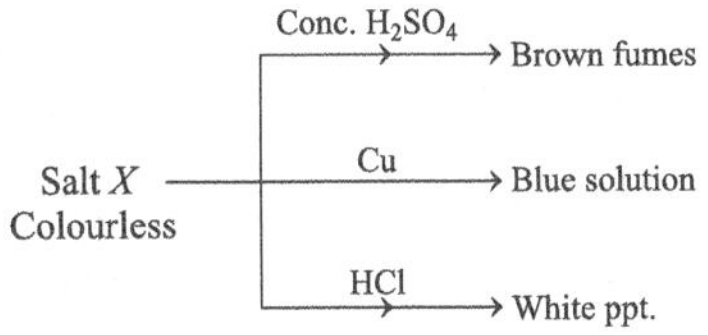

(a) $AgNO_3$ (b) $Cu(NO_3)_2$
(c) $Pb(NO_3)_2$ (d) $NaNO_3$

33. Benzoyl chloride on reaction with H_2 in presence of $Pd/BaSO_4$ produces a compound A. Which of the following will be the product, if A is treated with KCN in aqueous alcohol?

(a) 1- hydroxy-1,2-diphenylethan-1-one
(b) 2- hydroxy-1,2-diphenylethan-1-one
(c) 2- hydroxy-2,2-diphenylethan-2-one
(d) 2- hydroxy-1,1-diphenylethan-2-one

34. Given below are two statements.

Statement I Allyl isocyanide has 8σ and 3π-bonds.

Statement II The number of stereoisomers obtained by bromination of trans-2-butene is 1.

In the light of the above statements choose the correct answer from the options given below.

(a) Both Statement I and Statement II are correct.
(b) Both Statement I and Statement II are incorrect.
(c) Statement I is correct but Statement II is incorrect.
(d) Statement I is incorrect but Statement is correct.

35. Decreasing order of dehydration is

(A) 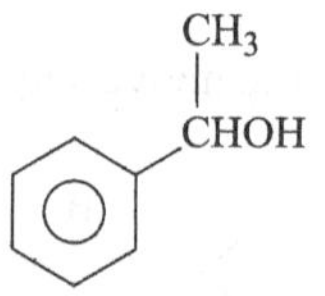 (B) 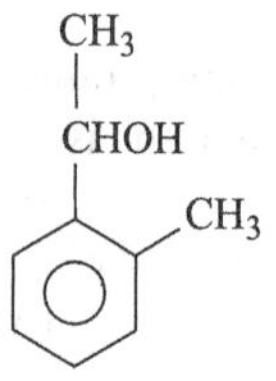

(C) 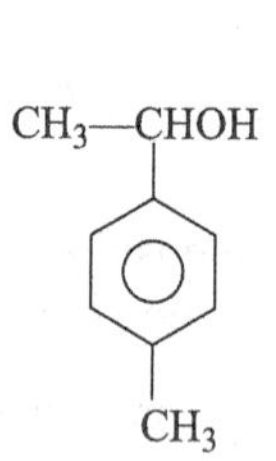 (D) 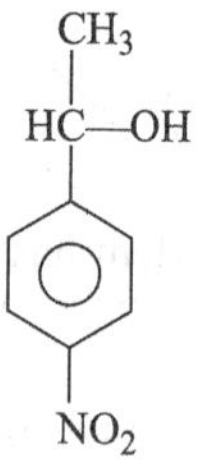

Choose the correct answer from the options given below.

(a) A > B > C > D (b) B > C > A > D
(c) D > A > C > B (d) D > A > B > C

36. Match List-I with List-II.

List-I (Complex)		List-II (Isomerism)	
A.	$Cr(NH_3)_6[Co(CN)_6]$	I.	Linkage isomerism
B.	$[Ni(NH_3)_3(NO_2)_3]$	II.	Solvate isomerism
C.	$[Fe(H_2O)_6]Cl_3$	III.	Co-ordination isomerism
D.	*cis*-$[CoCl_2(ox)_2]^{3-}$	IV.	Optical isomerism

Choose the correct answer from the options given below.

(a) A-II, B-I, C-III, D-IV
(b) A-IV, B-II, C-III, D-I
(c) A-III, B-I, C-II, D-IV
(d) A-I, B-II, C-III, D-IV

37. Consider the statements given below.

Statement I Enzyme activators can be considered as positive allosteric modulators for receptors.

Statement II When a vitamin is present along the enzyme, the catalytic activity is increased considerably.

In the light of the above statements choose the correct answer from the options given below.

(a) Both Statements I and Statement II are correct.
(b) Both Statement I and Statement II are incorrect.
(c) Statement I is correct but Statement I is incorrect.
(d) Statement I is incorrect but Statement II is correct.

38. Given below are oxides of some elements.

I. CuO II. Mn_2O_7 III. CO IV. SO_2 V. CrO_3

The total number of oxides that will be acidic

(a) 2 (b) 1
(c) 3 (d) 4

39. Choose the correct product formed during following reaction,

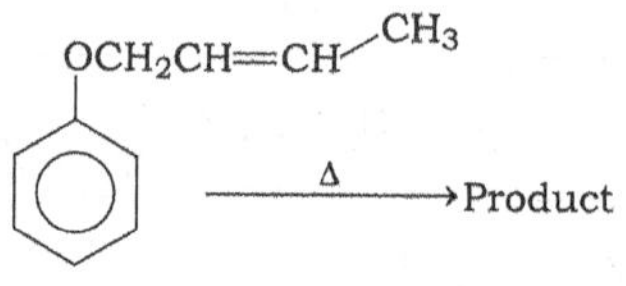

(a) 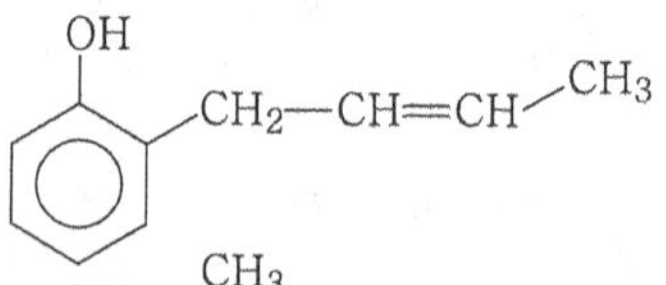

(b) 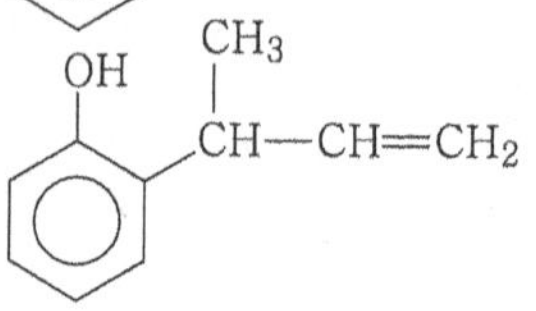

(c) 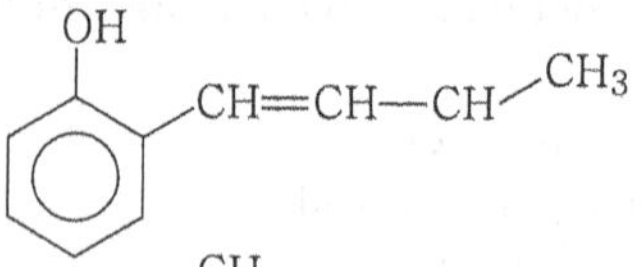

(d) 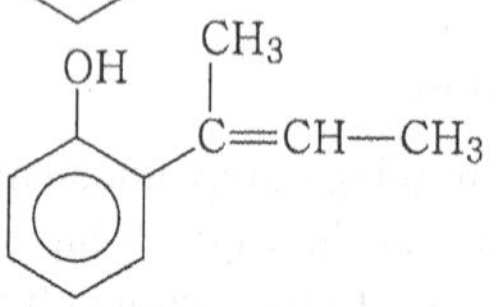

40. The maximum number of P—H bonds are contained in which of the following molecules?

(a) H_3PO_2 (b) H_3PO_3
(c) H_3PO_4 (d) $H_4P_2O_7$

41. Given below are two statements one is labelled as Assertion (A) and other is labelled as Reason (R).

Assertion (A) In water, orthoboric acid behaves as a weak monobasic acid.

Reason (R) In water, orthoboric acid acts as a proton donor.

In the light of the above statements choose the correct answer from the options given below.

(a) Both (A) and (R) are true and (R) is the correct explanation of (A).
(b) Both (A) and (R) are true but (R) is not the correct explanation of (A).
(c) (A) is true but (R) is false.
(d) (A) is false but (R) is true.

42. Match List-I with List-II.

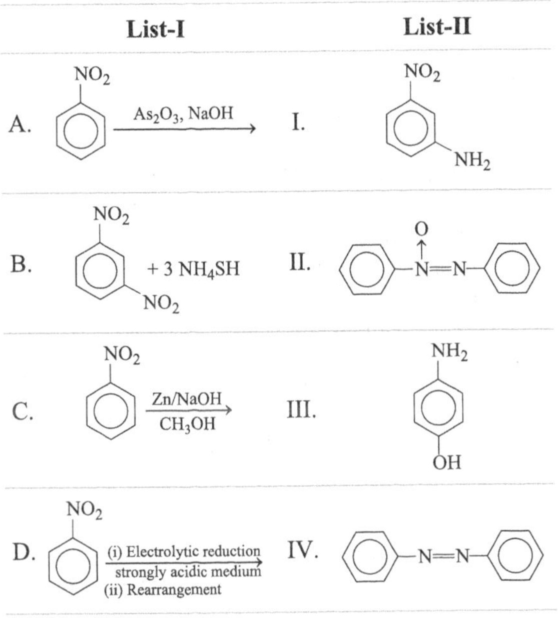

Choose the correct answer from the options given below.

(a) A-IV, B-I, C-II, D-III (b) A-II, B-I, C-IV, D-III
(c) A-III, B-II, C-I, D-IV (d) A-II, B-IV, C-III, D-I

43. Order of atomic radii,

A. Nd > Ce > Eu > Ho
B. Ho < Nd < Ce < Eu
C. Ce < Sm < Yb < Lu
D. Lu < Yb < Sm < Ce
E. Nd > Ce < Eu > Ho

Choose the correct answer from the given options below.

(a) B and D (b) A, D and E
(c) C and D (d) B, C and D

44. Equilibrium constants K_A and K_B for the following equilibrium

$NO(g) + \frac{1}{2}O_2 \overset{K_A}{\rightleftharpoons} NO_2\,(g)$ and

$2\,NO_2(g) \overset{K_B}{\rightleftharpoons} 2NO\,(g) + O_2\,(g)$ are related as,

(a) $K_A = \frac{1}{K_B}$ (b) $K_B = \frac{1}{\sqrt{K_A}}$
(c) $K_B = \frac{1}{K_A^2}$ (d) $K_A = \frac{1}{K_B^2}$

45. Which of the following statement is not correct?

(a) Absorption peak for $[Cr(NH_3)_6]^{3+}$ is 21680 cm^{-1}.
(b) Effective atomic no. of Pt in $[Pt\,Cl_6]^{2-}$ is 84.
(c) Crystal field stabilisation energy of d^2 in weak ligand field is $(-)0.8\Delta_0$.
(d) Example of weak ligand for d^5 configuration is $[MnF_6]^{4-}$.

Section B : Numerical Value Type Questions

46. 1 mole of CO_2 gas at 400 K is expanded under adiabatic conditions such that its volume becomes 16 times. The work done is ...... cal.
($\gamma = 1.50$ and $C_V = 6$ cal mol^{-1} for CO_2)

47. On mixing, heptane and octane form an ideal solution at 373 K, the vapour pressures of the two liquid components (heptane and octane) are 105 kPa and 45 kPa respectively. Vapour pressure of the solution obtained by mixing 25 g of heptane and 35 g of octane will be ......kPa. (Molar mass of heptane = 100 g mol^{-1} and of octane = 114 g mol^{-1})

48. A hydrogen atom in its ground state is exposed to radiation by the light of wavelength 970 Å. The ground state energy of hydrogen atom is −13.6 eV. The number of lines present in the emission spectrum is ....
[Given, $hc/e = 1237 \times 10^{-6}$ eV–m]

49. For the cell,
$Zn(s)|Zn^{2+}(aq, 1M)||Cu^{2+}(aq, 1M)|Cu(s)$ the standard emf of the cell is 1.70 V at 300 K. When the concentration of Zn^{2+} is changed to yM, the cell potential changes to 1.67 V. The value of y is ......M.
[Given, $F/R = 11500$ kV^{-1}, $\ln 10 = 2.30$]

50. Among the following the number of aromatic compounds is ....

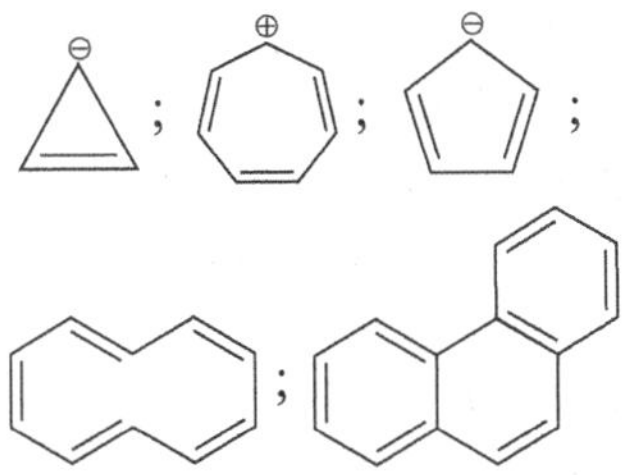

MATHEMATICS

Section A : Objective Type Questions

51. The angle between the lines whose direction cosines has the relation $l+m+n=0$ and $2l^2+2m^2-n^2=0$ is

(a) $\cos^{-1}\left(\frac{1}{3}\right)$ (b) $\cos^{-1}\left(\frac{-2}{3}\right)$

(c) $\cos^{-1}\left(\frac{-1}{3}\right)$ (d) $\cos^{-1}\left(\frac{\sqrt{3}}{2}\right)$

52. Let $p(x)$ be a polynomial with real coefficient such that $(x^3-\alpha^3)$ is a factor of $p(x)-p(\alpha)$ for each positive real number α, then $p(x)$ is the form of

(a) $f(x^3)$ (b) $xf(x^3)$

(c) $x^2f(x^3)$ (d) a constant

53. Three numbers are in AP with common difference d are removed from first n natural numbers and average of remaining numbers is found to be $\frac{43}{4}$, then ordered pair (n, d) can be

(a) (23, 5) (b) (19, 8) (c) (19, 5) (d) (23, 8)

54. If $A=\begin{bmatrix}\frac{\sqrt{3}}{2} & \frac{1}{2}\\ \frac{-1}{2} & \frac{\sqrt{3}}{2}\end{bmatrix}$, $B=\begin{bmatrix}1 & 1\\ 0 & 1\end{bmatrix}$ and $C=ABA^T$ and $x=A^TC^{2023}A$, then x is equal to

(a) $\begin{bmatrix}1 & 2023\\ 0 & 1\end{bmatrix}$

(b) $\begin{bmatrix}4+2023\sqrt{3} & 6150\\ 2025 & 4-2023\sqrt{3}\end{bmatrix}$

(c) $\frac{1}{4}\begin{bmatrix}2+\sqrt{3} & 1\\ -1 & 2-\sqrt{3}\end{bmatrix}$

(d) $\frac{1}{4}\begin{bmatrix}2023 & 2-\sqrt{3}\\ 2+\sqrt{3} & 2023\end{bmatrix}$

55. For $f(x)=\int_x^{x^2}\frac{du}{(\log u)^2}$, $x\neq 0, 1$, then consider the following statements for $f(x)$.

Statement I $f(x)$ is monotonically increasing in the interval $(2, \infty)$.

Statement II $f(x)$ is monotonically decreasing in the interval (0, 1).

(a) Statement I is true and Statement II is false
(b) Statement I is false and Statement II is true
(c) Statement I and Statement II both are true
(d) Statement I and Statement II both are false

56. Let $f(x)=\dfrac{2x(\sin x+\tan x)}{2\left[\frac{x+21\pi}{\pi}\right]-41}$, $x\neq n\pi$

(where, $[\cdot]$ represents greatest integer function), then f is

(a) an odd function
(b) an even function
(c) Both odd and even
(d) Neither odd nor even

57. If $4y=-3x$ and $\sqrt{3}y=x$ are the equations of a pair of conjugate diameters of an ellipse, then the eccentricity of the ellipse is

(a) $\sqrt{\frac{4+\sqrt{3}}{4}}$ (b) $\frac{\sqrt{4-\sqrt{3}}}{2}$

(c) $\sqrt{\frac{2-\sqrt{3}}{2}}$ (d) None of these

58. Let $\mathbf{a}=2\hat{\mathbf{i}}+8\hat{\mathbf{j}}+4\hat{\mathbf{k}}$, $\mathbf{b}=6\hat{\mathbf{i}}-4\hat{\mathbf{j}}+14\hat{\mathbf{k}}$ and $\mathbf{c}=4\hat{\mathbf{i}}-2\hat{\mathbf{j}}+8\hat{\mathbf{k}}$. If a vector $\mathbf{d}$ satisfies $\mathbf{d}\times\mathbf{b}=\mathbf{c}\times\mathbf{b}$ and $\mathbf{d}\cdot\mathbf{a}=32$, then $9\mathbf{d}$ is equal to

(a) $48\hat{\mathbf{i}}-26\hat{\mathbf{j}}+100\hat{\mathbf{k}}$ (b) $89\hat{\mathbf{i}}-48\hat{\mathbf{j}}+92\hat{\mathbf{k}}$

(c) $62\hat{\mathbf{i}}-72\hat{\mathbf{j}}+48\hat{\mathbf{k}}$ (d) $62\hat{\mathbf{i}}+88\hat{\mathbf{j}}-72\hat{\mathbf{k}}$

59. If A and B represents the complex numbers z_1 and z_2 such that $|z_1+z_2|=|z_1-z_2|$, then the circumcentre of ΔOAB, where O is origin, is

(a) $\frac{z_1+z_2}{2}$ (b) $\frac{z_1-z_2}{2}$

(c) $\frac{z_1+z_2}{3}$ (d) $\frac{z_1-z_2}{3}$

60. The parabola $y^2-px+12=0$ has directrix as the straight line $x=2$. Then, if p_1 and p_2 represent the two possible values of p (where $p_1, p_2\in R$, $p_1<p_2$), then $\left(\frac{p_1}{p_2}\right)$ is

(a) 3 (b) $\frac{1}{3}$ (c) -3 (d) $-\frac{1}{3}$

61. Let $\sqrt{d}$ be the distance between the point $A(2, 3, 4)$ and the intersecting point of line $L_1:\frac{x-1}{2}=\frac{y-2}{3}=\frac{z-3}{4}$ and line $L_2:\frac{x-4}{5}=\frac{y-1}{2}=\frac{z}{1}$ and $d=63-\sqrt{\mu}$. Then, μ must be equal to

(a) 246 (b) 169 (c) 244 (d) 269

62. The mean and variance of seven observations are 8 and 16, respectively. If 5 of the observations are 2, 4, 10, 12 and 14, then the product of the remaining two observations is

(a) 45 (b) 49
(c) 48 (d) 40

63. Let $< T_n >$ be an arithmetic progression with common difference $d (d \neq 0)$ and $< S_n >$ be a geometric progression with common ratio r, where r is a rational number and $0 < r < 1$. If $T_1 = d$, $S_1 = d^2$ and $\dfrac{T_1^2 + T_2^2 + T_3^2}{S_1 + S_2 + S_3}$ is a positive integer, then $6r$ is

(a) 2 (b) 3 (c) 6 (d) 9

64. The area of the greatest isosceles triangle that can be inscribed in a given ellipse having its vertex coincident with one end of the major axis, is

(a) $\dfrac{\sqrt{3}}{4} ab$ (b) $\dfrac{\sqrt{3}}{2} ab$
(c) $\dfrac{3\sqrt{3}}{4} ab$ (d) $\sqrt{3}ab$

65. The contents of 3 bags with respect to green and red marbles is as given in the table below.

Bag	*A*	*B*	*C*
Green	3	2	1
Red	1	2	3

A person randomly select one of the bags and draws a marble from it and retains it. If the marble is green, the persons draws the second marbles randomly from one of the two remaining bags.
If the first marble drawn is red the person draws one more marble from the same bag. The probability that the second drawn marble is green, is

(a) $\dfrac{73}{144}$ (b) $\dfrac{79}{140}$ (c) $\dfrac{63}{134}$ (d) $\dfrac{56}{137}$

66. Suppose $f(x) = ax + b$ and $g(x) = bx + a$, where a and b are positive integers.
If $f[g(20)] - g[f(20)] = 28$, then $a + b$ is

(a) 6 (b) 8
(c) 9 (d) 13

67. Let $f(x) = [\tan x (\cot x)]$, $x \in \left[\dfrac{\pi}{12}, \dfrac{\pi}{2}\right]$, where $[\cdot]$ denotes the greatest integer less than or equal to x. Then, the number of points, where $f(x)$ is discontinuous, is equal to

(a) 1 (b) 0 (c) 3 (d) ∞

68. The coefficient of x^{11} in the expansion of $(1 - 2x + 3x^2)(1 + x)^{11}$, is

(a) 196 (b) 64
(c) 121 (d) 144

69. If $\dfrac{a}{b}$ is the area of the region bounded by the curves $y = ex \ln x$ and $y = \dfrac{\ln x}{ex}$, where $\ln e = 1$, then $a + b$ is equal to

(a) $e^2 + 4e - 5$ (b) $e^2 + 4e + 5$
(c) $e^2 - 2e - 5$ (d) $e^2 - 4e + 5$

70. $f(x)$ is a continuous function for all real values of x and satisfies $\int_n^{n+1} f(x)\, dx = \dfrac{n^2}{2}$, $\forall\, n \in I$, then $\int_{-3}^{5} f(|x|)\, dx$ is equal to

(a) $\dfrac{19}{2}$ (b) $\dfrac{35}{2}$
(c) $\dfrac{17}{2}$ (d) None of these

Section B : Numerical Value Type Questions

71. The total number of even numbers with 3-digits such that, if 5 is one of the digits, then 7 is the next digit is equal to ....

72. If a function f satisfies the following conditions for all real values of x and y

(i) $f(x + y) = f(x) f(y)$
(ii) $f(x) = 1 + xg(x)$, where $\lim_{x \to 0} g(x) = 1$.

Then, the value of $\log f(10)$ is ....

73. If $T = \sum_{r=1}^{90} \tan^{-1}\left(\dfrac{2r}{2 + r^2 + r^4}\right)$, then $8190(\cot T - 1)$ is equal to ....

74. Line $y = 3x - b$ cuts the parabola $y = x^2 - 8x$ at points A and B. Then, the value of b for which the $\angle AOB$ is a right angle is equal to ....

75. Suppose $f(x)$ and $g(x)$ are differentiable functions such that $x \cdot g[f(x)] \cdot f'[g(x)] \cdot g'(x) = f[g(x)] \cdot g'[f(x)] \cdot f'(x)$ for all real x and $\int_0^a f[g(x)]\, dx = 1 - \dfrac{e^{-2a}}{2}$ for all real a. Given that $g[f(0)] = 1$. If the value of $g[f(4)] = e^{-4k}$, where $(k \in N)$, then k is equal to ....

DETAILED SOLUTIONS

Physics

1. *(d)* Energy of electron in ground state

$$= -13.6 \text{ eV.}$$

Final energy $= -13.6 + 12.75$

$$= -0.85 \text{ eV}$$

If n be the value of excited state, then

$$-\frac{13.6}{n^2} = -0.85$$

$$\Rightarrow \quad n^2 = 16$$

$$\Rightarrow \quad n = 4$$

Now, the number of spectral lines are

$$= \frac{n(n-1)}{2} = \frac{4 \times 3}{2} = 6 \text{ lines}$$

Study Tactics

Energy of the ground state is negative, since the electron absorbs total energy of the incident photon, hence, final energy of the electron is sum of ground state energy and energy of photon.

2. *(c)* Positive charge flow from higher potential to lower potential and also potential $\propto 1/r$ for sphere. So, smaller sphere have large potential.
During redistribution of charge, current will flow through the wire. Thus, electrical energy will be lost in the form of heat and light (spark).

3. *(a)* As for an ideal gas, $C_p - C_V = R$ and $\gamma = C_p / C_V$

So, $C_V = \dfrac{R}{(\gamma - 1)}$

For gas A,

$$\therefore \quad (C_V)_1 = \frac{R}{(5/3) - 1} = \frac{3}{2}R,$$

For gas B,

$$(C_V)_2 = \frac{R}{(7/5) - 1} = \frac{5}{2}R$$

and $(C_V)_{\text{mix}} = \dfrac{R}{(19/13) - 1} = \dfrac{13}{6}R$

Now, from conservation of energy,

$$\Delta U = \Delta U_1 + \Delta U_2$$

$$\Rightarrow (\mu_1 + \mu_2)(C_V)_{\text{mix}} \Delta T = [\mu_1 (C_V)_1 + \mu_2 (C_V)_2] \Delta T$$

i.e. $\quad (C_V)_{\text{mix}} = \dfrac{\mu_1 (C_V)_1 + \mu_2 (C_V)_2}{\mu_1 + \mu_2}$

We have, $\dfrac{13}{6}R = \dfrac{1 \times \frac{3}{2}R + \mu_2 \frac{5}{2}R}{1 + \mu_2} = \dfrac{(3 + 5\mu_2)R}{2(1 + \mu_2)}$

or $13 + 13\mu_2 = 9 + 15\mu_2$, i.e. $\mu_2 = 2$ gm mole

4. *(d)* (A) Energy density $= \dfrac{\text{Energy}}{\text{Volume}}$, so its unit $= \dfrac{\text{J}}{\text{m}^3} = \text{Jm}^{-3}$

(B) Young's modulus $= \dfrac{\text{Longitudinal stress}}{\text{Longitudinal strain}}$

$$= \frac{F/A}{\Delta l / l}, \text{ so its unit} = \frac{\text{N}}{\text{m}^2} = \text{Nm}^{-2}$$

(C) Permittivity of free space

Electric force, $F = \dfrac{1}{4\pi\varepsilon_0} \cdot \dfrac{q_1 q_2}{r^2} \Rightarrow \varepsilon_0 = \dfrac{q_1 q_2}{4\pi\varepsilon_0 F \cdot r^2}$

$\therefore$ So, unit of $\varepsilon_0 = \dfrac{\text{C}^2}{\text{N-m}^2}$

(D) Force constant, $k = \dfrac{F}{x} = \dfrac{\text{N}}{\text{m}} = \text{Nm}^{-1}$

5. *(d)* The magnetic field due to a long infinite wire at a perpendicular distance d is given by $|\mathbf{B}| = \dfrac{\mu_0 I}{2\pi d}$ and its direction is given by right hand thumb rule.
Magnetic field due to wire I_1 is pointed in inwards direction (according to right hand thumb rule)

$$\mathbf{B}_1 = \frac{\mu_0 I_1}{2\pi (d \cos\theta)} \cdot (-\hat{\mathbf{k}})$$

[$d\cos\theta$ gives perpendicular distance from first wire.]
Magnetic field due to wire I_2 will be pointed in upwards direction.

So, $\quad \mathbf{B}_2 = \dfrac{\mu_0 I_2}{2\pi (d \sin\theta)} \cdot (\hat{\mathbf{k}})$

Hence, net magnetic field is $\mathbf{B} = \mathbf{B}_1 + \mathbf{B}_2$

$$\Rightarrow \quad \mathbf{B} = \left[\frac{-\mu_0 I_1}{2\pi (d\cos\theta)} + \frac{\mu_0 I_2}{2\pi (d\sin\theta)}\right]\hat{\mathbf{k}}$$

$$= \frac{\mu_0}{2\pi d}\left[\frac{-I_1}{\cos\theta} + \frac{I_2}{\sin\theta}\right]\hat{\mathbf{k}}$$

Study Tactics

Use $B = \dfrac{\mu_0 I}{2\pi d}$ to find the individual magnetic field produced by I_1 and I_2. Here, d is the perpendicular distance of point P from the wire. Then, we use right hand formula to find the direction of magnetic field. Finally, add both the magnetic field using vector addition.

6. *(b)* Least count of screw gauge

$$= \frac{\text{Pitch}}{\text{Number of divisions on circular scale}}$$

$$= \frac{0.5\text{ mm}}{50} = 0.01 \text{ mm}$$

Least count of vernier callipers

$$= \frac{\text{Value of one division on main scale}}{\text{Number of divisions on sliding scale}}$$

$$= \frac{1\text{ mm}}{25} = 0.04\text{ mm}$$

Hence, option (b) is correct.

7. *(c)* Given, $p_1 = 1\,\text{atm}, n = 5\,\text{mol},$

$$T_1 = 293\,\text{K}$$

$$V_2 = \frac{V_1}{10}$$

For diatomic ideal gas, the degree of freedom, $f = 5$

The ratio of specific heat of gas, $\gamma = 1 + \frac{2}{f}$

$$\gamma = 1 + \frac{2}{5} = \frac{7}{5}$$

Using $T_1 V_1^{\gamma-1} = T_2 V_2^{\gamma-1}$

$$\Rightarrow \quad T_2 = T_1 \left(\frac{V_1}{V_2}\right)^{\gamma-1} = 293(10)^{0.4} = 736\,\text{K}$$

$$\text{Work done} = \frac{nR(T_1 - T_2)}{\gamma - 1}$$

$$= \frac{5 \times 8.3 \times (293 - 736)}{0.4} = -46\,\text{kJ}$$

$$\Delta U = \Delta Q - W = 0 - W = 46\,\text{kJ}$$

8. *(c)* Initially Ge and Si are both forward biased. So, current will effectively pass through Ge diode with a voltage drop of 0.3 V.

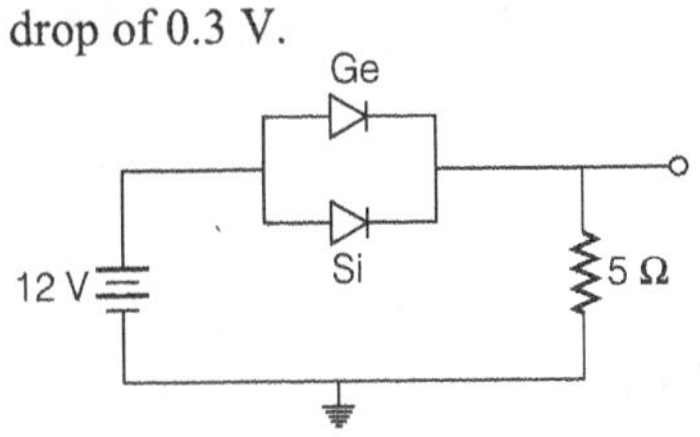

$\therefore$ Initial output voltage, $V_0 = 12 - 0.3 = 11.7$ V

If Ge is reversed biased, then only Si diode will work.

In this condition, output voltage, $V_0 = 12 - 0.7 = 11.3$ V

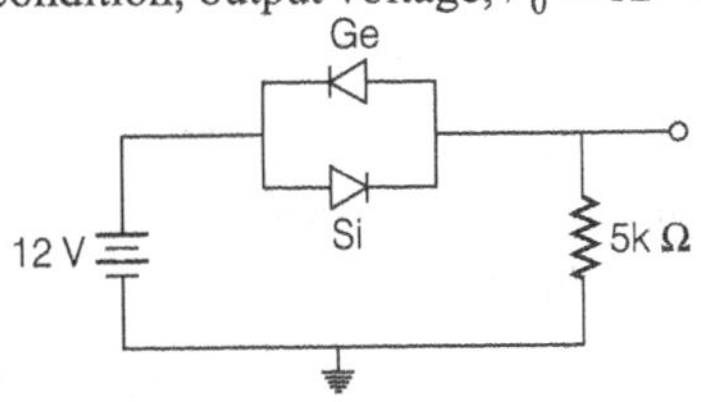

$\therefore$ Change in output voltage $= 11.7 - 11.3 = 0.4$ V

9. *(a)* $U(x) = k|x|^3 \Rightarrow F = -\frac{dU}{dx} = -3k|x|^2$ ...(i)

Also, for SHM,

$$x = A \sin \omega t \text{ and } \frac{d^2x}{dt^2} + \omega^2 x = 0$$

$$\Rightarrow \quad \text{Acceleration, } a = \frac{d^2x}{dt^2} = -\omega^2 x$$

$$\Rightarrow \quad F = ma = \frac{md^2x}{dt^2} = -m\omega^2 x \quad \text{...(ii)}$$

From Eqs. (i) and (ii), we get

$$\omega = \sqrt{\frac{3kx}{m}}$$

$$\because \quad T = \frac{2\pi}{\omega}$$

$$\Rightarrow \quad T = 2\pi\sqrt{\frac{m}{3kx}} = 2\pi\sqrt{\frac{m}{3k(A\sin\omega t)}} \Rightarrow T \propto \frac{1}{\sqrt{A}}$$

10. *(a)* Using the concept of conservation of angular momentum, the person contracts his hands, lowering his moment of inertia and increasing his angular velocity.

$$I\omega = \text{Constant}$$

The kinetic energy of an object increases as its angular velocity increases. It would also need the energy to pull the weights inwards from their current location.

Thus, option (a) is correct.

11. *(b)*

$$\mathbf{E} = (6\hat{\mathbf{i}} + 8\hat{\mathbf{j}})\,\text{N/kg}$$

$$\mathbf{r}_1 = 8\hat{\mathbf{i}} + 0\hat{\mathbf{j}}$$

$$V_1 = -\mathbf{E}\cdot\mathbf{r}_1 = -(6\hat{\mathbf{i}} + 8\hat{\mathbf{j}})\cdot 8\hat{\mathbf{i}}$$

$$V_1 = -48\,\text{N-m kg}^{-1}$$

$$\mathbf{r}_2 = 0\hat{\mathbf{i}} + 6\hat{\mathbf{j}}$$

$$V_2 = -\mathbf{E}\cdot\mathbf{r}_2 = -(6\hat{\mathbf{i}} + 8\hat{\mathbf{j}})\cdot(6\hat{\mathbf{j}})$$

$$= -48\,\text{N-mkg}^{-1}$$

$$\frac{V_1}{V_2} = \frac{-48}{-48} = 1$$

12. *(a)* Net process is

$$3\,{}^2\text{H}_1 \longrightarrow {}^4\text{He}_2 + p + n$$

Mass defect, $\Delta m = (3 \times 2.014) - 1.007 - 1.008 - 4.001$

$= 0.026$ amu

Energy released by fusion of 3 deuterons

$= 0.026 \times 931\,\text{MeV}$

$= 24.2\text{ MeV} = 3.87 \times 10^{-12}\,\text{J}$

Total energy that can be released by the star is

$$= \frac{10^{50}}{3} \times 3.87 \times 10^{-12}\,\text{J} \simeq 10^{38}\,\text{J}$$

$\therefore$ Time for which it can produce energy at the constant rate is $= \frac{10^{38}}{10^{16}} = 10^{22}$ s

13. *(c)* Impulse $= J =$ Change in momentum

$$F_{avg} = \frac{\Delta P}{\Delta t}$$

$\because \quad \Delta P_1 = \Delta P_2$

$\therefore \quad J_1 = J_2$

Given, $\quad \Delta t_1 = 4\text{ s}$ and $\Delta t_2 = 8\text{ s}$

Hence, F_{avg} in case (i) is more than F_{avg} in case (ii).

14. *(a)* The given zener diode circuit is given as,

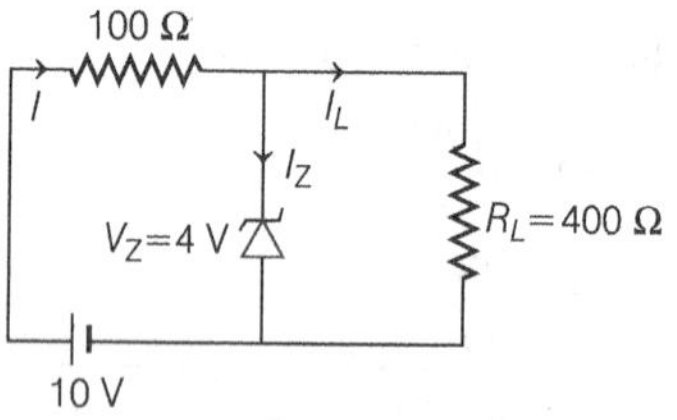

Potential drop across 400 Ω resistance,

$$V_L = V_Z = 4\text{V}$$

$\therefore$ Current through R_L,

$$I_L = \frac{V_L}{R_L} = \frac{4}{400} = 0.01\text{A}$$

Voltage drop across 100Ω resistance,

$$V' = 10 - 4 = 6\text{ V}$$

$\therefore \quad I = \frac{V'}{100} = \frac{6}{100} = 6 \times 10^{-2}\text{ A} = 0.06\text{A}$

$\therefore \quad I_Z = I - I_L$

$$= 0.06 - 0.01 = 0.05\text{A} = 50\text{ mA}$$

15. *(b)* Given, radius $r = 3\text{ m}$

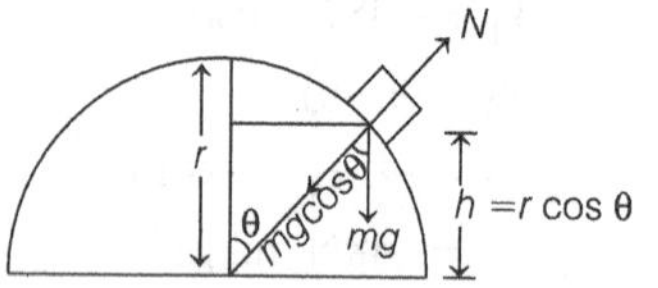

For the given free body diagram of the body,

$$mg\cos\theta - N = \frac{mv^2}{r}$$

When the body leaves the hemispherical surface

$\therefore \quad N = 0$

$\therefore \quad mg\cos\theta = \frac{mv^2}{r}$

$\therefore \quad v^2 = rg\cos\theta \quad \text{...(i)}$

Applying conservation of energy at height h at the detached point of hemisphere,

$$0 + mgr(1 - \cos\theta) = \frac{1}{2}mv^2 + 0$$

$\Rightarrow \quad v^2 = 2rg(1 - \cos\theta) \quad \text{...(ii)}$

From Eqs. (i) and (ii), we get

$$rg\cos\theta = 2rg(1 - \cos\theta)$$

$\Rightarrow \quad \cos\theta = \frac{2}{3}$

$\therefore \quad$ Height, $h = r\cos\theta = \frac{2r}{3} = \frac{2}{3} \times 3 = 2\text{m}$

16. *(b)* When the switch is open,

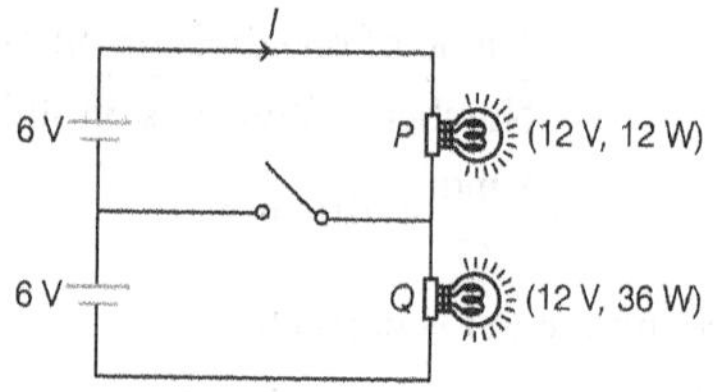

$\therefore \quad R_P = \frac{V^2}{P} = \frac{12 \times 12}{12} = 12\Omega$

$$R_Q = \frac{V^2}{P} = \frac{12 \times 12}{36} = 4\Omega$$

$\therefore \quad I = \frac{V}{R} = \frac{6+6}{12+4} = \frac{12}{16} = \frac{3}{4}\text{A}$

$\therefore$ Power drop across bulb P,

$$P_P = I^2 R_P = \frac{3}{4} \times \frac{3}{4} \times 12 = \frac{27}{4}\text{ W}$$

Power drop across bulb Q,

$$P_Q = I^2 R_Q = \frac{3}{4} \times \frac{3}{4} \times 4 = \frac{9}{4}\text{ W}$$

when the switch is closed,

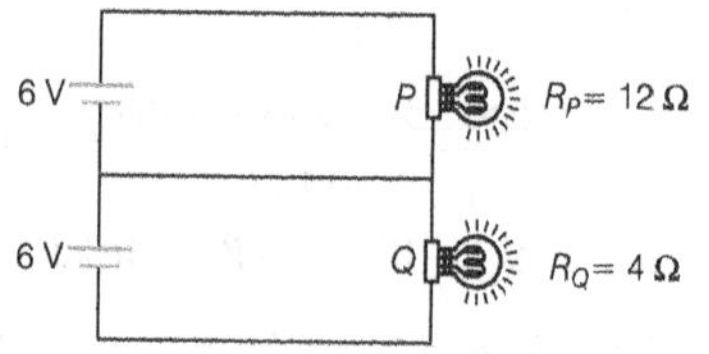

$\therefore \quad P_P' = \frac{V^2}{R_P} = \frac{6 \times 6}{12} = 3\text{ W}$

$$P_Q' = \frac{V^2}{R_Q} = \frac{6 \times 6}{4} = 9\text{ W}$$

$\therefore \quad P_P > P_P'$ and $P_Q < P_Q'$

Hence, intensity of bulb P decreases. Intensity of bulb Q increases.

17. *(b)* The energy density of EM wave is $u_B = \frac{1}{2} \cdot \frac{B_0^2}{\mu_0}$

$$\mu_0 = 4\pi \times 10^{-7}\text{ units},$$

$$u_B = 2.04 \times 10^{-8}\text{ Jm}^{-3}$$

$\therefore$ Amplitude of magnetic field,

$$B_0 = \sqrt{2\mu_0 u_B}$$

$$= \sqrt{2 \times 4\pi \times 10^{-7} \times 2.04 \times 10^{-8}}$$

$$= 22.64 \times 10^{-8}\text{T} = 226.4\text{ nT}$$

18. *(a)* Given, $V_A = V$

As, $\int_{V_A}^{V_B} dV = -\int E \cdot dr \cos 0°$

$$V_B - V_A = -E\int_0^{\theta} R\,d\theta$$

$$V_B = V_A - ER\theta$$

$\therefore \quad V_B = V - ER\theta$

19. *(d)* The saturation current in photoelectric experiment is directly proportional to intensity of light and is independent of the frequency of light.

$\therefore$ Photocurrent, $I \propto$ Intensity

$\therefore$ If intensity is doubled, then photocurrent is also doubled.

20. *(a)* For total internal reflection from glass-air interface, critical angle C must be less than the angle of incidence.

i.e. $\quad C < i$

or $\quad C < 45°$

But $\quad n = \dfrac{1}{\sin C}$

$\Rightarrow \quad C = \sin^{-1}\left(\dfrac{1}{n}\right)$

$\therefore \quad \sin^{-1}\left(\dfrac{1}{n}\right) < 45°$

$\Rightarrow \quad \dfrac{1}{n} < \sin 45°$

$\Rightarrow \quad n > \dfrac{1}{\sin 45°}$

$\Rightarrow \quad n > \dfrac{1}{1/\sqrt{2}} \Rightarrow n > \sqrt{2}$

21. (15) Given, $L = 15\text{ cm} = 0.15\text{ m}$

Stress $= 10^5\text{ N/m}^2$

Displacement, $\Delta x = 0.1\text{ cm} = 0.001\text{ m}$

The modulus of rigidity,

$$\eta = \frac{\text{Stress} \times L}{\Delta x} = \frac{10^5 \times 0.15}{0.001} = 15 \times 10^6 = p \times 10^6$$

Therefore, the value of p is 15.

22. (6) According to Cartesian sign convention,

$u = -40\text{ cm}, R = -20\text{ cm}$

Given, $\quad \mu_1 = 1$ and $\mu_2 = 1.33$

Applying equation for refraction through spherical surface, we get

$$\frac{\mu_2}{v} - \frac{\mu_1}{u} = \frac{\mu_2 - \mu_1}{R}$$

$$\frac{1.33}{v} - \frac{1}{-40} = \frac{1.33 - 1}{-20}$$

After solving, we get $v = -32\text{ cm}$

The magnification is, $m = \dfrac{h_2}{h_1} = \dfrac{\mu_1 v}{\mu_2 u}$

$\therefore \quad \dfrac{h_2}{1} = -\dfrac{1(32)}{1.33(-40)}$

or $\quad h_2 = 6\text{ mm}$

The positive sign shows that the image is erect.

23. (3) Escape velocity, $v_e = \sqrt{\dfrac{2GM}{R}}$

Velocity of projection, $v_P = \dfrac{1}{Z}\sqrt{\dfrac{2GM}{R}}$

Total initial energy $= \text{KE}_i + U_i$

$$= \frac{1}{2}mv_P^2 - \frac{GMm}{R}$$

$$= \frac{m}{2Z^2} \cdot \frac{2GM}{R} - \frac{GMm}{R}$$

$$= \frac{GMm}{R}\left[\frac{1}{Z^2} - 1\right]$$

At maximum height, $h_{\max} = \dfrac{R}{8}$, KE $= 0$

Total Final energy $= \text{KE}_f + U_f$

$$= 0 - \frac{GMm}{(R+h)}$$

$$\text{TE}_f = -\frac{GMm}{R + \frac{R}{8}} = -\frac{8GMm}{9R}$$

$\because \quad \text{TE}_i = \text{TE}_f$

$$\frac{GMm}{R}\left[\frac{1}{Z^2} - 1\right] = -\frac{8GMm}{9R}$$

$$\frac{1}{Z^2} - 1 = \frac{-8}{9}$$

$$\frac{1}{Z^2} = 1 - \frac{8}{9}$$

$$\frac{1}{Z^2} = \frac{1}{9} \Rightarrow Z = 3$$

24. (6) Given, condition, $h_1 = h_2$ $\quad$ ...(i)

Height is given as $H = \dfrac{u^2 \sin^2\theta}{2g}$ $\quad$ ...(ii)

From Eqs. (i) and (ii), we get

$\Rightarrow \quad u_1^2 \sin^2 45° = u_2^2 \sin^2\theta$

$\Rightarrow \quad \sin^2\theta = \dfrac{u_1^2}{u_2^2}\sin^2 45° = \dfrac{1}{3} \cdot \dfrac{1}{2} = \dfrac{1}{6}$

$\Rightarrow \quad \sin\theta = \dfrac{1}{\sqrt{6}}$

$\Rightarrow \quad \theta = \sin^{-1}\left(\dfrac{1}{\sqrt{6}}\right) \Rightarrow m = 6$

25. (5)

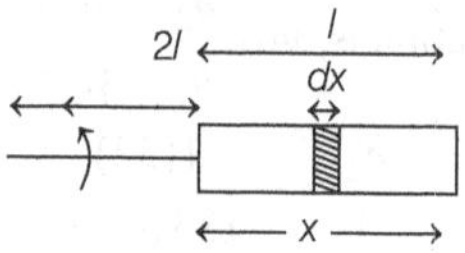

$(v = r\omega)$

$(\because e = Blv)$

$$e = \int_{2l}^{3l} \mathbf{B} \cdot (\omega x)\, dx$$

$$= B\omega \left[\frac{(3l)^2 - (2l)^2}{2} \right] = \frac{5B\omega l^2}{2}$$

$$= \frac{NB\omega l^2}{2} \text{ (given)}$$

On comparing, we get

$\therefore \quad N = 5$

Chemistry

26. (*d*) The reaction is represented as,

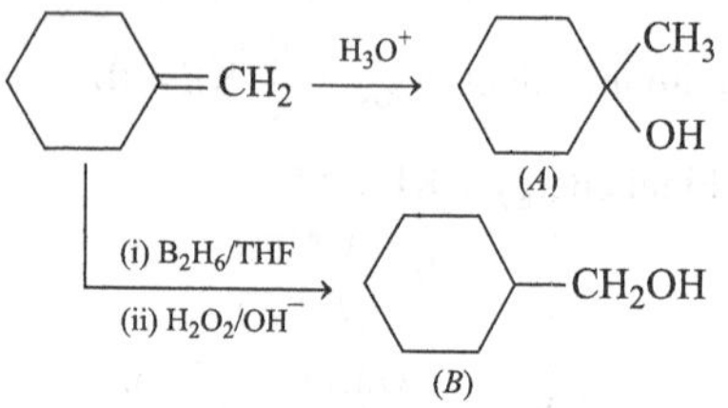

27. (*a*) Structure of glucose and fructose are as follows

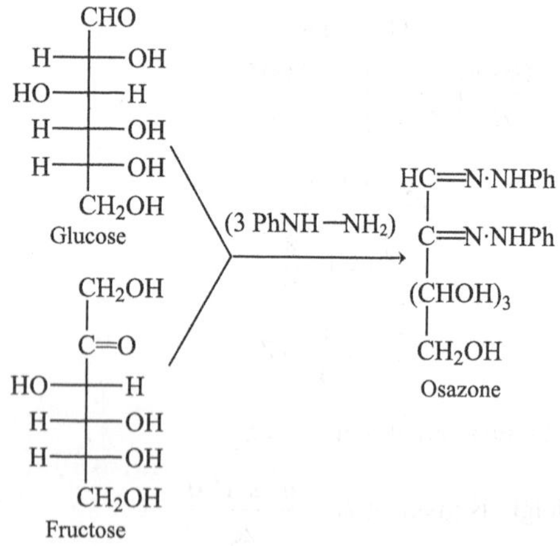

28. (*a*) The hybridisation of Xe in XeF_2 is sp^3d with 180° bond angle.

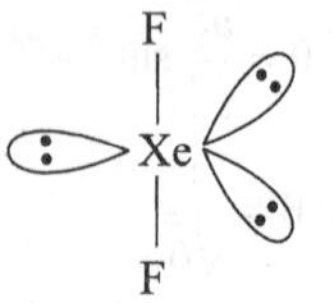

The geometry is linear and shape is trigonal bipyramidal with 3 equatorial lone pairs.

29. (*c*) $[Sc(H_2O)_6]^{3+}$ and $[Ti(H_2O)_6]^{4+}$ are colourless compounds due to $3d^0$ outer electronic configuration. No unpaired electron is present. Thus, option (c) is the incorrect statement.

30. (*c*)

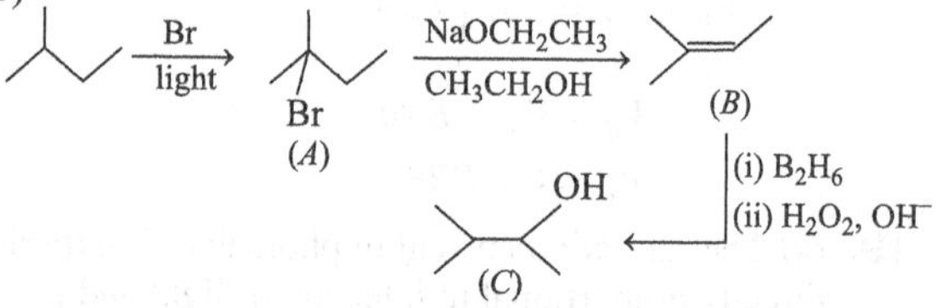

31. (*c*) As we know, $t_{1/2} \propto (P_0)^{1-n}$

$$\frac{(t_{1/2})_1}{(t_{1/2})_2} = \frac{(P_0)_1^{1-n}}{(P_0)_2^{1-n}}$$

$$\Rightarrow \quad \left(\frac{4}{2}\right) = \left(\frac{50}{100}\right)^{1-n}$$

$$2 = \left(\frac{1}{2}\right)^{1-n}$$

$$2 = (2)^{n-1}$$

$$n = 2$$

32. (*a*) $AgNO_3 + HCl \longrightarrow \underset{\text{White ppt.}}{AgCl} + HNO_3$

$$2AgNO_3 + Cu \longrightarrow 2Ag + \underset{\text{(Blue sol.)}}{Cu(NO_3)_2}$$

$$AgNO_3 + \underset{\text{(Conc.)}}{H_2SO_4} \longrightarrow Ag_2SO_4 + \underset{\text{(Brown fumes)}}{NO_2}$$

33. (*b*) **Rosenmund reaction** Benzoyl chloride on reaction with H_2 in presence of Pd / $BaSO_4$ produces benzaldehyde.

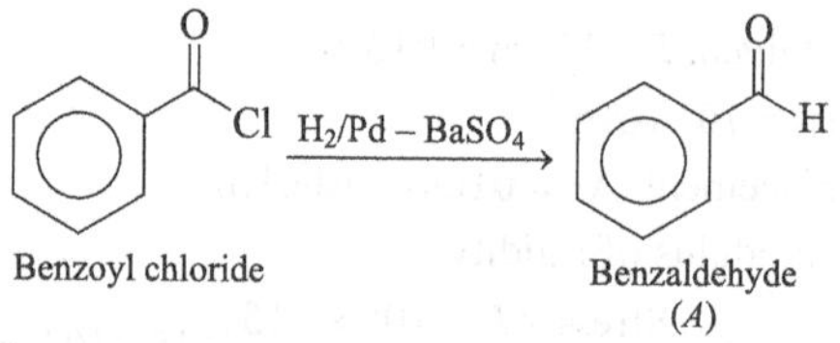

Benzoin condensation Benzoin condensation is a characteristic property of aromatic aldehyde. When benzaldehyde is treated with aqueous alcoholic solution of KCN it produce α-hydroxy ketone.

Necessary condition The aldehyde must be aromatic in nature.

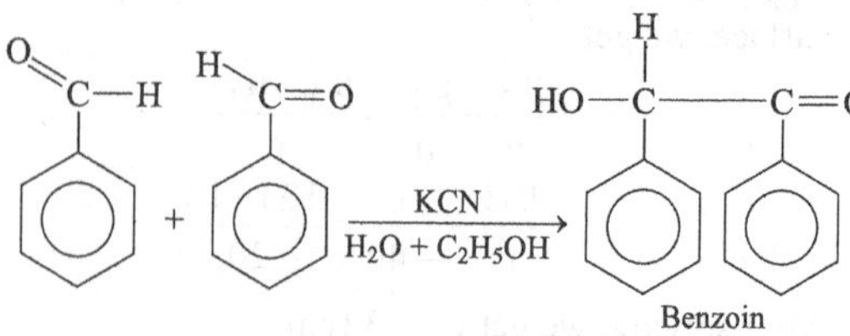

Nomenclature of benzoin

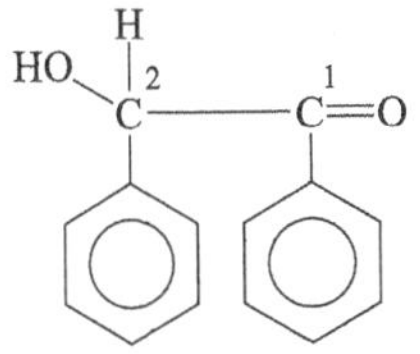

Parent chain → ethane

Functional group → ketone → one (suffix)

Substituents → 1, 2-diphenyl (prefix) → 2 hydroxy

IUPAC name = 2-hydroxy-1,2-diphenyl ethan-1-one

34. (*d*) Statement I is incorrect but Statement II is correct.

Its correct form is

Allyl isocyanide has 9 σ bonds, 3 π bonds and two non-bonded electrons.

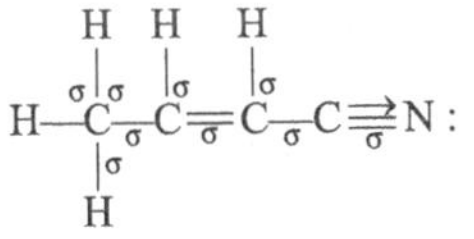

35. (*b*) In the given compounds, correct order of dehydration is

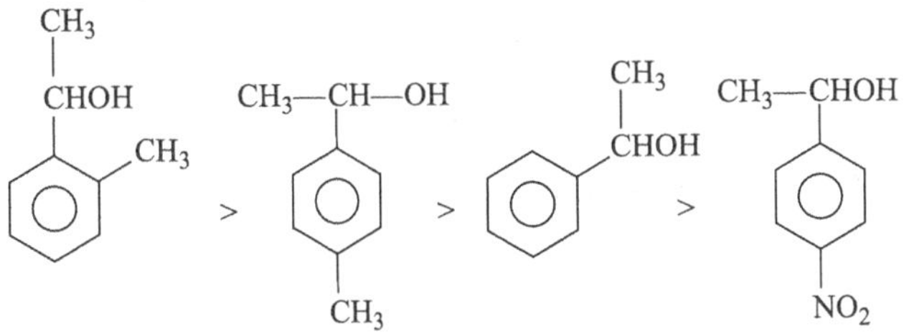

(B) (C) (A) (D)

36. (*c*) The correct match is A-III, B-I, C-II, D-IV.

- $[Cr(NH_3)_6][Co(CN)_6]$ shows coordination isomerism.
- $[Ni(NH_3)_3(NO_2)_3]$ shows linkage isomerism.
- $[Fe(H_2O)_6]Cl_3$ shows solvate isomerism.
- *cis*-$[CoCl_2(ox)_2]^{3-}$ shows optical isomerism.

37. (*a*) Enzyme activators should be allosteric since they allow the substrate to bind to the enzyme for the reaction to occur.

Hence, in this case, enzyme activators can be considered as positive allosteric modulators.

When a small non-protein like vitamin is present along the enzyme, the catalytic activity is increased considerably as the vitamin acts as a co-enzyme which enhances the enzymatic activity.

38. (*c*) $CrO_3 + H_2O \longrightarrow \underset{\text{Acid}}{H_2CrO_4}$

$SO_2 + 2H_2O \longrightarrow \underset{\text{Acid}}{H_2SO_3}$

$Mn_2O_7 + H_2O \longrightarrow \underset{\text{Acid}}{2HMnO_4}$

CO is neutral oxide, while CuO is basic in nature.

39. (*b*) **Fries rearrangement** Conversion of phenyl allyl ether into allyl phenol is known as Fries rearrangement.

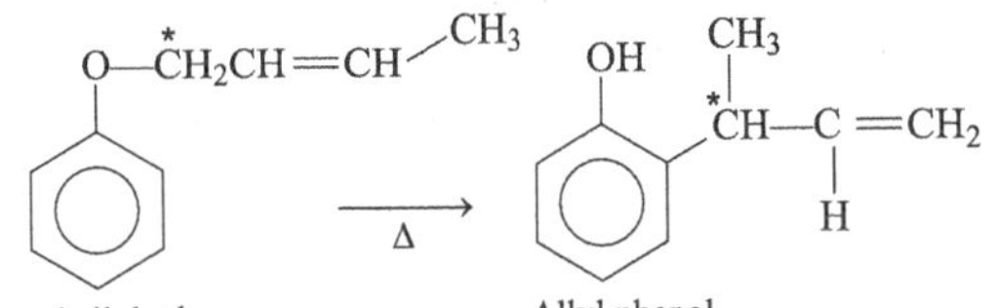

Phenyl allyl ether Allyl phenol

Remember in this reaction the C* attaches itself to benzene ring at *ortho*-position.

40. (*a*) The structures of the given oxyacids of phosphorus are as follows.

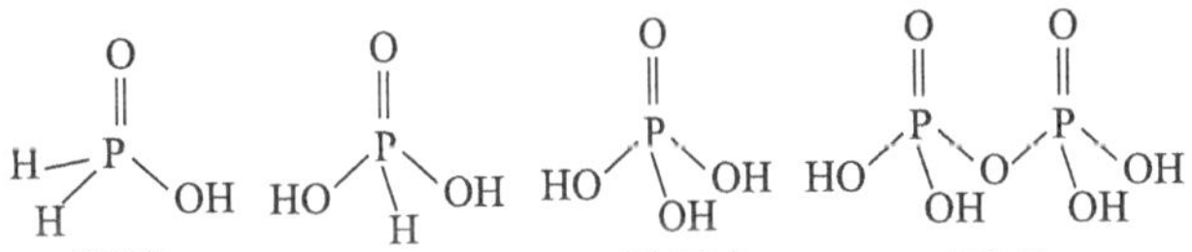

H_3PO_2	H_3PO_3	(H_3PO_4)	$H_4P_2O_7$
2P—H bonds	(1P—H bonds)	(No P—H bond)	(No P—H bond)

41. (*c*) (A) is true but (R) is false.

The structure of orthoboric acid is,

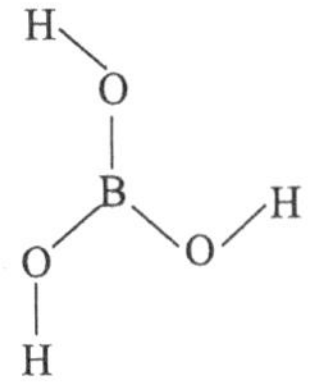

It cannot ionize in water to give a proton (H^+) whereas, it accepts lone pair of electrons and acts as a Lewis acid.

42. (*b*) The correct match is A-II, B-I, C-IV, D-III.

A. 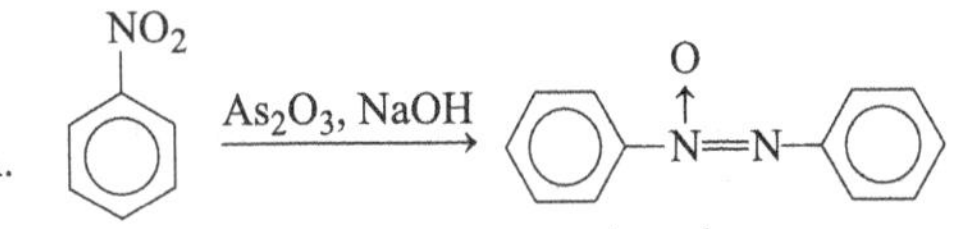

Nitrobenzene Azoxybenzene

B. 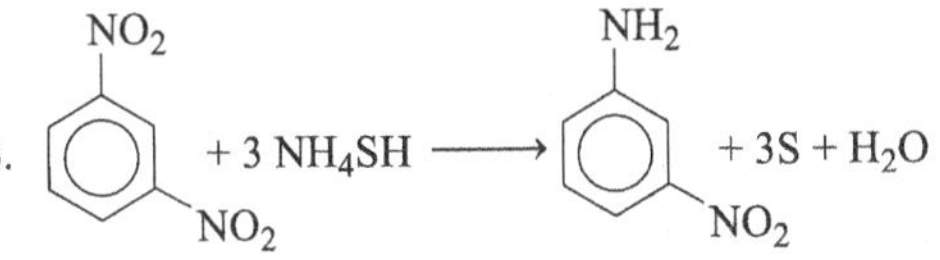

m-dinitrobenzene *m*-nitroaniline

C. 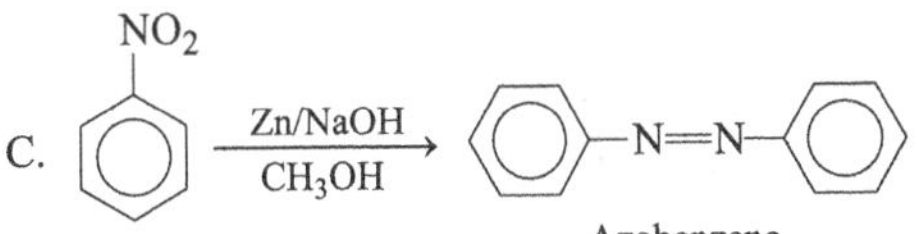

Nitrobenzene Azobenzene

D. 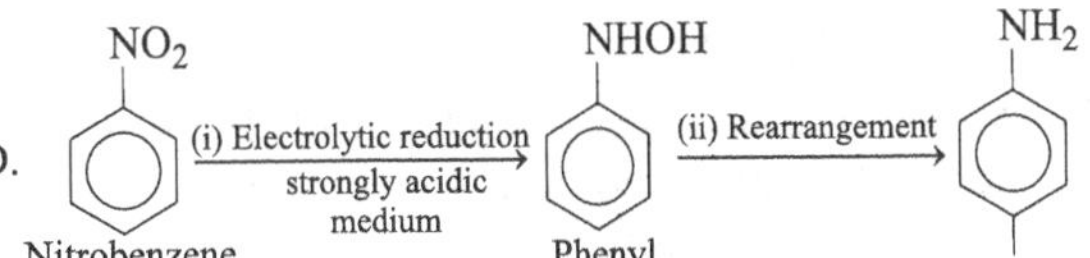

Nitrobenzene Phenyl hydroxylamine *p*-aminophenol

43. (*a*) Among the given order B and D are correct. For *f*-block elements, size decreases on going from left to right due to lanthanoid contraction.

Atomic radii follows the order

$$Ho < Nd < Ce < Eu$$
$$Ce > Sm > Yb > Lu$$

44. (*c*)
$$K_A = \frac{[NO_2]}{[NO][O_2]^{1/2}} \quad ...(i)$$
$$K_B = \frac{[NO]^2[O_2]}{[NO_2]^2} \quad ...(ii)$$
$$\frac{1}{K_B} = \frac{[NO_2]^2}{[NO][O_2]^{1/2}}$$
$$\frac{1}{\sqrt{K_B}} = \frac{[NO_2]}{[NO][O_2]^{1/2}} \quad ...(iii)$$

From Eqs. (i) and (iii)
$$K_A = \frac{1}{\sqrt{K_B}} \text{ and } K_B = \frac{1}{K_A^2}$$

45. (*b*) Among the given statements option (b) is not correct.

Effective atomic number of Pt in $[Pt\,Cl_6]^{2-}$ is 86.

46. (*1800*) For adiabatic condition,
$$\frac{T_2}{T_1} = \left(\frac{V_1}{V_2}\right)^{\gamma - 1}$$
$$\frac{T_2}{T_1} = \left(\frac{1}{16}\right)^{1.50-1} = \left(\frac{1}{16}\right)^{0.50} = \left(\frac{1}{16}\right)^{1/2} = \frac{1}{4}$$
$$T_2 = 400 \times \frac{1}{4} = 100 \text{ K}$$

Thus, $T_2 < T_1$, hence cooling takes place due to expansion under adiabatic condition.

$$\Delta E = q + W = W \ (\because q = 0 \text{ for adiabatic change.})$$

Sign of ΔE is negative because the gas expands.

$$W = -\Delta E = -C_V (T_2 - T_1)$$
$$= -6(100 - 400) = 1800 \text{ cal}$$

47. (*72*) $p_T = \chi_H \cdot p_H^\circ + \chi_O \cdot p_O^\circ$

$$\chi_H = \frac{\frac{25}{100}}{\frac{25}{100} + \frac{35}{114}} = 0.45$$

$\therefore \quad \chi_O = 1 - 0.45 = 0.55$

$$p_T = 0.45 \times 105 + 0.55 \times 45 = 72 \text{ kPa}$$

48. (*6*) The electron in the ground state of H-atom excited to the *n*th state after absorbing the radiation.

The wavelength of the radiation, $\lambda = 970$ Å

$$= 970 \times 10^{-10} \text{ m}$$

Energy gained by the electron, $E' = \dfrac{hc}{e\lambda}$

$$= \frac{1.237 \times 10^{-6}}{970 \times 10^{-10}}$$
$$= 12.75 \text{ eV}$$

49. (*10*) Given,

$$\underset{1\text{M}}{Zn(s) + Cu^{2+}(aq)} \longrightarrow \underset{y\text{M}}{Zn^{2+}(aq)} + Cu(s)$$
$$E_{cell} = 1.67 \text{ V}, E_{cell} = 1.70 \text{ V}$$
$$E_{cell} = E^\circ_{cell} - \frac{RT}{nF} \ln \frac{[Zn^{2+}]}{[Cu^{2+}]}$$
$$1.67 = 1.70 - \frac{RT}{2F} \ln y$$
$$\frac{RT}{2F} \ln y = 0.03$$
$$\ln y = \frac{2 \times 0.03}{300} \times \frac{F}{R}$$
$$\ln y = \frac{2 \times 0.03}{300} \times 11500$$
$$\ln y = 2.30$$
$$y = e^{2.30} = 10$$

50. (*3*) Among the given compounds 3 are aromatic

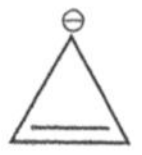

$4\pi e^-s$, planar (Anti-aromatic)

$6\pi e^-s$, conjugated, planar (Aromatic)

$6\pi e^-s$, conjugated, planar (Aromatic)

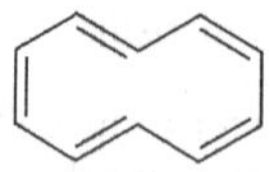

$10\pi e^-s$, non-planar, conjugated (non-Aromatic)

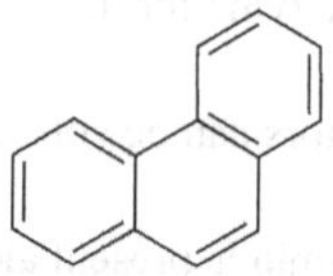

$14\pi e^-s$, planar, conjugated (Aromatic)

Study Tactics

To check the given compounds are aromatic or not, follow the following steps.

I. Molecule should be cyclic and planar.

II. It should be fully conjugated.

III. It should have $(4n + 2)\pi$ electron ($n = 0$ or any positive integer).

Mathematics

51. *(c)* We have, $l+m+n=0$

and $2l^2+2m^2-n^2=0$

Now, $2(l^2+m^2)-n^2=0$

$$\Rightarrow \quad 2(1-n^2)-n^2=0$$

$$\Rightarrow \quad 3n^2=2$$

$$\Rightarrow \quad n=\pm\sqrt{\frac{2}{3}}$$

Again, $2(l^2+m^2)=n^2$

$$\Rightarrow 2[(l+m)^2-2lm]=[-(l+m)]^2$$

$$l=m$$

$$\therefore \quad l+m=\pm\sqrt{\frac{2}{3}} \Rightarrow 2l=\pm\sqrt{\frac{2}{3}}$$

$$\Rightarrow \quad l=\pm\frac{1}{\sqrt{6}}=m$$

$\therefore$ Direction cosines are $\left(\frac{1}{\sqrt{6}},\frac{1}{\sqrt{6}},\sqrt{\frac{2}{3}}\right),\left(\frac{1}{\sqrt{6}},\frac{1}{\sqrt{6}},-\sqrt{\frac{2}{3}}\right)$

or $\left(-\frac{1}{\sqrt{6}},-\frac{1}{\sqrt{6}},\sqrt{\frac{2}{3}}\right)$ and $\left(-\frac{1}{\sqrt{6}},-\frac{1}{\sqrt{6}},-\sqrt{\frac{2}{3}}\right)$.

The angle between in both the case in $\cos^{-1}\left(\frac{-1}{3}\right)$.

52. *(a)* Let $p(x)=a_0+a_1x+a_2x^2+a_3x^3+\ldots$

$$=f_1(x^3)+x f_2(x^3)+x^2 f_3(x^3) \quad \ldots\text{(i)}$$

where, $f_1(x)$, $f_2(x)$ and $f_3(x)$ are polynomials with real coefficients.

As, $(x^3-\alpha^3)$ is a factor of $p(x)-p(\alpha)$.

$$\Rightarrow (x^3-\alpha^3)/(p(x)-p(\alpha))$$

$$\Rightarrow p(x)-p(\alpha)=(x^3-\alpha^3)f(x)$$

where, $f(x)$ is a polynomial with real coefficients.

Now, $p(x)-p(\alpha)=(x-\alpha)(x-\alpha\omega)(x-\alpha\omega^2)f(x)$ ...(ii)

where, ω is cube root of unity.

From Eqs. (i) and (ii), we get

$$f_1(\alpha^3)+\alpha f_2(\alpha^3)+\alpha^2 f_3(\alpha^3)=p(\alpha) \quad \ldots\text{(iii)}$$

$$f_1(\alpha^3)+\alpha\omega f_2(\alpha^3)+\alpha^2\omega^2 f_3(\alpha^3)=p(\alpha) \quad \ldots\text{(iv)}$$

$$f_1(\alpha^3)+\alpha\omega^2 f_2(\alpha^3)+\alpha^2\omega f_3(\alpha^3)=p(\alpha) \quad \ldots\text{(v)}$$

On adding above equations, we get

$$3f_1(\alpha^3)=3p(\alpha)$$

$$\Rightarrow \quad f_1(\alpha^3)=p(\alpha) \quad \ldots\text{(vi)}$$

From Eqs. (iii), (iv) and (vi), we get

$$f_2(\alpha^3)+\alpha f_3(\alpha^3)=0 \quad \ldots\text{(vii)}$$

$$f_2(\alpha^3)+\alpha\omega f_3(\alpha^3)=0 \quad \ldots\text{(viii)}$$

On solving Eqs. (vii) and (viii), we get

$$f_3(\alpha^3)=0,\ f_2(\alpha^3)=0, \forall \alpha>0$$

Thus, $f_2(x^3)=f_3(x^3)=0, \forall\, x\in R$

Hence, $p(x)=f(x^3)$ for some polynomial $f(x)$ with real coefficient.

53. *(c)* Let the removed numbers be $a-d, a, a+d$.

Given, $\frac{43}{4}=\frac{(1+2+\ldots+n)-(a-d+a+a+d)}{n-3}$

$$\Rightarrow \quad \frac{43}{4}=\frac{\frac{n(n+1)}{2}-3a}{n-3} \quad \ldots\text{(i)}$$

But $1+2+3\le a-d+a+a+d\le(n-2)+(n-1)+n$

$$\Rightarrow 6\le 3a\le 3n-3$$

$$\Rightarrow 3-3n\le -3a\le -6$$

$$\Rightarrow \frac{n(n+1)}{2}+3-3n\le\frac{n(n+1)}{2}-3a\le\frac{n(n+1)}{2}-6$$

$$\Rightarrow \quad \frac{n^2-5n+6}{2(n-3)}\le\frac{\frac{n(n+1)}{2}-3a}{n-3}\le\frac{n^2+n-12}{2(n-3)}$$

$$\Rightarrow \quad \frac{n^2-5n+6}{2(n-3)}\le\frac{43}{4}\le\frac{n^2+n-12}{2(n-3)} \quad \text{[from Eq. (i)]}$$

$$\Rightarrow \quad \frac{(n-2)(n-3)}{2(n-3)}\le\frac{43}{4}\le\frac{(n+4)(n-3)}{2(n-3)}$$

$$\Rightarrow \quad n-2\le\frac{43}{2}\le n+4$$

$$\Rightarrow \quad 2n-4\le 43\le 2n+8$$

$$\Rightarrow \quad 2n-4\le 43 \text{ and } 43\le 2n+8$$

$$\Rightarrow \quad n\le\frac{47}{2} \text{ and } \frac{35}{2}\le n$$

$$\therefore \quad \frac{35}{2}\le n\le\frac{47}{2}$$

$\Rightarrow n=18, 19, 20, 21, 22, 23$ [possible values]

If $n=19$ [from Eq. (i)]

$$\frac{43}{4}=\frac{190-3a}{16}$$

$$\Rightarrow a=6$$

If $d=5$, then $a-d, a, a+d$ are 1, 6, 11 [it is possible]

$(n,d)=(19,5)$

54. *(a)* Given, $A=\begin{bmatrix}\frac{\sqrt{3}}{2} & \frac{1}{2}\\ \frac{-1}{2} & \frac{\sqrt{3}}{2}\end{bmatrix}, B=\begin{bmatrix}1 & 1\\ 0 & 1\end{bmatrix}$,

$$C=ABA^T \text{ and } x=A^TC^{2023}A$$

$$\because \quad A^T=\begin{bmatrix}\frac{\sqrt{3}}{2} & \frac{-1}{2}\\ \frac{1}{2} & \frac{\sqrt{3}}{2}\end{bmatrix}$$

and $A^T A = \begin{bmatrix} \frac{\sqrt{3}}{2} & \frac{-1}{2} \\ \frac{1}{2} & \frac{\sqrt{3}}{2} \end{bmatrix} \begin{bmatrix} \frac{\sqrt{3}}{2} & \frac{1}{2} \\ \frac{-1}{2} & \frac{\sqrt{3}}{2} \end{bmatrix} = \begin{bmatrix} 1 & 0 \\ 0 & 1 \end{bmatrix}$

$\Rightarrow \quad A^T A = I = AA^T$

Now, $C^2 = (ABA^T) \cdot (ABA^T)$

$= AB(A^T A)BA^T$

$= AB(IB)A^T = AB^2 A^T$

$C^3 = C^2 \cdot C$

$= (AB^2 A^T) \cdot (AB \cdot A^T)$

$= AB^2 (A^T A) BA^T$

$= AB^2 (IB) A^T = AB^3 A^T$

Continuing in this way, we get

$C^{2023} = AB^{2023} A^T$

Now, $B^2 = \begin{bmatrix} 1 & 1 \\ 0 & 1 \end{bmatrix} \begin{bmatrix} 1 & 1 \\ 0 & 1 \end{bmatrix} = \begin{bmatrix} 1 & 2 \\ 0 & 1 \end{bmatrix}$

$B^3 = \begin{bmatrix} 1 & 2 \\ 0 & 1 \end{bmatrix} \begin{bmatrix} 1 & 1 \\ 0 & 1 \end{bmatrix} = \begin{bmatrix} 1 & 3 \\ 0 & 1 \end{bmatrix}$

Continuing in same manner, we have

$B^{2023} = \begin{bmatrix} 1 & 2023 \\ 0 & 1 \end{bmatrix}$

Now, $x = A^T C^{2023} A = A^T (AB^{2023} A^T) A$

$= (A^T A) B^{2023} (A^T A) = (IB^{2023}) I = B^{2023}$

$\Rightarrow \quad x = \begin{bmatrix} 1 & 2023 \\ 0 & 1 \end{bmatrix}$

55. *(c)* Given, $f(x) = \int_x^{x^2} \frac{du}{(\log u)^2}$

$\therefore \quad f'(x) = \frac{2x}{(\log x^2)^2} - \frac{1}{(\log x)^2}$

$= \frac{x}{2(\log x)^2} - \frac{1}{(\log x)^2} = \frac{1}{(\log x)^2}\left[\frac{x}{2} - 1\right]$

$\Rightarrow f'(x) > 0$, where $x > 2$ and $f'(x) < 0$ when $(0 < x < 1) \cup (1 < x < 2)$

$\therefore$ Both statements are true.

Study Tactics

To get the function $f'(x)$, use the Fundamental Theorem of calculus

$$F(x) = \int_{h(x)}^{g(x)} f(t)dt$$

$$\Rightarrow \quad \frac{dF}{dx} = f(g)g' - f(h)h'$$

And then check the value of x, where $f'(x)$ is increasing and decreasing.

56. *(a)* The denominator of $f(x)$ is

$2\left[\frac{x + 21\pi}{\pi}\right] - 41 = 2\left[\frac{x}{\pi} + 21\right] - 41$

$= 2\left\{21 + \left[\frac{x}{\pi}\right]\right\} - 41$

$= 2\left[\frac{x}{\pi}\right] + 1$

$\therefore \quad f(x) = \frac{x(\sin x + \tan x)}{\left[\frac{x}{\pi}\right] + \frac{1}{2}}$

$\Rightarrow \quad f(-x) = \frac{-x\{\sin(-x) + \tan(-x)\}}{\left[-\frac{x}{\pi}\right] + \frac{1}{2}}$

$= \frac{x(\sin x + \tan x)}{-1 - \left[\frac{x}{\pi}\right] + \frac{1}{2}}$

$= \frac{-x(\sin x + \tan x)}{\left[\frac{x}{\pi}\right] + \frac{1}{2}}$

$= -f(x)$

$\Rightarrow f(x)$ is an odd function.

57. *(b)* Let the equation of the ellipse be $\frac{x^2}{a^2} + \frac{y^2}{b^2} = 1$

Slope of the given diameters are

$m_1 = -\frac{3}{4}, m_2 = \frac{1}{\sqrt{3}}$

$\Rightarrow m_1 \cdot m_2 = \frac{-3}{4\sqrt{3}} = -\frac{b^2}{a^2}$ [by using the condition of conjugatey of two diameters]

$\Rightarrow \quad 3a^2 = 4\sqrt{3}b^2$

$\Rightarrow \quad 3a^2 = 4\sqrt{3}\,[a^2(1 - e^2)]$

$\Rightarrow \quad 3a^2 = 4\sqrt{3}a^2(1 - e^2)$

$\Rightarrow \quad 1 - e^2 = \frac{3}{4\sqrt{3}}$

$\Rightarrow \quad e^2 = 1 - \frac{3}{4\sqrt{3}}$

$\Rightarrow \quad e = \frac{\sqrt{4 - \sqrt{3}}}{2}$

58. *(a)* $\mathbf{a} = 2\hat{\mathbf{i}} + 8\hat{\mathbf{j}} + 4\hat{\mathbf{k}}$,

$\mathbf{b} = 6\hat{\mathbf{i}} - 4\hat{\mathbf{j}} + 14\hat{\mathbf{k}}$

and $\mathbf{c} = 4\hat{\mathbf{i}} - 2\hat{\mathbf{j}} + 8\hat{\mathbf{k}}$

Also, given $\mathbf{d} \cdot \mathbf{a} = 32$ and $\mathbf{d} \times \mathbf{b} = \mathbf{c} \times \mathbf{b}$

Then, $(\mathbf{d} - \mathbf{c}) \times \mathbf{b} = 0$

$\Rightarrow \quad \mathbf{d} - \mathbf{c} = \lambda\mathbf{b} \Rightarrow \mathbf{d} = \mathbf{c} + \lambda\mathbf{b}$

Now, $\mathbf{a}\cdot\mathbf{d}=\mathbf{a}\cdot(\mathbf{c}+\lambda\mathbf{b})=\mathbf{a}\cdot\mathbf{c}+\lambda(\mathbf{a}\cdot\mathbf{b})$

$$32=24+\lambda(36)$$
$$32-24=36\lambda$$
$$\frac{8}{36}=\lambda\Rightarrow\lambda=\frac{2}{9}$$

$\therefore$ $\mathbf{d}=\mathbf{c}+\lambda\mathbf{b}$

$$=(4\hat{\mathbf{i}}-2\hat{\mathbf{j}}+8\hat{\mathbf{k}})+\frac{2}{9}(6\hat{\mathbf{i}}-4\hat{\mathbf{j}}+14\hat{\mathbf{k}})$$
$$=(4\hat{\mathbf{i}}-2\hat{\mathbf{j}}+8\hat{\mathbf{k}})+\frac{12}{9}\hat{\mathbf{i}}-\frac{8}{9}\hat{\mathbf{j}}+\frac{28}{9}\hat{\mathbf{k}}$$
$$=\left(4+\frac{12}{9}\right)\hat{\mathbf{i}}-\left(2+\frac{8}{9}\right)\hat{\mathbf{j}}+\left(8+\frac{28}{9}\right)\hat{\mathbf{k}}$$
$$=\frac{48}{9}\hat{\mathbf{i}}-\frac{26}{9}\hat{\mathbf{j}}+\frac{100}{9}\hat{\mathbf{k}}$$
$$9\mathbf{d}=48\hat{\mathbf{i}}-26\hat{\mathbf{j}}+100\hat{\mathbf{k}}$$

59. *(a)* Given, $|z_1+z_2|=|z_1-z_2|$

$\Rightarrow$ $(z_1+z_2)(\bar{z}_1+\bar{z}_2)=(z_1-z_2)(\bar{z}_1-\bar{z}_2)$

$\Rightarrow |z_1|^2+|z_2|^2+z_1\bar{z}_2+z_2\bar{z}_1=|z_1|^2+|z_2|^2-z_1\bar{z}_2-z_2\bar{z}_1$

$\Rightarrow$ $z_1\bar{z}_2+z_2\bar{z}_1=0$

$\Rightarrow$ $\text{Re}(z_1\bar{z}_2)=0$

$\Rightarrow z_1\bar{z}_2$ is purely imaginary

$\Rightarrow \left(\frac{z_1}{z_2}\right)\cdot|z_2|^2$ is purely imaginary

$\Rightarrow \arg\left(\frac{z_1}{z_2}\right)=\pm\frac{\pi}{2}$

$\Rightarrow$ $\angle AOB=\frac{\pi}{2}$

$\Rightarrow$ Circumcentre of ΔOAB is mid-point of AB,

i.e. $\frac{z_1+z_2}{2}$.

60. *(c)* Given parabola has equation

$$y^2=px-12=p\left(x-\frac{12}{p}\right)$$

Thus, the directrix has equation $x-\frac{12}{p}=-\frac{p}{4}$

$\Rightarrow$ $x=\frac{12}{p}-\frac{p}{4}$

$\therefore$ $\frac{12}{p}-\frac{p}{4}=2$

$\Rightarrow$ $p^2+8p-48=0$

$\Rightarrow$ $(p-4)(p+12)=0$

$\therefore$ $p=4,-12$

So, $p_1=-12$ and $p_2=4$

$\therefore$ $\frac{p_1}{p_2}=-3$

61. *(b)* Given lines, $\frac{x-1}{2}=\frac{y-2}{3}=\frac{z-3}{4}$ ...(i)

and $\frac{x-4}{5}=\frac{y-1}{2}=\frac{z-0}{1}$ ...(ii)

Let P be any point on the line (i), then $P(2r_1+1, 3r_1+2, 4r_1+3)$.

Similarly, coordinates of any point Q on line (ii) are $Q(5r_2+4, 2r_2+1, r_2)$.

If lines (i) and (ii) intersect each other, then

$2r_1+1=5r_2+4\Rightarrow 2r_1-5r_2=3$ ...(iii)

$3r_1+2=2r_2+1\Rightarrow 3r_1-2r_2=-1$ ...(iv)

$4r_1+3=r_2\Rightarrow 4r_1-r_2=-3$ ...(v)

On solving Eqs. (iii) and (iv) we get

$r_1=-1$ amd $r_2=-1$

Hence, $P\equiv P(-1,-1,-1)$,

which is required intersection point.

$\Rightarrow \sqrt{d}$ is the distance between the point $A(2,3,4)$ and point $P(-1,-1,-1)$

$\Rightarrow$ $\sqrt{d}=\sqrt{(x_2-x_1)^2+(y_2-y_1)^2+(z_2-z_1)^2}$

$$\sqrt{d}=\sqrt{(2+1)^2+(3+1)^2+(4+1)^2}$$
$$\sqrt{d}=\sqrt{9+16+25}$$
$$d=50$$

Also, $d=63-\sqrt{\mu}$

$$50=63-\sqrt{\mu}$$
$$\sqrt{\mu}=13$$
$$\mu=169$$

62. *(c)* Let the remaining numbers be x and y.

$$\text{Mean}=\frac{2+4+10+12+14+x+y}{7}=8$$

$\Rightarrow x+y=14$ ...(i)

Variance

$$=\frac{2^2+4^2+10^2+12^2+14^2+x^2+y^2}{7}-8^2=16$$

$\Rightarrow$ $x^2+y^2=100$ ...(ii)

From Eqs. (i) and (ii), we get

$$(14-y)^2+y^2=100$$

$\Rightarrow$ $196+y^2-28y+y^2=100$

$\Rightarrow$ $y^2-14y+48=0$

$\Rightarrow$ $y=6,8$

$\therefore$ $x=8,6$

$$(x,y)=(6,8)\text{ or }(8,6)$$

Hence, product of the remaining two observations

$$=xy=48$$

63. *(3)* Here,

$$T_1 = d$$
$$T_2 = 2d$$
$$T_n = nd, \forall\, n \in N$$

and $S_1 = d^2$

$$S_2 = rd^2$$
$$S_n = r^{n-1}d^2, \forall n \in N$$

Now, $\dfrac{T_1^2 + T_2^2 + T_3^2}{S_1 + S_2 + S_3} = \dfrac{d^2(1^2 + 2^2 + 3^2)}{d^2(1 + r + r^2)}$

$= \dfrac{14}{1 + r + r^2}$

As, $\dfrac{14}{1 + r + r^2}$ is a positive integer say k,

we get

$$1 + r + r^2 = \frac{14}{k}$$

$\Rightarrow \quad r = -\dfrac{1}{2} \pm \sqrt{\dfrac{14}{k} - \dfrac{3}{4}}\ [\because 0 < r < 1]$

$\Rightarrow \quad \dfrac{1}{2} < \sqrt{\dfrac{14}{k} - \dfrac{3}{4}} < \dfrac{3}{2}$

$\Rightarrow \quad \dfrac{1}{4} < \dfrac{14}{k} - \dfrac{3}{4} < \dfrac{9}{4}$

$\Rightarrow \quad 1 < \dfrac{14}{k} < 3$

$\Rightarrow \quad 5 \le k \le 13$

For $k = 8$, $\dfrac{14}{k} - \dfrac{3}{4} =$ Perfect square of rational number

$\therefore$ For $k = 8$ and $r = \dfrac{1}{2} \Rightarrow 6r = 3$

Study Tactics

Write the T_1, T_2 and T_3 in the form of d (common difference), using arithmetic progression and S_1, S_2 and S_3 in the form of r (common ratio) using geometric progression. After that find the interval of k (as $5 \le k \le 13$).

64. *(c)*

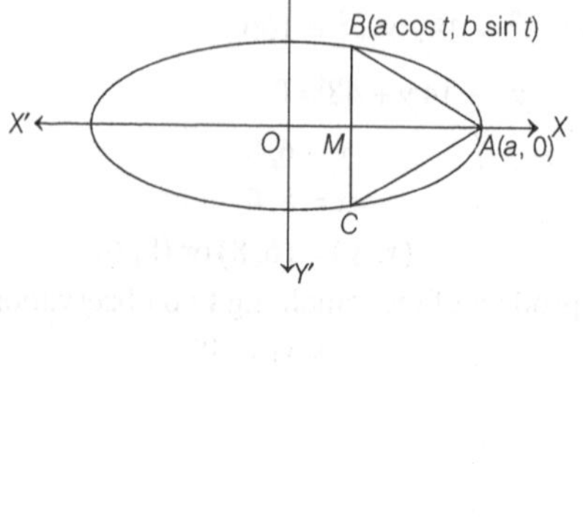

Let the equation of the ellipse be

$$\frac{x^2}{a^2} + \frac{y^2}{b^2} = 1$$

Consider an isosceles ΔABC, where $AB = AC$.

Let the coordinates of B be $(a\cos t, b\sin t)$.

$\Rightarrow$ The coordinates of C are $(a\cos t, -b\sin t)$.

Because BC is a vertical line and $BM = MC$.

Area of triangle $= 1/2(BC)(AM)$

$\Rightarrow \quad A = 1/2(2b\sin t)(a - a\cos t)$

$\Rightarrow \quad A(t) = ab(\sin t - \sin t\cos t)$

$A'(t) = ab(\cos t - \cos 2t)$

$A'(t) = 0$

$\Rightarrow \quad \cos t - \cos 2t = 0 \Rightarrow \cos t + 1 - 2\cos^2 t = 0$

$\Rightarrow \quad \cos t = 1, -1/2$

$\Rightarrow t = \cos^{-1}\left(\dfrac{-1}{2}\right) = \dfrac{2\pi}{3}$ is the relevant value.

$A''(2\pi/3) = ab\sqrt{3}/2\,(-1-2) < 0$

Hence, area is maximum for $t = \dfrac{2\pi}{3}$

Maximum area $= A\left(\dfrac{2\pi}{3}\right) = ab(\sin 2\pi/3 - \sin 2\pi/3 \cos 2\pi/3)$

$$= ab\left(\frac{\sqrt{3}}{2} + \frac{\sqrt{3}}{2}\frac{1}{2}\right) = \frac{3\sqrt{3}}{4}ab$$

65. *(a)*

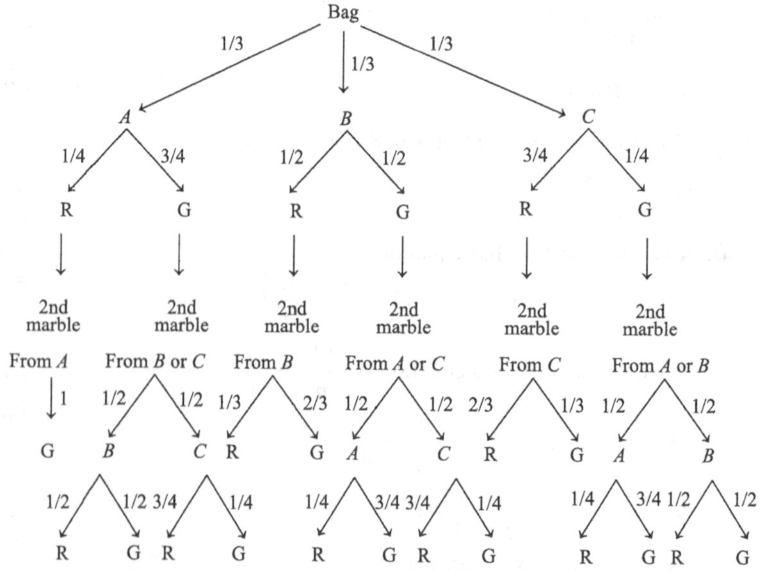

The possible situation with corresponding probabilities is explained in graphically.

The probability that second drawn marble is green, is

$$= \frac{1}{3}\left[\left\{\frac{1}{4}\cdot 1+\frac{3}{4}\cdot\frac{1}{2}\left(\frac{1}{2}+\frac{1}{4}\right)\right\}+\frac{1}{2}\left\{\frac{2}{3}+\frac{1}{2}\left(\frac{3}{4}+\frac{1}{4}\right)\right\} + \left\{\frac{3}{4}\cdot\frac{1}{3}+\frac{1}{4}\cdot\frac{1}{2}\left(\frac{3}{4}+\frac{1}{2}\right)\right\}\right]$$

$$= \frac{1}{3}\left[\frac{1}{4}+\frac{3}{8}\cdot\frac{3}{4}+\frac{1}{2}\left(\frac{2}{3}+\frac{1}{2}\right)+\left(\frac{3}{4}\cdot\frac{1}{3}+\frac{1}{8}\cdot\frac{5}{4}\right)\right]$$

$$= \frac{1}{3}\left[\frac{1}{4}+\frac{9}{32}+\frac{7}{12}+\frac{13}{32}\right]$$

$$= \frac{1}{3}\left[\frac{24+27+56+39}{96}\right]$$

$$= \frac{1}{3}\left[\frac{146}{96}\right] = \frac{1}{3}\left[\frac{73}{48}\right] = \frac{73}{144}$$

66. *(b)* $f[g(x)] = a(bx+a)+b$

$= abx + a^2 + b$ ...(i)

$g[f(x)] = b(ax+b)+a$

$= abx + b^2 + a$ ...(ii)

On subtracting Eq. (ii) from Eq. (i), we get

$f[g(20)] - g[f(20)] = a^2 - b^2 + b - a$

$\therefore\ (a^2 - b^2) + (b-a) = 28$

$\therefore\ (a-b)(a+b-1) = 28$

$= 1\times 28$ or 2×14 or 4×7

Let $a-b=1$ and $a+b-1=28$

$\Rightarrow\ a=15$ and $b=14$

If $a-b=2$ and $a+b-1=14$ (not possible)

If $a-b=4$ and $a+b-1=7$

$\Rightarrow a=6$ and $b=2$

$\therefore\ a+b=6+2=8$

67. *(c)* When $\cot x\in I$, $[\cot x]=\cot x$.

$\Rightarrow \tan x[\cot x]=1 \Rightarrow [\tan x(\cot x)]=1$

When $\cot x\notin I$, $[\cot x]<\cot x$.

$\Rightarrow \tan x[\cot x]<1 \Rightarrow [\tan x(\cot x)]=0$

$$\Rightarrow f(x)=[\tan x(\cot x)]=\begin{cases}1, & \cot x\in I\\ 0, & \cot x\notin I\end{cases}$$

So, $f(x)$ is discontinuous when $\cot x\in I$.

Now, $\frac{\pi}{12}\le x<\frac{\pi}{2}$

$\Rightarrow 0<\cot x\le 2+\sqrt{3}$

$\Rightarrow\ \cot x = 1, 2, 3$

Hence, number of points of discontinuity are $x=\cot^{-1}3$, $\cot^{-1}2$ and $\cot^{-1}1=\frac{\pi}{4}$

Thus, there are three points of discontinuity.

68. *(d)* $(1-2x+3x^2)(1+x)^{11} = (1-2x+3x^2)$

$(1+{}^{11}C_1x+{}^{11}C_2x^2+\ldots+{}^{11}C_9x^9+{}^{11}C_{10}x^{10}+x^{11})$

$\therefore$ Coefficient of x^{11}

$= 1-2\times{}^{11}C_{10}+3\times{}^{11}C_9$

$= 1-2\times 11+3\times\frac{11\times 10}{2\times 1} = 144$

69. *(a)* The given curves are

$y = ex\ln x$ ...(i)

and $y=\frac{\ln x}{ex}$ ...(ii)

Point of intersection of curves (i) and (ii), if

$$ex\ln x = \frac{\ln x}{ex}$$

$$\Rightarrow\ \left(ex-\frac{1}{ex}\right)\ln x = 0 \Rightarrow x=\frac{1}{e} \text{ or } x=1$$

At $x=\frac{1}{e}$ or $ex=1$, $y=-1$

At $x=1$, $y=0$

Point of intersection are $\left(\frac{1}{e},-1\right)$ and (1, 0).

Now, for $\frac{1}{e}<x<1$ or $\log\left(\frac{1}{e}\right)<\log x<\log 1$

$\Rightarrow -1<\log x<0$

i.e. $\log x$ is negative throughout.

$\therefore\ y_1 = ex\ln x$

$y_2=\frac{\ln x}{ex}$

Clearly, $y_1<y_2$ for interval $\frac{1}{e}<x<1$

The rough sketch of the two curves.

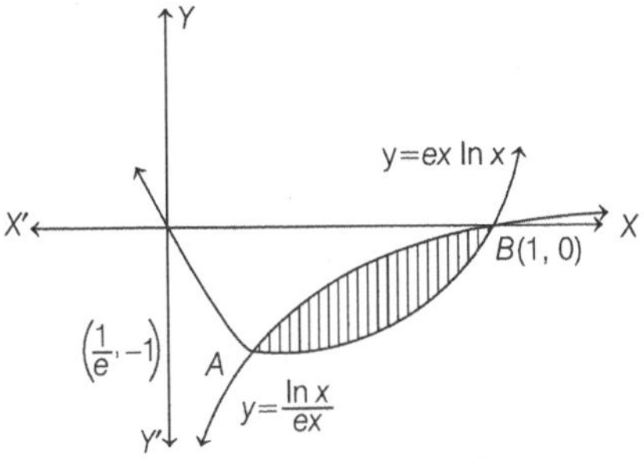

$\therefore$ Required area

$$=\left|\int_{1/e}^{1}(y_1-y_2)\,dx\right| = \left|\int_{1/e}^{1}\left(ex\ln x-\frac{\ln x}{ex}\right)dx\right|$$

$$=\left|e\int_{1/e}^{1}x\ln x\,dx-\frac{1}{e}\int_{1/e}^{1}\frac{\ln x}{x}\,dx\right|$$

$$=\left|e\left[\frac{x^2}{2}\ln x-\frac{x^2}{4}\right]_{1/e}^{1}-\frac{1}{e}\left[\frac{(\ln x)^2}{2}\right]_{1/e}^{1}\right|$$

$$= \left| e\left[\left(\frac{-1}{4}\right) - \left(\frac{-1}{2e^2} - \frac{1}{4e^2}\right)\right] - \frac{1}{e}\left[0 - \frac{1}{2}\right] \right|$$

$$= \left| e\left[-\frac{1}{4} + \frac{3}{4e^2}\right] + \frac{1}{2e} \right|$$

$$= \left| \frac{-e}{4} + \frac{3}{4e} + \frac{1}{2e} \right| = \left| \frac{5}{4e} - \frac{e}{4} \right|$$

$$= \left| \frac{5 - e^2}{4e} \right| = \frac{e^2 - 5}{4e} = \frac{a}{b}$$

$\therefore a + b = (e^2 - 5) + 4e$

Study Tactics

Find the intersection of both the curves, then roughly sketch the two curve after then, we find the area of required region (from 1/e to 1).

70. *(b)* $\int_{-3}^{5} f(|x|)\,dx = \int_{-3}^{3} f(|x|)\,dx + \int_{3}^{5} f(|x|)\,dx$

$$= 2\int_0^3 f(x)\,dx + \int_3^5 f(x)\,dx$$

$$= 2\left[\int_0^1 f(x)\,dx + \int_1^2 f(x)\,dx + \int_2^3 f(x)\,dx\right] + \left\{\int_3^4 f(x)\,dx + \int_4^5 f(x)\,dx\right\}$$

$$= 2\left(0 + \frac{1}{2} + \frac{2^2}{2}\right) + \left(\frac{9}{2} + \frac{16}{2}\right) = \frac{35}{2}$$

71. *(365)* For given condition, we can use 0, 2, 4, 6, 8 as unit place.

If 5 is at tens place, then as per the given condition, 7 should be at units place.

In such a situation, the number is not an even number. So, 5 cannot be at tens and ones places.

Hence, 5 can be only at hundreds place.

Here, two cases arises.

Case I When 5 is at hundreds place. If 5 is at hundreds place, then 7 will be at tens place. So, units place can be filled in 5 ways by using 0, 2, 4, 6, 8.

So, total number of even number $= 1 \times 1 \times 5 = 5$.

Case II When 5 is not a hundreds place. Now, hundreds place can be filled in 8 ways (0 and 5 cannot be used at hundred place). In tens place, we can use any one of the digits except 5.

So, this place can be filled in 9 ways.

At units place, we have to use one of the even digits. So, units place can be filled in 5 ways. So, total number of even numbers $= 8 \times 9 \times 5 = 360$.

Hence, the total numbers of required even numbers $= 360 + 5 = 365$

72. *(10)* On putting $x = y = 0$ in condition Eq. (i), we get

$$f(0) = f(0) \cdot f(0)$$

$\Rightarrow \quad f(0) = 0$ or $f(0) = 1$

If $f(0) = 0$, then putting $y = 0$ in Eq. (i), we get

$$f(x) = f(x + 0) = f(x)f(0) = f(x) \cdot 0 = 0$$

But by Eq. (ii) $\lim_{x \to 0} f(x) = 1 + \lim_{x \to 0} xg(x) = 1 + 0 = 1$,

which contradicts $f(x) = 0, \forall x \in R$

Hence, $f(0) = 1$

$$\lim_{h \to 0} \frac{f(x+h) - f(x)}{h} = \lim_{h \to 0} \frac{f(x)f(h) - f(x)}{h}$$

$$= f(x) \lim_{h \to 0} \frac{f(h) - 1}{h}$$

$$= f(x) \lim_{h \to 0} \frac{hg(h)}{h} \quad \text{[using Eq. (ii)]}$$

$$= f(x) \cdot \lim_{h \to 0} g(h) = f(x) \cdot 1 = f(x)$$

Thus, f is differentiable at every x and $f'(x) = f(x)$

$\Rightarrow \quad \frac{f'(x)}{f(x)} = 1$

$\Rightarrow \log f(x) = x + C$

put $\quad f(0) = 1$

$\Rightarrow \quad C = 0$

$\therefore \quad \log f(x) = x$

$\Rightarrow \log f(10) = 10$

Time Saver Tip

Gives, $f(x) = 1 + xg(x)$

$\Rightarrow \quad f'(x) = xg'(x) + g(x)$

$\Rightarrow \lim_{x \to 0} f'(x) = 0 + 1 = 1 \quad [\because \lim_{x \to 0} g(x) = 1]$

As, $f(x + y) = f(x)f(y)$ and $\lim_{x \to 0} f'(x) = 1$

$\therefore \quad f(x) = e^x$

Now, $\log f(10) = \log e^{10} = 10$

73. *(2)* We know that

$$\frac{2r}{2 + r^2 + r^4} = \frac{(1 + r + r^2) - (1 - r + r^2)}{1 + (1 + r + r^2)(1 - r + r^2)}$$

$$\therefore \quad T = \sum_{r=1}^{90} \tan^{-1}\left(\frac{2r}{2 + r^2 + r^4}\right)$$

$$= \sum_{r=1}^{90} \tan^{-1}\left[\frac{(1 + r + r^2) - (1 - r + r^2)}{1 + (1 + r + r^2)(1 - r + r^2)}\right]$$

$$= \sum_{r=1}^{90} [\tan^{-1}(1+r+r^2) - \tan^{-1}(1-r+r^2)]$$

$$= \sum_{r=1}^{90} [\tan^{-1}(1+r+r^2) - \tan^{-1}(1+(r-1)+(r-1)^2)]$$

$$= \tan^{-1}(1+90+90^2) - \tan^{-1} 1$$

$$= \tan^{-1} \frac{90+90^2}{90+90^2+2}$$

$$\Rightarrow \quad \tan T = \frac{8190}{8192}$$

$$\Rightarrow \quad 8190 \cot T = 8192$$

$$\Rightarrow 8190(\cot T - 1) = 2$$

74. *(23)* Given line $y = 3x - b$

$$\Rightarrow \quad 1 = \frac{3x - y}{b}$$

Homogenizing parabola with line

$$x^2 - 8x\left(\frac{3x-y}{b}\right) - y\left(\frac{3x-y}{b}\right) = 0$$

Since, $\angle AOB = 90°$

$\therefore$ Coefficient of x^2 + Coefficient of $y^2 = 0$

$$\Rightarrow \quad 1 - \frac{24}{b} + \frac{1}{b} = 0$$

$$\Rightarrow \quad \frac{b-24+1}{b} = 0 \Rightarrow b = 23$$

75. *(4)* Given, $x \cdot g[f(x)] \cdot f'(g(x))$

$$\cdot g'(x) = f[g(x)] \cdot g'[f(x)] \cdot f'(x)$$

$$\Rightarrow \quad x \cdot \frac{\frac{d}{dx} f[g(x)]}{f[g(x)]} = \frac{\frac{d}{dx} g[f(x)]}{g[f(x)]}$$

$$\Rightarrow x \cdot \frac{d}{dx}[\ln f(g(x))] = \frac{d}{dx} \ln [g(f(x)] \quad \ldots\text{(i)}$$

Now, $\int_0^a f[g(x)]\, dx = 1 - \frac{e^{-2a}}{2}$, differentiate w.r.t. a, we get

$$\Rightarrow \quad f[g(a)] = e^{-2a}$$

$$\Rightarrow \quad f[g(x)] = e^{-2x}$$

$$\Rightarrow \quad \log(f(g(x)) = -2x \quad \ldots\text{(ii)}$$

From Eqs. (i) and (ii), we get $-2x = \frac{d}{dx}\{\log(g(f(x)))\}$

$$\Rightarrow \quad \log[g(f(x))] = -x^2 + C$$

$$\Rightarrow \quad \log[g(f(0))] = C$$

$$\Rightarrow \quad C = 0 \quad [\because g(f(0)) = 1 \text{(given)}]$$

$$\Rightarrow \quad \log[g(f(x))] = -x^2$$

$$\Rightarrow \quad g[f(x)] = e^{-x^2}$$

$$\Rightarrow \quad g[f(4)] = e^{-16}$$

$$\Rightarrow \quad k = 4$$

JEE Main
TEST DRIVE

SECTION III

Achiever's Section

This section contains **2 Practice Sets** having a slightly higher difficulty level than JEE Main. These practice sets will help you achieve mastery over advanced concepts and give you an edge in the competitive landscape.

SECTION III

Achiever's Section

PRACTICE SET 14

- Total Marks : 300
- Total Time : 3 hrs
- No. of Questions : 75

Instructions

1. There are **three** parts in the questions paper consisting of **Physics**, **Chemistry** and **Mathematics** having 25 questions in each part of equal weightage. Each part (subject) has two sections.
2. **Section A** This section contains 20 multiple choice questions which have only one correct answer. Each question carries **4 marks** for correct answer and – **1 mark** for wrong answer.
3. **Section B** This section contains 5 questions. The answer to each of the questions is a numerical value. Each question carries **4 marks** for correct answer and – **1 mark** for wrong answer. For Section B, the answer should be rounded off to the nearest integer.

PHYSICS

Section A : Objective Type Questions

1. **Assertion** In adiabatic expansion, product of pV always decreases.

Reason In adiabatic expansion, work done by the gas is positive.

(a) Both Assertion and Reason are correct and Reason is the correct explanation of Assertion.
(b) Both Assertion and Reason are correct but Reason is not the correct explanation of Assertion.
(c) Assertion is correct but Reason is incorrect.
(d) Assertion is incorrect but Reason is correct.

2. A particle has an initial velocity $(4\hat{i}+6\hat{j})$ m/s and an acceleration of $(0.2\hat{i}+0.2\hat{j})$ m/s^2. Its speed after 10 s will be

(a) $10\sqrt{2}$ units (b) 10 units
(c) $6\sqrt{20}$ units (d) 10.8 units

3. Drift speed of electrons, when 1.6 A of current flows in a copper wire of cross-sectional area 4 mm^2 is v_d. If the electron density of copper is $9\times10^{28}/\text{m}^3$, then the value of v_d (in mm/s) is close to (charge of electron $=1.6\times10^{-19}$ C)

(a) 0.017 (b) 0.027
(c) 0.037 (d) 0.047

4. In a series L-C-R circuit, the voltage across R is 120 V and R is 1.2 kΩ with $C=1\,\mu$F. The resonant frequency ω is 200 rad/s. At resonance, the voltage across L is

(a) 500 V (b) 200 V
(c) 0.1×10^{-6} V (d) None of these

5. A constant force acting on a body of mass 5.0 kg changes its speed from 2.0 m/s to 3.5 m/s in time interval 25 s. The direction of motion of the body remains unchanged. What is the magnitude and direction of the force ?

(a) 0.18 N, along the direction of motion
(b) 0.3 N, along the direction of motion
(c) 0.18 N, opposite to direction of motion
(d) 0.3 N, opposite to direction of motion

6. The transition of hydrogen atom from different energy level is shown below.

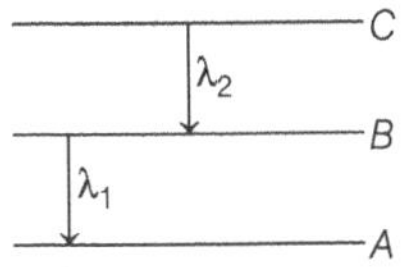

If the value of $\dfrac{\lambda_1}{\lambda_2}$ is found to be $\dfrac{7}{4n}$, then the value of n will be

(a) 2.5 (b) 9.8 (c) 4.9 (d) 19.6

7. The temperature of equal masses of three different liquids A, B and C are 12°C, 19°C and 28°C, respectively. The temperature when A and B are mixed is 16°C and when B and C are mixed is 23°C. The temperature when A and C are mixed, is

(a) 18.2°C (b) 22°C
(c) 20.2°C (d) 25.2°C

8. The length of a cylinder is measured with a meter rod having least count of 0.1 cm. Its diameter is measured with vernier calipers having least count 0.01 cm. Given that, the length is 5.0 cm and diameter is 2.00 cm. The percentage error in the calculated value of volume is

(a) 5% (b) 6% (c) 3% (d) 4%

9. The time period of revolution of electron in its ground state orbit in a hydrogen atom is 2.4×10^{-16} s. The frequency of revolution of the electron in its first excited state (in Hz) is

(a) 5.2×10^{14}
(b) 5.8×10^{14}
(c) 6.2×10^{14}
(d) None of the above

10. A hemispherical bowl is rotating with an angular speed ω. If there is no need of friction in revolving a box with bowl, which is placed inside the bowl as shown in the figure, then

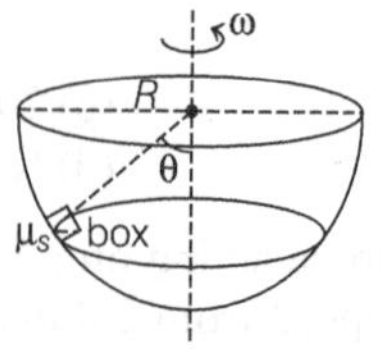

(a) $\cos\theta = \dfrac{g}{R\omega^2}$ (b) $\sin\theta = \dfrac{R\omega^2}{g}$

(c) $\tan\theta = \dfrac{g}{R\omega^2}$ (d) $\tan\theta = \dfrac{R\omega}{g}$

11. What is the instantaneous rate of change of voltage for displacement current x in the space between the parallel plate capacitor of capacity y?

(a) $x \cdot y$ (b) $\dfrac{x}{y}$

(c) $\dfrac{y}{x}$ (d) $x + y$

12. Which one of the following statements is correct ?

(a) The phenomenon of beat can take place for longitudinal wave only.
(b) Suppose two plane waves, one longitudinal and the other transverse having same frequency and amplitude are travelling in a medium in opposite directions with the same speed, by superposition of these waves, stationary waves cannot be obtained.
(c) When the sound wave refracted from air to water, then its wavelength remains unchanged.
(d) An open organ pipe contains longitudinal travelling wave.

13. The masses and radii of the earth and moon are M_1, R_1 and M_2, R_2, respectively. Their centres are at distance d apart. The minimum speed with which a particle of mass m should be projected from a point mid-way the line joining two centres, so that the particle will escape to infinity is

(a) $\sqrt{\dfrac{2G(M_1 + M_2)}{d}}$ (b) $\sqrt{\dfrac{4G(M_1 + M_2)}{d}}$

(c) $\sqrt{\dfrac{4GM_1M_2}{d}}$ (d) $\sqrt{\dfrac{4(M_1 + M_2)}{d}}$

14. Match the Column I (physical quantity) with Column II (dimension) and select the correct answer from the codes given below.

	Column I		Column II
A.	Magnetic field	1.	$[M^0L^0T^0]$
B.	Gravitational constant	2.	$[ML^0T^{-2}A^{-1}]$
C.	Universal gas constant	3.	$[ML^2T^{-2}K^{-1}mol^{-1}]$
D.	Refractive index	4.	$[M^{-1}L^3T^{-2}]$

Codes

	A	B	C	D		A	B	C	D
(a)	2	4	3	1	(b)	2	1	3	4
(c)	1	3	4	2	(d)	4	2	1	3

15. Which of the following statement(s) is/are correct ?

I. When a body slowly moves from one point to another point with some force, then the work done on the body is a non-zero constant.
II. The work done by all the forces (external and internal) on a system equals the change in total energy.
III. The negative of work done by a conservative force is equal to change in potential energy.

(a) Both I and II
(b) Both II and III
(c) Only III
(d) Only II

16. Given below are two statements.

Statement I When a disc is rotated counter-clockwise in a magnetic field (whose points perpendicularly into the plane of disc), rim of disc is more positive with respect to centre of disc.

Statement II Induced emf always opposes the change in magnetic flux responsible for its production.

In the light of above statements, choose the correct option given below.

(a) Both Statement I and Statement II are true.
(b) Both Statement I and Statement II are false.
(c) Statement I is true and Statement II is false.
(d) Statement I is false and Statement II is true.

17. If the potential at a point A (4, 0) is given by $V = (10x^2 + 12\log x - 24)$ J/C, then the electric field at this point is

(a) 83 N/C (b) 60 N/C (c) -60 N/C (d) -83 N/C

18. A steel bar of square cross-section, having area of cross-section A is subjected to a compressive force P as shown here.

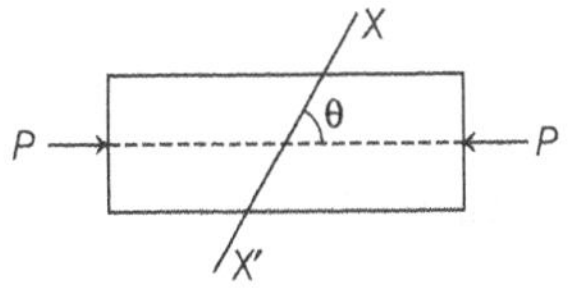

The intensity of shearing stress on the XX'-plane is

(a) $\frac{P}{A}\sin 2\theta$ (b) $\frac{P}{A}\cos 2\theta$

(c) $\frac{P}{2A}\sin 2\theta$ (d) $\frac{P}{2A}\cos 2\theta$

19. de-Broglie wavelength of an electron in the nth Bohr orbit of hydrogen atom is λ_n and the angular momentum is J_n, then

(a) $J_n \propto \lambda_n$ (b) $\lambda_n \propto 1/J_n$

(c) $J_n \propto \sqrt{\lambda_n}$ (d) None of these

20. The graph given below represents the I-V characteristics of a Zener diode. Which path of the characteristics curve is most relevant for its operation as a voltage regulator?

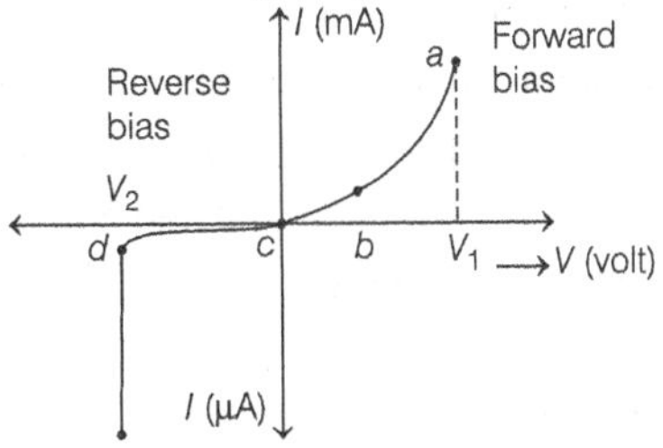

(a) ab
(b) bc
(c) cd
(d) de

Section B : Numerical Value Type Questions

21. A plane wave of wavelength 6250 Å is incident normally on a slit of width 2×10^{-2} cm. The width of the central maxima on a screen distant 50 cm will be $x\times10^{-3}$ m, then the value of x is ....

22. A ball is projected from the ground with velocity u at an angle of 60° with the horizontal surface. It reaches a maximum height of 160 m and returns to the ground. Upon hitting the ground for the first time, it losses half of its kinetic energy. Immediately after the bounce, the velocity of the ball makes an angle of 45° with the horizontal surface. The maximum height (in m), it reaches after the bounce is .... (nearest integer)

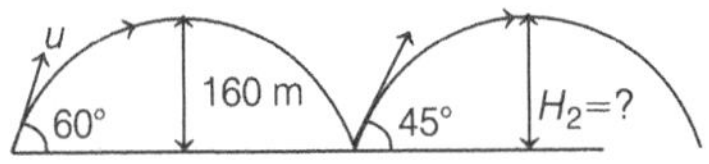

23. In a building, there are 20 bulbs of 15 W, 15 bulbs of 45 W, 15 LEDs of 9 W and 2 air conditioners of 1 kW. The voltage of electric main supply is 220 V.The minimum fuse capacity (rated value of fuse) of the building will be...... A.

24. A galvanometer coil has 600 turns and each turn has an average area of 4×10^{-4} m^2. If a torque of 1.6 N-m is required to keep this coil parallel to magnetic field when a current of 0.4 A is flowing through it, the strength of the field (in T) will be .... (nearest integer)

25. A ray of light AO of wavelength 670 nm in vacuum is incident on a glass slab at an angle of 60° and refracted at an angle of 30° along OB as shown in figure. The optical path length of light ray from A to B is .... m.

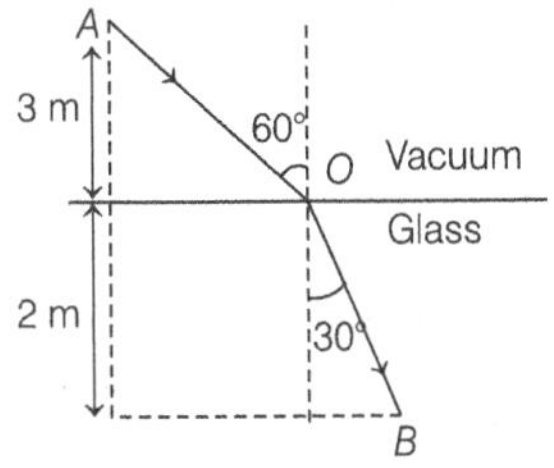

CHEMISTRY

Section A : Objective Type Questions

26. The correct statement(s) about the oxoacids, $HClO_4$ and HClO from the following is

(A) The central atom in both $HClO_4$ and HClO is sp^3-hybridised.

(B) $HClO_4$ is more acidic than HClO because of the resonance stabilisation of its anion.

(C) $HClO_4$ is formed in the reaction between Cl_2 and H_2O.

(D) The conjugate base of $HClO_4$ is weaker base than H_2O.

Choose the correct answer from the options given below.

(a) A, B and D
(b) B and D
(c) Only C
(d) A, B, C and D

27. XeF_4 reacts with SbF_5 to form

(a) XeF_6
(b) $[XeF_4]^+ [SbF_5]^-$
(c) $[XeF_3]^+ [SbF_6]^-$
(d) $Xe [SbF_4]$

28. Given below are two statements.

Statement I CHO–$(CHOH)_4$–CH_2OH with bromine water gives COOH–$(CHOH)_4$–CH_2OH.

CHO
|
(CHOH)$_4$
|
CH_2OH

COOH
|
(CHOH)$_4$
|
CH_2OH

Statement II CHO–$(CHOH)_4$–CH_2OH with HCN followed by hydrolysis and reduction with (HI/heat) gives $CH_3(CH_2)_5COOH$.

In the light of the above statements, choose the correct answer from the options given below.

(a) Statement I is false but Statement II is true.
(b) Both Statement I and Statement II are true.
(c) Both Statement I and Statement II are false.
(d) Statement I is true but Statement II is false.

29. What will be the final product obtained after reaction of aldohexose with bromine water followed by calcium salt and Fenton's reagent and methanol sequentially?

(a) 1, 2, 3, 4, 5-pentahydroxy hexanal
(b) 1, 2, 3, 4-tetrahydroxy pentanal
(c) 1, 2, 3, 4-tetramethoxy pentanal
(d) 1, 2, 3, 4, 5-pentahydroxy heptanal

30. Consider the following compounds.

I. $CH_3CH_2CH{=}CH_2$

II. $CH_2{=}C(CH_3){-}OH$

III. $CH_3CH_2CH(OH){-}CH_3$

IV. $CH_3{-}CH{=}CH{-}CH_2{-}OH$

V. $CH_3CH_2CH_2CH_2OH$

The total number of compounds that will be formed when butan-1-amine reacts with nitrous acid at 0°C.

(a) 2
(b) 3
(c) 4
(d) 1

31. Match List-I with List-II.

	List-I (Reagents)		List-II (Radicals)
A.	H_2S in presence of dil. HCl.	I.	Mg^{2+}
B.	NH_4OH in presence of NH_4Cl.	II.	Ni^{2+}
C.	H_2S in presence of NH_4OH.	III.	Al^{3+}
D.	Na_2HPO_4	IV.	Pb^{2+}

Choose the correct answer from the options given below.

(a) A-IV, B-III, C-II, D-I
(b) A-I, B-II, C-IV, D-III
(c) A-II, B-I, C-IV, D-III
(d) A-III, B-II, C-IV, D-I

32. *A*, *B* and *C* are hydroxy compounds of the elements *P*, *Q* and *R* respectively. *P*, *Q* and *R* are in the same period of the periodic table. *A* gives an aqueous solution of pH < 7. *B* reacts with both strong acids and strong alkalies. *C* gives an aqueous solution which is strongly alkaline.

Which of the following statement(s) is/are true?

I. All the three elements are metals.
II. The atomic radius decreases in the order *P*, *Q*, *R*.
III. The electronegativities decrease from *P* to *Q* to *R*.
IV. *P*, *Q* and *R* could be P, Al and Na, respectively.

Choose the correct option.

(a) I, II and IV
(b) III and IV
(c) II, III and IV
(d) I and III

33. What will be the product and degree of unsaturation present in the product for the following reaction?

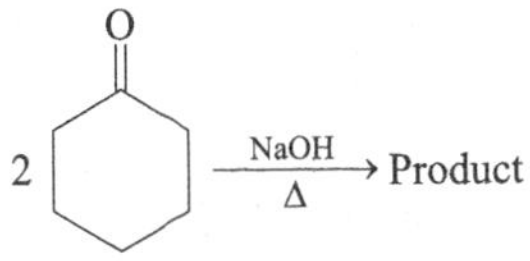

(a) 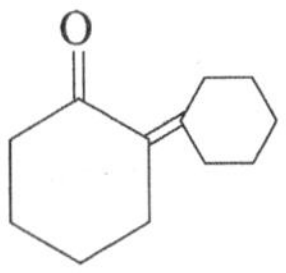 , 4

(b) 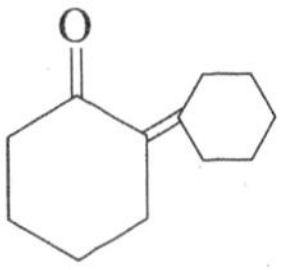 , 3

(c) 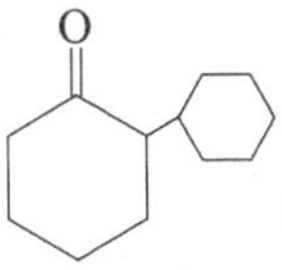 , 3

(d) 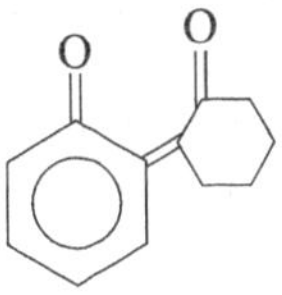 , 5

34. Which of the following is the best method for preparation of alkyl chloride?

(a) $ROH + SOCl_2 \longrightarrow$

(b) $ROH + PCl_5 \longrightarrow$

(c) $ROH + PCl_3 \longrightarrow$

(d) $ROH + HCl \xrightarrow{\text{Anhyd. } ZnCl_2}$

35. Which of the following has highest acidic strength?

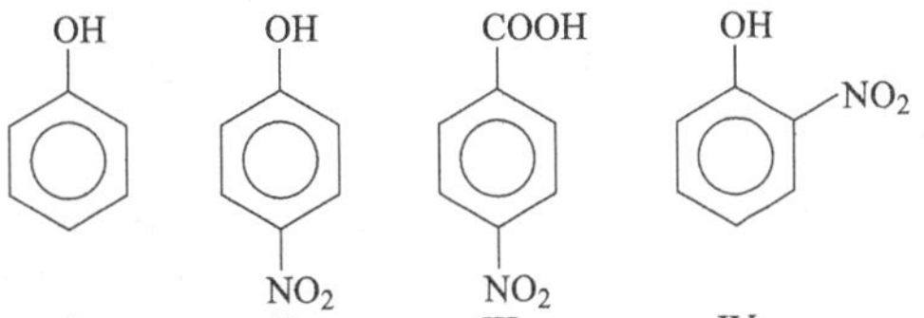

I II III IV

(a) II (b) III (c) IV (d) I

36. In reaction $A + 2B \rightleftharpoons 2C + D$, initial concentration of B was 1.5 times of $[A]$, but at equilibrium the concentrations of A and B became equal. The equilibrium constant for the reaction will be

(a) 1.4 (b) 4 (c) 3.5 (d) 8.6

37. The product obtained in following reaction is

$$CH_2{=}CH{-}CH_2{-}CH_2{-}C(=O)OH \xrightarrow[\text{Base}]{I{-}I}$$

(a) 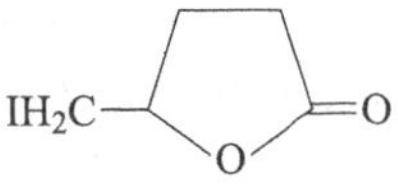

(b) 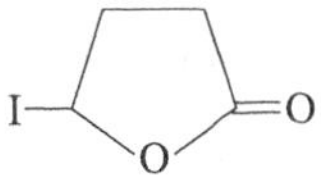

(c) 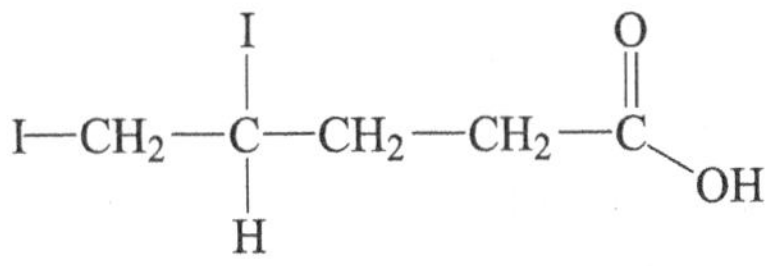

(d) 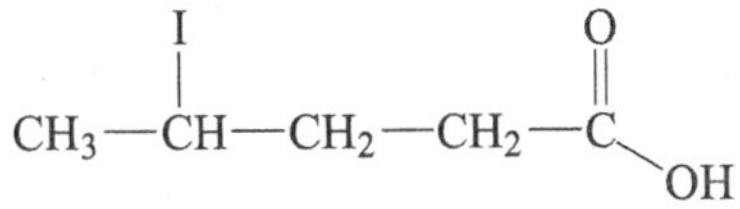

38. The number of paramagnetic species is

$$He_2^+, H_2, H_2^+, H_2^-$$

(a) 2 (b) 1 (c) 3 (d) 4

39. Which of the following Lewis structure does not contribute in resonance?

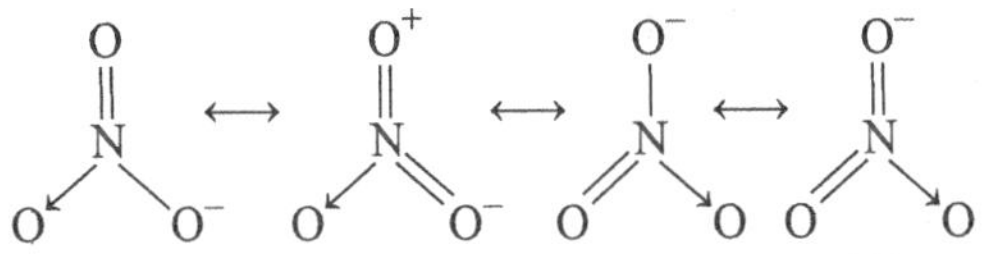

I II III IV

(a) I and II (b) II and IV

(c) III and I (d) IV only

40. A metal carbonate on being heated strongly gives a solid that forms a green solid with CoO (cobalt oxide). In which analytical group will the cation be precipitated and what will be the colour of precipitate.

(a) Group (IV), greenish white

(b) Group (III), red

(c) Group (II), yellow

(d) Group (I), greenish white

41. Match List-I with List-II and select the correct answer.

	List-I (Ion)		List-II (Shapes)
A.	ICl_2^-	I.	Linear
B.	BrF_2^+	II.	Angular
C.	ClF_4^-	III.	Tetrahedral
D.	$AlCl_4^-$	IV.	Square planar

Choose the correct option.

(a) A-I, B-II, C-III, D-IV (b) A-IV, B-I, C-II, D-III

(c) A-I, B-II, C-IV, D-III (b) A-II, B-I, C-III, D-IV

42. Given below are two statements.

Statement I For same volume expansion, work done by the gas is minimum in isobaric process and least in adiabatic process.

Statement II Work done is maximum in reversible processes.

In the light of the above statements, choose the correct answer from the options given below.

(a) Statement I is false but Statement II is true.
(b) Statement I is true but Statement II is false.
(c) Both Statement I and Statement II are true.
(d) Both Statement I and Statement II are false.

43. The correct order of basicities of the following compounds is

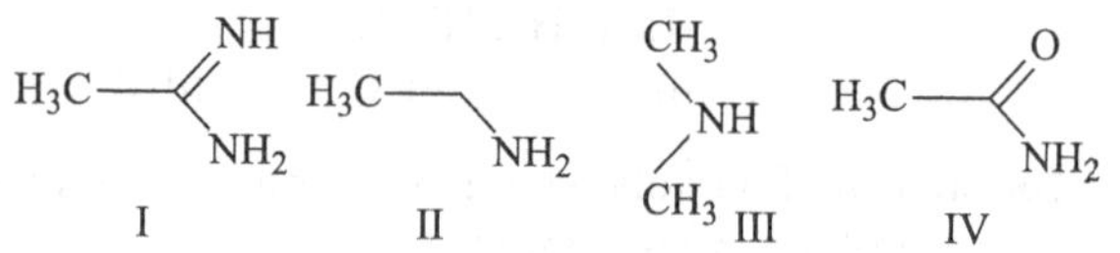

I II III IV

(a) II > I > III > IV
(b) I > III > II > IV
(c) III > I > II > IV
(d) I > II > III > IV

44. An organic compound with the formula $C_6H_{12}O_6$ forms a yellow crystalline solid with phenylhydrazine and gives a mixture of sorbitol and mannitol when reduced with sodium. Which among the following could be the compound?

(a) Fructose (b) Glucose
(c) Mannose (d) Sucrose

45. Given below are two statements one is labelled as Assertion (A) and other is labelled as Reason (R).

Assertion (A) $Fe_2(CO)_9$ is diamagnetic in nature.

Reason (R) In $Fe_2(CO)_9$, metal-metal bond between iron is present.

In the light of the above statements choose the correct answer from the options given below.

(a) Both (A) and (R) are correct but (R) is the correct explanation of (A).
(b) Both (A) and (R) are correct but (R) is not the correct explanation of (A).
(c) (A) is correct but (R) is incorrect.
(d) (A) is incorrect but (R) is correct.

Setion : B Numerical Value Type Questions

46. The standard emf of a cell, involving one electron change is found to be 0.591 V at 25°C. The equilibrium constant of the reaction is ($1F = 96500\ C\ mol^{-1}$)......$\times 10^{10}$.

47. The percentage of sulphur present in an organic compound if on analysis 0.53 g of this compound gives 1.158 g of $BaSO_4$......... .

48. A follows first order reaction, $A \longrightarrow$ Product
Concentration of A, changes from 0.1 M to 0.025 M in 40 min, when concentration of A is 0.01 M. The rate of reaction of A is ......$\times 10^{-6}$ mol L^{-1} min^{-1}.
[Nearest integer]

49. The difference between the radii of 3rd and 4th orbits of Li^{2+} is ΔR_1. The difference between the radii of 3rd and 4th orbits of He^+ is ΔR_2. The sum of ratio of ΔR_1 and ΔR_2 is ....

50. The highest value of the calculated spin only magnetic moment (in B.M) among all the transition metal complexes is .... **[Nearest integer]**

MATHEMATICS

Section A : Objective Type Questions

51. If $\log\left(\dfrac{dy}{dx}\right) = x + y$, $y(0) = 0$, then $y(-\log 2)$ is equal to

(a) $-\log 2$ (b) $\log \dfrac{3}{2}$ (c) $\log 2$ (d) $-\log \dfrac{3}{2}$

52. If $\int_0^{\pi} \dfrac{x\,dx}{1+\sin x} = k\pi$, then the value of k is

(a) 1 (b) 2 (c) 3 (d) 5

53. In ΔABC, the mid-point of the sides AB, BC and CA are respectively $(l, 0, 0)$, $(0, m, 0)$ and $(0, 0, n)$. Then, $\dfrac{AB^2 + BC^2 + CA^2}{l^2 + m^2 + n^2}$ is equal to

(a) 2 (b) 4 (c) 8 (d) 16

54. The value of $\lim\limits_{h \to 0} \dfrac{(1+h)^{\frac{1}{h}} - e\left(1 - \dfrac{h}{2}\right)}{h^2}$ is αe, then α is equal to

(a) $\dfrac{1}{24}$ (b) $\dfrac{11}{24}$
(c) $\dfrac{-11}{24}$ (d) $\dfrac{-1}{24}$

55. For $f(x) = xg(x)$ consider the following statements.

Statement I $f(x)$ is differentiable but f' is not continuous at $x = 0$.

Statement II Both f and g are differentiable at $x = 0$,

where $g(x) = \begin{cases} x \sin \dfrac{1}{x}, & x \neq 0 \\ 0, & x = 0 \end{cases}$

(a) Statement I is true and Statement II is false
(b) Statement I is false and Statement II is true
(c) Statement I and Statement II both are true
(d) Statement I and Statement II both are false

56. Let e_1 and e_2 are respectively the eccentricities of the hyperbola $\frac{x^2}{\operatorname{cosec}^2\theta} - \frac{y^2}{1} = 5\sin^2\theta$ and the ellipse $\frac{x^2}{1} + \frac{y^2}{\operatorname{cosec}^2\theta} = 5\sin^2\theta$. If $0 < \theta < \frac{\pi}{2}$ and $e_1 = \sqrt{7}\, e_2$, then the value of $2\tan^3\theta$ is equal to

(a) $6\sqrt{2}$ (b) $3\sqrt{2}$
(c) $2\sqrt{3}$ (d) $6\sqrt{3}$

57. If ${}^{m-1}C_k = (r^2 - 3)\,({}^mC_{k+1})$, then r belongs to

(a) $[-\sqrt{3}, \sqrt{3}]$ (b) $(\sqrt{3}, 2]$
(c) $(-\infty, 2)$ (d) $(2, \infty)$

58. Let O be the vertex and Q be any point on the parabola $x^2 = 16y$. If the point P divides the line segment OQ internally in the ratio $2:3$, then the locus of P is

(a) $8y = 5x^2$ (b) $y^2 = 2x$
(c) $32y = 5x^2$ (d) $x^2 = 2y$

59. The number of positive integral solution of the equation $\left[\frac{x}{[\pi^2]}\right] = \left[\frac{x}{\left[7\frac{1}{2}\right]}\right]$ is n, then the value of n is (where, $[\cdot]$ denotes greatest integer function)

(a) 5 (b) 15 (c) 8 (d) 20

60. The ordered pair (x, y) of real numbers with $0 < x < \frac{\pi}{2}$ such that $\frac{(\sin x)^{2y}}{(\cos x)^{\frac{y^2}{2}}} + \frac{(\cos x)^{2y}}{(\sin x)^{\frac{y^2}{2}}} = \sin 2x$, is

(a) $\left(\frac{\pi}{2}, 2\right)$ (b) $\left(\frac{\pi}{4}, 4\right)$ (c) $\left(\frac{\pi}{4}, 2\right)$ (d) $\left(\frac{\pi}{2}, 4\right)$

61. $\frac{\alpha^3}{2}\operatorname{cosec}^2\left(\frac{1}{2}\tan^{-1}\frac{\alpha}{\beta}\right) + \frac{\beta^3}{2}\sec^2\left(\frac{1}{2}\tan^{-1}\frac{\beta}{\alpha}\right)$ is equal to

(a) $(\alpha - \beta)(\alpha^2 + \beta^2)$
(b) $(\alpha + \beta)(\alpha^2 - \beta^2)$
(c) $(\alpha + \beta)(\alpha^2 + \beta^2)$
(c) None of the above

62. **Statement I** If A, B and C are acute positive angles such that $A + B + C = \pi$ and $\cot A \cot B \cot C = k$, then $k \le \frac{1}{3\sqrt{3}}$.

Statement II In ΔABC, $\tan A : \tan B : \tan C = 3:4:5$, then $\sin A \sin B \sin C$ is equal to $\frac{\sqrt{5}}{7}$.

(a) Only Statement I is true.
(b) Only Statement II is true.
(c) Both statements I and II are true.
(d) Both statements I and II are false.

63. If $\mathbf{p} = 3\mathbf{a} - 5\mathbf{b}$, $\mathbf{q} = 2\mathbf{a} + \mathbf{b}$, $\mathbf{r} = \mathbf{a} + 4\mathbf{b}$, $\mathbf{s} = -\mathbf{a} + \mathbf{b}$ are four vectors such that $\sin(\mathbf{p} \wedge \mathbf{q}) = 1$ and $\sin(\mathbf{r} \wedge \mathbf{s}) = 1$, then $\cos(\mathbf{a} \wedge \mathbf{b})$ is

(a) 0 (b) $\frac{\sqrt{43}}{5}$ (c) $-\frac{19}{5\sqrt{43}}$ (d) $\frac{19}{5\sqrt{43}}$

64. The coordinates of the point P on the line $\mathbf{r} = (\hat{\mathbf{i}} + \hat{\mathbf{j}} + \hat{\mathbf{k}}) + \lambda(-\hat{\mathbf{i}} + \hat{\mathbf{j}} - \hat{\mathbf{k}})$ which is nearest to the origin, is

(a) $\left(\frac{2}{3}, \frac{4}{3}, \frac{2}{3}\right)$ (b) $\left(\frac{-2}{3}, -\frac{4}{3}, \frac{2}{3}\right)$
(c) $\left(\frac{2}{3}, \frac{4}{3}, -\frac{2}{3}\right)$ (d) None of these

65. Let $f(x) = \max\{1 + \sin x, 1, 1 - \cos x\}$ for $x \in [0, 2\pi]$ and $g(x) = \max\{1, |x - 1|\}$ for $x \in R$, then $g(f(1))$ is equal to

(a) 0 (b) 1 (c) 2 (d) -1

66. If $I(n, m) = \int_0^{\pi} (n\sin x + m\cos x)^3\, dx$, where $n, m \in R$, $|n| \le 1, |m| \le 1$, then

(a) $I(n, m)$ is independent of n
(b) $I(n, m)$ is independent of m
(c) Maximum value of $I(n, 1)$ is $\frac{10}{3}$
(d) Minimum value of $I(n, 1)$ is $-\frac{10}{3}$

67. Let $A = \begin{bmatrix} 3 & -1 & -2 \\ 2 & 0 & k \\ 3 & -5 & 0 \end{bmatrix}$, where k is a real number let $B = [b_{ij}]$ such that $AB = \lambda I$, where λ is non-zero real number and I is the identity matrix of order 3. If $b_{23} = \frac{\lambda}{8}$ and $\det(B) = \frac{\lambda^2}{2}$, then $\lambda^3 : \det(B \operatorname{adj} A)$ is equal to

(a) $1:8$ (b) $8:1$ (c) $16:1$ (d) $1:16$

68. An ellipse and a hyperbola are confocal and the conjugate axis of the hyperbola is equal to the minor axis of the ellipse. If e_1 and e_2 are the eccentricities of the ellipse and the hyperbola respectively, then $\sqrt{e_1^2 + e_2^2}$ is equal to

(a) $2e_1 e_2$ (b) $e_1 e_2 \sqrt{2}$ (c) $\frac{e_1 e_2}{2}$ (d) $2e_1 e_2 + \sqrt{2}$

69. Let z and ω be two complex numbers such that $|z| \le 1, |\omega| \le 1, |z + i\omega| = |z - i\bar{\omega}| = 2$, then sum of all possible values of z, is

(a) 0 (b) 1 (c) 2 (d) None of these

70. Let the position vector of two points P and Q be $(4\hat{i} + 7\hat{j} + 8\hat{k})$ and $(2\hat{i} + 3\hat{j} + 4\hat{k})$. Let R and S be two point such that the direction ratios of the line PR and QS are (5, 9, 7) and (1, 1, −1), respectively. Let lines PR and QS intersect at T. If the vector TA is perpendicular to both PR and QS and the length of vector TA is $4\sqrt{26}$ units, $x > 0$, then the modulus of A is

(a) $\sqrt{420.5}$ units (b) $\sqrt{210.5}$ units
(c) $\sqrt{420.75}$ units (d) $\sqrt{420}$ units

Section B : Numerical Value Type Questions

71. If a, b, c, d and e are positive real numbers such that $a + b + c + d + e = 15$ and $ab^2c^3\, d^4e^5 = (120)^3\ (50)$. Then, $a^2 + b^2 + c^2 + d^2 + e^2$ is ....

72. A kite is moving horizontally at a height of 151.5 m. If the speed of the kite is 10 m/sec, how fast (in m/sec) is the string being let out, when the kite is 250 m from the boy who is flying the kite, the height of the boy being 1.5 m ....

73. Let A, B and C be 3 pair wise and mutually independent events each having probability P and if $P\left(\frac{\bar{C}}{A \cap (B \cup C)}\right) = \frac{x - P}{y - P}$, then $x^2 + y^2$ is equal to ....

74. If the standard deviation of $x_1, x_2, x_3, ..., x_4$ is 3, then standard deviation of $-2x_1 - 3, -2x_2 - 3, ..., -2x_n - 3$ is ....

75. A committee of 5 is to be chosen from a group of 9 people. Number of ways in which it can be formed if two particular persons either serve together or not at all and two other particular persons refuse to serve with each other, is equal to ....

DETAILED SOLUTIONS

Physics

1. *(b)* In adiabatic expansion, gas is cooled down.

$\because \quad T \propto pV$

Therefore, product of p and V will decrease.

Work done in adiabatic expansion, $W = \dfrac{p_1V_1 - p_2V_2}{\gamma - 1}$

As, $p_1V_1 > p_2V_2$

Hence, $W = +\text{ve}$

Hence, both Assertion and Reason are correct but Reason is not the correct explanation of Assertion.

2. *(b)* Using first equation of motion,

$$\mathbf{v} = \mathbf{u} + \mathbf{a}t$$
$$\mathbf{v} = 4\hat{\mathbf{i}} + 6\hat{\mathbf{j}} + (0.2\hat{\mathbf{i}} + 0.2\hat{\mathbf{j}}) \times 10$$
$$\mathbf{v} = 4\hat{\mathbf{i}} + 6\hat{\mathbf{j}} + 2\hat{\mathbf{i}} + 2\hat{\mathbf{j}}$$
$$\mathbf{v} = 6\hat{\mathbf{i}} + 8\hat{\mathbf{j}}$$
$$|\mathbf{v}| = \sqrt{6^2 + 8^2} = \sqrt{36 + 64} = 10\,\text{units}$$

3. *(b)* We know that,

$$I = neAv_d$$
$$v_d = \frac{I}{neA}$$
$$v_d = \frac{1.6}{9 \times 10^{28} \times 1.6 \times 10^{-19} \times 4 \times 10^{-6}}$$
$$v_d = 0.027 \times 10^{-3} = 0.027\,\text{mm/s}$$

4. *(a)* At resonance, $\omega L = \dfrac{1}{\omega C}$

Current flowing through the circuit,

$$I = \frac{V_R}{R} = \frac{120}{1200} = 0.1\,\text{A}$$

So, voltage across L is given by

$$V_L = IX_L = I\omega L$$

But $\omega L = 1/\omega C$

$$\therefore \quad V_L = \frac{I}{\omega C} = V_C = \frac{0.1}{200 \times 1 \times 10^{-6}} = 500\,\text{V}$$

5. *(b)* Mass of the body, $m = 5.0\,\text{kg}$

Initial speed, $u = 2.0\,\text{m/s}$

Final speed, $v = 3.5\,\text{m/s}$

Time, $t = 25\,\text{s}$

Force, $F = ?$

Using the first equation of motion,

$$v = u + at$$
$$\therefore \quad 3.5 = 2.0 + a \times 25$$
$$\text{or} \quad a = \frac{3.5 - 2.0}{25}\,\text{m/s}^2$$

Acceleration, $a = \dfrac{1.5}{25}\,\text{m/s}^2$

$\therefore$ Force acting on the body,

$$F = ma = 5.0 \times \frac{1.5}{25} = 0.3\,\text{N}$$

As direction of motion of the body remains unchanged, therefore the direction of force acting on the body is along the direction of motion.

6. *(b)* The given situation is shown in the figure.

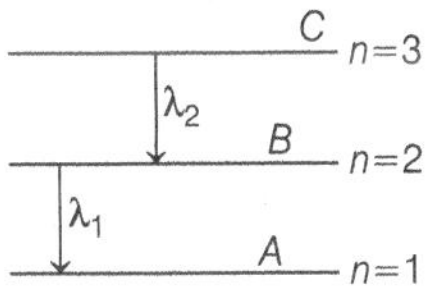

According to diagram,

$$E_B - E_A = \frac{hc}{\lambda_1}$$
$$\Rightarrow \quad \frac{-13.6}{2^2} - \frac{(-13.6)}{1^2} = \frac{hc}{\lambda_1}$$
$$\Rightarrow \quad 10.2 = \frac{hc}{\lambda_1} \Rightarrow \frac{hc}{\lambda_1} = 10.2 \quad \ldots \text{(i)}$$

Similarly, $E_C - E_B = \dfrac{hc}{\lambda_2}$

$$\Rightarrow \quad \frac{-13.6}{3^2} - \left(\frac{-13.6}{2^2}\right) = \frac{hc}{\lambda_2} \Rightarrow \frac{hc}{\lambda_2} = 1.89 \quad \ldots \text{(ii)}$$

On dividing Eq. (i) by Eq. (ii), we get

$$\frac{\dfrac{hc}{\lambda_1}}{\dfrac{hc}{\lambda_2}} = \frac{10.2}{1.89}$$
$$\Rightarrow \quad \frac{\lambda_1}{\lambda_2} = \frac{1}{5.6} = \frac{7}{4 \times 9.8} = \frac{7}{4n} \text{ (given)}$$
$$\therefore \quad n = 9.8$$

7. *(c)* By applying calorimetry method, we get

Heat gain = Heat lost

$$c_A\,(16 - 12) = c_B\,(19 - 16)$$
$$\Rightarrow \quad \frac{c_A}{c_B} = \frac{3}{4}$$

and $c_B\,(23 - 19) = c_C\,(28 - 23)$

$$\Rightarrow \quad \frac{c_B}{c_C} = \frac{5}{4}$$
$$\Rightarrow \quad \frac{c_A}{c_C} = \frac{c_A}{c_B} \times \frac{c_B}{c_C} = \frac{3}{4} \times \frac{5}{4} = \frac{15}{16} \quad \ldots \text{(i)}$$

If θ is the temperature when A and C are mixed, then

$c_A(\theta - 12) = c_C(28 - \theta)$

$$\Rightarrow \quad \frac{c_A}{c_C} = \frac{28-\theta}{\theta - 12} \quad \ldots(ii)$$

On solving Eqs. (i) and (ii), we get

$$\frac{28-\theta}{\theta-12} = \frac{15}{16} \Rightarrow 448 - 16\theta = 15\theta - 180$$

$$\therefore \quad \theta = 20.2°\text{ C}$$

8. *(c)* As we know that,

$$V = \pi r^2 h = \frac{\pi D^2 h}{4}$$

$$\frac{\Delta V}{V} = \frac{2\Delta D}{D} + \frac{\Delta h}{h}$$

$$\frac{\Delta V}{V} \times 100 = \left[2\left(\frac{0.01}{2.0}\right) + \left(\frac{0.1}{5.0}\right)\right] \times 100$$

$$= [0.01 + 0.02] \times 100 = 3\%$$

9. *(a)* Time period of revolution of an electron in its nth state (or nth orbit) of H-atom,

$$T_n = \frac{2\pi r_n}{v_n} = \frac{2\pi r_1 \cdot \left(\frac{n^2}{Z}\right)}{(v_1 Z/n)} = \frac{2\pi r_1}{v_1} \cdot \frac{n^3}{Z^2} = \frac{2\pi r_1}{v_1} \cdot n^3$$

where $Z = 1$ for H-atom r_1 and v_1 are radius and velocity of electron in ground state.

So, $\quad T_n = \frac{2\pi r_1}{v_1} \cdot n^3$ or $T_n = T_1 \cdot n^3$

where T_1 = time period of revolution in ground state.

Now, in first excited state $n = 2$, so time period,

$$T_2 = T_1 \times 2^3 = 8 \times 2.4 \times 10^{-16}\text{ s}$$

As, frequency $= \frac{1}{\text{time period}}$

So, frequency of revolution of electron in first excited state, $f_2 = \frac{1}{T_2} = \frac{1}{8 \times 2.4 \times 10^{-16}} = 5.2 \times 10^{14}\text{Hz}$

10. *(a)* Here, $N \sin\theta$ will act as F_c (centripetal force).

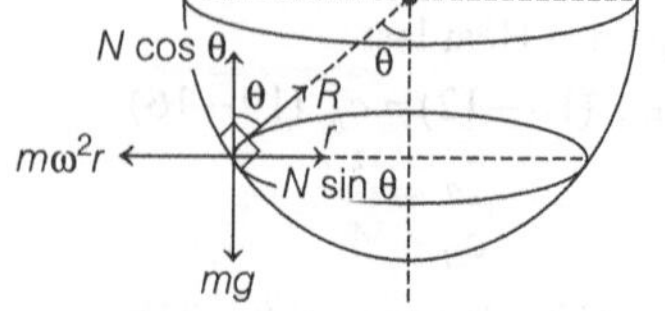

$$mr\omega^2 = N\sin\theta \quad \ldots(i)$$

$$mg = N\cos\theta \quad \ldots(ii)$$

Dividing Eq. (i) by Eq. (ii), we get

$$\tan\theta = \frac{r\omega^2}{g} \quad \ldots(iii)$$

Also, $\quad r = R\sin\theta \quad \ldots(iv)$

From Eqs. (iii) and (iv), we get

$$\tan\theta = \frac{\omega^2}{g} R\sin\theta \Rightarrow \cos\theta = \frac{g}{R\omega^2}$$

Study Tactics

When a body of mass m is kept on a revolving body in horizontal plane with angular velocity ω, then

Σ (component of all forces along the radius) $= m\omega^2 r$.

11. *(b)* The current between the plates of a capacitor,

$$i_d = \varepsilon_0 \cdot \frac{d\phi_E}{dt} = \varepsilon_0 \frac{d}{dt}(E \cdot A)$$

$$= \varepsilon_0 \frac{d}{dt}\left(\frac{V}{d} \cdot A\right) \quad \left[\because E = \frac{V}{d}\right]$$

$$= \frac{\varepsilon_0 A}{d} \cdot \left(\frac{dV}{dt}\right)$$

$$i_d = C\left(\frac{dV}{dt}\right) \quad \left[\because C = \frac{\varepsilon_0 A}{d}\right]$$

$$\therefore \quad \frac{dV}{dt} = \frac{i_d}{C} = \frac{x}{y}$$

12. *(b)* To obtain a stationary wave, two waves travelling in opposite directions, having same amplitude, same frequency are required. They must have same nature, means either both of the waves should be longitudinal or both of them should be transverse.

Beats takes place for both longitudinal and transverse waves.

An open organ pipe contains longitudinal stationary wave.

Hence, option (b) is correct.

13. *(b)*

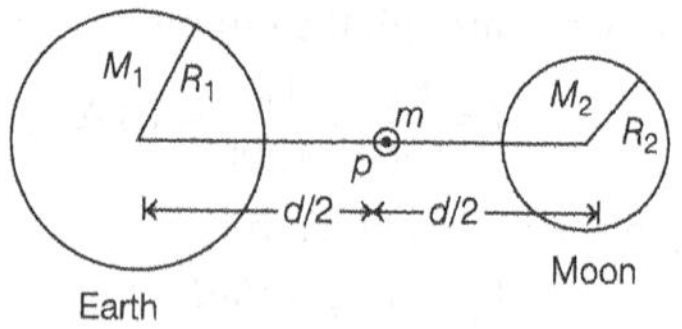

Gravitational potential of the system of earth and moon at mid-point P,

$$V = -\frac{GM_1}{d/2} - \frac{GM_2}{d/2} = -\frac{2G}{d}(M_1 + M_2)$$

$\therefore$ Potential energy of particle of mass m at mid-point P,

$$U = mV = -\frac{2Gm(M_1 + M_2)}{d}$$

So, when the particle is projected from mid-point to infinity.

$$|U| = \text{KE} \Rightarrow \frac{2Gm(M_1 + M_2)}{d} = \frac{1}{2}mv^2$$

$$\therefore \quad v = \sqrt{\frac{4G(M_1 + M_2)}{d}}$$

14. *(a)* Magnetic field (B)

$$B = \frac{F}{qv\sin\theta}$$

$$[\mathrm{B}] = \frac{[\mathrm{MLT^{-2}}]}{[\mathrm{AT}][\mathrm{LT^{-1}}]} = [\mathrm{ML^0T^{-2}A^{-1}}]$$

Gravitational constant $(G) = \dfrac{Fr^2}{m_1 m_2}$

$$\therefore \quad [\mathrm{G}] = \frac{[\mathrm{MLT^{-2}}]\cdot[\mathrm{L^2}]}{[\mathrm{M^2}]} = [\mathrm{M^{-1}L^3T^{-2}}]$$

Universal gas constant $(R) = \dfrac{pV}{nT} = \dfrac{\text{Work}}{nT}$

$$\therefore \quad [\mathrm{R}] = \frac{[\mathrm{ML^2T^{-2}}]}{[\mathrm{mol}][\mathrm{K}]} = [\mathrm{ML^2T^{-2}K^{-1}mol^{-1}}]$$

Refractive index is dimensionless.

$\therefore$ Refractive index $(\mu) = [\mathrm{M^0L^0T^0}]$

Hence, A $\rightarrow$ 2, B $\rightarrow$ 4, C $\rightarrow$ 3, D $\rightarrow$ 1.

15. *(c)* When the body slowly moves,

$$v_1 = v_2 = 0$$

$$\therefore \text{ Work done} = \Delta K = \frac{1}{2}mv_2^2 - \frac{1}{2}mv_1^2 = 0$$

By work-energy theorem,

$$(\mathrm{WD})_{\text{external}} + (\mathrm{WD})_{\text{internal}} = K_f - K_i = \Delta K$$

and $\Delta U = \text{Change in PE} = -W_C = -\int_a^b \mathbf{F_c}\cdot d\mathbf{r}$

16. *(a)*

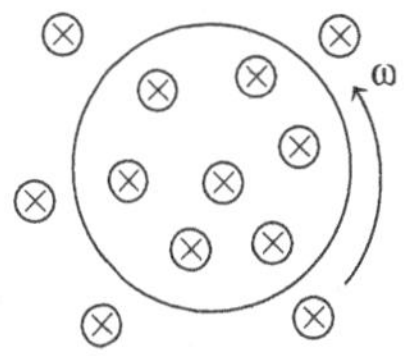

When a disc is rotating counter-clockwise, emf increases with increasing radius and rim is more positive with respect to the centre of the disc. Induced emf always opposes the change in magnetic flux which is responsible for its production.

17. *(d)* Electric field, $E = -\dfrac{dV}{dx}$

$$E = -20(x) - 12 \times \left(\frac{1}{x}\right) = -20x - \frac{12}{x}$$

At $\quad x = 4\text{ m}, E = -80 - 3 = -83\text{ N/C}$

Study Tactics

When potential, $V = f(x, y, z)$, then the electric field is given by

$$\mathbf{E} = \left(\frac{\partial v}{\partial x}\right)\hat{\mathbf{i}} + \left(\frac{\partial v}{\partial y}\right)\hat{\mathbf{j}} + \left(\frac{\partial v}{\partial z}\right)\hat{\mathbf{k}}$$

18. *(c)* Consider the equilibrium of cross-section XX'

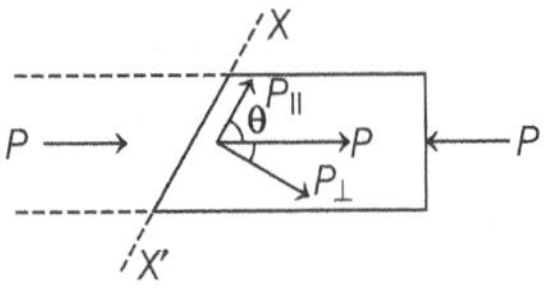

[force on both sides has to be P for a body in equilibrium]

Resolving the force in two components,

i.e. $P_{\parallel}$ (parallel to planes)

and $P_{\perp}$ (perpendicular to given plane XX')

Then, $P_{\parallel} = P\cos\theta$

Also, area of cross-section on XX' plane will be $\dfrac{A}{\sin\theta}$.

So, shearing stress

$$= \frac{P_{\parallel}}{\text{Area of cross-section}}$$

$$= \frac{P\cos\theta}{A/\sin\theta} = \frac{P\cos\theta\sin\theta}{A} = \frac{P}{2A}\sin 2\theta$$

19. *(a)* de-Broglie wavelength,

$$\lambda = \frac{h}{p} = \frac{h}{mv}$$

$$\Rightarrow \quad \lambda \propto \frac{1}{v}$$

Now, $\quad v \propto \dfrac{Z}{n}$ or $v \propto \dfrac{1}{n}$ [By Bohr atom relation]

Hence, $\lambda_n \propto n$ ...(i)

i.e. Wavelength in nth orbit λ_n is directly proportional to n (quantum number).

Now, $\quad J_n = n\dfrac{h}{2\pi}$

[J_n = angular momentum of electron in nth orbit]

i.e. $\quad J_n \propto n$ ...(ii)

Hence, from Eqs. (i) and (ii), we get

$$J_n \propto \lambda_n$$

20. *(d)* In region de, potential drop across diode remains constant for any value of current. This can be utilised for voltage regulation.

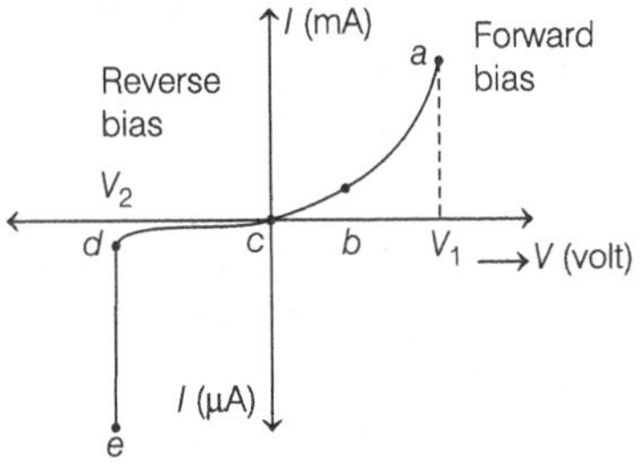

21. (3.12) Here, $\lambda = 6250\,\text{Å} = 6250\times10^{-10}$ m

$$a = 2\times10^{-2}\text{ cm} = 2\times10^{-4}\text{ m}$$
$$D = 50\text{ cm} = 0.5\text{ m}$$

The angular width of central maxima, $\beta = 2\theta = \dfrac{2\lambda}{a}$

$\therefore$ Linear width of central maxima,

$$L = \beta\times D = \frac{2\lambda}{a}\times D$$
$$= \frac{2\times6250\times10^{-10}\times0.5}{2\times10^{-4}}$$
$$= 3.12\times10^{-3}\text{ m}$$
$$= x\times10^{-3}\text{ m}$$

Hence, the value of x is 3.12.

22. (53) Height of object can be calculated by using formula

$$H = \frac{u^2\sin^2\theta}{2g}$$

First case

$$H_1 = \frac{u^2\sin^2 60^\circ}{2g} = 160$$
$$\frac{u^2}{2g} = \frac{640}{3} \quad \ldots\text{(i)}$$

On hitting ground, it losses half of its kinetic energy

$$\therefore \quad v = \frac{u}{\sqrt{2}}$$

Second case

$$H_2 = \left(\frac{u}{\sqrt{2}}\right)^2\frac{\sin^2 45^\circ}{2g}$$
$$H_2 = \frac{u^2}{4\times2g} \quad \ldots\text{(ii)}$$

From Eqs. (i) and (ii), we get

$$H_2 = \frac{1}{4}\times\frac{640}{3}$$
$$H_2 = 53.33\text{ m} \simeq 53\text{ m}$$

23. (15) Power of 20 bulbs (each of 15 W),

$$P_1 = 20\times15 = 300\text{ W}$$

Power of 15 bulbs (each of 45 W),

$$P_2 = 15\times45 = 675\text{ W}$$

Power of 15 LEDs (each of 9 W),

$$P_3 = 15\times9 = 135\text{ W}$$

Power of 2 air conditioner (each of 1 kW)

$$P_4 = 2\times1000 = 2000\text{ W}$$

Total power of the building,

$$P = P_1 + P_2 + P_3 + P_4$$
$$P = 300+675+135+2000 = 3110\text{ W}$$

Using, $P = VI$, Current drawn from main supply of 220V will be

$$I = \frac{P}{V} = \frac{3110}{220} = 14.136\text{ A}$$

$\therefore$ Minimum fuse capacity required is 15 A.

24. (17) Torque, $\tau = BINA\sin\theta$

Given, $\theta = 90^\circ \Rightarrow \sin 90^\circ = 1$

$$\tau = 1.6\text{ N-m}$$
$$N = 600$$
$$I = 0.4\text{ A}$$
$$A = 4\times10^{-4}\text{ m}^2$$
$$B = \frac{\tau}{INA\sin\theta}$$
$$= \frac{1.6}{0.4\times600\times4\times10^{-4}\times1}$$
$$B \simeq 17\text{ T}$$

25. (10) From the figure,

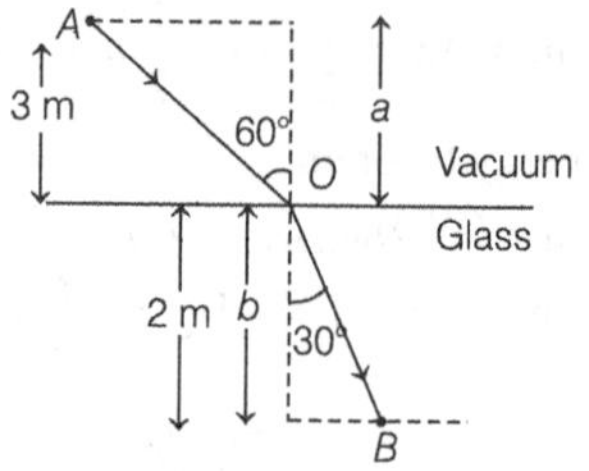

$$\cos 60^\circ = \frac{a}{AO}$$
$$AO = \frac{a}{\cos 60^\circ} = 2a \quad \ldots\text{(i)}$$

and

$$\cos 30^\circ = \frac{b}{BO}$$
$$BO = \frac{b}{\cos 30^\circ} = \frac{2}{\sqrt{3}}b \quad \ldots\text{(ii)}$$

Optical path length of light ray

$$= AO + \mu(BO) \quad \ldots\text{(iii)}$$

Here, μ can be determined using Snell's law,

i.e., $$\mu = \frac{\sin 60^\circ}{\sin 30^\circ} = \frac{\sqrt{3}/2}{1/2} = \sqrt{3} \quad \ldots\text{(iv)}$$

Substituting values from Eqs. (i), (ii) and (iv) in Eq. (iii), we get

$$\therefore \text{Optical path} = 2a + \left(\sqrt{3}\times\frac{2}{\sqrt{3}}b\right)$$
$$= 2a + 2b$$
$$= 2(3+2) = 10\text{ m}$$

Chemistry

26. (*a*) The correct statements are (A), (B) and (D).

(A)

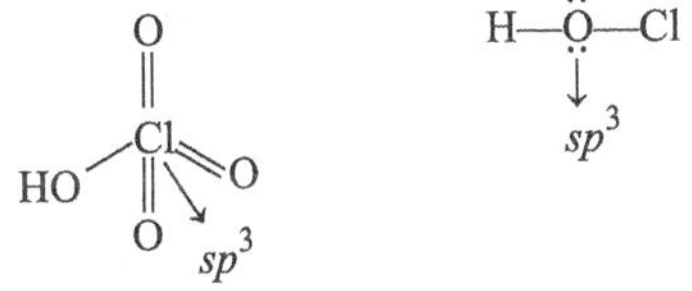

Both the central atoms are sp^3 hybridised.

(B) ClO_4^- is resonance stabilised anion

$$HClO_4 \longrightarrow ClO_4^- + H^+$$

$$HClO \longrightarrow ClO^- + H^+$$

Hence, $HClO_4$ is more acidic than HClO.

(C) is the incorrect statement.

The correct statement is as follows:

$$Cl_2 + H_2O \longrightarrow HCl + HOCl$$

(D) $HClO_4 + H_2O \longrightarrow ClO_4^- + H_3O^+$

Since, H_2O is accepting H^+ from $HClO_4$ so H_2O is stronger base compared to ClO_4^-.

27. (*c*) $\underset{\text{Lewis base}}{XeF_4} + \underset{\text{Lewis acid}}{SbF_5} \longrightarrow \underset{\text{Adduct}}{[XeF_3]^+ [SbF_6]^-}$

28. (*b*) Both Statement I and Statement II are true. The reactions are as follows

- Bromine water is a mild oxidising agent.

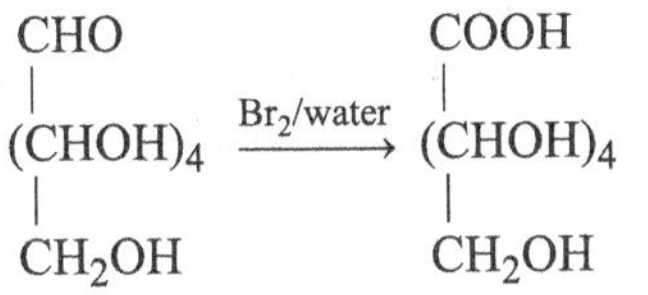

Glucose Gluconic acid

- Reaction with HCN

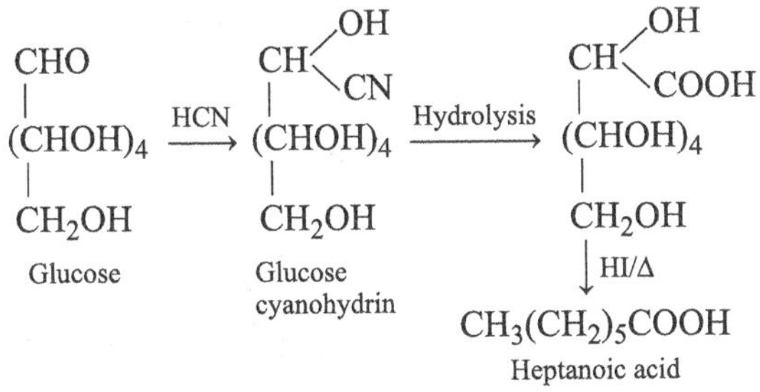

29. (*c*) **Ruff's degradation** The process of conversion of aldose to another aldose having one carbon atom less than former aldose by using following two sequential step i.e.

(i) Br_2 / H_2O and

(ii) Ca salt $+ H_2O_2 / Fe^{2+}$ (Fenton's reagent), CH_3OH

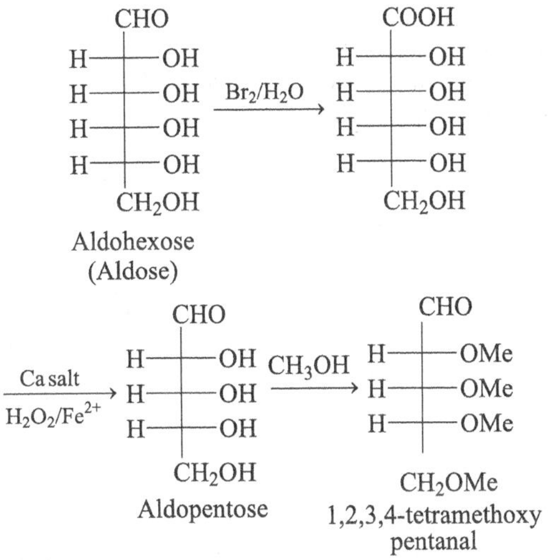

30. (*b*) Among the given compounds only I, III and V will formed when butan-1-amine reacts with nitrous acid at 0°C.

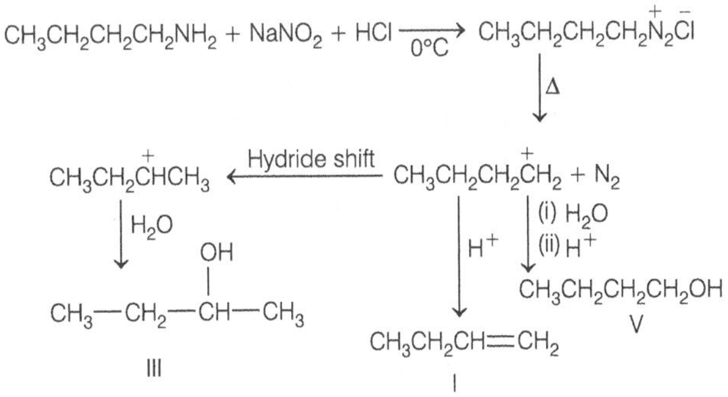

31. (*a*) The correct match is A-IV, B-III, C-II, D-I.

32. (*b*) [Keep in mind about all the elements in 3rd period and their electronegativity order]

I. All three elements are not metals, Al and Na are metals while P is non-metal.

II. Atomic radius decrease as $R > Q > P$.

III. Electronegativity of P(P), Al (Q) and Na(R) decreases as $P > Q > R$.

IV. Na, P and Al all the three belong to 3rd period.

Hydroxy compound of phosphorus = H_3PO_4 (A) which is acidic so pH < 7.

Hydroxide of aluminium is $Al(OH)_3$ is amphoteric in nature.

Hydroxide of sodium is NaOH (C) which is strongly alkaline in nature.

Study Tactics

Students are advised to solve the problem by using the concept of nature of hydroxides of element and electronegativity of elements.

33. (*a*) **Aldol condensation** Aldehyde or ketone having α H-atom undergo aldol condensation *via* enolate ion formation in presence of base to produce α, β-unsaturated product.

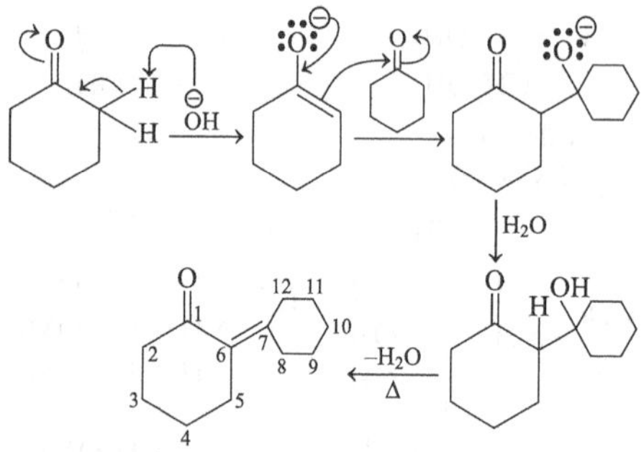

Unsaturation in organic compound Unsaturation in organic compound can be calculated by using following mathematical relation,

$$\text{Unsaturation} = (\text{C}+1) - \frac{\text{H}}{2} - \frac{X}{2} + \frac{\text{N}}{2}$$

where, C = number of carbons

H = number of hydrogens

X = number of halogens

N = number of nitrogens.

The product formed in the reaction has molecular formula

$= C_{12}H_{18}O$

$$\text{Unsaturation} = (12+1) - \frac{18}{2}$$

$= 13 - 9 = 4$

Hence, number of unsaturation in the product formed is 4.

34. (*a*) The chlorination of alcohol by using $SOCl_2$ (thionyl chloride) is the best method for the preparation of alkyl halide because in this method all other products formed are gaseous and thus halides are obtained in pure state.

$$ROH + SOCl_2 \xrightarrow{\Delta} R\text{—Cl} + HCl\uparrow + SO_2\uparrow$$

35. (*b*) Presence of EWG (electron withdrawing group) on ring withdraw electron density from the ring and hence increase the acidity of compound. The correct order of electron withdrawing nature are as follows

$$NO_2 > COOH > OH$$

The —COOH group shows acidity by loosing H^+ from more polar O— H bond, hence (III) is most acidic among all.

Study Tactics

First compare the acidity of carboxylic acid and phenol then compare the substituted phenol and phenol, finally carboxylic acid and substituted carboxylic acid.

36. (*b*) Let the degree of dissociation = x

	A	+ $2B$ ⇌	$2C$	+ D
Initial concentration	1	1.5	0	0
At equilibrium	$(1-x)$	$(1.5-2x)$	$2x$	x

Given, $(1-x) = (1.5-2x)$

$1 = 1.5 - 2x + x = 1.5 - x$

$x = 1.5 - 1 = 0.5$

Equilibrium constant for the reaction

$$K_C = \frac{[C]^2[D]}{[A][B]^2} = \frac{(2x)^2}{(1-x)} \cdot \frac{(x)}{(1.5-2x)^2}$$

$\because \quad x = 0.5$

$$K_C = \frac{(2\times 0.5)^2\,(0.5)}{(1-0.5)(1.5-2\times 0.5)^2}$$

$$= \frac{(1)\times(0.5)}{(0.5)(0.5)^2}$$

$$= \frac{0.5}{0.5\times 0.25}$$

$$= \frac{0.5}{0.125} = 4$$

37. (*a*) **Iodolactonisation** The γ, δ unsaturated carboxylic acid undergo reaction with I_2 in presence of a base to produce lactone as a product. This process is known as iodolactonisation.

The above given reaction is also an example of iodolactonisation reaction which sequentially occurs as

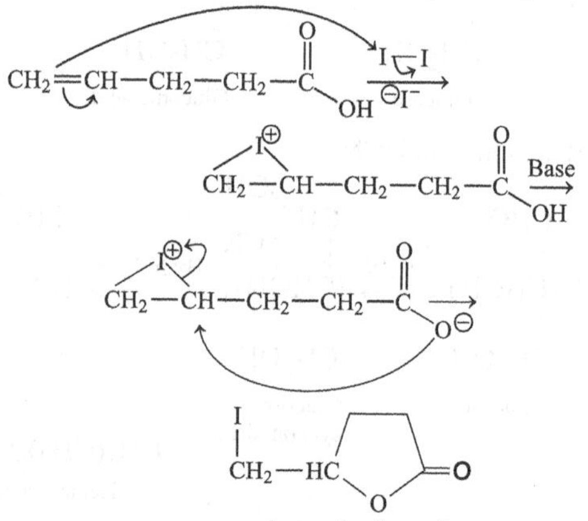

lactone (cyclic ester)
5 membered ring

Baeyer strain theory, Baeyer proposed a theory in order to explain stability of cyclic ring. According to this theory, lesser the value of angle strain, more stable is the ring and more easy is its preparation.

$$\text{Angle strain} = \frac{1}{2}(109°28' - \theta)$$

where, θ = internal angle between two bonds.

For 5 membered ring,

$$\text{Angle strain} = \frac{1}{2}(109°28' - 108°) = 0.64°$$

For 6 membered ring,

$$\text{Angle strain} = \frac{1}{2}(109°28' - 120°) = -5°36'$$

Hence, the five membered lactone (cyclic ester) is more stable than the six membered cyclic ester.

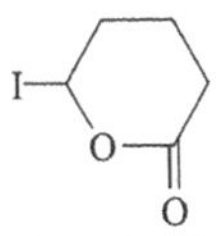

6-membered (less stable)

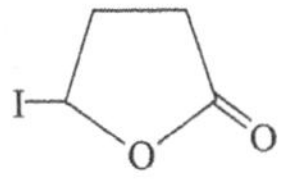

5-membered (more stable)

38. (*c*) He_2^+ and H_2^- have three electrons, one must be unpaired. H_2^+ has one unpaired electron whereas H_2 has two electrons and thus diamagnetic.

39. (*b*) Structure II is non-contributing as more electronegative atom carries a positive charge. Thus, it is not stable structure. Similarly, structure IV does not contribute as O atom carrying –ve charge violates octet rule, hence not stable.

40. (*a*) The formation of a green solid with CoO indicates Zn^{2+} (charcoal cavity test) and Zn^{2+} group precipitated in group IV.

$$ZnCO_3 \xrightarrow{\text{Heat}} ZnO \xrightarrow[\text{Heat}]{\text{CoO}} \underset{\text{Green solid}}{CoZnO_2}$$

41. (*c*) The correct match is A-I, B-II, C-IV, D-III

Ions	**BP**	**LP**	**Shape**
ICl_2^-	2	3	Linear
BrF_2^+	2	2	Angular
ClF_4^-	4	2	Square planar
$AlCl_4^-$	4	0	Tetrahedral

42. (*a*) Work done by the gas is maximum in isobaric process and least in adiabatic process.

43. (*b*) The resonance structures of $CH_3—\underset{NH_2}{\underset{|}{C}}=NH$ contains – ve charge on one of the two nitrogen atoms. Hence, it is most basic.

- $CH_3—CH_2NH_2$ is 1° amine which is less basic than 2° amine i.e. $(CH_3)_2NH$ due the presence of $+I$ effect in $(CH_3)_2NH$.
- $CH_3—\overset{O}{\overset{||}{C}}—NH_2$ will be least basic due to the delocalisation of electrons on nitrogen atom through resonance.

44. (*a*)

(i) Since, the compound form a yellow crystalline solid, i.e. osazone with phenyl hydrazine, it may be aldohexose or a ketohexose.

(ii) Since, on reduction, compound forms a mixture of sorbitol and mannitol it must be a ketohexose, i.e. fructose.

(iii) Glucose on reduction gives only sorbitol.

45. (*a*) Both (A) and (R) are true but (R) is the correct explanation of (A).

46. (*1*) Relation between K_{eq} and $E°_{cell}$ is

$$E°_{cell} = \frac{2.303RT}{nF}\log K_{eq}$$

$$E°_{cell} = \frac{0.0591}{n}\log K_{eq} \text{ (at 298K)}$$

$$0.591 = \frac{0.0591}{1}\log K_{eq}$$

$$\therefore \log K_{eq} = 10$$

$$\therefore K_{eq} = 1 \times 10^{10}$$

47. (*30*) Percentage of sulphur

$$= \frac{32}{233} \times \frac{\text{mass of } BaSO_4}{\text{mass of organic compound}} \times 100$$

$$= \frac{32}{233} \times \frac{1.158}{0.53} \times 100 = 30\%$$

48. (*347*) $A \longrightarrow$ Product (first order reaction)

For first order reaction,

$$\text{Rate constant } (k) = \frac{2.303}{t}\log_{10}\frac{[A]_0}{[A]_t}$$

At $t = 40$ min,

$$= \frac{2.303}{40}\log_{10}\frac{0.1}{0.025} = \frac{2.303}{40}\log_{10}4$$

$$= \frac{2.303}{40} \times 2\log_{10}2 = \frac{2.303}{40} \times 2 \times 0.3010$$

$$k = 0.0347 \text{ min}^{-1}$$

At concentration of $A = 0.01\text{M} = [A]$

$$\text{Rate} = \left(\frac{dx}{dt}\right) = k[A]$$

$$\frac{dx}{dt} = 0.0347 \times 0.01$$

$$= 347 \times 10^{-6} \text{ mol L}^{-1}\text{min}^{-1}$$

49. *(5)* As we know, $r = 0.529\frac{n^2}{Z}$ Å

For Li^{2+}, $(r_{Li^{2+}})_{n=4} - (r_{Li^{2+}})_{n=3} = \frac{0.529}{3}(16-9) = \Delta R_1$

For He^+, $(r_{He^+})_{n=4} - (r_{He^+})_{n=3} = \frac{0.529}{2}(16-9) = \Delta R_2$

$$\frac{\Delta R_1}{\Delta R_2} = \frac{\frac{0.529}{3}(16-9)}{\frac{0.529}{2}(16-9)} = \frac{2}{3}$$

Hence, the ratio of $\Delta R_1 : \Delta R_2$ is 2 : 3 and the sum of their ratio is 5.

50. *(6)* The spin only magnetic moment (μ) (in BM) is given by $\mu = \sqrt{n(n+2)}$

where, n = number of unpaired electrons.

The highest value of n in transition metal complex is 5 in its d^5-configuration.

$\therefore \quad \mu = \sqrt{5(5+2)}$ BM $= 5.916$ BM $\simeq 6$ BM

Mathematics

51. *(d)* We have, $\log\left(\frac{dy}{dx}\right) = x + y$

$\Rightarrow \quad \frac{dy}{dx} = e^{x+y}$

$\Rightarrow \quad \frac{dy}{dx} = e^x e^y$

$\Rightarrow \quad e^{-y}dy = e^x dx$

$\Rightarrow \quad -e^{-y} = e^x + C$

At $x = 0$, $y = 0$

$\therefore \quad -1 = 1 + C$

$\Rightarrow \quad C = -2$

$\therefore \quad -e^{-y} = e^x - 2$

$\Rightarrow \quad e^y = \frac{1}{2-e^x}$

$\Rightarrow \quad y = \log\left(\frac{1}{2-e^x}\right) = -\log(2-e^x)$

Then, $y(-\log 2) = -\log(2 - e^{-\log 2})$

$= -\log\left(2 - \frac{1}{2}\right) = -\log\frac{3}{2}$

52. *(a)* Let $I = \int_0^{\pi} \frac{x}{1+\sin x}dx$ ...(i)

Then, $I = \int_0^{\pi} \frac{\pi - x}{1+\sin(\pi - x)}dx$

$\left[\because \int_0^a f(x)dx = \int_0^a f(a-x)dx\right]$

$\Rightarrow \quad I = \int_0^{\pi} \frac{\pi - x}{1+\sin x}dx$ $[\because \sin(\pi - \theta) = \sin\theta]$ ...(ii)

On adding Eqs. (i) and (ii), we get

$2I = \int_0^{\pi} \frac{\pi}{(1+\sin x)}dx = \pi\int_0^{\pi} \frac{1}{(1+\sin x)}dx$

$= \pi\int_0^{\pi} \frac{1-\sin x}{(1+\sin x)(1-\sin x)}dx$

[multiply numerator and denominator by $(1-\sin x)$]

$\Rightarrow 2I = \pi\int_0^{\pi} \frac{1-\sin x}{1-\sin^2 x}dx$

$= \pi\int_0^{\pi} \frac{1}{\cos^2 x}dx - \pi\int_0^{\pi} \frac{\sin x}{\cos^2 x}dx$

$[\because \sin^2\theta + \cos^2\theta = 1]$

$\Rightarrow 2I = \pi\int_0^{\pi} \sec^2 x\,dx - \pi\int_0^{\pi} \sec x \cdot \tan x\,dx$

$\Rightarrow 2I = \pi[\tan x - \sec x]_0^{\pi}$

$\Rightarrow 2I = \pi[\tan\pi - \sec\pi - (\tan 0 - \sec 0)]$

$\Rightarrow 2I = \pi[0 + 1 - 0 + 1]$

$\Rightarrow 2I = 2\pi \Rightarrow I = \pi \Rightarrow K = 1$

53. *(c)*

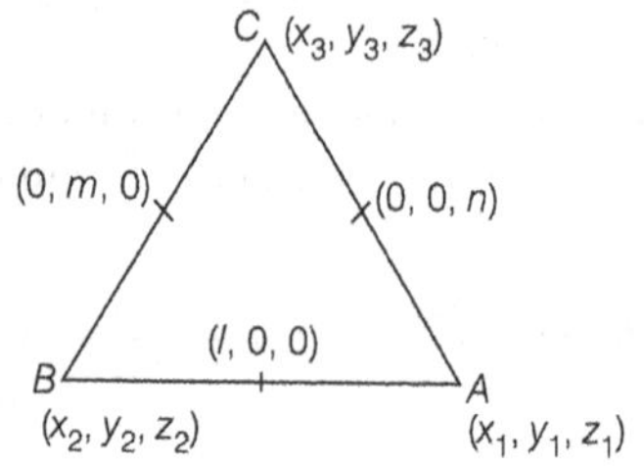

From the figure,

$x_1 + x_2 = 2l,\ y_1 + y_2 = 0,\ z_1 + z_2 = 0$

$x_2 + x_3 = 0,\ y_2 + y_3 = 2m,\ z_2 + z_3 = 0$

and $x_1 + x_3 = 0,\ y_1 + y_3 = 0,\ z_1 + z_3 = 2n$

On solving, we get

$x_1 = l, x_2 = l$ and $x_3 = -l$

$y_1 = -m,\ y_2 = m$ and $y_3 = m$

and $z_1 = n,\ z_2 = -n,\ z_3 = n$

$\because$ Coordinates are $A(l,-m,n), B(l,m,-n)$ and $C(-l,m,n)$.

$\therefore \frac{AB^2 + BC^2 + CA^2}{l^2 + m^2 + n^2}$

$= \frac{4m^2 + 4n^2 + 4l^2 + 4n^2 + 4l^2 + 4m^2}{l^2 + m^2 + n^2} = 8$

54. *(b)* Let $x = (1+h)^{\frac{1}{h}}$

On taking log both sides, we get

$\log x = \frac{1}{h}\log(1+h) = \frac{1}{h}\left[h - \frac{h^2}{2} + \frac{h^3}{3} - \frac{h^4}{4} + \dots\right]$

$$\log x = 1 - \frac{h}{2} + \frac{h^2}{3} - \frac{h^3}{4} + \ldots$$

$$x = e^{1-\frac{h}{2}+\frac{h^2}{3}-\frac{h^3}{4}+\ldots} \Rightarrow ee^{\left(-\frac{h}{2}+\frac{h^2}{3}-\frac{h^3}{4}+\ldots\right)}$$

$$= e\left[1 + \left(-\frac{h}{2}+\frac{h^2}{3}-\frac{h^3}{4}+\ldots\right) + \frac{1}{2!}\left(-\frac{h}{2}+\frac{h^2}{3}-\frac{h^3}{4}+\ldots\right)^2 + \ldots\right]$$

$$\Rightarrow x - e + \frac{eh}{2} = eh^2\left[\frac{1}{3} + 0(h) + \frac{1}{2}\left(-\frac{1}{2}+0(h)\right)^2 + \ldots\right]$$

where, $0(h)$ is terms containing h.

$$\Rightarrow \lim_{h\to 0} \frac{(1+h)^{\frac{1}{h}} - e\left(1-\frac{h}{2}\right)}{h^2} = e\left[\frac{1}{3}+\frac{1}{8}\right] = \frac{11}{24}e = \alpha e$$

$$\Rightarrow \qquad \alpha = \frac{11}{24}$$

Study Tactics

Assume $x = (1+h)^{\frac{1}{h}}$

Take log both side and apply expansion of $\log(1+h)$

i.e. $\log(1+h) = h - \frac{h^2}{2} + \frac{h^3}{3} - \frac{h^4}{4} + \ldots\ldots$

Now, use above in given expression and solve.

55. *(a)* Since, $g(x) = \begin{cases} x\sin\frac{1}{x}, & x \neq 0 \\ 0, & x = 0 \end{cases}$

and $f(x) = xg(x)$

$\therefore$ $f(x) = \begin{cases} x^2\sin\frac{1}{x}, & x \neq 0 \\ 0, & x = 0 \end{cases}$

and for $x \neq 0$, $f'(x) = x^2\cos\frac{1}{x}\left(-\frac{1}{x^2}\right) + 2x\sin\frac{1}{x}$

$$= -\cos\frac{1}{x} + 2x\sin\frac{1}{x}$$

which exists for all $x \neq 0$.

For $x = 0$

$$f'(0) = \lim_{x\to 0}\frac{f(x)-f(0)}{x-0}$$

$$= \lim_{x\to 0}\frac{x^2\sin\frac{1}{x} - 0}{x-0} = \lim_{x\to 0} x\sin\frac{1}{x} = 0$$

$$\Rightarrow f'(x) = \begin{cases} -\cos\frac{1}{x} + 2x\sin\frac{1}{x}, & x \neq 0 \\ 0, & x = 0 \end{cases}$$

At $x = 0$, $\cos\frac{1}{x}$ is not continuous, therefore $f'(x)$ is not continuous at $x = 0$.

Now, $Lg'(0) = \lim_{x\to 0}\frac{0-(-x)\sin\left(-\frac{1}{x}\right)}{x} = \lim_{x\to 0} -\sin\frac{1}{x}$

Limit does not exist.

g is not differentiable at $x = 0$

$\therefore$ Only Statement I is true.

56. *(d)* Given equation of hyperbola is

$$\frac{x^2}{\text{cosec}^2\theta} - \frac{y^2}{1} = 5\sin^2\theta$$

$$\Rightarrow \frac{x^2}{5} - \frac{y^2}{5\sin^2\theta} = 1$$

$$\therefore \quad e_1^2 = \frac{5+5\sin^2\theta}{5}$$

$$\Rightarrow \quad e_1^2 = 1 + \sin^2\theta \qquad \ldots\text{(i)}$$

Also, given the equation of ellipse is

$$\frac{x^2}{1} + \frac{y^2}{\text{cosec}^2\theta} = 5\sin^2\theta$$

$$\Rightarrow \frac{x^2}{5\sin^2\theta} + \frac{y^2}{5} = 1$$

$$\therefore \quad e_2^2 = \frac{5-5\sin^2\theta}{5}$$

$$\Rightarrow \quad e_2^2 = \cos^2\theta \qquad \ldots\text{(ii)}$$

$$\because \quad e_1 = \sqrt{7}e_2 \qquad \text{[given]}$$

$$\Rightarrow \quad e_1^2 = 7e_2^2$$

$$\Rightarrow \quad 1 + \sin^2\theta = 7\cos^2\theta \qquad \text{[using Eqs. (i) and (ii)]}$$

$$\Rightarrow 1 + \sin^2\theta + 7\sin^2\theta - 7 = 0$$

$$\Rightarrow \quad 8\sin^2\theta - 6 = 0 \Rightarrow \sin\theta = \frac{\sqrt{3}}{2}$$

$$\therefore \quad \cos\theta = \frac{1}{2}$$

$$\therefore 2\tan^3\theta = 2\left[\frac{\sqrt{3}}{2}\times\frac{2}{1}\right]^3 = 6\sqrt{3}$$

57. *(b)* ${}^{m-1}C_k = (r^2-3)({}^mC_{k+1})$

$$\Rightarrow \frac{(m-1)!}{k!(m-1-k)!} = (r^2-3)\frac{(m)!}{(k+1)!(m-k-1)!}$$

$$\Rightarrow \frac{(m-1)!}{k!(m-1-k)!} = (r^2-3)\frac{m(m-1)!}{(k+1)k!(m-1-k)!}$$

$$\Rightarrow \quad r^2 - 3 = \frac{k+1}{m} \qquad \ldots\text{(i)}$$

But $0 \leq k \leq m-1$

$$\Rightarrow \quad 1 \leq k+1 \leq m$$

$\Rightarrow \quad \frac{1}{m} \le \frac{k+1}{m} \le 1$

$\Rightarrow \quad \frac{1}{m} \le r^2 - 3 \le 1$ [using Eq. (i)]

$\Rightarrow \quad \frac{1}{m} + 3 \le r^2 \le 4 \Rightarrow 3 < r^2 \le 4$ $\left[\because \frac{1}{m} \ge 0\right]$

$\Rightarrow \quad -2 \le r < -\sqrt{3}$ or $\sqrt{3} < r \le 2$

$\Rightarrow \quad r \in (\sqrt{3}, 2]$

58. *(c)* Any point on the parabola $x^2 = 16y$ is $(2at, at^2)$ i.e. $(8t, 4t^2)$.

$\therefore$ Point P divides the line segment joining $O(0,0)$ and $Q(8t, 4t^2)$ in the ratio 2 : 3.

Now, by section formula for internal division.

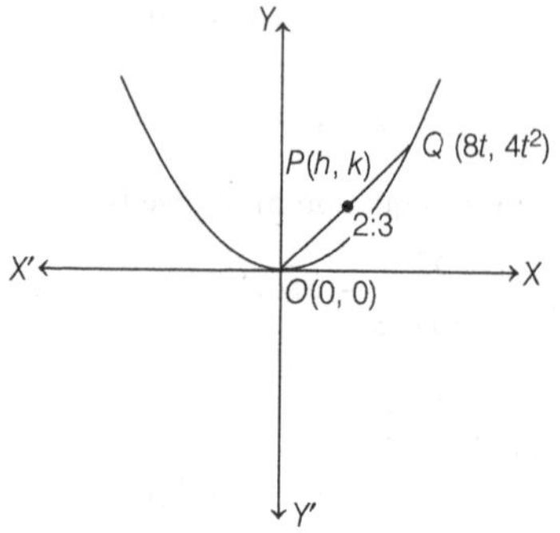

$\therefore \quad h = \frac{2 \times 8t + 3 \times 0}{5} \Rightarrow h = \frac{16t}{5}$

and $\quad k = \frac{2 \times 4t^2 + 3 \times 0}{5} \Rightarrow k = \frac{8t^2}{5}$

$\Rightarrow \quad k = \frac{5}{32}h^2 \Rightarrow 32k = 5h^2$

$\therefore \quad 32y = 5x^2$

which is the required locus.

59. *(b)* Given, $\left[\frac{x}{9}\right] = \left[\frac{x}{7}\right]$

Case I $0 \le \frac{x}{7} < 1$ and $0 \le \frac{x}{9} < 1$

$\Rightarrow \quad 0 \le x < 7$ and $0 \le x < 9$

$\Rightarrow x = 1, 2, 3, 4, 5, 6$

Case II $1 \le \frac{x}{7} < 2$ and $1 \le \frac{x}{9} < 2$

$\Rightarrow \quad 7 \le x < 14$ and $9 \le x < 18$

$\Rightarrow \quad x = 9, 10, 11, 12, 13$

Case III $2 \le \frac{x}{7} < 3$ and $2 \le \frac{x}{9} < 3$

$\Rightarrow \quad 14 \le x < 21$ and $18 \le x < 27$

$\Rightarrow \quad x = 18, 19, 20$

Case IV $3 \le \frac{x}{7} < 4$ and $3 \le \frac{x}{9} < 4$

$\Rightarrow \quad x = 27$

$n = 6 + 5 + 3 + 1 = 15$

60. *(c)* Using AM $\ge$ GM inequality,

$$\frac{(\sin x)^{2y}}{(\cos x)^{\frac{y^2}{2}}} + \frac{(\cos x)^{2y}}{(\sin x)^{\frac{y^2}{2}}} \ge 2\sqrt{\frac{(\sin x \cos x)^{2y}}{(\cos x \sin x)^{\frac{y^2}{2}}}}$$

$$= 2(\sin x \cos x)^{y - \frac{y^2}{4}}$$

$\Rightarrow \quad \sin 2x \ge 2(\sin x \cos x)^{y - \frac{y^2}{4}}$

$\Rightarrow \quad \sin x \cos x \ge (\sin x \cos x)^{y - \frac{y^2}{4}}$

$\Rightarrow \quad y - \frac{y^2}{4} \ge 1$

$\Rightarrow y^2 - 4y + 4 \le 0 \Rightarrow (y-2)^2 \le 0$

$\Rightarrow y = 2$ and $\sin x = \cos x \Rightarrow x = \frac{\pi}{4}$

$\therefore \left(\frac{\pi}{4}, 2\right)$ is the ordered pair.

61. *(c)* $\frac{\alpha^3}{2} \operatorname{cosec}^2\left(\frac{1}{2}\tan^{-1}\frac{\alpha}{\beta}\right) + \frac{\beta^3}{2}\sec^2\left(\frac{1}{2}\tan^{-1}\frac{\beta}{\alpha}\right)$

$$= \alpha^3 \frac{1}{1 - \cos\left(\tan^{-1}\frac{\alpha}{\beta}\right)} + \beta^3 \frac{1}{1 + \cos\left(\tan^{-1}\frac{\beta}{\alpha}\right)}$$

$$= \alpha^3 \frac{1}{1 - \cos\left(\cos^{-1}\frac{\beta}{\sqrt{\alpha^2+\beta^2}}\right)} + \beta^3 \frac{1}{1 + \cos\left(\cos^{-1}\frac{\alpha}{\sqrt{\alpha^2+\beta^2}}\right)}$$

$$= \alpha^3 \frac{1}{1 - \frac{\beta}{\sqrt{\alpha^2+\beta^2}}} + \beta^3 \frac{1}{1 + \frac{\alpha}{\sqrt{\alpha^2+\beta^2}}}$$

$$= \sqrt{\alpha^2+\beta^2}\left(\frac{\alpha^3}{\sqrt{\alpha^2+\beta^2} - \beta} + \frac{\beta^3}{\sqrt{\alpha^2+\beta^2} + \alpha}\right)$$

$$= \sqrt{\alpha^2+\beta^2}\left[\alpha^3 \frac{(\sqrt{\alpha^2+\beta^2} + \beta)}{\alpha^2} + \beta^3 \frac{(\sqrt{\alpha^2+\beta^2} - \alpha)}{\beta^2}\right]$$

$$= \sqrt{\alpha^2+\beta^2}\,(\alpha+\beta)\sqrt{\alpha^2+\beta^2}$$

$$= (\alpha+\beta)(\alpha^2+\beta^2)$$

62. *(a)* **Statement I** Given, $A + B + C = \pi$

$\Rightarrow \tan A + \tan B + \tan C = \tan A \tan B \tan C$

Now, AM $\geq$ GM

$\Rightarrow \dfrac{\tan A + \tan B + \tan C}{3} \geq (\tan A \cdot \tan B \cdot \tan C)^{1/3}$

$\Rightarrow (\tan A \tan B \tan C)^{2/3} \geq 3$

$\Rightarrow \left(\dfrac{1}{k}\right)^{2/3} \geq 3 \Rightarrow k \leq \dfrac{1}{3\sqrt{3}}$

Hence, Statement I is true.

Statement II We have,

$\tan A : \tan B : \tan C = 3:4:5$

$\Rightarrow \dfrac{\tan A}{3} = \dfrac{\tan B}{4} = \dfrac{\tan C}{5} = \lambda$ (say)

$\Rightarrow \tan A = 3\lambda, \tan B = 4\lambda$ and $\tan C = 5\lambda$

Now, we have

$\tan A + \tan B + \tan C = \tan A \tan B \tan C$

$\Rightarrow 3\lambda + 4\lambda + 5\lambda = 60\lambda^3$

$\Rightarrow \lambda = \dfrac{1}{\sqrt{5}}$ [as λ cannot be negative]

$\Rightarrow \tan A = \dfrac{3}{\sqrt{5}}, \tan B = \dfrac{4}{\sqrt{5}}$ and $\tan C = \sqrt{5}$

$\Rightarrow \sin A = \dfrac{3}{\sqrt{14}}, \sin B = \dfrac{4}{\sqrt{21}}$ and $\sin C = \dfrac{\sqrt{5}}{\sqrt{6}}$

$\therefore \sin A \sin B \sin C = \dfrac{3 \times 4 \times \sqrt{5}}{\sqrt{14} \times \sqrt{21} \times \sqrt{6}} = \dfrac{2\sqrt{5}}{7}$

So, Statement II is false.

63. *(d)* $\mathbf{p} = 3\mathbf{a} - 5\mathbf{b}, \mathbf{q} = 2\mathbf{a} + \mathbf{b}$

$\mathbf{r} = \mathbf{a} + 4\mathbf{b}, \mathbf{s} = -\mathbf{a} + \mathbf{b}$

$\because \mathbf{p} \cdot \mathbf{q} = 0$

$\therefore 6|\mathbf{a}|^2 - 5|\mathbf{b}|^2 - 7\mathbf{a} \cdot \mathbf{b} = 0$ ...(i)

$\because \mathbf{r} \cdot \mathbf{s} = 0$

$\Rightarrow -|\mathbf{a}|^2 - 3\mathbf{a} \cdot \mathbf{b} + 4|\mathbf{b}|^2 = 0$ ...(ii)

$\Rightarrow \dfrac{6|\mathbf{a}|^2 - 5|\mathbf{b}|^2}{7} = \dfrac{4|\mathbf{b}|^2 - |\mathbf{a}|^2}{3}$ [from Eqs. (i) and (ii)]

$\Rightarrow 25|\mathbf{a}|^2 = 43|\mathbf{b}|^2$

$\Rightarrow |\mathbf{a}| = \dfrac{\sqrt{43}}{5}|\mathbf{b}|$

$\therefore \cos\theta = \dfrac{1}{7}\left(6\left|\dfrac{\mathbf{a}}{\mathbf{b}}\right| - 5\left|\dfrac{\mathbf{b}}{\mathbf{a}}\right|\right)$

$= \dfrac{1}{7}\left(\dfrac{6\sqrt{43}}{5} - 5 \cdot \dfrac{5}{\sqrt{43}}\right)$

$= \dfrac{1}{7}\left(\dfrac{6 \times 43 - 125}{5\sqrt{43}}\right) = \dfrac{19}{5\sqrt{43}}$

64. *(a)* Let the point P be (x, y, z), then the vector $x\hat{\mathbf{i}} + y\hat{\mathbf{j}} + z\hat{\mathbf{k}}$ will lie on the line.

$\Rightarrow (x-1)\hat{\mathbf{i}} + (y-1)\hat{\mathbf{j}} + (z-1)\hat{\mathbf{k}} = -\lambda\hat{\mathbf{i}} + \lambda\hat{\mathbf{j}} - \lambda\hat{\mathbf{k}}$

$\Rightarrow x = 1-\lambda, y = 1+\lambda$ and $z = 1-\lambda$

Now, point P is nearest to the origin.

$\Rightarrow D = (1-\lambda)^2 + (1+\lambda)^2 + (1-\lambda)^2$

$\dfrac{dD}{d\lambda} = -4(1-\lambda) + 2(1+\lambda) = 0 \Rightarrow \lambda = \dfrac{1}{3}$

The point is $\left(\dfrac{2}{3}, \dfrac{4}{3}, \dfrac{2}{3}\right)$.

65. *(b)* $f(x) = \max\{1+\sin x, 1, 1-\cos x\}$ for $x \in [0, 2\pi]$

$$f(x) = \begin{cases} 1+\sin x, & 0 \leq x \leq \dfrac{3\pi}{4} \\ 1-\cos x, & \dfrac{3\pi}{4} \leq x \leq \dfrac{3\pi}{2} \\ 1, & \dfrac{3\pi}{2} \leq x \leq 2\pi \end{cases}$$

and $g(x) = \max\{1, |x-1|\}$ for $x \in R$

$$g(x) = \begin{cases} 1-x, & x \leq 0 \\ 1, & 0 \leq x \leq 2 \\ x-1, & x \geq 2 \end{cases}$$

$f(1) = 1 + \sin 1$

$\Rightarrow g(f(1)) = g(1+\sin 1) = \max\{1, |1+\sin 1 - 1|\}$

$\Rightarrow g(f(1)) = 1$

66. *(c)* We have, $I(n,m) = \int_0^\pi (n\sin x + m\cos x)^3\, dx$

$= \int_0^\pi n^3 \sin^3 x\, dx + \int_0^\pi 3n^2 m \sin^2 x \cos x\, dx$

$+ \int_0^\pi 3nm^2 \sin x \cos^2 x\, dx + \int_0^\pi m^3 \cos^3 x\, dx$

$= 2n^3 \int_0^{\frac{\pi}{2}} \sin^3 x\, dx + 3mn^2 \int_0^\pi \sin^2 x \cos x\, dx$

$+ 3nm^2 \int_0^\pi \cos^2 x \sin x\, dx + m^3 \int_0^\pi \cos^3 x\, dx$

$= 2n^3 I_1 + 3mn^2 I_2 + 3nm^2 I_3 + m^3 I_4$

$I_1 = \int_0^{\frac{\pi}{2}} \sin^3 x\, dx$

By using Wallis' formula,

$\int_0^{\frac{\pi}{2}} \sin^n x\, dx = \dfrac{2 \cdot 4 \cdot 6 \ldots (n-1)}{1 \cdot 3 \cdot 5 \ldots n}$, where n is odd

$\therefore I_1 = \dfrac{2}{1 \cdot 3} = \dfrac{2}{3}$

$I_2 = \int_0^\pi \sin^2 x \cos x\, dx$

Let $\sin x = t \Rightarrow \cos x\, dx = dt$

when $x = 0, t = 0$

when $x = \pi, t = \pi \Rightarrow I_2 = 0$

$$I_3 = \int_0^{\pi} \cos^2 x \sin x \, dx$$

Let $\cos x = t \Rightarrow \sin x \, dx = -dt$

when $x = 0, t = 1$

when $x = \pi, t = -1$

$$\therefore \quad I_3 = \int_1^{-1} -t^2 dt = \int_{-1}^{1} t^2 dt = \left[\frac{t^3}{3}\right]_{-1}^{1} = \frac{2}{3}$$

$$I_4 = \int_0^{\pi} \cos^3 x \, dx \quad \text{...(i)}$$

$$\Rightarrow I_4 = \int_0^{\pi} \cos^3 (\pi - x) \, dx = -\int_0^{\pi} \cos^3 x \, dx \quad \text{...(ii)}$$

On adding Eqs. (i) and (ii), we get

$$2I_4 = 0 \Rightarrow I_4 = 0$$

$$\therefore \quad I(n, m) = 2n^3 \times \frac{2}{3} + 3m^2 n \left(\frac{2}{3}\right)$$

$$\Rightarrow I(n, m) = \frac{4}{3} n^3 + 2m^2 n$$

which is dependent on both n and m.

Now, $I(n, 1) = \frac{4}{3} n^3 + 2n \le \frac{4}{3} + 2 = \frac{10}{3}$

$\therefore$ Maximum value of $I(n, 1) = \frac{10}{3}$

67. *(a)* Given, $AB = \lambda I$

$$\Rightarrow \quad B = (\lambda I) A^{-1} = \lambda A^{-1}$$

$\because \det(B) = \frac{\lambda^2}{2} \ne 0$

So, A is invertible.

Cofactors of A

$C_{11} = 5k, \quad C_{12} = 3k, \quad C_{13} = -10,$

$C_{21} = 10, \quad C_{22} = 6, \quad C_{23} = 12,$

$C_{31} = -k, \quad C_{32} = -3k - 4, \quad C_{33} = 2,$

$$\det(A) = 3C_{11} + (-1)C_{12} + (-2)C_{13}$$
$$= 15k - 3k + 20 = 12k + 20$$

Now, $b_{23} = -\lambda \times \frac{1}{12k + 20}$,

$$C_{32} = \frac{-\lambda(-3k - 4)}{12k + 20} = \frac{\lambda}{8}$$

$\Rightarrow 12k + 20 = 24k + 32$

$\Rightarrow \quad k = -1$

$$\therefore \quad \det(B) = \lambda^3 \det A^{-1} = \frac{\lambda^3}{\det(A)}$$

Now, $\det(B \operatorname{adj} A) = (\det B)[\det(\operatorname{adj} A)]$

$$= (\det B)(\det A)^2$$

$$= \frac{\lambda^3}{(\det A)} (\det A)^2$$

$$= \lambda^3 \det A = \lambda^3 (12k + 20)$$

$$= \lambda^3 (-12 + 20) \quad [\because k = -1]$$

$$= 8\lambda^3$$

$$\Rightarrow \quad \frac{\lambda^3}{\det(B \operatorname{adj} A)} = \frac{1}{8} = 1:8$$

68. *(b)* Let the foci of the ellipse be $(\pm ae_1, 0)$ and those of the hyperbola be $(\pm Ae_2, 0)$.

According to the question,

$$ae_1 = Ae_2 \quad \text{...(i)}$$

Also, it is given that the conjugate axis of the hyperbola is equal to the minor axis of the ellipse.

Therefore, $a^2(1 - e_1^2) = A^2(e_2^2 - 1)$ ...(ii)

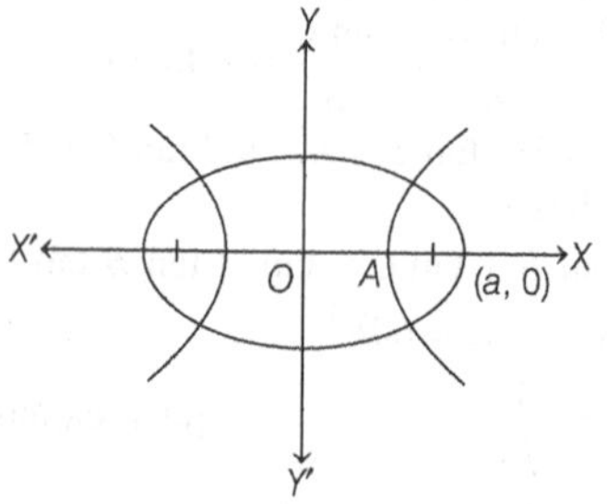

From Eqs. (i) and (ii), we get

$$\left(\frac{e_2}{e_1}\right)^2 = \frac{e_2^2 - 1}{1 - e_1^2}$$

$$\Rightarrow \quad 1 = \frac{e_1^2(e_2^2 - 1)}{e_2^2(1 - e_1^2)}$$

$$\Rightarrow \quad e_1^2 e_2^2 - e_1^2 = e_2^2 - e_1^2 e_2^2$$

$$\Rightarrow \quad e_1^2 + e_2^2 = 2e_1^2 e_2^2$$

On taking square root both sides, we get

$$\sqrt{e_1^2 + e_2^2} = e_1 e_2 \sqrt{2}$$

69. *(a)* $\because |z| \le 1, |\omega| \le 1$

$\Rightarrow z, \omega, -i\omega, i\overline{\omega}$ lies inside or on the boundary of $|z| \le 1$.

Since, $|z + i\omega| = |z - i\overline{\omega}| = 2$

$\Rightarrow -i\omega$ and $i\overline{\omega}$ are end points of diameter having one end point as z.

$\Rightarrow \quad -i\omega = i\omega = -z$

$\therefore \quad |z + z| = 2 \Rightarrow |z| = 1$

Now, $\quad -z = -i\omega = i\overline{\omega}$

$\Rightarrow \omega = -iz$ and $\overline{\omega} = iz$

$\Rightarrow \overline{\omega} = i\overline{z}$ and $\overline{\omega} = iz$

$\Rightarrow i\overline{z} = iz$

$\Rightarrow \quad z = \overline{z}$

$\therefore$ z is real and since $|z| = 1$

$\Rightarrow \quad z = 1, -1$

$\therefore$ Sum of values of $z = 0$

70. *(c)* $P = (4\hat{i} + 7\hat{j} + 8\hat{k})$ and $Q = (2\hat{i} + 3\hat{j} + 4\hat{k})$

$V_{PR} = (5, 9, 7)$ and $V_{QS} = (1, 1, -1)$

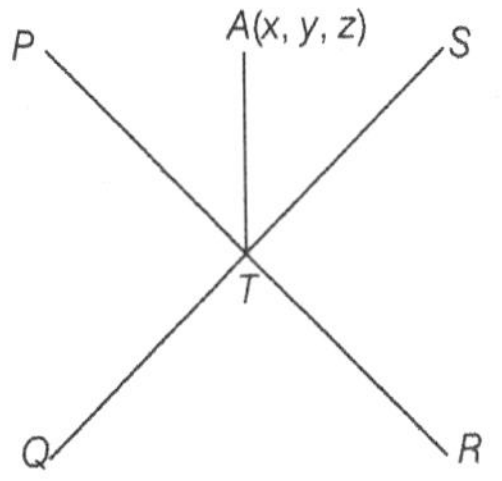

Equation of line PR

$$PR = (4\hat{i} + 7\hat{j} + 8\hat{k}) + \lambda(5\hat{i} + 9\hat{j} + 7\hat{k})$$

Equation of line QS

$$QS = (2\hat{i} + 3\hat{j} + 4\hat{k}) + \mu(\hat{i} + \hat{j} - \hat{k})$$

Let T be the point of intersection.

$$T \equiv (4 + 5\lambda, 7 + 9\lambda, 8 + 7\lambda)$$
$$T \equiv (2 + \mu, 3 + \mu, 4 - \mu)$$

$\therefore \quad 4 + 5\lambda = 2 + \mu$

$\Rightarrow 5\lambda - \mu + 2 = 0 \quad$ ...(i)

$\therefore \quad 7 + 9\lambda = 3 + \mu$

$\Rightarrow 9\lambda - \mu + 4 = 0 \quad$ ...(ii)

From Eqs. (i) and (ii), we get

$$\lambda = -\frac{1}{2} \text{ and } \mu = -\frac{1}{2}$$

So, the point of intersection of two lines, $T \equiv \left(\frac{3}{2}, \frac{5}{2}, \frac{9}{2}\right)$

Now, direction cosines of TA will be $V_{PR} \times V_{QS}$

$$\Rightarrow \begin{vmatrix} \hat{i} & \hat{j} & \hat{k} \\ 5 & 9 & 7 \\ 1 & 1 & -1 \end{vmatrix} = -16\hat{i} + 12\hat{j} - 4\hat{k}$$

$$\therefore L_{TA} = \left(\frac{3}{2}\hat{i} + \frac{5}{2}\hat{j} + \frac{9}{2}\hat{k}\right) + x(-16\hat{i} + 12\hat{j} - 4\hat{k})$$

Let $A \equiv \left(\frac{3}{2} - 16x, \frac{5}{2} + 12x, \frac{9}{2} - 4x\right)$

$$\Rightarrow TA = \sqrt{\left(\frac{3}{2} - \frac{3}{2} + 16x\right)^2 + \left(\frac{5}{2} - \frac{5}{2} - 12x\right)^2 + \left(\frac{9}{2} - \frac{9}{2} + 4x\right)^2}$$
$$= 4\sqrt{26}$$

$\Rightarrow (16x)^2 + (12x)^2 + (4x)^2 = 416$

[on squaring both sides]

$\Rightarrow \quad 416x^2 = 416$

$\Rightarrow \quad x^2 = 1$

$\therefore \quad x = \pm 1 \quad$ [$x = -1$, rejected as $x > 0$]

$\therefore \quad A \equiv \left(\frac{-29}{2}, \frac{29}{2}, \frac{1}{2}\right)$

$$|A| = \sqrt{\left(\frac{-29}{2}\right)^2 + \left(\frac{29}{2}\right)^2 + \left(\frac{1}{2}\right)^2}$$
$$= \sqrt{\frac{841}{4} + \frac{841}{4} + \frac{1}{4}}$$
$$= \sqrt{420.75} \text{ units}$$

71. *(55)* Given, $a + b + c + d + e = 15$

$$\Rightarrow a + \frac{b}{2} + \frac{b}{2} + \frac{c}{3} + \frac{c}{3} + \frac{c}{3} + \frac{d}{4} + \frac{d}{4} + \frac{d}{4} + \frac{d}{4} + \frac{e}{5} + \frac{e}{5} + \frac{e}{5} + \frac{e}{5} + \frac{e}{5} = 15$$

$$\Rightarrow \text{AM} = \frac{\left[a + \frac{b}{2} + \frac{b}{2} + \frac{c}{3} + \frac{c}{3} + \frac{c}{3} + \frac{d}{4} + \frac{d}{4} + \frac{d}{4} + \frac{d}{4} + \frac{e}{5} + \frac{e}{5} + \frac{e}{5} + \frac{e}{5} + \frac{e}{5}\right]}{15} = 1$$

$$\text{GM} = \left(a \cdot \frac{b^2}{2^2} \cdot \frac{c^3}{3^3} \cdot \frac{d^4}{4^4} \cdot \frac{e^5}{5^5}\right)^{\frac{1}{15}}$$

$$= \left(\frac{(120)^3\, 50}{2^2\, 3^3\, 4^4\, 5^5}\right)^{\frac{1}{15}} = 1$$

$\because$ GM = AM

Hence, $a = \frac{b}{2} = \frac{c}{3} = \frac{d}{4} = \frac{e}{5}$

$\Rightarrow \quad a = 1, b = 2, c = 3, d = 4$ and $e = 5$

$\therefore \quad a^2 + b^2 + c^2 + d^2 + e^2 = 1^2 + 2^2 + 3^2 + 4^2 + 5^2$

$$= \frac{5 \times 6 \times 11}{6} = 55$$

72. *(8)* Let the position of the kite at time t be at C.

$\therefore BC = 151.5$ m.

Let AD be the boy who is flying the kite.

Let $AB = x = DE$ and $DC = y$

$\therefore \quad CE = BC - BE = 151.5 - 1.5 = 150$ m.

Therefore, from the right angled ΔCDE,

$$y^2 = x^2 + (150)^2 \quad \text{...(i)}$$

$$\therefore 2y\frac{dy}{dt} = 2x\frac{dx}{dt} + 0$$

$$\therefore \quad \frac{dy}{dt} = \frac{x}{y}\frac{dx}{dt} = \frac{x}{y} \times 10 \quad \left[\because \frac{dx}{dt} = 10 \text{ m/sec}\right]$$

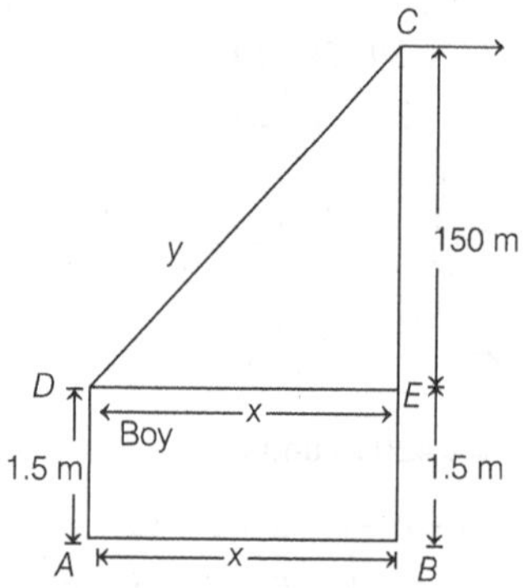

$$= \frac{10x}{y} = \frac{10\sqrt{(y)^2 - (150)^2}}{y} \quad \text{[from Eq. (i)]}$$

When $y = 250\,\text{m}$

Then, $\frac{dy}{dt} = \frac{10\sqrt{(250)^2 - (150)^2}}{250}$

$$= \frac{10 \times 200}{250} = 8\,\text{m/sec}$$

Hence, the string is being let out at the rate of 8m/sec when 250 m of string is out.

73. *(5)* $P[A \cap (B \cup C)] = P[(A \cap B) \cup (A \cap C)]$

$$= P(A \cap B) + P(A \cap C) - P(A \cap B \cap C)$$

$$= P^2 + P^2 - P^3 = P^2(2 - P)$$

$$P[\overline{C} \cap \{A \cap (B \cup C)\}] = P(A \cap B \cap \overline{C})$$

$$= P^2 - P^3 = P^2(1 - P)$$

$$\therefore P\left(\frac{\overline{C}}{A \cap (B \cup C)}\right) = \frac{P[\overline{C} \cap \{A \cap (B \cup C)\}]}{P[A \cap (B \cup C)]}$$

$$= \frac{P^2\,(1 - P)}{P^2\,(2 - P)}$$

$$= \frac{1 - P}{2 - P} = \frac{x - P}{y - P}$$

$\Rightarrow \quad x = 1$ and $y = 2$

$\therefore x^2 + y^2 = 5$

74. *(6)* We know that standard deviation is independent of change of origin

i.e. if $d_i = \frac{x - A}{h}$, then $\sigma_x = |h| \sigma_d$

In this case, $-2x_i - 3 = \frac{x_i + \frac{3}{2}}{-\frac{1}{2}}$

So, $\quad h = -\frac{1}{2}$

Thus, $\sigma_d = \frac{1}{|h|} \sigma_x = \frac{1}{\left|-\frac{1}{2}\right|} \times 3 = 6$

75. *(41)* Committee of 5 to be formed from 9 people 2 particular person always together ;

2 other are refuse to be together ;

So, 5 people have no condition

= 2 particular who want to be always together and 3 from remaining 5(two of them who refuse together are eliminate)

+ 2 particular person always together and one of the two who does not want to together remaining 2 from remaining 5

+ one of them who does not want be together and 4 from remaining five

+ five from remaining five

$$= 1 \times {}^5C_3 + 2 \times {}^5C_2 + 2 \times {}^5C_4 + {}^5C_5$$

$$= 10 + 20 + 2 \times 5 + 1$$

$$= 41$$

PRACTICE SET 15

Total Marks : 300	Total Time : 3 hrs	No. of Questions : 75

Instructions

1. There are **three** parts in the questions paper consisting of **Physics**, **Chemistry** and **Mathematics** having 25 questions in each part of equal weightage. Each part (subject) has two sections.
2. **Section A** This section contains 20 multiple choice questions which have only one correct answer. Each question carries **4 marks** for correct answer and – **1 mark** for wrong answer.
3. **Section B** This section contains 5 questions. The answer to each of the questions is a numerical value. Each question carries **4 marks** for correct answer and – **1 mark** for wrong answer. For Section B, the answer should be rounded off to the nearest integer.

PHYSICS

Section A : Objective Type Questions

1. When a certain metallic surface is illuminated with monochromatic light of wavelength 2.8×10^{-7} m, the stopping potential for photoelectric current is 3 V and when the same surface is illuminated with light of wavelength 5.6×10^{-7}m, the stopping potential is 1 V. The threshold wavelength of this surface for photoelectric effect is

(a) 2.8×10^{-7} m (b) 1.12×10^{-7} m
(c) 1.12×10^{-6} m (d) 2.8×10^{-6} m

2. Statement I By Gauss's law of magnetism, the net magnetic flux through any closed surface is zero.

Statement II The net magnetic field inside a closed surface is zero.

In the light of above statements, choose the most appropriate answer from the options given below.

(a) Both Statement I and Statement II are true.
(b) Both Statement I and Statement II are false.
(c) Statement I is true, Statement II is false.
(d) Statement I is false, Statement II is true.

3. Radius of an air bubble at the bottom of the lake is r and it becomes $2r$ when the air bubble rises to the top surface of the lake. If p cm of water be atomospheric pressure, then the depth of lake is

(a) $2p$ (b) $8p$ (c) $4p$ (d) $7p$

4. A particle which is constrained to move along the X-axis, is subjected to force in the same direction, which varies with the distance x of the particle from the origin as $F(x) = -Kx + ax^3$. Here, K and a are positive constants. For $x \geq 0$, the functional form of the potential energy $U(x)$ of the particle is

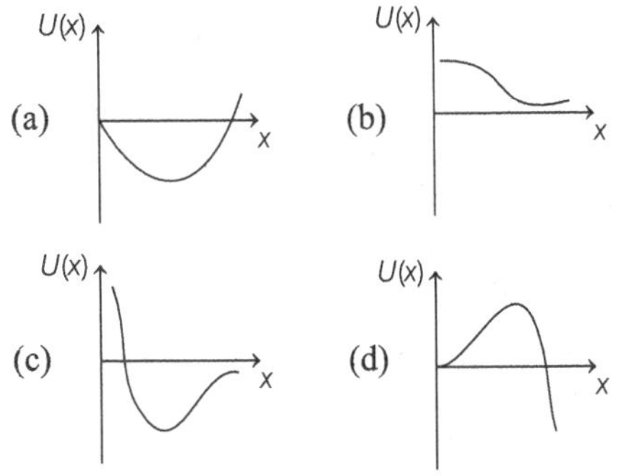

5. A point source of electromagnetic radiation has an average power output of 1000 W. The maximum value of electric field at a distance 4.0 m from the source is

(a) 32.5 V/m
(b) 61.2 V/m
(c) 57.8 V/m
(d) 64.72 V/m

6. **Assertion** The amplitude of a wave determines its energy.

Reason Waves with higher amplitude have higher energy.

In the light of above statements, choose the most appropriate answer from the options given below.

(a) Both Assertion and Reason are correct and Reason is the correct explanation of Assertion.
(b) Both Assertion and Reason are correct but Reason is not the correct explanation of Assertion.
(c) Assertion is correct but Reason is incorrect.
(d) Assertion is incorrect but Reason is correct.

7. A pair of coils of turns n_1 and n_2 are kept close together. Current passing through the first coil is reduced at rate r and emf 3 mV is developed across the other coil. If the second coil carries current which is then reduced at rate $r/2$, the emf produced across the first coil will be

(a) 1.5 mV (b) $1.5 \times \frac{n_1}{n_2}$ mV
(c) $6 \frac{n_2}{n_1}$ mV (d) $1.5 \frac{n_2}{n_1}$ mV

8. A wide vessel with a small hole in the bottom is filled with water and kerosene. If the thickness of water layer is $h_1 = 30$ cm and kerosene is $h_2 = 20$ cm, then the velocity of flow of water is

(a) $\sqrt{2g\left(\frac{h_1}{\rho_2} + \frac{h_2}{\rho_1}\right)}$ (b) $\sqrt{2g\left(h_1 + \frac{h_2\rho_2}{\rho_1}\right)}$
(c) $\sqrt{2g\left(\frac{h_1\rho_2}{\rho_1} + h_2\right)}$ (d) $\sqrt{2g\left(\frac{h_1 + h_2}{\rho_1 + \rho_2}\right)}$

9. If 59 divisions of the main scale coincide with 60 divisions of the vernier scale and the smallest division of the main scale is 0.5°, then the least count of the instrument is

(a) 1 min (b) 30 s (c) 20 s (d) 2 min

10. Five rods of same dimensions are arranged as shown in figure. They have thermal conductivities K_1, K_2, K_3, K_4 and K_5. When points A and B are maintained at different temperatures, no heat would flow through central rod, if

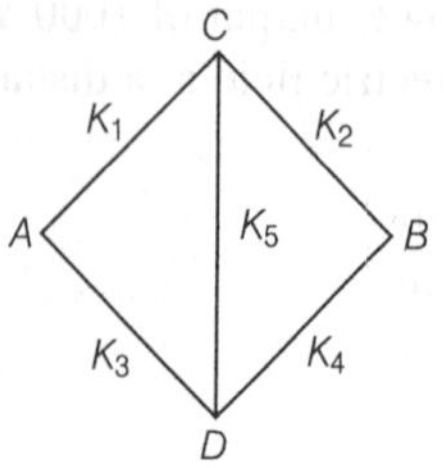

(a) $K_1K_4 = K_2K_3$ (b) $K_1 = K_4$ and $K_2 = K_3$
(c) $\frac{K_1}{K_4} = \frac{K_2}{K_3}$ (d) $K_1K_2 = K_3K_4$

11. The ratio of escape velocity at the earth to the escape velocity at a planet whose radius and mean density are four times as that of earth is

(a) 1 : 4 (b) 4 : 1 (c) 1 : 8 (d) 8 : 1

12. Four resistors of resistances $10\,\Omega, 16\,\Omega, 8\,\Omega$ and $3\,\Omega$ are connected in cyclic order to form a wheatstone bridge. The resistance (in Ω) that should be connected in series with $3\,\Omega$ resistor to balance the wheatstone bridge is

(a) 6Ω (b) 8Ω (c) 2Ω (d) 10Ω

13. A particle of mass $m = 1.6 \times 10^{-27}$ kg and charge $q = 1.6 \times 10^{-19}$ C moves at a speed of 1.0×10^7 m/s. It enters a region of uniform magnetic field at a point E, as shown in figure. The field has a strength of 1.0 T. The magnetic field is directed into the plane of the paper. The particle leaves the region of the field at the point F, then distance EF and the angle θ will be

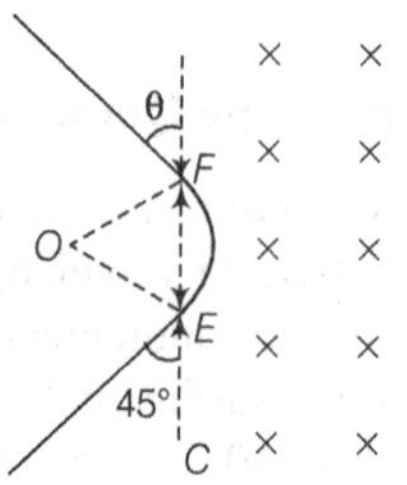

(a) 14 cm and 55° (b) 15 cm and 45°
(c) 14 cm and 45° (d) 15 cm and 55°

14. The acceleration of a particle varies with displacement as $a = -4x$. At $x = 4$ m and $t = 0$, particle is at rest. Select the correct alternative.

(a) The maximum speed of particle is 8 m/s.
(b) The maximum speed of particle is 2 m/s.
(c) The maximum speed of particle is 4 m/s.
(d) The maximum speed of particle is 6 m/s.

15. Which of the following statement is correct ?

(a) Work done in isothermal process is zero.
(b) Isothermal process follows Boyle's law.
(c) By second law of thermodynamics, the efficiency of heat engine is equal to 1.
(d) $\Delta Q - \Delta W$ is path dependent.

16. I. The mass of any gas of volume 22.4 L is equal to its molecular weight.

II. The collisions of the molecules of an ideal gas are partially elastic.

III. If γ is the ratio of specific heats and R is the universal gas constant, then molar specific heat at constant volume C_V is $\frac{\gamma R}{\gamma - 1}$.

Which of the following statement(s) is/are correct ?

(a) Both I and III (b) Both II and III
(c) Only III (d) Only I

17. An electron is revolving in its 3rd excited state (radius = 5.75Å), then de-Broglie wavelength of this electron is

(a) 8.25 Å (b) 3.5 Å (c) 9.02 Å (d) 12.9 Å

18. Match the Column I (symbol) with Column II (their respective logic gate) and select the correct answer from the codes given below.

	Column I		Column II
A.	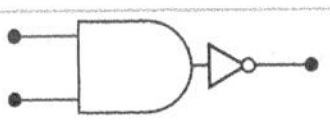	1.	NOR gate
B.	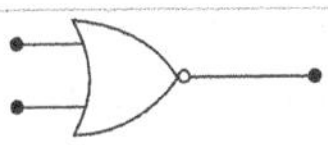	2.	OR gate
C.	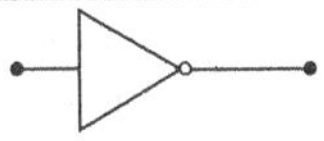	3.	NOT gate
D.	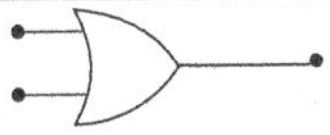	4.	NAND gate

Codes

	A	B	C	D		A	B	C	D
(a)	1	3	4	2	(b)	4	1	3	2
(c)	4	2	3	1	(d)	2	4	3	1

19. Young's modulii of two wires P and Q are in the ratio of 3 : 2. Wire P is 4 m long and has radius R. Wire Q is 3 m long and has radius of 2 mm. If the two wires stretch by the same length for a given load, then the value of R is close to

(a) 2 mm (b) 23 mm (c) 23 m (d) 1.88 mm

20. Obtain the binding energy of the nuclei ${}_2He^4$, ${}_{26}Fe^{56}$ and ${}_{83}Bi^{209}$ in unit of MeV from the following data.

$m_H = 1.007825$ amu, $m_n = 1.008665$ amu,

$m({}_2He^4) = 4.002604$ amu,

$m({}_{26}Fe^{56}) = 55.934939$ amu,

$m({}_{83}Bi^{209}) = 208.980388$ amu.

Which nucleus have greater binding energy per nucleon? Choose from the given options.

(a) ${}_2He^4$ (b) ${}_{26}Fe^{56}$
(c) ${}_{83}Bi^{209}$ (d) All have same

Section B : Numerical Value Type Questions

21. In Young's double slit experiment, the length of band AB is 1 mm. The fringe width is 0.021 mm. Find the number of fringes.

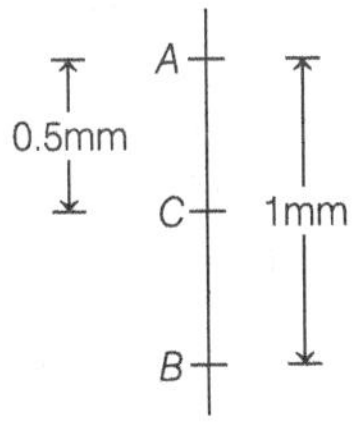

22. When a car is at rest, its driver see rain drops falling on it vertically. When driving the car with speed v, he see that rain drops are coming at an angle 45° to vertical. On further increasing the speed of car to nv, this angle changes to 60°, then the value of n^2 is ....

23. Variation of heat supplied Q with gas volume V is as shown for an isothermal process.

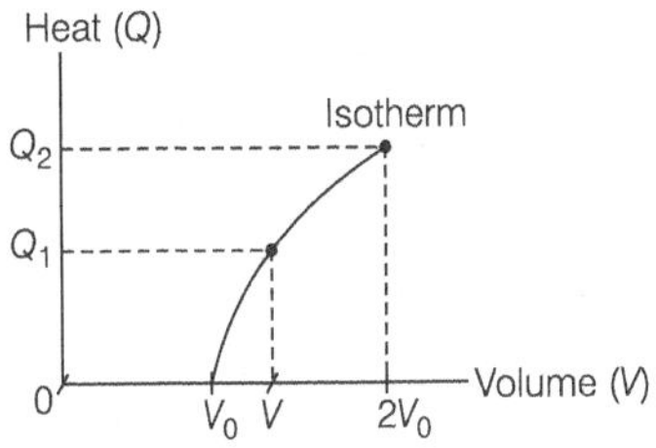

If $\frac{Q_2}{Q_1} = 2$, then $\frac{V}{V_0}$ is $\sqrt{x}$. The value of x is ....

24. The value of time constant for given circuit is ......s.

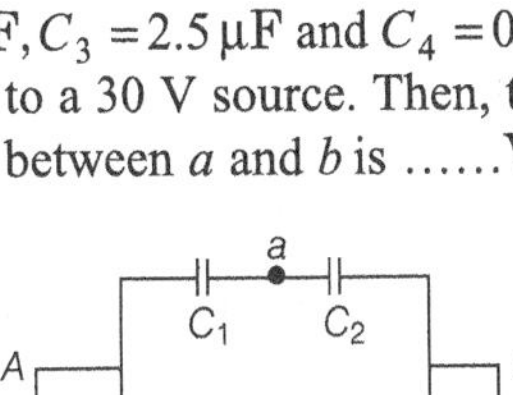

25. Four capacitors with capacitances $C_1 = 1\ \mu F$, $C_2 = 1.5\ \mu F$, $C_3 = 2.5\ \mu F$ and $C_4 = 0.5\ \mu F$ are connected to a 30 V source. Then, the potential difference between a and b is ......V.

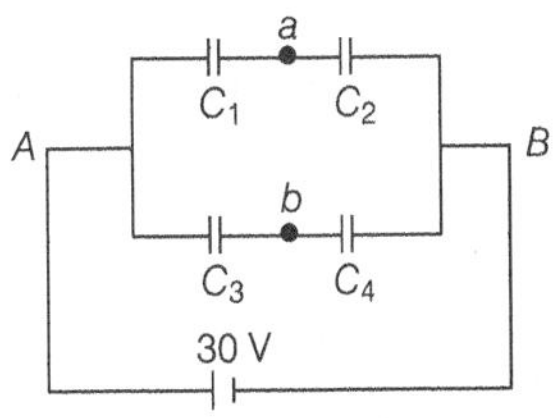

CHEMISTRY

Section A : Objective Type Questions

26. Fixed volume of 0.1 M $C_6H_5COOH(pK_a = 4.4)$ solution is added into 0.2 M sodium benzoate solution and forms a 300 mL resultant acidic buffer solution. If pH of the buffer solution is 4.7, then the volume of benzoic acid will be

(a) 450 mL (b) 250 mL (c) 300 mL (d) 150 mL

27. When a protic acid is added to NH_3, it forms X and when a strong base is added to NH_3, it forms Y. Consider $NH_3 = Z$ and arrange them in correct order of their bond angle.

(a) $Y > Z > X$ (b) $X > Z > Y$
(c) $Z > Y > X$ (d) $Z > X > Y$

28. Which of the following statement(s) is/are correct?

A. Hydrolysis of sucrose with dilute acid yields an equimolar mixture of D-(+)-glucose and D-(+)-fructose.
B. Acid hydrolysis of sucrose is accompanied by a change in optical rotation.
C. In sucrose C_1 of glucose and C_2 of fructose are linked by glycosidic linkage.
D. Aqueous solution of sucrose exhibit mutarotation.

Choose the correct answer from the options given below.

(a) A, B and C (b) A and D
(c) B and C (d) B and D

29. Match List-I and List-II.

	List-I		List-II
A.	$R—\overset{O}{\overset{\|\|}{C}}—Cl \longrightarrow R—CHO$	I.	$Br_2/NaOH$
B.	$R—CH_2—COOH \longrightarrow R—\underset{Cl}{\underset{\|}{CH}}—COOH$	II.	$H_2/Pd\text{-}BaSO_4$
C.	$R—\overset{O}{\overset{\|\|}{C}}—NH_2 \longrightarrow R—NH_2$	III.	Cu, HCl
D.	$ArN_2^+Cl^- \longrightarrow ArCl + N_2$	IV.	Cl_2/Red P, H_2O

Choose the correct answer from the options given below.

(a) A-III, B-IV, C-I, D-II
(b) A-III, B-I, C-IV, D-II
(c) A-II, B-IV, C-I, D-III
(d) A-II, B-I, C-IV, D-III

30. The frequency of one of the lines in Paschen series of the hydrogen atom is 2330×10^{14} Hz. The quantum number n_2 which produced this transition is

(a) 6 (b) 5 (c) 4 (d) 3

31. Match List-I with List-II.

	List-I		List-II
A.	V_2O_5	I.	Oxidation of ethene to ethanal
B.	$TiCl_4 + Al(CH_3)_3$	II.	Hydrogenation
C.	$PdCl_2$	III.	Oxidation of SO_2 in the manufacture of H_2SO_4
D.	Ni-complex	IV.	Polymerisation of ethylene

Choose the correct answer from the options given below.

(a) A-II, B-III, C-I, D-IV (b) A-III, B-IV, C-I, D-II
(c) A-III, B-I, C-IV, D-II (d) A-IV, B-III, C-II, D-I

32. Among the group 16 elements, one having highest value of negative electron gain enthalpy is

(a) O (b) S (c) Se (d) Te

33. Which of the following will be obtained by keeping ether in contact with air for a long time.

(a) $C_2H_5—O—CH(CH_3)—O—OH$
(b) $C_2H_5—OCH_2—OH$
(c) $C_2H_5—O—C_2H_5OH$
(d) $CH_3—OCH(CH_3)—O—OH$

34. In the reaction shown below the six membered ring is generated by shifting which bond.

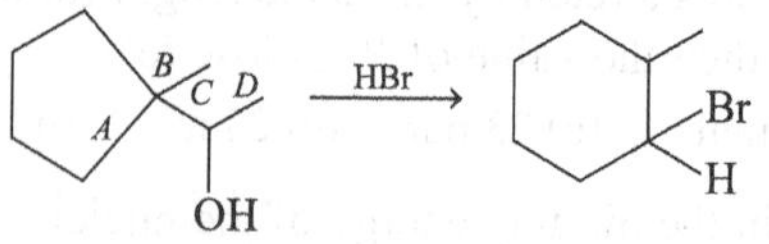

(a) A (b) B (c) C (d) D

35. Order of covalent bond:

A. NaF > NaI; NaCl < NaI
B. NaF < NaI; LiF > NaF
C. $PbCl_2 < PbCl_4$; CuCl > KCl
D. KF < KCl; $MgCl_2$ > NaCl
E. LiF > KF; NaCl > $MgCl_2$

Choose the correct answer from the options given below:

(a) A, C and D (b) B and C
(c) B, C and D (d) A, C and E

36. In which of the following processes the bond order increases and paramagnetic character changes to diamagnetic ones?

(a) $O_2 \longrightarrow O_2^{2-}$ (b) $N_2 \longrightarrow N_2^+$
(c) $NO \longrightarrow NO^+$ (d) $O_2 \longrightarrow O_2^+$

37. Consider the following statements.

Statement I $[Co(NH_3)_6][Co(CN)_6]$ do not exhibit coordination isomerism.

Statement II Coordination isomerism arises from the interchange of ligands between cationic and anionic entities present in a complex.

In the light of the above statements, choose the most appropriate answer from the options given below.

(a) Both Statement I and Statement II are incorrect.
(b) Statement I is correct but Statement II is incorrect.
(c) Statement I is incorrect but Statement II is correct.
(d) Both Statement I and Statement II are correct.

38. The reagent '*X*' in the given reaction is

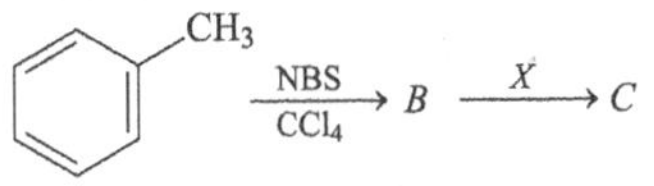

A

The product *C* can also be obtained by treatment of *A* with SeO_2.

(a) *aq.* H_2SO_4 (b) $LiAlH_4$
(c) $ZnCl_2/HCl$ (d) *aq.* KOH

39. Correct order of wavelength of absorption for the following complexes is

A. $[Fe(CN)_6]^{4-}$ B. $[Fe(H_2O)_6]^{2+}$ C. $[PtCl_4]^{2-}$

Choose the correct answer from the options given below.

(a) A < B < C (b) B < A < C
(c) C < B < A (d) C < A < B

40. Which of the following is incorrect statement?

(a) Dimethylamine reacts with KOH and phenol to form an azo dye.
(b) Primary amines show intermolecular hydrogen bonding.
(c) Amines have lower boiling points as compared to those of alcohols.
(d) Aromatic amines are generally less basic than alkylamine.

41. When 6, 6-dimethyl cyclohex- 2-en-1-ol react with MnO_2, it produces *A*. What will be the product *B* when *A* is treated with CH_3Li in presence of CuCl and its optical activity. Choose the correct option.

(a) *B* is 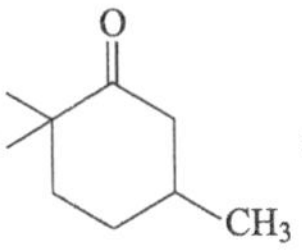

and optically inactive.

(b) *B* is 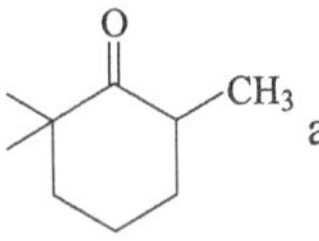

and optically active.

(c) *B* is 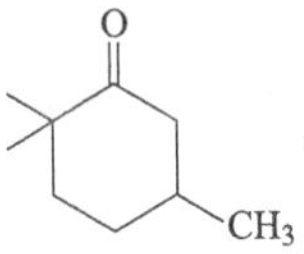

and optically active.

(d) *B* is 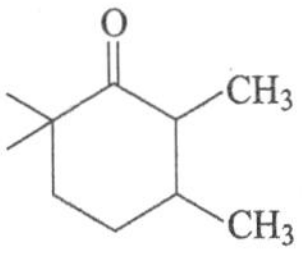

and optically active.

42. If the radius of the first orbit of hydrogen atom is $r_0/2$, then the de Broglie's wavelength of electron in 4th orbit is

(a) $\frac{\pi r_0}{8}$ (b) $\frac{\pi r_0}{4}$ (c) $8\pi r_0$ (d) $4\pi r_0$

43. Which among the following has unequal bond lengths?

(a) XeF_4 (b) BF_4^- (c) SiF_4 (d) SF_4

44.

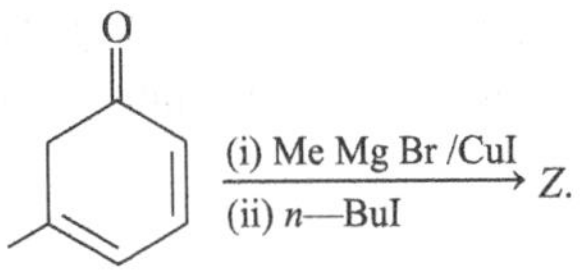

Z is

(a) 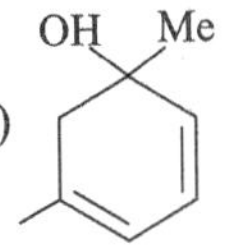

(b) 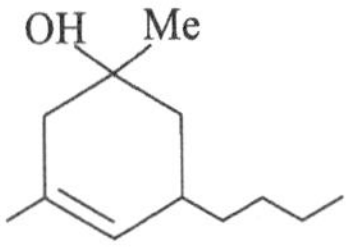

(c) 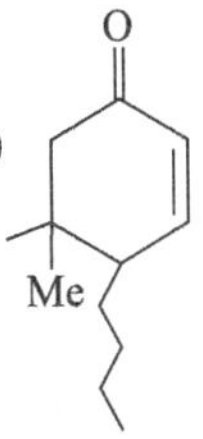

(d) 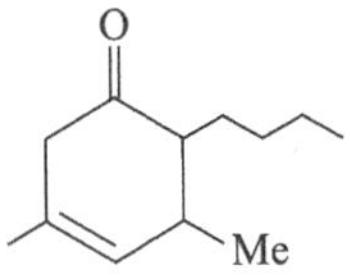

45. Consider the following carbocations.

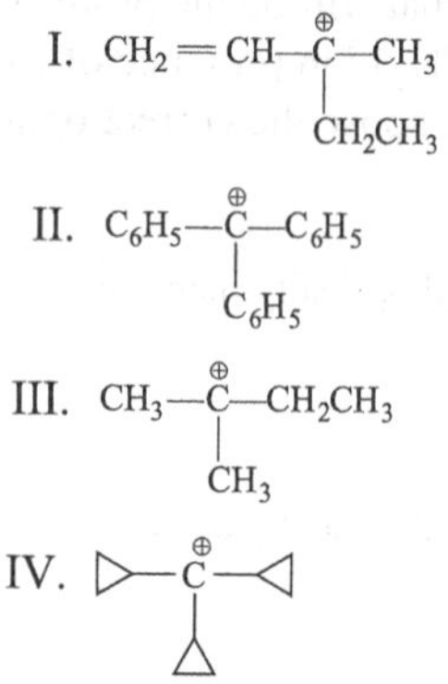

The correct increasing order of hydride affinity for following carbocation is

(a) III < IV < I < II (b) IV < II < I < III
(c) I < II < III < IV (d) II < III < IV < I

Section B : Numerical Value Type Questions

46. Consider the following cell reaction,

$2\,Fe(s) + O_2(g) + 4\,H^+(aq) \longrightarrow 2\,Fe^{2+}(aq) + 2H_2O(l);\ E° = 1.67\ V$

At $[Fe^{2+}] = 10^{-3}$ M, $p_{(O_2)} = 0.1$ atm and pH = 3, the cell potential at 25°C is ...... $\times 10^{-1}$ V.

47. The rate constant of a zero order reaction is 2.0×10^{-2} mol L^{-1} s^{-1}. If the concentration of the reactant after 25 s is 0.5 M, the initial concentration is ......M.

48. A system is allowed to move form state P to Q following path PRQ by absorbing 80 J of heat energy. The work done by the system is 30J. The work done by the system in reaching state Q from P through path PSQ is 10J.

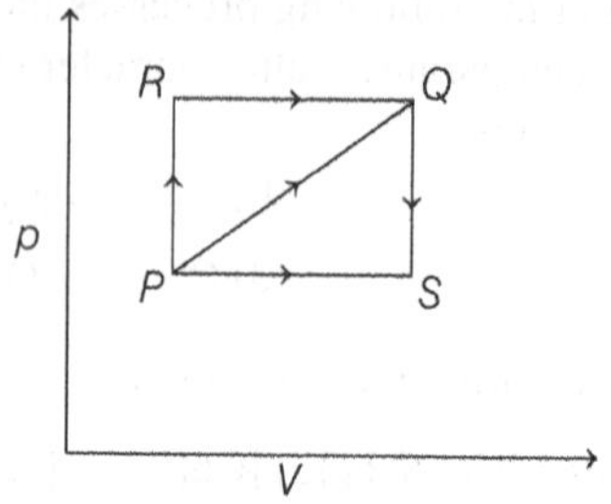

Consider these statements:

(A) Increase in internal energy form state P to state Q is 50J.

(B) If path PSQ is followed to reach state Q, $\Delta U = 50$ J.

(C) If work done by the system is path PQ is 20J, the heat absorbed during path PQ is 70J.

(D) The value $U_R - U_P$ is equal to $U_S - U_Q$.

(E) Heat absorbed by the system to reach Q from P through path PSQ is 60 J.

The number of correct statements from the above is.......

49. The number of correct statement(s) from the following for copper (At. no. 29) is/are ....

A. Cu(I) is easily oxidized.
B. Cu(I) complexes are generally colourless.
C. Cu(II) complexes are generally diamagnetic.
D. Cu(II) complexes are generally paramagnetic.

50. During the estimation of nitrogen in organic compound by Kjeldahl's method, the ammonia evolved by 0.5 g of the compound neutralised 10 ml of 1 M H_2SO_4. The percentage of nitrogen in the compound is ....

MATHEMATICS

Section A : Objective Type Questions

51. Let $z_1 = 10 + 6i$ and $z_2 = 4 + 6i$. If z is any complex number such that amplitude of $\frac{z - z_1}{z - z_2}$ is $\frac{\pi}{4}$ such that $|z - 7 - 9i| = \alpha$, then α is equal to

(a) $3\sqrt{2}$ (b) $2\sqrt{3}$
(c) $\sqrt{2}$ (d) $3\sqrt{3}$

52. $\int_0^\infty f(x)\,dx = \frac{\pi}{2}$ and $f(x)$ is an even function, then $\int_0^\infty f\left(x - \frac{1}{x}\right) dx$ is equal to

(a) $\frac{\pi}{4}$

(b) $\frac{\pi}{2}$

(c) π

(d) None of the above

53. Three numbers are drawn from the set {1, 2, 3, ..., n} with replacement. The probability that their sum is $2n$, is

(a) $\frac{1}{2n}$ (b) $\frac{1}{2}$
(c) $\frac{(n-1)(n+4)}{2n^3}$ (d) $\frac{(n+1)(n+2)}{n^3}$

54. $A=\begin{bmatrix}1&0&0\\0&1&1\\0&-2&4\end{bmatrix}$ and $I=\begin{bmatrix}1&0&0\\0&1&0\\0&0&1\end{bmatrix}$ and $A^{-1}=\left[\frac{1}{6}(aA^2+bA+cI)\right]$, then the value of $a+b+c$ is

(a) 6 (b) 4 (c) 12 (d) 8

55. If total numbers of runs scored in r matches is $\left(\frac{r+1}{4}\right)(2^{r+1}-r-2)$, where $r>1$ and the runs scored in the kth match are given by $k\,2^{r+1-k}$, where $1\le k\le r$, then r is equal to

(a) 6 (b) 7 (c) 8 (d) 9

56. Consider the following statements

Statement I $C_0^2-2C_1^2+3C_2^2-\ldots+(-1)^n(n+1)C_n^2=(n+1)(-1)^n$ when n is odd.

Statement II $C_0^2-2C_1^2+3C_2^2-\ldots+(-1)^n(n+1)C_n^2=(-1)^{n/2}\left(\frac{n}{2}+1\right)\frac{n!}{\left(\frac{n}{2}\right)!\left(\frac{n}{2}\right)!}$, if n is even, then

(a) Statement I is true and Statement II is false
(b) Statement I is false and Statement II is true
(c) Statement I and Statement II both are true
(d) Statement I and Statement II both are false

57. Let three matrices $A=\begin{bmatrix}2&1\\4&1\end{bmatrix}$, $B=\begin{bmatrix}3&4\\2&3\end{bmatrix}$ and $C=\begin{bmatrix}3&-4\\-2&3\end{bmatrix}$, then the value of $\text{tr}(A)+\text{tr}\left[\frac{A(BC)}{2}\right]+\text{tr}\left[\frac{A(BC)^2}{4}\right]+\text{tr}\left[\frac{A(BC)^3}{8}\right]+\ldots\infty$ is ....

(a) 6 (b) 0 (c) 16 (d) 8

58. Let f and g be two differentiable functions defined from $R\to R^+$. If $f(x)$ has local maximum at $x=c$ and $g(x)$ has a local minimum at $x=c$, then $h(x)=\frac{f(x)}{g(x)}$

(a) has a local maximum at $x=c$
(b) has a local minimum at $x=c$
(c) is monotonic at $x=c$
(d) has a point of inflection at $x=c$

59. Let the major axis of a standard ellipse equals the transverse axis of a standard hyperbola and the director circles have radius equal to $2R$ and R, respectively. If e_1 and e_2 are the eccentricities of the ellipse and hyperbola respectively, then the correct relation is

(a) $4e_1^2-e_2^2=6$ (b) $2e_1^2-4e_2^2=3$
(c) $4e_2^2-e_1^2=6$ (d) $2e_1^2-e_2^2=4$

60. The line $x+y=k$ meets the X and Y-axes at P and Q, respectively. A ΔPAB is inscribed in the ΔOPQ, O being the origin, with right angled at B. A and B lie respectively on OQ and PQ. If the area of the ΔPAB is $\frac{3}{8}$ of the area of ΔOPQ, then $\frac{PB}{QB}$ is equal to

(a) 4 (b) 1 (c) 2 (d) 3

61. If $f(x)=\begin{cases}\dfrac{ae^{\frac{1}{|x+2|}}-1}{2-e^{\frac{1}{|x+2|}}}, & -3<x<-2\\ b, & x=-2\\ \sin\left(\dfrac{x^4-16}{x^5+32}\right), & -2<x<0\end{cases}$ is continuous at $x=-2$, then (a,b) is equal to

(a) $(\sin 5, -\sin 5)$ (b) $\left(\sin\frac{2}{5}, -\sin\frac{2}{5}\right)$
(c) $\left(-\sin\frac{2}{5}, \sin\frac{2}{5}\right)$ (d) $\left(-\sin\frac{2}{5}, -\sin\frac{2}{5}\right)$

62. Number of 4-digit numbers of the from $N=abcd$ which satisfy following three conditions?

(i) $4000\le N<6000$
(ii) N is a multiple of 5
(iii) $3\le b<c\le 6$ is equal to

(a) 12 (b) 18 (c) 24 (d) 48

63. The direction cosines of a vector **r** which is equally inclined to OX, OY and OZ. If $|\mathbf{r}|$ is given, then the total number of such vectors are

(a) 2 (b) 8 (c) 6 (d) 12

64. The value of $\int_{-\pi}^{\pi}\frac{2x(1+\sin x)}{1+\cos^2 x}dx$ is equal to

(a) $-\sqrt{2}\pi^2$ (b) π^2
(c) $\frac{\pi^2}{\sqrt{2}}$ (d) $-\frac{\pi^2}{\sqrt{2}}$

65. If $A > 0$, $B > 0$ and $A + B = \frac{\pi}{6}$, then the maximum value of $\tan A \cdot \tan B$ is $(a - \sqrt{b})^2$, then the value of $a + b$ is

(a) 5 (b) 3 (c) 7 (d) 2

66. If $y = f(x)$ is a solution of $xy_1 - y = x \tan \frac{y}{x}$, $y(1) = \frac{\pi}{2}$, then $\frac{1}{\pi}\int_0^1 y\,dx$ is equal to

(a) $\frac{1}{8}$ (b) $\frac{5}{8}$ (c) $\frac{3}{8}$ (d) $-\frac{1}{8}$

67. Vertices B and C of ΔABC lie along the line $\frac{x+2}{2} = \frac{y-1}{1} = \frac{z-0}{4}$. If A has coordinates $(1, -1, 2)$ and line segment BC has length 8 units, then the area of triangle is

(a) $\sqrt{\frac{4544}{7}}$ sq units (b) $\sqrt{\frac{284}{7}}$ sq units

(c) $\sqrt{\frac{71}{7}}$ sq units (d) $\sqrt{\frac{1136}{7}}$ sq units

68. The line $y = mx$ bisects the area enclosed by the curve $y = 1 + 4x - x^2$ and the lines $x = 0$, $x = \frac{3}{2}$ and $y = 0$. Then, the value of $6m$ is

(a) 17 (b) 13
(c) 22 (d) 9

69. For the given two data sets

	Variance	Mean	Number of observation
X	4	2	5
Y	5	4	5

Then, the variance of the combined data set is

(a) $\frac{5}{2}$ (b) $\frac{7}{2}$

(c) $\frac{11}{2}$ (d) $\frac{13}{2}$

70. Let $f: R \to R$ be a continuous function such that $f(3x) - f(x) = x$. If $f(8) = 6$, then $f(18)$ is equal to

(a) 11 (b) 10 (c) 9 (d) 8

Section B : Numerical Value Type Questions

71. Let the position vectors of the points P, A and B be $\mathbf{r}$, $\hat{\mathbf{i}} + \hat{\mathbf{j}} + \hat{\mathbf{k}}$ and $-\hat{\mathbf{i}} + \hat{\mathbf{k}}$. If **PA** is perpendicular to **PB** but $\mathbf{r}$ is not perpendicular to $\mathbf{r} - (\hat{\mathbf{j}} + 2\hat{\mathbf{k}})$, then $|\mathbf{r}|^2$ is equal to ......

72. Let $f: R \to \left(0, \frac{2\pi}{3}\right]$ defined as $f(x) = \cot^{-1}(x^2 - 4x + \alpha)$. Then, the smallest integral value of α such that $f(x)$ is into function is ....

73. If $(\mathbf{a} + 3\mathbf{b}) \cdot (7\mathbf{a} - 5\mathbf{b}) = 0$ and $(\mathbf{a} - 4\mathbf{b}) \cdot (7\mathbf{a} - 2\mathbf{b}) = 0$. Then, the angle between $\mathbf{a}$ and $\mathbf{b}$ (in degree) is ....

74. Let $x^2 = 8y$ be the given parabola and $P(6, 2)$, $Q(1, 2)$ be the end points of a line segment. If R lies on the line segment such that $PR : QR = 1 : \beta$ and R is an interior point of the parabola, then the number of integral values of 2β is ....

75. If the straight line joining the origin to the points of intersection of the straight line $69x + 25y = 3450$ and the circle $(x - 25)^2 + (y - 69)^2 = c^2$ are at right angles, then $c^2 - (69)^2 = r^2$, where r is equal to ....

DETAILED SOLUTIONS

Physics

1. *(c)* Given, $\lambda_1 = 2.8\times 10^{-7}$ m

$$(V_s)_1 = 3\text{V}$$
$$\lambda_2 = 5.6\times 10^{-7}\text{ m}$$
$$(V_s)_2 = 1\text{V}$$

As we know that,

$$\frac{hc}{\lambda} - \frac{hc}{\lambda_0} = (eV_s)_1$$

For the first case,

$$\frac{hc}{\lambda_1} - \frac{hc}{\lambda_0} = 3\text{eV}$$

$$\Rightarrow \quad \frac{hc}{2.8\times 10^{-7}} - \frac{hc}{\lambda_0} = 3\times 1.6\times 10^{-19} \quad \ldots\text{(i)}$$

For the second case,

$$\frac{hc}{\lambda_2} - \frac{hc}{\lambda_0} = (eV_s)_2$$

$$\frac{hc}{5.6\times 10^{-7}} - \frac{hc}{\lambda_0} = 1\text{ eV}$$

$$\frac{hc}{5.6\times 10^{-7}} - \frac{hc}{\lambda_0} = 1\times 1.6\times 10^{-19} \quad \ldots\text{(ii)}$$

On dividing Eq. (i) by Eq. (ii), we get

$$\frac{1}{2.8\times 10^{-7}} - \frac{1}{\lambda_0} = 3\left(\frac{1}{5.6\times 10^{-7}} - \frac{1}{\lambda_0}\right)$$

$$\frac{3}{\lambda_0} - \frac{1}{\lambda_0} = \frac{3}{5.6\times 10^{-7}} - \frac{1}{2.8\times 10^{-7}}$$

$$\lambda_0 = 11.2\times 10^{-7}$$

$$\lambda_0 = 1.12\times 10^{-6}\text{ m}$$

Hence, threshold wavelength of surface is 1.12×10^{-6} m.

2. *(a)* By Gauss's law of magnetism, $\phi = \oint \mathbf{B}\cdot\mathbf{A} = 0$

For a closed surface,

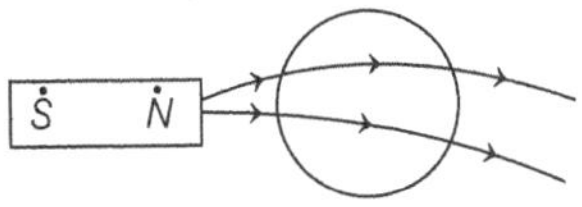

Total number of entering field lines is equal to total number of leaving field lines. So, net magnetic field inside the closed surface is zero.

3. *(d)* We know that, $p_1V_1 = p_2V_2$

According to the question

$$(p+h)\rho g.\frac{4}{3}\pi r^3 = p\rho g.\frac{4}{3}\pi(2r)^3 \quad \text{or} \quad p+h = 8p$$

$$\therefore \quad h = 7p$$

4. *(d)* The potential energy of the particle is given by

$$U = -\int F\cdot dx$$
$$= -\int(-Kx + ax^3)\,dx$$

or, $$U = K\frac{x^2}{2} - a\frac{x^4}{4} = \frac{x^2}{4}(2K - ax^2) \ldots\text{(i)}$$

From Eq. (i), it follows that $U = 0$ at two values of x which are $x = 0$ and $x = \sqrt{\frac{2K}{a}}$. Hence, graphs (b) and (c) are not possible. Also, U is maximum or minimum at a value of x given by $\frac{dU}{dx} = 0$ i.e.,

$$0 = \frac{d}{dx}\left(\frac{Kx^2}{2} - \frac{ax^4}{4}\right)$$
$$= Kx - ax^3 = x(K - ax^2)$$

or, $$x = \sqrt{\frac{K}{a}}$$

At this value of x, U is maximum, if $\frac{d^2U}{dx^2} < 0$

Now, $$\frac{d^2U}{dx^2} = \frac{d}{dx}(Kx - ax^3)$$
$$= K - 3ax^2$$

At $x = \sqrt{\frac{K}{a}} \Rightarrow \frac{d^2U}{dx^2} = K - 3a.\frac{K}{a} = -2K$ which is negative.

Hence, U is maximum at $x = \sqrt{\frac{K}{a}}$.

So, graph (a) is not possible.

Also, U is negative for $x > \sqrt{\frac{2K}{a}}$.

Therefore, the correct graph is (d).

5. *(b)* Given, $P = 1000$ W and $R = 4$ m

Intensity of EM wave is given by

$$I = \frac{P}{4\pi R^2}$$

$$V_{\text{avg}}c = \frac{1}{2}\varepsilon_0 E_0^2\cdot c \qquad [\because I = V_{\text{avg}}\cdot c]$$

$$\Rightarrow \quad \frac{P}{4\pi R^2} = \frac{1}{2}\varepsilon_0 E_0^2 c \Rightarrow E_0 = \sqrt{\frac{P}{2\pi R^2\varepsilon_0 c}}$$

Hence, by putting values

$$E_0 = \sqrt{\frac{1000}{2\times 3.14\times 4^2\times 8.85\times 10^{-12}\times 3\times 10^8}}$$

$$\Rightarrow \quad E_0 = 61.2\text{ V/m}$$

6. *(a)* The amplitude of a wave represents the maximum displacement of particles in the medium from their equilibrium position. A wave with higher amplitude carries more energy because it causes greater disturbances in the medium.

7. *(a)* $M_{21} = M_{12}$, i.e. mutual inductance in first coil due to second coil is the same as the mutual inductance in second coil due to first coil. So, different currents in two coils don't effect the value of mutual inductance.

Now, induced emf $\propto$ rate of charge of current. So, if rate is reduced to half, induced emf shall also reduce to half.

Hence, induced emf will be $\frac{3}{2}$ mV = 1.5 mV.

8. *(b)*

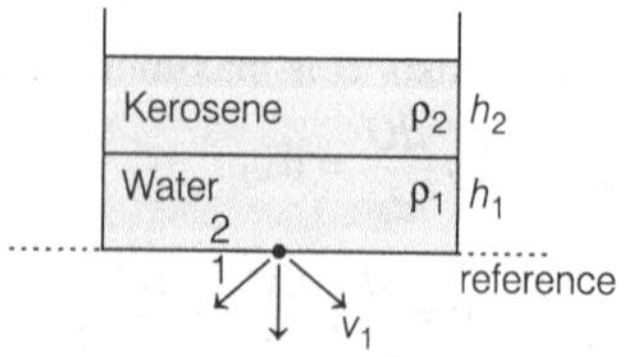

Let A is the area of the vessel and a is area of the hole. Apply Bernoulli's theorem just outside and inside the hole,

$$\frac{1}{2}\rho_1 v_1^2 + p_{atm} = 0 \quad \ldots\text{(i)}$$

$$\frac{1}{2}\rho_1 v_2^2 + p_{atm} + h_2\rho_2 g + h_1\rho_1 g = 0 \quad \ldots\text{(ii)}$$

and
$$av_1 = Av_2 \quad \ldots\text{(iii)}$$

Substitute value of p_{atm} in Eq. (ii) from Eq. (i). Then, substitute value of v_2 from Eq. (iii) and we obtain the following equation

$$\frac{1}{2}\rho_1 v_1^2\left[1-\frac{a^2}{A^2}\right] = \left[h_1 + \frac{h_2\rho_2}{\rho_1}\right]\rho_1 g$$

$$\Rightarrow \quad v_1^2\left[1-\frac{a^2}{A^2}\right] = \left[h_1 + \frac{h_2\rho_2}{\rho_1}\right]2g$$

$$\because \quad \frac{a^2}{A^2} << 1$$

$$\therefore \quad v_1^2 = 2g\left(h_1 + \frac{h_2\rho_2}{\rho_1}\right) \Rightarrow v_1 = \sqrt{2g\left(h_1 + \frac{h_2\rho_2}{\rho_1}\right)}$$

9. *(b)* If n divisions of vernier scale conicides with $(n-1)$ divisions of main scale, then

$$\text{LC} = \left(1 - \frac{n-1}{n}\right) \times \text{MSD reading}$$

Given, $n = 60$, MSD reading $= 0.5° = 0.5 \times 60 \times 60$ s

$$\therefore \quad \text{LC} = \left(1 - \frac{59}{60}\right) \times 0.5 \times 60 \times 60 = 30\text{ s}$$

10. *(a)* The equivalent electrical circuit figure in this case is of Wheatstone bridge. No current would flow through central rod CD when bridge is balanced.

The condition for balanced bridge is $\frac{P}{Q} = \frac{R}{S}$ (in terms of resistance).

$$\frac{1/K_1}{1/K_2} = \frac{1/K_3}{1/K_4}$$

$$\Rightarrow \quad \frac{K_2}{K_1} = \frac{K_4}{K_3} \Rightarrow K_1K_4 = K_2K_3$$

11. *(c)* The escape velocity is given by

$$v = \sqrt{\frac{2GM}{R}} = \sqrt{\frac{2G}{R} \times \frac{4}{3}\pi R^3 \times \rho} \quad \left(\because M = \frac{4}{3}\pi R^3 \times \rho\right)$$

$$\therefore \quad v \propto R\sqrt{\rho}$$

$$\frac{v_e}{v_p} = \left(\frac{R_e}{R_p}\right)\sqrt{\frac{\rho_e}{\rho_p}}$$

$$= \frac{1}{4}\sqrt{\frac{1}{4}} = \frac{1}{8} \quad [\because R_e : R_p = 1:4, \rho_e : \rho_p = 1:4]$$

12. *(c)*

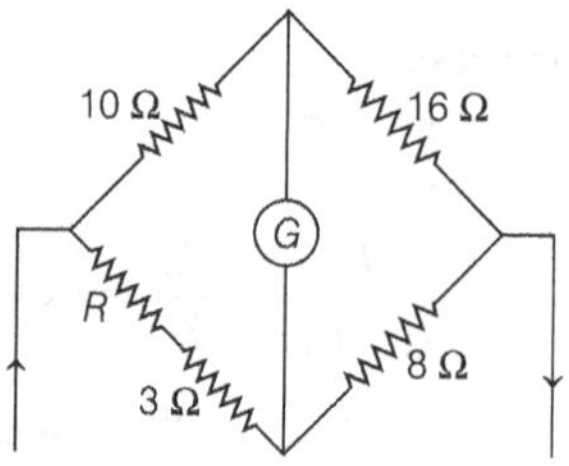

For balanced wheatstone bridge, $\frac{10}{R+3} = \frac{16}{8}$

$$\Rightarrow \quad R = 2\Omega$$

13. *(c)*

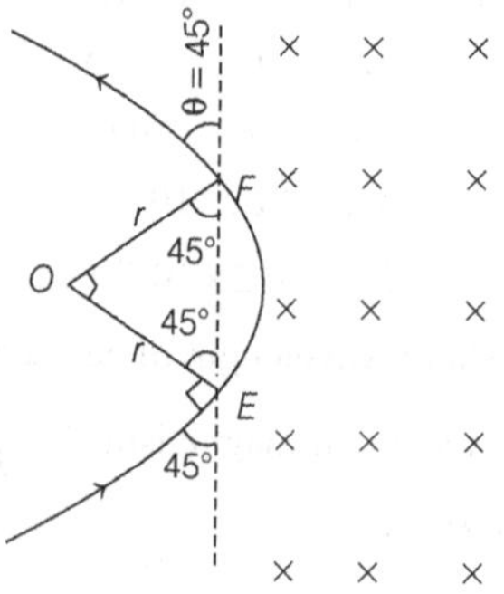

By symmetry, $\theta = 45°$

$$r = \frac{mv}{Bq}$$

$$EF = \sqrt{r^2 + r^2}$$

$$= \sqrt{2}r = \sqrt{2} \times \frac{mv}{Bq}$$

$$= \sqrt{2} \times \frac{1.6 \times 10^{-27} \times 1.0 \times 10^7}{1 \times 1.6 \times 10^{-19}}$$

$$= 14 \text{ cm}$$

14. *(a)* Acceleration of particle is $a = -4x$

$$\Rightarrow \quad v \cdot \frac{dv}{dx} = -4x \Rightarrow \int v \cdot dv = -\int 4x\,dx$$

$$\Rightarrow \quad \frac{v^2}{2} = -2x^2 + C'$$

$$\therefore \quad v = \sqrt{C' - 4x^2} \quad \ldots(i)$$

Now, for v to be maximum, $\frac{dv}{dt} = a = 0$

$$\therefore \quad -4x = 0$$

$$\therefore \quad x = 0$$

From Eq. (i), at $x = 4$ m, $v = 0$

$$0 = C' - 4 \times 4^2$$

$$\therefore \quad C' = 64$$

$$\therefore \quad v = \sqrt{64 - 4x^2}$$

$$\therefore \quad v_{\max} = (\text{Velocity at } x = 0) = \sqrt{64} = 8 \text{ m/s}$$

15. *(b)* Work done in isochoric process is zero.

In isothermal process,

Temperature, $T =$ constant

$$\therefore \quad pV = nRT = \text{constant}$$

$\Rightarrow p \propto \frac{1}{V}$, which is Boyle's law

The second law of thermodynamics states that, it is impossible for heat engine to convert all the heat it receives from a thermal energy reservior to work completely.

As, $\Delta Q - \Delta W = \Delta U$, which is independent of path.

16. *(d)* Molar volume of a gas = 22.4 L

∴ Mass of the gas of volume 22.4 L at STP is equal to its molecular weight. Collisions of the molecules of an ideal gas are perfectly elastic.

$$\text{As,} \quad C_p - C_V = R \quad \ldots(i)$$

$$\frac{C_p}{C_V} = \gamma \Rightarrow C_p = \gamma C_V \quad \ldots(ii)$$

From Eqs. (i) and (ii), we get

$$\gamma C_V - C_V = R$$

$$\therefore \quad C_V = \frac{R}{\gamma - 1}$$

17. *(c)* By Bohr's postulate for revolving electron,

Angular momentum, $L = \frac{nh}{2\pi} \Rightarrow mvr_n = \frac{nh}{2\pi}$

Momentum of the electron, $p = mv = \frac{nh}{2\pi r_n}$

∴ de-Broglie wavelength associated with the electron is

$$\lambda_n = \frac{h}{p} = \frac{2\pi r_n}{n}$$

$$\therefore \quad n = 4, r_n = 5.75 \text{ Å}$$

$$\therefore \quad \lambda_n = \frac{2\pi \times 5.75}{4} = 9.02 \text{ Å}$$

18. *(b)* The representation of NOR gate is

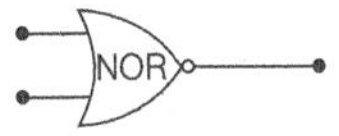

The representation of OR gate is

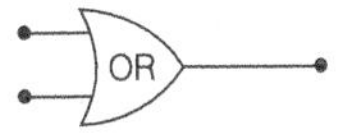

The representation of NOT gate is

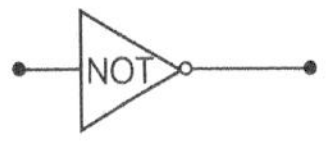

The representation of NAND gate is

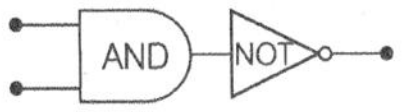

OR

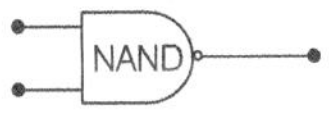

19. *(d)* When a wire is stretched, the change in length of wire is $\Delta l = \frac{Fl}{\pi r^2 Y}$, where Y is its Young's modulus.

Here, for wires P and Q,

$$l_P = 4 \text{ m}, l_Q = 3 \text{ m}$$

$$\frac{Y_P}{Y_Q} = \frac{3}{2}, r_Q = 2 \text{ mm} = 2 \times 10^{-3} \text{ m and } \frac{F_P}{F_Q} = 1$$

As, it is given that $\Delta l_P = \Delta l_Q$

$$\Rightarrow \frac{F_P l_P}{\pi r_P^2 Y_P} = \frac{F_Q l_Q}{\pi r_Q^2 Y_Q}$$

$$r_P^2 = \frac{F_P}{F_Q} \cdot \frac{l_P}{l_Q} \cdot \frac{Y_Q}{Y_P} \cdot r_Q^2 = 1 \times \frac{4}{3} \times \frac{2}{3} \times 4 \times 10^{-6} \text{ m}$$

$$= \frac{32}{9} \times 10^{-6} \text{ m} = 3.55 \times 10^{-6} \text{ m}$$

$$r_P = 1.88 \times 10^{-3} \text{ m}$$

$$r_P = 1.88 \text{ mm}$$

20. *(b)* Mass defect in $_2He^4$

$$= [2.01565 + 2.01733 - 4.002604]$$

$$= 0.030376 \text{ amu}$$

Binding energy = (0.030376×931) MeV = 28.28 MeV

∴ Binding energy per nucleon = $\frac{28.28}{4} = 7.07$ MeV

Mass defect in

$$_{26}Fe^{56} = 30 \times 1.008665 + 26 \times 1.007825 - 55.934939$$

$$= 0.528461 \text{ amu}$$

∴ Total binding energy = 0.528461×931.5 MeV

$$= 492.26 \text{ MeV}$$

∴ BE per nucleon = $\frac{492.26}{56} = 8.790$ MeV

Mass defect in $_{83}Bi^{209}$ nucleus

$= 83 \times 1.007825 + 126 \times 1.008665 - 208.980388 = 1.760872$

Total BE $= 1.760872 \times 931.5 = 1640.26$ MeV

$\therefore$ BE per nucleon $= \dfrac{1640.26}{209} = 7.848$ MeV

Therefore, $_{26}Fe^{56}$ have greater binding energy per nucleon.

21. (49) The number of fringes on either side of C of screen is

$$n_1 = \left[\frac{AC}{\beta}\right] = \left[\frac{0.5}{0.021}\right] = [23.8] \approx 24$$

Total number of fringes $= 2n_1 +$ Fringe at centre
$= 2n_1 + 1 = 2 \times 24 + 1$
$= 48 + 1 = 49$

22. (3) Let, $\mathbf{v}_{rg} = -a\hat{\mathbf{j}}$, $\mathbf{v}_{mg} = v\hat{\mathbf{i}}$

$\therefore \quad \mathbf{v}_{rm} = \mathbf{v}_{rg} - \mathbf{v}_{mg} = \mathbf{v}_{rg} + (-\mathbf{v}_{mg})$

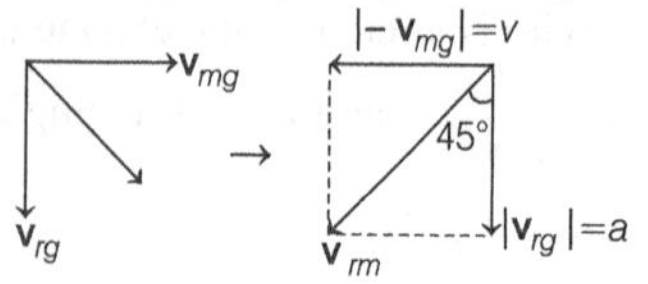

$\therefore \quad \tan 45° = \dfrac{v}{a} \Rightarrow a = v$

Again, when driver increases his speed to nv in same direction, then

$$\mathbf{v}'_{mg} = nv\hat{\mathbf{i}}$$

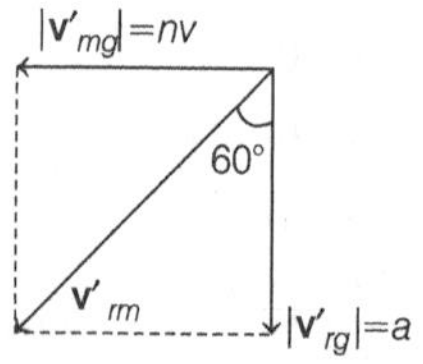

$\therefore \quad \tan 60° = \dfrac{nv}{a}$

$\Rightarrow \quad \sqrt{3} = \dfrac{nv}{a} = \dfrac{nv}{v} = n$ [using Eq. (i)]

$n = \sqrt{3}$

$\therefore \quad n^2 = 3$

23. (2) Heat supplied to gas in isothermal process is given by

$$Q = nRT \ln \frac{V_{\text{final}}}{V_{\text{initial}}}$$

So, for increase in volume from V_0 to V

$$Q_1 = nRT \ln \frac{V}{V_0} \quad \dots \text{(i)}$$

while for increase in volume from V_0 to $2V_0$

$$Q_2 = nRT \ln \frac{2V_0}{V_0}$$

$$\Rightarrow 2Q_1 = nRT \ln 2 \quad \dots \text{(ii)}$$

Dividing Eq. (i) by Eq. (ii), we get

$$\frac{1}{2} = \frac{\ln \frac{V}{V_0}}{\ln 2} \Rightarrow \ln \frac{V}{V_0} = \ln \sqrt{2}$$

$$V = \sqrt{2}V_0 \Rightarrow \frac{V}{V_0} = \sqrt{2} = \sqrt{x} \quad \text{(given)}$$

$\therefore \quad x = 2$

24. (4) After short circuiting, the battery in DC inductor offers zero resistance.

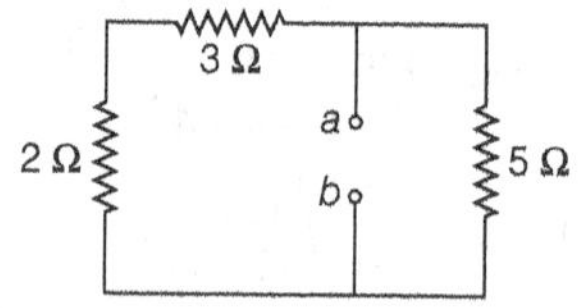

2 Ω and 3 Ω are in series and this combination is in parallel with 5 Ω.

$$R_{ab} = \frac{(2+3) \times 5}{(2+3+5)} = 2.5\,\Omega$$

Now, time constant, $\tau = \dfrac{L}{R_{ab}} = \dfrac{10}{2.5} = 4$ s

25. (13)

Study Tactics

In this problem, first calculate the equivalent resistance across the points a and b and then apply the formula τ (time constant) $= L/R$

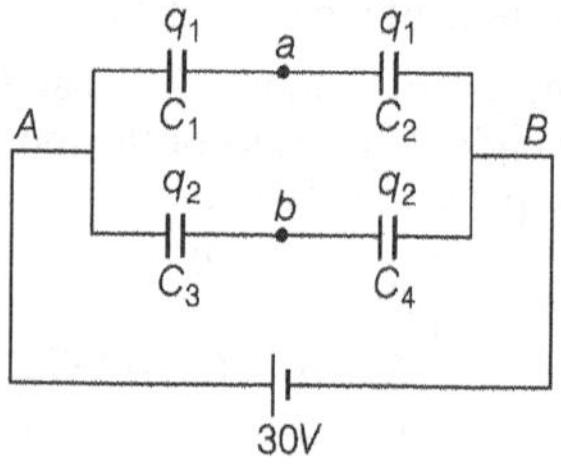

Let q_1 be the charge on capacitors C_1 and C_2 and q_2 be the charge on capacitors C_3 and C_4.

$\therefore \quad V_A - V_B = (V_A - V_a) + (V_a - V_B)$

$\Rightarrow \quad 30 = \dfrac{q_1}{C_1} + \dfrac{q_1}{C_2} = \dfrac{q_1}{1} + \dfrac{q_1}{1.5}$

$\therefore \quad q_1 = 18\,\mu C$

$V_A - V_B = (V_A - V_b) + (V_b - V_B)$

Also, $\quad 30 = \dfrac{q_2}{C_3} + \dfrac{q_2}{C_4} = \dfrac{q_2}{2.5} + \dfrac{q_2}{0.5}$

$\therefore \quad q_2 = 12.5\,\mu C$

$\therefore \quad V_A - V_a = \dfrac{q_1}{C_1} = \dfrac{18}{1} = 18$ V

$V_A - V_b = \dfrac{q_2}{C_3} = \dfrac{12.5}{2.5} = 5$ V

$\therefore \quad V_b - V_a = 18 - 5 = 13$ V

Chemistry

26. (*d*) $\text{pH} = \text{p}K_a + \log\frac{[\text{Salt}]}{[\text{Acid}]}$

$$\text{pH} = \text{p}K_a + \log\frac{[C_6H_5\ COO^-]}{[C_6H_5COOH]}$$

$$4.7 = 4.4 + \log\frac{[C_6H_5COO^-]}{[C_6H_5COOH]}$$

$$0.3 = \log\frac{[C_6H_5\ COO^-]}{[C_6H_5\ COOH]}$$

$$\frac{[C_6H_5COO^-]}{[C_6H_5COOH]} = 1.995 \simeq 2$$

Let, volume of acid $= V$ mL

$$\frac{0.2\times(300-V)}{0.1\times V} = 2 \Rightarrow V = 150 \text{ mL}$$

27. (*b*) When NH_3 is treated with acid it produces X.

$$\underset{[Z]}{NH_3} \xrightarrow{H^+} \underset{[X]}{NH_4^+}$$

When NH_3 is treated with strong base.

$$\underset{[Z]}{NH_3} \xrightarrow{\text{Base}} \underset{[Y]}{NH_2^{\ominus}}$$

Bond angle of molecule or ion can easily be understood in terms of structure and bent rule.

Structure of $\underset{(X)}{NH_4^+}$, $\underset{(Z)}{NH_3}$, $\underset{(Y)}{NH_2^{\ominus}}$

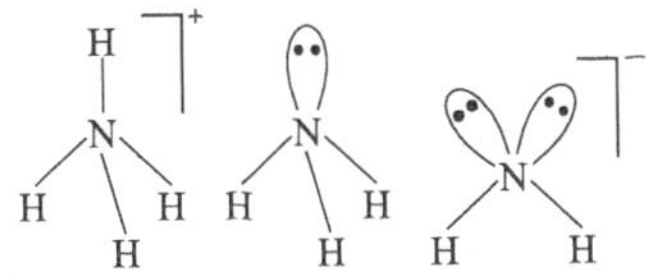

X Z Y

4 *bp* 3*bp* + 1*lp* 2 *bp* + 2*lp*

Bond angle decreases as the number of lone pair increases. Hence, the correct order is $X > Z > Y$.

28. (*c*) Among the given statement B and C are correct while A and D are incorrect.

Hydrolysis of sucrose On hydrolysis it produces equimolar mixture of D- (+)- glucose and D- (−)- fructose.

$$C_{12}H_{22}O_{11} + H_2O \longrightarrow \underset{\text{D-(+)-glucose}}{C_6H_{12}O_6} + \underset{\text{D-(−)-fructose}}{C_6H_{12}O_6}$$

Glycosidic linkage In sucrose glycosidic linkage exist between C_1 of glucose and C_2 of fructose

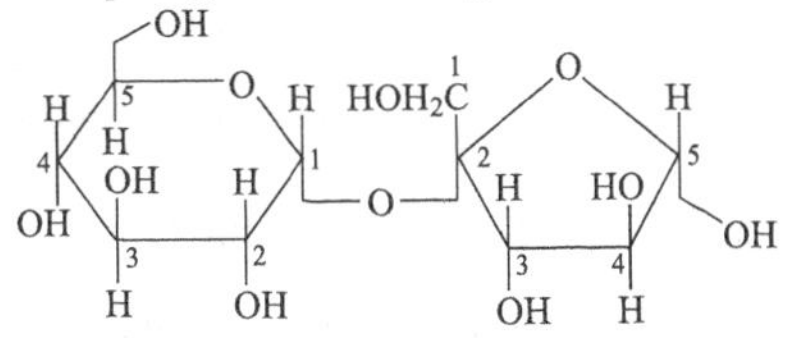

α-D-glucose β-D-fructose

Solution of sucrose is dextrorotatory (specific rotation = ± 66.7°). When sucrose is hydrolysed it forms equimolar quantity of D-glucose and D-fructose.

The solution of formed products is found to be laevorotatory (specific rotation = − 20.0°). This change in optical properties of sucrose is called inversion of cane sugar.

29. (*c*) The correct match is A-II, B-IV, C-I, D-III.

A. H_2/Pd-$BaSO_4$ is used to reduce acyl chlorides to their corresponding aldehydes.

B. Cl_2/Red P, H_2O is used to convert carboxylic acid to monochloro substituted acid.

C. Br_2/NaOH is used to convert amides to amine.

D. Cu, HCl is used to convert diazonium salt into aryl chloride.

30. (*b*) $\lambda = \frac{c}{\nu}$

$$\lambda = \frac{3\times10^8}{2.33\times10^{14}} = 1.288\times10^{-6}\ \text{m}$$

For paschen series $n_1 = 3$, $n_2 = n$

$$\frac{1}{\lambda} = \left(\frac{1}{(3)^2} - \frac{1}{n^2}\right)$$

$$776397.52 = 10967700\left(\frac{n^2-9}{9n^2}\right)$$

$$9 = n^2(0.363)$$

$$n^2 = 24.815$$

$$n \simeq 5$$

Therefore, option (b) is correct.

31. (*b*) The correct match is A-III, B-IV, C-I, D-II.

A. V_2O_5 is used as catalyst for oxidation of SO_2 in the manufacture of H_2SO_4.

B. $TiCl_4 + Al(CH_3)_3$ is used in the polymerisation of ethylene.

C. $PdCl_2$ is used in wacker process, in which alkene changed into aldehyde *via* catalytic cyclic process initiated by $PdCl_2$.

D. Nickel based catalysts are widely used for selective hydrogenation of polyunsaturated compounds formed during steam creating.

32. (*b*) Due to smaller size of oxygen addition of electron is not easy. Hence, the order of electron gain enthalpy is

$$S > Se > Te > Po > O$$

33. (*a*) Ethers on prolonged contact with air or oxygen form peroxides which are explosives and should be handled with care.

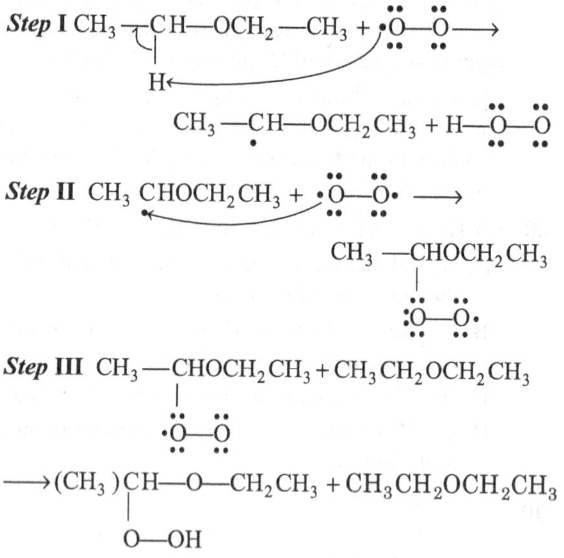

34. (*a*) Five membered ring is being converted into six membered ring. Hence, 6th bond must come from within the ring. Therefore, bond A is responsible for the formation of 6-membered ring.

35. (*c*) B, C and D are correct.

(i) For NaX salts, order of covalent character is as follows:

$$NaF < NaCl < NaBr < NaI$$

(ii) Lithium has small size compared to sodium. Hence, it is less likely for Li to donate an electron to form ionic bond. Hence Li salts are more covalent.

(iii) Due to inert pair effect $PbCl_2$ is more ionic while $PbCl_4$ being in +4 oxidation state shows more covalent character due to smaller size of Pb (IV).

(iv) As we move left to right in a period, ionic character decreases, hence $MgCl_2 > NaCl$.

36. (*c*) (i) $O_2 \longrightarrow O_2^{2-}$

Bond order of $O_2 \longrightarrow 2$ (Paramagnetic)
Bond order of $O_2^{2-} \longrightarrow 1$(Diamagnetic)

B.O. decreases by 1.

(ii) $N_2 \longrightarrow N_2^+$

B.O. of $N_2 \longrightarrow 3$ (Diamagnetic)
B.O. of $N_2^+ \longrightarrow 2.5$ (Paramagnetic)
B.O. decreases by 0.5.

(iii) $NO \longrightarrow NO^+$

B.O. of $NO \longrightarrow 2.5$ (Paramagnetic)
B.O. of $NO^+ \longrightarrow 3$ (Diamagnetic)
B.O. increases by 0.5.

(iv) $O_2 \longrightarrow O_2^+$

B.O. of $O_2 \longrightarrow 2$ (Paramagnetic)
B.O. of $O_2^+ \longrightarrow 2.5$ (Paramagnetic)
B.O. increases by 0.5.
Thus, option (c) is correct.

37. (*c*) Coordination isomerism is observed in the coordination compounds having both cationic and anionic complex ions. Hence, $[Co(NH_3)_6][Co(CN)_6]$ exhibit coordination isomerism because, exchange of ligands between cation and anion occurs between them. Therefore, statement I is incorrect but Statement II is correct.

38. (*d*) **Benzylic bromination** Benzylic bromination of toluene can be done by its treatment with NBS in CCl_4 occurs as follows

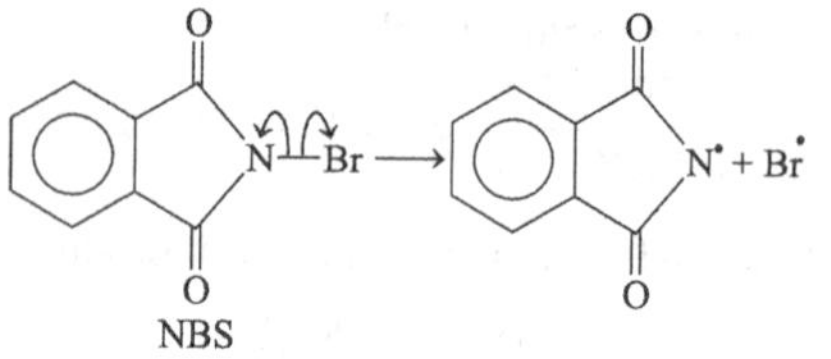

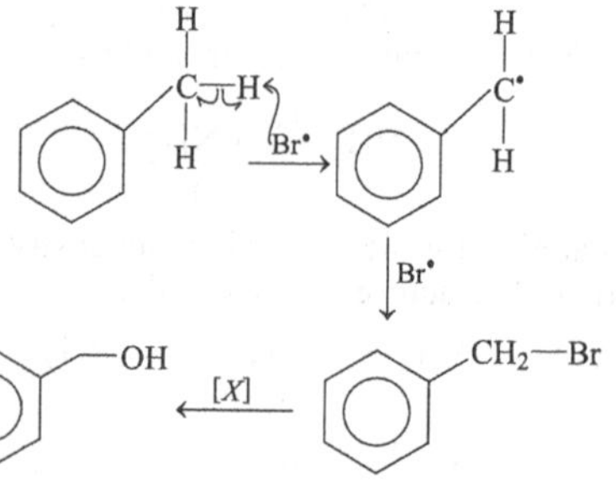

Benzylic oxidation This can be done by treating toluene with SeO_2.

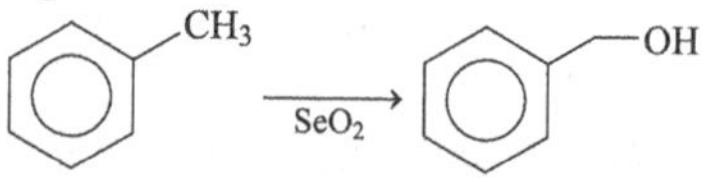

Hence, the best reagent used for above conversion is *aq.* KOH.

$\therefore \quad [X] = Aq.\ KOH$

39. (*a*) Crystal field splitting energy is directly proportional to strength of ligand.
Order of strength of ligand is

$$CN^- > H_2O > Cl^-$$

For octahedral complex, splitting is greater, hence it has more energy of absorption while tetrahedral complex has less splitting hence less energy of absorption.

Since, energy $\propto \dfrac{1}{\text{wavelength}}$, correct order is

$[Fe(CN)_6]^{4-} < [Fe(H_2O)_6]^{2+} < [PtCl_4]^{2-}$.

40. (*a*) Azodyes are not formed by secondary amine i.e. dimethylamine.

41. (*c*) To complete the reaction, these steps are followed. Complete reaction involved in preparation of Gilman reagent.

Gilman's reagent When CuCl and CH_3 Li mixed together, it produces Gilman's reagent.

$$CH_3Li + CuCl \longrightarrow (CH_3)_2CuLi$$

When Gilman reagent is treated with α, β unsaturated aldehyde or ketone, it produces product through 1, 4 addition.
Structural formula of 6, 6-dimethyl cyclohex-2-en-1-ol is

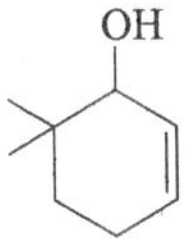

This molecule on reaction with MnO_2 undergoes oxidation to produce corresponding ketone as shown

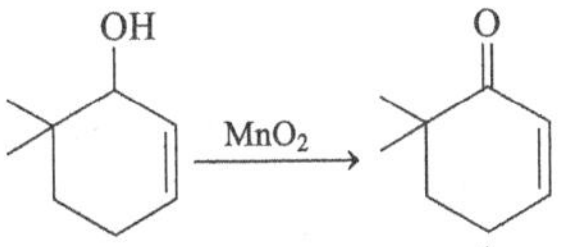

A

A further undergo reaction with $(CH_3)_2CuLi$ to give *C* through 1,4-addition.

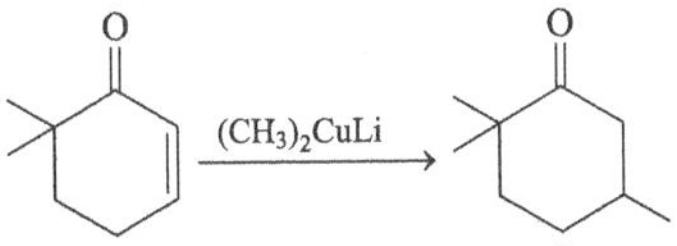

A *B*

The formation of *B via* 1, 4-addition is due to transfer of CH_3 from less polar Cu—CH_3 bond to less electropositive position.

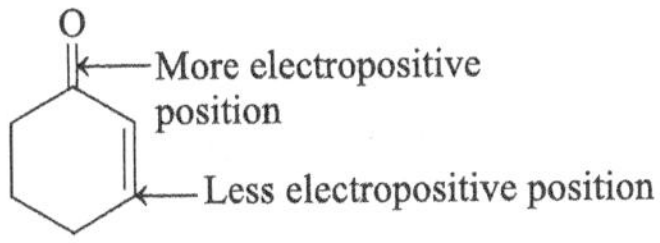

The product *B* contains asymmetric carbon hence product *B* is optically active.

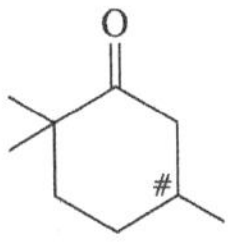

42. (*d*) From de Broglie's relationship,

$$2\pi r = n\lambda$$

$$\lambda = \frac{2\pi r}{n}$$

where, $r = \frac{n^2}{Z} a_0$

$$\lambda = \frac{2\times\pi}{n} \times \frac{n^2}{Z} a_0$$

Given, $a_0 = \frac{r_0}{2}$, $n = 4$

$$\lambda = 2\times\pi\times\frac{n}{Z}\times\frac{r_0}{2} = \frac{\pi\times 4\times r_0}{1} = 4\pi r_0$$

43. (*d*)

(a) XeF_4 is sp^3d^2-hybridised with 4 bond pair and 2 lone pair. Structure is square planar. So, all bond length are equal.

(b) $BF_4^- \longrightarrow 4\,bp$, sp^3-hybridised. So, structure is tetrahedral. All bond lengths are equal.

(c) $SiF_4 \longrightarrow sp^3$-hybridised with tetrahedral geometry. So, bond lengths are equal.

(d) $SF_4 \longrightarrow sp^3d$-hybridised with 4 *bp* and 1 *lp* and have distorted TBP geometry or *see-saw* shape.

So, axial bonds are longer than equatorial bond. Hence, option (d) is correct.

44. (*c*)

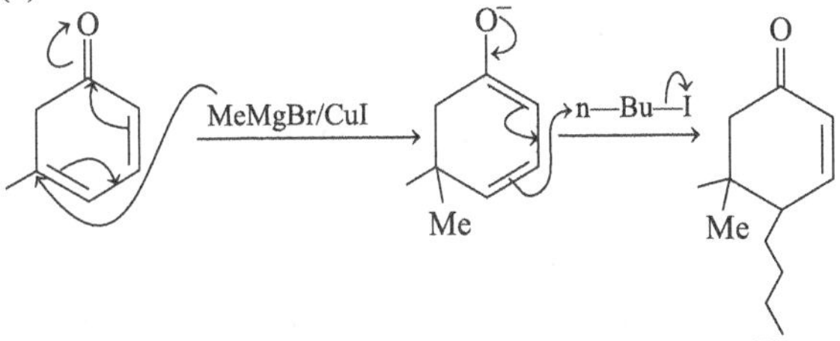

(*Z*)

45. (*b*) The increasing order of hydride affinity for the given carbocation is IV < II < I < III.

The correct hydride affinity order of carbocations will be decided by the stability of carbocation.

Higher the stability of carbocation, lower will be the hydride affinity.

46. (*16*) In the reaction, Fe is oxidised and O_2 is reduced.

$$2Fe \longrightarrow 2Fe^{2+} + 4e^-$$

$$4e^- + O_2 + 4H^+ \longrightarrow 2\,H_2O$$

$$\therefore\ E_{Cell} = E^\circ_{OP_{Fe}} - \frac{0.059}{4}\log[Fe^{2+}]^2 + E^\circ_{RP_{O_2}} + \frac{0.059}{4}\log p_{O_2}\times[H^+]^4$$

$$= E^\circ_{Cell} + \frac{0.059}{4}\log\frac{p_{O_2}\times[H^+]^4}{[Fe^{2+}]^2}$$

$$= 1.67 + \frac{0.059}{4}\log\frac{0.1\times(10^{-3})^4}{(10^{-3})^2}$$

$$= 1.67 + \frac{0.059}{4}\log 10^{-7} = 1.67 + \frac{0.059\times(-7)}{4}$$

$$= 1.67 - 0.103 = 1.567\text{ V}$$

$$\approx 1.6 \approx 16\times 10^{-1}\text{ V}$$

47. (1) For zero order reaction,

rate of reaction = rate constant

$$= 2\times10^{-2} \text{ mol L}^{-1}\text{s}^{-1}$$

Also, $$k = \frac{[A_0]-[A]}{t}$$

$$2\times10^{-2} = \frac{[A_0]-0.5}{25}$$

$$[A_0] = 50\times10^{-2} + 0.5$$
$$= 1\text{M}$$

48. (4) $PRQ = PR + QR$

$PSQ = PS + QS$

(A) ΔU is a state function, hence does not depend on path. So, for P to Q,

$$\Delta U = q + w$$
$$\Delta U = 80\text{ J} - 30\text{ J}$$
$$= 50\text{ J}$$

(B) As the state change is same $(P-Q), \Delta U = 50\text{ J}$.

(C) For path PQ,

$$\Delta U = q + w$$
$$50 = q - 20$$
$$q = 70\text{ J}$$

Heat absorbed = 70 J

(D) Since, ΔU is a state function and PR and SQ are different states hence they are not equal.

(E) $\Delta U = q + w$

$$50 = q - 10$$
$$q = 60\text{ J}$$

Hence, A, B, C and E are correct.

49. (3) Statement given in A, B and D are correct while statement C is incorrect. Its correct form is

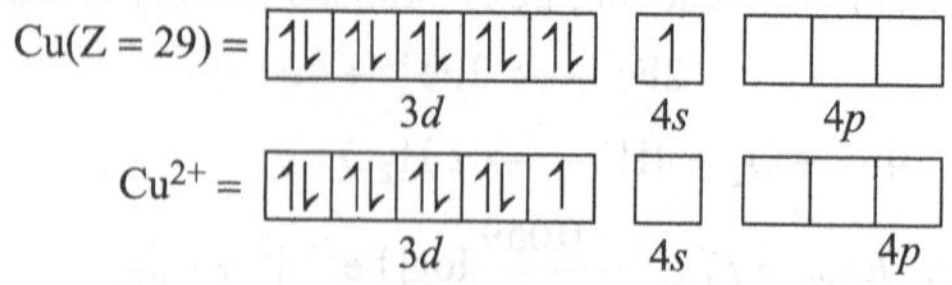

As the Cu^{2+} contains one unpaired electron so the magnetic nature would be paramagnetic.

50. (56) 1 M of 10 mL H_2SO_4 = 1M of 20 mL of NH_3

1000 mL of 1 M ammonia contains = 14 g of N

20 mL of 1 M ammonia contains $= \frac{14\times20}{1000}$ g of N

$\therefore$ % of N $= \frac{14\times20}{1000\times0.5}\times100$

$= 56\ \%$

Mathematics

51. *(a)* Let $z = a + ib$

$$\frac{z-z_1}{z-z_2} = \frac{a+ib-10-6i}{a+ib-4-6i}$$

$$= \frac{(a-10)+i(b-6)}{(a-4)+i(b-6)} \times \frac{(a-4)-i(b-6)}{(a-4)-i(b-6)}$$

$$= \frac{(a-10)(a-4)+(b-6)^2 + i[(a-4)(b-6) - (a-10)(b-6)]}{(a-4)^2+(b-6)^2}$$

$$\frac{z-z_1}{z-z_2} = \frac{(a-10)(a-4)+(b-6)^2 + i\,6(b-6)}{(a-4)^4+(b-6)^2}$$

Given, $\text{amp}\left(\frac{z-z_1}{z-z_2}\right) = \frac{\pi}{4}$

$$\Rightarrow \frac{6(b-6)}{(a-10)(a-4)+(b-6)^2} = 1$$

$$\Rightarrow (a-10)(a-4)+(b-6)^2 = 6b-36$$

$$\Rightarrow a^2+b^2-14a-18b+112 = 0 \quad \text{...(i)}$$

Now, $\alpha = |z-7-9i| = |(a+ib)-7-9i|$

$$= |(a-7)+i(b-9)|$$
$$= \sqrt{(a-7)^2+(b-9)^2}$$
$$= \sqrt{a^2+b^2-14a-18b+130}$$
$$= \sqrt{(a^2+b^2-14a-18b+112)+18}$$
$$= \sqrt{18} \quad \text{[using Eq. (i)]}$$
$$= 3\sqrt{2}$$

$\therefore$ $\alpha = 3\sqrt{2}$

Study Tactics

Assume $z = a + ib$ and put it in given expression $\frac{z-z_1}{z-z_2}$ and simplify.

Use formula of amplitude, $\theta = \tan^{-1}\frac{y}{x}$

Now, use above in given expression $|z-7-9i| = \alpha$

52. *(b)* $\int_0^\infty f(x)\,dx = \frac{\pi}{2}$

Put $x = t - \frac{1}{t} \Rightarrow dx = \left(1+\frac{1}{t^2}\right)dt$

$$\therefore \int_0^\infty f(x)\,dx = \int_1^\infty f\left(t-\frac{1}{t}\right)\cdot\left(1+\frac{1}{t^2}\right)dt$$

$$= \int_1^\infty f\left(t-\frac{1}{t}\right)dt + \int_1^\infty f\left(t-\frac{1}{t}\right)\cdot\frac{1}{t^2}\,dt$$

Put $\frac{1}{t}=y \Rightarrow -\frac{1}{t^2}dt = dy$ in second integral.

$$=\int_1^{\infty} f\left(t-\frac{1}{t}\right)dt + \int_1^0 f\left(\frac{1}{y}-y\right)\cdot(-dy)$$

$$=\int_1^{\infty} f\left(t-\frac{1}{t}\right)dt - \int_1^0 f\left(y-\frac{1}{y}\right)dy$$

$$=\int_1^{\infty} f\left(t-\frac{1}{t}\right)dt + \int_0^1 f\left(t-\frac{1}{t}\right)dt$$

$$=\int_0^{\infty} f\left(t-\frac{1}{t}\right)dt = \int_0^{\infty} f\left(x-\frac{1}{x}\right)dx$$

$$=\int_0^{\infty} f(x)\,dx = \frac{\pi}{2}$$

53. *(c)* Given numbers $=\{1, 2, 3, \ldots, n\}$

Total number of ways of selecting three numbers with replacement $= n\times n\times n = n^3$.

Now, calculate the number of favourable cases.

Let $x+y+z=2n$, where $1\le x\le n$, $1\le y\le n$ and $1\le z\le n$.

Clearly, favourable number of cases in the number of solution of above equation.

$=$ Coefficient of x^{2n} in $(x+x^2+\ldots+x^n)^3$

$=$ Coefficient of x^{2n} in $x^3(1+x+x^2+\ldots+x^{n-1})^3$

$=$ Coefficient of x^{2n-3} in $(1+x+x^2+\ldots+x^{n-1})^3$

$=$ Coefficient of x^{2n-3} in $\left(\frac{1-x^n}{1-x}\right)^3$

$=$ Coefficient of x^{2n-3} in $(1-x^n)^3(1-x)^{-3}$

$=$ Coefficient of x^{2n-3} in $(1-x^{3n}-3x^n+3x^{2n})(1-x)^{-3}$

$=$ Coefficient of x^{n-3} in $(1-x)^{-3} = \frac{(n-1)(n+4)}{2}$

$\therefore$ Required probability $= \frac{\frac{(n-1)(n+4)}{2}}{n^3} = \frac{(n-1)(n+4)}{2n^3}$

54. *(a)* Given a matrix $A=\begin{bmatrix}1 & 0 & 0\\ 0 & 1 & 1\\ 0 & -2 & 4\end{bmatrix}$

$\therefore$ Characteristic equation of a matrix A is

$$|A-\lambda I|=0$$

$$\begin{vmatrix}1-\lambda & 0 & 0\\ 0 & 1-\lambda & 1\\ 0 & -2 & 4-\lambda\end{vmatrix}=0$$

$$(1-\lambda)[(1-\lambda)(4-\lambda)+2]=0$$

$\Rightarrow (1-\lambda)(4+\lambda^2-4\lambda-\lambda+2)=0$

$\Rightarrow \quad (1-\lambda)(\lambda^2-5\lambda+6)=0$

$\Rightarrow \lambda^2-5\lambda+6-\lambda^3+5\lambda^2-6\lambda=0$

$\Rightarrow \quad -\lambda^3+6\lambda^2-11\lambda+6=0$

$\Rightarrow \quad \lambda^3-6\lambda^2+11\lambda-6=0$

Also, by Cayley Hamilton theorem, every square matrix satisfies its characteristic equation.

$\therefore \quad A^3-6A^2+11A-6I=0$

$$6I = A^3-6A^2+11A$$

On multiplying by A^{-1}, we get

$$6A^{-1}=A^2-6A+11I$$

$$A^{-1}=\frac{1}{6}(A^2-6A+11I)$$

$\because \quad A^{-1}=\frac{1}{6}(aA^2+bA+cI)$

On comparing, we get

$$a=1, b=-6 \text{ and } c=11$$

$\therefore a+b+c=1-6+11=6$

55. *(b)* Given, runs scored in kth match $=k2^{r+1-k}$, $1\le k\le r$

and runs scored in r matches $=\left(\frac{r+1}{4}\right)(2^{r+1}-r-2)$

$\therefore \quad \sum_{k=1}^{r} k2^{r+1-k} = \left(\frac{r+1}{4}\right)(2^{r+1}-r-2)$

$\Rightarrow \quad 2^{r+1}\left[\sum_{k=1}^{r} k2^{-k}\right] = \left(\frac{r+1}{4}\right)(2^{r+1}-r-2)$

$\Rightarrow 2^{r+1}\left[\frac{1}{2}+\frac{2}{2^2}+\frac{3}{2^3}+\ldots+\frac{r}{2^r}\right]$

$$=\left(\frac{r+1}{4}\right)(2^{r+1}-r-2) \quad \ldots(i)$$

Let $\quad T=\frac{1}{2}+\frac{2}{2^2}+\frac{3}{2^3}+\ldots+\frac{r}{2^r} \quad \ldots(ii)$

$\Rightarrow \quad \frac{1}{2}T=\frac{1}{2^2}+\frac{2}{2^3}+\ldots+\frac{r}{2^{r+1}} \quad \ldots(iii)$

On subtracting Eq. (iii) from Eq. (ii), we get

$$\frac{1}{2}T=\frac{1}{2}+\frac{1}{2^2}+\frac{1}{2^3}+\ldots+\frac{1}{2^r}-\frac{r}{2^{r+1}}$$

$\Rightarrow \quad \frac{1}{2}T=\frac{\frac{1}{2}\left(1-\frac{1}{2^r}\right)}{1-\frac{1}{2}}-\frac{r}{2^{r+1}} \Rightarrow T=2\left[1-\frac{1}{2^r}-\frac{r}{2^{r+1}}\right]$

Now, Eq. (i) becomes,

$$2\cdot 2^{r+1}\left[1-\frac{1}{2^r}-\frac{r}{2^{r+1}}\right]=\left(\frac{r+1}{4}\right)[2^{r+1}-r-2]$$

$\Rightarrow \quad 2[2^{r+1}-r-2]=\left(\frac{r+1}{4}\right)(2^{r+1}-r-2)$

$\Rightarrow \quad \frac{r+1}{4}=2 \Rightarrow r=7$

56. *(b)* When n is odd, let $n = 2p-1$, so

$$S = C_0^2 - 2C_1^2 + 3C_2^2 - \ldots + (-1)^{2p-1}(2p)C_{2p-1}^2 \quad \ldots(i)$$

Using ${}^nC_r = {}^nC_{n-r}$, Eq. (i) can be written as,

$$S = -2pC_0^2 + (2p-1)C_1^2 - (2p-2)C_2^2 + \ldots + C_{2p-1}^2 \quad \ldots(ii)$$

On adding Eqs. (i) and (ii), we get

$$2S = (1-2p)[C_0^2 - C_1^2 + C_2^2 - \ldots + (-1)^{2p-1}C_{2p-1}^2]$$

$$= (1-2p)[(C_0^2 - C_{2p-1}^2) - (C_1^2 - C_{2p-2}^2) + \ldots + (-1)^p(C_p^2 - C_{p-1}^2)]$$

$$= (1-2p)(0) \quad [\because {}^nC_r = {}^nC_{n-r}]$$

$\Rightarrow S = 0$

Statement I is false.

When n is even, let $n = 2p$, then

$$S = C_0^2 - 2C_1^2 + 3C_2^2 - \ldots + (-1)^{2p-1}(2p)C_{2p-1}^2 + (-1)^{2p}(2p+1)C_{2p}^2 \quad \ldots(iii)$$

$$\Rightarrow S = (2p+1)C_0^2 - (2p)C_1^2 + (2p-1)C_2^2 - \ldots + (-1)^{2p-1}C_{2p-1}^2 + (-1)^{2p}C_{2p}^2 \quad \ldots(iv)$$

On adding Eqs. (iii) and (iv), we get

$$2S = (2p+2)[C_0^2 - C_1^2 + C_2^2 - C_3^2 + \ldots + (-1)^{2p}C_{2p}^2]$$

$$\Rightarrow S = (p+1)[C_0^2 - C_1^2 + C_2^2 - C_3^2 + \ldots + (-1)^{2p}C_{2p}^2] \quad \ldots(v)$$

$\because C_0^2 - C_1^2 + C_2^2 - C_3^2 + \ldots + (-1)^{2p}C_{2p}^2$

= Coefficient of constant term in $[C_0 - C_1x + C_2x^2 - \ldots + (-1)^{2p}C_{2p}x^{2p}]$

$$\times\left[C_0 + C_1\frac{1}{x} + C_2\frac{1}{x^2} + \ldots + C_{2p}\frac{1}{x^{2p}}\right]$$

= Coefficient of contstant term in $(1-x)^{2p}\left(1+\frac{1}{x}\right)^{2p}$

= Coefficient of x^{2p} in $(1-x)^{2p}(1+x)^{2p}$

= Coefficient of x^{2p} in $(1-x^2)^{2p}$

$= (-1)^p\, {}^{2p}C_p = (-1)^{n/2} \cdot {}^nC_{n/2}$

From Eq. (v), we get

$$S = (p+1)(-1)^p\, {}^{2p}C_p = (p+1)(-1)^{n/2}\, {}^nC_{n/2}$$

$$= (-1)^{n/2}\left(\frac{n}{2}+1\right)\frac{n!}{\left(\frac{n}{2}\right)!\left(\frac{n}{2}\right)!}$$

Statement II is true.

57. *(a)* We have, $A = \begin{bmatrix} 2 & 1 \\ 4 & 1 \end{bmatrix}$,

$$B = \begin{bmatrix} 3 & 4 \\ 2 & 3 \end{bmatrix}, C = \begin{bmatrix} 3 & -4 \\ -2 & 3 \end{bmatrix}$$

$$BC = \begin{bmatrix} 3 & 4 \\ 2 & 3 \end{bmatrix}\begin{bmatrix} 3 & -4 \\ -2 & 3 \end{bmatrix}$$

$$BC = \begin{bmatrix} 1 & 0 \\ 0 & 1 \end{bmatrix}$$

$$A(BC) = A$$

$$A(BC)^2 = A$$

Then, $\text{tr}(A) + \text{tr}\left(\frac{A(BC)}{2}\right) + \text{tr}\left(\frac{A(BC)^2}{4}\right) + \text{tr}\left(\frac{A(BC)^3}{8}\right) + \ldots\infty$

$$= \text{tr}(A) + \text{tr}\left(\frac{A}{2}\right) + \text{tr}\left(\frac{A}{4}\right) + \text{tr}\left(\frac{A}{8}\right) + \ldots\ldots\infty$$

$= S$ [let]

$$S = 3 + \frac{3}{2} + \frac{3}{4} + \frac{3}{8} + \ldots$$

$$S = \frac{3}{1-\frac{1}{2}} \Rightarrow S = \frac{6}{1} = 6 \quad \left[\text{sum of infinite GP} = \frac{a}{1-r}\right]$$

Time Saver Tip

$$\because BC = \begin{bmatrix} 3 & 4 \\ 2 & 3 \end{bmatrix}\begin{bmatrix} 3 & -4 \\ -2 & 3 \end{bmatrix} = \begin{bmatrix} 1 & 0 \\ 0 & 1 \end{bmatrix} = I$$

$$\therefore \text{tr}(A) + \text{tr}\left(\frac{A(BC)}{2}\right) + \text{tr}\left(\frac{A(BC)^2}{4}\right) + \text{tr}\left(\frac{A(BC)^3}{r}\right) + \ldots\ldots$$

$$= \text{tr}(A) + \text{tr}\left(\frac{A}{2}\right) + \text{tr}\left(\frac{A}{2^2}\right) + \ldots\ldots\infty$$

$$= \text{tr}(A) + \frac{1}{2}\text{tr}(A) + \frac{1}{2^2}\text{tr}(A) + \ldots\ldots\infty$$

$$= \frac{tr(A)}{1-\frac{1}{2}} = 2\,\text{tr}(A) = 2(2+1) = 6$$

58. *(a)* We have, at $x = c$, $f(x)$ and $g(x)$ has local maximum and local minimum, respectively.

$$h(x) = \frac{f(x)}{g(x)}$$

$$h'(x) = \frac{f'(x)g(x) - g'(x)f(x)}{[g(x)]^2}$$

$$h'(c^+) = \frac{(-\text{ve})(+\text{ve}) - (+\text{ve})(+\text{ve})}{(+\text{ve})^2} = -\text{ve (negative)}$$

$\Rightarrow h'(c^+) < 0$

and $h'(c^-) = \dfrac{(+\text{ve})(+\text{ve}) - (-\text{ve})(+\text{ve})}{(+\text{ve})^2} = +\text{ve (positive)}$

$h'(c^-) > 0$

$\Rightarrow h(x)$ has local maximum at $x = c$.

59. *(c)* We have,

$$a_e = a_h \quad \ldots(i)$$
$$a_e^2 + b_e^2 = 4R^2 \quad \ldots(ii)$$
and $a_h^2 - b_h^2 = R^2 \quad \ldots(iii)$

Now, $a_e^2 e_1^2 = a_e^2 - (4R^2 - a_e^2)$

and $a_h^2 e_2^2 = a_h^2 + b_h^2$

$$= a_e^2 + a_e^2 - R^2 = 2a_e^2 - R^2$$

$$\Rightarrow \quad e_1^2 = 2 - \frac{4R^2}{e^2} \text{ and } e_2^2 = 2 - \frac{R^2}{e^2}$$

$$\Rightarrow e_1^2 - 4e_2^2 = -6 \Rightarrow 4e_2^2 - e_1^2 = 6$$

60. *(d)* Let $\frac{PB}{QB} = x$, then the coordinates of B are $\left(\frac{k}{1+x}, \frac{kx}{1+x}\right)$. [by section formula]

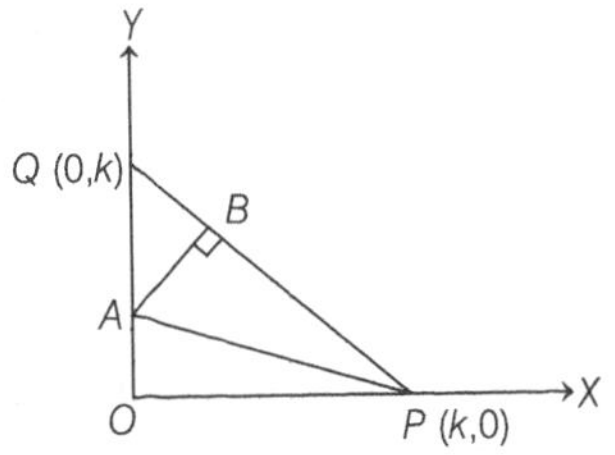

Slope of $AB = \frac{-1}{\text{Slope of } PQ} = \frac{-1}{-1} = 1$

Now, equation of AB is

$$y - \frac{kx}{1+x} = x - \frac{k}{1+x}$$

$$\Rightarrow \quad x - y = \frac{k}{1+x} - \frac{kx}{1+x} \Rightarrow x - y = k\frac{(1-x)}{1+x}$$

So, the coordinates of A are $\left(0, \left(\frac{x-1}{x+1}\right)k\right)$.

$$AB = \frac{(\sqrt{2}k)}{1+x} \text{ and } PB = \frac{\sqrt{2}\,kx}{1+x}$$

Therefore, the area of ΔPAB

$$= \frac{1}{2} \times \frac{kx}{1+x} \times \frac{k}{1+x} \times (\sqrt{2})^2 = \frac{k^2 x}{(1+x)^2}$$

Area of $\Delta OPQ = \frac{k^2}{2}$

According to the question,

$$\frac{k^2 x}{(1+x)^2} = \frac{3}{8} \cdot \frac{k^2}{2}$$

$$\Rightarrow \quad 3x^2 - 10x + 3 = 0$$

$$\Rightarrow \quad (x-3)\left(x - \frac{1}{3}\right) = 0 \Rightarrow x = 3, x = \frac{1}{3}$$

If $x = \frac{1}{3}$, A lies outside the segment OQ and hence the required value of $x = 3$.

$$\therefore \quad \frac{PB}{QB} = 3$$

61. *(b)* At $x = -2$,

$$\text{LHL} = \lim_{x \to -2^-} f(x) = \lim_{x \to -2^-} \frac{ae^{\frac{1}{|x+2|}} - 1}{2 - e^{\frac{1}{|x+2|}}}$$

$$= \lim_{x \to -2^-} \frac{a - e^{-\frac{1}{|x+2|}}}{2e^{-\frac{1}{x+2}} - 1}$$

As $x \to -2$, $\frac{1}{|x+2|} \to \infty$

$$\lim_{x \to -2^-} f(x) = -a$$

$\therefore \quad \text{LHL} = -a$

At $x = -2$,

$$\text{RHL} = \lim_{x \to -2^+} f(x) = \lim_{x \to -2^+} \sin\left(\frac{x^4 - 16}{x^5 + 32}\right)$$

$$= \lim_{x \to -2^+} \sin\left[\frac{(x-2)(x^2+4)}{(x^4 - 2x^3 + 4x^2 - 8x + 16)}\right]$$

$$= \sin \frac{(-4)(8)}{16+16+16+16+16}$$

$$= \sin\left(-\frac{2}{5}\right)$$

$$\text{RHL} = \sin\left(-\frac{2}{5}\right) = -\sin\frac{2}{5}$$

$\because f(x)$ is continuous at $x = -2$.

Then, LHL = RHL = $f(-2)$

$$\Rightarrow \quad -a = -\sin\frac{2}{5} = b$$

$$\therefore \quad a = \sin\frac{2}{5}, b = -\sin\frac{2}{5}$$

$$(a, b) = \left(\sin\frac{2}{5}, -\sin\frac{2}{5}\right)$$

62. *(c)* $N = abcd$

At Ist position = 4 or 5 $\Rightarrow$ 2 options

At IVth place = 0 or 5

$\Rightarrow$ 2 options (multiple of 5)

At IInd and IIIrd place.

$\Rightarrow$ If $b = 3$, $c = 4, 5, 6$

If $\quad b = 4$, $c = 5, 6$

If $b = 5, c = 6$

For b and c

$\Rightarrow$ 6 cases

Total $= 2 \times 2 \times 6 = 24$

63. *(a)* Let l, m and n be the direction cosines of $\mathbf{r}$.

Since, $\mathbf{r}$ is equally inclined to OX, OY and OZ.

$\therefore \quad l = m = n$

We know that $l^2 + m^2 + n^2 = 1$

$\Rightarrow \quad 3l^2 = 1 \quad [\because l = m = n]$

$\Rightarrow \quad l = \pm \frac{1}{\sqrt{3}}$

$\therefore$ Direction cosines of $\mathbf{r}$ are either

$$\frac{1}{\sqrt{3}}, \frac{1}{\sqrt{3}}, \frac{1}{\sqrt{3}} \text{ or } -\frac{1}{\sqrt{3}}, -\frac{1}{\sqrt{3}}, -\frac{1}{\sqrt{3}}$$

Now, $\mathbf{r} = |\mathbf{r}|(l\hat{\mathbf{i}} + m\hat{\mathbf{j}} + n\hat{\mathbf{k}})$

$$\mathbf{r} = |\mathbf{r}|\left(\frac{1}{\sqrt{3}}\hat{\mathbf{i}} + \frac{1}{\sqrt{3}}\hat{\mathbf{j}} + \frac{1}{\sqrt{3}}\hat{\mathbf{k}}\right)$$

or $$\mathbf{r} = |\mathbf{r}|\left(-\frac{1}{\sqrt{3}}\hat{\mathbf{i}} - \frac{1}{\sqrt{3}}\hat{\mathbf{j}} - \frac{1}{\sqrt{3}}\hat{\mathbf{k}}\right)$$

Thus, there are 2 such vectors.

64. *(b)* Let $I = \int_{-\pi}^{\pi} \frac{2x(1+\sin x)}{1+\cos^2 x}\,dx$

$$= \int_{-\pi}^{\pi} \frac{2x}{1+\cos^2 x}\,dx + \int_{-\pi}^{\pi} \frac{2x\sin x}{1+\cos^2 x}\,dx = I_1 + I_2$$

where, $I_1 = \int_{-\pi}^{\pi} \frac{2x}{1+\cos^2 x}\,dx = 0$

$\because \quad \int_{-a}^{a} f(x)dx = 0$, if $f(-x) = -f(x)$

and $\quad I_2 = \int_{-\pi}^{\pi} \frac{2x\sin x}{1+\cos^2 x}\,dx$

$$I_2 = 4\int_0^{\pi} \frac{x\sin x}{1+\cos^2 x}\,dx$$

$$= 4\int_0^{\pi} \frac{(\pi - x)\sin(\pi - x)}{1+\cos^2(\pi - x)}\,dx$$

$$= 4\pi\int_0^{\pi} \frac{\sin x}{1+\cos^2 x}\,dx - 4\int_0^{\pi} \frac{x\sin x}{1+\cos^2 x}\,dx$$

$$= 4\pi\int_0^{\pi} \frac{\sin x}{1+\cos^2 x}\,dx - I_2$$

$\Rightarrow \quad 2I_2 = 4\pi\int_0^{\pi} \frac{\sin x\,dx}{1+\cos^2 x}$

$$I_2 = -2\pi[\tan^{-1}(\cos x)]_0^{\pi}$$

$$= -2\pi\left(-\frac{\pi}{4} - \frac{\pi}{4}\right) = \pi^2$$

$\therefore \quad I = \pi^2$

65. *(a)* $A > 0, B > 0$ and $A + B = \frac{\pi}{6}$

$$\tan(A + B) = \tan\frac{\pi}{6}$$

$$\frac{\tan A + \tan B}{1 - \tan A \cdot \tan B} = \frac{1}{\sqrt{3}}$$

$$1 - \tan A \cdot \tan B = (\tan A + \tan B) \cdot \sqrt{3}$$

$$\tan A \cdot \tan B = 1 - (\tan A + \tan B)\sqrt{3}$$

$\tan A \cdot \tan B$ will be maximum, when $\tan A + \tan B$ will be minimum.

Now, $\tan A + \tan B = \frac{\sin A}{\cos A} + \frac{\sin B}{\cos B}$

$$= \frac{\sin(A+B)}{\cos A\cos B} = \frac{\sin 30^\circ}{\cos A\cos B}$$

$$= \frac{1}{2\cos A\cos B}$$

$$= \frac{1}{\cos(A+B) + \cos(A-B)}$$

$$= \frac{1}{\frac{\sqrt{3}}{2} + \cos(A-B)}$$

For $\tan A + \tan B$ to be minimum, $\cos(A - B)$ should be maximum

i.e. $\quad \cos(A - B) = 1$

$\Rightarrow \quad A = B = \frac{\pi}{12} \quad \left[\because A + B = \frac{\pi}{6}\right]$

$\therefore$ The maximum value of $\tan A \cdot \tan B$

$$= \tan\frac{\pi}{12} \cdot \tan\frac{\pi}{12} = (2 - \sqrt{3})^2$$

$\because \quad a = 2, b = 3$

$\therefore \quad a + b = 5$

66. *(a)* Given differential equation is

$$x\frac{dy}{dx} - y = x\tan\frac{y}{x}$$

$\Rightarrow \quad \frac{dy}{dx} - \frac{y}{x} = \tan\frac{y}{x} \quad \ldots(i)$

Let $\quad \frac{y}{x} = u$

$\Rightarrow \quad y = ux$

$\Rightarrow \quad \frac{dy}{dx} = u + x\frac{du}{dx}$

Eq. (i) becomes,

$$u + x\frac{du}{dx} - u = \tan u$$

$\Rightarrow \quad \cot u\,du = \frac{dx}{x}$

$\Rightarrow \quad \log|\sin u| = \log|x| + \log C$

$\Rightarrow \quad \log\left|\sin\frac{y}{x}\right| = \log C|x|$

$\Rightarrow \quad \left|\sin\frac{y}{x}\right| = C|x|$

At $x = 1,\ y = \frac{\pi}{2}$

$\therefore \quad \left|\sin\frac{\pi}{2}\right| = C|1|$

$\Rightarrow \quad C = 1$

So, $\quad \left|\sin\frac{y}{x}\right| = |x|$

$\Rightarrow \quad \sin\frac{y}{x} = \pm x$

For $0 < x < 1$

$$\sin\frac{y}{x} = x \Rightarrow y = x\sin^{-1}x$$

Now, $\int_0^1 y\,dx = \int_0^1 x\sin^{-1}x\,dx$

$$= \left(\frac{x^2}{2}\sin^{-1}x\right)_0^1 + \frac{1}{2}\int_0^1 \frac{1-x^2-1}{\sqrt{1-x^2}}dx$$

$$= \frac{1}{2}\cdot\frac{\pi}{2} + \frac{1}{2}\int_0^1 \sqrt{1-x^2}\,dx - \frac{1}{2}\int_0^1 \frac{dx}{\sqrt{1-x^2}}$$

$$= \frac{\pi}{4} - \frac{1}{2}(\sin^{-1}x)_0^1 + \frac{1}{2}\left(\frac{x}{2}\sqrt{1-x^2} + \frac{1}{2}\sin^{-1}x\right)_0^1$$

$$= \frac{\pi}{8}$$

$$\Rightarrow \quad \frac{1}{\pi}\int_0^1 y\,dx = \frac{1}{8}$$

67. *(d)* Clearly, height h of ΔABC is the length of perpendicular from $A(1, -1, 2)$ to the line $\frac{x+2}{2} = \frac{y-1}{1} = \frac{z-0}{4}$ which passes through $P(-2, 1, 0)$ and is parallel to $\mathbf{b} = 2\hat{\mathbf{i}} + \hat{\mathbf{j}} + 4\hat{\mathbf{k}}$.

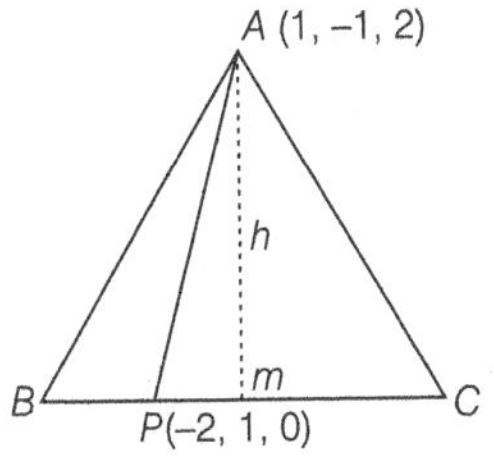

$$\therefore \quad h = \frac{|\mathbf{PA} \times \mathbf{b}|}{|\mathbf{b}|}$$

Now, $\mathbf{PA} = -3\hat{\mathbf{i}} + 2\hat{\mathbf{j}} - 2\hat{\mathbf{k}}$

and $\quad \mathbf{b} = 2\hat{\mathbf{i}} + \hat{\mathbf{j}} + 4\hat{\mathbf{k}}$

$$\therefore \quad \mathbf{PA} \times \mathbf{b} = \begin{vmatrix} \hat{\mathbf{i}} & \hat{\mathbf{j}} & \hat{\mathbf{k}} \\ -3 & 2 & -2 \\ 2 & 1 & 4 \end{vmatrix} = 10\hat{\mathbf{i}} + 8\hat{\mathbf{j}} - 7\hat{\mathbf{k}}$$

$$\therefore \quad |\mathbf{PA} \times \mathbf{b}| = \sqrt{(10)^2 + (8)^2 + (7)^2} = \sqrt{213}$$

$$|\mathbf{b}| = \sqrt{(2)^2 + (1)^2 + (4)^2} = \sqrt{21}$$

$$\because \quad h = \frac{|\mathbf{PA} \times \mathbf{b}|}{|\mathbf{b}|} = \frac{\sqrt{213}}{\sqrt{21}} = \sqrt{\frac{71}{7}}$$

It is given that the length of BC is 8 units.

$$\therefore \text{ Area of } \Delta ABC = \frac{1}{2} \times (BC \times h)$$

$$= \frac{1}{2} \times 8 \times \sqrt{\frac{71}{7}} = \sqrt{\frac{1136}{7}} \text{ sq units.}$$

68. *(b)* We have, curves $y = 1 + 4x - x^2$

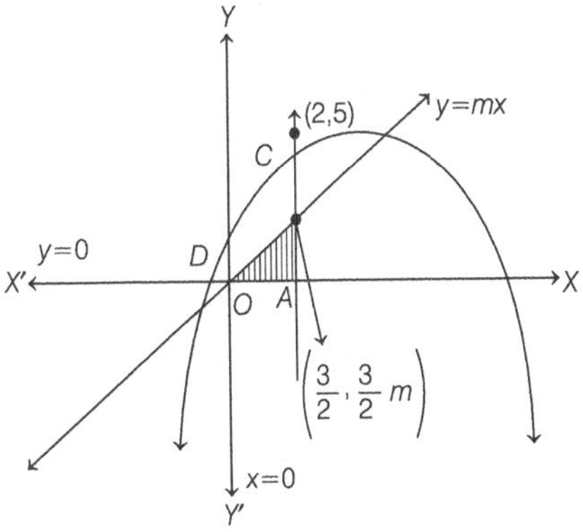

and lines $x = 0,\ x = \frac{3}{2}$

and $\quad y = 0$

$\quad y = 1 + 4x - x^2$

$\Rightarrow (x-2)^2 = -(y-5)$

Vertex is (2, 5).

Area of $OACD = \int_0^{3/2} (1 + 4x - x^2)dx$

$$= \left[x + \frac{4x^2}{2} - \frac{x^3}{3}\right]_0^{3/2}$$

$$= \frac{39}{8} \qquad \ldots\text{(i)}$$

and area of $\Delta OAC = \frac{1}{2} \times \frac{3}{2} \times \frac{3}{2} \times m$

$$= \frac{9}{8}m \qquad \ldots\text{(ii)}$$

From Eqs. (i) and (ii), we get

$$\frac{9}{8}m = \frac{1}{2}\times\frac{39}{8}$$

$$\Rightarrow \quad m = \frac{8}{9}\times\frac{1}{2}\times\frac{39}{8}$$

$$\Rightarrow \quad m = \frac{13}{6} \Rightarrow 6m = 13$$

69. *(c)* Given, var $(X) = 4$, var $(Y) = 5$

$$\overline{X} = 2, \overline{Y} = 4$$

$$n_X = 5 \text{ and } n_Y = 5$$

$$\therefore \quad \text{var}(X) = \frac{\Sigma X_i^2}{n_X} - (\overline{x})^2$$

$$4 = \frac{1}{5}\Sigma x^2 - 2^2 \Rightarrow \Sigma x^2 = 40$$

and $\text{var}(Y) = \dfrac{\Sigma Y_i^2}{n_Y} - (\overline{y})^2$

$$5 = \frac{1}{5}\Sigma y^2 - 4^2$$

$$\Rightarrow \quad \Sigma y^2 = 105$$

$$\therefore \quad \Sigma x^2 + \Sigma y^2 = 105 + 40 = 145$$

and $\quad \overline{X} = \dfrac{\Sigma x}{n_X}$

$$\Rightarrow \quad 2 = \frac{\Sigma x}{5}$$

$$\Rightarrow \quad \Sigma x = 10$$

and $\quad \overline{Y} = \dfrac{\Sigma y}{n_Y} \Rightarrow 4 = \dfrac{\Sigma y}{5}$

$$\Rightarrow \quad \Sigma y = 20$$

$$\therefore \quad \Sigma x + \Sigma y = \Sigma(x+y) = 30$$

Variance of combined data

$$= \frac{1}{10}\Sigma(x^2+y^2) - \left[\frac{1}{10}\Sigma(x+y)\right]^2$$

$$= \frac{1}{10}\times 145 - \left(\frac{1}{10}\times 30\right)^2 = \frac{145}{10} - 9 = \frac{11}{2}$$

70. *(a)* Given, $f(3x) - f(x) = x$ $\quad\ldots$(i)

Replacing x by $\frac{x}{3}$, we get

$$f(x) - f\left(\frac{x}{3}\right) = \frac{x}{3} \quad \ldots\text{(ii)}$$

Again, replacing x by $\frac{x}{3}$ in Eq. (ii), we get

$$f\left(\frac{x}{3}\right) - f\left(\frac{x}{9}\right) = \frac{x}{3^2}$$

On continuing in this way, we get

$$f\left(\frac{x}{3^{n-2}}\right) - f\left(\frac{x}{3^{n-1}}\right) = \frac{x}{3^{n-1}} \quad \ldots\text{(iii)}$$

On adding all these and taking $n \to \infty$,

$$\lim_{n\to\infty}\left(f(3x) - f\left(\frac{x}{3^{n-1}}\right)\right) = x\left(1 + \frac{1}{3} + \frac{1}{3^2} + \ldots\right)$$

$$\Rightarrow \quad f(3x) - f(0) = \frac{3x}{2} \quad \ldots\text{(iv)}$$

On putting $x = \frac{8}{3}$ in Eq. (iv), we get

$$f(8) - f(0) = 4$$

$$\Rightarrow \quad f(0) = 6 - 4 = 2$$

On putting $x = \frac{18}{3}$ in Eq. (iv), we get

$$f(18) - 2 = \frac{3}{2}\times\frac{18}{3}$$

$$f(18) = 9 + 2 = 11$$

71. *(5)* Let $\mathbf{a} = \hat{\mathbf{i}} + \hat{\mathbf{j}} + \hat{\mathbf{k}}$

$$\mathbf{b} = -\hat{\mathbf{i}} + \hat{\mathbf{k}}$$

$$\therefore \quad \mathbf{PA} \perp \mathbf{PB}$$

$$\Rightarrow \quad (\mathbf{a} - \mathbf{r})\cdot(\mathbf{b} - \mathbf{r}) = 0$$

$$\Rightarrow \quad \mathbf{r}^2 - \mathbf{r}\cdot(\mathbf{a} + \mathbf{b}) + \mathbf{a}\cdot\mathbf{b} = 0 \quad \ldots\text{(i)}$$

$$\mathbf{a}\cdot\mathbf{b} = -1 + 0 + 1 = 0$$

and $\mathbf{a} + \mathbf{b} = \hat{\mathbf{j}} + 2\hat{\mathbf{k}}$

From Eq. (i), we get

$$\mathbf{r}^2 - \mathbf{r}\cdot(\hat{\mathbf{j}} + 2\hat{\mathbf{k}}) = 0$$

$$\Rightarrow \quad \mathbf{r}\cdot[\mathbf{r} - (\hat{\mathbf{j}} + 2\hat{\mathbf{k}})] = 0$$

$\because$ **r** is not perpendicular to $r - (\hat{\mathbf{j}} + 2\hat{\mathbf{k}})$.

$$\Rightarrow \quad \mathbf{r} = \hat{\mathbf{j}} + 2\hat{\mathbf{k}}$$

and $|\mathbf{r}|^2 = \mathbf{r}\cdot\mathbf{r} = (\hat{\mathbf{j}} + 2\hat{\mathbf{k}})\cdot(\hat{\mathbf{j}} + 2\hat{\mathbf{k}}) = 5$

72. *(4)* Clearly, $\cot^{-1}(x^2 - 4x + \alpha) > \dfrac{2\pi}{3}$

or $x^2 - 4x + \alpha > \dfrac{-1}{\sqrt{3}}, \forall x \in R$

$$\Rightarrow \quad x^2 - 4x + \alpha + \frac{1}{\sqrt{3}} > 0, \forall x \in R$$

So, discriminant < 0

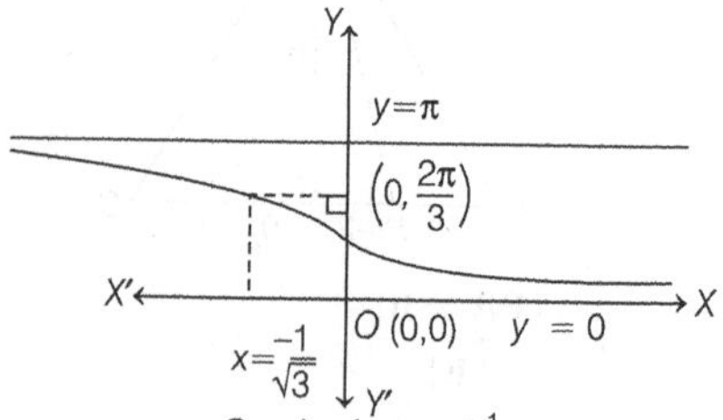

Graph of $y = \cot^{-1}x$

$$\Rightarrow \quad 16-4\left(\alpha+\frac{1}{\sqrt{3}}\right)<0$$

$$4-\alpha-\frac{1}{\sqrt{3}}<0$$

$$\Rightarrow \quad \alpha>4-\frac{1}{\sqrt{3}}$$

$$\therefore \quad \alpha \in\left(4-\frac{1}{\sqrt{3}}, \infty\right)$$

Hence, the minimum integral $\alpha = 4$.

73. *(60)* We have,

$$(\mathbf{a}+3\mathbf{b})\cdot(7\mathbf{a}-5\mathbf{b})=0$$

$$\Rightarrow 7|\mathbf{a}|^2+16\,\mathbf{a}\cdot\mathbf{b}-15|\mathbf{b}|^2=0 \quad \ldots(\text{i})$$

and $\quad (\mathbf{a}-4\mathbf{b})\cdot(7\mathbf{a}-2\mathbf{b})=0$

$$\Rightarrow 7|\mathbf{a}|^2-30\,\mathbf{a}\cdot\mathbf{b}+8|\mathbf{b}|^2=0 \quad \ldots(\text{i})$$

From Eqs. (i) and (ii), we get

$$\mathbf{a}\cdot\mathbf{b}=\frac{15|\mathbf{b}|^2-7|\mathbf{a}|^2}{16} \text{ and}$$

$$\mathbf{a}\cdot\mathbf{b}=\frac{1}{30}(7|\mathbf{a}|^2+8|\mathbf{b}|^2)$$

$$\therefore \quad \frac{15|\mathbf{b}|^2-7|\mathbf{a}|^2}{8}=\frac{1}{15}(7|\mathbf{a}|^2+8|\mathbf{b}|^2)$$

$$\Rightarrow 15(15|\mathbf{b}|^2-7|\mathbf{a}|^2)=8(7|\mathbf{a}|^2+8|\mathbf{b}|^2)$$

$$\Rightarrow \quad 225|\mathbf{b}|^2-105|\mathbf{a}|^2=56|\mathbf{a}|^2+64|\mathbf{b}|^2$$

$$\Rightarrow \quad 161|\mathbf{b}|^2=161|\mathbf{a}|^2$$

$$\Rightarrow \quad |\mathbf{b}|^2=|\mathbf{a}|^2$$

From Eqs. (i), we get

$$16\,\mathbf{a}\cdot\mathbf{b}=15|\mathbf{b}|^2-7|\mathbf{a}|^2=15|\mathbf{b}|^2-7|\mathbf{b}|^2$$

$$\Rightarrow 16\,\mathbf{a}\cdot\mathbf{b}=8|\mathbf{b}|^2$$

$$\Rightarrow \quad \mathbf{a}\cdot\mathbf{b}=\frac{1}{2}|\mathbf{b}|^2$$

$$\Rightarrow |\mathbf{a}||\mathbf{b}|\cos\theta=\frac{1}{2}|\mathbf{b}|^2$$

$$\Rightarrow \quad \cos\theta=\frac{1}{2}$$

$$\therefore \quad \theta=60^\circ$$

74. *(2)* On applying section formula, the coordinates of

$$R \equiv\left(\frac{1\times 1+6\beta}{1+\beta}, \frac{1\times 2+2\beta}{1+\beta}\right) \equiv\left(\frac{1+6\beta}{1+\beta}, 2\right)$$

Since, R lies within the parabola, so

$$\left(\frac{1+6\beta}{1+\beta}\right)^2-8\cdot 2<0$$

$$\Rightarrow \quad (1+6\beta)^2-16(1+\beta)^2<0$$

$$\Rightarrow \quad (1+6\beta+4+4\beta)(1+6\beta-4-4\beta)<0$$

$$\Rightarrow \quad (10\beta+5)(2\beta-3)<0$$

$$\Rightarrow \quad (2\cdot\beta+1)(2\cdot\beta-3)<0$$

$$\Rightarrow \quad \left(\beta+\frac{1}{2}\right)\left(\beta-\frac{3}{2}\right)<0$$

$$\therefore \quad -\frac{1}{2}<\beta<\frac{3}{2}$$

But since, R lies on the line segment, so β cannot be negative.

$$\therefore \quad 0<\beta<\frac{3}{2}$$

$$\Rightarrow \quad 0<2\beta<3$$

$\therefore$ Number of integral values of $2\beta = 2$

75. *(25)* Making the equation of the circle homogeneous, with the help of equation of line, we get

$$x^2+y^2-2(25x+69y)\frac{(69x+25y)}{2\times 69\times 25}+[(69)^2+(25)^2-c^2]\left[\frac{69x+25y}{2\times 69\times 25}\right]^2=0$$

$$\Rightarrow \quad (2\times 69\times 25)^2(x^2+y^2)-4\times 69\times 25^2x(69x+25y)$$
$$-4\times 69^2\times 25y(69x+25y)+[(69)^2+(25)^2-c^2]$$
$$[(69)^2x^2+25^2y^2+2\times 69\times 25xy]=0$$

which is the equation of required lines at right angled.

Coefficient of x^2 + Coefficient of $y^2 = 0$

$$\Rightarrow \quad (69^2+25^2-c^2)(69^2+25^2)=0$$

$$\Rightarrow \quad c^2=69^2+25^2$$

$$\Rightarrow \quad r=25$$